The Last Interglacial-Glacial Transition in North America

Edited by

Peter U. Clark
Department of Geosciences
Oregon State University
Corvallis, Oregon 97331

and

Peter D. Lea
Department of Geology
Bowdoin College
Brunswick, Maine 04011

270

1992

Published by The Geological Society of America, Inc.
3300 Penrose Place, P.O. Box 9140, Boulder, Colorado 80301

Printed in U.S.A.

GSA Books Science Editor Richard A. Hoppin

Library of Congress Cataloging-in-Publication Data

The Last interglacial-glacial transition in North America / edited by
Peter U. Clark and Peter D. Lea.
p. cm. — (Special paper / Geological Society of America ;
270)
Includes bibliographical references and index.
ISBN 0-8137-2270-5
1. Glacial epoch—North America. 2. Drift—North America. 3. Ice
sheets—North America. 4. Pluvial lakes—North America. I. Clark,
Peter U., 1956- . II. Lea, Peter D. III. Series: Special papers
(Geological Society of America) ; 270.
QE697.L294 1992
551.7'92'097—dc20 92-18020
CIP

Cover photo: Exposure of Quaternary soils and sediments in Illinois. From the base, Paleozoic limestone, Illinoian till, Sangamon Geosol, Roxana Silt, Robein Silt, Farmdale and Indian Point Geosols, lower (organic) Peoria Loess, upper Peoria Loess. Photograph taken at the Athens North Quarry, Illinois, by B. Brandon Curry.

10 9 8 7 6 5 4 3 2 1

Contents

Preface

This volume is an outgrowth of a symposium, Last Interglaciation/Glaciation Transition (122–64 ka) in North America, held at the 1988 annual meeting of the Geological Society of America in Denver, Colorado. Fourteen of the seventeen papers presented at the symposium appear in this volume. Several additional papers not presented at the symposium are also included.

In addressing the last interglacial-glacial transition in North America, we emphasize records of former ice sheets, glaciers, and pluvial lakes. Although a wealth of additional available information on this transition is thus not included in this volume, our primary purpose is to focus on the response of North American ice sheets and glaciers to climate change following the last interglaciation. To evaluate and describe this response, chapters in this volume offer geographically significant coverage that provides a continental perspective of former North American ice sheets and glaciers.

In developing a comprehensive model of the ice ages, it is equally important to understand the geologic record of an interglacial-glacial transition as it is to understand the transition from a glaciation to an interglaciation. The transition from the last glacial maximum ca. 21,000 yr to the present interglaciation has been intensively studied and represents one of the best documented events of Earth's history. By comparison, the geologic record of the last interglacial-glacial transition is poorly preserved in discontinuous sediments, and chronometric methods for accurately dating this interval remain elusive. Despite these limitations, there have been a number of major advances in our understanding of this transition over the last 10–15 years. Chapters in this volume represent a current and balanced description of these advances.

We were saddened to hear of the passing of Dr. R. P. Goldthwait while this volume was in press. A true Friend of the Pleistocene, Dr. Goldthwait was a preeminent glacial geologist whose contributions strongly influenced the direction of Quaternary studies, not the least of which was the subject of this volume. We dedicate this volume to his memory.

We wish to thank the many people who have been involved in the development of this volume. First and foremost, our thanks go to the 41 authors and coauthors who contributed papers. Because all contributors are active and productive scientists with busy time schedules, such a remarkable voluntary response is testimony to the interest, significance, and vitality of the subject. We regret the four-year delay from the time of the symposium to the time of publication; however, we credit not only those authors who submitted papers in a timely matter, but also those who, while somewhat tardy, persisted under gentle prodding in getting their papers in, thus ensuring the breadth of coverage that we deemed necessary to have a successful and nearly complete representation.

The rigor and clarity of these chapters are also a reflection of hard work by the many scientists who formally reviewed these chapters. Their participation in this volume is greatly appreciated.

Peter U. Clark
Corvallis, Oregon

Peter D. Lea
Brunswick, Maine

Geological Society of America
Special Paper 270
1992

The last interglacial-glacial transition in North America: Introduction

Peter U. Clark
Department of Geosciences, Oregon State University, Corvallis, Oregon 97331

INTRODUCTION

During the Quaternary Period, large ice sheets and smaller ice masses repeatedly grew and decayed in response to climate change. The record of ice-volume variations in the deep-sea oxygen-isotope record identifies more than 20 glacial cycles during the Quaternary. In contrast, much of the continental record of glaciations older than that of the last glacial cycle has been eroded, and what remains is difficult to date accurately. The record of ice-sheet growth and decay on land is therefore much more fragmentary than the record in deep-sea sediments (e.g., Sibrava and others, 1986), and the oxygen-isotope stratigraphy is commonly used as a reference for continental glaciation (e.g., Kukla, 1977; Andrews and Barry, 1978; Fulton, 1984; Richmond and Fullerton, 1986).

The transition from the last glacial maximum (ca. 18 ka) to the present interglaciation has been intensively studied and represents one of the best documented events of Earth's history. Of equal importance is understanding the transition from the last interglaciation to the last glaciation. Continuous records of climate change in deep-sea sediments and polar ice suggest that peak interglaciation ca. 125 ka (oxygen-isotope substage 5e; see oxygen-isotope time scale in Fig. 2E; Peteet and others, Chapter 5, this volume) was followed by periods of global cooling and increased ice volume during isotope stage 5, culminating with a major glaciation during isotope stage 4. This transition is poorly represented in discontinuous continental sediments, and chronometric methods for accurately dating this interval remain elusive. Continental records of glaciation are of particular interest, however, because they provide direct evidence of when and where ice accumulated and how extensive glaciation was. Continental records of glaciation following the last interglaciation are thus of crucial significance for (1) linking climate forcing functions (orbital variations, CO_2) to response of the cryosphere, (2) providing boundary conditions for numerical ice-sheet and climate modeling, and (3) evaluating the ice-volume signal in sea level and marine oxygen-isotope records.

UNDERSTANDING THE LAST INTERGLACIAL-GLACIAL TRANSITION IN NORTH AMERICA

The frequency of glacial cycles identified in deep-sea oxygen-isotope records corresponds to variations in insolation received on Earth caused by changes in the geometry of Earth's orbit about the sun, indicating a causal relation (Hays and others, 1976; Imbrie and others, 1984; Ruddiman and others, 1986). It is generally recognized, however, that insolation variations alone can not explain the magnitude of the climatic response (Peteet and others, this volume), particularly the 100 ka cycle of ice volume that has dominated the climate signal for the past 700 ka but has no power at this period in the orbital forcing (Hays and others, 1976; Kominz and Pisias, 1979; Imbrie, 1985).

Hays and others (1976) argued that nonlinear responses in the climate system must have amplified orbital forcings in order to explain the relation between insolation variations and growth and decay of the Northern Hemisphere ice sheets (cf. Wigley, 1976; Imbrie and Imbrie, 1980). Many models have examined the physical response of ice sheets to insolation variations by including such nonlinear feedback processes as glacial isostatic rebound, accumulation/ablation interactions, and calving (e.g., Oerlemans, 1980; Pollard and others, 1980; Imbrie and Imbrie, 1980; Budd and Smith, 1981; Birchfield and others, 1981; Denton and Hughes, 1983; Hyde and Peltier, 1985). Other models have examined the role of the oceans in ice-sheet inception (Lamb and Woodruffe, 1970; Johnson and Andrews, 1979; Ruddiman and McIntyre, 1979, 1981; Broecker and Denton, 1989). Several models have investigated the effect of changing atmospheric gases and aerosols in amplifying insolation forcing (Manabe and Broccoli, 1985; Rind and others, 1989; Peteet and others, this volume).

As recorded in the deep-sea oxygen-isotope stratigraphy, the last glacial cycle corresponds to the time since the last period of minimum ice volume (the last interglaciation, isotope substage 5e; CLIMAP Project Members, 1984; see Muhs, this volume) to the present interglaciation (isotope stage 1).

Clark, P. U., 1992, The last interglacial-glacial transition in North America: Introduction, *in* Clark, P. U., and Lea, P. D., eds., The Last Interglacial-Glacial Transition in North America: Boulder, Colorado, Geological Society of America Special Paper 270.

The most complete and best-dated segment of this cycle in North America is that of the last deglaciation (e.g., Ruddiman and Wright, 1987a). Well-preserved sediments and landforms are within the range of ^{14}C dating, and the major patterns of ice retreat can thus be identified and followed through time in some detail (Dyke and Prest, 1987). As a result, the North American terrestrial record of the transition from the last glaciation to the present interglaciation contains a large body of spatial and temporal data that provide important insights into the response of the Earth to forcings and feedbacks involved in deglaciation (Ruddiman and Duplessy, 1985; Ruddiman and Wright, 1987b; COHMAP Members, 1988; Broecker and Denton, 1989).

In contrast, there is a relative dearth of comparable information about the transition from the last interglaciation to the last glaciation in North America. The stratigraphic record of ice-sheet inception during the last glacial cycle is limited by poor preservation and, more important, by a paucity of dating techniques that are applicable to sediments and fossils from this interval.

As with the record of deglaciation, however, spatial and temporal information on the inception of North American ice sheets and glaciers is critical to understanding and modeling the response of the Earth to climatic forcings (Budd and Smith, 1981, 1987; Rind and others, 1989; Peteet and others, this volume). Therefore, in order to develop a model that explains a complete glacial cycle with respect to insolation variations, internal feedbacks, and ice-sheet response, it is necessary to have as complete an understanding as possible of the location and timing of ice sheet and glacier growth as well as disintegration.

This discussion raises four important questions in evaluating the response of the climate system, as measured by changes in ice volume, to forcing mechanisms: (1) where and when did ice sheets and smaller ice masses begin to grow following the last ice-volume minimum (isotope substage 5e) in the Northern Hemisphere; (2) how large did the ice masses grow during the early phases of the last glaciation; (3) can the continental stratigraphic record of changes in ice volume delimit the ice-volume signal in deep-sea oxygen-isotope records; and (4) how do models which incorporate various forcing functions of climate change, including ice-sheet, ocean, and atmosphere models, compare with the continental stratigraphic record of ice growth.

Chapters in this volume address these questions from the perspective of the stratigraphic record of the transition from the last interglaciation to the last glaciation in North America. We define this transition as the interval from the last period of minimum global ice volume (isotope stage 5e) to isotope stage 4. Oxygen-isotope records indicate that global ice volume reached as much as 70% of that of the last glacial maximum (isotope stage 2) during this interval (Mix, this volume).

The first four chapters provide a framework for the remainder of the volume. Goldthwait reviews the early ideas, many of them his own, that led to the recognition in the North American continental record of a major glacial event (early Wisconsinan) preceding the last glacial maximum and postdating the last interglaciation. Mix discusses the deep-sea oxygen-isotope record of isotope stages 4 and 5, emphasizing the ice-volume signal (magnitude and timing) in the isotope record. Muhs reviews sea-level fluctuations during the last interglacial-glacial transition that are recorded by marine terraces on North America and on islands close to North America. Peteet and others review climate records from ice cores, ocean sediments, and terrestrial environments bearing on the question of climate response to forcings during the last interglacial-glacial transition. The authors then summarize the major results of Rind and others (1989); the effects of insolation variations on ice-sheet growth are investigated in a general circulation model.

The remaining 16 chapters in this volume, summarized below, emphasize our current understanding of the stratigraphic record of North American ice sheets, glaciers, and pluvial lakes during the last interglacial-glacial transition. These chapters address the nature of the lithostratigraphic records from many of the key regions where this transition is best documented, and thus represent an excellent spatial sampling of most of glaciated North America (Fig. 1). In addition, these chapters focus on the chronology of the transition, which is the critical factor that must be addressed in order to resolve many of the issues concerning climate forcings and ice-sheet response. Many chapters present important new numerical-age data bearing on the timing and nature of the stratigraphic record.

The purpose of the remainder of this chapter is to present an overview of the volume with respect to the critical issues concerning the timing and nature of the last interglacial-glacial transition and implications of the North American stratigraphic record to this transition.

ICE-SHEET INCEPTION

Broecker and van Donk (1970) proposed that the sawtooth shape of the deep-sea oxygen-isotope record represented long (ca. 90 ka) and gradual but continuous periods of ice accumulation that ended abruptly with rapid (ca. 10 ka) deglaciations ("terminations"). This pattern indicates a significantly longer response of ice sheets to climatic forcings that lead to their growth in comparison to their decay, and has been emphasized in a number of models that attempt to explain the 100 ka ice-volume signal, particularly with respect to nonlinear feedbacks that may have amplified the insolation forcing during deglaciation (e.g., Imbrie and Imbrie, 1980; Pollard, 1983; Peltier, 1987).

Although the sawtooth pattern reflects a longer time period involved in growth of ice sheets to their full-glacial maximum than in their decay, a number of studies have addressed the secondary punctuations in the oxygen-isotope record that indicate rates of ice growth comparable to those associated with deglaciation during terminations. Ruddiman and McIntyre (1982) described one such event that occurred during isotope substage 7b (ca. 230 ka). They estimated that during this "interglaciation" (stage 7), continental ice grew to 30%–55% of full glacial volumes within 10 ka.

The deep-sea oxygen-isotope record and records of sea-level

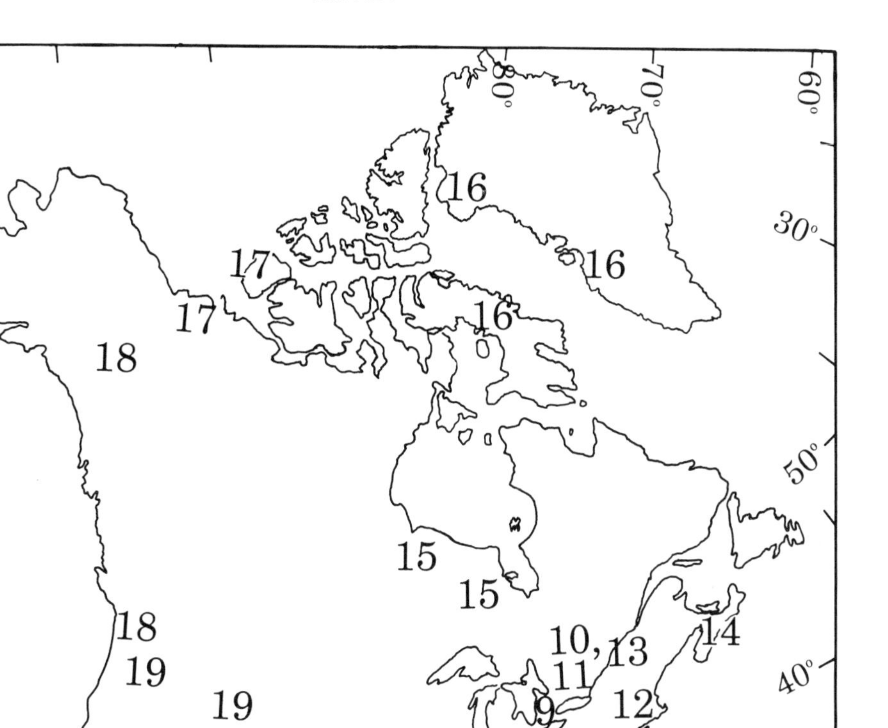

Figure 1. Areas in North America where the last interglacial-glacial transition is discussed in this volume. Numbers refer to chapters in volume (6 = Curry and Follmer; 7 = Miller and others; 8 = Szabo; 9 = Dreimanis; 10 = Eyles and Williams; 11 = Hicock and Dreimanis; 12 = Oldale and Colman; 13 = Lamothe and others; 14 = Stea and others; 15 = Thorleifson and others; 16 = Miller and others; 17 = Vincent; 18 = Clague and others; 19 = Colman and Pierce; 20 = Oviatt and McCoy; 21 = Bachhuber).

change from marine terraces also identify periods of rapid ice growth during the last glacial cycle. Estimates vary, however, as to the magnitudes and rates of ice-volume increase.

In first interpreting Quaternary deep-sea oxygen-isotope records, Emiliani (1955) emphasized the paleotemperature signal over that of ice volume. Subsequent work has shown that ice-volume related changes in ocean chemistry were the primary signal in the oxygen-isotope record (Shackleton, 1967; Shackleton and Opdyke, 1973). Delimiting the ice-volume signal in deep-sea oxygen-isotope records remains elusive, however, because the role of other contributing factors to the isotope signal cannot be readily quantified (Duplessy, 1978; Mix, 1987, this volume).

Mix identifies temperature, atmospheric vapor transport, and changes in ice-sheet isotopic composition as important variables that, in addition to ice volume, determine the oxygen-isotope ratios measured in fossil foraminifera. By evaluating the likely contribution of each of these variables, Mix estimates the magnitude of ice-volume changes in the deep-sea oxygen-isotope record during the last interglacial-glacial transition. He argues that substages 5d and 5b as well as stage 4 were times of large and rapid ice-sheet growth.

Muhs reviews evidence for changes in sea level during the last interglacial-glacial transition recorded by raised marine terraces. Uplifted terraces on Barbados, Haiti, and New Guinea indicate that sea level ca. 105 and 80 ka (substages 5c and 5a) was considerably lower than present. In contrast, terraces on the east and west coasts of North America, Bermuda, and the Bahamas suggest sea level at these times was nearly the same as present (cf. Hollin and Hearty, 1990).

The New Guinea marine-terrace record identifies a sea-level drop of ca. 60 m by 112 ka (isotope substage 5d) following the last interglacial sea-level high ca. 120 ka (Chappell and Shackleton, 1986), or 50% of the full-glacial sea-level drop (121 ± 5 m; Fairbanks, 1989). The Barbados sea-level record (Bard and others, 1990) identifies a sea-level drop to −70 m below present during isotope stage 4 following a sea level at ~−15 to −19 m during substage 5a (Muhs, this volume). These estimates of sea-level change during substage 5d and stage 4 agree with the estimates of ice-volume interpreted from the oxygen-isotope record (Mix, this volume). According to the New Guinea and Barbados records, sea level never fully recovered to the present level (cf. Muhs, this volume), but instead was punctuated by rapid fluctuations that culminated with a drop of 30–70 m to the last glacial maximum sea-level lowstand.

Peteet and others review evidence in high-resolution marine, terrestrial, and ice-core records for rapid climatic change following the last interglaciation that correspond to changes in ice volume. The timing of the climate signal corresponds to insolation variations, but the magnitude of the response to insolation forcing is unrelated, indicating nonlinear feedbacks.

If ice-volume change is faithfully recorded by the oxygen-isotope record and marine terraces, these data indicate that development of the Northern Hemisphere ice sheets to their last glacial maximum was characterized by several periods of rapid and large fluctuations of ice volume that are in phase with orbital forcing. Moreover, rates of ice growth ($\geq$1 m/100 yr of sea-level change) are comparable to the rate of decay during terminations (Fairbanks, 1989), although volume changes are of lesser magnitude.

Many workers have proposed that ice-sheet inception occurred when lowering of regional snowline intersected extensive upland plateaus of northern Canada and Fennoscandinavia ("instantaneous glacierization") (Brooks, 1926; Ives, 1957; Lamb and Woodruffe, 1970; Barry and others, 1975; Ives and others, 1975).

Using the Goddard Institute for Space Studies (GISS) general circulation model (GCM), Peteet and others investigated the role of insolation forcing in the formation of the Laurentide Ice Sheet at 116 ka, when summer solar insolation was reduced and sea-level and oxygen-isotope records indicate a period of rapid and major ice growth. Despite enhanced reduction in summer and fall insolation as well as in CO_2 concentrations, their model failed to maintain ice on land (cf. Rind and others, 1989). When sea-surface temperatures were lowered to full-glacial values, the model maintained ice only on northern Baffin Island. Reconstructed sea-surface temperatures in the northwest Atlantic (Ruddiman and McIntyre, 1979, 1981), Baffin Bay and nearshore Baffin Island (Miller and others, this volume), however, suggest relatively warm oceans during ice-sheet inception. Peteet and others conclude that either some as-yet unidentified forcing factor must be involved in ice-sheet inception, or the GCM is too insensitive to insolation variations.

Records of ice volume indicate a more complex structure to a glacial cycle than the simple sawtooth pattern described by Broecker and van Donk (1970). Instead, development of the ice sheets to their full-glacial extent should be regarded as a series of step-like "glacial buildup periods" (Andrews and Barry, 1978). The GCM experiments by Rind and others (1989) and Peteet and others indicate that unknown but probably complex nonlinear feedbacks are required to amplify orbital forcing in the growth of ice sheets.

In examining the structure of the last glacial cycle, it is thus equally important to address those times of rapid ice growth as well as rapid ice decay. This is best done from the perspective of the land record, where information on the timing, location, and magnitude of ice-volume change can define oxygen-isotope and sea-level records as well as provide critical input for and validation of climate models.

Within this context, North America is significant because it hosted the Laurentide Ice Sheet, or the largest Northern Hemisphere ice mass, as well as the Cordilleran Ice Sheet. In addition, the large number of smaller North American ice masses provide a more sensitive record of glacier response to climate change than do the large ice sheets. The record of large pluvial lakes in the southwestern United States contains paleoclimatic information that is independent of but complementary to the glacial record.

LAST INTERGLACIAL-GLACIAL TRANSITION IN NORTH AMERICA

The stratigraphic perspective of the last interglacial-glacial transition in North America has evolved significantly since it was first examined by T. C. Chamberlin and Frank Leverett nearly 100 years ago. Our current understanding of this transition, and thus of the structure of the last glacial cycle, has been strongly influenced by this early work. We briefly trace the development of this thinking, identifying in particular those advances that have had a continuing effect on interpretation of the glacial record.

Early ideas

Prior to the application of radiocarbon dating to continental records and oxygen-isotope analyses to deep-sea sediments, the last glacial cycle was characterized as a single major glaciation, but with many ice-margin fluctuations of unknown duration and extent. Chamberlin (1895) introduced the term "Wisconsin formation" for well-expressed end moraines deposited by the Lake Michigan Lobe of the Laurentide Ice Sheet associated with this glaciation.

Leverett (1898) introduced the term "Sangamon" for the soil developed in glacial deposits ("Illinoian till sheet"; Leverett, *in* Chamberlin, 1896) underlying deposits of the Wisconsin formation. Leverett (1898) recognized the stratigraphic significance of the Sangamon Soil as representing a significant break in the glacial record, or an interglaciation.

Deposits of the "Wisconsin formation" were initially subdivided into stratigraphic units based on major end moraines that represented their maximum limit. Discordances in the alignment of moraines were interpreted as a result of age differences. Chamberlin (1893) mentioned that Leverett identified two divisions of the deposits that Chamberlin (1895) later classified as Wisconsin. In discussing the "Wisconsin stage"[1], Leverett (1899) formally

[1]Note differing meanings of the term "stage" as defined by the 1933 and 1961 Stratigraphic Codes and as applied to Quaternary deposits.

defined the "Early Wisconsin" and "Late Wisconsin" from moraines and associated deposits at the southern end of the Lake Michigan Lobe. Leverett (1929) later recognized an additional subdivision of moraines and deposits, which he referred to as "Middle Wisconsin." Chamberlin (1893, 1895) and Leverett (1899) interpreted the stratigraphic units as representing significant "shifting of the ice lobes . . . and a retreat of some consequence between the . . . ice advances" (Leverett, 1899, p. 317).

Leighton (1933; Kay and Leighton, 1933) revised the classification of the Wisconsin, replacing the terms "Early, Middle, and Late" with "Tazewell," "Cary," and "Mankato," respectively, which were derived from local geographic areas, as well as included the "Iowan" substage at the base of the Wisconsin stage (Fig. 2A). Leighton and Willman (1950) and Leighton (1957) subsequently added the "Farmdale" and "Valders" substages (Fig. 2B).

Frye and others (1948) first added the adjectival ending "an" to Wisconsin to identify the stage as a time-stratigraphic unit.

Identification of early Wisconsin glaciation in North America

As of 1950, Wisconsin sediments were thought to represent a single major expansion of the Laurentide Ice Sheet following the last (Sangamon) interglaciation, although substages of the Wisconsin were interpreted as representing readvances of the ice margin. Subsequent radiocarbon dating has confirmed that the substages were associated with one glaciation, except for the Iowan substage, which was identified as pre-Wisconsin (Ruhe, 1969).

After 1950, three significant events occurred that dramati-

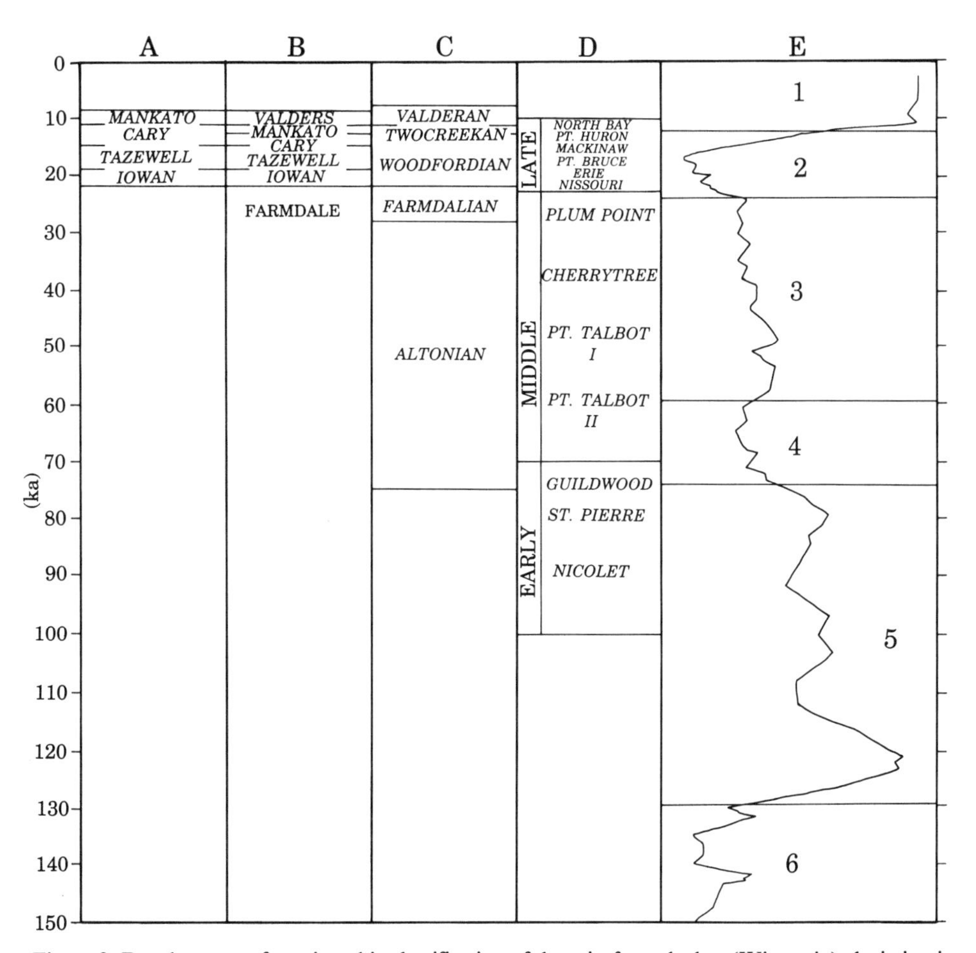

Figure 2. Development of stratigraphic classification of deposits from the last (Wisconsin) glaciation in the midcontinent of North America. A: Classification of deposits in Illinois by Leighton (1933). Time subdivision after Willman and Frye (1970). B: Classification of deposits in Illinois by Leighton (1957). Time subdivision after Willman and Frye (1970). C: Classification of deposits in Illinois by Willman and Frye (1970). D: Classification of deposits in the eastern Great Lakes and St. Lawrence Valley by Dreimanis and Karrow (1972), with time scale revised by Terasmae and Dreimanis (1976). E: Oxygen isotope curve through isotope stage 6 (after Martinson and others, 1987).

cally changed our understanding of the last glacial cycle (see Goldthwait, this volume). Radiocarbon dating was applied to fossiliferous sediments associated with glacial sediments, thus establishing a numerical chronology of late Quaternary events (Flint and Rubin, 1955; Flint, 1956). Oxygen-isotope ratios were measured on foraminifera in deep-sea cores, thus identifying a nearly continuous record of Quaternary climate change (Emiliani, 1955). Detailed investigations of the stratigraphic record of the Laurentide Ice Sheet were conducted in areas other than the classical area of Illinois, thus providing a more representative history of the ice sheet (e.g., Dreimanis, 1957; Goldthwait, 1959; McDonald, 1969; Gadd, 1971; Skinner, 1973; Miller and others, 1977).

Two important interpretations bearing on the structure of the last glacial cycle were drawn from these various studies. First, radiocarbon dating demonstrated that the oldest Wisconsin sediments identified up to 1950 by Leighton and his predecessors were deposited after 25,000 yr B.P. (Fig. 2), and the Laurentide Ice Sheet reached its maximum extent 18,000–20,000 yr B.P. (Flint and Rubin, 1955). Second, radiocarbon dating identified an extensive glaciation that preceded the last glacial maximum but apparently postdated the last interglaciation. Weak soil development in deposits of this glaciation and overlying organic-rich deposits with radiocarbon ages ⩾30,000 yr B.P. indicated glacial retreat prior to the readvance of the ice sheet ca. 18,000 yr B.P. Evidence of an intervening glacial event also came from the deep-sea oxygen-isotope record (Emiliani, 1955) and pluvial lake deposits in the western United States (Flint and Gale, 1958).

Identification of this penultimate glacial event in the stratigraphic record of the Ohio River basin (Forsyth, 1957, 1958, 1965; Forsyth and LaRoque, 1956; Goldthwait, 1959) and the Erie basin (Dreimanis, 1957, 1960) led to subdivision of the Wisconsin glaciation into predominantly glacial early and late Wisconsin substages, separated by a cool, partly glacial interstadial complex (middle Wisconsin) (Dreimanis and Karrow, 1972; Dreimanis and Goldthwait, 1973) (Fig. 2D).

Radiocarbon dating of deposits in Illinois associated with Leverett's early, middle, and late substages of the Wisconsin glaciation (Leighton's Tazewell, Cary, and Mankato) indicated they were associated with the younger part of the last glacial stage (<25,000 yr B.P.) (Fig. 2B). Frye and Willman (1960) redefined these deposits as belonging to the Woodfordian substage (Fig. 2C). Loess and glacial deposits were identified in Illinois, however, with radiocarbon ages ⩾28,000 yr B.P., but overlying the Sangamon Soil (Frye and Willman, 1960; Kempton and Hackett, 1963; Frye and others, 1968). Glacial and eolian (loess) deposits from this period, defined as the Altonian substage (Frye and Willman, 1960), are separated from glacial deposits of the Woodfordian substage by organic-rich silts and peaty deposits belonging to the Farmdalian substage (Frye and Willman, 1960) (Fig. 2C).

Although the stratigraphic record of the last glacial stage in the eastern Great Lakes region (Dreimanis and Karrow, 1972) differs in detail from that described in the Lake Michigan basin (Frye and Willman, 1973) (Fig. 2), both records recognized a significant period of glaciation following the last interglaciation but separated from the last glacial maximum by a significant interstadial. This tripartite subdivision of the last glacial cycle is also represented in the ice-volume signal of the oxygen-isotope deep-sea record (isotope stages 4, 3, and 2) (Shackleton and Opdyke, 1973; Mix, this volume) (Fig. 2E).

Reappraisal of early Wisconsin glaciation

A number of new developments have occurred since the subdivision of the Wisconsin(an) stage in the benchmark publications by Willman and Frye (1970), Dreimanis and Karrow (1972), and Dreimanis and Goldthwait (1973). Improvements and new developments in dating methods, reinvestigation of key stratigraphic relations, and description and dating of stratigraphic records from other sectors of the Laurentide Ice Sheet indicate that the concept of early Wisconsin glaciation may be significantly different and more complex than that suggested up to the early 1970s (Clark and Lea, 1986).

Most chapters in this volume address these new developments from the classic areas of the Lake Michigan, Ohio River, Erie, and Ontario basins, as well as from a number of other areas where there is a stratigraphic record of the last interglacial-glacial transition. Overall, these chapters provide a detailed overview of the structure of this transition in North America. In particular, these chapters are of critical importance to establishing the temporal and spatial patterns of growth and development of ice sheets, glaciers, and pluvial lakes at the close of the last interglacial.

The purpose of the following discussion is to summarize the major results of these chapters as they bear on the question of the timing and nature of the last interglacial-glacial transition in North America.

Laurentide Ice Sheet

Chapters addressing the area covered by the Laurentide Ice Sheet provide a continental-scale perspective of events involving its growth and development. Thorleifson and others present stratigraphic evidence and amino-acid data for alternating glacial and interglacial environments in the Hudson Bay lowland, near the former center of the Laurentide Ice Sheet. An interglacial marine episode, correlated to isotope substage 5e, was followed by west-northwest glacial flow, attributed to ice-sheet inception in Quebec during stage 5. If the thermoluminescence (TL) chronology is correct, this advance is recorded by the Rocksand and Amery tills. A subsequent interval of deglaciation, marine incursion, and isostatic recovery led to ice-free conditions in the lowlands until isotope stage 3. If the TL ages are too young, then glaciation during stage 5 is recorded by the Long Spruce, Sachigo, and Adam tills, and the Hudson Bay lowlands remained ice-covered throughout the remainder of the last glacial (stages 4, 3, and 2), although flow shifted to the southwest and south in response to the migration of an ice divide across the Hudson Bay region.

The most extensive advance of the ice sheet following the last interglacial occurred in the high latitudes during stage 5. Miller and others summarize the stratigraphy of the land and marine records of the Baffin Bay region. Sediments from Baffin Island (Ayr Lake till) and northwest Greenland indicate an extensive ice-sheet advance onto the continental shelf during early isotope stage 5, and restricted ice extent throughout the remainder of the last glacial cycle (isotope stages 4, 3, and 2). Nearshore marine records identify warm marine conditions prior to and following this glacial advance. Records of glacial erosion from sediment cores support extensive ice surrounding Baffin Bay during isotope stage 5.

Vincent summarizes the stratigraphic, paleoecologic, and geochronologic record of ice-sheet glaciation of the western Canadian arctic. He concludes that an extensive advance of the northwestern margin of the Laurentide Ice Sheet, beyond late Wisconsinan[2] limits, occurred during isotope stage 5 or 4, depositing the Bar Harbour, Mercy, Jesse, and Sachs tills. Vincent notes that various workers have proposed different chronologies for glacial advances in this region, and suggests that the apparent differences might be reconciled if Keewatin sector ice persisted more or less stably in northwestern Canada during much of isotope stages 5 through 2.

Stratigraphic records from Quebec and from Nova Scotia indicate development of an independent ice cap over the Appalachian uplands during isotope stage 4. Lamothe and others identify evidence in the northeastern Appalachians for an ice cap that developed during the early Wisconsinan and was correlated to isotope stage 4. This event is recorded by the Chaudiere till, which overlies sediments of the last interglacial (Massawippi Formation).

Stea and others present evidence for a pre–Late Wisconsinan history of glaciation, high sea levels, and vegetation in Nova Scotia, using U/Th, amino-acid, and electron-spin-resonance age estimates for shells and wood. Formation of a raised marine platform and deposition of forest beds during isotope stage 5 (Sangamonian) was followed by at least four phases of ice flow. Stea and others associate the first post-interglacial advance in Nova Scotia with an ice cap over the Appalachians. This advance, correlated to isotope stage 4, deposited the Red Head and East Milford tills as well as various shelly diamictons, and may have extended to the edge of the continental shelf, where it deposited Scotian Shelf drift.

The Laurentide Ice Sheet subsequently advanced into the St. Lawrence Valley, depositing the Levrard till, and coalesced with the ice cap over the Appalachians (Lamothe and others), while independent ice caps remained active over Nova Scotia throughout the last glaciation (Stea and others).

Stratigraphic records of the last glacial cycle from the southern area of the Laurentide Ice Sheet suggest that the ice-sheet margin remained retracted until the late Wisconsin, when it advanced to its maximum position. Many tills formerly interpreted as recording an early Wisconsin glaciation are now thought to be older.

Oldale and Colman critically examine long-held assumptions of extensive early Wisconsinan (isotope stage 4) ice-sheet glaciation of New England. They conclude that the penultimate glacial advance into southern New England occurred during the Illinoian (isotope stage 6), and that ice advanced no farther than northern New England during isotope stages 4 and 3. Lower tills of southern New England, including the Montauk till member, therefore probably predate the last interglaciation.

Eyles and Williams summarize recent work from the glacial sedimentary record exposed in the Toronto, Ontario, area. Sediments and fossils from the Don Beds and Scarborough Formation record climatic deterioration during the last interglaciation, which they correlate to isotope stage 5. Early Wisconsin (isotope stage 4) glaciation is represented by glaciolacustrine sediments, including the Sunnybrook "Till." Eyles and Williams argue that the Laurentide Ice Sheet remained north of Toronto until the late Wisconsin, when it overrode the area and advanced to its last glacial maximum extent.

Hicock and Dreimanis present the results of their detailed reinvestigation of Sunnybrook drift in the Toronto area. They agree with Eyles and Williams that the unit includes glaciolacustrine members, but argue that clast fabrics and small-scale erosional and deformational features also define a till member formed by lodgement and by deformation of glaciolacustrine sediments below a grounded lobe of the Laurentide Ice Sheet. Hicock and Dreimanis thus contest the claim of Eyles and Williams that ice remained north of Toronto until the late Wisconsin. Note, however, that Dreimanis's reinterpretation of the glacial stratigraphy in the Erie basin (this volume) reduces the difference between alternative early Wisconsin ice margins to only about 100 km.

Dreimanis proposes that the Bradtville drift, which had long been considered to represent early Wisconsinan glaciation in the Erie basin, is instead pre-Sangamonian. Early Wisconsinan glaciation, correlated to isotope stage 4, is indicated by varved glaciolacustrine sediments (Member B of the Tyrconnel Formation), which Dreimanis attributes to damming of the eastern outlet of Lake Erie by the Laurentide Ice Sheet.

Szabo reevaluates the stratigraphic record of the last interglacial-glacial transition from northern Ohio. On the basis of recent surficial mapping and new geochronologic data, Szabo argues that glacial sediments (Titusville, Garfield Heights, and Millbrook tills) previously interpreted as early Wisconsinan are pre-Wisconsinan, and evidence of early Wisconsinan glaciation is absent.

Miller and others discuss new geochronological data

[2]Because there are several schemes used in naming Quaternary chronostratigraphic units (e.g., Frye and others, 1968; Fulton, 1984; Richmond and Fullerton, 1986), the editors of this volume have not tried to enforce any one scheme over another in any of the chapters. In summarizing each of the chapters in this volume, I follow the specific usage of the terms as used in each chapter. The editors strongly support, however, attempts to reconcile the stratigraphic problems unique to the Quaternary (e.g., Johnson, 1989; Johnson and others, 1991).

that indicate the Sydney Soil, which was one of the original cornerstones in identifying early Wisconsin glaciation (Forsyth, 1965; Goldthwait, this volume), may be equivalent to the Sangamon Soil in Illinois. These new data also indicate that glacial sediments in southwestern Ohio (Fairhaven till) and southeastern Indiana (Whitewater till) previously considered middle and early Wisconsinan are pre-Wisconsinan.

Curry and Follmer review sedimentological, pedological, and paleontological records of the last interglacial-glacial transition in Illinois. The earliest record of the Laurentide Ice Sheet is indicated by the Roxanna silt (loess), which first was deposited by 50 ka. The Laurentide Ice Sheet, however, did not advance into Illinois until 25 ka.

Cordilleran Ice Sheet

Clague and others review lithostratigraphic and geochronologic evidence for the extent of the Cordilleran Ice Sheet in western Canada and northwestern Washington during the early Wisconsinan (correlated to isotope stage 4). Available data suggest moderately extensive early Wisconsinan ice in northwestern Washington and British Columbia. In contrast, data from the southwestern Yukon Territory indicate that drift sheets beyond the late Wisconsinan ice limit there probably correlate with isotope stage 6. The extent of early Wisconsinan ice behind late Wisconsinan limits in the Yukon, however, is not well known. Clague and others also summarize paleoecological data from Sangamonian sediments in Yukon Territory, British Columbia, and Washington. All areas include evidence for climates warmer and drier than those of today.

Mountain glacier and pluvial lakes of the western United States

Important data on the nature of the last interglacial-glacial transition south of ice-sheet limits in the western United States are provided by the record of glacier advances in diverse mountain ranges and pluvial-lake fluctuations in intervening basins. Colman and Pierce summarize weathering-rind data that indicate variable response of mountain glaciers during the early Wisconsinan (correlated to isotope stage 4). Some ranges record early Wisconsinan glacial advances beyond late Wisconsinan limits, whereas others have no record of extensive early Wisconsinan glaciation. Colman and Pierce point out that small mountain glaciers may have responded more quickly and more variably to short but intense climatic excursions than did large ice sheets, and hypothesize that interregional differences may relate to local climatic and/or hypsometric effects.

Oviatt and McCoy review pluvial records from the Great Basin of the western United States. Where the chronology is well determined, expansion of pluvial lakes appears to be approximately coincident with glaciations. There is some geochronologic evidence for lake-level rise in the Bonneville basin during the early Wisconsin, probably equivalent to isotope stage 4. In addition, there may have been an early Wisconsin pluvial lake in the Lahontan basin, but the geochronologic control is equivocal. Relatively large early Wisconsin lakes may have formed in the Searles and Panamint basins.

Oviatt and McCoy also review glacial records from mountain ranges in or near the Great Basin (cf. Colman and Pierce, this volume). Only moraines from the Sierra Nevada may record an early Wisconsin glaciation as extensive as the late Wisconsin in the region surrounding the Great Basin.

Bachhuber uses sedimentological and paleoecological data to present a detailed record of pre–late Wisconsinan paleolimnologic change from the Estancia basin of New Mexico. Unlike the deep, freshwater lake associated with full-pluvial conditions of the late Wisconsinan, conditions during the early and/or middle Wisconsinan were characterized by alternating wet and dry playas and shallow saline lakes. Bachhuber notes that climatic excursions recorded in the pre–late Wisconsinan sediments were minor, and that the saline-lake complexes probably resulted from cold/dry conditions rather than cold/wet conditions inferred for full pluvials.

DISCUSSION

A significant problem acknowledged by many chapters in this volume addressing the stratigraphy of the last interglacial-glacial transition centers on dating methods. In particular, the dearth of suitable material which, when available, is subject to errors in accuracy and precision presents a major obstacle to developing a well-delimited chronology. Nevertheless, recent application of new dating methods (e.g., amino-acid racemization, TL, weathering rinds) has significantly revitalized and improved our understanding of this critical time period.

The stratigraphic record of the last interglacial-glacial transition in North America provides important information toward understanding when and where ice sheets, glaciers, and pluvial lakes developed in response to climatic forcings involved in the initiation of the last glaciation. Stratigraphic records from the Hudson Bay lowlands indicate that the Laurentide Ice Sheet developed in Quebec east of Hudson Bay during isotope stage 5, perhaps as early as substage 5d. Along its northeastern margin on Baffin Island, the ice sheet advanced to its maximum position of the last glaciation early during isotope stage 5. The ice sheet also reached its maximum extent along its northwestern margin early during the last glaciation, perhaps during isotope stage 5.

In contrast, stratigraphic records from the southern areas of the ice sheet do not record evidence of glaciation during isotope stage 5, although Lamothe and others suggest that the ice sheet may have advanced into the St. Lawrence Valley during stage 5. An independent ice cap developed in the Appalachians during isotope stage 4 and advanced as far as Nova Scotia before the ice sheet coalesced with it in southeastern Quebec. Stratigraphic records from Maine to Illinois are consistent in suggesting that the ice-sheet margin did not advance to its maximum position until the late Wisconsin (isotope stage 2).

These data indicate that fluctuations of the northern margin of the Laurentide Ice Sheet were out of phase with fluctuations of its southern margin. The ice sheet developed and reached its maximum extent along its northern margin during isotope stage 5, and subsequently remained at some position behind that margin throughout the remainder of the last glacial cycle. In contrast, the southern margin of the ice sheet remained retracted until it advanced to its maximum position during isotope stage 2.

Stratigraphic records of the Cordilleran Ice Sheet indicate (1) no evidence for the ice sheet until isotope stage 4, and (2) that fluctuations of its margins were out of phase with those of the Laurentide Ice Sheet. During the last glaciation, therefore, the northern margin of the ice sheet did not reach its maximum extent until the late Wisconsin (isotope stage 2), whereas its southern margin was possibly as extensive during the early Wisconsin (isotope stage 4) as during the late Wisconsin. Because the position of the early Wisconsin margin cannot be precisely identified, however, the absolute differences between its positions and those of the late Wisconsin are not known.

Stratigraphic records of mountain glaciers from the western United States indicate a variable response to climate change during the last glacial cycle; some glaciers developed to their maximum extent during the early Wisconsin (isotope stage 4), whereas others remained retracted until the late Wisconsin. Similarly, pluvial lake records identify variable responses, although the largest lakes (Bonneville and Lahontan) did not reach their maximum sizes until the late Wisconsin.

These records provide important spatial and temporal data on the development of North American ice sheets and glaciers that are required for constructing modeling studies of ice-sheet growth in response to climatic forcing (Budd and Smith, 1981, 1987; Rind and others, 1989; Peteet and others, this volume). Thus, evidence for growth of the Laurentide Ice Sheet during isotope stage 5 underscores the complex but unknown nonlinear feedbacks necessary to amplify orbital forcing of ice-sheet growth during this time, as discussed by Peteet and others (this volume).

Of the Northern Hemisphere ice masses, the Laurentide Ice Sheet was the most important contributor to glacio-eustatic fluctuations; therefore, stratigraphic records of this ice sheet are also important to delimiting sea-level records, including the oxygen-isotope ice-volume signal. Evidence for significant periods of ice growth during isotope stages 5 and 4, but maximum ice volume probably being reached during isotope stage 2, support the ice-volume estimates made by Mix (this volume). Conflicting sea-level records for the last interglacial-glacial transition, discussed by Muhs (this volume), must be reconciled with the stratigraphic record of the Laurentide Ice Sheet.

ACKNOWLEDGMENTS

Peter Lea provided summaries of chapters by Hicock and Dreimanis, Oldale and Colman, Stea and others, Vincent, Thorleifson and others, Clague and others, Colman and Pierce, and Bachhuber.

REFERENCES CITED

Andrews, J. T., and Barry, R. G., 1978, Glacial inception and disintegration during the last glaciation: Annual Reviews of Earth and Planetary Sciences, v. 6, p. 205–228.

Bard, E., Hamelin, B., Fairbanks, R. G., and Zindler, A., 1990, Calibration of the ^{14}C timescale over the past 30,000 years using mass spectrometric U-Th ages from Barbados corals: Nature, v. 345, p. 405–410.

Barry, R. G., Andrews, J. T., and Mahaffy, M. A., 1975, Continental ice sheets: Conditions for growth: Science, v. 190, p. 979–981.

Birchfield, G. E., Weertman, J., and Lunde, A. T., 1981, A paleoclimate model of the Northern Hemisphere ice sheets: Quaternary Research, v. 15, p. 126–142.

Broecker, W. S., and Denton, G. H., 1989, The role of ocean-atmosphere reorganizations in glacial cycles: Geochimica et Cosmochimica Acta, v. 53, p. 2465–2501.

Broecker, W. S., and van Donk, J., 1970, Insolation changes, ice volumes, and the O^{18} record in deep-sea cores: Reviews of Geophysics and Space Physics, v. 8, p. 169–198.

Brooks, C.E.P., 1926, Climate through the ages: New Haven, Connecticut, Yale University Press, 439 p.

Budd, W. F., and Smith, I. N., 1981, The growth and retreat of ice sheets in response to orbital radiation changes: International Association of Hydrological Sciences Publication 131, p. 369–409.

——, 1987, Conditions for growth and retreat of the Laurentide Ice Sheet: Geographie Physique et Quaternaire, v. 41, p. 279–290.

Chamberlin, T. C., 1893, The diversity of the glacial period: American Journal of Science, v. 45, p. 171–200.

——, 1895, The classification of American glacial deposits: Journal of Geology, v. 3, p. 270–277.

——, 1896, Nomenclature of glacial formations: Journal of Geology, v. 4, p. 872–876.

Chappell, J., and Shackleton, N. J., 1986, Oxygen isotopes and sea level: Nature, v. 324, p. 137–138.

Clark, P. U., and Lea, P. D., 1986, Reappraisal of early Wisconsin glaciation in North America: Geological Society of America Abstracts with Programs, v. 18, p. 565.

CLIMAP Project Members, 1984, The last interglacial ocean: Quaternary Research, v. 21, p. 123–224.

COHMAP Members, 1988, Climatic changes of the last 18,000 years: Observations and model simulations: Science, v. 241, p. 1043–1052.

Denton, G. H., and Hughes, T. J., 1983, Milankovitch theory of ice ages: Hypothesis of ice-sheet linkage between regional insolation and global climate: Quaternary Research, v. 20, p. 125–144.

Dreimanis, A., 1957, Stratigraphy of the Wisconsin glacial stage along the northwestern shore of Lake Erie: Science, v. 126, p. 166–168.

——, 1960, Pre-classical Wisconsin in the eastern portion of the Great Lakes region, North America: International Geological Congress, 21st, Copenhagen, Report Session 4, p. 108–119.

Dreimanis, A., and Goldthwait, R. P., 1973, Wisconsin glaciation in the Huron, Erie, and Ontario lobes, *in* Black, R. F., Goldthwait, R. P., and Willman, H. B., eds., The Wisconsinan Stage: Geological Society of America Memoir 136, p. 71–106.

Dreimanis, A., and Karrow, P. F., 1972, Glacial history of the Great Lakes–St. Lawrence region, the classification of the Wisconsin(an) Stage, and its correlatives: International Geological Congress, 24th, Montreal, p. 5–15.

Duplessy, J. C., 1978, Isotope studies, *in* Gribben, J., ed., Climatic change: Cambridge, England, Cambridge University Press, p. 46–67.

Dyke, A. S., and Prest, V. K., 1987, Late Wisconsinan and Holocene history of the Laurentide Ice Sheet: Geographie Physique et Quaternaire, v. 41, p. 237–264.

Emiliani, C., 1955, Pleistocene temperatures: Journal of Geology, v. 63, p. 538–578.

Fairbanks, R. G., 1989, A 17,000-year glacio-eustatic sea level record: Influence of glacial melting rates on the Younger Dryas event and deep-ocean circulation: Nature, v. 342, p. 637–642.

Flint, R. F., 1956, New radiocarbon dates and late Pleistocene stratigraphy: American Journal of Science, v. 254, p. 265–287.

Flint, R. F., and Gale, W. A., 1958, Stratigraphy and radiocarbon dates at Searles Lake, California: American Journal of Science, v. 256, p. 689–714.

Flint, R. F., and Rubin, M., 1955, Radiocarbon dates of pre-Mankato events in eastern and central North America: Science, v. 121, p. 649–658.

Forsyth, J. L., 1957, "Early" Wisconsin in Ohio: Geological Society of America Bulletin, v. 68, p. 1728.

——, 1958, New exposure of the buried "Sidney-type" soil along upper Brush Creek, Ohio: Geological Society of America Bulletin, v. 69, p. 1565.

——, 1965, Age of the buried soil in the Sidney, Ohio area: American Journal of Science, v. 263, p. 251–297.

Forsyth, J. L., and LaRocque, J.A.A., 1956, Age of the buried soil at Sidney, Ohio: Geological Society of America Bulletin, v. 67, p. 1696.

Frye, J. C., and Willman, H. B., 1960, Classification of the Wisconsinan Stage in the Lake Michigan glacial lobe: Illinois Geological Survey Circular 285, 16 p.

——, 1973, Wisconsinan climatic history interpreted from Lake Michigan Lobe deposits and soils, *in* Black, R. F., Goldthwait, R. P., and Willman, H. B., eds., The Wisconsinan Stage: Geological Society of America Memoir 136, p. 135–152.

Frye, J. C., Swineford, A., and Leonard, A. B., 1948, Correlation of Pleistocene deposits of the central Great Plains with the glacial section: Journal of Geology, v. 56, p. 501–525.

Frye, J. C., Willman, H. B., Rubin, M., and Black, R. F., 1968, Definition of Wisconsinan Stage: U.S. Geological Survey Bulletin 1274-E, p. E1–E22.

Fulton, R. J., 1984, Summary: Quaternary stratigraphy of Canada: Geological Survey of Canada Paper 84-10, p. 1–5.

Gadd, N. R., 1971, Pleistocene geology of the St. Lawrence Lowland: Geological Survey of Canada Memoir 359, 153 p.

Goldthwait, R. P., 1959, Scenes in Ohio during the last Ice Age: Ohio Journal of Science, v. 59, p. 193–216.

Hays, J. D., Imbrie, J., and Shackleton, N. J., 1976, Variations in the Earth's orbit: Pacemaker of the ice ages: Science, v. 194, p. 1121–1132.

Hollin, J. T., and Hearty, P. J., 1990, South Carolina interglacial sites and stage 5 sea levels: Quaternary Research, v. 33, p. 1–17.

Hyde, W. T., and Peltier, W. R., 1985, Sensitivity experiments with a model of the ice age cycle: The response to harmonic forcing: Journal of Atmospheric Science, v. 44, p. 1351–1374.

Imbrie, J., 1985, A theoretical framework for the Pleistocene ice ages: Geological Society of London Journal, v. 142, p. 417–432.

Imbrie, J., and Imbrie, J. Z., 1980, Modeling the climatic response to orbital variations: Science, v. 207, p. 943–953.

Imbrie, J., and eight others, 1984, The orbital theory of Pleistocene climate: Support from a revised chronology of the marine ^{18}O record, *in* Berger, A. L., Imbrie, J., Hays, J. D., Kukla, G., and Saltzman, B., eds., Milankovitch and climate, Part 1: Dordrecht, D. Riedel, p. 269–305.

Ives, J. D., 1957, Glaciation of the Torngat Mountains: Geographical Bulletin, v. 10, p. 67–87.

Ives, J. D., Andrews, J. T., and Barry, R. G., 1975, Growth and decay of the Laurentide Ice Sheet and comparisons with Fenno-Scandinavia: Die Naturwissenschaten, v. 62, p. 118–125.

Johnson, R. G., and Andrews, J. T., 1979, Rapid ice-sheet growth and initiation of the last glaciation: Quaternary Research, v. 12, p. 119–134.

Johnson, W. H., 1989, Quaternary glacial stratigraphic classifications: Chamberlin's legacy, their evolution, and the need for reform: Geological Society of America Abstracts with Programs, v. 21, no. 6, p. A123.

Johnson, W. H., Hansel, A. K., Follmer, L. R., and Curry, B. B., 1991, Late Quaternary temporal classification in Illinois: Geochronologic or diachronic? Geological Society of America Abstracts with Programs, v. 23, no. 3, p. 19–20.

Kay, G. F., and Leighton, M. M., 1933, Eldoran epoch of the Pleistocene Period: Geological Society of America Bulletin, v. 44, p. 669–674.

Kempton, J. P., and Hackett, J. E., 1963, Radiocarbon dates from the pre-Woodfordian Wisconsinan of northern Illinois: Geological Society of America Special Paper 76, p. 91.

Kominz, M. A., and Pisias, N. G., 1979, Pleistocene climate: Deterministic or stochastic?: Science, v. 204, p. 171–173.

Kukla, G. J., 1977, Pleistocene land-sea correlations: I. Europe: Earth Science Reviews, v. 13, p. 307–374.

Lamb, H. H., and Woodruffe, A., 1970, Atmospheric circulation during the last Ice Age: Quaternary Research, v. 1, p. 29–58.

Leighton, M. M., 1933, The naming of the subdivisions of the Wisconsin glacial age: Science, v. 77, p. 168.

——, 1957, The Cary-Mankato-Valders problem: Journal of Geology, v. 58, p. 599–623.

Leighton, M. M., and Willman, H. B., 1950, Loess formations of the Mississippi Valley: Journal of Geology, v. 58, p. 599–623.

Leverett, F., 1898, The weathered zone (Sangamon) between the Iowan loess and Illinoian till sheet: Journal of Geology, v. 6, p. 171–181.

——, 1899, The Illinois glacial lobe: U.S. Geological Survey Monograph 38, 817 p.

——, 1929, Moraines and shorelines of the Lake Superior region: U.S. Geological Survey Professional Paper 154, p. 1–72.

Manabe, S., and Broccoli, A. J., 1985, The influence of continental ice sheets on the climate of an Ice Age: Journal of Geophysical Research, v. 90, p. 2167–2190.

Martinson, D. G., Pisias, N. G., Hays, J. D., Imbrie, J., Moore, T. C., Jr., and Shackleton, N. J., 1987, Age dating and the orbital theory of the ice ages: development of a high-resolution 0 to 300,000-year chronostratigraphy: Quaternary Research, v. 27, p. 1–29.

McDonald, B. G., 1969, Glacial and interglacial stratigraphy, Hudson Bay Lowland: Geological Survey of Canada Paper 68-53, p. 78–99.

Miller, G. H., Andrews, J. T., and Short, S. K., 1977, The last interglacial-glacial cycle, Clyde foreland, Baffin Island, N.W.T.: Stratigraphy, biostratigraphy and chronology: Canadian Journal of Earth Sciences, v. 14, p. 2824–2857.

Mix, A. C., 1987, The oxygen-isotope record of glaciation, *in* Ruddiman, W. F., and Wright, H. E., Jr., eds., North America and adjacent oceans during the last deglaciation: Boulder, Colorado, Geological Society of America, The Geology of North America, v. K-3, p. 111–135.

Oerlemans, J., 1980, Model experiments on the 100,000-year glacial cycle: Nature, v. 287, p. 430–432.

Peltier, W. R., 1987, Glacial isostasy, mantle viscosity, and Pleistocene climatic change, *in* Ruddiman, W. F., and Wright, H. E., Jr., eds., North America and adjacent oceans during the last deglaciation: Boulder, Colorado, Geological Society of America, The Geology of North America, v. K-3, p. 155–182.

Pollard, D., 1983, Ice-age simulation with a calving ice sheet model: Quaternary Research, v. 20, p. 30–48.

Pollard, D., Ingersoll, A. P., and Lockwood, J. G., 1980, Response of a zonal climate-ice sheet model to the orbital perturbations during the Quaternary ice ages: Tellus, v. 32, p. 301–319.

Richmond, G. M., and Fullerton, D. S., 1986, Introduction to Quaternary glaciations in the United States of America: Quaternary Science Reviews, v. 5, p. 3–10.

Rind, D., Peteet, D., and Kukla, G., 1989, Can Milankovitch orbital variations initiate the growth of ice sheets in a general circulation model?: Journal of Geophysical Research, v. 94, p. 12,851–12,871.

Ruddiman, W. F., and Duplessy, J. C., 1985, Conference on the last deglaciation: Timing and mechanism: Quaternary Research, v. 23, p. 1–17.

Ruddiman, W. F., and McIntyre, A., 1979, Warmth of the subpolar North Atlantic ocean during Northern Hemisphere ice-sheet growth: Science, v. 204, p. 173–175.

——, 1981, Oceanic mechanisms for the amplification of the 23,000-year ice-volume cycle: Science, v. 212, p. 617–627.

——, 1982, Severity and speed of Northern Hemisphere glaciation pulses: The limiting case?: Geological Society of America Bulletin, v. 93, p. 1273–1279.

Ruddiman, W. F., and Wright, H. E., Jr., eds., 1987a, North America and adjacent oceans during the last deglaciation: Boulder, Colorado, Geological Society of America, The Geology of North America, v. K-3, 501 p.

——, 1987b, Introduction, *in* Ruddiman, W. F., and Wright, H. E., Jr., eds., North America and adjacent oceans during the last deglaciation: Boulder, Colorado, Geological Society of America, The Geology of North America, v. K-3, p. 1–12.

Ruddiman, W. F., Raymo, M., and McIntyre, A., 1986, Matuyama 41,000-year cycles: North Atlantic Ocean and Northern Hemisphere ice sheets: Earth and Planetary Science Letters, v. 80, p. 117–129.

Ruhe, R. V., 1969, Quaternary landscapes in Iowa: Ames, Iowa State University Press, 255 p.

Shackleton, N. J., 1967, Oxygen isotope analyses and Pleistocene temperatures reassessed: Nature, v. 215, p. 15–17.

Shackleton, N. J., and Opdyke, N. D., 1973, Oxygen isotope and paleomagnetic stratigraphy of equatorial Pacific core V28-238: Oxygen isotope temperatures and ice volumes on a 10^5 and 10^6 year scale: Quaternary Research, v. 3, p. 39–55.

Sibrava, V., Bowen, D. Q., and Richmond, G. M., eds., 1986, Quaternary glaciations in the Northern Hemisphere: Quaternary Science Reviews, v. 5, 510 p.

Skinner, R. G., 1973, Quaternary stratigraphy of the Moose River basin, Ontario: Geological Survey of Canada Bulletin 225, 77 p.

Terasmae, J., and Dreimanis, A., 1976, Quaternary stratigraphy of southern Ontario, *in* Mahaney, W. C., ed., Quaternary stratigraphy of North America: Stroudsburg, Pennsylvania, Dowden, Hutchinson and Ross, Inc., p. 51–63.

Wigley, T.M.L., 1976, Spectral analysis and the astronomical theory of climatic change: Nature, v. 264, p. 629–631.

Willman, H. B., and Frye, J. C., 1970, Pleistocene stratigraphy of Illinois: Illinois State Geological Survey Bulletin 94, 204 p.

Manuscript Accepted by the Society September 6, 1991

Printed in U.S.A.

Geological Society of America
Special Paper 270
1992

Historical overview of early Wisconsin glaciation

Richard P. Goldthwait*
Department of Geology and Mineralogy, Ohio State University, Columbia, Ohio 43210

ABSTRACT

"Early Wisconsin" is a time term introduced by T. C. Chamberlin more than a century ago. Right or wrong, it has been applied to almost 100 units of drift in North America. These embrace any unit too young to bear a true weathering gumbotil or deep pebble rotting (pre-Wisconsin), and yet too weathered and covered with moderately thick loess to be fresh "classical" Wisconsin (now Wisconsinan). Several such drifts, like the long-argued Iowan, have been denied a separate place in time by showing that they are younger soil on older eroded drift like "Kansan."

The revolution in dating generated by radiocarbon measurements of organic matter, from 1950 onward, eliminated a few other "early Wisconsin" cases, including those of Chamberlin. However, from 1950 to 1970 many >35,000 or >40,000 yr B.P. radiocarbon ages reinforced an early Wisconsin possibility. In Ohio we found six kinds of evidence for early Wisconsin glaciation, but only three proved to be good. The greatest confirmation during these two decades came from sea-floor coring; sea-surface temperatures and oxygen isotopes showed one cool long Wisconsinan stage 75 to 18 ka. By 1970 more precise radiocarbon dating confirmed some actual glacial sections on land, as well, and the records of the majority agreed there were climaxes in extent of ice at 70 ka and 20 ka.

ORIGIN OF IDEAS AND TERMINOLOGY

The idea of an ice age (Pleistocene) began more than 150 years ago, but it took one-half century before the last stage (Wisconsinan) was distinguished, and another 50 years before the early Wisconsin (Altonian) concept was really brewing.[1] The ice-age concept was championed first in 1837 by Louis Agassiz, who grew up in the Alps and got interested in the erratic- and striation-studded extensions of Alpine glaciers into the Jura Mountains (Agassiz, 1840). Others like Bernhardi (1832) had toyed with the idea as to how Scandinavian ice had spread. Agassiz proceeded to convince eminent British geologists of an independent Scottish ice sheet that spread over most of England. James Geikie of Scotland took up the cudgel and discovered "several intervening periods of less Arctic conditions . . . (during which) . . . the ice gradually melted away" (Geikie, 1871, p. 552). When in the late 1840s Agassiz moved to North America, he had already influenced Edward Hitchcock to the extent that Hitchcock (1841) found evidence of an ice sheet over New England. The idea spread rapidly to the midwestern United States, where clear interglacial organic exposures are common. Whittlesey (1866) and others even recognized an "older glacial drift" by its subdued topography and weathering. It extended farther south. The discovery of interglacial soils ("gumbotil") in Iowa, Ilinois, and Indiana allowed a four-fold division like that in Europe (Newberry, 1874; and others). T. C. Chamberlin (1883) referred to the latest stage as "East Wisconsin" drift because he lived in Madison, Wisconsin, then, but the name was not formalized to "Wisconsin" until Geikie (1894) wrote American subdivisions and Chamberlin (1895) published his famous classification paper.

Early Wisconsin is another story that developed slowly. One by one, drifts appeared that did not fit the simple four-fold classification. One, which was named Iowan by Chamberlin (1895), had been discovered in 1891 by McGee and was much studied by Kay and Leighton (1933). As late as 1947, Flint's (p. 246, footnote) much-used text said: "There is now no doubt that Iowan is an integral part of Wisconsin stage." Not until 1965 did Ruhe

*Deceased.

[1]Throughout this chapter the latest, most-used substage names, as applied in Illinois now, are in parentheses.

Goldthwait, R. P., 1992, Historical overview of early Wisconsin glaciation, *in* Clark, P. U., and Lea, P. D., eds., The last interglacial-glacial transition in North America: Boulder, Colorado, Geological Society of America Special Paper 270.

and others finally dispose of the type Iowan as a much older "Kansan-Yarmouth" drift—a soil truncated by overriding late Wisconsin ice. However, the precarious and false placing of Iowan for 60 years helped to generate most ideas of early Wisconsin.

STRATIGRAPHY OF UPPER PLEISTOCENE

All across North America, from 1930 on, there was great emphasis upon detailed stratigraphy, and the number of workers multiplied many times. One very early candidate for early Wisconsin was Farmdale loess at Farm Creek in west-central Illinois (Lake Michigan lobe; Leighton, 1926). It is a dark loess just under the Tazewell-Woodfordian loess (then called Peorian, now Morton), but overlying other loesses and the Sangamonian profile. By 1962 those "other" loesses had been delineated by mineralogy all across Illinois and called "Roxana" (Frye and others, 1962). It was an even better candidate—but no till. An older till sheet (Winnabago) without Illinoian-Sangamon character was found projecting westward from under classical Tazewell-Shelbyville moraine (Shaffer, 1956) and later was explored under Woodfordian drift (Kempton, 1963).

In southeastern Indiana (Miami lobe; Gooding, 1963), a lithologically distinct Whitewater till was overridden by later Wisconsin (Woodfordian) tills. At nearby Smith Farm, these tills overlie cold-weather (spruce-oak) organic silts (Kapp and Gooding, 1964), which in turn overlie deeply leached Sangamon soils. At Sidney, in western Ohio (Miami lobe; Forsyth, 1957, 1965), a thin Miami-type soil in well-drained till was overlain by the regional till plain. Nearby at Brush Creek it also overlies a cool-climate silt and peat (pine pollen) on clay-rich middle Wisconsin accretion gley. Hamilton, Ohio, near the south end of the same lobe, once exhibited a fine section of Illinoian till over a lake bed, deeply weathered by Sangamon soils and covered by two distinct Wisconsinan till sheets. At Gahanna, in central Ohio (Scioto lobe; Goldthwait, 1965), a long-studied sandwich of till–sandy gravel–till had a faint patchy Fox-type soil on the sand (Fig. 1).

The best-studied exposures were along the north shore of Lake Erie (Erie lobe; Dreimanis, 1958, 1960). Here the buried Bradtville till is overlain by organic-rich sands and thick Catfish Creek tills. Perhaps the Bradtville till is correlative with the Sunnybrook till (Karrow, 1967) exposed in Scarborough Bluffs east of Toronto, Ontario. These have been studied for more than a century, together with famous Don (Sangamon) interglacial beds nearby. Note that, in all of these cases, the "classical" latest tills (Woodfordian) lie on top of and extend much farther south than early Wisconsin tills. By 1965 it began to seem as though the early Wisconsin (Altonian) ice sheet did not push nearly as far as later ice, and that there was a long cool middle Wisconsin interval.

Balanced against this hypothesis were unsuccessful candidates for older (Altonian) tongues of till projecting farther south than the late Wisconsin Shelbyville-Cuba-Kent terminal moraines. In all cases the topography was "subdued," the weathering was intermediate between Woodfordian and Sangamonian, and the loess cover was relatively deep. Most eastern was Olean drift in upstate New York and Pennsylvania (Salamanca reentrant; MacClintock and Apfel, 1944). White and Totten's (1965) Titusville till in northeastern Ohio projected from under the fresher Woodfordian tills. For a time (1952–1970), we believed that the distinctive Boston till with deep Russell soils in southern Ohio represented an exposed tongue of early Wisconsin age (Goldthwait, 1952; Goldthwait and Rosengreen, 1969). However, not one of these stood the test of time as radiocarbon ages were determined; most became early late Wisconsin (Farmdalian) (Quinn, 1974; Quinn and Goldthwait, 1985). Some, like Whitewater and Winnebago, became Illinoian or older.

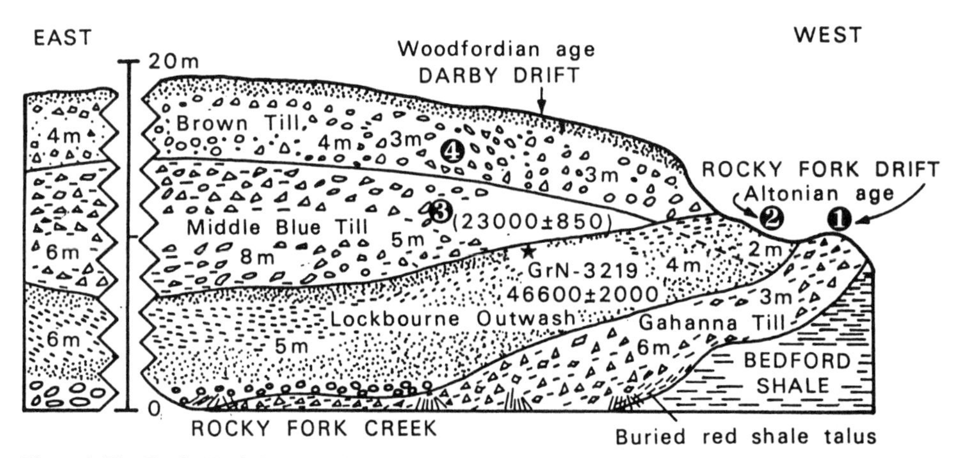

Figure 1. The Rocky Fork Cut at Gahanna in central Ohio as depicted in 1965 for INQUA Field Trip G. Elliptical imbricate boulders of Berea sandstone showed up at the base of the left (east) end soon thereafter. The sandy gravel (2) was interpreted then as Lockbourne outwash, named from widespread outcrops to the right (west) and lower down under till plains from Columbus (south). The ^{14}C age in layer 3, Middle Blue till, came from wood drilled out of a nearby farm well. It is one of many 21,000 to 23,000 yr B.P. ages found in till of this mid-lobe (Scioto).

There were many mountain glacier extensions in the western states in which some moderately fresh looking moraines were deposited farther downvalley than those of the last major glaciation; surfaces were subdued and pebbles more rotted than those upvalley. The earliest was a classic study in the Sierra Nevada (Blackwelder, 1931) that defined the Tahoe stage. Subsequently, extensive studies from southern Colorado through Wyoming identified the Bull Lake stage (Richmond, 1957, 1965). This was later extended through Yellowstone Park and the northern Rocky Mountains (Richmond and others, 1965). Outwash-terrace connections enhanced many correlations.

The Puget Sound ice lobe from north of Washington made slightly greater and older advances northwest and southeast of Olympic Mountains during Salmon Springs glaciation. In most places it was superseded by the Vashon stade of Fraser glaciation, so it was thought that "Salmon Springs Glaciation may represent Early to Middle Wisconsin time" (Crandell, 1965, p. 345). Between Wisconsin glaciations, stratigraphic sands of the Olympia interglaciation contained pollen of a cool, moist time. All this was denied as early Wisconsin record when, years later, the Salmon Springs drift was shown to be much older (<1 Ma tephra) (Westgate and others, 1987). However, another drift, Possession, in northern Puget Sound, is in early Wisconsin position (Easterbrook, 1986).

On Cascade peaks southward from Mount Ranier, radiating mountain glaciers deposited Hayden Creek drift, which was at first correlated to the Salmon Springs advance (Crandell and Miller, 1974). Even the mountain glaciers that spread far across Alaska recorded two great expansions here and there within fresh-drift time; the Knik stage around Cook Inlet was the earlier (Péwé and others, 1965). Thus, by 1965 to 1970, a major early Wisconsin equivalent in western mountains seemed evident, and unlike the mid-continent it was slightly more extensive than later ice. All of this was, from 1960 to 1970, more or less supported by numerical dating.

REVOLUTION BY RADIOMETRIC DATING

Not only did stratigraphy become the dominant approach to glacial geology by the mid-century, but means of numerical dating suddenly became available. From before 1900 until the 1920s, the age for the conclusion of the last glaciation was estimated by the rate of recession of Niagara Falls and other erosional-weathering processes (Leverett, 1909). I was taught that the Wisconsin stage ended 25 to 30 ka. Interglacials were compared induration by relative depths of gumbotil (Kay, 1931). However, a limited but precise dating method has been introduced to this continent in 1920 (Antevs, 1922); i.e., measuring the thickness of clay-silt annual couplets in ice-contact lakes and then matching sections. Up and down the Connecticut River Valley, Antevs accounted for more than 4,000 late-glacial years. Then, by continuing up the Ottawa River Valley, with obvious no-overlap gaps, he estimated enough additional years to partially confirm a 20 ka plus post-glacial figure (Antevs, 1928).

After World War II, as an outgrowth of atomic research, there were "solid carbon" dates based on the decay of carbon-14 in logs (Libby, 1952). By 1952, most of the estimates of last vestige of ice had been cut in half—a veritable revolution in dating! In Ohio, where we had many buried logs and organic masses in till, we could define an early maximum of ice near 22,000 yr B.P. and continuing terminal oscillations until 18,000 yr B.P. (Goldthwait, 1959; Dreimanis and Goldthwait, 1973). Maximum ice push of Wisconsin time proved to be not at all simultaneous along the ice margin, for the Des Moines lobe invaded Minnesota to central Iowa about 14,500 yr B.P.—as suggested by overlapping moraines three decades earlier. Now many tills and most intertill organic beds could be checked by liquid ^{14}C, but only reliably back to 30,000 yr B.P. That was the confidence limit of radiocarbon counting in the early 1950s. By 1960, better sample treatment and longer counting time brought dates to >40,000 and >50,000 yr B.P. Gahanna till in Ohio was >37,000 yr B.P.; Whitewater till in Indiana was >41,000 yr B.P.; the Winnebago Formation in Illinois was >38,000 yr B.P.; shells in Roxana loess dated >38,000 to >45,000 yr B.P.; and Wadena lower till in Minnesota was >40,000 yr B.P. Port Talbot peat just over Bradtville till in Ontario was >40,000 yr B.P. and shells above Sunnybrook till in Ontario were >52,000 yr B.P. Scattered, disconnected organic masses under tills from Maine to North Dakota were >40,000 yr B.P. The Farmdalian stage in Illinois had been well dated to 22,000 to 28,000 yr B.P., which presumably placed all >40,000 situations in middle or early Wisconsinan time.[2]

In the early to middle 1960s, by long-repeated counts on preconcentrated samples, the Groëningen Radiocarbon Lab in Holland began to come up with some ages back to 70,000 yr B.P. These were mistrusted by some, and they had large statistical errors (± 2,000 or more). Most surprising was the great antiquity of a post-Sangamon peat between fresh tills south of Quebec City near Trois Rivière, Quebec; this was ca. 65,000 yr B.P. by three samples (Gadd, 1971). A till-covered peat at Otto, New York, confirmed that there might have been a very early ice invasion in the Northeast (Muller, 1965). However, the seven dates from Port Talbot beds just over Bradtville till in Ontario showed that till came between 45,500 yr B.P. above it but after >60,000 yr B.P. (date below it). Together with a log above Gahanna till in Ohio at 46,600 yr B.P., these ages indicated the major thrust of early Wisconsinan (Altonian) ice probably was about 55 ka. This was the evidence and thinking as the 1970s began.

All the stratigraphic deduction was based upon the assumption that a "greater than" date associated with an underlying till was early Wisconsin, not Illinoian, providing that there was no Sangamonian deep soil profile or weathering evident on that

[2]Use of "Wisconsinan" for "Wisconsin" was promulgated by the Illinois Geological Survey in the 1960s. Where found in older literature (1965 and before) the older term "Wisconsin" is used here; both are intended to refer to the same time span.

underlying till. From 1950 through the 1960s, it was pointed out that many of these early Wisconsinan buried till candidates contained abundant chlorite, which is rapidly weathered away in time. However, today the trend, even in papers to follow, is to question whether this negative weathering evidence is valid. Was the Sangamonian soil eroded away by the advance of late Wisconsinan (Woodfordian) ice? Or was the older pre-Wisconsin till protected from weathering (soil) by late glacial/interglacial sediments lying over it? One by one many of the >40,000 yr B.P. tills are being assigned or reassigned by some study in the 1980s to the Illinoian stage: the Winnabago Formation in Illinois, the Bradtville till in Ontario, the Whitewater till in Indiana-Ohio, and perhaps the Gahanna till in Ohio. Only a younger (early Wisconsin) age sample within the till itself, will satisfy all. I have seen many excellent Sangamonian soil profiles under Wisconsin till while on field trips conducted in southern Ohio, Indiana, and Illinois, and even in Toronto. They have lost most chlorite and are not so thick as still-earlier interglacials, yet these paleosols remain today in spite of the overriding. For this reason I favor the early Wisconsin probability in most of these ">40,000 yr B.P." cases.

OCEAN FLOOR TO THE RESCUE

None of these dating methods went back far enough with certainty to pin down the very beginning of last (Wisconsinan) glaciation on land. Nevertheless, parallel studies of unrelated sediments on the adjacent sea floors evolved a temperature-controlled framework that did fit these terrestrial glacial age estimates. Ocean-floor radiocarbon ages on shells were supplemented and confirmed roughly by the limit of two other isotopic age measurements: uranium-thorium and potassium-argon. Species assemblages of foraminifera indicated surface-water temperature fluctuation from warm to cold several times in the Atlantic and Gulf of Mexico. Sea ice was continuous around the North Pole from 80,000 yr B.P. until today (Beard, 1973). Fairly cool but fluctuating temperatures persisted 50 to 28 ka, but very cold climaxes occurred 75 to 65 ka and 22 to 14 ka. For a period in the late 1960s, oxygen isotopes ($^{16}O/^{18}O$) were applied to the same temperature fluctuations, but then it was realized (Shackelton and Opdyke, 1973) that these related mostly to ice volume. Oxygen isotopes in Greenland ice cores also showed main cooling at 73 to 59 ka and 32 to 13 ka (Langway and others, 1973). Thus after volumes of literature and much controversy, there was consensus about ocean temperature regimes that showed the Wisconsinan began shortly after 80 ka.

The chief question remaining in 1970 was not whether there *was* an early Wisconsin ice advance, but how far and where did ice sheets expand in the early and middle Wisconsin (Altonian). The long interval between the early and late Wisconsinan was certainly warmer than the glacial-arctic phases, but not warmer than today. Pollen from peat in the St. Lawrence lowland suggested temperatures 2 to 4 °C cooler (Terasmae, 1960); in upstate New York, the climate was 5 °C cooler (Otto section; Muller, 1965); just east of Toronto, summer was cooler by 5 °C (Terasmae, 1960); and the north shore of Lake Erie may have been 12 °C cooler. Pine trees grew in Sidney, Ohio. Although these deposits are not all from the same millennium, they confirm that the middle Wisconsin was not a full interglacial event and lasted 20,000 to 30,000 years.

BEST EARLY WISCONSIN CANDIDATES TODAY

The longest and most-intense studies of glacial sediments that are purported to be between the Sangamon and late Wisconsin have been made in eastern Canada. The best-dated and traced are the Levrard till and Deschaillons varves of the Nicolet stade (proposed to be 75 to 65 ka) just under the St. Pierre sediments south of Quebec City (Lamothe, 1989). The Becancour till (Gadd, 1971) is relegated to Illinoian age. The studies began with the classic studies of Coleman (1901) on the Don Brickyards in Toronto. These led to careful studies and tracing by Karrow (1967) at Scarborough Bluffs on the north shore of Lake Ontario. Now questions have been raised about the ice-contact identity of Sunnybrook "till" by the sedimentary studies of Eyles and Eyles (1983), but they are countered by Hicock and Dreimanis (1989). Long, careful studies at many Lake Erie cliffs have been done by Dreimanis and associates. Even there, some interpretations have changed; for example, the Dunwich till is believed to predate 45 to 50 ka (seven Groeningen dates just above it averaged 45,500 yr B.P.) because it is in the midst of Port Talbot beds. However, does it demonstrate a middle Wisconsin ice invasion? For a time Dreimanis thought not; it might not be "in situ" till, but rather a reworked and slumped diamicton. Now it becomes the only early Wisconsin candidate as the Bradtville till is relegated to the Illinoian stage (Dreimanis, this volume). However, these questions all were asked long after the historic period up to 1970 reported here.

About 300 km (200 mi) to the south the Gahanna till at Rocky Fork Creek cut (Scioto sublobe) suffered similar interpretive changes through this century (Fig. 1). Hubbard (1911) said the lower till (Gahanna 1) was Illinoian age because it was the second till down. J. E. Carman had doubts that he never published, because there was thick sand with some gravel beds rather than Sangamon age soil between the two tills. The cut remained open and well trimmed until 1977, so I studied it with classes annually for 30 years and in three dimensions as it receded back 65 m (200 ft) (Goldthwait, 1965). Little by little the upper Wisconsinan till was divided into two tills (brown 4 and blue 3), two advances, which were differentiated by color, stone counts, pebble fabric, and by discontinuous silt lenses between. Then a very nice boulder pavement with uniform striae on the boulders appeared near the blue-brown contact. More pertinent was the rusty clay-enriched pod of Fox-type soil on the bedded sand-gravel between the lower tills. Horizontally this soil extended as a faint yellow zone. Because the overlying till was compact and 9 m (28 ft) thick, and not visibly jointed, this was interpreted as the vestige of a former surface Fox soil. Near it and 45 cm (1.5 ft) below the upper till–sand contact were the water-worn short sticks of wood radiocarbon dated at 46,600 yr B.P. This debris could have been washed in from some outside source, probably from the Appalachian low plateau on the east, as suggested by crossbedding in the sand. As such, the age of the till below may be 50,000, 60,000, or even 70,000 yr B.P.—at least older than 46,000 yr B.P. Up until 1975 the interpreted age was early Wisconsin.

Not all of the supporting kinds of evidence demonstrated on Friends of Pleistocene excursions and in guides of 1952 (15th Eastern Reunion) and 1962 (13th Midwest Reunion) have stood the test of time. The pod of Fox-type soil, as seen at 45 other localities, which involved leaching and clay enrichment atop Lockbourne Gravels, is in question. It was challenged by Gooding and others (1959); they called it surface soil, translocated down till joints in Holocene time. This is true probably in about 20 of the localities, where till cover is less than 3 m (10 ft) thick and "soil" is discontinuous beneath obvious joints. In Rocky Fork cut and another 20 sites where overlying till is 3 to 15 m (10 to 50 ft) thick, or soil extends continuously more than 25 m along the contact, there is more doubt. It appears that in all 12 southwest-central Ohio counties involved, the outwash over the lower (Gahanna) till was very widespread, and some soil on its surface represents a middle Wisconsin weathering (see Gooding and others, 1959).

Doubt was raised about the 16 km (10 mi) correlation from Rocky Fork to the type Lockbourne Gravel because of the following. (1) Stone count lithologies differed significantly. (2) This Rocky Fork exposure is more than 15 m (50 ft) above any nearby valley occurrence of Lockbourne Gravels. (3) At the northeast end of the Rocky Fork cut, exposed by one flooding cloudburst, was a 1.5 m basal layer of huge (15 to 100 cm) rounded boulders of local sandstone (Berea member) imbricated from east to west. Because this site is in a west-trending valley, and is cut into Mississippian "Olentangy red shale" on the Appalachian low scarp, the gravels may be reinterpreted as middle Wisconsinan alluvial fan built over a sharp eroded contact of the Gahanna till and onto Lockbourne outwash surface to the west.

The two cuts south of Sidney, Ohio, 100 m (65 mi) west of Gahanna, seem to support the early Wisconsin hypothesis, but for a different sublobe (Miami). Forsyth (1965) described an extensive, well-structured, Miami-type soil on top of lower till along the railroad cut. An upper till, 5 m thick, covered it 23,000 yr B.P., as recorded by logs of young spruce forest. Gooding (1975) misinterpreted this (J. Forsyth, 1978, personal commun. to D. P. Stewart) to indicate a short exposure interval correlative with his St. Paris interval in silt under till farther south (Preble County, Ohio). Gooding neglected the Brush Creek site 2 km south, where the same upper till lay on depression-type soil with clayey organic colluvial accumulations and peat dated variously from 22,430 to 50,000 yr B.P. No Groëningen date was ever obtained for this peat, but certainly some till below both localities must be "Early" Wisconsin (J. Forsyth, personal commun. to D. P. Stewart). None of the clay alteration (chlorite loss) of a true Sangamon gumbotil could be found in the clayey organic material, so it does not appear to be Sangamonian.

In conclusion, many of the best eastern prospects for an early Wisconsin glaciation have been under critical review since 1970, but they appear to be valid. Perhaps some of the following chapters can pick between the 55,000(?), 65,000(?), or 70,000 year age possibilities. Perhaps all three are valid. The evidence is mostly buried and scattered, so a solution may take a long time.

REFERENCES CITED

Agassiz, F. J., 1840, Étude sur les glaciérs: Neuchatel, 347 p.

Antevs, E., 1922, The recession of the last ice sheet in New England: American Geographical Society Research Series, no. 11, 120 p.

——, 1928, The last glaciation: American Geographical Society Research Series, no. 17, 292 p.

Beard, J. H., 1973, Pleistocene-Holocene boundary and Wisconsinan substages, Gulf of Mexico, *in* Black, R. F., Goldthwait, R. P., and Willman, H. B., eds., The Wisconsinan Stage: Geological Society of America Memoir 136, p. 277–316.

Bernhardi, R., 1832, A hypothesis of extensive glaciation in prehistoric time (translation): Jahrbuch für Mineralogie, Geognosie, und Petrofactenkund (Heidelberg), ser. III, p. 257–267.

Blackwelder, E., 1931, Pleistocene glaciation in the Sierra Nevada and Basin Ranges: Geological Society of America Bulletin, v. 42, p. 865.

Chamberlin, T. C., 1883, Preliminary paper on the terminal moraine of the second glacial epoch: U.S. Geological Survey, 3rd Annual Report, p. 291–402.

——, 1895, The classification of American glacial deposits: Journal of Geology, v. 3, p. 270–277.

Coleman, A. P., 1901, Glacial and interglacial beds near Toronto: Journal of Geology, v. 9, p. 285–310.

Crandell, D. R., 1965, The glacial history of western Washington and Oregon, *in* Wright, H. E., Jr., and Frey, D. G., eds., The Quaternary of the United States: Princeton, New Jersey, Princeton University Press, p. 341–353.

Crandell, D. R., and Miller, R. D., 1974, Quaternary stratigraphy and the extent of glaciation in the Mount Ranier region, Washington: U.S. Geological Survey Professional Paper 847, 59 p.

Dreimanis, A., 1958, Wisconsin stratigraphy at Port Talbot on the north shore of Lake Erie, Ontario: Ohio Journal of Science, v. 58, p. 65–84.

——, 1960, Pre-classical Wisconsin on the eastern portion of the Great Lakes region, North America: International Geological Congress, XXI, Norden (Copenhagen), v. 4, p. 108–119.

Dreimanis, A., and Goldthwait, R. P., 1973, Wisconsin glaciation in the Huron, Erie, and Ontario lobes, *in* Black, R. F., Goldthwait, R. P., and Willman, H. B., eds., The Wisconsinan Stage: Geological Society of America Memoir 136, p. 71–106.

Easterbrook, D. J., 1986, Stratigraphy and chronology of Quaternary deposits of the Puget Lowland and Olympic Mountains of Washington and the Cascade Mountains of Washington and Oregon: Quaternary glaciations in the Northern Hemisphere: Quaternary Science Reviews, v. 5, p. 145–159.

Eyles, C. H., and Eyles, N., 1983, Sedimentation in a large lake; a reinterpretation of the late Pleistocene stratiagraphy at Scarborough Bluffs, Ontario, Canada:

Geology, v. 11, p. 146–152.
Flint, R. F., 1947, Glacial geology and the Pleistocene Epoch: New York, John Wiley & Sons, 589 p.
Forsyth, J. L., 1957, "Early" Wisconsin drift in Ohio [abs.]: Geological Society of America Bulletin 68, p. 1728.
—— , 1965, Age of the buried soil in the Sidney, Ohio area: American Journal of Science, v. 263, p. 521–597.
Frye, J. C., Glass, H. D., and Willman, H. B., 1962, Stratigraphy and mineralogy of the Wisconsinan loesses of Illinois: Illinois Geological Survey Circular 334, 55 p.
Gadd, N. R., 1971, Pleistocene geology of the central St. Lawrence Lowland: Geological Survey of Canada Memoir 359, 153 p.
Geikie, J., 1871, On changes of climate during the Glacial Epoch: Geological Magazine, v. 8, p. 545–553.
—— , 1894, The Great ice age and its relation to the antiquity of man (third edition): London, Stanford, 850 p.
Goldthwait, R. P., 1952, The 1952 field conference of Friends of the Pleistocene: Columbus, Ohio State University, mimeographed guide for 15th meeting (also *in* Science, v. 116, p. 244).
—— , 1959, Scenes in Ohio during the last Ice Age: Ohio Journal of Science, v. 59, p. 193–216.
—— , 1965, *in* INQUA 7th Congress, Guidebook for field conference G: Nebraska Academy of Science, p. 64–82.
Goldthwait, R. P., and Rosengreen, T., 1969, Till stratigraphy from Columbus southwest to Highland County, Ohio (Geological Society of America, North Central Section, Columbus, Ohio, 3rd Annual Meeting guidebook, Field Trip 2): Geological Society of America, p. 2-1–2-17.
Gooding, A., 1963, Illinoian and Wisconsin glaciations in the Whitewater Basin, southeastern Indiana and adjacent areas: Journal of Geology, v. 71, p. 665–682.
Gooding, A. M., 1975, The Sidney interstadial and late Wisconsin history in Indiana and Ohio: American Journal of Science, v. 275, p. 993–1011.
Gooding, A. M., Thorp, J., and Gamble, E., 1959, Leached, clay-enriched zones in post-Sangamon drift in southwestern Ohio and southeastern Indiana: Geological Society of America Bulletin, v. 70, p. 921–926.
Hicock, S. R., and Dreimanis, A., 1989, Sunnybrook drift indicates a grounded early Wisconsin glacier in the Lake Ontario basin: Geology, v. 17, p. 169–172.
Hitchcock, E., 1841, First anniversary address before the Association of American Geologists at their 2nd annual meeting in Philadelphia, April 5, 1841: American Journal of Science, v. 41, p. 232–275.
Hubbard, G. D., 1911, Physiography or surficial geology: Part II, *in* Geology of the Columbus Quadrangle: Columbus, Geological Survey of Ohio, ser. 4, Bulletin 14, p. 51–110.
Kapp, R., and Gooding, A., 1964, Pleistocene vegetational studies in the Whitewater Basin, southeastern Indiana: Journal of Geology, v. 72, p. 307–326.
Karrow, P. F., 1967, Pleistocene geology of the Scarborough area: Ontario Department of Mines Geological Report 46, 108 p.
Kay, G. F., 1931, Classification and duration of the Pleistocene period: Geological Society of America Bulletin, v. 42, p. 425–466.
Kay, G. F., and Leighton, M. M., 1933, Eldoran epoch of the Pleistocene period: Geological Society of America Bulletin, v. 49, p. 669–674.
Kempton, J. P., 1963, Subsurface stratigraphy of the Pleistocene deposits of central northern Illinois: Illinois Geological Survey Circular 356, 43 p.
Lamothe, M., 1989, A new framework of the Pleistocene stratigraphy of the central St. Lawrence Lowland, southern Quebec: Geographie Physique et Quaternaire, v. 43, p. 119–129.
Langway, C. C., Jr., Dansgaard, W., Johnsen, S. J., and Clausen, H., 1973, Climatic fluctuations during the late Pleistocene, *in* Black, R. F., Goldthwait, R. P., and Willman, H. B., eds., The Wisconsinan Stage: *in* Geological Society of America Memoir 136, p. 317–321.
Leighton, M. M., 1926, A notable type Pleistocene section; the Farm Creek exposure near Peoria, Illinois: Journal of Geology, v. 34, p. 167–174.
Leverett, F., 1909, Weathering and erosion as time measures: American Journal of Science, v. 177, p. 349–368.
Libby, W. F., 1952, Radiocarbon dating: Chicago, Illinois, University of Chicago Press, 174 p.
MacClintock, P., and Apfel, E. T., 1944, Correlation of the drifts of the Salamanca re-entrant, New York: Geological Society of America Bulletin, v. 55, p. 1143–1164.
McGee, W. I., 1891, The Pleistocene history of northwestern Iowa: U.S. Geological Survey, 11th Annual Report, p. 189–577.
Muller, E. H., 1965, Quaternary geology of New York, *in* Wright, H. E., Jr., and Frey, D. G., eds., Princeton, New Jersey, Princeton University Press, p. 99–112.
Newberry, J. S., 1874, Geology of Ohio; surface geology: Geological Survey of Ohio Report 2, part 1, p. 1–80.
Péwé, T. L., Hopkins, D. M., and Giddings, J. L., 1965, The Quaternary geology and archaeology of Alaska, *in* Wright, H. E., Jr., and Frey, D. G., eds., The Quaternary of the United States: Princeton, New Jersey, p. 355–374.
Quinn, M. J., 1974, Glacial geology of Ross County, Ohio [Ph.D. thesis]: Columbus, Ohio State University, 162 p.
Quinn, M. J., and Goldthwait, R. P., 1985, Glacial geology of Ross County, Ohio: Division of the Ohio Geological Survey, Report of Investigation No. 127, 42 p.
Richmond, G. M., 1957, Correlation of Quaternary deposits in the Rocky Mountain region, U.S.A.: INQUA, International Congress V (Madrid-Barcelona) Résumées des Communications, p. 157.
—— , 1965, Glaciation of the Rocky Mountains, *in* Wright, H. E., Jr., and Frey, D. G., eds., Princeton, New Jersey, Princeton University Press, p. 217–230.
Richmond, G. M., Fryxell, R., Neff, G. E., and Weis, P. L., 1965, The Cordilleran ice sheet of the northern Rocky Mountains, and related Quaternary history of the Columbia Plateau, *in* Wright, H. E., Jr., and Frey, D. G., eds., The Quaternary of the United States: Princeton, New Jersey, Princeton University Press, p. 231–242.
Ruhe, R. V., Dietz, W. P., Fenton, T. E., and Hall, G. F., 1968, The Iowan problem: Iowa Geological Survey Report of Investigation No. 7, 40 p.
Shackleton, N. J., and Opdyke, N. D., 1973, Oxygen isotope and paleomagnetic stratigraphy of equatorial Pacific core V 28-238: Oxygen isotope temperatures and ice volumes on a 10^5 and 10^6 year scale: Quaternary Research, v. 3, p. 39–55.
Shaffer, P. R., 1956, Farmdale drift into northwestern Illinois: Illinois Geological Survey Report of Investigation 198, 25 p.
Terasmae, J., 1960, A palynological study of the Pleistocene interglacial beds at Toronto, Ontario: Canadian Geological Survey Bulletin 56, p. 23–41.
Westgate, J. A., Easterbrook, D. J., Naeser, N. D., and Carson, R. J., 1987, Lake Tapps tephra; an early Pleistocene stratigraphic marker in the Puget Lowland, Washington: Quaternary Research, v. 28, p. 340–355.
White, G. W., and Totten, S. M., 1965, Wisconsinan age of the Titusville Till (formerly called "Inner Illinoian"), northwest Pennsylvania: Science, v. 148, p. 234–235.
Whittlesey, C., 1866, On the fresh-water glacial drift of the northwestern states: Smithsonian Contributions to Knowledge no. 197, 32 p.

Manuscript Received by the Society September 6, 1991

Printed in U.S.A.

Geological Society of America
Special Paper 270
1992

The marine oxygen isotope record: Constraints on timing and extent of ice-growth events (120–65 ka)

Alan C. Mix
College of Oceanography, Oregon State University, Corvallis, Oregon 97331

ABSTRACT

Oxygen-isotope analyses of foraminifera record the timing and magnitude of ice-growth events during ice ages. The timing of the major isotopic shifts during the late Pleistocene is relatively well determined, although small improvements continue to be made. In the SPECMAP marine oxygen-isotope time scale, the major isotopic transitions due to ice-growth events of the last glacial inception are centered at ca. 118–112 ka (substage 5e/5d), ca. 98–92 ka (substage 5c/5b), and ca. 75–68 ka (substage 5a/4). The ice volume at each isotopic event remains only partly known. This is due to difficulties in isolating local temperature effects on $\delta^{18}O$, changes in the isotopic composition of ice, and variations in ocean circulation and atmospheric vapor transport. Considering likely temperature, water-mass effects, and analytical errors, we can not discern any size difference in the part of the $\delta^{18}O$ shifts due to ice growth at the substage 5e/5d and the stage 5/4 transitions. Both were roughly the same size, about 0.6‰. Assuming that the mean isotopic composition of glacier ice was –30‰ to –35‰, ice-volume changes are about 25×10^6 km^3 across these transitions. The substage 5c/5d isotopic transition, the smallest of the three major ice-growth events of stage 5, was about a 0.3‰ shift, or 13×10^6 km^3 of ice, using similar assumptions. In spite of the uncertainties in translating isotope measurements quantitatively into ice volumes, it is clear that the isotopic shifts between 120 and 65 ka require relatively large and rapid changes in global ice volume.

INTRODUCTION

Oxygen-isotope ratios from fossil foraminifera (expressed as $\delta^{18}O$) delimit the timing and pattern of global glaciation. Qualitatively, this information has been critical for defining the history of the ice ages. Quantitative interpretation of this record in terms of ice volume, however, is not straightforward. Here I discuss the marine oxygen-isotope record and focus on the timing and extent of glaciation associated with the major ice-growth transitions of the last ice age (i.e., isotopic stages 5 and 4, ca. 122 to 65 ka). More complete reviews of the complexities of the oxygen-isotope record appeared elsewhere (Duplessy, 1978; Mix, 1987).

Early work on $\delta^{18}O$ of foraminifera centered on its use as a paleothermometer (Emiliani, 1955). Emiliani knew that $\delta^{18}O$ of calcite depends both on the temperature and the isotopic composition of the local water. He concluded that past temperature changes were more important than water-composition changes in Pleistocene $\delta^{18}O$ records from planktic foraminifera. After analyzing $\delta^{18}O$ in benthic foraminifera, Shackleton (1967) concluded the opposite, that water isotopic composition effects driven by ice-volume changes dominated the down-core record. Because the deep sea is relatively cold, and has such a large volume, Shackleton (1967) assumed that it must have been thermally stable during the Pleistocene ice ages. He noted that down-core $\delta^{18}O$ records from benthic and planktic foraminifera roughly covary, and thus most of the observed variations in $\delta^{18}O$ must reflect ice-volume change.

The field of isotopic paleoceanography developed through the 1970s using Shackleton's assumption. Until recently most workers believed that the glacial-interglacial contrast in $\delta^{18}O$ due to ice-volume change was at least 1.5‰. Some estimates ranged up to 1.8‰ (Broecker, 1978). Extrapolating modern ice-sheet isotopic compositions, most workers have assumed a $\delta^{18}O$ composition of mean glacier ice of –30‰ to –35‰ vs. standard

Mix, A. C., 1992, The marine oxygen isotope record: Constraints on timing and extent of ice-growth events (120–65 ka), *in* Clark, P. U., and Lea, P. D., eds., The Last Interglacial-Glacial Transition in North America: Boulder, Colorado, Geological Society of America Special Paper 270.

mean ocean water (SMOW) (e.g., Dansgaard and Tauber, 1969). Such estimates imply eustatic sea-level changes of at least 150 m, which suggests very large ice sheets. This inference fueled the ongoing debate among glacial geologists on the extent and thickness of glacier ice (e.g., Andrews, 1973, 1982; Hughes and others, 1981).

$\delta^{18}O$ AND DEEP-SEA TEMPERATURES

Recent work on oxygen isotopes and sea levels has rekindled the original controversy about the temperature and ice-volume effects. If ice-volume change dominates the marine $\delta^{18}O$ record, and if the isotopic composition of glacier ice is roughly constant, then variations in $\delta^{18}O$ should parallel sea-level variations. Chappell and Shackleton (1986) found significant discrepancies between benthic $\delta^{18}O$ data from the deep Pacific Ocean and a detailed sea-level reconstruction from New Guinea (Fig. 1). They suggested total glacial-interglacial sea-level changes of ~130 m since the last glacial maximum. More recent sea-level reconstructions based on corals of the past ca. 17 ka from Barbados (Fairbanks, 1989) refine this estimate, with 121 ± 5 m of sea-level change.

Chappell and Shackleton (1986) concluded that the deep Pacific was as warm as present only briefly during the last interglacial maximum (substage 5e). In their view, throughout most of the last glaciation it was at least 1.5 °C colder than now. This would reduce the estimate for the maximum glacial-interglacial ice-volume effect on $\delta^{18}O$ to about 1.3‰. In their preferred temperature reconstruction for the deep sea, most of the abyssal cooling occurred during the initiation of the last ice age (the 5e/5d transition, ca. 115 ka). Warming occurred at the glacial termination (the stage 2/1 transition, ca. 10 ka). If so, the 5e/5d isotopic transition in benthic foraminifera at most sites, if taken without temperature corrections, would overestimate ice-volume change. Changes in $\delta^{18}O$ within the last ice age, such as the stage 5/4 transition near 70 ka, should record changes in ice volume reasonably well.

The search for more reliable estimates of the ice-volume effect on oxygen-isotope records moved toward finding places where little or no temperature changes have occurred. Labeyrie and others (1987) and Duplessy and others (1988) compiled a $\delta^{18}O$ record from benthic foraminifera in the Norwegian Sea, where deep water is now near the freezing point (–1.9 °C). Because deep water here could not have been much colder, they inferred that its temperature was constant throughout the last ice age. Unfortunately, the Norwegian Sea compilation lacks benthic foraminifera in the glacial maximum section. To get around this problem, Labeyrie and others (1987) and Duplessy and others (1988) included some Pacific benthic data in the compilation, after adjusting scales to give similar values in overlapping sections. This composite record reduces the estimate of the total glacial-interglacial ice-volume effect on $\delta^{18}O$ further, to about 1.1‰ (Fig. 2). If correct, it expands the role of deep-ocean temperature change to more than one-third of the total isotopic signal from benthic foraminifera in typical open-ocean cores. It would require that most of the deep ocean below 2,000 m depth cooled dramatically, to below –0.5 °C during glaciation.

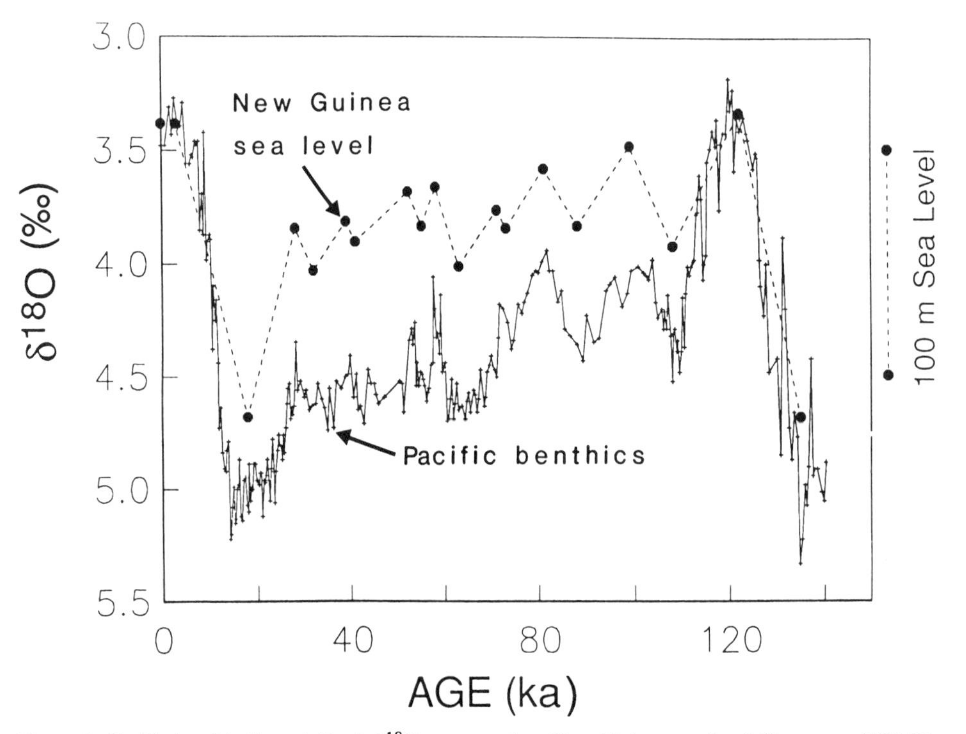

Figure 1. Pacific benthic foraminiferal $\delta^{18}O$ compared to New Guinea sea level (from core V19-30; Shackleton and Pisias, 1985).

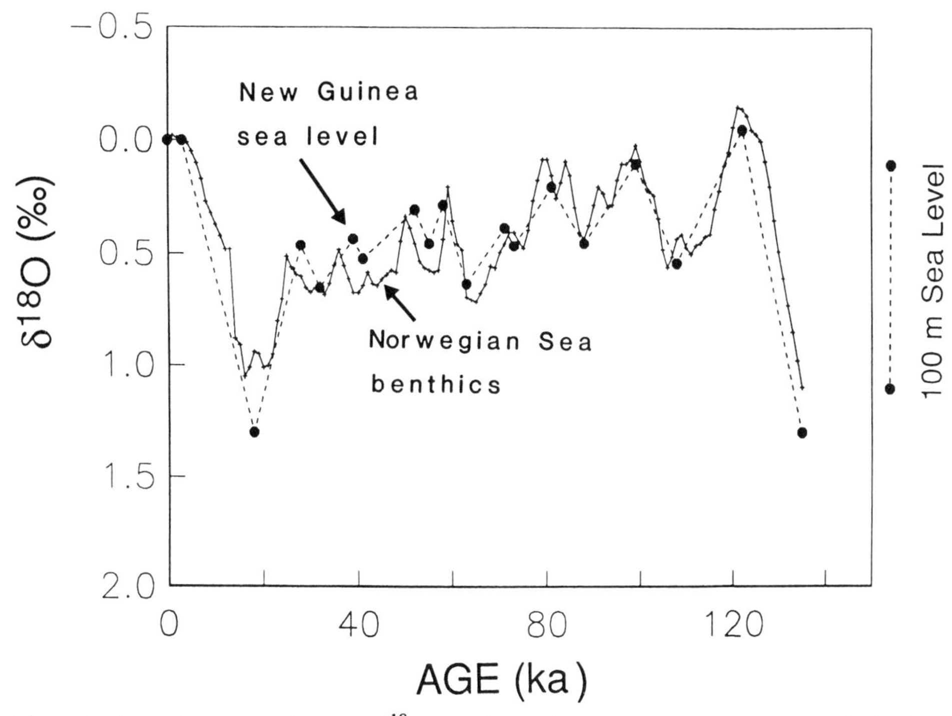

Figure 2. Estimated temperature-free $\delta^{18}O$ record based on Norwegian Sea and Pacific benthic foraminiferal data (composite record; from Duplessy and others, 1988) compared to New Guinea sea level.

Mix and Pisias (1988) considered the physical consequences of the interpretation of Labeyrie and others (1987) and Duplessy and others (1988) in a simple model of the deep-ocean heat budget. They concluded that major changes in deep-ocean circulation would be needed to maintain such cold deep-sea temperatures. Even forming all deep water at the freezing point would not be enough to cool the mean deep ocean to below –0.5 °C. In addition, the heat supplied to the deep sea by cross-thermocline mixing and geothermal heating must have been removed more efficiently than it is now. Mix and Pisias (1988) showed that this could be done by forming deep water much faster at the glacial maximum than at present. Recent reconstructions of paleoventilation rates of the deep sea, based on accelerator radiocarbon dating of benthic and planktic foraminifera, seems to preclude this possibility (Shackleton and others, 1988; Broecker and others, 1988). Considering this problem, Mix and Pisias (1988) preferred a model that had total glacial-interglacial ice-volume effects on $\delta^{18}O$ of about 1.4‰. This would permit modest ice-age cooling of the deep sea, without requiring radical changes in the heat budget. The issue remains unresolved.

$\delta^{18}O$ AND STABLE TROPICAL SEA-SURFACE TEMPERATURES

Following Matthews and Poore (1981), Shackleton (1987) considered the tropical surface oceans to have been thermally stable. He used planktic foraminiferal $\delta^{18}O$ data from the western tropical Pacific as a proxy for ice-volume changes. The planktic isotope data match the New Guinea sea-level record much better than do the benthic data (Fig. 3). Using the difference between the planktic and benthic isotope records, Shackleton (1987) inferred deep-sea temperature changes of more than 2 °C. This would make the total glacial-interglacial $\delta^{18}O$ change due to ice volume equal 1.2‰. As in Chappell and Shackleton (1986), most of the deep-sea cooling occurred at the substage 5e/5d transition.

A planktic foraminiferal record from the tropical Atlantic (Mix, 1987) has larger changes in $\delta^{18}O$ than that from the Pacific (Fig. 4); the total glacial-interglacial range is 1.8‰. The $\delta^{18}O$ shifts for interstadial events within stage 5, however, agree reasonably well with the sea-level record, being significantly lower than the Pacific benthic values. The largest offsets between the Atlantic planktic $\delta^{18}O$ record and the sea-level data are within isotopic stages 2–4 (ca. 10–70 ka), where the Atlantic planktic record gives the largest $\delta^{18}O$ departures from modern of all the records cited here. This suggests either cooling of the tropical Atlantic surface waters by ~2 °C, or increases in surface water $\delta^{18}O$ of 0.4‰ to 0.6‰ during stages 2–4. Some ice-age cooling of this area is consistent with faunal estimates of temperature change (CLIMAP, 1981 and 1984; McIntyre and others, 1989), but the pattern is not identical. Both temperature effects and watermass effects are likely contributors to the tropical Atlantic $\delta^{18}O$ record.

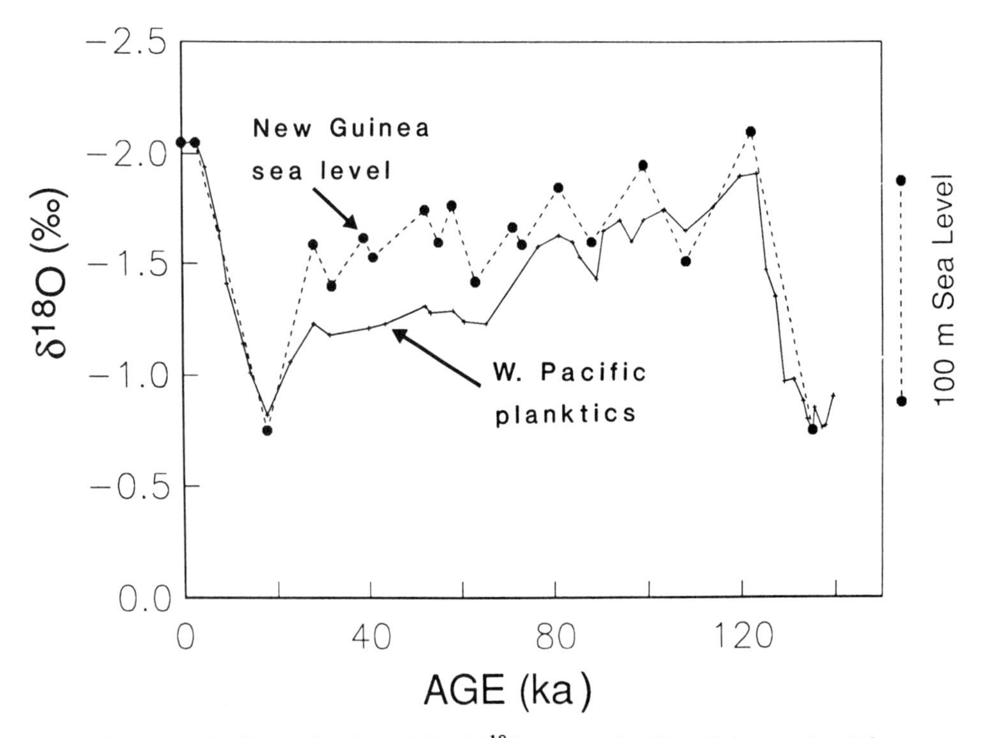

Figure 3. Western Pacific planktic foraminiferal $\delta^{18}O$ compared to New Guinea sea level (from core RC17-177; Shackleton, 1987).

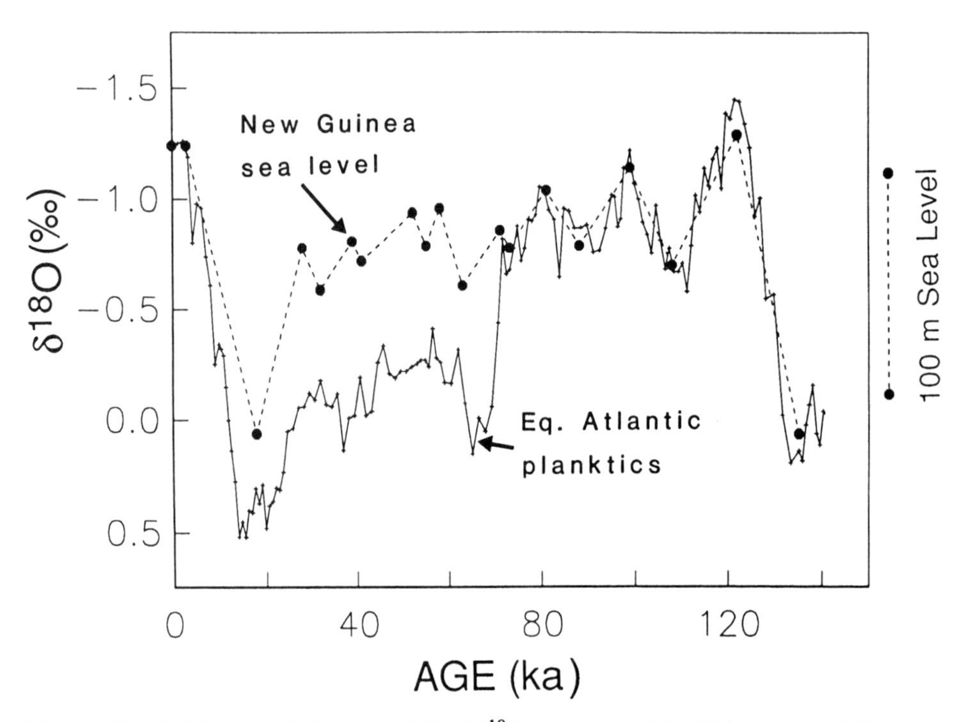

Figure 4. Tropical Atlantic planktic foraminiferal $\delta^{18}O$ compared to New Guinea sea level (from core V30-40; data in Appendix).

$\delta^{18}O$ AND ATMOSPHERIC VAPOR TRANSPORT

There is a potential pitfall in using the $\delta^{18}O$ record from warm surface waters as an ice-volume proxy. Emphasizing temperature change to be a contaminant to the $\delta^{18}O$ record of ice volume, most studies implicitly assume no change in water-mass $\delta^{18}O$, other than the global budget adjustments due to ice-volume change. At present, however, surface waters in the polar regions have lower $\delta^{18}O$ than in the tropics because of the net transport of ^{18}O-depleted water vapor from low to high latitudes. Variations in this vapor transport would change the $\delta^{18}O$ of the warm surface ocean. This in turn would change foraminiferal $\delta^{18}O$, even in the absence of temperature changes.

Broecker (1989) considered the effects of vapor transport from the Atlantic Ocean to the Pacific Ocean, across Panama. It may be possible to minimize this effect simply by averaging representative planktic records from both oceans (weighted to the volume of near-surface waters in the each ocean, roughly 3:1 for Indo-Pacific: Atlantic). A more difficult task may be to remove the possible effects of changing amounts of vapor transport from low to high latitudes. This effect, noted conceptually by Broecker (1986), and quantified here, can cause systematic differences between average planktic and benthic $\delta^{18}O$ records, even in the absence of temperature changes.

The poleward vapor-transport effect is illustrated in a simple three-box model of the ocean, which contains a warm-surface and a cold-surface ocean (both 100 m thick) and a 3,900-m-deep ocean (Fig. 5). Water fluxes approximate the modern ocean, with advective overturn (f_o) of 45 Sv (Sv = Sverdrups, 1 Sv = 10^6 $m^3 \cdot s^{-1}$). Vapor transport from the low- to high-latitude box (f_v) occurs with a $\delta^{18}O$ value 25‰ lower than that of warm surface water. This simplifies reality somewhat, but is sufficient to illustrate the effect. Modern vapor fluxes are not well known (J. Kutzbach, 1989, personal commun.). I estimate the net vapor flux by averaging precipitation minus evaporation (*P-E*) for all areas poleward of lat 60°. Kutzbach and Guetter's (1986) simulation of the modern world in a general circulation model gives *P-E* of 1.9 mm/day for the polar area. The net vapor flux from the low latitudes required to support this excess precipitation is 1.5 Sv. This compares well with the value used by Broecker (1986) of 1.0 Sv.

Diffusion between boxes in the model is set to physically reasonable values that yield acceptable values for the modern distribution of $\delta^{18}O$ in the ocean. At present, the $\delta^{18}O$ value of average deep water is about −0.1‰ (vs. SMOW). Warm surface water $\delta^{18}O$ averages about +0.5‰, and polar surface water averages about −0.2‰ to −0.3‰ (Craig and Gordon, 1965, with 0.10‰ correction by Craig, 1987). To fit these values, cross-thermocline (warm surface to deep water) diffusion (K_{zw}) is set here to be equivalent to two-way (vertical) transfer of 4.5 Sv of water. Diffusion between cold surface water and deep water (K_{zc}) is higher (11 Sv) due to lower density contrast. Between the cold and warm surface boxes, diffusion (K_h) is equal to two-way horizontal water transfers of 5 Sv. The total advective and diffusive water transfers between surface- and deep-water boxes, 59.5 Sv in this box model, compare reasonably well with box-model values of 55 Sv (Stuiver and others, 1983), 52 Sv (Broecker and Peng, 1986), and 76 Sv (Keir, 1988) designed to fit radiocarbon and other geochemical distributions in the modern world oceans.

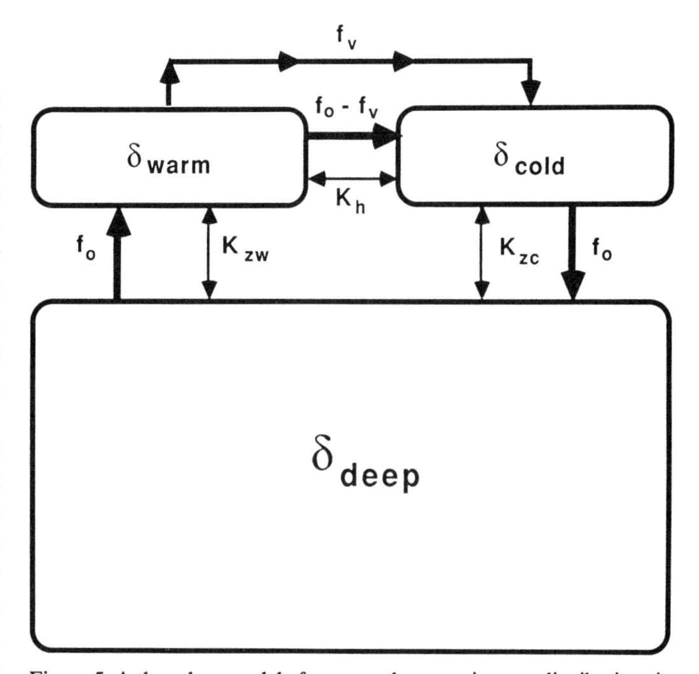

Figure 5. A three-box model of water and oxygen isotope distributions in the world ocean. δ_{warm}, δ_{cold}, and δ_{deep} refer to the oxygen isotope compositions (vs. SMOW) of water masses in the warm surface, cold surface, and mean deep ocean, respectively. The *K* terms represent diffusion between boxes (see text), and the *f* terms are advective water transfers (see text). f_o is thermohaline overturn, and f_v is the net flux of water vapor through the atmosphere.

Figure 6 illustrates sensitivity of water-mass $\delta^{18}O$ to changing poleward vapor flux, while all other water transfers are constant. If there were no net vapor flux, there would be no transfer of low $\delta^{18}O$ water to the cold surface box. In this case, all water masses of the ocean would have the same $\delta^{18}O$ value of −0.1‰ (the global average value used as input). As the vapor flux increases, ^{18}O is enriched greatly in the warm surface waters (to about 1.2‰ for a vapor transport of 3 Sv), and depleted slightly in the cold surface and deep oceans. The smaller changes in these cold boxes reflect the larger volume of the deep sea and more effective diffusive mixing between the deep and the cold-surface ocean. The largest changes in water-mass $\delta^{18}O$ are in the warm surface ocean. This means that tropical planktic foraminiferal $\delta^{18}O$ measurements will record ice volume with fidelity only if vapor fluxes are constant.

To estimate the possible variations in ice-age vapor fluxes, the modern and 18 ka general circulation model simulations of Kutzbach and Guetter (1986) are used. Although atmospheric general circulation models are imperfect in their simulation of the water cycle, this comparison is useful as a sensitivity test. As

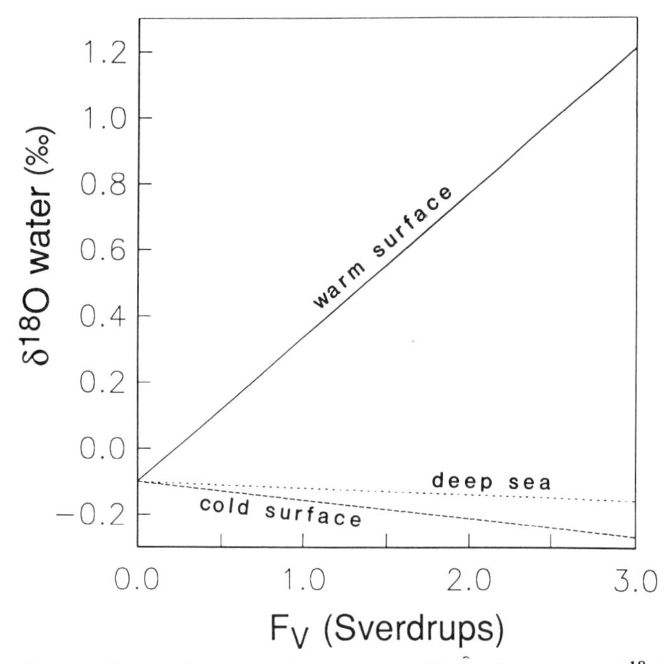

Figure 6. Box model output, illustrating sensitivity of water-mass $\delta^{18}O$ to varying vapor flux from low to high latitudes. The oxygen isotope value of warm surface water, δ_{warm}, increases with increasing poleward vapor flux. A modern value for f_v of 1.5 Sv gives δ_{warm} = 0.55‰. Ice-age reduction of f_v to 1.0 Sv gives δ_{warm} = 0.34‰. The changes in cold surface water and deep water are in the opposite sense, but much smaller. The exact values are a function of the model simplifications, but the sense of change and approximate scaling are robust.

noted above, the modern simulation gives average *P-E* of 1.9 mm/day for areas poleward of lat 60°. For the glacial world, the same model gives *P-E* for the high latitudes of 1.4 mm/day, lower by nearly one-third.

Reducing poleward vapor flux from the modern value of 1.5 Sv to a glacial value of 1.0 Sv would lower $\delta^{18}O$ in warm surface water from 0.55‰ to 0.34‰ (Fig. 6). Thus, without other changes, $\delta^{18}O$ in tropical planktic foraminifera would underestimate the glacial-interglacial ice-volume change by about 0.2‰. After correcting for the vapor-transport effect, Shackleton's (1987) inferred glacial-interglacial $\delta^{18}O$ range due to ice volume of 1.3‰ from western Pacific planktic foraminifera would increase to about 1.5‰.

This is only an approximation, because the model is very simple. As with any box-model study, a legitimate question is how sensitive the model is to slight changes in its simplifying assumptions. One model assumption that may affect the sensitivity of the response is the mixed-layer thickness of 100 m. To test this assumption, the model was rerun with a mixed-layer thickness of 300 m. This changes the results by only about 10% (i.e., $\delta^{18}O$ of warm surface water = 1.1 ‰ instead of 1.2‰ for vapor flux of 3 Sv). Another simplification is the treatment of diffusion, which is poorly understood in the modern ocean. In the limit, setting all diffusion values to zero increases the sensitivity of warm surface waters to vapor flux. For a vapor flux of 3 Sv, the $\delta^{18}O$ of warm surface water would be 1.5‰ instead of 1.2‰. Even with this extreme assumption, however, the difference between the interglacial and glacial case for warm surface water $\delta^{18}O$ would be 0.3‰, similar to the standard run above.

The box-model result is relatively sensitive to the rate of thermohaline overturn in the ocean. If thermohaline overturn were lower than at present, the sensitivity of $\delta^{18}O$ distributions to changes in vapor transport would be increased. For example, if advective overturn is set to 22.5 Sv instead of 45 Sv, a poleward vapor flux of 3 Sv yields warm surface-water $\delta^{18}O$ of 1.9‰ instead of 1.2‰ in the standard case. Smaller but opposite effects occur in the cold-surface and deep oceans. Thus, thermohaline overturn could be nearly as important as vapor flux on the distribution of $\delta^{18}O$ in the oceans. If lower poleward vapor fluxes were linked to higher thermohaline overturn (perhaps due to less input of low-density fresh water to deep-water source areas), the sensitivity of the warm surface ocean $\delta^{18}O$ to the coupled effects would be dramatically increased, by as much a factor of 3. If the opposite were true, that lower vapor fluxes co-occurred with lower overturn, there would be little or no isotopic response. Although there are some preliminary indications that radiocarbon ages of deep waters were greater in the glacial maximum than at present (Shackleton and others, 1988; Broecker and others, 1988), we can not yet fully determine what circulation change caused this, so we cannot yet say how $\delta^{18}O$ in tropical surface oceans would be affected.

Thus, although other effects of changing atmospheric and oceanic circulation (especially rates of thermohaline overturn) could modify the above result slightly, the concept of higher $\delta^{18}O$ in the tropical oceans associated with higher poleward vapor transport shown in Figure 6 is robust. If temperature effects could be reliably removed from both planktic and benthic $\delta^{18}O$, the difference between the average tropical planktic record and average benthic records would record both poleward vapor transport and thermohaline overturn. The box model suggests that likely changes in vapor flux would cause relatively small change in warm surface water $\delta^{18}O$, on the order of a few tenths per mille; however, these are large enough to affect critical interpretations of climatic history. They are averages for the entire warm surface ocean, and local effects could be larger. To isolate ice-volume signals quantitatively from oxygen isotopes in planktonic foraminifera, we must also understand global-scale atmospheric and oceanic circulation changes.

CONSTRAINTS ON THE SIZE OF ICE-GROWTH EVENTS

The focus here is the size of the isotopic effects associated with the major ice-growth transitions of substages 5e/5d (ca. 115 ka), 5c/5b (ca. 95 ka), and 5a/4 (ca. 75 ka). Specifically, Which of these isotopic events was largest? Can the size of these isotopic events in different areas help to isolate the contributions of ice volume, temperature, or water-mass changes to the isotopic record of these events?

TABLE 1. $\delta^{18}O$ AT EXTREME EVENTS*

Data Type	$\delta^{18}O$ Event							
	5e (‰)	5d (‰)	5c (‰)	5b (‰)	5a (‰)	4 (‰)	2 (‰)	1 (‰)
Pacific benthic	0.0	0.9	0.6	1.0	0.6	1.2	1.6	0.0
Norwegian Sea benthic	-0.1	0.5	0.1	0.4	0.1	0.7	1.0	0.0
Equatorial Atlantic planktic	-0.1	0.6	0.2	0.5	0.3	1.2	1.8	0.0
West Pacific planktic	0.1	0.4	0.3	0.6	0.4	0.8	1.3	0.0

*Departures from modern values.

Table 1 summarizes the isotopic departures from modern values at the $\delta^{18}O$ maxima and minima associated with each event. Data from the four records illustrated in Figures 1–4 are included. The deep Pacific record comes from benthic foraminifera in core V19-30, from the eastern tropical Pacific at lat 3.4°S, long 83.5°W, 3,091 m depth (Shackleton and Pisias, 1985). The Norwegian Sea benthic record is a composite from several cores (including some Pacific data for the glacial maximum). Because Duplessy and others (1988) believed that this composite was not contaminated by temperature changes, they referred to it as "mean ocean water $\delta^{18}O$." The equatorial Atlantic planktic record comes from analyses of *Globigernoides sacculifer* (with terminal chamber, 355–415 μm) in core V30-40 at lat 0.2°S, long 23.2°W, 3,706 m depth. These are the highest-resolution planktic data available from the tropical Atlantic (Appendix). The western Pacific planktic data come from analyses of *G. sacculifer* in core RC17-177, from lat 1.8°N, long 159.5°E, 2,600 m depth (Shackleton, 1987). This core has a relatively low sedimentation rate (ca. 1.5 cm/1,000 yr). Shackleton (1987) noted that the signal amplitude in this record may have been reduced due to bioturbation.

Because the different cores used here were analyzed at different resolution, the $\delta^{18}O$ values for events in Table 1 were smoothed from the original data with a gaussian filter 6,000 years wide. This smoothing reduces random errors, and ensures that all the records are examined at the same resolution. The errors for each value in Table 1 reflect random measurement errors, random variations within the sample population of foraminifera, and potentially nonrandom site-specific problems. We can estimate the errors for the first two items. Most modern mass-spectrometer laboratories now report long-term errors on carbonate standards of less than 0.1‰. For example, the standard deviation of $\delta^{18}O$ values of a local carbonate powder in the laboratory at Oregon State University is ± 0.09‰ over a 1-year period (1989). The average short-term standard deviation (within daily runs) of the same standard over the same period is ±0.04‰. To estimate errors in the foraminiferal data (including both mass-spectrometry errors and random population errors), I use the data in the Appendix, where 59 samples have replicate analyses. Within these samples, the average deviation of an analysis from the mean at each sample is 0.07‰. For this core, at least, population effects do not appear to add to the variability induced by measurement error. Because these random errors are less than 0.1‰, and because the records are smoothed, values in Tables 1 and 2 are reported to a precision of 0.1‰. Site-specific errors, which may be nonrandom, cannot be assessed in detail. For example, the possible problem of bioturbation in core RC17-177 cannot be resolved. In the future, this type of analysis should be done with an average of many cores. For now, however, I have used what I believe are the best available data from each area at the time of writing (May, 1989).

The size of the isotopic events varies in the different records (Table 1). The peak interglacial event (substage 5e) is the same as the modern in the Pacific benthics, lower by 0.1‰ in Norwegian Sea benthics and Atlantic planktics, but higher by 0.1‰ in western Pacific planktics. Although coral terrace and oxygen-isotope records are commonly cited as evidence that sea level was higher (and ice volume lower) during substage 5e than now, the evidence for this from the foraminiferal $\delta^{18}O$ records discussed here is equivocal.

Table 2 records the isotopic shifts at each of the ice-growth boundaries, measured at the four different sites. Taken without corrections, the Pacific benthic foraminiferal record suggests that the earliest transition (5e/5d) was the largest. It has a 0.9‰ $\delta^{18}O$ shift, relative to a 0.6‰ shift at the stage 5/4 transition. The Norwegian Sea benthic record has equal shifts of 0.6‰ at the 5e/5d and 5/4 transitions. In the Atlantic planktic record, a 0.7‰ shift occurs at the 5e/5d boundary, and a 0.9‰ shift is at the 5/4 boundary. The western Pacific planktic data have $\delta^{18}O$ shifts smaller than in the Atlantic, about 0.3‰ for the 5e/5d boundary and 0.4‰ for the 5/4 boundary. The four different

TABLE 2. ICE GROWTH AND TOTAL GLACIAL-INTERGLACIAL OFFSETS IN $\delta^{18}O$

Data Type	Oxygen-Isotopic Shifts ($\delta^{18}O$ ‰)			
	5e/5d	5c/5b	5a/4	2/1
Pacific benthic	0.9	0.4	0.6	1.6
Norwegian Sea benthic	0.6	0.3	0.6	1.0
Equatorial Atlantic planktic	0.7	0.3	0.9	1.8
West Pacific planktic	0.3	0.3	0.4	1.3

records agree that the size of the substage 5c/5b isotopic shift was about 0.3‰ to 0.4‰ (Table 2), identifying it as the smallest of the ice-growth isotopic transitions of stage 5.

At the substage 5d glacial event, the largest isotopic departure from modern values comes from the Pacific benthic record, with a value of 0.9‰ above modern (Table 1). The other values are smaller, averaging 0.5‰ for this event. The anomalously large benthic isotopic shift at the 5e/5d boundary is consistent with significant cooling of the deep sea at this time. If the ice-volume effect on $\delta^{18}O$ at substage 5d was 0.5‰ greater than modern, it implies an increase of ice volume of more than 20×10^6 km^3 (for ice $\delta^{18}O$ of −35‰). This is less ice than predicted by the model of Budd and Smith (1979), which was able to grow more than 30×10^6 km^3 of ice at the 5e/5d transition. In their model, however, substage 5d has the most ice of the past 120 ka. It has no 5c/5b ice-growth event, and the 5a/4 event is smaller than that at 5e/5d. In contrast, the model of glacial inception developed by Andrews and Mahaffy (1976) accounted for 2.7 to 7.0×10^6 km^3 of ice in North America, less than predicted by the isotope record. If this mismatch is explained by deep-sea cooling, then the deep Pacific would have to be cooler than at present by more than 3 °C, bringing its potential temperature below −1.5 °C. This is unlikely, because it approaches the freezing point of seawater. Rind and others (1989) and Peteet and others (this volume) find no ice growth in an atmospheric general circulation model (GCM). If this were correct, and the entire $\delta^{18}O$ record at this time reflected temperature, it would put the deep-ocean temperature below the freezing point, a physical impossibility.

Shackleton (1987) inferred cooling of about 2 °C for the deep Pacific at the 5e/5d boundary, and essentially no temperature change at the substage 5c/5b or the stage 5/4 boundaries. If this is correct, the isotopic shift due to ice-volume change at both the 5e/5d and 5a/4 boundaries was about 0.6‰.

Although the Atlantic and Pacific planktic records do not give the same values for the isotopic shifts, they agree that the 5/4 transition was the largest event. Ice-age vapor transport to the high latitudes that was lower than modern would make the planktic records underestimate true ice-volume change. The model results discussed above suggest a signal of a few tenths per mille for this effect between interglacial and glacial maxima. It is not certain whether significant reductions in vapor flux would have occurred at the 5e/5d boundary, the 5/4 boundary, or both. Ice-core stable-isotope records from Antarctica and Greenland, which are generally interpreted as a function of rayleigh distillation of water out of clouds, should reflect this large-scale vapor transport. If so, the largest change in poleward vapor transport occurred at the 5e/5d transition, where the largest ice-growth shifts in deuterium/hydrogen ratios (D/H) or $\delta^{18}O$ occur at both sites (Dansgaard and others, 1985; Jouzel and others, 1987). A 0.2‰ vapor transport effect at the 5e/5d transition would offset the difference in amplitude between the 5e/5d and 5a/4 events in the planktic $\delta^{18}O$.

If we make this correction to the planktic $\delta^{18}O$ record at the 5e/5d transition, and make a volume-weighted average of the Atlantic and Pacific planktic records to minimize the effects of Atlantic-Pacific vapor transport (Broecker, 1989), there would be roughly equal $\delta^{18}O$ shifts of about 0.5‰ to 0.6‰ across both the 5e/5d and 5/4 transitions due to ice volume. This is consistent with the Norwegian Sea benthic record, and also with the Pacific deep-water record, if there is some deep-sea cooling here at the 5e/5d transition and none at the 5/4 transition. Thus, within the limits of analytical errors in the data (about 0.1‰ on all data), possible isotopic effects of deep-sea cooling (about 0.2‰ to 0.3‰ on benthics), and vapor transport (about 0.2‰ on planktics), we cannot discern any differences in the size of the ice-growth isotopic events at 5/4 and 5e/5d. This contrasts with insolation forcing, which has a larger departure from modern at the 5e/5d transition than at 5/4 (Berger, 1978). The stage 5/4 event started with larger ice volumes, and thus probably reached greater ice volumes, as isotopic values in stage 4 (ca. 65 ka) reach about 70% of their glacial maximum values.

If we assume for simplicity a mean glacier ice $\delta^{18}O$ value of −30‰ to −35‰, and ice-volume contributions to the 5e/5d and 5a/4 ice-growth transitions of 0.6‰, then ice volume increased by about 25×10^6 km^3 at these times. The $\delta^{18}O$ shift, and presumably also the ice-volume change at the 5c/5b transition, was about half that, 0.3‰ or 13×10^6 km^3.

It is probably not appropriate to assume a constant $\delta^{18}O$ value for glacier ice, however (Mix and Ruddiman, 1984). The sensitivity of mean ocean water $\delta^{18}O$ change to ice-volume change is directly proportional to the isotopic offset between average ocean water and average ice. If early ice was less depleted in ^{18}O than later ice (because the ice sheets were smaller), the substage 5e/5d ice-volume transition would be underrepresented in the isotopic record.

The sensitivity of the ocean's isotope budget to ice-volume change would also depend on where the early ice grew. Many studies (e.g., Ives and others, 1975; Andrews and Mahaffy, 1976) cited Labrador and Baffin Island as the centers for initial glacierization. Mean $\delta^{18}O$ of modern precipitation here is nearly −20‰ (International Atomic Energy Agency, 1981). Ice growth at this isotopic composition would yield a relatively small change in the ocean's mean $\delta^{18}O$. In contrast, a typical $\delta^{18}O$ value of snow in Antarctica is −40‰ (Morgan, 1982). Ice growth in Antarctica, with this highly ^{18}O-depleted snow, would be overrepresented in the marine record by a factor of two relative to early ice growth in Europe or North America. It is important to remember that the marine oxygen-isotope record is not simply an average of all ice-volume changes on the globe. It is a *weighted* average, controlled by the oxygen-isotope composition of each parcel of ice.

CONSTRAINTS ON THE TIMING OF ICE-GROWTH EVENTS

Because direct evidence for early glaciation on land is difficult to date, the marine record is especially important for establishing the timing of the last glacial inception. Much effort has gone into developing a numerical time scale for marine $\delta^{18}O$ variations in the Quaternary. Early attempts used radiometrically dated magnetic boundaries and assumed constant sedimentation rates between these few control points (Shackleton and Opdyke, 1973; Emiliani and Shackleton, 1974). Mesollela and others (1969) and Broecker and Van Donk (1970) correlated the marine $\delta^{18}O$ record to coral terraces dated by uranium-series isotopes, but these ages were relatively imprecise. Until recently, more detailed dating required indirect methods.

Hays and others (1976) first identified orbital periodicities in the marine $\delta^{18}O$ data and made the first attempt to refine the time scale by tuning $\delta^{18}O$ variations to orbital variations. Through many subsequent iterations, this tuning effort, combined with stacking of many $\delta^{18}O$ records, has yielded the current SPECMAP time scales for benthic and planktic foraminifera (Imbrie and others, 1984; Martinson and others, 1987). The primary assumption is that isotopic variations are phase locked with orbital variations within the dominant orbital frequency bands of 1/41 ka^{-1}, and 1/23 ka^{-1}. The assumption is *not* that ice volume is a directly forced response to orbital changes. Although it could be such, tuning would work equally well if orbital changes acted as a pacemaker, setting the phase and frequency of a naturally varying climate without directly driving the variations.

According to the SPECMAP time scale, the ages of major events within stage 5 are the peak interglaciation (substage 5e) at 122 ka, the initial glacial stade (substage 5d) at 107 ka, and the youngest interstade (substage 5a) at 80 ka. The major ice-growth events discussed earlier center on 115 ± 4 ka (substage 5e/5d), 94 ± 4 ka (substage 5c/5b), and 72 ± 4 ka (stage 5/4). The first full glacial event of the most recent ice age (stage 4) dates at ca. 65 ka.

One test of the accuracy of the time scale is high coherency between the orbital variations and the data in both the tuned bands, and in the untuned 1/100 ka^{-1} band (Imbrie and others, 1984). High coherency means that there is a linear relation between the time-varying amplitudes of the orbital forcing and the varying amplitudes of the isotopic signal within each frequency band. Pisias (1983) showed that this result of high coherency would not be possible by artificial tuning of random variations to systematically varying orbital configurations. In addition, nonisotopic signals in the same samples are coherent with orbital variations after the oxygen-isotopic time scale is applied (Martinson and others, 1987; Imbrie and others, 1989). Note that the nonisotopic signals are not tuned to the orbital changes. Only $\delta^{18}O$ is tuned.

An assumption of the SPECMAP tuning strategy is that the phase offset between the orbital changes and the $\delta^{18}O$ data is constant within each tuned frequency band. Amplitude information is not used for tuning. Pisias and others (1990) relaxed this assumption of fixed phase and considered both the amplitude modulation and phase for tuning. Time-scale adjustments made by this strategy are small, typically less than 4 ka. The largest adjustments occur on major deglaciations. The effect on the last interglacial substage 5e is to move it from 122 ka (Imbrie and others, 1984) to ca. 126 ka (Pisias and others, 1990).

A challenge to the marine isotopic time scale comes from $\delta^{18}O$ variations in a calcite vein from Devils Hole, Nevada (Winograd and others, 1988). This records ground-water $\delta^{18}O$ variations linked mostly to precipitation temperature. The pattern of $\delta^{18}O$ change here visually resembles the marine record, but is opposite the sense of change. Dating was done with the ^{230}Th-^{234}U and ^{234}U-^{238}U methods, and error bars average 13 ka. At Devils Hole the apparent interglacial maximum falls at ca. 140 ka, nearly 20 ka older than the inferred age of the marine interglacial event 5e. Because of this large difference, Winograd and others (1988) questioned the dating of the marine record, and the conclusions drawn from it regarding orbital pacing of Pleistocene climate change.

Recent radiometric ages from the coral sea-level record support the tuned marine $\delta^{18}O$ time scales. New techniques of uranium-series dating by mass spectrometry yield very precise dates on coral terraces. Bard and others (1990) correct the radiocarbon time scale and increase glacial maximum ages by a few thousand years. The uranium-series age for the last interglacial sea-level highstand correlated to $\delta^{18}O$ substage 5e is 126 ± 3 ka (n = 10). The radiometric age for the late stage 5 interstade (5a) is 87.7 ± 0.4 ka (n = 2) (Edwards and others, 1987). Although these ages are a few thousand years older than those given by the SPECMAP time scale, they are significantly younger than those of Winograd and others (1988). Edwards and others (1987, p. 1552) stated "We have found no evidence for high sea level 140 kyr ago."

SUMMARY AND CONCLUSIONS

The marine oxygen-isotope record from foraminifera yields extremely valuable information about the timing and magnitude of glaciation on a global scale. The timing of the major isotopic shifts is relatively well determined. In the SPECMAP $\delta^{18}O$ time scale, the major ice-growth events of the last glacial inception are centered at ca. 118–112 ka (5e/5d), ca. 98–92 ka (5c/5b), and ca. 75–68 ka (5a/4). Refinements continue, but are small. Precise radiometric dates from coral terraces suggest that the SPECMAP time scale gives ages that are too young by at most a few thousand years. The ice volume at each isotopic event remains poorly known, however, due to difficulties in isolating local temperature and water-mass effects on $\delta^{18}O$, and changes in the isotopic composition of ice.

Many workers now believe that major cooling of the deep sea occurred during the last glacial inception (substage 5e/5d). If this is true, the uncorrected $\delta^{18}O$ record from benthic foraminifera overestimates the size of this ice-growth transition. In contrast, the $\delta^{18}O$ record from average low-latitude planktic foraminifera could underestimate the size of the substage 5e/5d ice-volume shift, if a decrease in vapor transport from low to high latitudes occurred at this time. Given the limits of analytical errors in the $\delta^{18}O$ data, incomplete data on deep-sea cooling, and poorly understood vapor transport and circulation effects, we cannot discern from the oxygen isotope data any differences in the size of the ice-growth events at the stage 5/4 and substage 5e/5d boundaries. It appears from the isotope records cited here that the substage 5c/5b ice-growth event was the smallest of the three major ice-growth events of stage 5.

Analysis of spatial variations of down-core $\delta^{18}O$ records in terms of water-mass effects, and attempts to account for their cause, is just beginning. Much work remains to be done. In any case, the isotopic shifts at the ice-growth transitions imply relatively large and rapid changes in global ice volume within isotopic stage 5 (ca. 120–75 ka). These events are not yet adequately accounted for by three-dimensional ice-sheet models.

ACKNOWLEDGMENTS

This research was supported by the National Science Foundation. I thank W. F. Ruddiman, an anonymous reviewer, and the editors for very useful comments.

REFERENCES CITED

Andrews, J. T., 1973, The Wisconsin Laurentide ice sheets. Dispersal centers, problems of rates of retreat, and climatic implications: Arctic and Alpine Research, v. 5, p. 185–199.

——, 1982, On the reconstruction of Pleistocene ice sheets: Quaternary Science Reviews, v. 1, p. 1–30.

Andrews, J. T., and Mahaffy, M.A.W., 1976, Growth rate of the Laurentide ice sheet and sea level lowering (with emphasis on the 115,000 PB sea level low): Quaternary Research, v. 6, p. 167–183.

Bard, E., Hamelin, B., Fairbanks, R. G., and Zindler, A., 1990, Calibration of the ^{14}C timescale over the past 30,000 years using mass spectrometric U-Th ages for Barbados corals: Nature, v. 345, p. 405–410.

Berger, A. L., 1978, Long-term variations of caloric insolation resulting from the Earth's orbital elements: Quaternary Research, v. 9, p. 139–167.

Broecker, W. S., 1978, The cause of glacial to interglacial climate change, *in* Evolution of planetary atmospheres and climatology of the Earth: Nice, France, Centre National d'Etudes Spatiales, p. 165–177.

——, 1986, Oxygen isotope constraints on surface ocean temperatures: Quaternary Research, v. 26, p. 121–134.

APPENDIX. V30-40 355-425 μm *G. sacculifer*

Depth	$\delta^{18}O$	$\delta^{13}C$	Depth	$\delta^{18}O$	$\delta^{13}C$	Depth	$\delta^{18}O$	$\delta^{13}C$	Depth	$\delta^{18}O$	$\delta^{13}C$	Depth	$\delta^{18}O$	$\delta^{13}C$
0.0	-1.22	1.98	84.0	0.30	1.86	159.0	-0.24	2.01	249.0	-0.99	2.04	309.0	-1.01	1.94
0.0	-1.29	2.02	87.0	0.31	1.80	165.0	-0.27	1.95	249.0	-0.83	1.89	309.0	-0.67	1.78
3.0	-1.22	1.89	90.0	0.37	1.64	168.0	-0.27	2.01	252.0	-0.63	1.85	312.0	-0.79	1.93
3.0	-1.28	1.96	90.0	0.09	1.63	171.0	-0.24	1.92	252.0	-0.66	1.96	312.0	-0.72	1.86
6.0	-1.26	2.05	93.0	0.05	2.02	174.0	-0.42	1.87	255.0	-0.96	1.72	315.0	-0.94	2.02
9.0	-1.19	2.12	96.0	0.07	1.69	174.0	-0.41	1.75	258.0	-0.92	1.84	315.0	-1.01	1.82
12.0	-0.80	1.92	96.0	0.00	1.75	177.0	-0.28	1.76	258.0	-0.97	1.91	318.0	-0.62	1.62
15.0	-0.98	2.04	99.0	-0.02	1.89	180.0	-0.26	1.69	261.0	-0.86	1.81	318.0	-1.01	2.04
18.0	-0.96	2.11	99.0	-0.09	1.80	183.0	-0.17	1.67	261.0	-0.88	1.97	321.0	-0.68	1.70
21.0	-0.74	1.84	102.0	-0.06	1.66	186.0	-0.27	1.80	264.0	-0.94	1.75	324.0	-0.78	1.65
24.0	-0.61	1.93	105.0	-0.12	1.92	186.0	-0.06	1.69	264.0	-0.80	2.04	327.0	-0.67	1.80
27.0	-0.25	1.84	108.0	-0.09	1.88	189.0	-0.32	1.58	267.0	-0.94	1.74	330.0	-0.67	1.77
30.0	-0.34	1.64	111.0	-0.18	1.84	192.0	-0.08	1.85	267.0	-0.83	1.99	333.0	-0.71	1.88
33.0	-0.32	1.58	114.0	-0.07	2.06	192.0	-0.07	1.82	270.0	-0.79	1.66	336.0	-0.58	1.81
36.0	-0.42	1.68	117.0	-0.06	1.85	195.0	0.15	1.93	270.0	-0.73	1.94	339.0	-0.82	1.59
36.0	-0.16	1.69	120.0	-0.12	2.01	198.0	-0.01	1.80	273.0	-0.77	1.98	339.0	-0.76	1.77
39.0	-0.15	1.71	123.0	0.10	1.90	201.0	0.05	1.93	276.0	-0.87	1.72	342.0	-1.02	1.75
42.0	0.00	1.73	123.0	0.17	1.82	204.0	-0.06	2.01	279.0	-1.02	1.81	345.0	-0.94	1.79
48.0	0.27	1.78	126.0	-0.01	1.82	210.0	-0.82	2.15	282.0	-1.01	1.76	348.0	-1.14	1.94
51.0	0.52	1.79	129.0	-0.02	1.92	213.0	-0.66	2.19	285.0	-0.85	1.87	351.0	-1.06	1.84
54.0	0.45	1.97	132.0	-0.19	2.00	216.0	-0.68	2.06	285.0	-0.90	2.08	354.0	-1.18	1.62
54.0	0.45	1.86	132.0	-0.20	1.75	219.0	-0.78	2.22	288.0	-0.98	1.66	357.0	-1.23	1.74
57.0	0.52	1.92	135.0	-0.02	1.94	222.0	-0.88	2.20	288.0	-0.84	1.83	360.0	-1.13	1.70
60.0	0.40	1.98	138.0	-0.04	2.02	225.0	-0.72	2.09	291.0	-1.13	1.79	360.0	-0.96	1.73
63.0	0.41	1.77	141.0	-0.22	2.13	228.0	-0.78	2.05	291.0	-1.15	1.99	363.0	-1.39	2.02
66.0	0.30	1.78	141.0	-0.30	1.93	231.0	-0.91	2.33	294.0	-1.14	1.86	366.0	-1.36	1.88
69.0	0.37	1.75	144.0	-0.27	1.92	234.0	-0.90	2.26	297.0	-1.22	1.87	369.0	-1.45	2.02
72.0	0.30	1.61	144.0	-0.40	1.87	237.0	-0.93	2.05	300.0	-0.94	1.70	372.0	-1.44	1.87
72.0	0.27	1.79	147.0	-0.21	2.04	240.0	-1.02	2.23	300.0	-1.20	2.09	375.0	-1.20	1.71
75.0	0.48	1.70	150.0	-0.19	2.04	240.0	-1.09	2.30	303.0	-1.00	1.94	375.0	-1.48	2.10
78.0	0.38	1.95	153.0	-0.22	2.02	243.0	-1.05	2.04	306.0	-0.77	1.83	378.0	-1.06	1.63
81.0	0.36	1.76	156.0	-0.22	2.00	246.0	-0.95	2.01	306.0	-1.02	2.05	378.0	-1.40	1.64

——, 1989, The salinity contrast between the Atlantic and Pacific oceans during glacial time: Paleoceanography, v. 4, p. 207–212.

Broecker, W. S., and Peng, T. H., 1986, Carbon cycle: 1985: Radiocarbon, v. 28, p. 309–327.

Broecker, W. S., and Van Donk, J., 1970, Insolation changes, ice volumes and the O-18 record in deep-sea cores: Reviews of Geophysics and Space Physics, v. 8, p. 169–198.

Broecker, W. S., Andree, M., Bonani, G., Wolfli, W., Oeschger, H., Klas, M., Mix, A., and Curry, W., 1988, Preliminary estimates for the radiocarbon age of deep water in the glacial ocean: Paleoceanography, v. 3, p. 659–669.

Budd, W. F., and Smith, I. N., 1979, The growth and retreat of ice sheets in response to orbital radiation changes, *in* Allison, I., ed., Sea level, ice, and climatic change: International Association of Hydrological Sciences Publication 131, p. 369–409.

Chappell, J., and Shackleton, N. J., 1986, Oxygen Isotopes and sea level: Nature, v. 324, p. 137–140.

CLIMAP, 1981, Seasonal reconstructions of the Earth's surface at the last glacial maximum: Geological Society of America Map and Chart Series MC-36.

CLIMAP, 1984, The last interglacial ocean: Quaternary Research, v. 21, p. 123–224.

Craig, H., 1987, Stable isotopes of water: ^{2}H (= D) and ^{18}O, *in* Ostlund, H. G., Craig, H., Broecker, W. S. and Spencer, D., eds., GEOSECS Atlantic, Pacific, and Indian Ocean expeditions, 7: Washington, D.C., National Science Foundation, p. 6–7.

Craig, H., and Gordon, L. I., 1965, Deuterium and oxygen 18 variations in the ocean and marine atmosphere, *in* Tongiorgi, E., ed., Stable isotopes in oceanic studies and paleotemperatures (Third SPOLETO Conference on Nuclear Geology: Pisa, Consiglio Nazionale Delle Richerche, Laboratorio di Geologia Nucleare, p. 9–130.

Dansgaard, W., and Tauber, H., 1969, Glacier oxygen-18 content and Pleistocene ocean temperatures: Science, v. 166, p. 499–502.

Dansgaard, W., Clausen, H. B., Gundestrup, N., Johnsen, S. J., and Rygner, C., 1985, Dating and climatic interpretation of two deep Greenland ice cores, *in* Langway, C. C., Jr., Oeschger, H., and Dansgaard, W., eds., Greenland ice core: Geophysics, geochemistry, and the environment: American Geophysical Union Geophysical Monograph 33, p. 71–76.

Duplessy, J. C., 1978, Isotope studies, *in* Gribben, J., ed., Climatic change: Cambridge, Cambridge University Press, p. 46–67.

Duplessy, J. C., Labeyrie, L., and Blanc, P. L., 1988, Norwegian Sea deep water variations over the last climatic cycle: Paleo-oceanographical implications, *in* Wanner, H. and Siegenthaler, U., eds., Long and short term variability of climate: Berlin, Springer Verlag, p. 83–116.

APPENDIX. V30-40 355-425 μm *G. sacculifer* (continued)

Depth	$\delta^{18}O$	$\delta^{13}C$	Depth	$\delta^{18}O$	$\delta^{13}C$	Depth	$\delta^{18}O$	$\delta^{13}C$	Depth	$\delta^{18}O$	$\delta^{13}C$	Depth	$\delta^{18}O$	$\delta^{13}C$
381.0	-0.92	1.73	456.0	0.09	1.61	522.0	-0.51	1.70	600.0	-1.29	2.30	672.0	-0.50	1.97
384.0	-1.01	1.66	459.0	0.09	1.68	525.0	-0.51	1.82	600.0	-1.38	2.19	675.0	-0.44	1.85
387.0	-0.42	1.51	459.0	0.28	1.71	528.0	-0.39	1.73	603.0	-1.22	2.37	678.0	-0.45	1.61
378.0	-0.67	1.63	462.0	0.28	1.88	531.0	-0.47	1.45	606.0	-1.37	2.26	678.0	-0.64	1.67
390.0	-0.57	1.63	462.0	0.41	1.63	531.0	-0.33	1.68	609.0	-1.32	2.10	681.0	-0.67	1.94
393.0	-0.02	1.52	465.0	0.22	1.71	534.0	-0.44	1.66	612.0	-1.49	2.02	684.0	-0.75	1.69
396.0	0.19	1.47	468.0	0.10	1.54	537.0	-0.47	1.63	612.0	-1.30	1.88	687.0	-0.90	1.78
399.0	0.03	1.54	468.0	0.28	1.91	540.0	-0.63	1.58	615.0	-1.03	1.69	690.0	-0.98	1.76
399.0	0.24	1.52	471.0	-0.03	1.67	543.0	-0.44	1.55	618.0	-1.00	1.92	693.0	-0.98	1.99
302.0	0.15	1.27	474.0	0.12	1.62	546.0	-0.39	1.71	621.0	-0.98	1.88	696.0	-1.09	1.70
402.0	0.21	1.46	477.0	-0.31	1.92	549.0	-0.19	1.78	624.0	-0.58	1.91	699.0	-1.18	1.98
405.0	0.02	1.54	477.0	0.12	1.55	552.0	-0.34	1.79	624.0	-0.85	1.90	702.0	-1.18	1.89
408.0	-0.07	1.42	480.0	-0.20	1.77	555.0	-0.22	1.79	627.0	-0.77	1.81	705.0	-1.30	1.63
411.0	-0.16	1.65	480.0	0.20	1.49	558.0	-0.24	1.76	627.0	-0.87	1.89	708.0	-0.98	1.64
414.0	0.06	1.51	483.0	-0.17	1.84	561.0	-0.51	1.88	630.0	-0.99	2.04	711.0	-0.68	1.23
417.0	0.11	1.42	486.0	-0.09	1.70	561.0	-0.45	1.69	633.0	-1.06	1.89	714.0	-0.42	1.44
420.0	-0.01	1.56	489.0	0.01	1.49	564.0	-0.36	1.65	636.0	-0.80	1.86	717.0	-0.10	1.67
420.0	-0.07	1.61	489.0	-0.04	1.46	567.0	-0.53	1.66	636.0	-0.79	1.88	720.0	-0.21	1.68
423.0	-0.03	1.32	492.0	-0.26	1.71	570.0	-0.69	1.68	639.0	-1.07	1.75	723.0	-0.23	1.43
423.0	0.00	1.73	495.0	-0.27	1.74	573.0	-0.76	1.77	642.0	-1.07	2.03	726.0	-0.20	1.36
426.0	0.05	1.52	498.0	-0.21	1.85	576.0	-0.80	1.73	645.0	-1.09	1.98	729.0	-0.18	1.69
429.0	0.07	1.53	498.0	-0.25	1.58	579.0	-0.99	1.92	648.0	-0.95	1.87	732.0	-0.21	1.67
432.0	-0.11	1.46	501.0	-0.43	1.60	582.0	-1.09	1.89	651.0	-0.96	1.82	735.0	-0.19	1.49
435.0	-0.07	1.53	504.0	-0.39	1.47	585.0	-1.13	2.05	654.0	-0.84	1.57	738.0	-0.35	1.76
438.0	0.04	1.52	504.0	-0.36	1.46	588.0	-1.09	1.90	657.0	-0.62	1.54	741.0	-0.41	1.74
441.0	-0.07	1.58	507.0	-0.45	1.85	588.0	-1.11	2.39	660.0	-0.47	1.74	744.0	-0.42	1.60
444.0	0.21	1.46	510.0	-0.58	1.82	291.0	-0.95	2.09	663.0	-0.44	1.71	747.0	-0.28	1.58
447.0	0.12	1.47	513.0	-0.59	1.69	591.0	-1.18	2.18	666.0	-0.36	1.70	750.0	-0.36	1.69
450.0	0.02	1.60	516.0	-0.48	1.63	594.0	-1.15	2.16	669.0	-0.42	1.87	753.0	-0.70	2.11
453.0	0.33	1.64	519.0	-0.35	1.64	597.0	-1.17	2.06						

Edwards, R. L., Chen, J. H., Ku, T.-L., and Wasserburg, G. J., 1987, Precise timing of the last interglacial period from mass spectrometric determination of thorium-230 in corals: Science, v. 236, p. 1547–1553.

Emiliani, C., 1955, Pleistocene temperatures: Journal of Geology, v. 63, p. 538–578.

Emiliani, C., and Shackleton, N. J., 1974, The Brunhes Epoch: Isotopic paleotemperatures and geochronology: Science, v. 183, p. 511–514.

Fairbanks, R. G., 1989, A 17,000-year glacio-eustatic sea level record: Influence of glacial melting rates on the Younger Dryas event and deep-ocean circulation: Nature, v. 342, p. 637–642.

Hays, J. D., Imbrie, J., and Shackleton, N. J., 1976, Variations in the Earth's orbit: Pacemaker of the ice ages: Science, v. 194, p. 1121–1132.

Hughes, T. J., Denton, G. H., Andersen, B. G., Schilling, D. H., Fastook, J. L., and Lingle, C. S., 1981, The last great ice sheets, a global view, *in* Denton, G. H., and Hughes, T. J., eds., The last great ice sheets: New York, Wiley Interscience, p. 263–317.

Imbrie, J., and eight others, 1984, The orbital theory of Pleistocene climate: Support from a revised chronology of the marine $\delta^{18}O$ record, *in* Berger, A., Imbrie, J., Hays, J., Kukla, G., and Saltzman, B., eds., Milankovitch and climate, Part 1: Dordrecht, Reidel, p. 269–305.

Imbrie, J., McIntyre, A., and Mix, A., 1989, Oceanic response to orbital forcing in the late Quaternary: Observational and experimental strategies, *in* Berger, A., Schneider, S., and Duplessy, J. C., eds., Climate and the geosciences: Dordrecht, Kluwer, p. 121–164.

International Atomic Energy Agency, 1981, Statistical treatment of environmental isotope data in precipitation: IAEA Technical Report Series 206, p. 1–255.

Ives, J. D., Andrews, J. T., and Barry, R. G., 1975, Growth and decay of the Laurentide Ice Sheet and comparisons with Fenno-Scandinavia: Naturwissenschaften, v. 62, p. 118–125.

Jouzel, J., Lorius, C., Petit, J. R., Genthon, C., Barkov, N. I., Kotlyakov, V. M., and Petrov, V. M., 1987, Vostok ice core: A continuous isotope temperature record over the last climatic cycle (160,000 years): Nature, v. 329, p. 403–408.

Keir, R. S., 1988, On the late Pleistocene ocean geochemistry and circulation: Paleoceanography, v. 3, p. 413–446.

Kutzbach, J. E., and Guetter, P., 1986, The influence of changing orbital parameters and surface boundary conditions on climate simulations for the past 18,000 years: Journal of Atmospheric Science, v. 43, p. 1726–1759.

Labeyrie, L. D., Duplessy, J. C., and Blanc, P. L., 1987, Variations in mode of formation and temperature of oceanic deep waters over the past 125,000 years: Nature, v. 327, p. 477–482.

Martinson, D. G., Pisias, N. G., Hays, J. D., Imbrie, J., Moore, T. C., and Shackleton, N. J., 1987, Age dating and the orbital theory of the ice ages: Development of a high-resolution 0 to 300,000-year chronostratigraphy: Quaternary Research, v. 27, p. 1–30.

Matthews, R. K., and Poore, R. Z., 1981, Tertiary $\delta^{18}O$ record and glacio-eustatic sea-level fluctuations: Geology, v. 8, p. 501–504.

McIntyre, A., Ruddiman, W. F., Karlin, K., and Mix, A. C., 1989, Surface water responses of the equatorial Atlantic to orbital forcing: Paleoceanography, v. 4, p. 19–55.

Mesolella, K. J., Matthews, R. K., Broecker, W. S., and Thurber, D. L., 1969, The astronomical theory of climatic change: Barbados data: Journal of Geology, v. 77, p. 250–274.

Mix, A. C., 1987, The oxygen-isotope record of glaciation, *in* Ruddiman, W. F., and Wright, H., eds., North America and adjacent oceans during the last deglaciation: Boulder, Colorado, Geological Society of America, The Geology of North America, v. K-3, p. 111–135.

Mix, A. C., and Pisias, N. G., 1988, Oxygen isotope analyses and deep-sea temperature changes: Implications for rates of oceanic mixing: Nature, v. 331, p. 249–251.

Mix, A. C., and Ruddiman, W. F., 1984, Oxygen-isotope analyses and Pleistocene ice volumes: Quaternary Research, v. 21, p. 1–20.

Morgan, V. I., 1982, Antarctic ice sheet surface oxygen isotope values: Journal of Glaciology, v. 28, p. 315–323.

Pisias, N. G., 1983, Geologic time series from deep-sea sediments: Time scales and distortion by bioturbation: Marine Geology, v. 51, p. 99–113.

Pisias, N. G., Mix, A. C., and Zahn, R., 1990, Non-linear response in the global climate system: Evidence from benthic oxygen isotopic record in core RC13-110: Paleoceanography, v. 5, p. 147–160.

Rind, D., Peteet, D., and Kukla, G., 1989, Can Milankovitch orbital variations initiate the growth of ice sheets in a general circulation model?: Journal of Geophysical Research, v. 94, p. 12,851–12,871.

Shackleton, N. J., 1967, Oxygen Isotope analyses and Pleistocene temperatures, reassessed: Nature, v. 215, p. 15–17.

——, 1987, Oxygen isotopes, ice volume, and sea level: Quaternary Science Reviews, v. 6, p. 183–190.

Shackleton, N. J., and Opdyke, N. D., 1973, Oxygen isotope and palaeomagnetic stratigraphy of equatorial Pacific core V28-238: Oxygen isotope temperatures and ice volumes on a 10^5 and 10^6 year scale: Quaternary Research, v. 3, p. 39–55.

Shackleton, N. J., and Pisias, N. G., 1985, Atmospheric carbon dioxide, orbital forcing, and climate, *in* Sundquist, E. T., and Broecker, W. S., eds., The carbon cycle and atmospheric CO_2: Natural variations Archean to present: American Geophysical Union Geophysical Monograph 32, p. 303–318.

Shackleton, N. J., Duplessy, J. C., Arnold, M., Maurice, P., Hall, M. A., and Cartlidge, J., 1988, Radiocarbon age of last glacial Pacific deep water: Nature, v. 335, p. 708–711.

Stuiver, M., Quay, P. D., and Ostlund, H. G., 1983, Abyssal water carbon-14 distribution and the age of the world ocean: Science, v. 219, p. 849–851.

Winograd, I., Szabo, B. J., Coplen, T. B., and Riggs, A. C., 1988, A 250,000-year climatic record from Great Basin vein calcite: Implications for Milankovitch theory: Science, v. 242, p. 1275–1280.

Manuscript Submitted June 1989
Manuscript Accepted by the Society September 6, 1991

Printed in U.S.A.

Geological Society of America
Special Paper 270
1992

The last interglacial-glacial transition in North America: Evidence from uranium-series dating of coastal deposits

Daniel R. Muhs
U.S. Geological Survey, MS 424, Box 25046, Denver Federal Center, Denver, Colorado 80225

ABSTRACT

Considerable uncertainty exists as to whether the last interglacial was relatively "short" (~10 ka) or "long" (~20–60 ka), although most investigators generally agree that the last interglacial correlates with all or part of deep-sea oxygen-isotope stage 5. A compilation of reliable U-series ages of marine terrace corals from deposits that have been correlated with isotope stage 5 indicates that there were three relatively high sea-level stands at ca 125–120 ka, ca. 105 ka, and ca. 85–80 ka, and these ages agree with the times of high sea level predicted by the Milankovitch orbital-forcing theory. At a number of localities, however, there are apparently reliable coral ages of ca. 145–135 ka and ca. 70 ka, and the Milankovitch theory would not predict high sea levels at these times. These ages are at present unexplained and require further study.

The issue of whether the last interglacial was "short" or "long"can be addressed by examining the evidence for how high sea level was during the stands at ca. 125 ka, ca. 105 ka, and ca. 80 ka, because sea level is inversely proportional to global ice volume. In tectonically stable areas such as Bermuda, the Bahamas, the Yucatan peninsula, and Florida, there is clear evidence that sea level at ca. 125 ka was +3 to +10 m higher than present. During the ca. 105 ka and ca. 80 ka high sea-level stands, there is conflicting evidence for how high sea levels were. Studies of uplifted terraces on Barbados and Haiti and most studies of terraces on New Guinea indicate sea levels considerably lower than present. Studies of the terraces and deposits on the east and west coasts of North America, Bermuda, and the Bahamas, however, indicate sea levels close to, or only slightly below, the present at these times. Thus, data from Barbados, Haiti, and New Guinea indicate a "short" last interglacial centering ca. 125 ka, but data from the other localities indicate that sea level was high during much of the period from 125 to 80 ka, and that there were two minor ice advances in that period.

If it is accepted that the last interglacial period was relatively "long" and ended sometime after ca. 80 ka, then coastal deposits on the California Channel Islands record a shift in the nature of sedimentation at the interglacial/glacial transition. Marine terraces that are ca. 80 ka are overlain by two eolianite units separated by paleosols. U-series ages of the terrace corals and carbonate rhizoliths indicate that eolian sedimentation occurred between ca. 80 and 49 ka, and again between ca. 27 and 14 ka. Eolian sands were apparently derived from carbonate-rich shelf sediments during glacially-lowered sea levels, because there are not sufficient beach sources for calcareous sediment at present. The times of eolian sedimentation agree well with times of glaciation predicted by the Milankovitch model of climatic change.

Muhs, D. R., 1992, The last interglacial-glacial transition in North America: Evidence from uranium-series dating of coastal deposits, *in* Clark, P. U., and Lea, P. D., eds., The Last Interglacial-Glacial Transition in North America: Boulder, Colorado, Geological Society of America Special Paper 270.

INTRODUCTION

Accurate information about major climatic shifts of the past is critical to the prediction of future climates because these can be used to test alternative models of climate change. The last interglacial/glacial transition is an important interval in the late Quaternary history of North America. Less is known about the last interglacial/glacial transition than the last glacial/Holocene transition, because the former occurred at least 75 ka ago and perhaps as much as 115 ka. The record of the last interglacial/glacial transition is not as well preserved as the last glacial/Holocene transition, there are fewer dating methods available for developing an accurate chronology, and fewer investigators have studied this time interval.

There seems to be agreement among a great many investigators that the last interglacial is found in various geologic records, but there is considerable disagreement as to the chronology of this interval of time. In the midcontinent, the last interglacial is represented stratigraphically by the Sangamon soil (Follmer, 1978). Many investigators in the midcontinent have suggested that the period during which this well-developed soil formed was warmer than the present, but perhaps no longer than the Holocene, or about 10 ka (see review in Boardman [1985]). Fulton and Prest (1987) and St.-Onge (1987) suggest that the "Sangamonian Stage" in Canada lasted from 130 to 80 ka, and it is considered to be correlative with all of deep-sea oxygen-isotope stage 5. However, they regard the "Sangamon interglaciation" to be a time-transgressive term for geologic-climate or event units. Richmond and Fullerton (1986) also correlated the glacial record with the deep-sea oxygen-isotope record, but restricted the term "Sangamon" to substage 5e of the deep-sea record (ca. 132–122 ka) and referred to substages 5a–5d as the "Eowisconsin." In describing the Clear Lake, California, pollen record, Adam (1988) correlated a general period of high oak pollen with all of deep-sea isotope stage 5, but pointed out that the early part of this period, which he correlated with substage 5e, is the only time period that is truly "interglacial" in temperature terms. In their major study of the marine record from deep-sea cores, the CLIMAP Project Members (1984) restricted their definition of the "last interglacial" to substage 5e on the grounds that this was the last time that glacial ice volume was as low as or lower than present.

There are several major challenges to the concept that the last interglacial had a duration of only about 10 ka and is correlative only with isotope substage 5e, as suggested by the studies cited above. From the pedologic record, Ruhe (1974) and Ruhe and others (1974) proposed that the well-developed (compared to modern soils) Sangamon soils in the midwestern United States are the result of a warmer, wetter climate in that region during the period of soil formation. In contrast, Follmer (1982), Muhs (1983a), and Boardman (1985) argued that characteristics of the Sangamon soil could also be explained by a period of pedogenesis significantly in excess of 10 ka. Speleothem growth periods in Canadian mountains and Iowa, dated by U-series methods, suggest that the last interglacial period may have had a duration of 60 ka or more (Harmon and others, 1977; Lively, 1983). Oxygen-isotope data from an ice core taken from Devon Island off northern Canada also suggest a last interglacial much longer than 10 ka (Koerner and Fisher, 1985). Thus, two schools of thought have emerged from studies of proxy climate data: those paleoenvironmental records suggesting a "short" (ca 10 ka) last interglacial and those suggesting a "long" (ca 20–60 ka) last interglacial.

In this chapter I review age estimates and paleo-sea levels of the last interglacial derived from marine terraces found on the east and west coasts of North America and on islands close to North America such as Bermuda, the Bahamas, Barbados, and Haiti, particularly with regard to the record of the timing of the last interglacial. Terrace ages on the Huon peninsula of New Guinea are also examined, because this area has been studied carefully and has given one of the most detailed sea-level records of the late Quaternary. Finally, some preliminary results of studies of marine and eolian sediments on the Channel Islands of California and their implications for the record of the last interglacial/glacial transition are presented.

AGE ESTIMATES OF DEEP-SEA OXYGEN-ISOTOPE STAGE 5

A common ground in almost all of the studies cited above is correlation to the oxygen-isotope record in deep-sea cores, which is considered to be a much more complete record than those of glacial deposits, buried soils, pollen, speleothems, or ice. It follows, therefore, that one of the most important questions concerning the record of the last interglacial in deep-sea cores is the method(s) by which it has been dated. Most investigators agree that all or part of isotope stage 5 represents the last interglacial, so it is pertinent to examine the chronological data for this interval of time.

A compilation of data from some of the better-studied deep-sea cores reveals a range of age estimates for both the beginning and end of oxygen-isotope stage 5 (Table 1). The age estimates, derived by a variety of methods, are in general agreement, but there are significant differences when they are examined in detail. Both the ^{230}Th excess and Al accumulation methods used for core V28-238 indicate that stage 5 could have begun significantly earlier (by 10–15 ka) than the "orbital tuning" and ^{231}Pa/^{230}Th/sedimentation-rate age estimates suggest for core RC11-120. It is also possible to compare two different cores (V28-238 and CH72-02) dated by the same method (^{230}Th excess). These two cores differ in their starting dates for stage 5 by 20 ka and in their ending dates for stage 5 by almost 13 ka. Given all of the uncertainties in the dating methods that have been used on deep-sea cores, it is not possible, at the present time, to favor one chronology over another. Collectively, these data suggest that further refinements in independent dating of deep-sea cores is desirable.

Because of the problems in dating cores directly, a number of investigators have attempted to develop time scales for the deep-sea record by "tuning" the oxygen-isotope record to the

TABLE 1. ESTIMATES OF THE DEEP-SEA OXYGEN ISOTOPE STAGE 4/5 AND STAGE 5/6 BOUNDARIES

Core	4/5 Boundary (ka)	5/6 Boundary (ka)	Method(s) of Age Determination	Reference*
V28-238	77.9†	133.9†	Paleomagnetism/ sedimentation rate	1
V28-238	81.6	144.7	^{230}Th excess	2
V28-238	80.4	138	Aluminum accumulation	2
CH72-02	68.7	124	^{230}Th excess	3
RC11-120	73	127	^{231}Pa/^{230}Th/sedimentation rate	4
RC11-120	73.9	129.8	Orbital tuning	5
RC14-37	73.5	132	K/Ar/sedimentation rate	6

*1 = Shackleton and Opdyke (1973); 2 = Kominz and others (1979); 3 = Southon and others (1987); 4 = Hays and others (1976); 5 = Martinson and others (1987); 6 = Ninkovich and others (1978).
†Calculated by Muhs using core data from Shackleton and Opdyke (1973) and 730 ka as the time of the Brunhes-Matuyama boundary (Mankinen and Dalrymple, 1979).

time scale of changes in the distribution of solar radiation over the surface of the Earth (Morley and Hays, 1981; Johnson, 1982; Martinson and others, 1987). This approach assumes that the major factors driving climate changes are changes in the Earth's orbital parameters, as outlined by Milankovitch (1941). Unfortunately, this method is somewhat circular in its reasoning: it was the coincidence of independent U-series age estimates of deep-sea cores (using the ^{230}Th excess and ^{231}Pa excess methods) and uplifted coral reefs (using the ^{230}Th/^{234}U method) that was the original basis of support for the Milankovitch orbital-forcing theory (Broecker and others, 1968). This concern is magnified in light of recent climatic modeling evidence of an uncertain role for orbital factors in the growth and maintenance of continental ice sheets (Rind and others, 1989, this volume).

Until new methods for dating cores become available, some insight can be gained by examining the sea-level highstands recorded by marine terraces and dated by U-series analysis of fossil corals associated with these terraces. Marine terraces can yield data about both the timing of interglacial periods and the amount of global glacial ice, as recorded by the magnitude of sea-level rise.

MARINE TERRACES AS INDICATORS OF THE TIMING OF INTERGLACIAL SEA-LEVEL HIGHSTANDS

The importance of marine terraces as indicators for the timing of interglacial periods stems from the fact that heights of sea level have an inverse relation to amounts of global ice. Investigators generally agree that emergent marine terraces (whether coral reefs or cliff-platform types) represent highstands of sea level. On coasts undergoing tectonic uplift, terraces may also form during lowstands of sea level, but such terraces are not now exposed and many that formed were probably largely eroded away by higher stands of sea level. However, terraces that formed during highstands on tectonically active coasts can be uplifted sufficiently during the succeeding intervals of lowstands that they are elevated and hence preserved above sea level during subsequent highstands. For both coral-reef growth and cliff-platform cutting on tectonically emergent coasts, the optimum time of formation will be when the rate of sea-level rise matches the rate of uplift. This will occur during stage 1, 2, or 3 (as shown in Fig. 1), depending on whether the coast has a high, moderate, or low rate of uplift, respectively. On stable coasts, terrace formation will occur when sea level is not changing, shown as stage 4 in Figure 1. On both uplifting and stable coasts, terraces will be abandoned when sea level falls (stage 5 in Fig. 1). Thus, on coasts where tectonic uplift has been occurring throughout the Quaternary, a series of emergent terraces may be present which record the major highstands of sea level that correspond to times of minimal global ice, or interglacial and interstadial intervals (Fig. 2).

Under the assumption that, on uplifting coasts, marine terraces correspond to some part of a sea-level highstand, it follows that the intervals between times of successive terrace formation must represent lowstands that occur during buildup and maintenance of glacial ice. Thus, flights of marine terraces have a record of ice-volume fluctuations. The best method for determining the

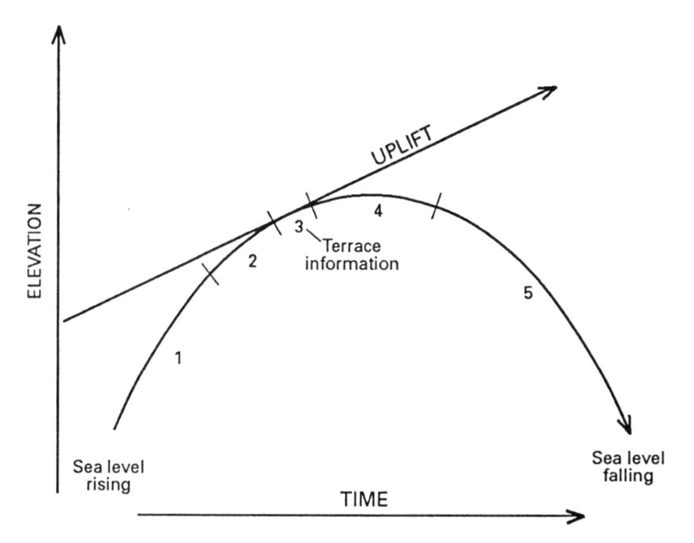

Figure 1. Model of marine terrace formation as a function of sea-level change and tectonic uplift. The diagonal line marked "uplift" represents a constant rate of uplift through time on a tectonically active coastline. The curved line represents the rise and fall of sea level during an interglacial sea-level highstand. The numbered segments of the sea-level curve represent different time periods during the history of sea-level rise and fall. On a tectonically active coast, terrace formation begins in stage 2 (rising sea) and is completed in stage 3, when the rate of uplift matches the rate of sea-level rise (shown here by the constant uplift rate line as a tangent to the sea-level curve in stage 3). The terrace is abandoned during stages 4 and 5. On a tectonically stable coast, terrace formation would take place during stage 4, when sea level is high, but not changing. Modified from Bradley and Griggs (1976).

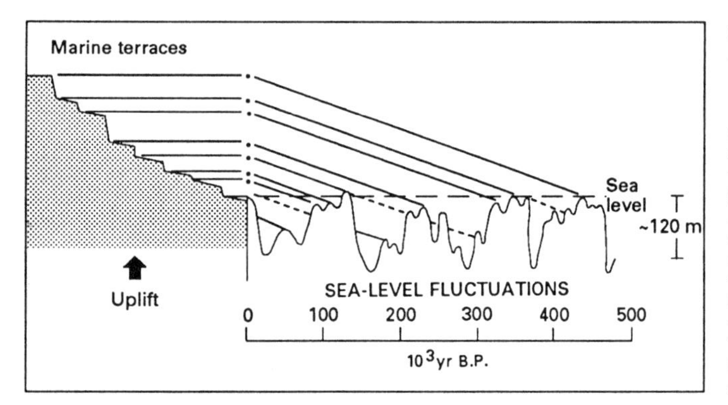

Figure 2. Model of successive marine-terrace formation on a tectonically active coast. Terraces form during highstands of sea level and are uplifted and preserved during succeeding lowstands of sea level. Slopes of diagonal lines represent uplift rates. Modified from Lajoie (1986).

time of terrace formation comes from uranium-series (^{230}Th/^{234}U, ^{234}U/^{238}U, and ^{231}Pa/^{235}U) dating of unaltered, aragonitic corals, either colonial or solitary forms.

URANIUM-SERIES AGES OF MARINE TERRACES

Marine terrace record

Detailed studies on the tectonically emergent islands of Barbados, New Guinea, and Haiti have identified three uplifted coral-reef terraces that record three highstands of sea level correlated, by many investigators, with some or all of deep-sea oxygen-isotope stage 5. The terraces are listed in order of ascending elevation and increasing age: on Barbados, the Worthing, Ventnor, and Rendezvous Hill terraces (Bender and others, 1979; also called Barbados I, II, and III, respectively, by Mesolella and others, 1969); on New Guinea, reef complexes V, VI, and VII (Chappell, 1974; Bloom and others, 1974); and on Haiti, the Mole, Saint, and Nicolas terraces (Dodge and others, 1983). These terraces have been dated by U-series methods as ca. 80 ka, 105 ka, and 125 ka, and have been correlated with, from youngest to oldest, isotope substages 5a, 5c, and 5e, respectively, by Shackleton and Opdyke (1973), Shackleton and Matthews (1977), Fairbanks and Matthews (1978), Bender and others (1979), Aharon (1983), Dodge and others (1983), Matthews (1985), and Chappell and Shackleton (1986). Even with the uncertainties in the age assignments for isotope stage 5 (Table 1). these are probably reasonable correlations. On the tectonically active west coast of North America, two and sometimes three terraces have been correlated with the Barbados, New Guinea, and Haiti terraces described above by Ku and Kern (1974), Kern (1977), Kennedy and others (1982), Muhs and Szabo (1982), Muhs and others (1987b, 1988, 1990), and Rockwell and others (1989). Terraces on the west coast of North America are not emergent coral reefs, but rather are uplifted shore platforms that are overlain by marine sands and gravels. These marine-terrace deposits frequently contain solitary corals or colonial hydrocorals that allow U-series dating.

On coastlines thought to be tectonically stable (i.e., not undergoing significant uplift or subsidence), such as the Bahamas, Bermuda, the Yucatan peninsula, and the east coast of the United States, there is a far less-clear picture of the sea-level record. Marine deposits on these coasts generally do not form geomorphically distinct terraces as they do on uplifting coasts, and there is considerably more controversy concerning their interpretation. However, a number of these emergent marine deposits contain corals, and U-series ages have been reported by Broecker and Thurber (1965), Osmond and others (1965), Neumann and Moore (1975), Szabo and others (1978), Harmon and others (1983), Szabo (1985), and Szabo and Halley (1988).

Uranium-series systematics

The reliability of a U-series age of a fossil coral is contingent on that organism having incorporated a small amount of U (but no Th) from sea water during growth, and having maintained a closed system with respect to U and its daughter products after death. The most rigorous test for closed-system conditions is agreement between ^{230}Th/^{234}U and ^{231}Pa/^{235}U ages, but few laboratories analyze samples for ^{231}Pa abundances. However, a ^{230}Th/^{234}U age can be interpreted to be reliable if the fossil coral passes the following tests: (1) the sample is 95%–100% aragonite; (2) the ^{230}Th/^{232}Th value is high (>20), indicating no incorporation of Th-bearing minerals; (3) the U concentration is 2–3 ppm (typical for colonial corals; solitary corals appear to be higher), indicating no post-mortem gains or losses of U; and (4) the ^{234}U/^{238}U value is consistent with the ^{230}Th/^{234}U age, assuming that the initial ^{234}U/^{238}U value in sea water at the time of coral growth was 1.14–1.15, and that this ratio has decreased with ^{234}U decay toward an equilibrium value of 1.00. Measurements of modern sea water have given ^{234}U/^{238}U values of 1.14–1.15 (Ku and others, 1977; Chen and others, 1986).

In Table 2, I have compiled all published U-series ages of corals of which I am aware from the study areas described above that are from marine deposits or emergent reefs that have either been correlated by various investigators with oxygen-isotope stage 5 or have ages that are in the estimated age ranges for isotope stage 5 given in Table 1. In making this compilation I have included only those samples that meet the closed-system criteria described above. Because of methodological differences between laboratories (Harmon and others, 1979; Ivanovich and others, 1984), I have recalculated all of the ages for the samples in Table 2 using half-lives of 75,200 yr and 244,000 yr for ^{230}Th and ^{234}U, respectively, as well as giving the original ages reported by the authors. All uncertainties given in Table 2 and in the text are 1 sigma.

Uranium-series ages of terrace corals: A general picture

When all closed-system U-series ages of terrace corals that have been correlated with isotope stage 5 are plotted as a frequency histogram, there is evidence of two distinct highstands of sea level ca. 125–120 ka and 85–80 ka and a less-distinct highstand ca. 105 ka (Fig. 3). The three age modes agree well with the peaks of summer insolation at high latitudes as calculated by Berger (1978) and therefore it can be said that in general the sea-level record supports the Milankovitch orbital-forcing theory of climatic change (Fig. 3). However, when this plot is examined in some detail, there are a number of problematic ages. There are several coral ages between 120 and 105 ka, although most of these permit agreement with the 125–120 ka or 105 ka ages when their analytical errors are considered. There are also several ages in the range 145–135 ka that are analytically distinguishable from the ca. 125–120 ka ages. Some workers have argued that there may be two distinct highstands of sea level in the time interval from 120 to 140 ka (Chappell and Veeh, 1978; Moore, 1982). Only a single ice-volume minimum in this time period is recorded in deep-sea cores, but it is interesting that ages of ca. 145–135 ka for terrace corals are in agreement with some of the age estimates for the start of substage 5e based on direct analyses of the cores (Table 1). Winograd and others (1988) cited the ca. 140 ka ages as supportive evidence for a start of substage 5e that is considerably earlier than that used by most investigators. The 145–135 ka ages for corals and for the start of substage 5e in cores present a challenge to the Milankovitch orbital-forcing theory, because insolation during summer at high latitudes was relatively low around this time (Fig. 3). Further discussion of this problem can be found in Chappell and Veeh (1978), Moore (1982), Kaufman (1986), and Pillans (1987).

In addition to uncertainties about when isotope stage 5 began, there are problems in interpreting U-series data from terraces to determine when isotope stage 5 ended. As discussed earlier, some workers regard the whole of deep-sea stage 5 as the last interglacial period; others would argue that only substage 5e can be considered "last interglacial." If the latter interpretation is used, then the data plotted in Figure 3 imply that the last interglacial ended ca. 120–115 ka. If the former interpretation is used, then the close of the last interglacial period, on the basis of what is portrayed in Figure 3, is much harder to define. There are a number of terrace corals in Table 2 and Figure 3 with ages ranging from 75 to 60 ka from deposits that have been correlated with isotope substage 5a. As with the 145–135 ka corals, these ages represent a challenge to the Milankovitch theory, because summer insolation at high latitudes was quite low at this time. Thus, there are a number of problems and inconsistencies in the data given in Table 2 and plotted in Figure 3, even though all these samples passed the geochemical and isotopic screening described above. It is useful to discuss these problems on a study area by study area basis.

Barbados

The coral terraces on Barbados have probably received more attention for U-series dating than any other terrace sequence in the world, and as a result, the data tabulated in Table 2 allow for detailed comparisons. Most of the ages for the Worthing terrace cluster around 80 ka. This age was used by Broecker and others (1968) and Mesolella and others (1969) to provide support for the orbital-forcing theory of climatic change, because this postdates a time (ca. 85 ka) of high summer insolation at high latitudes (Fig. 3). However, two recent suites of analyses of corals from this terrace disagree with the ca. 80 ka age estimates. Radtke and others (1988) analyzed three corals from the Worthing terrace and obtained ages of 66 ± 2 ka, 70 ± 2 ka, and 72 ± 2 ka (Table 2). These dates are problematic because they indicate a time of formation of the Worthing terrace when insolation was very low at high latitudes (Fig. 3). A second problematic set of ages was reported by Edwards and others (1987). These workers obtained mass-spectrometric ages of 87 ka (with an uncertainty of only 200–300 yr) for the Worthing terrace, which would indicate that sea level was high at a time just *before* the high-latitude insolation maximum (Fig. 3). This would require that melting of most glacial ice occurred well before summer radiation was at a maximum at high latitudes. It is perhaps more reasonable to think that sea level would be at a maximum sometime after the solar-radiation maximum.

A similar problem can be identified with the mass spectrometric ages of 111–112 ka for the Ventnor terrace reported by Edwards and others (1987). These ages precede, by a significant amount, an insolation maximum that occurred at about 103 ka. Other workers who have analyzed corals from this terrace have obtained ages that average about 105 ka, but which range from 100 to 111 ka, with an uncertainty of 2–6 ka (Table 2). Edwards and others (1987) concluded that factors other than orbital forcing could have caused a high sea level at 111–112 ka, but the other ages of the Ventnor terrace would still permit an orbitally-induced sea-level highstand.

There are also problems associated with the highstand of about 120–129 ka recorded by the Rendezvous Hill terrace on Barbados. Recent U-series ages of coral from this terrace reported by Ku and others (1990) are mostly 120–125 ka, in agreement with previous analyses, but three of their samples gave ages of 99 ± 3 ka, 103 ± 2 ka, and 106 ± 2 ka. These ages are in the general range of ages for the younger Ventnor terrace (Table 2). Whereas it is possible for older terrace corals to be reworked onto a younger terrace surface, it is difficult to envision how younger terrace corals could be reworked onto an older, topographically higher terrace. In addition, these corals do not show any evidence of significant recrystallization, secondary U additions, or other open-system effects. Therefore, it is difficult to explain the younger-than-expected ages in these samples. All of the age discrepancies discussed for the Worthing, Ventnor, and Rendezvous Hill terrace corals could be the result of subtle diagenetic processes, as suggested by Ku and others (1990), but there are at present no clues as to what these processes might be.

A final problem with the terrace sequence on Barbados concerns the newly-named Maxwell terrace reported by Ku and others (1990). This terrace actually appeared on the reef trend map of Mesolella and others (1969, Fig. 6), but was not mapped as a separate reef crest by Bender and others (1979, Fig. 2). Ku

TABLE 2. ISOTOPIC ACTIVITY RATIOS AND URANIUM-SERIES AGES OF CORALS FROM MARINE DEPOSITS THAT ARE THOUGHT TO BE CORRELATE WITH ISOTOPE STAGE 5

Sample Number	$^{234}U/^{238}U$ Activity Ratios	$^{230}Th/^{234}U$ Activity Ratios	Age* (ka)	Age† (ka)	Ref.§
North Carolina and Virginia (Norfolk and Kempsville Formations)					
C-2	1.10 ± 0.02	0.50 ± 0.02	74 ± 4	74 ± 4	1
C-18	1.11 ± 0.02	0.52 ± 0.02	79 ± 5	78 ± 5	1
C-36A	1.10 ± 0.02	0.48 ± 0.02	69 ± 4	70 ± 4	1
C-36B	1.09 ± 0.02	0.46 ± 0.02	67 ± 4	66 ± 4	1
C-15	1.10 ± 0.02	0.49 ± 0.02	72 ± 4	73 ± 4	1
C-34	1.12 ± 0.02	0.49 ± 0.02	72 ± 4	72 ± 4	1
South Carolina ("Late Wando" Formation)					
	1.09 ± 0.02	0.56 ± 0.01	87 ± 4	88 ± 3	1
Florida (Key Largo Limestone)					
KL1	1.13 ± 0.01	0.75 ± 0.02	145 ± 14	144 ± 8	2
765A	1.12 ± 0.01	0.61 ± 0.03	95 ± 9	100 ± 8	3
801D	1.09 ± 0.02	0.71 ± 0.04	130 ± 20	131 ± 15	3
801F	1.09 ± 0.01	0.71 ± 0.03	130 ± 15	131 ± 11	3
801C	1.10 ± 0.02	0.75 ± 0.03	140 ± 15	145 ± 13	4
KL-5	1.14 ± 0.03	0.69 ± 0.04	125 ± 15	122 ± 14	4
KL-8	1.09 ± 0.02	0.68 ± 0.03	125 ± 11	121 ± 10	4
KL-10	1.08 ± 0.03	0.70 ± 0.05	133 ± 19	128 ± 18	4
KL-10	1.15 ± 0.08	0.74 ± 0.05	145 ± 14	139 ± 21	4
Bermuda (Southampton Formation)					
790515-6	1.08 ± 0.02	0.54 ± 0.02	85 ± 6	83 ± 5	5
PG-218	1.08 ± 0.01	0.53 ± 0.02	83 ± 5	81 ± 5	5
Bermuda (Spencer's Point Formation)					
75012	1.11 ± 0.02	0.60 ± 0.03	97 ± 6	97 ± 8	5
780524-14	1.11 ± 0.03	0.62 ± 0.02	106 ± 6	103 ± 5	5
780522-5	1.05 ± 0.01	0.63 ± 0.02	108 ± 6	107 ± 6	5
Bermuda (Devonshire Formation)					
75014	1.10 ± 0.02	0.67 ± 0.02	118 ± 6	117 ± 7	5
PG-321	1.06 ± 0.02	0.67 ± 0.04	118 ± 11	118 ± 14	5
PG-4	1.09 ± 0.01	0.67 ± 0.02	118 ± 6	118 ± 6	5
780522-6	1.10 ± 0.02	0.68 ± 0.02	121 ± 6	120 ± 7	5
75013	1.11 ± 0.02	0.69 ± 0.02	124 ± 6	123 ± 7	5
016	1.10 ± 0.03	0.70 ± 0.03	124 ± 8	127 ± 11	5
75020	1.09 ± 0.02	0.69 ± 0.02	124 ± 6	124 ± 7	5
PG-324	1.11 ± 0.03	0.69 ± 0.03	124 ± 8	123 ± 11	5
PG-322	1.10 ± 0.02	0.69 ± 0.04	124 ± 12	124 ± 14	5
75015	1.09 ± 0.01	0.68 ± 0.01	125 ± 6	121 ± 3	5
790513-2	1.06 ± 0.02	0.70 ± 0.03	127 ± 9	128 ± 11	5
PG-304	1.07 ± 0.02	0.70 ± 0.03	127 ± 9	128 ± 11	5
780521	1.11 ± 0.01	0.70 ± 0.02	127 ± 6	127 ± 7	5
PG-220	1.07 ± 0.01	0.71 ± 0.02	131 ± 7	131 ± 8	5
PG-210	1.12 ± 0.02	0.72 ± 0.02	134 ± 8	133 ± 7	5
Northern Bahamas					
717F	1.14 ± 0.02	0.56 ± 0.04	80 ± 3	87 ± 10	3
20	1.08 ± 0.02	0.68 ± 0.02	120 ± 6	121 ± 7	6
21	1.10 ± 0.02	0.70 ± 0.02	128 ± 6	126 ± 7	6
5-7	1.13 ± 0.03	0.67 ± 0.02	125 ± 6	117 ± 6	6
5-8	1.13 ± 0.03	0.67 ± 0.02	125 ± 6	117 ± 6	6
6-3	1.13 ± 0.03	0.71 ± 0.03	140 ± 9	129 ± 11	6
36-C	1.11 ± 0.03	0.74 ± 0.03	146 ± 9	141 ± 12	6
43-1	1.09 ± 0.03	0.68 ± 0.02	121 ± 6	121 ± 7	6
43-3	1.06 ± 0.03	0.70 ± 0.02	121 ± 6	128 ± 7	6
46-2B	1.13 ± 0.03	0.69 ± 0.02	132 ± 7	123 ± 6	6
47-D	1.07 ± 0.03	0.66 ± 0.02	112 ± 6	115 ± 6	6
San Salvador, Bahamas					
GB-3	1.10 ± 0.02	0.69 ± 0.02	123 ± 9	124 ± 7	7
Haiti (Mole Terrace)					
E12	1.10 ± 0.02	0.53 ± 0.01	80 ± 3	81 ± 3	8
02	1.12 ± 0.02	0.54 ± 0.01	82 ± 2	82 ± 3	8
K1	1.10 ± 0.02	0.51 ± 0.01	77 ± 3	76 ± 2	8
P3	1.10 ± 0.02	0.54 ± 0.01	83 ± 3	83 ± 2	8
G1	1.09 ± 0.02	0.70 ± 0.02	128 ± 6	127 ± 7	8
Haiti (Saint Terrace)					
Q5	1.10 ± 0.01	0.64 ± 0.02	108 ± 5	108 ± 6	8
Haiti (Nicolas Terrace)					
B2	1.10 ± 0.02	0.71 ± 0.02	130 ± 6	130 ± 8	8
B11	1.12 ± 0.02	0.71 ± 0.01	132 ± 5	130 ± 4	8
C1 (α-spec)	1.12 ± 0.02	0.71 ± 0.02	130 ± 6	130 ± 7	8
C4 (α-spec)	1.12 ± 0.02	0.70 ± 0.02	126 ± 6	126 ± 7	8
C1 (mass-spec)	1.107 ± 0.002	0.684 ± 0.002	121.9 ± 0.5	121.5 ± 0.7	9
C4#1 (mass-spec)	1.109 ± 0.004	0.690 ± 0.002	123.6 ± 0.8	123.4 ± 0.7	9
Barbados (Worthing Terrace)					
AGA-1	1.11 ± 0.02	0.52 ± 0.02	79 ± 4	78 ± 5	10
OC-26	1.13 ± 0.01	0.53 ± 0.01	82 ± 2	80 ± 2	10
AEH-1	1.12 ± 0.01	0.53 ± 0.02	82 ± 4	80 ± 5	10
FS-3	1.13 ± 0.02	0.54 ± 0.02	84 ± 4	83 ± 5	10
1152C	1.11 ± 0.01	0.52 ± 0.01	79 ± 2	78 ± 3	11
B-4	1.11 ± 0.01	0.46 ± 0.01	67 ± 2	66 ± 2	12
B-5	1.13 ± 0.02	0.48 ± 0.01	70 ± 3	70 ± 2	12
B-7	1.12 ± 0.02	0.49 ± 0.03	73 ± 6	72 ± 6	12
OC-51A	1.126 ± 0.003	0.560 ± 0.001	87.5 ± 0.3	87.3 ± 0.3	13
OC-51B	1.126 ± 0.003	0.562 ± 0.001	87.9 ± 0.4	87.8 ± 0.2	13
FS-50A	1.127 ± 0.008	0.519 ± 0.006	78 ± 1	78 ± 1	14
FS-51	1.134 ± 0.008	0.508 ± 0.006	76 ± 1	76 ± 2	14
OC-50	1.126 ± 0.007	0.551 ± 0.006	85 ± 1	85 ± 2	14
OC-51	1.131 ± 0.007	0.542 ± 0.006	83 ± 1	83 ± 1	14
OC-53	1.13 ± 0.01	0.543 ± 0.007	83 ± 2	83 ± 2	14
Barbados (Ventnor Terrace)					
1144C	1.13 ± 0.01	0.63 ± 0.01	105 ± 3	105 ± 3	11
FT-1	1.11 ± 0.02	0.62 ± 0.02	104 ± 4	103 ± 6	10
FT-1	1.12 ± 0.02	0.61 ± 0.02	100 ± 4	100 ± 5	10
FT-1	1.11 ± 0.02	0.61 ± 0.02	100 ± 4	100 ± 5	10
AFZ-1	1.10 ± 0.01	0.62 ± 0.02	104 ± 4	103 ± 5	10
AFZ-1	1.11 ± 0.01	0.62 ± 0.02	104 ± 4	103 ± 6	10
AFK-1	1.10 ± 0.01	0.62 ± 0.02	104 ± 6	103 ± 5	10
AEG-2	1.10 ± 0.02	0.65 ± 0.02	111 ± 6	111 ± 6	10
X-5	1.13 ± 0.02	0.65 ± 0.02	111 ± 6	111 ± 6	10
B59a	1.13 ± 0.02	0.63 ± 0.01	100+6/-3	105 ± 3	12
FT-50A	1.126 ± 0.003	0.653 ± 0.001	112.0 ± 0.5	111.6 ± 0.3	13
FT-50B	1.127 ± 0.002	0.653 ± 0.002	111.8 ± 0.6	111.6 ± 0.6	13
FT-50B	1.124 ± 0.003	0.654 ± 0.001	112.3 ± 0.5	112.0 ± 0.3	13
ANM-21	1.116 ± 0.007	0.620 ± 0.007	102 ± 2	102 ± 2	14

TABLE 2. ISOTOPIC ACTIVITY RATIOS AND URANIUM-SERIES AGES OF CORALS FROM MARINE DEPOSITS THAT ARE THOUGHT TO BE CORRELATE WITH ISOTOPE STAGE 5 (continued)

Sample Number	$^{234}U/^{238}U$ Activity Ratios	$^{230}Th/^{234}U$ Activity Ratios	Age* (ka)	Age† (ka)	Ref.§
BARBADOS (VENTNOR TERRACE) continued					
AM-22	1.10 ± 0.01	0.615 ± 0.008	102 ± 2	102 ± 3	14
FT-51(1)	1.117 ± 0.006	0.65 ± 0.01	110 ± 2	111 ± 3	14
BAB-10	1.125 ± 0.008	0.623 ± 0.007	103 ± 2	103 ± 2	14
BARBADOS (MAXWELL TERRACE)					
AEJ-5	1.12 ± 0.01	0.69 ± 0.02	124 ± 6	123 ± 7	10
AEJ-21	1.105 ± 0.007	0.702 ± 0.007	128 ± 2	128 ± 3	14
AEJ-22(2)	1.12 ± 0.01	0.683 ± 0.008	121 ± 3	121 ± 3	14
AGP-10	1.111 ± 0.007	0.661 ± 0.007	114 ± 2	114 ± 2	14
AGP-11	1.099 ± 0.008	0.662 ± 0.008	118 ± 2	118 ± 2	14
AGP-12(1)	1.10 ± 0.01	0.662 ± 0.008	115 ± 2	115 ± 3	14
BARBADOS (RENDEZVOUS HILL TERRACE)					
1152E	1.10 ± 0.01	0.69 ± 0.01	125 ± 5	125 ± 4	11
AFS-1	1.12 ± 0.03	0.69 ± 0.02	124 ± 6	123 ± 7	10
AFM-T-2	1.11 ± 0.01	0.70 ± 0.02	127 ± 6	127 ± 7	10
AFM-B-1	1.09 ± 0.02	0.70 ± 0.02	127 ± 6	127 ± 7	10
ADR-1	1.11 ± 0.02	0.70 ± 0.02	127 ± 6	127 ± 7	10
B66	1.13 ± 0.02	0.70 ± 0.02	128 ± 7	125 ± 7	12
AFM-20	1.111 ± 0.002	0.707 ± 0.002	129.2 ± 0.7	129.0 ± 0.7	13
R-52	1.115 ± 0.003	0.704 ± 0.002	128.1 ± 0.8	127.9 ± 0.7	13
AFS-10	1.110 ± 0.002	0.696 ± 0.001	125.7 ± 0.6	125.4 ± 0.4	13
AFS-11	1.114 ± 0.003	0.687 ± 0.001	122.6 ± 0.7	122.3 ± 0.7	13
AFS-12A	1.109 ± 0.002	0.685 ± 0.001	122.1 ± 0.5	121.8 ± 0.3	13
AFS-12B	1.111 ± 0.003	0.687 ± 0.001	122.7 ± 0.6	122.4 ± 0.3	13
AFS-12C	1.109 ± 0.003	0.692 ± 0.001	124.5 ± 0.6	124.1 ± 0.4	14
AFM-20A(1)	1.099 ± 0.009	0.673 ± 0.009	118 ± 3	118 ± 3	14
AFM-20A(2)	1.114 ± 0.009	0.680 ± 0.008	120 ± 3	120 ± 3	14
AFM-22A(1)	1.103 ± 0.008	0.619 ± 0.007	102 ± 2	102 ± 2	14
AFM-22A(2)	1.111 ± 0.009	0.608 ± 0.008	99 ± 2	99 ± 3	14
AFM-23(1)	1.116 ± 0.009	0.659 ± 0.008	114 ± 2	114 ± 3	14
AFM-23(2)	1.11 ± 0.01	0.631 ± 0.008	106 ± 2	106 ± 2	14
R-51(2)	1.111 ± 0.008	0.688 ± 0.007	123 ± 2	123 ± 3	14
AFS-10	1.097 ± 0.009	0.685 ± 0.008	122 ± 3	122 ± 3	14
AFS-11	1.096 ± 0.009	0.690 ± 0.008	124 ± 3	124 ± 3	14
AFS-12	1.12 ± 0.01	0.682 ± 0.009	121 ± 3	120 ± 3	14
AFM3#1	1.108 ± 0.004	0.694 ± 0.002	125.1 ± 0.8	124.8 ± 0.7	9
YUCATAN PENINSULA					
355-11	1.10 ± 0.01	0.68 ± 0.02	121 ± 6	120 ± 7	15
355-12	1.10 ± 0.01	0.69 ± 0.02	123 ± 6	124 ± 7	15
355-14	1.10 ± 0.01	0.68 ± 0.02	120 ± 6	120 ± 7	15
293-9	1.11 ± 0.01	0.69 ± 0.02	120 ± 6	123 ± 7	15
C-22	1.10 ± 0.01	0.68 ± 0.02	121 ± 6	120 ± 7	15
OREGON (WHISKY RUN TERRACE)					
M2798	1.10 ± 0.01	0.54 ± 0.02	83 ± 5	83 ± 5	16
CALIFORNIA (POINT ARENA)					
M7824	1.12 ± 0.02	0.51 ± 0.02	76 ± 4	76 ± 4	16
CALIFORNIA (POINT LOMA, NESTOR TERRACE)					
2577	1.11 ± 0.01	0.64 ± 0.02	109 ± 6	108 ± 6	17
2577	1.12 ± 0.02	0.71 ± 0.02	131 ± 8	130 ± 7	17
2577	1.12 ± 0.01	0.69 ± 0.02	124 ± 7	123 ± 7	17
BAJA CALIFORNIA (PUNTA BANDA, LIGHTHOUSE TERRACE)					
PB-8	1.11 ± 0.01	0.52 ± 0.01	78 ± 3	78 ± 3	18
BAJA CALIFORNIA (PUNTA BANDA, SEA CAVE TERRACE)					
PB-3	1.11 ± 0.01	0.68 ± 0.01	120 ± 3	120 ± 3	18
PB-10	1.10 ± 0.01	0.69 ± 0.01	124 ± 4	124 ± 4	18
BAJA CALIFORNIA (MAGDALENA TERRACE)					
AO-3	1.08 ± 0.02	0.66 ± 0.02	116 ± 8	115 ± 6	19
AO-6	1.14 ± 0.03	0.66 ± 0.04	118 ± 12	113 ± 13	19
BAJA CALIFORNIA (MULEGE TERRACE)					
3258	1.10 ± 0.03	0.75 ± 0.02	144 ± 7	145 ± 8	20
3261	1.11 ± 0.03	0.69 ± 0.02	124 ± 5	123 ± 7	20
BAJA CALIFORNIA (LA PAZ AREA)					
C-A	1.12 ± 0.02	0.75 ± 0.02	146 ± 9	144 ± 7	21
C-B	1.12 ± 0.02	0.72 ± 0.02	135 ± 6	133 ± 7	21
NEW GUINEA (REEF COMPLEX V)					
8	1.13 ± 0.02	0.43 ± 0.02	61 ± 4	60 ± 4	22
12	1.10 ± 0.02	0.54 ± 0.02	84 ± 4	83 ± 5	22
38	1.11 ± 0.02	0.55 ± 0.02	86 ± 4	85 ± 5	22
NEW GUINEA (REEF COMPLEX VI)					
14b	1.12 ± 0.02	0.63 ± 0.03	107 ± 9	105 ± 9	22
20	1.09 ± 0.02	0.63 ± 0.02	107 ± 6	106 ± 6	22
NEW GUINEA (REEF COMPLEX VIIB)					
NG618	1.11 ± 0.01	0.67 ± 0.02	116 ± 7	117 ± 6	23
NG618	1.11 ± 0.01	0.68 ± 0.02	119 ± 7	120 ± 7	23
NEW GUINEA (REEF COMPLEX VIIA)					
NG616	1.10 ± 0.01	0.74 ± 0.03	140 ± 10	141 ± 12	23
NG616	1.12 ± 0.02	0.72 ± 0.03	133 ± 10	133 ± 11	23
15	1.08 ± 0.02	0.74 ± 0.02	142 ± 8	142 ± 8	22

*Age reported by authors in reference cited.

†Age calculated by Muhs using half-lives of ^{230}Th and ^{234}U of 75,200 and 244,000 years, respectively.

§1 = Szabo (1985); 2 = Szabo and Halley (1988); 3 = Broecker and Thurber (1965); 4 = Osmond and others (1965); 5 = Harmon and others (1983); 6 = Neumann and Moore (1975); 7 = Szabo and others (1988); 8 = Dodge and others (1983); 9 = Bard and others (1990); 10 = Mesolella and others (1969); 11 = Ku (1968); 12 = Radtke and others (1988); 13 = Edwards and others (1987); 14 = Ku and others (1990); 15 = Szabo and others (1978); 16 = Muhs and others (1990); 17 = Ku and Kern (1974); 18 = Rockwell and others (1989); 19 = Omura and others (1979); 20 = Ashby and others (1987); 21 = Szabo and others (1990); 22 = Bloom and others (1974); 23 = Veeh and Chappell (1970).

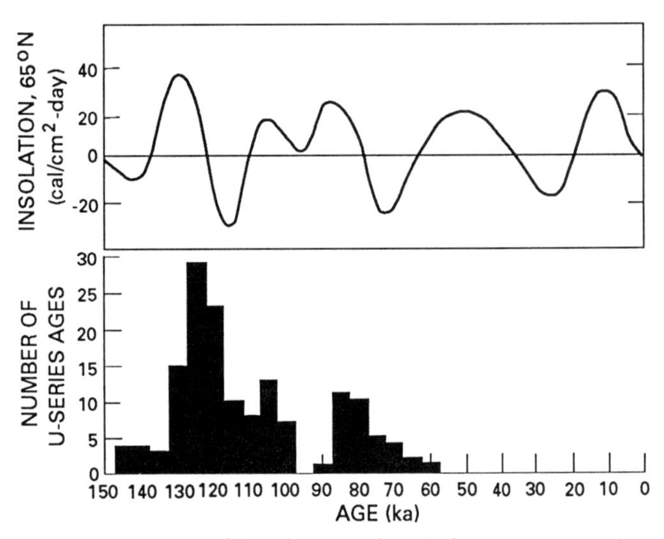

Figure 3. Histogram of U-series ages of corals from terraces that have been correlated with deep-sea oxygen-isotope stage 5. Ages plotted are those calculated by Muhs from data in Table 2, rounded off to the closest 5,000 years. Also shown for comparison is average insolation received at the top of the atmosphere at lat 65°N for the summer half year, expressed as deviations from the present (A.D. 1950) value. Insolation data are from Berger (1978).

and others (1990) tested the hypothesis that these two terraces could represent the "dual peak" proposed by Chappell and Veeh (1978), Aharon and others (1980), and Moore (1982) and alluded to earlier in this chapter. They confirmed the earlier results of Mesolella and others (1969) that there is no significant age difference between these two terraces and thus they do not support the hypothesis of bipartite sea levels in the time period 120–140 ka. However, what is not explained by these data is why there are *two* geomorphically distinct terraces, both with *Acropora palmata* reef-crest corals, that are the same age.

East Coast of the United States

At localities in Virginia and North Carolina, several corals from the Norfolk and Kempsville formations (or their equivalents), meeting all of the closed-system requirements, give ages ranging from 66 to 78 ka and averaging about 72 ka (Szabo, 1985; Table 2). On the basis of their fossil assemblages (equivalent to modern ones), and their estimated paleo-sea level (+4 to +10 m), Mixon and others (1982) thought that these deposits should be about 125 ka, and speculated that the corals giving ages of 66–78 ka had undergone postdepositional loss of ^{230}Th. Given the relative immobility of Th in most near-surface environments, ^{230}Th loss does not seem very likely, and Szabo (1985) later showed that the ^{230}Th/^{234}U ages were reliable, on the basis of concordant ^{231}Pa/^{235}U ages. As for the ca. 70 ka corals from the Worthing terrace on Barbados, these ages suggest a very high sea level at a time when summer insolation at high latitudes was quite low (Fig. 3), and thus they are in conflict with the orbital-forcing theory.

Localities with 145–135 ka terrace corals

Certain terrace corals from New Guinea, Baja California, Florida, and the Bahamas have given ages in the range of 145–135 ka (Table 2). As discussed earlier, these ages are problematic in that they are in a time period of relatively low summer insolation at high latitudes (Fig. 3), yet there is no apparent evidence for open-system conditions. In Baja California, Florida, and the Bahamas, at least some of these corals are apparently found in the same deposits that yielded corals with 120–125 ka ages (Table 2). These data suggest that subtle diagenetic processes, for which we have little or no geochemical or isotopic evidence, may be responsible for bringing about open-system conditions. Geomorphic, stratigraphic, U-series, and stable-isotope data from the rapidly uplifting Huon Peninsula of New Guinea, however, indicate that reef complex VIIa is distinct from reef complex VIIb; the former has been dated at 142–133 ka and the latter has been dated as 120–117 ka (Veeh and Chappell, 1970; Chappell, 1974; Bloom and others, 1974; Aharon, 1983). It is possible that on tectonically stable coasts or coasts with very low uplift rates such as Baja California Sur, Florida, and the Bahamas), one would expect a mixture of corals of two different ages if there were two highstands of sea level of about the same magnitude, not greatly separated in time. It is possible that future high-precision mass-spectrometric U-series analyses of terrace corals will shed further light on this issue.

MAGNITUDE OF SEA-LEVEL RISE DURING ISOTOPE STAGE 5

Sea level at ca. 125 ka

In tectonically stable areas distant from plate boundaries, U-series dating of corals indicates that sea level ca. 125 ka was about 5–6 m higher than present. Patch reefs and marine calcarenites on Bermuda dated ca. 125 ka indicate a former sea level 5 m higher than present (Harmon and others, 1983). In the northern Bahamas, coral-bearing marine conglomerates and in situ reefs that have been dated ca. 125 ka occur at 0.2 to 2.3 m above current sea level; bioerosional notches that are thought to have formed during the same high sea-level stand occur at 5 to 6 m above sea level (Neumann and Moore, 1975). Elsewhere in the Bahamas, ooids in beach deposits dated ca. 125 ka indicate a possible paleo-sea level of up to 10 m higher than present sea level (Garrett and Gould, 1984; Muhs and others, 1987a). In Florida, the Key Largo limestone has a maximum elevation of 6 m above sea level on Windley Key (Szabo and Halley, 1988), and ages of ca. 125 and 145 ka have been reported for corals from this locality (Broecker and Thurber, 1965; Osmond and others, 1965; Szabo and Halley, 1988; Table 2). Corals dated ca. 125 ka on the Yucatan Peninsula are associated with beach deposits that indicate a paleo-sea level of 3 to 6 m higher than present (Szabo and others, 1978). Thus, there is convincing evidence that sea level ca. 125 ka stood at 6 ± 4 m above the present, and there is general agreement among most investigators that this marks a time of significantly lower global ice volume compared to the present.

Sea levels at ca. 105 and 80 ka

Whether the relatively high sea-level stands ca. 80 and 105 ka can be regarded as "interglacial" depends in part on how high the sea stood at these times, which in turn has an inverse relation to the volume of global ice. Studies of terraces on the tectonically emergent islands of Barbados, New Guinea, and Haiti all suggest that sea levels were significantly below the present at these times, based on elevations of the ca. 80, 105, and 125 ka terraces, an assumption of a +6 m highstand of sea level ca. 125 ka, and an assumption of a constant rate of uplift in the past 125 ka. The rate of uplift is calculated from the present elevation of the 125 ka terrace (–6 m) divided by its age. Using the ages and present elevations of the 80 and 105 ka terraces, it is then possible to calculate the paleo-sea levels at the time of formation of these terraces. Using the latest measurements of the elevations of reefs reported by Chappell and Shackleton (1986) for New Guinea, Bender and others (1979) for Barbados, and Dodge and others (1983) for Haiti, sea level is estimated to have been at –16 to –18 m at 105 ka and –15 to –16 m at 80 ka, relative to present. By the use of an alternative method, Bloom and Yonekura (1985) recalculated paleo-sea levels on New Guinea with the surprising conclusion that sea level at 80 ka is estimated to have been ~7 m lower relative to present and sea level at 105 ka is estimated to have been about the same as the present (paleo-sea levels estimated for terraces younger than ca. 80 ka gave about the same results by both methods). Unfortunately, there are too few elevation data for terraces on Barbados and Haiti to make comparative paleo-sea-level estimates using the Bloom and Yonekura (1985) method. It is not known why the two methods should yield such significantly different results, but it is clear that the elevations of the 80 and 105 ka sea-level highstands are not yet firm.

Data from the tectonically stable coasts of Bermuda, the Bahamas, and the eastern United States and from the tectonically uplifting coast of western North America, indicate sea levels near to, or even higher than, present at 80 and 105 ka. On Bermuda, there are several patchy, coral-bearing marine conglomerates that Harmon and others (1983) have dated ca. 80 and 105 ka. These investigators interpreted these marine conglomerates as storm deposits and said that currently submerged speleothems (also dated by U series) on Bermuda indicate that sea level was significantly lower than present at 80 and 105 ka. Vacher and Hearty (1989) pointed out that the conglomerates with the 80 ka corals were from localities on the protected side of Bermuda. They questioned the storm origin of both deposits, and suggested instead that the 105 ka ages represented scatter about true ages of ca. 125 ka, but the 80 ka corals represented a different highstand, at about 1 m higher than present, at that time.

Neumann and Moore (1975) reported U-series ages of corals from low-elevation marine deposits in the tectonically stable northern Bahamas. Among these were two samples of *Diploria* from the Berry Islands, collected from marine conglomerates found at elevations of 1.0–1.5 m above sea level. The ages of these samples are reported to be 103 ± 5 ka and 105 ± 5 ka. Unfortunately, isotopic data for these samples were not given, so it is not possible to evaluate whether these ages are reasonable. If they have experienced closed-system conditions, the data indicate that sea level ca. 105 ka was as high or higher than present. Supporting data for this idea are found on New Providence Island in the northern Bahamas, where there is a complex sequence of oolitic eolianites described by Garrett and Gould (1984). Muhs and others (1987a) analyzed some of the six eolianite units described by Garrett and Gould (1984) and reported preliminary U-series ages of ca. 105 ka for one of the units. Note that the U-series ages of ooids date the time of ooid formation, not the time of eolian deposition, although it is possible that these two events may not be greatly separated in time. In any case, the times of ooid formation on the Bahamas platform are significant for sea-level studies. Ooids form only in very shallow waters on the Bahamas platform today; most of this platform is less than about 6 m deep at present. If sea level dropped more than about 6 m below the present, most of the platform would be subaerially exposed, and ooid formation would not take place. Hence, U-series ages of ooids of ca. 105 ka imply a sea level at that time within about 6 m of the present. A sea-level stand higher than about –10 to –15 m, relative to present, ca. 100 ka is also implied by high-precision U-series ages of a submerged speleothem on Grand Bahama Island (Li and others, 1989). Broecker and Thurber (1965) reported a U-series age of 80 ± 8 ka (recalculated to be 87 ± 10 ka; see Table 2) for a coral found on a marine terrace cut into oolitic eolianite from the Berry Islands of the northern Bahamas. The elevation of this terrace is about 1 to 2 m above present sea level (N. Newell, 1990, personal commun.) and it clearly implies a sea level higher than present at the time of its formation. Thus, in the northern Bahamas, a tectonically stable setting, there is evidence for sea levels close to the present ca. 105 ka and possibly also ca. 80 ka.

The east coast of the United States is apparently undergoing a rather low long-term rate of uplift, ~0.02 m/1000 yr (Cronin, 1981), so that for the late Pleistocene, it can be considered to be relatively stable. The ages of certain marine deposits on the east coast of the United States remain controversial (Cronin and others, 1981; McCarten and others, 1982; Szabo, 1985; Corrado and others, 1986; Wehmiller and others, 1988; Hollin and Hearty, 1990), but ages of other marine deposits are agreed upon by most workers who have studied them. In Virginia and North Carolina, the Norfolk and Kempsville formations have corals in them that have been analyzed by Szabo (1985). He reported U-series ages of corals from these deposits that ranged from 66 ± 4 ka to 78 ± 5 ka (see Table 2), and correlated these units with oxygen-isotope substage 5a. Amino-acid data reported by Wehmiller and others (1988) support this correlation. These formations are currently 4–10 m above sea level (Cronin and others, 1981). Thus, if the correlation of these deposits with substage 5a is accepted, then the data imply a sea level higher than present from 66 ± 4 ka to 78 ±

5 ka. The major uncertainty here is the correlation of the deposits with substage 5a when the ages of the corals could be as young as 66 ka (see earlier discussion in this paper on the timing of sea-level highstands). In the Charleston, South Carolina, area, Szabo (1985) analyzed several corals from what he described as "late Wando" deposits. The top of this formation is currently 3–10 m above sea level (McCarten and others, 1982). Corals from this unit have relatively low $^{230}Th/^{232}Th$ values, but Szabo (1985) used several corals to derive an isochron plot that showed excellent linearity and gave an age of 88 ± 3 ka (Table 2). Szabo (1985) correlated this unit with isotope substage 5c, but a correlation with substage 5a may be more appropriate. Wehmiller and others (1988) used amino-acid ratios in shells from the same deposits to estimate an age of ca. 200 ka; they thought that the disagreement with Szabo's (1985) U-series ages was due to problems in the U-series systematics of the corals. A simpler explanation is that the shells with apparent amino-acid age estimates of ca. 200 ka are reworked from an older deposit, such as those for which Szabo (1985) reported U-series ages of coral of around 200 ka. If the ca. 80 ka age and correlation to isotope substage 5a are accepted, the elevation of the late Wando deposits implies a higher-than-present sea level at 88 ka.

The west coast of North America is tectonically active, and paleo-sea-level estimates must be made in a manner similar to that used for Barbados, New Guinea, and Haiti. Marine terraces in this area contain solitary corals or hydrocorals that have been dated by the U-series method by Ku and Kern (1974), Muhs and Szabo (1982), Muhs and others (1987b, 1988, 1990), and Rockwell and others (1989). Three localities (Point Loma and San Nicolas Island, California, and Punta Banda, Baja California, Mexico) have confidently dated 80 ka and 125 ka terraces. In addition, Punta Banda has a terrace of intermediate elevation (between the 80 ka and 125 ka terraces) that Rockwell and others (1989) interpreted to be equivalent to the 105 ka terrace found on Barbados, New Guinea, and Haiti. Assuming a sea level 6 m higher than present at 125 ka and a constant uplift rate, the elevations of the shoreline angles of these terraces indicate sea levels ~1 m lower than present at 105 ka and ~4 m lower than present at 80 ka.

Thus, data from two tectonically stable areas off North America (Bermuda and the Bahamas) and the east and west coasts of North America indicate that sea levels may have stood relatively high at both 80 ka and 105 ka, when compared to the New Guinea, Barbados, and Haiti terrace data. If sea levels at these times were in fact close to the present, the implication is that isotope substages 5a and 5c were really periods of an "interglacial" character, at least in terms of global ice volume. With relatively high sea levels (and thus low ice volumes) during isotope substages 5a, 5c, and 5e, isotope stage 5 may be viewed as a rather long interglacial period that was interrupted by two short-lived and spatially limited glaciations during substages 5d and 5b. These two short periods of ice growth are recorded as the sea-level lowstands that produced initial emergence of the 125 ka and 105 ka terraces. What is not clear, however, is why paleo-sea levels calculated from Barbados, New Guinea, and Haiti differ so significantly from those calculated in the same manner from terrace data on the west coast of North America. Resolution of this issue will require more studies on both stable and tectonically active coasts.

IS THE LAST INTERGLACIAL-GLACIAL TRANSITION FOUND IN THE COASTAL SEDIMENTARY RECORD?

Coastal stratigraphy on the California Channel Islands

Recent studies conducted on the California Channel Islands (Fig. 4) indicate that the last interglacial/glacial transition may be recorded by a shift from marine to eolian sedimentation in some coastal areas. Marine terraces on the California Channel Islands are locally overlain by calcareous eolian sand, some of which has been weakly lithified into eolianite (Fig. 5) (Orr, 1960; Vedder and Norris, 1963; Johnson, 1977; Muhs, 1983b). This cemented, carbonate-rich sand commonly occurs in areas where there is no present sand source. Even where potential beach sources exist, the eolianite has a much higher carbonate content than the adjacent beaches (Vedder and Norris, 1963; Johnson, 1972). Instead, carbonate content of the eolian sand and eolianite is closer to that of modern, sandy, insular shelf sediments of the Channel Islands. On San Nicolas Island, for example, modern shelf sediments average 49% $CaCO_3$, whereas modern beach sediments on the island average only 13% $CaCO_3$ (Vedder and Norris, 1963). In addition, the eolian deposits do not grade laterally into marine-terrace deposits, but are found on top of them, separated by a paleosol. This stratigraphic relation is significant, because it indicates that eolian sedimentation occurred after sea level was lowered. Radiocarbon ages of charcoal found in paleosols (locations 178 and 180 of Johnson [1977]) that underlie and overlie eolianite on San Miguel Island (Fig. 4) indicate that eolian sedimentation took place close to the last glacial maximum, between about 18 and 20 ka (Johnson, 1977). Collectively, these observations indicate that eolian sands were deposited during glacial periods, when sea level was significantly lower and calcareous shelf sediments were exposed to the wind (Fig. 6). The best estimate of the magnitude of sea-level lowering during the last glacial maximum comes from a recent study of submerged corals found off the island of Barbados reported by Fairbanks (1989). His data indicate that the last glacial maximum ca. 21–22 ka (Bard and others, 1990) had a sea level about 121 m lower than present. I mapped the extent of San Nicolas and San Clemente islands at the time of the last glacial maximum using the present 120 m isobath. The results indicate that large insular shelf areas would have been exposed 21–22 ka (Figs. 7 and 8).

On San Clemente Island, Muhs (1983b) mapped two eolianite units (in addition to modern, active dune sand) that can be distinguished on the basis of stratigraphic relations and degree of soil development (Fig. 9). The older unit, mapped as "Qeo" in Figure 9, is not found on the first terrace (ca. 80 ka) or the second

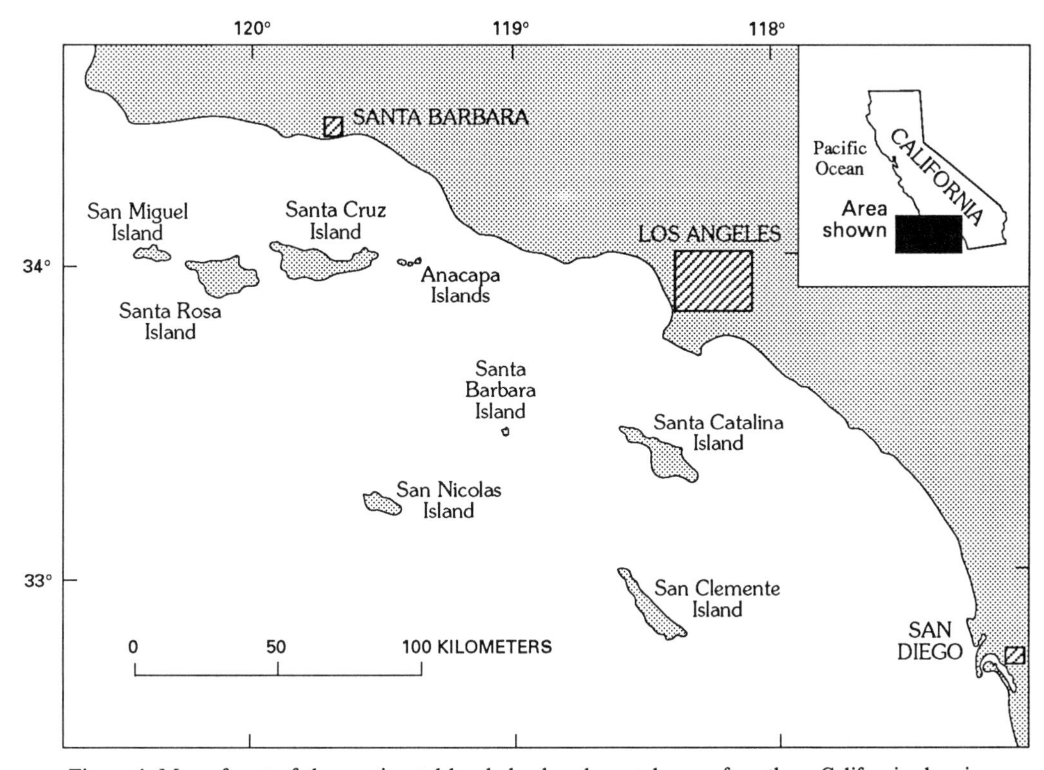

Figure 4. Map of part of the continental borderland and coastal area of southern California showing localities referred to in the text.

Figure 5. Cross-bedded eolianite exposed on Vizcaino Point, San Nicolas Island, California.

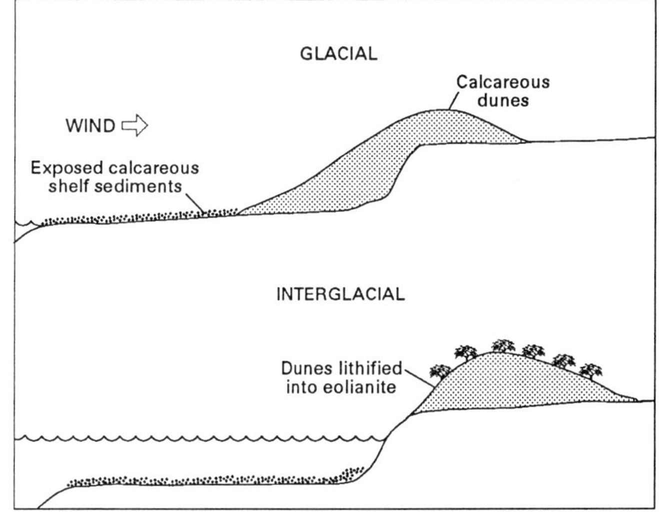

Figure 6. Model of eolianite formation during glacial periods on coasts lacking sandy beaches but which have abundant calcareous, sandy shelf sediments.

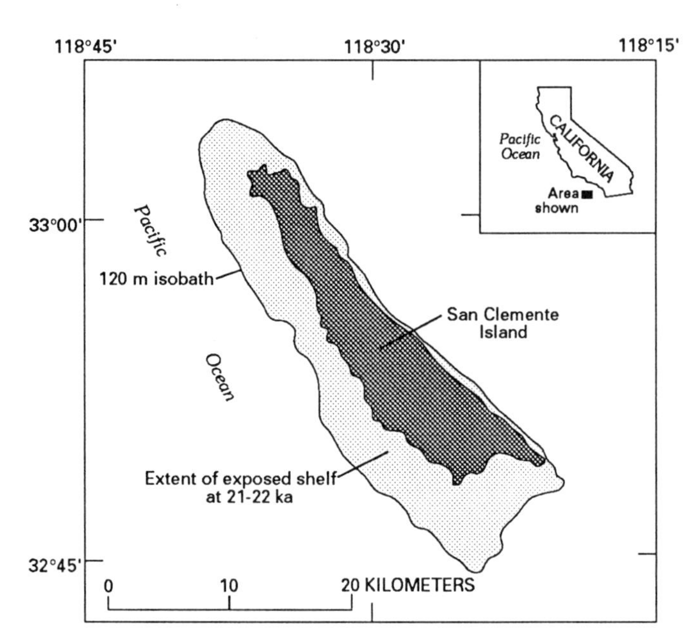

Figure 7. Shelf area on San Clemente Island, California, that would have been exposed during the last glacial maximum at 21–22 ka.

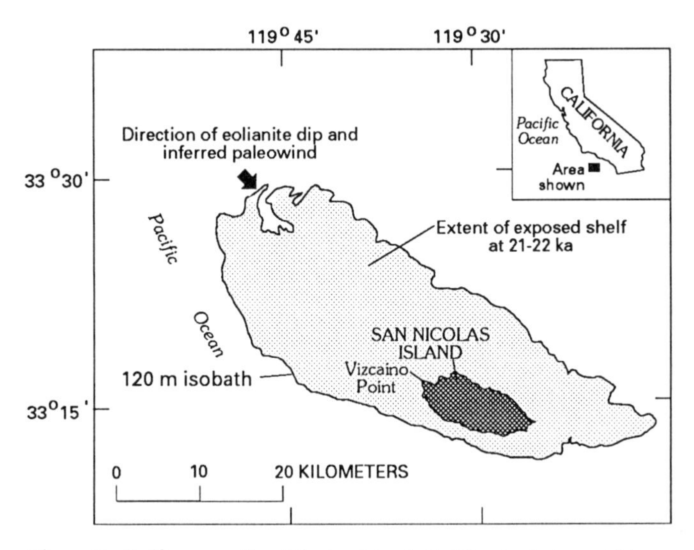

Figure 8. Shelf area on San Nicolas Island, California, that would have been exposed during the last glacial maximum at 21–22 ka.

terrace (ca. 127 ka) (Muhs, 1983b; Muhs and Szabo, 1982), but is found on many of the higher marine terraces, and is characterized by strongly developed soils with red, argillic B horizons. Thus, the stratigraphy and soil data suggest that this eolianite is older than 127 ka, perhaps significantly older. The younger eolianite (mapped as "Qey" in Fig. 9) is best exposed in modern sea cliffs on the west coast of San Clemente Island, where it is underlain by deposits of the first terrace, estimated by amino-acid ratios on shells to be about 80 ka (Muhs, 1983b). The Qey unit is subdivided into a number of members separated by paleosols (Fig. 10). The major members are shown as E2 and E3 (hereafter referred to as "lower" and "upper" members, respectively) in Figure 10. Where surface soils on the Qey unit are found, they are characterized by minimally developed B horizons that lack red colors and have significantly lower clay contents than the soils of the Qeo unit. Muhs (1983b) reported a ^{14}C age of ca. 22,000 yr B.P. for land snails found in the P3 paleosol, but this is probably a minimum age. Sands from all these eolian units were collected for thermoluminescence (TL) dating, but only the sediments from unit E3 had a sufficient amount of 4–11 μm material for analysis. This sample (Alpha #2611) yielded the following values: U, 1.3 ppm; Th, 1.0 ppm, K_2O, 0.37%; "a" value, 0.117; total dose rate, 0.95 rad/yr; equivalent dose, 1.168 krad; age estimate (using 75% of the laboratory measured saturation water content), 12,200 ± 1800 yr B.P. It is not known whether this silt-sized material was deposited at the same time as the more abundant sand-sized sediment, or if it is a secondary eolian silt blown in from a more distant source, such as the California mainland (Muhs, 1983c).

Calcareous eolian sands, locally lithified into eolianite, are also exposed in cliff sections around Vizcaino Point on San Nicolas Island (Fig. 8). Two eolian units overlie a paleosol developed on the first marine terrace (Fig. 11). On the basis of the similar stratigraphy, I correlate the two eolian units on San Nicolas Island with the two major eolianite units overlying the first marine terrace on San Clemente Island (units E2 and E3 in Fig. 10).

Uranium-series ages of rhizoliths found in California eolianites

On both San Nicolas Island and San Clemente Island, secondary carbonates are found that have a similar relation to the eolian sands. Near the upper part of each eolianite or eolian sand unit, rhizoliths (carbonate root casts) and horizontal pedogenic calcretes mark intervals of eolian sand stabilization and pedogenesis (Fig. 12). Eolian sedimentation should have ceased when sea-level rise progressed far enough to flood the shelf source areas. Because rhizoliths record times of vegetation stabilization when source areas had been cut off, they may record minimum ages for the underlying eolian sands.

Because rhizoliths on the California Channel Islands are relatively pure (up to 99% $CaCO_3$) secondary carbonates and are very dense, they are potentially suitable for U-series dating. X-ray diffraction analyses showed that the interiors of all but one of the rhizoliths studied are composed of calcite that lack primary (shell-derived) aragonite. This is significant because it indicates that, in all but one case, the rhizoliths are completely secondary and U-series ages should reflect the time of calcite precipitation. Only data for those rhizoliths that completely lack aragonite are reported here. Rhizoliths were collected for U-series dating from the three eolianite units on San Clemente Island (Qeo, and Qey, lower and upper) and the two eolian units that overlie the first marine terrace on San Nicolas Island. In addition, solitary corals (*Balanophyllia elegans*) were found in the first terrace deposits at

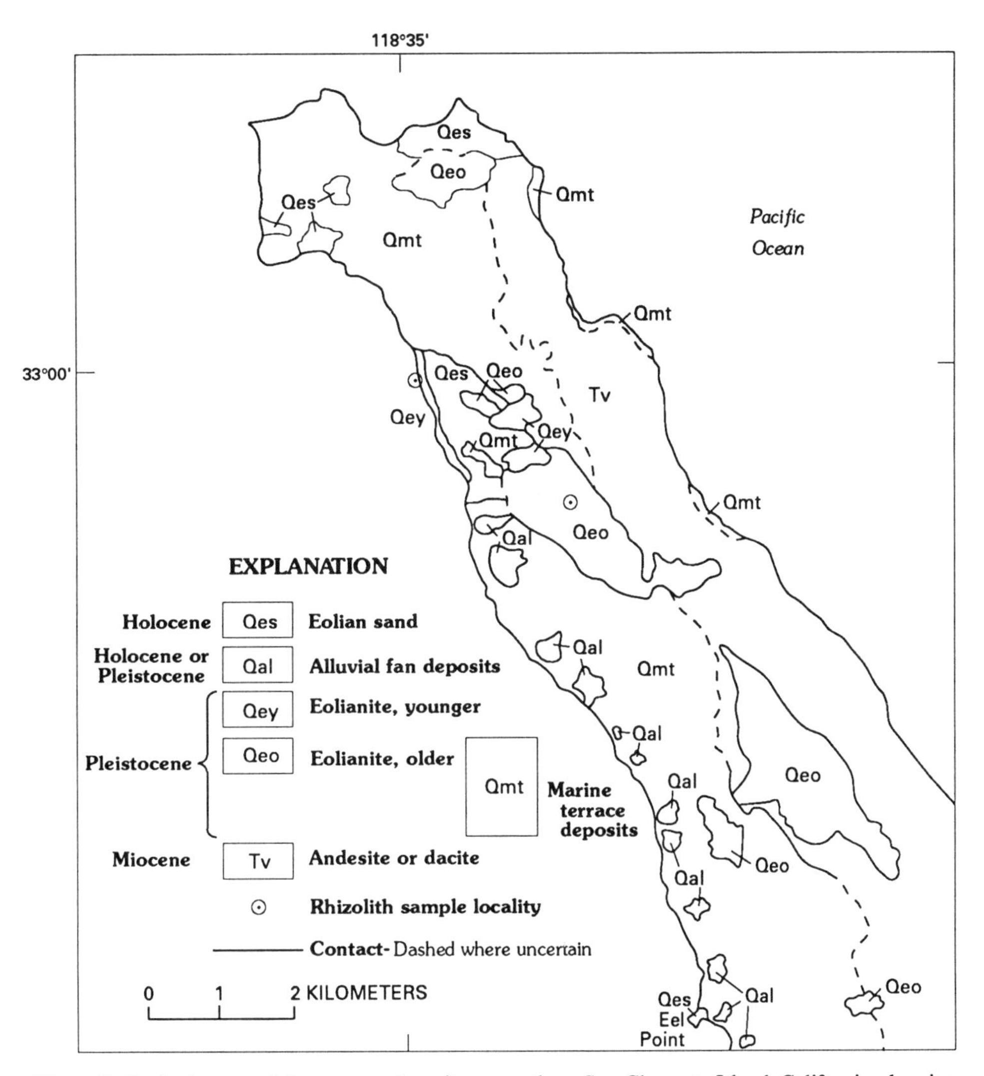

Figure 9. Geologic map of Quaternary deposits on northern San Clemente Island, California, showing rhizolith sample localities. Geologic map from Muhs (1983b).

Vizcaino Point on San Nicolas Island. No corals were found in the first terrace deposits on San Clemente Island.

Uranium-series ages were determined for the marine terrace corals and eolianite rhizoliths by isotope-dilution alpha spectrometry. Only the relatively pure interiors of rhizoliths were sampled for U-series analyses. All samples were cleaned mechanically prior to analysis to remove adhering detrital mineral grains. Samples were then powdered and heated to 900 °C for about 8 hrs to destroy organic matter and convert the $CaCO_3$ to CaO. The coral samples and some of the rhizoliths were then dissolved in 6*N* HCl and allowed to equilibrate with a combined ^{229}Th-^{236}U spike. After equilibration, a ferric nitrate carrier was added and the isotopes of U and Th were coprecipitated with NH_4OH. The precipitates were centrifuge washed, dissolved in 6*N* HCl, and loaded onto an ion-exchange column in chloride form to separate U from Th. The U and Th separates were taken to dryness and then dissolved in 8*N* HNO_3 and loaded onto ion-exchange columns in nitrate form for further purification. The purified U and Th separates were then taken to dryness. The thorium residue was then dissolved with about 0.2 ml of 0.1*N* HNO_3 and 0.1 ml of thenoyltrifluoroacetone (dissolved in benzene) was added to the solution. The solution was then agitated to transfer the Th into the organic phase; this was then transferred dropwise onto a heated stainless-steel disk. The U residue was dissolved in a drop of 6*N* HCl, mixed with about 1.5 ml of NH_4Cl buffer and adjusted to a pH of about 6 with NH_4OH. The solution containing the U was then added to a Teflon plating apparatus and electroplated onto a stainless-steel disk over a period of about 0.5 h at a current of 1.8 A. The disks containing the U and Th samples were then placed in an alpha spectrometer for counting. After some samples had been counted sufficiently, it was concluded that some detrital materials had remained in the samples after cleaning; this was apparent from $^{230}Th/^{232}Th$ activity ratios lower than about 20 (Table 3). In such cases, some of

Figure 10. Cliff exposure of eolianite at Vizcaino Point, San Nicolas Island, showing horizontal calcrete underlain by carbonate rhizoliths.

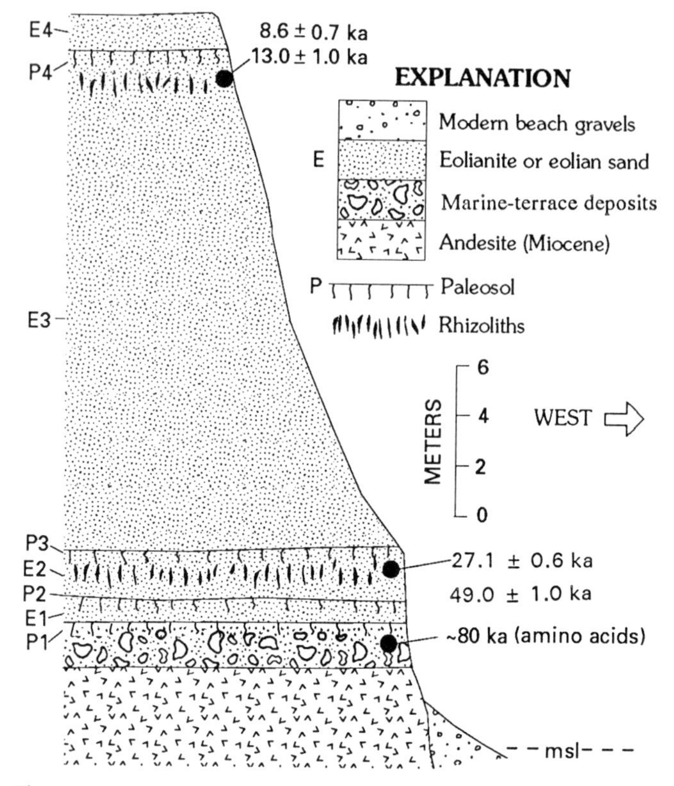

Figure 11. Cliff section exposed on the northwest coast of San Clemente Island, California, showing U-series ages of rhizoliths (see Table 3). Stratigraphy from Muhs (1983b).

the activity of ^{230}Th is contributed by the detrital minerals rather than all of the ^{230}Th activity being due to in situ radioactive decay of ^{234}U in the carbonate. This problem was minimized on all subsequent samples by gradual dissolution of the carbonate using dilute (0.25*N*) HNO_3. However, even with dilute HNO_3 leaching, some U and Th can still be leached from the detrital minerals. In the event that the $^{230}Th/^{232}Th$ ratios are still less than about 20 after dilute acid leaching, a correction can be made (using the method outlined by Ku and Liang, 1984) by a separate analysis of the residue (if a sufficient amount is present) or the host eolian sediment, which is the obvious source of the contaminating material. In the case of the San Clemente Island rhizoliths, only the samples from the youngest unit required corrections. Hence, samples of the appropriate host eolian sediment were analyzed, using a complete decomposition with an HNO_3-$HClO_4$-HF mixture; corrections were made using the mixing-line plot method of Ku and Liang (1984) with the isotopic composition of the host sediment as a proxy for the residue (Table 4). For all samples reported in Tables 3 and 4, errors for activity ratios and ages are based on counting statistics and are 1 sigma.

Uranium-series analysis of the fossil corals indicate that the first terrace on San Nicolas Island is 80 ± 2 ka and thus correlates with ca. 80 ka terraces reported from Barbados, New Guinea, Haiti, northern Baja California, and other areas discussed earlier. The 80 ka age is reasonable because these corals have (1) U concentrations that are typical for this species (Ku and Kern, 1974; Rockwell and others, 1989; Muhs and others, 1990); (2) $^{230}Th/^{232}Th$ ratios that are significantly greater than 20, thus requiring no corrections for detrital materials; and (3) $^{234}U/^{238}U$ ratios that are consistent with an 80 ka age and initial $^{234}U/^{238}U$ ratios in sea water of 1.14–1.15, within limits of analytical uncertainty.

Uranium-series ages of rhizoliths are in agreement with the stratigraphy and relative ages of the deposits and indicate that rhizoliths are suitable for U-series dating (Table 3 and Figs. 10 and 11). Within a given eolian unit, there is a significant range of rhizolith ages, but this is expected because a rhizolith can form during any time period that a surface is stabilized with vegetation. The oldest mapped eolianite on San Clemente Island is the Qeo unit, which is not found on the first (ca. 80 ka) or second (ca. 127 ka) marine terraces, but is found on most of the higher terraces. Rhizoliths B and D from this unit give ages of 152 ± 5 ka and 166

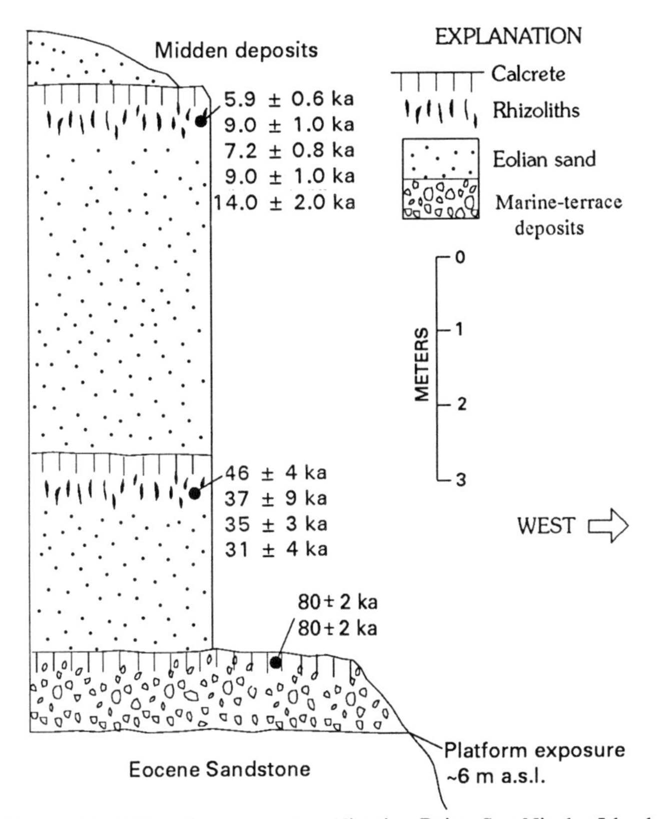

Figure 12. Cliff section exposed at Vizcaino Point, San Nicolas Island, California, showing U-series ages of rhizoliths and marine terrace corals (see Table 3).

± 10 ka (Table 3), in agreement with the stratigraphic relation to the marine terraces. Rhizolith A has an apparent age of 21 ka, but it also has a U concentration that is a factor of 4 to 5 greater than the other rhizoliths from the Qeo unit, and up to an order of magnitude higher than that found in the younger rhizoliths. Hence, it is likely that this rhizolith has undergone some recent, secondary U uptake. More U-series ages are needed from this unit, but these preliminary data indicate that eolian deposition could have taken place during the lowstands of sea level preceding the formation of the second terrace (ca. 127 ka) or the third terrace (ca. 200 ka) (Muhs and Szabo, 1982). The younger Qey eolianite on San Clemente Island is subdivided into "lower" and "upper" members on the basis of the stratigraphy exposed on the west coast (Fig. 10). U-series ages of rhizoliths support this relative age relation (Table 3 and Fig. 10). The lower member was apparently deposited between ca. 80 ka (the estimated age of the underlying marine terrace deposits) and 49 ± 1 ka, the age of the oldest rhizolith in this unit. The surface of this eolianite was apparently stabilized by vegetation at least until 27 ka, the age of the youngest rhizolith. Some time after 27 ka, the upper member of Qey (shown as E3 in Fig. 10) was deposited and stabilized by 13 ± 1 ka, the age of the oldest rhizolith from this unit, in agreement with the TL age estimate of 12 ± 2 ka of the 4–11 μm fraction of the deposit.

On San Nicolas Island, U-series ages of rhizoliths support the correlation of the two eolian sand units overlying the 80 ka marine terrace with the upper and lower members of the Qey eolianite on San Clemente Island (Table 3 and Fig. 11). The lower member on San Nicolas Island was deposited between 80 ± 2 ka and 46 ± 4 ka, or about the same time as the lower, E2 eolianite on San Clemente Island. The surface of this unit was stabilized by vegetation at least until about 31 ± 4 ka, the age of the youngest rhizolith. Eolian sedimentation began again sometime after ca. 31 ka and continued until 14 ± 2 ka, the age of the youngest rhizolith in the upper eolian unit.

Inferred late Quaternary sea-level history on the California Channel Islands

Uranium-series dating of these eolianites indicates a sequence of events that agrees with sea-level history recorded elsewhere. After the relatively high sea-level stand ca. 80 ka, sea level lowered and what is now the first marine terrace on both islands emerged. After initial emergence, and while sea level was still dropping, soils formed on the marine-terrace deposits (Figs. 10 and 11). When sea level had lowered sufficiently to expose calcareous shelf sands, eolian sedimentation began and resulted in the deposition of the lower eolian members. Sea level apparently rose sufficiently to cut off the source of the sands sometime around 46–49 ka, because this is when rhizolith formation, indicating vegetation stabilization, began on both islands. This period of stabilization continued until about 27–31 ka, the age of the youngest rhizoliths in the lower eolian members. On New Guinea, reef complexes IV (ca. 60 ka) IIIa (ca. 40–50 ka), IIIb (ca. 40 ka), and II (ca. 28 ka) record four relatively high sea-level stands during the interval from 60 to 28 ka. Calculations of the position of sea level at these times are in broad agreement by both the Bloom and Yonekura (1985) and Chappell and Shackleton (1986) methods, and indicate sea levels as high as –24 to –28 m, relative to present, ca. 60 ka, and as low as –35 to –44 m, relative to present, ca. 28 ka. If sea levels were this high, significant portions of the shelf areas of both islands would be submerged, and it is probable that most eolian sand sources would be cut off. Sometime after 27–31 ka, sea level dropped again, and eolian sedimentation was renewed and continued until 13–14 ka. These observations agree with those made by Johnson (1977) of an eolianite deposited between 18 and 20 ka on San Miguel Island, California. Recent studies by Fairbanks (1989) of submerged reefs off Barbados indicate that sea level was around 121 m lower than present at the last glacial maximum. The last glacial maximum is now estimated to be 21–22 ka by high-precision U-series dating of corals (Bard and others, 1990). The bracketing U-series ages of rhizoliths of 27–31 ka and 13–14 ka from San Nicolas and San Clemente islands are in broad agreement with these new age estimates of the last glacial maximum from Barbados, and indicate that major eolian sedimentation took place during the last glacial maximum, when sea level was lowered ~121 m and broad shelf areas were exposed on both islands (Figs. 7 and 8).

TABLE 3. URANIUM CONCENTRATIONS, ISOTOPIC ACTIVITY RATIOS, AND U-SERIES AGES OF CALIFORNIA CHANNEL ISLANDS RHIZOLITHS AND CORALS

Locality, Geologic Unit (Sample)*	U (ppm)	$^{234}U/^{238}U$	$^{230}Th/^{232}Th$	$^{230}Th/^{234}U$	Uncorrected Age† (ka)	Corrected Age§ (ka)
		Activity Ratios				
San Clemente Island, Qey, upper						
R, A (HNO_3 leach)	0.646 ± 0.008	1.07 ± 0.01	7.7 ± 0.4	0.110 ± 0.002	12.6 ± 0.3	8.6 ± 0.7
R, B (HNO_3 leach)	0.715 ± 0.009	1.08 ± 0.01	17 ± 1	0.127 ± 0.002	14.7 ± 0.3	13 ± 1
San Nicolas Island, upper eolian sand						
R, A (HCl leach)	0.86 ± 0.01	1.10 ± 0.01	2.2 ± 0.1	0.130 ± 0.004	15.1 ± 0.5	5.9 ± 0.6
R, A (HNO_3 leach)	0.82 ± 0.01	1.12 ± 0.01	3.4 ± 0.3	0.129 ± 0.007	15.0 ± 0.9	9 ± 1
R, B (HCl leach)	0.85 ± 0.01	1.08 ± 0.01	3.0 ± 0.2	0.114 ± 0.004	13.1 ± 0.5	7.2 ± 0.8
R, B (HNO_3 leach)	0.82 ± 0.01	1.10 ± 0.02	3.8 ± 0.2	0.121 ± 0.004	14.0 ± 0.5	9 ± 1
R, C (HNO_3 leach)	0.577 ± 0.008	1.09 ± 0.02	8 ± 1	0.142 ± 0.006	16.6 ± 0.8	14 ± 3
San Clemente Island, Qey, lower						
R, A (HNO_3 leach)	0.687 ± 0.008	1.128± 0.009	23 ± 2	0.222 ± 0.004	27.1 ± 0.6	27.1 ± 0.6
R, B (HCl leach)	0.726 ± 0.009	1.16 ± 0.01	21.6 ± 0.8	0.367 ± 0.005	49 ± 1	49 ± 1
San Nicolas Island, lower eolian sand						
R, A (HCl leach)	0.667 ± 0.008	1.11 ± 0.01	4.2 ± 0.1	0.417 ± 0.006	58 ± 1	46 ± 4
R, A (HNO_3 leach)	0.597 ± 0.008	1.13 ± 0.01	17 ± 2	0.304 ± 0.008	39 ± 1	37 ± 9
R, C (HCl leach)	0.86 ± 0.01	1.13 ± 0.01	7.1 ± 0.2	0.316 ± 0.005	41 ± 1	35 ± 3
R, C (HNO_3 leach)	0.77 ± 0.01	1.11 ± 0.01	17 ± 2	0.262 ± 0.006	33 ± 1	31 ± 4
San Nicolas Island, first marine terrace						
Coral, A (B. e.)	4.78 ± 0.07	1.14 ± 0.01	64 ± 7	0.530 ± 0.009	80 ± 2	80 ± 2
Coral, B (B.e.)	4.19 ± 0.06	1.12 ± 0.01	26 ± 1	0.529 ± 0.009	80 ± 2	80 ± 2
San Clemente Island, Qeo						
R, A (HNO_3 leach)	6.45 ± 0.08	1.046± 0.004	81 ± 5	0.178 ± 0.002	21.2 ± 0.3	21.2 ± 0.3
R, B (HNO_3 leach)	1.64 ± 0.02	1.042± 0.008	92 ± 6	0.76 ± 0.01	152 ± 5	152 ± 5
R, D (HNO_3 leach)	1.34 ± 0.02	1.12 ± 0.01	>800	0.80 ± 0.02	166 ±10	166 ±10

*R = Rhizolith. Corals are *Balanophyllia elegans* (B.e.); A was collected from Los Angeles County Museum of Natural History locality number 10621 and B was collected from locality number 11009.
†Calculated using half-lives of ^{230}Th and ^{234}U of 75,200 and 244,000 years, respectively.
§Corrected for inherited ^{230}Th using isotopic composition of the host sediment (Table 4) and two-point mixing-line plot (Ku and Liang, 1984) when ^{230}Th ratio is less than 20.

TABLE 4. URANIUM AND THORIUM ISOTOPIC COMPOSITION OF WHOLE-SEDIMENT EOLIAN SANDS USED FOR AGE CORRECTIONS

Locality, Geologic Unit U (ppm)	Th (ppm)	$^{234}U/^{238}U$ Activity Ratios	$^{230}Th/^{232}Th$	$^{230}Th/^{234}U$
San Clemente Island, Qey, upper eolianite				
0.82 ± 0.01	0.96 ± 0.03	1.03 ± 0.1	2.59 ± 0.09	0.95 ± 0.02
San Nicolas Island, upper eolian sands				
1.07 ± 0.01	1.72 ± 0.04	1.04 ± 0.01	1.45 ± 0.04	0.73 ± 0.02
San Nicolas Island, lower eolian sands				
0.90 ± 0.01	1.71 ± 0.03	1.07 ± 0.01	1.30 ± 0.02	0.74 ± 0.01

Thus, eolian sands on the southern Channel Islands record two periods of glaciation, an advance sometime between ca. 80 and 49 ka, and an advance sometime between 27–31 ka and 13–14 ka. Additional U-series dating of rhizoliths may narrow these age ranges significantly. Inferred times of low sea level (glacials) from the Channel Islands eolianite record agree well with times of lowered summer insolation in high latitudes, whereas times of marine-terrace formation (interglacials) agree well with periods of higher summer insolation in high latitudes (Fig. 13).

The marine-terrace record indicates that sea levels were low ca. 115 and 95 ka (Fig. 3). Were sea levels low enough at these times to expose the insular shelves of the Channel Islands and provide a source of sediment for eolianite formation? This question cannot be answered with the available data, but I have observed two eolianite units overlying the second, ca. 125 ka terrace at Vizcaino Point on San Nicolas Island. These two eolianites may correlate with the two eolian units overlying the ca. 80 ka terrace, or they may record earlier (ca. 115 and 95 ka) lowstands of sea level. Future U-series analyses should answer this question.

If the ca. 80 ka highstand of sea level is considered to be part of a "long interglacial" that includes all of deep-sea isotope stage 5, then the shift from marine to eolian sedimentation recorded by the coastal deposits on the California Channel Islands reflects the last interglacial/glacial transition. Such a shift in the type of sedimentation should be recorded on other coasts as well and invites comparison to the California Channel Islands. The model of eolianite formation during glacial periods is supported by observations in other localities, where eolianites overlie marine-terrace deposits and/or eolianites extend below sea level. These relations have been observed in South Africa (Marker, 1976), Australia (Fairbridge and Johnson, 1978; Sprigg, 1979), Hawaii (Stearns, 1978), Mallorca (Butzer and Cuerda, 1962; Butzer, 1975), Lebanon (Wright, 1962), and Puerto Rico (Kaye, 1959). There are, however, other localities where eolianites grade laterally into beach or marine deposits, such as in Bermuda and the Bahamas (Land and others, 1967; Garrett and Gould, 1984), indicating formation of eolianite during highstands of sea level. As Gardner (1983) pointed out, it is probably unwise to postulate a single relation of eolianite formation to sea level for all areas.

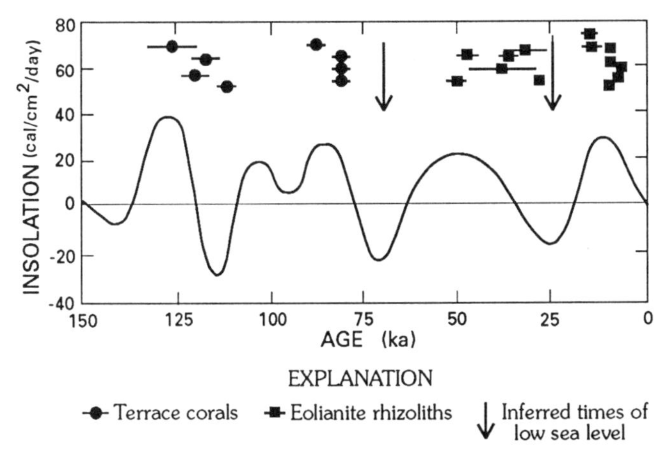

Figure 13. Plot of U-series ages of eolianite rhizoliths and marine-terrace corals from San Nicolas and San Clemente Islands, California (data from Table 3; Muhs and Szabo [1982]; and updated from Muhs and others [1987b]). Vertical arrows show times of inferred low sea level (and also continental glaciation) based on the distribution of U-series ages. Shown for comparison is a plot of average insolation received at the top of the atmosphere at lat 65°N for the summer half year, expressed as deviations from the present (A.D. 1950) value. Insolation data are from Berger (1978).

Nevertheless, the results presented here suggest that with careful field work and a rigorous dating program, a valuable record of the last interglacial/glacial transition could be found on many subtropical and tropical coasts.

CONCLUSIONS

Compilation of reliably dated corals from terraces that have been correlated with deep-sea oxygen-isotope stage 5 indicate, in general, three relatively high stands of sea level ca. 125–120 ka, 105 ka, and 85–80 ka. These times of high sea level are in agreement with calculated periods of high summer insolation at high latitudes, which would have favored global ice melting and resulted in relatively high sea levels. On both Barbados and the east coast of the United States, there are a number of corals from marine deposits that are apparently reliable and indicate a high sea level ca. 70 ka. Studies from both relatively stable coasts and tectonically emergent coasts indicate that there may have been a high sea level at 145–135 ka, but data are sparse and stratigraphic evidence is lacking, except on New Guinea. However, high sea levels at both 70 ka and 145–135 ka represent potential challenges to the Milankovitch orbital-forcing theory of climatic change, because summer insolation at high latitudes was low at these times, and would have favored glaciation and lowered sea levels. What is clearly needed to help resolve these issues are high-precision mass-spectrometric U-series analyses of corals for dating and careful studies of corals to investigate possible subtle diagenesis and U migration, using scanning-electron microscopy and fission-track mapping.

There is disagreement among investigators from different fields concerning the length of the last interglacial. Many glacial geologists and most marine scientists favor a "short" last interglacial (correlated to deep-sea substage 5e only), whereas some pedologists, speleothem investigators, and ice-core workers favor a "long" interglacial (most or all of deep-sea stage 5). The marine terrace record has added to the confusion, because whether the last interglacial was long or short depends on how high sea level was ca. 105 and 80 ka; records from different coasts disagree on the elevations reached by the sea at these times. Paleo-sea-level elevations generated by an assumption of constant uplift rate during the past 125 ka on Barbados, New Guinea, and Haiti all indicate sea levels significantly lower than present at 105 ka and 80 ka. New Guinea paleo-sea levels calculated by an alternative method indicate sea levels close to the present at these times. On the east and west coasts of North America and on Bermuda and the Bahamas, there is also evidence for sea levels close to the present or even higher ca. 105 and 80 ka. Taken as a whole, these results indicate that more studies are needed on other coasts before we can make confident statements about sea-level elevations during the last interglacial/glacial cycle.

Coastal deposits on the California Channel Islands record a shift from marine to eolian sedimentation at the last interglacial/glacial transition if it is assumed that the last interglacial was "long," and ended sometime after ca. 80 ka. Marine terrace deposits contain solitary corals that have been dated ca. 80 ka by U-series methods or have shells that give amino-acid ratios indicative of an age of ca. 80 ka. A paleosol has developed on these deposits, and this soil is in turn overlain by two eolian sand or eolianite units, separated by another paleosol. Carbonate rhizoliths in the eolian units give U-series ages that indicate two episodes of sedimentation, one between ca. 80 ka and 49–46 ka, and another between 31–27 ka and 14–13 ka. These eolian sands could only have been deposited when sea level was low and shelf sands were exposed, because there are not sufficient beach sediments at present to account for the volume of eolian sand. Greatly lowered sea levels imply glacial conditions, and the times of lowered sea level on the California Channel Islands, as indicated by the bracketing U-series ages, agree well with times of lower summer insolation at high latitudes. These observations indicate that marine terrace/eolianite sequences may provide a valuable record of the last interglacial/glacial transition, and studies should be initiated on other coasts where such deposits exist.

ACKNOWLEDGMENTS

This study was supported by the Global Change and Climate History Program of the U.S. Geological Survey. I thank the U.S. Navy for access to San Nicolas and San Clemente Islands. Ron Dow and Andy Yatsko of the U.S. Navy provided logistical support and helpful discussions on the islands. Tracy Rowland, George Kennedy, Don Johnson, John Rosholt, and Tom Rockwell assisted with field work on various trips. Lisa Hu prepared some of the rhizolith samples for dating, and Chuck Bush assisted with processing of the U-series data. Art Bloom, Ken Pierce, Barney Szabo, and an unknown reviewer read earlier drafts of this paper; I thank them for their comments. I also appreciate the suggestions of the volume editors, Peter Clark and Peter Lea. Pat Wilber drafted the figures and Marge Henneck typed the tables.

REFERENCES CITED

Adam, D. P., 1988, Palynology of two upper Quaternary cores from Clear Lake, Lake County, California: U.S. Geological Survey Professional Paper 1363, p. 1–86.

Aharon, P., 1983, 140,000 year isotope climate record from raised coral reefs in New Guinea: Nature, v. 304, p. 720–723.

Aharon, P., Chappell, J., and Compston, W., 1980, Stable isotope and sea-level data from New Guinea supports Antarctic ice-surge theory of ice ages: Nature, v. 283, p. 649–651.

Ashby, J. R., Ku, T. L., and Minch, J. A., 1987, Uranium series ages of corals from the upper Pleisotcene Mulege terrace, Baja California Sur, Mexico: Geology, v. 15, p. 139–141.

Bard, E., Hamelin, B., Fairbanks, R. G., and Zindler, A., 1990, Calibration of the ^{14}C timescale over the past 30,000 years using mass spectrometric U-Th ages from Barbados corals: Nature, v. 345, p. 405–410.

Bender, M. L., Fairbanks, R. G., Taylor, F. W., Matthews, R. K., Goddard, J. G., and Broecker, W. S., 1979, Uranium-series dating of the Pleistocene reef tracts of Barbados, West Indies: Geological Society of America Bulletin, v. 90, p. 577–594.

Berger, A. L., 1978, Long-term variations of caloric insolation resulting from the earth's orbital elements: Quaternary Research, v. 9, p. 139–167.

Bloom, A. L., and Yonekura, N., 1985, Coastal terraces generated by sea-level change and tectonic uplift, *in* Woldenberg, M. J., ed., Models in geomorphology: Boston, Allen and Unwin, p. 139–154.

Bloom, A. L., Brocker, W. S., Chappell, J.M.A., Matthews, R. K., and Mesolella, K. J., 1974, Quaternary sea level fluctuations on a tectonic coast: New $^{230}Th/^{234}U$ dates from the Huon Peninsula, New Guinea: Quaternary Research, v. 4, p. 185–205.

Boardman, J., 1985, Comparison of soils in midwestern United States and western Europe with the interglacial record: Quaternary Research, v. 23, p. 62–75.

Bradley, W. C., and Griggs, G. B., 1976, Form, genesis, and deformation of central California wave-cut platforms: Geological Society of America Bulletin, v. 87, p. 433–449.

Broecker, W. S., and Thurber, D. L., 1965, Uranium-series dating of corals and oolites from Bahaman and Florida Key limestones: Science, v. 149, p. 58–60.

Broecker, W. S., Thurber, D. L., Goddard, J., Ku, T. L., Matthews, R. K., and Mesolella, K. J., 1968, Milankovitch hypothesis supported by precise dating of coral reefs and deep-sea sediments: Science, v. 159, p. 297–300.

Butzer, K., 1975, Pleistocene littoral-sedimentary cycles of the Mediterranean Basin: A Mallorquin view, *in* Butzer, K. W. and Isaac, G. L., eds., After the Australopithecines: Stratigraphy, ecology and culture change in the middle Pleistocene: Chicago, Aldine, p. 25–71.

Butzer, K., and Cuerda, J., 1962, Coastal stratigraphy of southern Mallorca and its implications for the Pleistocene chronology of the Mediterranean Sea: Journal of Geology, v. 70, p. 398–416.

Chappell, J., 1974, Geology of coral terraces, Huon Peninsula, New Guinea: A study of Quaternary tectonic movements and sea-level changes: Geological Society of America Bulletin, v. 85, p. 553–570.

Chappell, J., and Shackleton, N. J., 1986, Oxygen isotopes and sea level: Nature, v. 324, p. 137–140.

Chappell, J., and Veeh, H. H., 1978, Late Quaternary tectonic movements and sea-level changes at Timor and Atauro Island: Geological Society of America Bulletin, v. 89, p. 356–358.

Chen, J. H., Edwards, R. L., and Wasserburg, G. J., 1986, ^{238}U, ^{234}U and ^{232}Th in seawater: Earth and Planetary Science Letters, v. 80, p. 241–251.

CLIMAP Project Members, 1984, The last interglacial ocean: Quaternary Research, v. 21, p. 123–224.

Corrado, J. C., Weems, R. E., Hare, P. E., and Bambach, R. K., 1986, Capabilities and limitations of applied aminostratigraphy, as illustrated by analyses of *Mulinia lateralis* from the late Cenozoic marine beds near Charleston, South Carolina: South Carolina Geology, v. 30, p. 19–46.

Cronin, T. M., 1981, Rates and possible causes of neotectonic vertical crustal movements of the emerged southeastern United States Atlantic Coastal Plain: Geological Society of America Bulletin, v. 92, p. 812–833.

Cronin, T. M., Szabo, B. J., Ager, T. A., Hazel, J. E., and Owens, J. P., 1981, Quaternary climates and sea levels of the U.S. Atlantic coastal plain: Science, v. 211, p. 233–240.

Dodge, R. E., Fairbanks, R. G., Benninger, L. K., and Maurrasse, F., 1983, Pleistocene sea levels from raised coral reefs of Haiti: Science, v. 219, p. 1423–1425.

Edwards, R. L., Chen, J. H., Ku, T. L., and Wasserburg, G. J., 1987, Precise timing of the last interglacial period from mass spectrometric determination of thorium-230 in corals: Science, v. 236, p. 1547–1553.

Fairbanks, R. G., 1989, A 17,000-year glacio-eustatic sea level record: Influence of glacial melting rates on the Younger Dryas event and deep-ocean circulation: Nature, v. 342, p. 637–642.

Fairbanks, R. G., and Matthews, R. K., 1978, The marine oxygen isotope record in Pleistocene coral, Barbados, West Indies: Quaternary Research, v. 10, p. 181–196.

Fairbridge, R. W., and Johnson, D. L., 1978, Eolianite, *in* Fairbridge, R. W., and Bourgeois, J., eds., The encyclopedia of sedimentology: Stroudsburg, Dowden, Hutchinson and Ross, p. 279–282.

Follmer, L. R., 1978, The Sangamon Soil in its type area—A review, *in* Mahaney, W. C., ed., Quaternary soils: Toronto, York University, p. 125–165.

——, 1982, The geomorphology of the Sangamon surface: Its spatial and temporal attributes, *in* Thorn, C. E., ed., Space and time in geomorphology: London, George Allen and Unwin, p. 117–146.

Fulton, R. J., and Prest, V. K., 1987, The Laurentide ice sheet and its significance: Géographie Physique et Quaternaire, v. 41, p. 181–186.

Gardner, R.A.M., 1983, Aeolianite, *in* Goudie, A. S., and Pye, K., eds., Chemical sediments and geomorphology: London, Academic Press, p. 265–300.

Garrett, P., and Gould, S. J., 1984, Geology of New Providence Island, Bahamas: Geological Society of America Bulletin, v. 95, p. 209–220.

Harmon, R. S., Ford, D. C., and Schwarcz, H. P., 1977, Interglacial chronology of the Rocky and Mackenzie mountains based upon ^{230}Th-^{234}U dating of calcite speleothems: Canadian Journal of Earth Sciences, v. 14, p. 2543–2552.

Harmon, R. S., Ku, T. L., Matthews, R. K., and Smart, P. L., 1979, Limits of U-series analysis: Phase 1 results of the Uranium-Series Intercomparison Project: Geology, v. 7, p. 405–409.

Harmon, R. S., and eight others, 1983, U-series and amino-acid racemization geochronology of Bermuda: Implications for eustatic sea-level fluctuation over the past 250,000 years: Palaeogeography, Palaeoclimatology, Palaeoecology, v. 44, p. 41–70.

Hays, J. D., Imbrie, J., and Shackleton, N. J., 1976, Variations in the Earth's orbit: Pacemaker of the Ice Ages: Science, v. 194, p. 1121–1132.

Hollin, J. T., and Hearty, P. J., 1990, South Carolina interglacial sites and stage 5 sea levels: Quaternary Research, v. 33, p. 1–17.

Ivanovich, M., Ku, T. L., Harmon, R. S., and Smart, P. L., 1984, Uranium Series Intercomparison Project (USIP): Nuclear Instruments and Methods in Physics Research, v. 223, p. 466–471.

Johnson, D. L., 1972, Landscape evolution on San Miguel Island, California [Ph.D. thesis]: Lawrence, University of Kansas, 391 p.

——, 1977, The late Quaternary climate of coastal California: Evidence for an Ice Age refugium: Quaternary Research, v. 8, p. 154–179.

Johnson, R. G., 1982, Brunhes-Matuyama magnetic reversal dated at 790,000 yr B.P. by marine-astronomical correlations: Quaternary Research, v. 17, p. 135–147.

Kaufman, A., 1986, The distribution of ^{230}Th/^{234}U ages in corals and the number of last interglacial high-sea stands: Quaternary Research, v. 25, p. 55–62.

Kaye, C. A., 1959, Shoreline features and Quaternary shoreline changes, Puerto Rico: U.S. Geological Survey Professional Paper 317B, p. 49–140.

Kennedy, G. L., Lajoie, K. R., and Wehmiller, J. F., 1982, Aminostratigraphy and faunal correlations of late Quaternary marine terraces, Pacific Coast, USA: Nature, v. 299, p. 545–547.

Kern, J. P., 1977, Origin and history of upper Pleistocene marine terraces, San Diego, California: Geological Society of America Bulletin, v. 88, p. 1553–1566.

Koerner, R. M., and Fisher, D. A., 1985, The Devon Island ice core and the glacial record, *in* Andrews, J. T., ed., Quaternary environments, Eastern Canadian Arctic, Baffin Bay, and Western Greenland: Boston, Allen and Unwin, p. 309–327.

Kominz, M. A., Heath, G. R., Ku, T. L., and Pisias, N. G., 1979, Brunhes time scales and the interpretation of climatic change: Earth and Planetary Science Letters, v. 45, p. 394–410.

Ku, T. L., 1968, Protactinium 231 method of dating coral from Barbados Island: Journal of Geophysical Research, v. 73, p. 2271–2276.

Ku, T. L., and Kern, J. P., 1974, Uranium-series age of the upper Pleistocene Nestor terrace, San Diego, California: Geological Society of America Bulletin, v. 85, p. 1713–1716.

Ku, T. L., and Liang, Z. C., 1984, The dating of impure carbonates with decay-series isotopes: Nuclear Instruments and Methods in Physics Research, v. 223, p. 563–571.

Ku, T. L., Knauss, K. G., and Mathieu, G. G., 1977, Uranium in open ocean: Concentration and isotopic composition: Deep-Sea Research, v. 24, p. 1005–1017.

Ku, T. L., Ivanovich, M., and Luo, S., 1990, U-series dating of last interglacial high sea stands: Barbados revisited: Quaternary Research, v. 33, p. 129–147.

Lajoie, K. R., 1986, Coastal tectonics, *in* Active tectonics: Washington, D.C., National Academy Press, p. 95–124.

Land, L. S., Mackenzie, F. T., and Gould, S. J., 1967, The Pleistocene history of Bermuda: Geological Society of America Bulletin, v. 78, p. 993–1006.

Li, W. X., Lundberg, J., Dickin, A. P., Ford, D. C., Schwarcz, H. P., McNutt, R., and Williams, D., 1989, High-precision mass-spectrometric uranium-series dating of cave deposits and implications for palaeoclimate studies: Nature, v. 339, p. 534–536.

Lively, R. S., 1983, Late Quaternary U-series speleothem growth record from southeastern Minnesota: Geology, v. 11, p. 259–262.

Mankinen, E. A., and Dalrymple, G. B., 1979, Revised geomagnetic polarity time scale for the interval 0-5 m.y. B.P.: Journal of Geophysical Research, v. 84, p. 615–626.

Marker, M. E., 1976, Aeolianite: Australian and South African deposits compared: Annals of the South African Museum, v. 71, p. 115–124.

Martinson, D. G., Pisias, N. G., Hays, J. D., Imbrie, J., Moore, T. C., Jr., and Shackleton, N. J., 1987, Age dating and the orbital theory of the ice ages: Development of a high-resolution 0 to 300,000-year chronostratigraphy:

Quaternary Research, v. 27, p. 1–29.
Matthews, R. K., 1985, The $\delta^{18}O$ signal of deep-sea planktonic foraminifera at low latitudes as an ice-volume indicator: South African Journal of Science, v. 81, p. 274–275.
McCartan, L., Owens, J. P., Blackwelder, B. W., Szabo, B. J., Belknap, D. F., Kriausakul, N., Mitterer, R. M., and Wehmiller, J. F., 1982, Comparison of amino acid geochronometry with lithostratigraphy, biostratigraphy, uranium-series coral dating, and magnetostratigraphy in the Atlantic coastal plain of the southeastern United States: Quaternary Research, v. 18, p. 337–359.
Mesolella, K. J., Matthews, R. K., Broecker, W. S., and Thurber, D. L., 1969, The astronomical theory of climatic change: Barbados data: Journal of Geology, v. 77, p. 250–274.
Milankovitch, M. M., 1941, Canon of insolation and the Ice Age problem: Beograd, Koniglich Serbische Akademie (English translation: Jerusalem, Israel, Israel Program for Scientific Translations, 1969), 484 p.
Mixon, R. B., Szabo, B. J., and Owens, J. P., 1982, Uranium-series dating of mollusks and corals, and age of Pleistocene deposits, Chesapeake Bay area, Virginia and Maryland: U.S. Geological Survey Professional Paper 1067-E, p. E1–E-18.
Moore, W. S., 1982, Late Pleistocene sea-level history, *in* Ivanovitch, M., and Harmon, R. S., eds., Uranium-series disequilibrium: Applications to environmental problems: Oxford, Clarendon Press, p. 481–496.
Morley, J. J., and Hays, J. D., 1981, Towards a high-resolution, global, deep-sea chronology for the last 750,000 years: Earth and Planetary Science Letters, v. 53, p. 279–295.
Muhs, D. R., 1983a, An estimate of the duration of the Sangamon interglacial in the midwestern United States using sea-level curves and ice-margin paleogeography: Geological Society of America Abstracts with Programs, v. 15, no. 4, p. 211.
——, 1983b, Quaternary sea-level events on northern San Clemente Island, California: Quaternary Research, v. 20, p. 322–341.
——, 1983c, Airborne dust fall on the California Channel Islands, U.S.A.: Journal of Arid Environments, v. 6, p. 223–238.
Muhs, D. R., and Szabo, B. J., 1982, Uranium-series age of the Eel Point terrace, San Clemente Island, California: Geology, v. 10, p. 23–26.
Muhs, D. R., Bush, C. A., and Rowland, T. R., 1987a, Uranium-series age determinations of Quaternary eolianites and implications for sea-level history, New Providence Island, Bahamas: Geological Society of America Abstracts with Programs, v. 19, no. 7, p. 780.
Muhs, D. R., Kennedy, G. L., and Miller, G. H., 1987b, New uranium-series ages of marine terraces and late Quaternary sea level history, San Nicolas Island, California: Geological Society of America Abstracts with Programs, v. 19, no. 7, p. 780–781.
Muhs, D. R., Kennedy, G. L., and Rockwell, T. K., 1988, Uranium-series ages of corals from marine terraces on the Pacific coast of North America: Implications for the timing and magnitude of late Pleistocene sea level changes: American Quaternary Association Program and Abstracts, 10th Biennial Meeting, p. 140.
Muhs, D. R., Kelsey, H. M., Miller, G. H., Kennedy, G. L., Whelan, J. F., and McInelly, G. W., 1990, Age estimates and uplift rates for late Pleistocene marine terraces: Southern Oregon portion of the Cascadia forearc: Journal of Geophysical Research, v. 95, p. 6685–6698.
Neumann, A. C., and Moore, W. S., 1975, Sea level events and Pleistocene coral ages in the northern Bahamas: Quaternary Research, v. 5, p. 215–224.
Ninkovich, D., Shackleton, N. J., Abdel-Monem, A. A., Obradovich, J. D., and Izett, G., 1978, K-Ar age of the late Pleistocene eruption of Toba, north Sumatra: Nature, v. 276, p. 574–577.
Omura, A., Emerson, W. K., and Ku, T. L., 1979, Uranium-series ages of echinoids and corals from the upper Pleistocene Magdalena terrace, Baja California Sur, Mexico: The Nautilus, v. 94, p. 184–189.
Orr, P. C., 1960, Late Pleistocene marine terraces on Santa Rosa Island, California: Geological Society of America Bulletin, v. 71, p. 1113–1120.
Osmond, J. K., Carpenter, J. R., and Windom, H. L., 1965, $^{230}Th/^{234}U$ age of the Pleistocene corals and oolites of Florida: Journal of Geophysical Research, v. 70, p. 1843–1847.
Pillans, B., 1987, Quaternary sea-level changes: Southern Hemisphere data, *in* Devoy, R.J.N., ed., Sea surface studies: A global view: London, Croom Helm, p. 264–293.
Radtke, U., Grun, R., and Schwarcz, H. P., 1988, Electron spin resonance dating of the Pleistocene coral reef tracts of Barbados: Quaternary Research, v. 29, p. 197–215.
Richmond, G. M., and Fullerton, D. S., 1986, Introduction to Quaternary glaciations in the United States of America, *in* Sibrava, V., Bowen, D. Q., and Richmond, G. M., eds., Quaternary glaciations in the Northern Hemisphere: Oxford, Pergamon Press, p. 3–10.
Rind, D., Peteet, D., and Kukla, G., 1989, Can Milankovitch orbital variations initiate the growth of ice sheets in a general circulation model?: Journal of Geophysical Research, v. 94, p. 12,851–12,871.
Rockwell, T. K., Muhs, D. R., Kennedy, G. L., Hatch, M. E., Wilson, S. H., and Klinger, R. E., 1989, Uranium-series ages, faunal correlations and tectonic deformation of marine terraces within the Agua Blanca fault zone at Punta Banda, northern Baja California, Mexico, *in* Abbott, P. L., ed., Geologic studies in Baja California: Los Angeles, Pacific Section, Society of Economic Paleontologists and Mineralogists, p. 1–16.
Ruhe, R. V., 1974, Sangamon paleosols and Quaternary environments in midwestern United States, *in* Mahaney, W. C., ed., Quaternary environments: Proceedings of a symposium: Toronto, York University Geographical Monographs no. 5, p. 153–167.
Ruhe, R. V., Hall, R. D., and Canepa, A. P., 1974, Sangamon paleosols of southwestern Indiana, U.S.A.: Geoderma, v. 12, p. 191–200.
Shackleton, N. J., and Matthews, R. K., 1977, Oxygen isotope stratigraphy of late Pleistocene coral terraces in Barbados: Nature, v. 268, p. 618–620.
Shackleton, N. J., and Opdyke, N. D., 1973, Oxygen isotope and paleomagnetic stratigraphy of equatorial Pacific core V28-238: Oxygen isotope temperatures and ice volumes on a 10^5 and 10^6 year scale: Quaternary Research, v. 3, p. 39–55.
Southon, J. R., Ku, T. L., Nelson, D. E., Reyss, J. L., Duplessy, J. C., and Vogel, J. S., 1987, ^{10}Be in a deep-sea core: Implications regarding ^{10}Be production changes over the past 420 ka: Earth and Planetary Science Letters, v. 85, p. 356–364.
Sprigg, R. C., 1979, Stranded and submerged sea-beach systems of southeast South Australia and the aeolian desert cycle: Sedimentary Geology, v. 22, p. 53–96.
St.-Onge, D. A., 1987, The Sangamonian Stage and the Laurentide ice sheet: Géographie Physique et Quaternaire, v. 41, p. 189–198.
Stearns, H. T., 1978, Quaternary shorelines in the Hawaiian Islands: Bernice P. Bishop Museum Bulletin 237, p. 1–57.
Szabo, B. J., 1985, Uranium-series dating of fossil corals from marine sediments of southeastern United States Atlantic coastal plain: Geological Society of America Bulletin, v. 96, p. 398–406.
Szabo, B. J., and Halley, R. B., 1988, $^{230}Th/^{234}U$ ages of aragonitic corals from the Key Largo limestone of south Florida: American Quaternary Association Program and Abstracts, 10th Biennial Meeting, p. 154.
Szabo, B. J., Ward, W. C., Weidie, A. E., and Brady, M. J., 1978, Age and magnitude of late Pleistocene sea-level rise on the eastern Yucatan Peninsula: Geology, v. 6, p. 713–715.
Szabo, B. J., Hattin, D. E., and Warren, V. L., 1988, Age of fossil reef at Grotto Beach, San Salvador, Bahamas, and its implications regarding sea level during the last interglacial high stand: American Quaternary Association, Program and Abstracts, 10th Biennial Meeting, p. 155.
Szabo, B. J., Hausback, B. P., and Smith, J. T., 1990, Relative inactivity during the last 140,000 years of a portion of the La Paz fault, southern Baja California Sur, Mexico: Environmental Geology and Water Sciences, v. 15, p. 119–122.
Vacher, H. L., and Hearty, P., 1989, History of stage 5 sea level in Bermuda: Review with new evidence of a brief rise to present sea level during substage 5a: Quaternary Science Reviews, v. 8, p. 159–168.

Vedder, J. G., and Norris, R. M., 1963, Geology of San Nicolas Island, California: U.S. Geological Survey Professional Paper 369, p. 1–65.

Veeh, H. H., and Chappell, J., 1970, Astronomical theory of climatic change: Support from New Guinea: Science, v. 167, p. 862–865.

Wehmiller, J. F., Belknap, D. F. Boutin, B. S., Mirecki, J. E., Rahaim, S. D., and York, L. L., 1988, A review of the aminostratigraphy of Quaternary mollusks from United States Atlantic Coastal Plain sites, *in* Easterbrook, D. J., ed., Dating Quaternary sediments: Geological Society of America Special Paper 227, p. 69–110.

Winograd, I. J., Szabo, B. J., Coplen, T. B., and Riggs, A. C., 1988, A 250,000-year climatic record from Great Basin vein calcite: Implications for Milankovitch theory: Science, v. 242, p. 1275–1280.

Wright, H. E., Jr., 1962, Late Pleistocene geology of coastal Lebanon: Quaternaria, v. 6, p. 525–539.

Manuscript Accepted by the Society September 6, 1991

Geological Society of America
Special Paper 270
1992

Wisconsinan ice-sheet initiation: Milankovitch forcing, paleoclimatic data, and global climate modeling

D. Peteet and D. Rind
Goddard Space Flight Center, Institute for Space Studies, 2880 Broadway, New York, NY 10025
G. Kukla
Lamont-Doherty Geological Observatory of Columbia University, Palisades, NY 10964

ABSTRACT

Questions concerning the initiation and growth of the continental ice sheets at the onset of the last glaciation are explored by comparing Milankovitch forcing, paleoclimatic records, and the results of global circulation model experiments. Decreases in the solar radiation at lat 50° and 60°N over the past 140 ka provide an apparent reason for the cooling intervals, which, in accord with paleoclimatic evidence, started with a dramatic and rapid temperature decrease ca. 115 ka. Subsequent insolation decreases, although in phase with paleoclimatic evidence of cooling, appear to be unrelated to the magnitude of the response, particularly during marine oxygen-isotope stages 4 and 2. The gradual and relatively small insolation changes during stage 3 could not have alone caused the rapid climatic oscillations evident in ice cores, pollen records, and deep-sea sediments. The magnitude of the Holocene warming appears to be greater than that generated by an equivalent insolation regime 80 ka.

A series of experiments performed using the Goddard Institute for Space Studies (GISS) general circulation model driven by modified insolation input corresponding to the time of the initiation of the last glaciation failed to maintain snow cover at high latitudes during the summer, despite reduced summer and fall insolation at 116 ka. Even when the differences between the insolation fields at 116 and 114 ka were magnified by a factor of five, snow did not accumulate. When 10-m-thick ice was added in the model to locations where continental ice sheets existed during the last glacial maximum, the model failed to maintain this ice; melting rates over North America were such that the ice would have disappeared within five years. Only when the sea-surface temperatures and carbon-dioxide concentrations were reduced to the full glacial values was the model capable of maintaining the ice, but only in a restricted region of northern Baffin Island.

Modeling results and evaluation of the paleoclimatic records imply that many key questions concerning the response of the atmosphere-ocean-biosphere-cryosphere to orbital forcing remain unresolved. Mechanisms and feedbacks involved in the general pattern and details of global climate change can be defined only as chronological resolution of the paleoclimatic record is improved.

INTRODUCTION

Although climatic variation over the past million years, as evidenced by faunal deep-sea records, has been linked to variations in the Earth's orbit (Hays and others, 1976), the forcings involved in past transitions from warm to cold climates, and from interglacial to glacial modes, are not understood. Much consideration has been given the idea that changes in insolation alone

Peteet, D., Rind, D., and Kukla, G., 1992, Wisconsinan ice-sheet initiation: Milankovitch forcing, paleoclimatic data, and global climate modeling, *in* Clark, P. U., and Lea, P. D., eds., The Last Interglacial-Glacial Transition in North America: Boulder, Colorado, Geological Society of America Special Paper 270.

cannot explain the observed paleoclimatic shifts, and that nonlinear responses of the climate system operate as amplifiers of the effects of direct radiative forcing. However, such nonlinear processes are difficult to sort out and to test.

Most studies of the deep-sea records and orbital forcing have focused on the changes over the past 2.4 m.y. (Ruddiman, 1987). Imbrie and Imbrie (1980) found that part of the shape and enigmatic strength of the 100 ka signal in the paleoclimatic record could be reproduced by assuming that ablation was a much faster process than accumulation. However, ice-sheet models (Birchfield and others, 1981; Pollard, 1982) suggest that this explanation accounts for only part of the nonlinear response. Ruddiman and Wright (1987) summarize internal feedback processes that have been proposed to accelerate deglaciations. These include bedrock rebound (Oerlemans, 1980; Birchfield and others, 1981; Peltier, 1987), iceberg calving (Denton and Hughes, 1983; Pollard, 1984; Andrews, 1987), and moisture feedback (Ewing and Donn, 1966; Ruddiman, 1987). Forces internal to the Earth system have also been proposed to explain the onset of glaciations, including CO_2 decreases and dust increases. Ice-core stratigraphy indicates that these processes are correlative with atmospheric cooling (Barnola and others, 1987; De Angelis and others, 1987). Broecker and Denton (1989) proposed that glacial cycles are driven by abrupt reorganizations in the ocean-atmosphere system, linked to orbitally induced changes in fresh-water transports that drive the glacial cycles. These reorganizations thus synchronize the timing of glacial events in the Northern and Southern hemispheres.

Processes and feedbacks involved in climatic change are complex; marine, terrestrial, and ice-core records of climate change must be compared in order to understand how the climate system operates. At no time is this complexity more evident than in the transition from the last interglacial to the last glacial maximum and then to the Holocene. In particular, major questions persist: When and where did the initiation of the Laurentide ice sheet take place? What was the rate and magnitude of ice-sheet growth and climatic change at various intervals from the last interglacial to the last glacial maximum (LGM), ca. 18 ka? What caused the greatest buildup of ice sheets at the last glacial maximum, a time of relatively weak insolation decrease? Do the marine, terrestrial, and ice-core records of climatic change provide a converging stratigraphic record that can be interpreted as a direct response to Milankovitch orbital forcing?

In addressing these questions, we first discuss the Milankovitch variations of the past 140 ka. We then examine paleoenvironmental records spanning the last interglacial/glacial cycle and summarize the results of recent Goddard Institute for Space Studies (GISS) climate modeling efforts (Rind and others, 1989) designed to simulate the initiation and maintenance of an ice sheet under past insolation regimes.

MILANKOVITCH ORBITAL VARIATIONS (140–0 ka)

The record of insolation variations due to orbital forcing at selected northern latitudes over the past 140 ka (Berger, 1978) (Fig. 1) gives an apparent reason for ice-sheet initiation starting about 116 ka. The 10 ka that followed the last interglacial maximum at 125 ka (Bard and others, 1989) was accompanied by a rapid, large drop in summer half-year insolation in the high latitudes of the Northern Hemisphere. This drop in insolation is the greatest decrease that has occurred within the past 140 ka. The insolation shifts centered at about 90, 70, and 23 ka also demonstrate summer negative departures at high latitudes, progressively less pronounced than the decrease centered at 116 ka. Judging from the paleoclimatic evidence, however, the magnitude of ice buildup increased from 116 to 23 ka, and thus is in inverse relation to the magnitude of insolation decreases. Furthermore, the magnitude of the insolation increase at 10 ka appears to have been less than the increases at 100 and 80 ka.

ENVIRONMENTAL CHANGE DURING THE INTERGLACIAL-GLACIAL TRANSITION

The peak of the last interglacial is marked by raised coral reefs dated by U/Th at 125 + 6 ka (Mesolella and others, 1969; Bloom and others, 1974; Dodge and others, 1983). Recent improvement in mass-spectrometry techniques has resulted in high analytical precision for U/Th ages on several coral reefs of the last interglacial, ages that range from 122 ± 1 ka to 130 ± 1 ka (Edwards and others, 1987; Bard and others, 1989). Kaufman (1986) estimated the duration of the last interglacial as not more than 12 ka, based upon the narrow temporal distribution of the 80 most-reliable coral analyses. This estimate agrees with the ±10 ka duration of the last interglacial determined from the annually laminated diatomites in Germany (Muller, 1965). All of these data support the dominant role of Milankovitch forcing in the timing of the last interglaciation. A conflicting record that supports an earlier interglacial peak (Winograd and others, 1988) suggests that the relation is possibly more complex.

Pollen, foraminifera, and ice-core data reveal that the last interglacial ended with a major environmental change toward colder conditions ca. 115 ka. Palynological records from northwestern Europe indicate that the last interglacial was characterized by hardwood forests with a species composition similar to that of modern deciduous forests in the same region. The shift from a warm, temperate environment to colder conditions ca. 115 ka is reflected in a number of pollen records throughout western Europe (Wijmstra, 1969; Woillard and Mook, 1982; Turon, 1984; de Beaulieu and Reille, 1984; Behre, 1989) (Figs. 2 and 3). Pollen records from annually laminated deposits in Germany show that the hardwood forests were continuously present for ~10 to 11 ka during the last interglacial (Muller, 1965; Turner, 1975). They were then replaced by a taiga (pine-birch assemblage) at about 115 ka (Woillard and Mook, 1982; Turner, 1975; Kukla, 1980), which was in turn shortly replaced by tundra. In the central European loess belt, temperate grasslands were covered by a thin dust layer ("marker band"; Kukla, 1977) during the same transition to colder conditions. An episode of intense hillwash and formation of pellet sands by torrential downpours

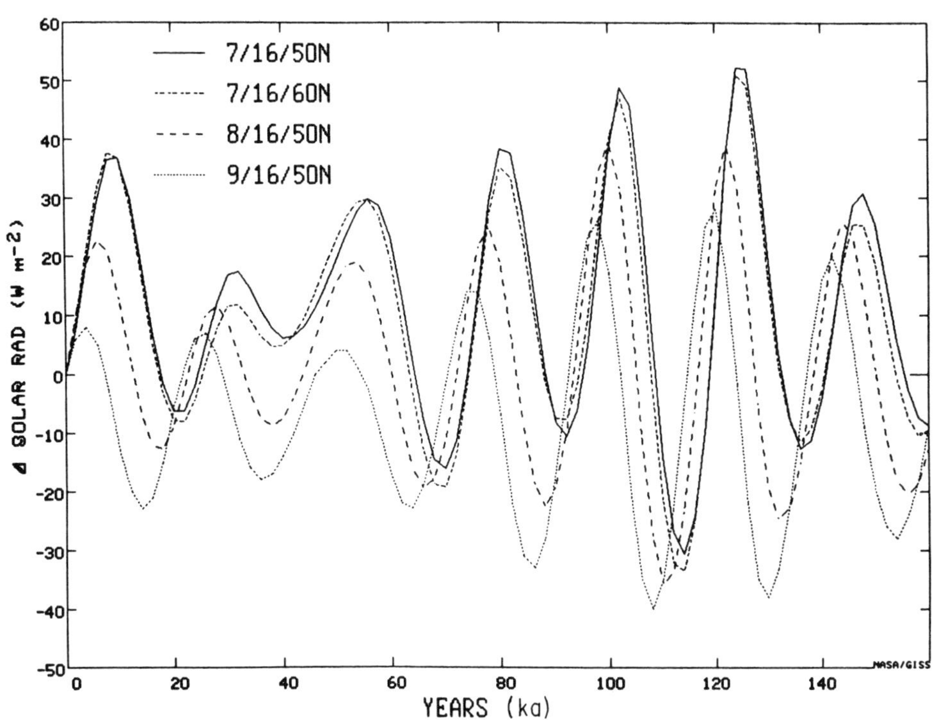

Figure 1. Change of incident solar radiation during the past 160 ka in July, August, and September at lat 50°N, and during July at 60°N. (Berger, 1978).

followed, and in Czechoslovakia, woolly rhinoceros and mammoths replaced forest elephants (Kukla and Lozek, 1961; Kukla and Koci, 1972).

Zagwijn (1983) estimated a drop of sea level of at least 32 m at the Eemian/Weichselian boundary in the Netherlands, and from glacial evidence Mangerud (1990) concluded that eustatic sea level dropped between 10 and 20 m before any significant ice growth occurred in Scandinavia. However, the ice growth eventually resulted in major ice-sheet formation in Scandanavia in substage 5d, which subsequently melted during the warming of substage 5c (Mangerud, 1990).

Records of summer sea-surface temperature in the North Atlantic, as indicated by planktonic foraminifera in core V23-82 (Fig. 2A) (Sancetta and others, 1973; Ruddiman, 1977; Ruddiman and others, 1980), show large and rapid changes toward colder temperatures ca. 115 ka. Similarly, the record of sea-surface temperature for RC11-120 in the southern Indian Ocean also demonstrates a drop in temperature at the 5e/5d transition to values close to those reached during the LGM. The oxygen-isotope change, however, suggests that the ice volume grew to only about half that of the LGM (Hays and others, 1976). The records of pollen and oxygen isotopes from the northeastern Pacific Ocean (Heusser and Shackleton, 1979) (Fig. 4) show a similar difference ca. 115 ka, with a dramatic shift toward colder conditions on land while the marine isotopic shift is relatively small. From a suite of North Atlantic deep-sea cores, Ruddiman (1977) and Heinrich (1988) demonstrated that total input rates of ice-rafted sediment to the North Atlantic and Norwegian Sea increased only slightly at 115 ka (Fig. 5).

The record of climatic change from the Vostok ice core, shown as profiles of deuterium, ^{18}O, and CO_2 spanning the past 160 ka (Fig. 6) (Jouzel and others, 1987; Barnola and others, 1987), indicates a rapid rate of cooling at 116 ka, which is in phase with the Northern Hemisphere high-latitude change in incident solar radiation (Fig. 1). The magnitudes of the inferred temperature and CO_2 decreases are greater than or equal to any of those in the subsequent record up to the Holocene. The duration of the last interglacial in the Vostok record, about 20 ka (Fig. 6), was determined by an ice-flow model (Jouzel and others, 1987). It is substantially greater than that evidenced by pollen or ocean faunal records discussed above. However, recent revision of this Vostok model makes the duration and age of the last interglacial comparable with the marine stratigraphy (Petit and others, 1990).

These records for North Atlantic sea-surface temperature (SST), pollen, ice-rafted debris, and ice-core isotopic data (Figs. 2–6) indicate a relatively great environmental shift at 115 ka that is larger than that expected from the amplitude of the marine oxygen-isotope change (Fig. 7). Was the buildup of ice on land perhaps delayed behind the other environmental changes? It is

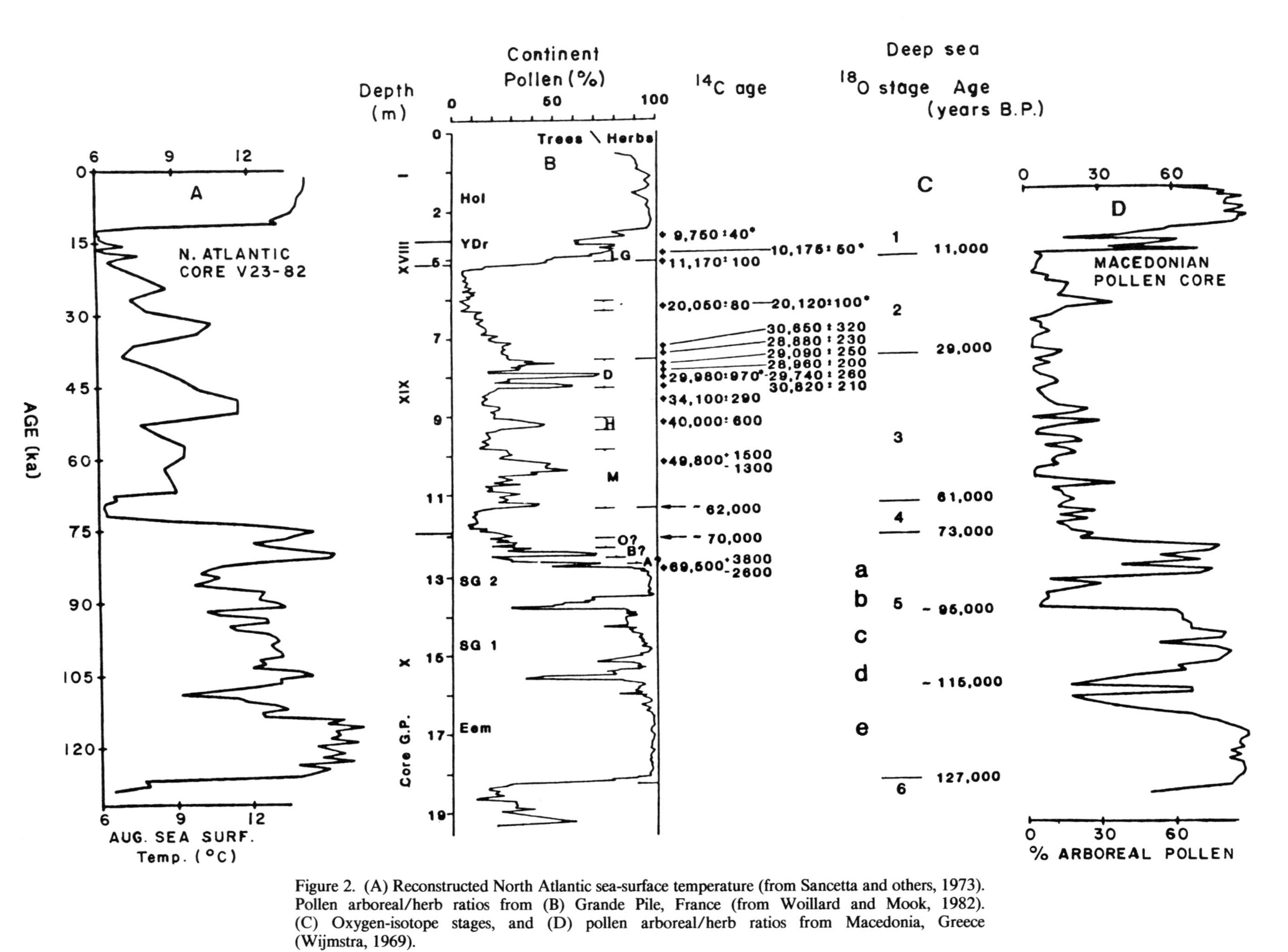

Figure 2. (A) Reconstructed North Atlantic sea-surface temperature (from Sancetta and others, 1973). Pollen arboreal/herb ratios from (B) Grande Pile, France (from Woillard and Mook, 1982). (C) Oxygen-isotope stages, and (D) pollen arboreal/herb ratios from Macedonia, Greece (Wijmstra, 1969).

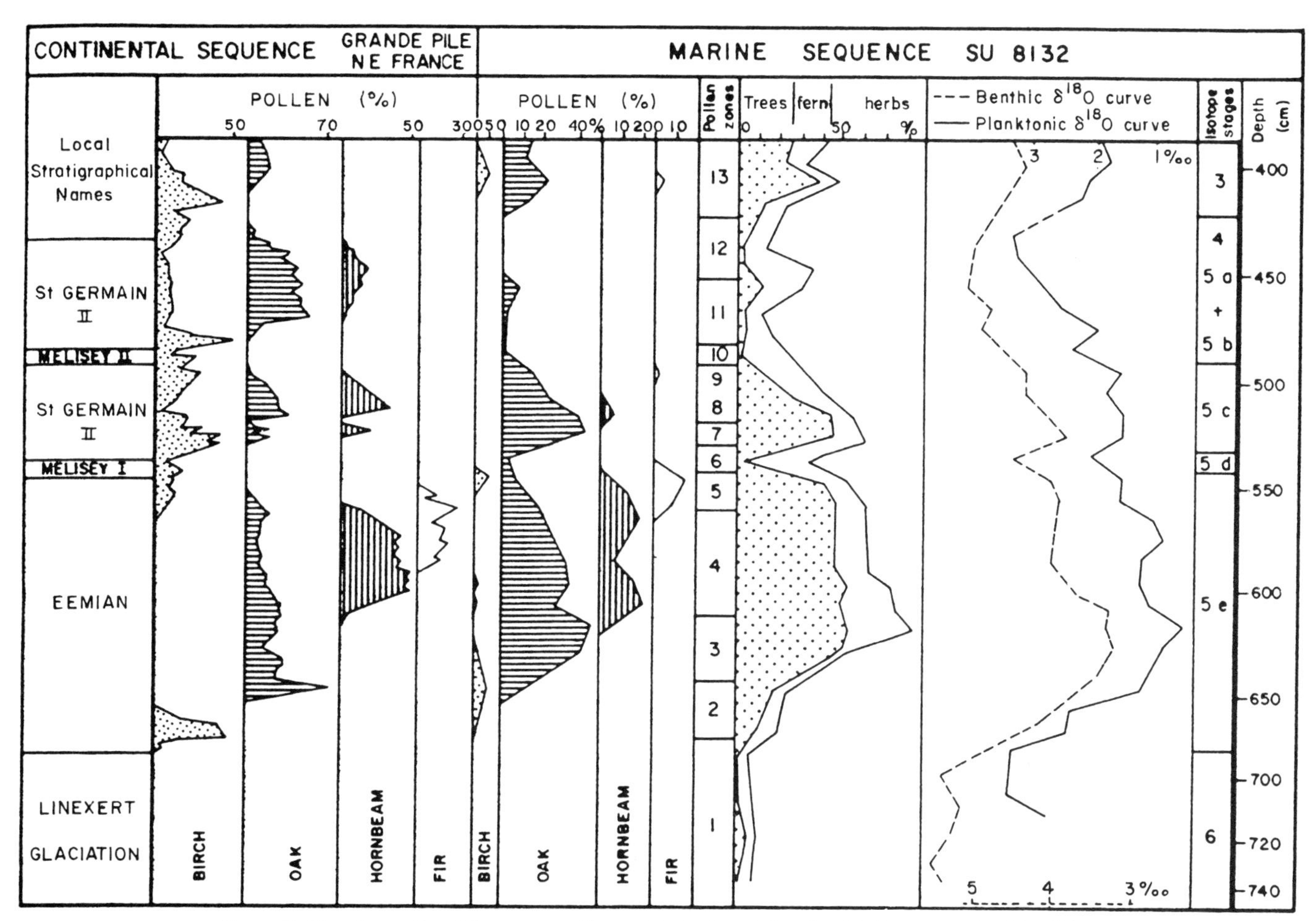

Figure 3. French continental sequence (from Woillard and Mook, 1982) compared with offshore marine sequence of pollen and isotopes in core SU 8132 (from Turon, 1984).

not currently possible to answer this question because of the complexity of distinguishing temperature from ice volume within the isotopic signal.

The volume of continental ice buildup over the past 125 ka cannot be derived from isotopic records of benthic foraminifera for three major reasons: (1) the isotope record reflects a changing mixture of ice-volume and deep-ocean-temperature effects (Chappell and Shackleton, 1986; Broecker, 1986; Labeyrie and others, 1987); (2) the isotopic composition of ice sheets has varied with time (Mix and Ruddiman, 1984); and (3) chronological uncertainties arise due to problems with bioturbation and limitations of dating methods. Oxygen-isotope records from planktonic foraminifera are further complicated by differences in seasonal growth, depth habitat, and salinity (Epstein and Mayeda, 1953; Berger and others, 1978; Chappell and Shackleton, 1986; Mix, this volume).

An independent sea-level curve is needed to determine the ice-volume component of the oxygen-isotope curve, but until very recently (Fairbanks, 1989) the only available sea-level records were from raised coral reefs, which do not indicate the sea-level lowstands corresponding to the ice maxima. However, the coral-reef highstands at 125, 105, and 82 ka, separated by sea-level drops, do identify sea-level maxima in phase with summer insolation peaks at high latitudes (Mesolella and others, 1969).

The stacked isotope record of Martinson and others (1987) (Fig. 7) identifies maxima of the oxygen-isotope ratio that correspond to peak glaciations at 18 and 135 ka. It also shows substantial ice volume at 90 and 110 ka. According to this oxygen-isotope record, the most rapid rate of temperature decrease and/or ice-volume increase occurred from 123–110 ka, and is in phase with the greatest summer-insolation decrease in the Northern Hemisphere. However, the isotope record itself is "tuned" to the orbital periodicities (with an error of ±5 ka) (Martinson and others, 1987), so that a detailed independent comparison of the timing of the oxygen-isotope changes with insolation changes is not possible.

If interpreted as a qualitative measure of land-based ice volume, the oxygen-isotope record (Fig. 7) indicates an ice-volume change between substage 5e and substages 5c and 5a of approximately one-third of the full glacial/interglacial amplitude. This implies a drop in sea level of at least 40 m, based on the

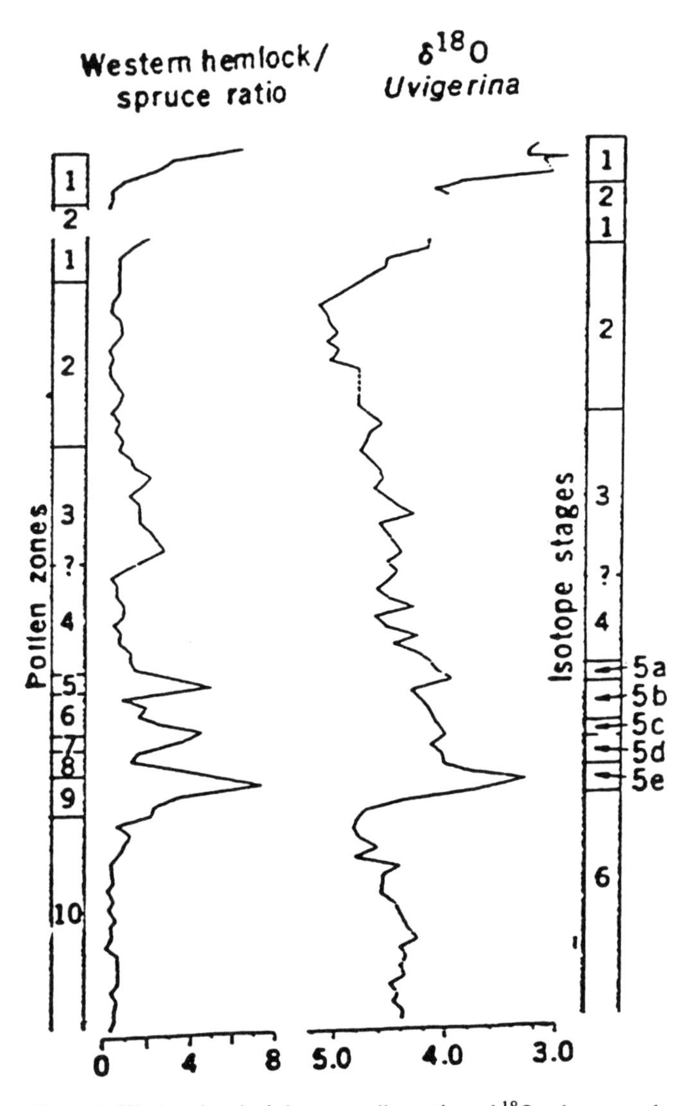

Figure 4. Western hemlock/spruce pollen ratio and ^{18}O values over the last glacial/interglacial cycle in a core from the northeastern Pacific Ocean (from Heusser and Shackleton, 1979).

130 m glacial-interglacial estimate by Chappell and Shackleton (1986) and the recent calculation of 121 ± 5 m by Fairbanks (1989). However, the coral reefs in New Guinea indicate a drop of only 25 m (Chappell and Shackleton, 1986), confirming the complexity in interpreting the isotopic record.

Difficulties in estimating the ice volume and geographic extent of continental ice sheets during the 115–110 ka interval arise because of the few reliable chronometric methods available for this interval and because of the removal of the glacial stratigraphic record by subsequent ice advances. Stratigraphic evidence for early Wisconsinan (marine isotope stage 4 and/or 5) glaciation is cited from the Erie basin (Dreimanis, 1988), the St. Lawrence lowland (Lamothe and others, 1988), Baffin Island (Andrews and others, 1985; Miller and Andrews, 1988), the western Canadian Arctic (Vincent, 1988), and the Hudson Bay lowland (Thorleifson and others, 1988). In western North America, evidence is cited from northwestern Washington (Clague and others, 1988) and several alpine areas (Colman and Pierce, 1988). In most of these areas, accurate dating methods were not applicable and the age estimates are approximations. Westgate (1987) noted that the best hope for developing a well-dated glacial history would be in volcanic regions, where the K/Ar, $^{40}Ar/^{39}Ar$, and tephrochronological techniques can be used. As yet, however, no detailed and continuous stratigraphic record has been retrieved from such areas.

CLIMATIC CHANGE 115–70 ka

Milankovitch insolation curves for high latitudes (Fig. 1) indicate two warm peaks following the last interglacial, at 100 and 80 ka. Such peaks are visible in continuous pollen (Woillard and Mook, 1982; Wijmstra, 1969; Turon, 1984; Heusser and Shackleton, 1979), foraminiferal (Sancetta and others, 1973), and ice-core records (Jouzel and others, 1987; Chappellaz and others, 1990) (Figs. 2–7), and appear to be in phase with the Milankovitch forcing. However, as evident in the figures, the magnitude of the warming response inferred from the pollen and sea-surface temperatures is far greater than that shown in the ice cores or in the marine isotopes.

A dramatic cooling at 75 ka (the marine oxygen-isotope stage 5/4 boundary, Fig. 7) is recognized by a variety of paleoclimatic indices, implying that the environmental change at this time was greater than at the 5e/5d boundary. The extreme nature of this cooling, however, is not uniquely explained by the orbital forcing, which shows a smaller insolation decrease than at 116 ka (Fig. 1). The extent of the Laurentide Ice Sheet at this time is not well known due to the poor dating control and erosion of glacial deposits during younger ice advances. However, it was probably less extensive along its southern margin and more extensive along its northern margin during stage 4 than during stage 2 (papers in this volume). Mangerud (1990) concluded that the last Scandinavian ice sheet developed at this time. Seret and others (1990) provided evidence for larger glaciation in the French Vosges sometime between 70 and 29 ka than during the last glacial maximum. Moraines in the Southern Hemisphere suggest that the ice advance in stage 4 was greater than that in stage 2 (Mercer, 1983).

Although the greatest change in temperature in the Vostok core occurred at 115 ka, the temperature minimum reached at ca. 64 ka is lower (Jouzel and others, 1987) (Fig. 6). The CO_2 values at 64 ka are ~40 ppm lower than at 115 ka (Barnola and others, 1987; Genthon and others, 1987) (Fig. 6). Aluminum concentration (a continental indicator) in the Vostok core also increases dramatically (De Angelis and others, 1987) (Fig. 8). This evidence suggests that decreasing CO_2 and increasing atmospheric dust in the atmosphere may have amplified the global cooling. In addition, CO_2 and dust fluctuations may provide a link for the similar timing of glaciation in the Northern and Southern hemispheres.

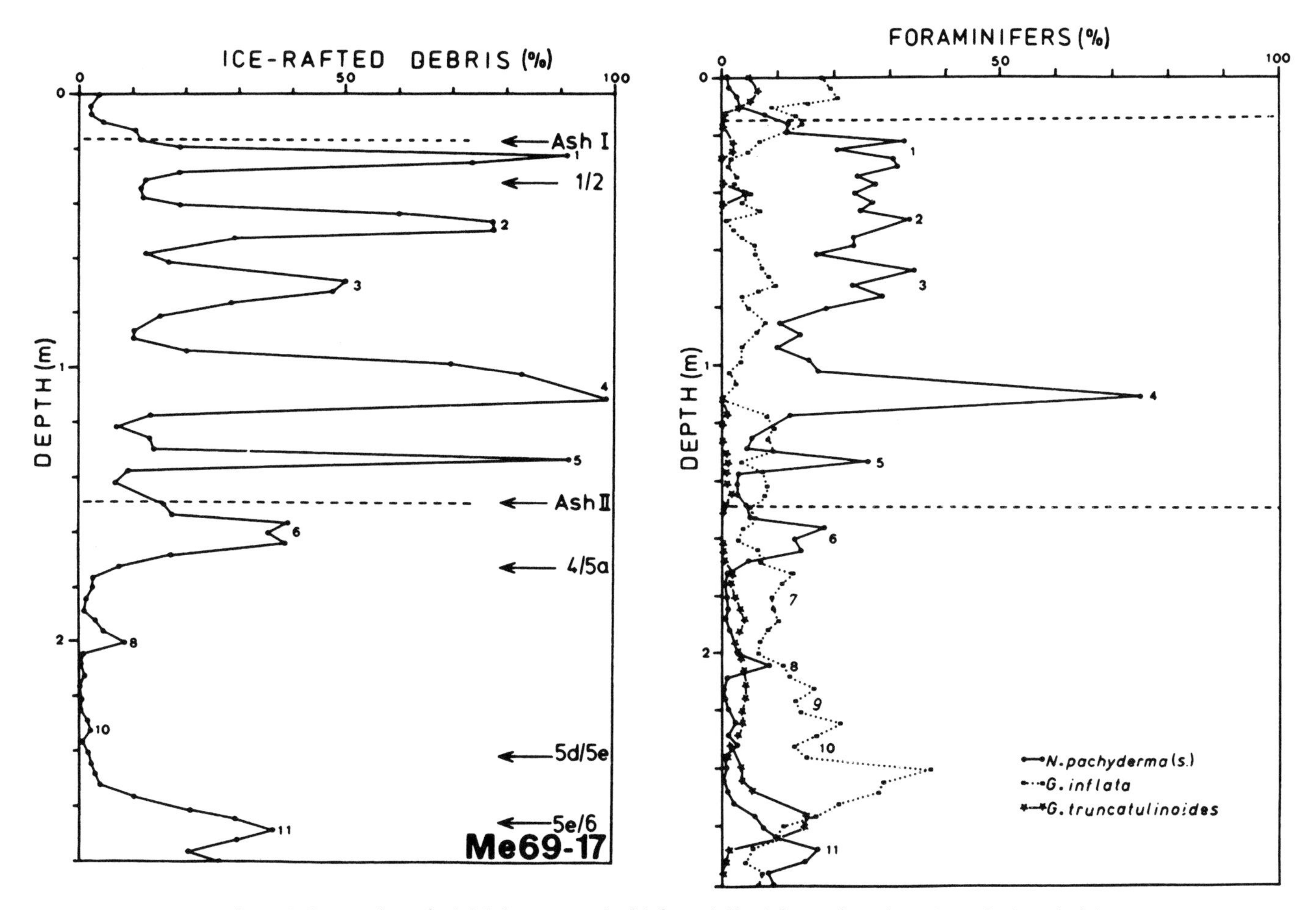

Figure 5. Percent ice-rafted debris compared with foraminiferal fluctuations throughout the last glacial cycle in North Atlantic core Me69-17 (from Heinrich, 1988).

^{13}C data from the North Atlantic (Mix and Fairbanks, 1985; Duplessy and others, 1988), Caribbean, and Southern Ocean (D. Oppo, 1990, personal commun.) show that the deep Atlantic circulation was similar in isotope stage 4 to that of full glacial isotope stage 2.

RAPID CLIMATIC OSCILLATIONS 70–25 ka

The insolation record exhibits two relatively warm peaks (ca. 50 and 30 ka) during marine isotope stage 3. Each peak is followed by a relatively minor and gradual decrease, culminating in the decrease to the LGM (18 ka) (Fig. 1). This record is paralleled in the loess sequences from China (Kukla and An, 1989).

In contrast to the gradual insolation changes, most paleoclimatic records from this interval indicate rapid alternations of warm and cold events, which in frequency and timing appear to be unrelated to the Milankovitch forcing. The European pollen stratigraphy (Woillard and Mook, 1982; Wijmstra, 1969; van der Hammen and others, 1967; Behre, 1989) suggests that temperature increases were at least one-half to three-fourths as large as those that took place during the last two terminations, i.e., at the 6/5 and 2/1 stage boundaries (Figs. 2 and 3). Coastal areas in Scandinavia were ice-free several times between 75 and 30 ka, and the final expansion of Weichselian ice toward its maximum did not take place until after 28 ka (Mangerud, 1990). Well-developed soils indicating warmth in midwestern North America (e.g., Berti, 1975; Gruger, 1972) and Alaska (Hopkins, 1982) parallel these European changes, although the chronology in America is not well developed.

Dansgaard and others (1984) showed that rapid and extreme fluctuations in stable isotopes (signifying air-temperature differences of 5 °C) from two sites in Greenland appear to correlate over the past 50 ka. These fluctuations are in phase with changes in CO_2 (Stauffer and others, 1985) and dust content (Hammer and others, 1985). Although the high-frequency CO_2 oscillations have not been documented in the Vostok record (Fig. 6), this Southern Hemisphere record suggests that a minimum of four oscillations took place, with temperature shifts of least 3 °C. Peaks in the concentration of atmospheric methane in Antarctic

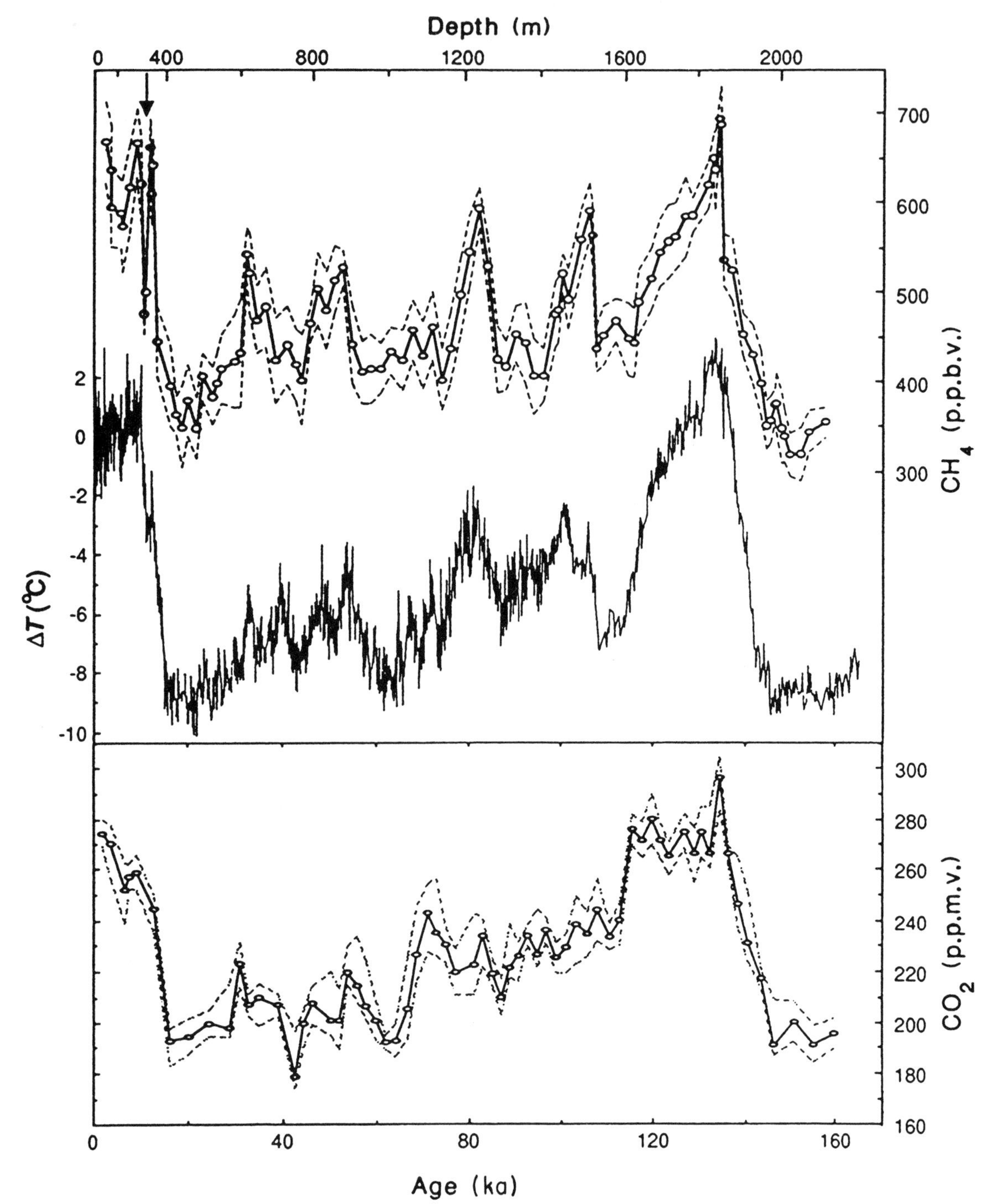

Figure 6. Vostok ice-core record of methane (top), deuterium (middle), and CO_2 (lower) changes over the past 160 ka (from Jouzel and others, 1987; Barnola and others, 1987; Chapellaz and others, 1990). ppbv is parts per billion volume; ppmv is parts per million volume.

ice cores parallel the warming intervals (Chappellaz and others, 1990). A CO_2 minimum at about 40 ka marks the lowest value of the past 160 ka. The Vostok dust record shows major peaks at 60–70 and 30–20 ka, with minor fluctuations in the interim (Fig. 8).

Rapid and substantial temperature fluctuations recorded in ice cores during stage 3 correspond with oscillations in the North Atlantic marine sediment record of species abundance and ice rafting (Ruddiman, 1977; Ruddiman and McIntyre, 1981; Heinrich, 1988) (Fig. 5). Using accelerator radiocarbon ages on a handpicked planktonic polar species (*Neogloboquadrina pachyderma*), Broecker and others (1988) identified four rapid climatic

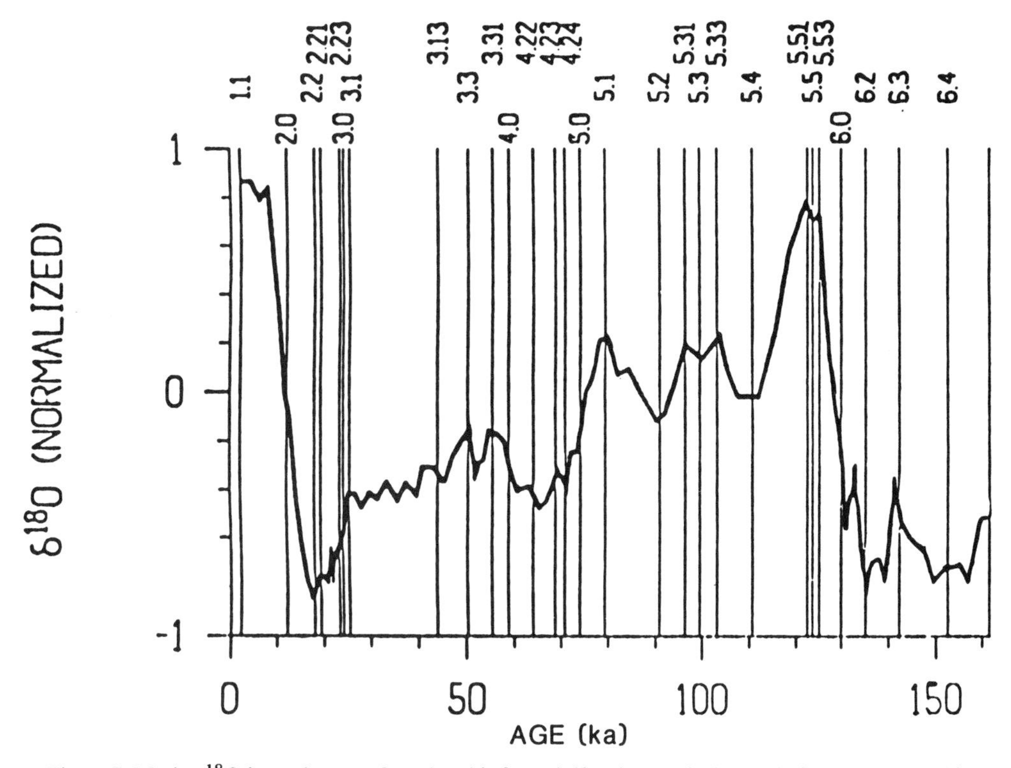

Figure 7. Marine ^{18}O isotopic curve from benthic foraminifera is a stacked record of many cores and is tuned to the orbital frequencies (from Martinson and others, 1987).

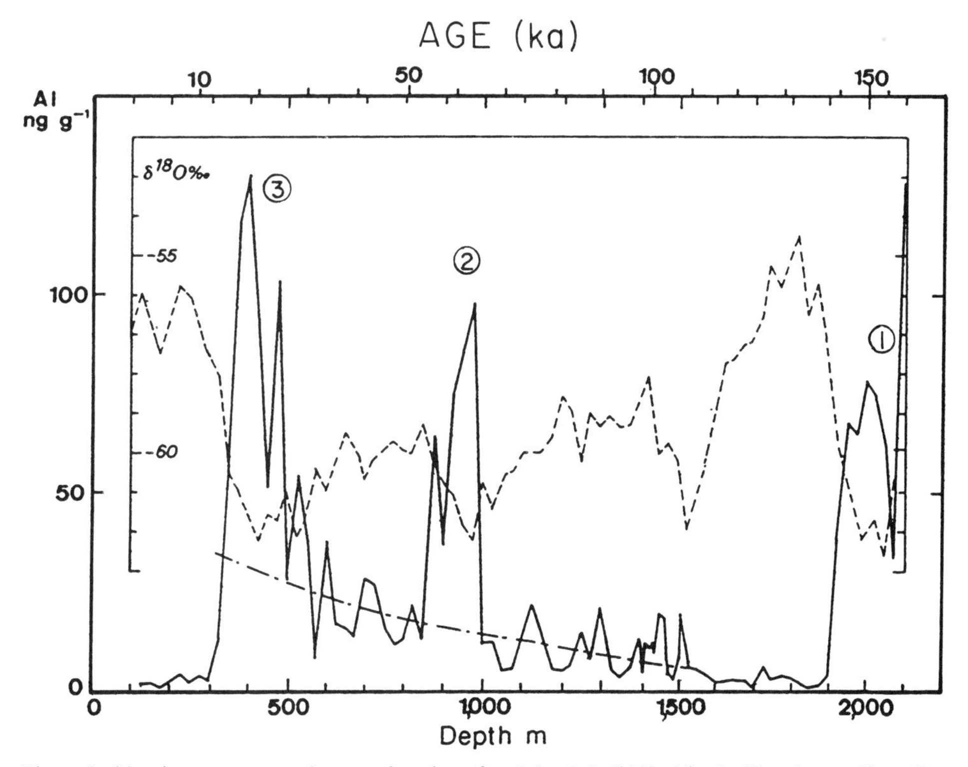

Figure 8. Aluminum concentration as a function of real depth (solid line) in the Vostok core (from De Angelis and others, 1987). The climatic reference is given by ^{18}O (broken line) from Lorius and others (1985). The lower dot-dash line represents the estimated background. Points 1, 2, and 3 are major dust peaks.

oscillations between 40 and 22 ka in the North Atlantic. Paterne and others (1986) found the same number of rapid oscillations during this interval in the Tyrrhenian Sea, using tephrochronology to date a planktonic foraminiferal sequence. Unfortunately, lack of radiocarbon chronology of similar oscillations in ice cores and poor chronological control of pollen records precludes correlation amongst all records.

LAST GLACIAL MAXIMUM (LGM)

Although the insolation minimum at 20 ka is relatively minor (Fig. 1), paleoclimatic evidence demonstrates severe climatic conditions worldwide during the LGM. Pollen, foraminiferal, and ice-core data point to local temperature minima and dust maxima. The reason for the intensity of this climatic response to a relatively minor orbital perturbation is not apparent. Possibly the lack of a major insolation maximum for the preceding 50 ka allowed for the magnitude of ice-sheet growth, but the forcing is not obvious.

Fairbanks (1989) documented the first continuous and detailed ^{14}C dated record of sea-level change during the last deglaciation from coral reefs drilled off the shore of Barbados. This record indicates sea-level rise as early as 17 ka, which suggests an immediate response to a minor insolation increase.

HOLOCENE WARMING

The insolation peak at 10 ka appears to be responsible for the deglaciation and rapid warming that is evident throughout various environmental records. However, the magnitude of the response, including the major loss of ice sheets, is large when compared to the magnitude of the insolation peak, which is less than or equal to any of the peaks in stage 5. If the marine ^{18}O is used as an indication of ice volume, it is not clear why the continental ice sheets completely disappeared during the Holocene, but only partially melted during substages 5c or 5a (see Figs. 1 and 7). Although possible feedbacks include the increases in CO_2 and CH_4 (Barnola and others, 1987; Stauffer and others, 1988; Raynaud and others, 1988; Chappellaz and others, 1990), it is also possible that the large decrease of the meridional gradient associated with the peak obliquity strengthened the impact of the high summer half-year insolation to the high latitudes during the last deglaciation, ca. 13 to 9 ka (Kukla and others, 1981).

GISS GENERAL CIRCULATION MODEL STUDIES

Can Milankovitch variations actually produce the required conditions for ice-sheet growth? In a recent study (Rind and others, 1989), we used the GISS global climate model (GCM) to investigate whether variations in solar radiation would have initiated model growth of ice sheets, or, failing that, sustain snow or thin ice sheets over a large geographic region. The results showed that even under the radiative conditions most favorable for ice-sheet growth, ca. 116 ka (stage 5d), ice sheets were neither initiated nor maintained by the model. We review the results of this study, and conclude that either Milankovitch variations alone could not produce conditions required for ice-sheet growth, or that the GISS model incompletely represents the feedbacks operating in the climate system.

Solar-Radiation Changes

Insolation fields of 116 and 106 ka were used in the model input. In order to examine the potential impact of the insolation change that took place at the time of the end of the last interglacial, we also used a modified insolation field (MIF), which represents an extreme insolation drop that is five times the decrease from 116–114 ka. The characteristics of the temporal and spatial changes in solar radiation relative to today for these insolation fields are shown in Figure 9. At 116 ka, summer insolation was reduced compared to the present at high northern latitudes, with increases in other seasons (Fig. 9A). Temperature decreases in summer are not large, however, compared with estimates of the cooling needed for ice growth to take place. However, land ice may have been growing, as implied by the marine isotopic evidence (Fig. 7). Near 106 ka, when the marine isotopic record indicates that ice volume reached a peak, solar insolation in late summer and fall was reduced relative to 116 ka at high northern latitudes, and increased in spring (Fig. 9B). This scenario suggests that radiation conditions for maximum cooling and ice growth occurred with a shift toward a mid-summer insolation minimum.

Because we are not able to observe the impact of gradually changing geographic shifts in insolation through the progression of geologic time, i.e., in a time-transgressive model, we adjusted our methodology to the limitations of the equilibrium-model approach. The insolation shift from an interglacial to a colder climate was most rapid at about 115 ka. Changes in insolation distribution in high latitudes of the Northern Hemisphere at this time included fall decreases and spring increases. In order to magnify the impacts of those shifts in the model, we took the difference in radiation values from 116 to 114 ka, multiplied the difference five times, and fit orbital parameters to the resultant insolation field. As a result, we decreased late summer and fall insolation up to 25% (Fig. 9C) (in comparison with the changes of 10% at 106 ka produced by the real orbital variations; Fig. 9B) and increased spring insolation correspondingly. These altered radiation fields (MIF) were then used in the GCM with both specified and model-calculated sea-surface temperatures, to determine whether they would favor changes in atmospheric circulation conducive to ice growth.

For ice to accumulate, a positive annual snow budget must develop. This means that some snow must remain throughout the summer. Even with altered radiation fields, the model could not maintain snow cover through July over much of the North American continent (Fig. 10); snow-cover values are not much different from model-simulated snow cover for the current climate. In other words, the model was unable to generate ice-covered ground with the insolation input that existed at the onset of the

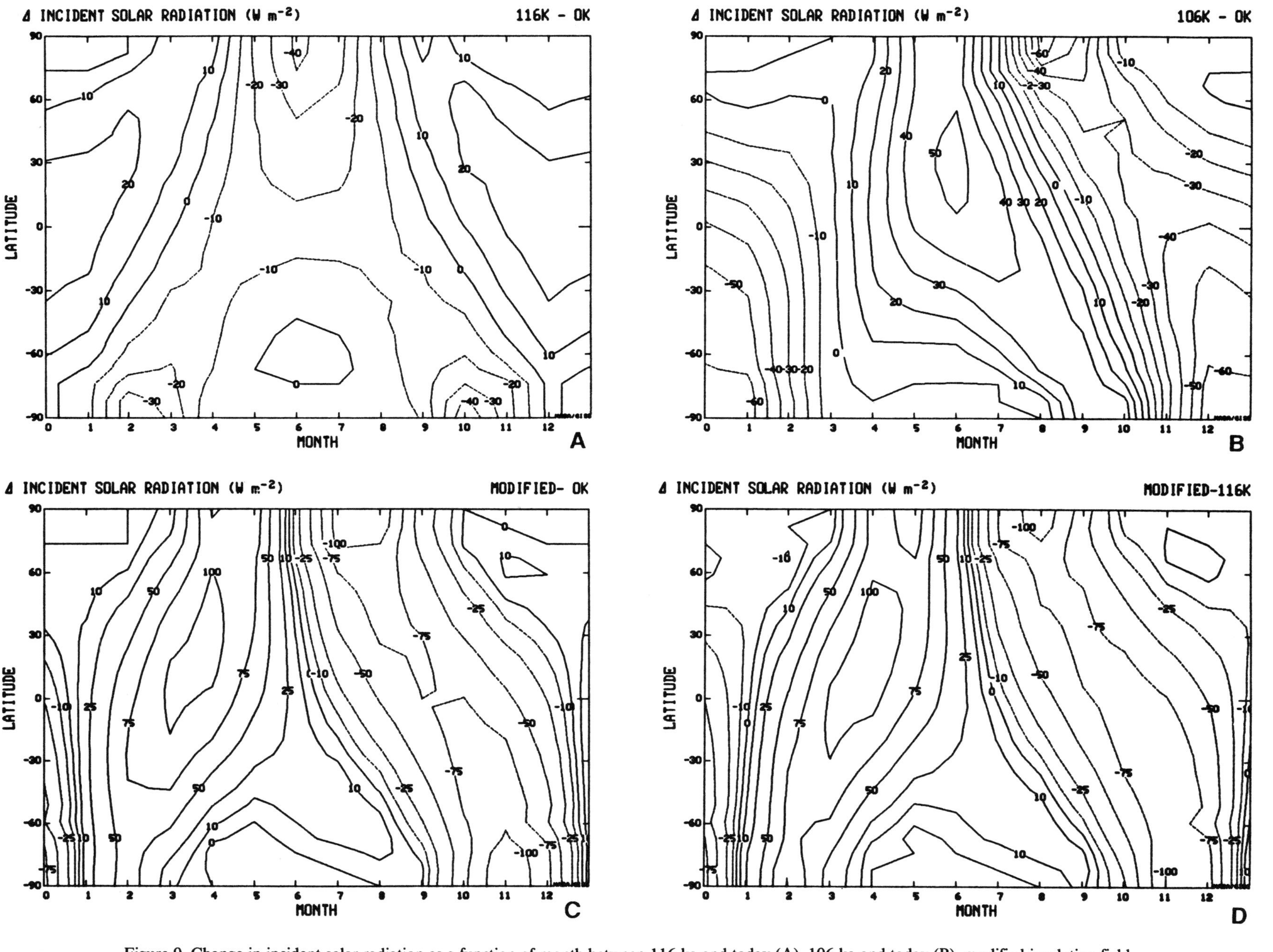

Figure 9. Change in incident solar radiation as a function of month between 116 ka and today (A), 106 ka and today (B), modified insolation field (MIF) and today (C), and the MIF and 116 ka (D) (from Rind and others, 1989).

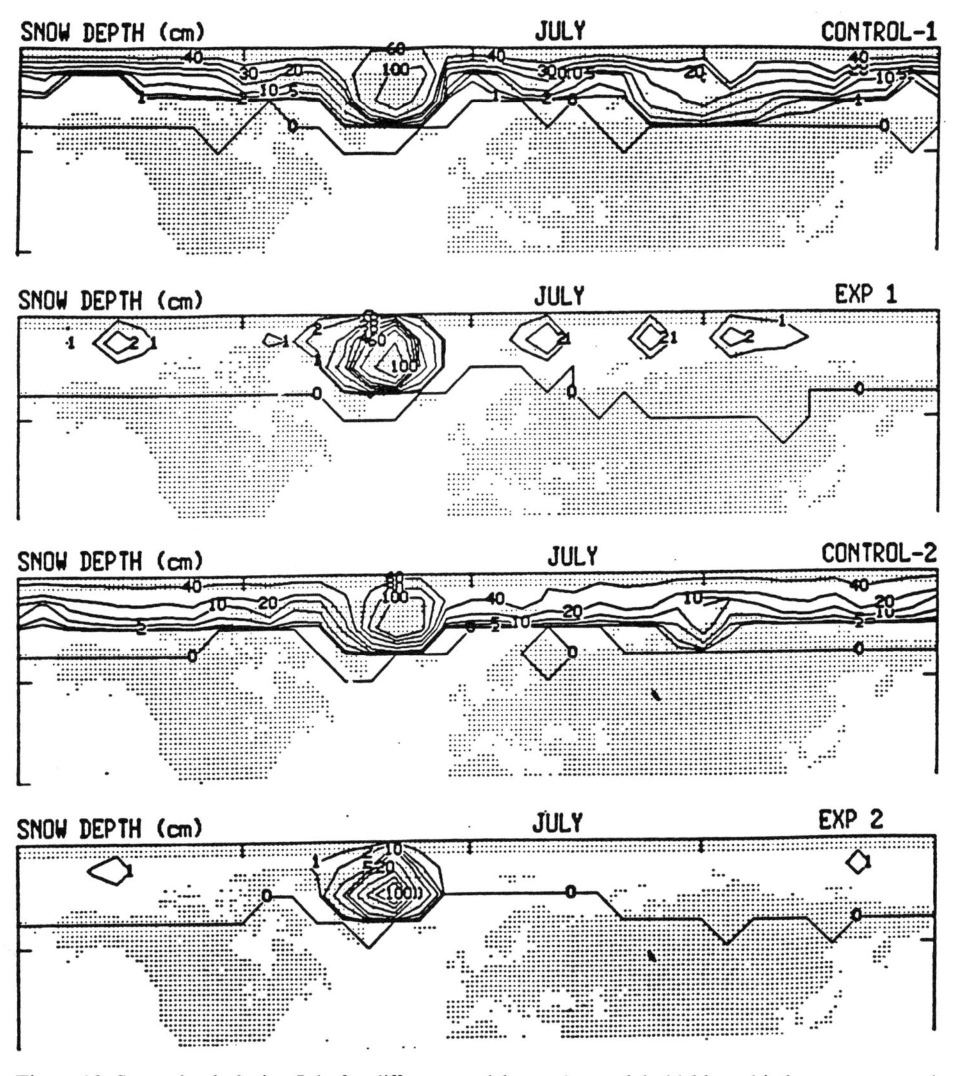

Figure 10. Snow depth during July for different model runs (control-1, 116 ka orbital parameters and current SSTs; Experiment (EXP) 1, modified insolation field (MIF) and current SSTs; control-2, 116 ka orbital parameters and calculated SSTs; EXP, MIF and calculated SSTs). Values less than 1 cm generally indicate that snow cover is not continuous during the month (from Rind and others, 1989).

last glaciation. However, it is possible that the coarse resolution of the model did not properly simulate isolated snow-covered areas on higher terrain or in shaded regions, which—at least conceptually—could have advanced and spread over surrounding regions. Nor does the model include glacier dynamics, which might allow ice to grow at some very high latitude and spread as a high ice sheet southward, where it would then influence climate through positive feedbacks. We therefore tried additional experiments to see if ice would be at least maintained by the model, if not generated.

10 m ice sheet

A 10-m-thick ice layer was put into grid box locations where ice existed during the LGM (CLIMAP Project Members, 1981). We then tested whether the altered solar-radiation fields would allow the model to maintain that ice. In the MIF experiment, the model did not sustain the 10 m of ice. Melting rates over much of North America were such that the ice would have disappeared within five years. Results are similar using the actual 106 ka radiation field, indicating that the rapid radiative melting in the model is not very dependent upon the details of seasonal distribution of the solar-radiation fields.

Reduced CO_2 and the LGM cold ocean

Observations of the CO_2 levels in the Vostok ice core show values reduced by a maximum of 70 ppm (300–230 ppm) during the 116–106 ka interval (Fig. 6). We thus reduced the atmospheric CO_2 by this amount. The result indicates that global air temperature in the model would have cooled to approximately full glacial conditions.

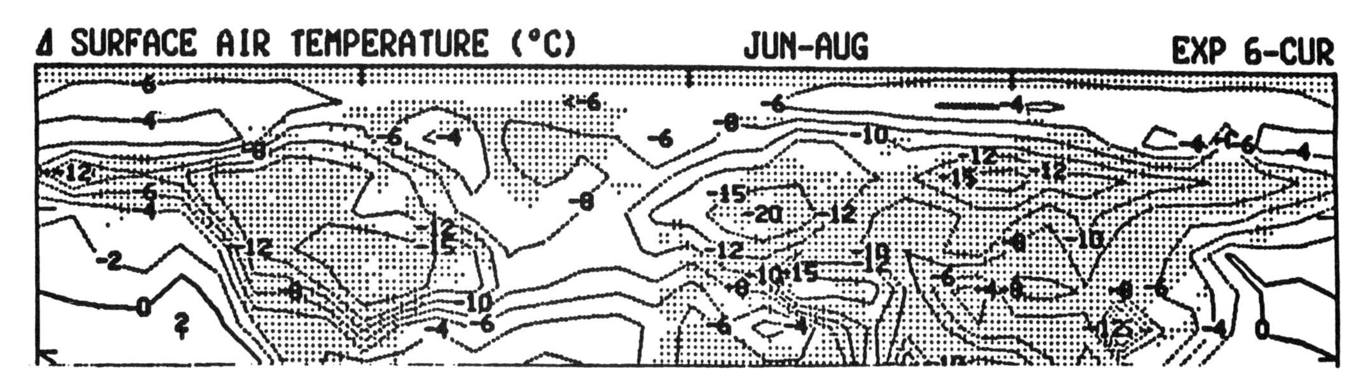

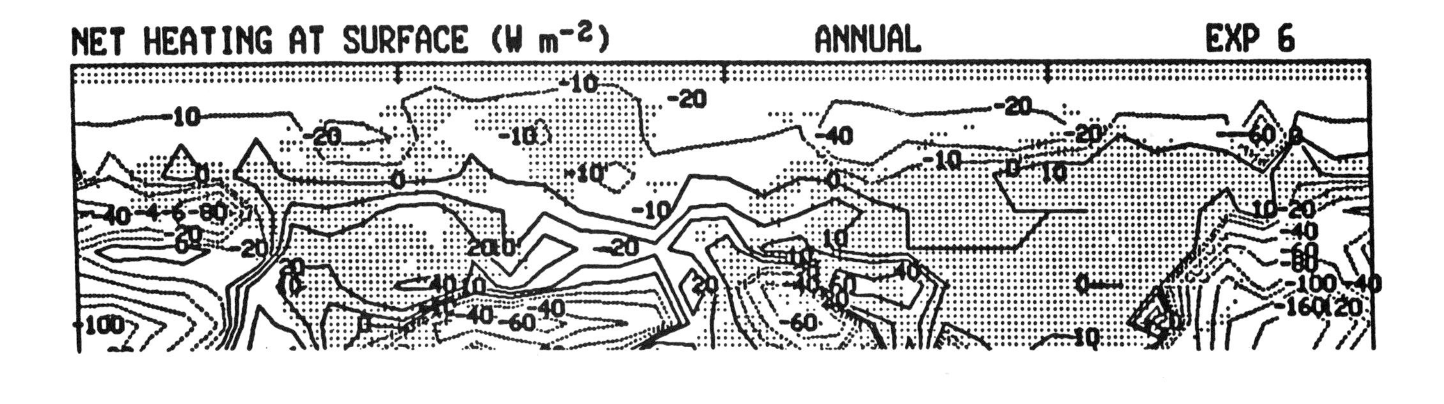

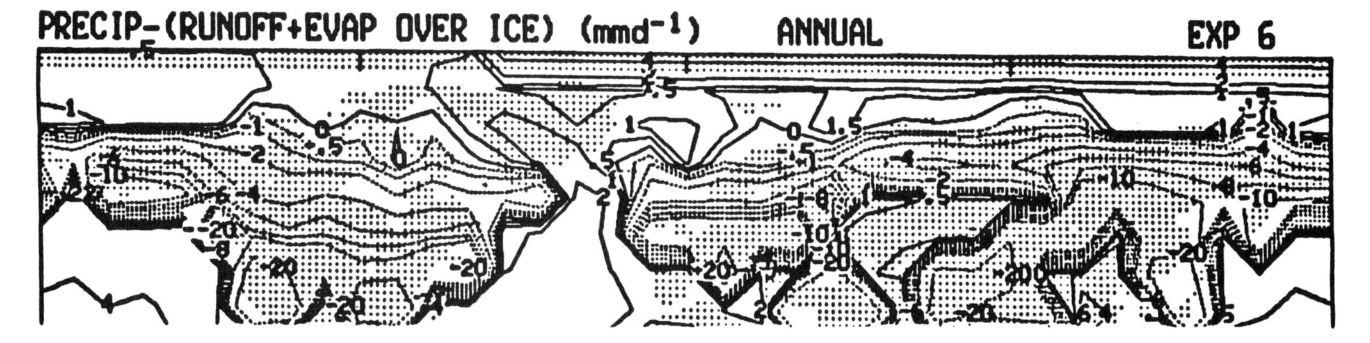

Figure 11. Change in summer temperature, Experiment (EXP) 6 (CLIMAP ice-age ocean, sea ice, land ice, and lowered CO_2) minus model current climate values for summer (top), annual net surface heating in EXP 6 (middle), and annual net mass balance over land ice in EXP 6 (bottom) (from Rind and others, 1989).

We then repeated the experiment, reconstructing the full ice-age ocean with cold SSTs and extensive sea ice. The resulting net heating at the surface and precipitation minus runoff and evaporation are shown in Figure 11. Although cooling intensified somewhat at the highest latitudes, ice would have been maintained only in very restricted areas (e.g., northern Baffin Island).

Despite the limited success of our MIF studies, the phasing of paleoclimatic signals suggests that this experiment may be unrealistic. Ruddiman and others (1980) have shown that ocean cooling *followed* ice-sheet growth in the North Atlantic region, based upon a comparison of the oxygen-isotope signal and SSTs. Note, however, that a cooling of the deep ocean may negate some of the isotopic offset (Chappell and Shackleton, 1986). In addition, in the Vostok ice-core record, the CO_2 decreases appear to lag temperature depression (Barnola and others, 1987) (Fig. 6). Thus, the CO_2 shift seems to have been a result of cooler oceans rather than a trigger of the cooling. Major increase of atmospheric dust in the ice core and ice rafting in the North Atlantic appear at about the same time as the lowest values of CO_2 in Vorstok, ca. 70 ka.

Perhaps major continental ice sheets did not build in the Northern Hemisphere until the middle Wisconsinan (64 ka), but until the terrestrial chronology is improved, this question may remain unanswered. Mix (this volume) notes that the marine-isotopic transitions for stage 5e/5d and 5/4 were roughly the same size, and that these data are in possible conflict with the coral terrace data from New Guinea (Chappell and Shackleton, 1986), due to uncertainties in interpretation of the isotopic records.

Last Glacial Maximum SST reduced by 2 °C

As discussed by Rind and Peteet (1985), there is evidence that the LGM climate was colder than is simulated by this model using the SSTs of CLIMAP Project Members (1981). When SSTs were uniformly reduced by 2 °C below the CLIMAP values, global radiation balance of the 18 ka simulation was improved. Using the reduced SST values as input to the present experiment, the net radiation and mass balances became slightly more favorable for ice-sheet maintenance over Baffin Island (Fig. 11). In eastern North America, however, ice sheets would still have melted in all but very restricted regions (Rind and others, 1989).

Summary of the GCM results

The GISS GCM is unable to grow ice with insolation values of stage 5 as input. If the model could not initiate or sustain ice at about 110 ka, it is unlikely that it would do so at any other time between then and now, as no other radiation shift was apparently more favorable (Fig. 1). Only if ice sheets grew high enough could the model have supported them, because the greater thickness results in less atmosphere above the ice and lower total concentration of greenhouse gases between the ice-sheet surface and the top of the atmosphere. This allows long-wave radiation to escape more easily, producing a less-positive energy balance.

DISCUSSION

Although the initiation of ice ages is associated with Milankovitch forcing, enhanced ice growth in marine-isotope stage 4 and subsequently in stage 2 does not seem in proportion to the insolation reductions during these stages. Nor does the Milankovitch forcing explain the rapid oscillations in stage 3 or the magnitude of the Holocene ice melt.

From the modeling results of the GISS GCM, it is clear that Milankovitch orbital forcing alone is not strong enough to initiate ice-sheet growth in the model, or to sustain it. Neeman and others (1988, p. 11,176) suggested that climate and ice-sheet modelers "Have usually avoided a confrontation with the possibility that the orbital forcing alone, without an amplifying mechanism which is not yet understood, is too weak to cause the ice sheet fluctuations." In the model experiments with additional lowered CO_2, colder SSTs, and an initial low ice sheet in place, the problem of growing or maintaining ice sheets remains. Either the model is not as sensitive to orbital variations as the real world climate, or some important unknown or poorly known feedback process is left out (e.g., deep water formation or atmospheric dust, which we did not model due to the difficulty in simulating realistically in the model). Rind and others (1989) discussed the possible problems in initiation of ice-sheet growth in the model, including uncertainties in clouds, oceans, and a limited geographic resolution. However, none of these uncertainties appears likely to have had a substantial impact. If the problem in initiation of ice sheets is due to model deficiency, this has implications for model predictions of future climate change.

We return to the questions posed in the Introduction, to assess our state of knowledge about the possible feedbacks which might be involved.

1. When did initiation of the Laurentide ice sheet take place? This question remains unanswered. Records of North Atlantic SST and ice rafting, as well as Vostok dust records and pollen stratigraphy, suggest that continental ice sheets may not have grown substantially until ca. 64 ka (oxygen-isotope stage 4). ^{13}C data also suggests a major shift in deep-water formation sometime near stage 4. Although most paleoclimatic data record large temperature and precipitation changes within stage 5, it is not clear how much ice formed in this interval. The 5e/5d insolation shift is more extreme than that at the stage 5/4 boundary (see Fig. 1). Because the climatic response was apparently different, we must conclude that the relation between the insolation forcing and the paleoclimatic record is not yet understood.

2. What was the rate and magnitude of ice-sheet growth at various intervals from the last interglaciation to the present?

Although the marine, land, and ice-core records do not have a common chronology, it appears that temperature shifts were often rapid and dramatic (i.e., pollen stratigraphy, SSTs, ice rafting, and Vostok and Greenland ice-core stratigraphy). A major change in climate apparently occurred at the stage 5/4 transition, possibly concurrent with changes in deep-water circulation. As demonstrated in the Vostok record, this change was closely related to a decrease in CO_2 and to an increase in dust influx. Rapid climatic oscillations in stage 3 are not explained by Milankovitch forcing, and their relation to the magnitude of ice advance and retreat during this stage is unknown. An independent sea-level curve is needed to ascertain the ice-volume component of the deep-sea isotope record.

3. What caused the buildup of ice sheets that characterized the last glacial maximum?

The reason for the maximum buildup of ice sheets about 20 ka is not obvious from the Milankovitch forcing, which shows less reason for cooling than at 116 ka. Paleoclimatic information recovered from the marine, ice, and terrestrial records also do not suggest an obvious reason for this major ice buildup. Modeling studies failed to initiate the ice increase by insolation decrease alone, and even lowered CO_2 and the replacement of global SSTs with LGM values did not sustain ice-sheet growth. The Vostok record suggests a possible feedback through increased atmospheric dust, presumably due to increased wind speed, a drier climate, and/or greater exposure of continental shelves. Ocean circulation changes could also trigger conditions leading to ice growth. However, each of these changes could only have occurred after a causative climatic shift.

4. Do the marine, terrestrial, and ice-core records of climatic change provide a converging stratigraphic record that can be interpreted as a response to orbital forcing?

Although the environmental response to orbital forcing appears to be related in phase to the orbital forcing, the magnitude of the response is not linear. During isotopic stage 5, the coral-reef sequence of sea-level maxima during substages 5e, 5c, and 5a is paralleled by temperature maxima, as recorded by pollen and North Atlantic sea-surface temperature records. The expected cooling in stage 4 appears to be magnified, possibly by changes in CO_2 and the dust increases. Rapid oscillations during stage 3 do not appear to be related to the gradual insolation changes during this interval. The magnitude of ice-sheet growth during stage 2 is not explained by Milankovitch forcing, just as the ice melt 10 ka seems larger than that expected by the insolation regime. Variability in regional responses to climate forcing and interpretation of these paleoclimatic records, as well as poor chronological resolution control, may play a contributing role to gaps in our understanding of the mechanisms involved in the climate system.

ACKNOWLEDGMENTS

This research was supported by NASA and NSF. We note that this manuscript was written in 1989, and more recent publications render some of the analysis outdated due to the lag in publication.

REFERENCES CITED

Andrews, J. T., 1987, The late Wisconsin glaciation and deglaciation of the Laurentide ice sheet, *in* Ruddiman, W. F. and Wright, Jr., H. E. eds., North Ameria and adjacent oceans during the last deglaciation: Boulder, Colorado, Geological Society of America, The Geology of North America, v. K-3, p. 13–37.

Andrews, J. T., Aksu, A., Kelly, M., Klassen, R., Miller, G. H., Mode, W. N., and Mudie, P., 1985, Land/ocean correlations during the last interglacial/glacial transition, Baffin Bay, northwestern North Atlantic: A review: Quaternary Science Reviews, v. 4, p. 333–355.

Bard, E., Hamelin, B., Lao, Y., Anderson, R. F., and Fairbanks, R. G., 1989, Dating of the last interglacial period by U/Th mass spectrometry of Caribbean corals [abs.]: Cambridge, England, 3rd International Conference on Paleoceanography, p. 49.

Barnola, J. M., Raynaud, D., Korotkevich, Y. S., and Lorius, C., 1987, Vostok ice core provides 160,000-year record of atmospheric CO_2: Nature, v. 329, p. 408–418.

Behre, K. E., 1989, Biostratigraphy of the last glacial period in Europe: Quaternary Science Reviews, v. 8, p. 25–44.

Berger, A. L., 1978, Long-term variations of daily insolation and Quaternary climatic changes: Journal of Atmospheric Science, v. 3, p. 2362–2367.

Berger, W. H., Killingly, J. S., and Vincent, E., 1978, Stable isotopes in deep-sea carbonates; Box core ERDC-92, West Equatorial Pacific: Oceanologica Acta, v. 1, p. 221–230.

Berti, A. A., 1975, Paleobotany of Wisconsinan interstadials, eastern Great Lakes region, North America: Quaternary Research, v. 5, p. 591–620.

Birchfield, G. E., Weertman, J., and Lunde, A. T., 1981, A paleoclimate model of Northern Hemisphere ice sheets: Quaternary Research, v. 15, p. 126–142.

Bloom, A. L., Broecker, W. S., Chappell, J., Matthews, R. K., and Mesollela, K. S., 1974, Quaternary sea level fluctuations on a tectonic coast: New $^{230}Th/^{234}U$ dates from Huon Peninsula, New Guinea: Quaternary Research, v. 4, p. 185–205.

Broecker, W. S., 1986, Oxygen isotope constraints on surface ocean temperature: Quaternary Research, v. 26, p. 121–134.

Broecker, W. S., and Denton, G. H., 1989, The role of ocean-atmosphere reorganizations in glacial cycles: Geochimica et Cosmochimica Acta, v. 53, p. 2465–2501.

Broecker, W. S., Andree, M., Bonani, G., Wolfli, W., Oeschger, H., and Klas, M., 1988, Can the Greenland climatic jumps be identified in records from ocean and land?: Quaternary Research, v. 30, p. 1–6.

Chappell, J., and Shackleton, N. J., 1986, Oxygen isotopes and sea level: Nature, v. 324, p. 137–140.

Chappellaz, J., Barnola, J. M., Raynaud, D., Korotkevich, Y. S., and Lorius, C., 1990, Ice core record of atmospheric methane over the past 160,000 years: Nature, v. 345, p. 127–131.

Clague, J. J., Easterbrook, D. J., Hughes, O. L., and Matthews, J. V., Jr., 1988, Was there an ice sheet in the Canadian cordillera during the early Wisconsinan?—Reexamination of the geological record: Geological Society of America Abstracts with Programs, v. 20, no. 7, p. A4.

CLIMAP Project Members, 1981, Seasonal reconstruction of the Earth's surface at the last glacial maximum: Geological Society of America Map and Chart Series MC-36.

Colman, S. M., and Pierce, K. L., 1988, Variable records of early Wisconsin alpine glaciation in the northwestern United States: Geological Society of America Abstracts with Programs, v. 20, no. 7, p. A5.

Dansgaard, W., Johnson, S. J., Clausen, H. B., Dahl-Jensen, D., Gundestrup, N., and Hammer, C. U., 1984, North Atlantic climatic oscillations revealed by deep Greenland ice cores, *in* Hansen, J. E., and Takahashi, T., eds., Climate processes and climate sensitivity: Washington, D.C., American Geophysical Union, p. 288–298.

De Angelis, M., Barkov, N. I., and Petrov, V. N., 1987, Aerosol concentrations over the last climatic cycle (160 kyr) from an ice core: Nature, v. 325, p. 318–321.

de Beaulieu, J. L., and Reille, M., 1984, A long upper Pleistocene pollen record from Les Echets near Lyon, France: Boreas, v. 13, p. 111–132.

Denton, G. H., and Hughes, T., 1983, Milankovitch theory of ice age; hypothesis of ice-sheet linkage between regional insolation and global climates: Quaternary Research, v. 20, p. 125–144.

Dodge, R. E., Fairbanks, R. G., Benninger, L. K., and Maurrasse, F., 1983, Pleistocene sea levels from raised coral reefs of Haiti: Science, v. 219, p. 1423–1425.

Dreimanis, A., 1988, The early Wisconsin in the Erie basin: Geological Society of America Abstracts with Programs, v. 20, no. 7, p. A3.

Duplessy, J. C., Shackleton, N. J., Fairbanks, R. G., Labeyrie, L., Oppo, D., and Kallel, N., 1988, Deepwater source variations during the last climatic cycle and their impact on the global deepwater circulation: Paleoceanography, v. 3, p. 343–360.

Edwards, R. L., Chen, J. H., and Wasserburg, G. J., 1987, ^{238}U-^{234}U-^{230}Th-^{232}Th systematics and the precise measurement of time over the past 500,000 years: Earth and Planetary Science Letters, v. 81, p. 175–192.

Epstein, S., and Mayeda, T., 1953, Variation of O-18 content of water from natural sources: Geochimica et Cosmochimica Acta, v. 4, p. 213–224.

Ewing, W. M., and Donn, W. L., 1966, A theory of ice ages: Science, v. 23, p. 1061–1066.

Fairbanks, R., 1989, Glacio-eustatic sea level record 0–17,000 years before present: Influence of glacial melting rates on Younger Dryas "event" and deep ocean circulation: Nature, v. 342, p. 637–642.

Genthon, C., Barnola, J. M., Raynaud, D., Lorius, C., Jouzel, J., Barkov, N. I., Korotkevich, Y. S., and Kotlyakov, V. M., 1987, Vostok ice core: Climatic response to CO_2 and orbital forcing changes over the last climatic cycle: Nature, v. 329, p. 414–418.

Gruger, E., 1972, Pollen and seed studies of Wisconsinan vegetation in Illinois, U.S.A.: Geological Society of America Bulletin, v. 83, p. 2715–2734.

Hammer, C. U., Clausen, H. B., Dansgaard, W., Neftel, A., Kristindottir, P., and Johnson, E., 1985, Continuous impurity analysis along the Dye 3 deep core, *in* Langway, C. C., Oeschger, H., and Dansgaard, W., eds., Greenland ice core: Geophysics, geochemistry, and the environment: American Geophysical Union Monograph 33, p. 90–94.

Hays, J. D., Imbrie, J., and Shackleton, N. J., 1976, Variations in Earth's orbit: Pacemaker of the ice ages: Science, v. 194, p. 1121–1132.

Heinrich, H., 1988, Origin and consequences of cyclic ice rafting in the northeast Atlantic Ocean during the past 130,000 years: Quaternary Research, v. 29, p. 142–152.

Heusser, L. E., and Shackleton, N. J., 1979, Direct marine-continental correlation: 150,000-year oxygen isotope record from the North Pacific: Science, v. 204, p. 837–839.

Hopkins, D. M., 1982, Aspects of the paleogeography of Beringia during the late Pleistocene, *in* Hopkins, D., Matthews, J. V., Jr., Schweger, C., and Young, S., eds., Paleoecology of Beringia: New York, Academic Press, p. 3–28.

Imbrie, J., and Imbrie, J. Z., 1980, Modeling the climatic response to orbital variations: Science, v. 207, p. 943–953.

Jouzel, J., Lorius, C., Petit, J. R., Genthon, C., Barkov, N. I., Kotlyakov, V. M., and Petrov, V. M., 1987, Vostok ice core: A continuous isotope temperature record over the last climatic cycle (160,000 years): Nature, v. 329, p. 403–408.

Kaufman, A., 1986, The distribution of $^{230}Th/^{234}U$ ages in corals and the number of last interglacial high-sea stands: Quaternary Research, v. 25, p. 55–62.

Kukla, G., 1977, Pleistocene land-sea correlations. I, Europe: Earth-Science Reviews, v. 13, p. 307–374.

——, 1980, End of the last interglacial: A predictive model of the future?, *in* Van Zinderen Bakker, E. M., Sr., and Coetzee, J. A., eds., Paleoecology of Africa and the surrounding islands: Rotterdam, Balkema, p. 395–408.

Kukla, G., and An, Z., 1989, Loess stratigraphy in central China: Palaeogeography, Palaeoceanography, Palaeoecology, v. 72, p. 203–205.

Kukla, G., and Koci, A., 1972, End of the last interglacial in the loess record: Quaternary Research, v. 2, p. 374–383.

Kukla, G., and Lozek, V., 1961, Loess and related deposits. Survey of Czechoslovak Quaternary (Czwartorzed Europy Srodkowej i Wschodniej: INQUA, VIth International Congress): Prace, Ton XXXIV, WARSZAWA, Institut Geologiczny, v. 34, p. 11–28.

Kukla, G., Berger, A., Lotti, R., and Brown, J., 1981, Orbital signature of interglacials: Nature, v. 290, p. 295–300.

Labeyrie, L. D., Duplessy, J. C., and Blanc, P. L., 1987, Variations in mode of formation and temperature of oceanic deep waters over the past 125,000 years: Nature, v. 327, p. 477–482.

Lamothe, M., Parent, M., and Shilts, W. W., 1988, Early Wisconsinan events in the St. Lawrence lowland and in the Appalachians of southern Quebec: Geological Society of America Abstracts with Programs, v. 20, no. 7, p. A3.

Lorius, C., Jouzel, J., Ritz, C., Merlivat, L., Barkov, N. I., Korotkevich, Y. S., and Kotlyakov, V. M., 1985, A 150,000-year climatic record from Antarctic ice: Nature, v. 316, p. 591–596.

Mangerud, J., 1990, The Scandinavian ice sheet through the last interglacial/glacial cycle; *in* Frenzel, B., ed., Klimageschichtliche Problem des Holozans and des Letzten interglazials: Palacklimaforschung 1: Stuttgart, New York, Gustav Fischer, p. 142–154.

Martinson, D. G., Pisias, N. G., Hays, J. D., Imbrie, J., Moore, T. C., Jr., and Shackleton, N., 1987, Age dating and the orbital theory of the ice ages: Development of a high-resolution 0 to 300,000-year chronostratigraphy: Quaternary Research, v. 27, p. 1–29.

Mercer, J. H., 1983, Late Cainozoic glacial variations in South America south of the equator, *in* Rabassa, J., ed., Quaternary of South America and Antarctic Peninsula: Rotterdam, Balkema, p. 45–58.

Mesolella, K. J., Matthews, R. K., Broecker, W. S., and Thurber, D. L., 1969, The astronomical theory of climate change: Journal of Geology, v. 77, p. 250–274.

Miller, G. H., and Andrews, J. T., 1988, Timing and character of early Wisconsin ice sheet growth in Canada: Geological Society of America Abstracts with Programs, v. 20, no. 7, p. A3.

Mix, A., and Fairbanks, R. G., 1985, North Atlantic surface-ocean control of Pleistocene deep-ocean circulation: Earth and Planetary Science Letters, v. 73, p. 231–243.

Mix, A., and Ruddiman, W. F., 1984, Oxygen isotope analyses and Pleistocene ice volumes: Quaternary Research, v. 21, p. 1–20.

Muller, H., 1965, Eine pollenanalytische Neubearbeitung des Interglazialprofils von Bilshausen (Unter-Eichsfeld): Geologisches Jahrbuch, v. 83, p. 327–352.

Neeman, B. U., Ohring, G., and Joseph, J., 1988, The Milankovitch theory and climate sensitivity 2. Interaction between the Northern Hemisphere ice sheets and the climate system: Journal of Geophysical Research, v. 93, p. 11,175–11,191.

Oerlemans, J., 1980, Model experiments on the 100,000-year glacial cycle: Nature, v. 287, p. 430–432.

Paterne, M., Guichard, F., Labeyrie, J., Gillot, P. Y., and Duplessy, J. C., 1986, Tyrrhenian Sea tephrochronology of the oxygen isotope record for the past 60,000 years: Marine Geology, v. 72, p. 259–285.

Peltier, W. R., 1987, Glacial isostasy, mantle viscosity, and Pleistocene climatic change, *in* Ruddiman, W. F., and Wright, H. E., eds., North America and adjacent oceans during the last deglaciation: Boulder, Colorado, Geological Society of America, The Geology of North America, v. K-3, p. 155–182.

Petit, J. R., Mounier, L., Jouzel, J., Korotkevich, Y. S., Kotlyakov, V. I., and Lorius, C., 1990, Paleoclimatological and chronological implications of the Vostok core dust record: Nature, v. 343, p. 56–58.

Pollard, D., 1982, A simple ice-sheet model yields realistic 100-kyr glacial cycles: Nature, v. 296, p. 334–338.

——, 1984, Some ice-age aspects of a calving ice sheet model, *in* Berger, A., Imbrie, J., Hays, J., Kukla, G., and Saltzman, B., eds., Milankovitch and climate (NATO ASI Series): Dordrecht, Netherlands, D. Reidel, p. 591–564.

Raynaud, D., Chapellaz, J., Barnola, J. M., Korotkevich, K. S., and Lorius, C., 1988, Climatic and CH_4 cycle implications of glacial-interglacial CH_4 changes in the Vostok core: Nature, v. 333, p. 655–657.

Rind, D., and Peteet, D., 1985, Terrestrial conditions at the last glacial maximum and CLIMAP sea-surface temperature estimates: Are they consistent?: Quaternary Research, v. 24, p. 1–22.

Rind, D., Peteet, D., and Kukla, G., 1989, Can Milankovitch orbital variations initiate the growth of ice sheets in a general circulation model?: Journal of Geophysical Research, v. 94, no. D10, p. 12,851–12,871.

Ruddiman, W. F., 1977, Late Quaternary deposition of ice-rafted sand in the subpolar North Atlantic (lat 40° to 65°): Geological Society of America Bulletin, v. 88, p. 1813–1827.

——, 1987, Northern Oceans, *in* Ruddiman, W. F., and Wright, H. E., Jr., North America and adjacent oceans during the last deglaciation: Boulder, Colorado, Geological Society of America, The Geology of North America, v. K-3, p. 137–154.

Ruddiman, W. F., and McIntyre, A., 1981, North Atlantic during the last deglaciation: Palaeogeography, Palaeoceanography, Palaeoecology, v. 35, p. 145–214.

Ruddiman, W. F., and Wright, H. E., Jr., 1987, Introduction, *in* Ruddiman, W. F., and Wright, H. E., Jr., eds., North America and adjacent oceans during the last deglaciation: Boulder, Colorado, Geological Society of America, The Geology of North America, v. K-3, p. 1–12.

Ruddiman, W. F., McIntyre, A., Niebler-Hunt, B., Durazzi, J. T., 1980, Oceanic evidence for the mechanism of rapid Northern Hemisphere glaciation: Quaternary Research, v. 13, p. 33–64.

Sancetta, C., Imbrie, J., and Kipp, N. G., 1973, Climatic record of the past 130,000 years in North Atlantic deep-sea core V23-82: Correlation with the terrestrial record: Quaternary Research, v. 3, p. 110–116.

Seret, G., Dricot, E., and Wansard, G., 1990, Evidence for an early glacial maximum in the French Vosqes during the last glacial maximum: Nature, v. 346, p. 453–456.

Stauffer, B., Neftel, A., Oeschger, H., and Schwander, J., 1985, CO_2 concentrations in air extracted from Greenland ice samples, *in* Langway, C. C., Oeschger, H., and Dansgaard, H., eds., Greenland ice core: Geophysics, geochemistry, and the environment: American Geophysical Union Monograph 33, p. 85–89.

Stauffer, B., Lochbronner, E., Oeschger, H., and Schwander, J., 1988, Methane concentrations in the glacial atmosphere was only half that of the pre-industrial Holocene: Nature, v. 332, p. 812–814.

Thorleifson, L. H., Shilts, W. W., Wyatt, P. H., and Neilsen, E., 1988, Early Wisconsin glaciation of the Hudson Bay lowland: Geological Society of America Abstracts with Programs, v. 20, no. 7, p. A4.

Turner, C., 1975, The correlation and duration of middle Pleistocene interglacial periods in Northwest Europe, *in* Butzer, K. W., and Isaac, G. L., eds., After the Australopithecines: The Hague, Mouton, p. 259–308.

Turon, J. L., 1984, Direct land/sea correlations in the last interglacial complex: Nature, v. 309, p. 673–676.

van der Hammen, T., Maarleveld, G. C., Vogel, J. C., and Zagwijn, W. H., 1967, Stratigraphic, climatic succession and radiocarbon dating of the last glacial in The Netherlands: Geologie en Mijnbouw, v. 46, p. 80–95.

Vincent, J. S., 1988, The Sangamonian and early Wisconsinan record in the western Canadian Arctic: Geological Society of America Abstracts with Programs, v. 20, no. 7, A4.

Westgate, J. A., Easterbrook, D. J., Naeser, N. D., and Carson, R. J., 1987, Lake Tapps tephra: An early Pleistocene stratigraphic marker in the Puget lowland, Washington: Quaternary Research, v. 28, p. 340–355.

Wijmstra, T. A., 1969, Palynology of the first 30 meters of a 120 m deep section in northern Greece: Acta Botanica Neerlandica, v. 18, p. 511–527.

Winograd, I. J., Szabo, B. J., Coplen, T. B., and Riggs, A. C., 1988, A 250,000-year climatic record from Great Basin Vein Calcite: Implications for Milankovitch theory: Science, v. 242, p. 1275–1280.

Woillard, G., and Mook, W. G., 1982, Carbon-14 dates at Grande Pile: Correlation of land and sea chronologies: Science, v. 215, p. 159–161.

Zagwijn, W. H., 1983, Sea-level changes in the Netherlands during the Eemian: Geologie en Mijnbouw, v. 62, p. 437–450.

Manuscript Accepted by the Society September 6, 1991

Geological Society of America
Special Paper 270
1992

The last interglacial-glacial transition in Illinois: 123–25 ka

B. Brandon Curry and Leon R. Follmer
Illinois State Geological Survey, 615 East Peabody Drive, Champaign, Illinois 61820

ABSTRACT

Sediments, soils, and fossils are used to interpret paleoenvironmental conditions during the last interglacial-glacial transition in Illinois. The sediments include the classic Sangamonian-Wisconsinan and Wisconsinan/Farmdalian-Woodfordian successions; both consist of loess overlying pedogenically modified or generally colluviated sediment. Although the contacts between the loess and soil are defined as isochronous, they are diachronous. The age of basal Wisconsinan deposits spans from at least 50 to 22 ka, and the age of basal Woodfordian sediments spans from 25 to 20 ka.

Recognition of the last interglacial episode, the Sangamonian Age, is based on Mollisols, Alfisols, and Ultisols that formed in Illinoian glacigenic deposits. Sangamonian flora and fauna, although sparsely preserved, include pollen well represented by deciduous trees, grasses, and *Ambrosia.* Sangamonian vertebrates included giant tortoise, mastodon, giant beaver, snapping turtle, and short-nosed gar. The Wisconsinan Age (Altonian Subage) began as Roxana Silt (loess) was deposited under periglacial conditions. The vegetation was characterized by coniferous trees that grew in weakly developed, often organic-rich, cryoturbated soils. Periglacial conditions persisted during the Altonian and Farmdalian Subages, from before 50 ka to about 25 ka, after which loess, outwash, and till were deposited under glacial conditions during the Woodfordian Subage and much of the remainder of the Wisconsinan Age. The vegetation that grew during the last glacial episode includes plants that are now found at and north of the treeline in Canada.

INTRODUCTION

Willman and Frye (1970) defined chronostratigraphic boundaries within the late Pleistocene glacial drift in Illinois on the basis of lithostratigraphic units. They assumed that the interval of time beween the last interglacial and glacial episodes was short and that it could be considered an isochron. The last glacial episode began as loess (Roxana Silt) was deposited along the Illinois and Mississippi river valleys. Willman and Frye attributed this loess to an ice margin that advanced as far south as about 41.5°N in Illinois. Following Willman and Frye's (1970) concept of an extensive early Wisconsinan glaciation, others throughout the midcontinent (e.g., Dreimanis and Goldthwait, 1973; Gooding, 1975; Fullerton, 1986) attributed pre–late Wisconsinan glacial deposits to the early Wisconsinan interval, although a reliable chronology for most deposits was lacking (Clark and Lea, 1986).

Ongoing studies in Illinois have refined some of Willman and Frye's (1970) concepts. Kempton and others (1985) and Curry (1989) showed that several tills thought to be early Wisconsinan in northern Illinois are actually pre-Sangamonian (Illinoian); evidence for the primary glacial source of the earliest Wisconsinan loess, probably north of Illinois, has not been found (Winters and others, 1988; Johnson and Follmer, 1989; Leigh, 1991). We agree with McKay (1979b), who suggested that Willman and Frye's (1970) age estimate of 75 ka for the base of the Roxana in the type area was excessive, and that an age of about 50 ka is more reasonable.

The base of the Wisconsinan succession is recognized by the lowermost occurrence of loess that is partly composed of unweathered silt grains and pedogenically mixed with older weathered sediments (Frye and others, 1974). This criterion often requires detailed field and laboratory data to sort out the effects of pedogenic and geogenic processes, namely pedoturbation and addition of loess, respectively.

Curry, B. B., and Follmer, L. R., 1992, The last interglacial-glacial transition in Illinois: 123–25 ka, *in* Clark, P. U., and Lea, P. D., eds., The Last Interglacial-Glacial Transition in North America: Boulder, Colorado, Geological Society of America Special Paper 270.

This chapter reviews the sediment, soil, and fossil records of the last interglacial-glacial transition in Illinois; in addition we present data from the Athens Quarry, Hopwood Farm, Pittsburg Basin, and Lomax localities (Fig. 1). We interpret the last interglacial-glacial transition to span the time from the last interglacial maximum to the last glacial maximum. There are no isotopic or faunal records from Illinois to estimate changes in past temperatures. Instead, a climatic interpretation is derived by combining interpretations from the sediment, soil, and fossil pollen and plant macrofossils.

STRATIGRAPHIC BACKGROUND

In Illinois we recognize the following chronostratigraphic and lithostratigraphic units deposited from about 150 to 22 ka (Willman and Frye, 1970; Fig. 2): Illinoian till and Sangamonian

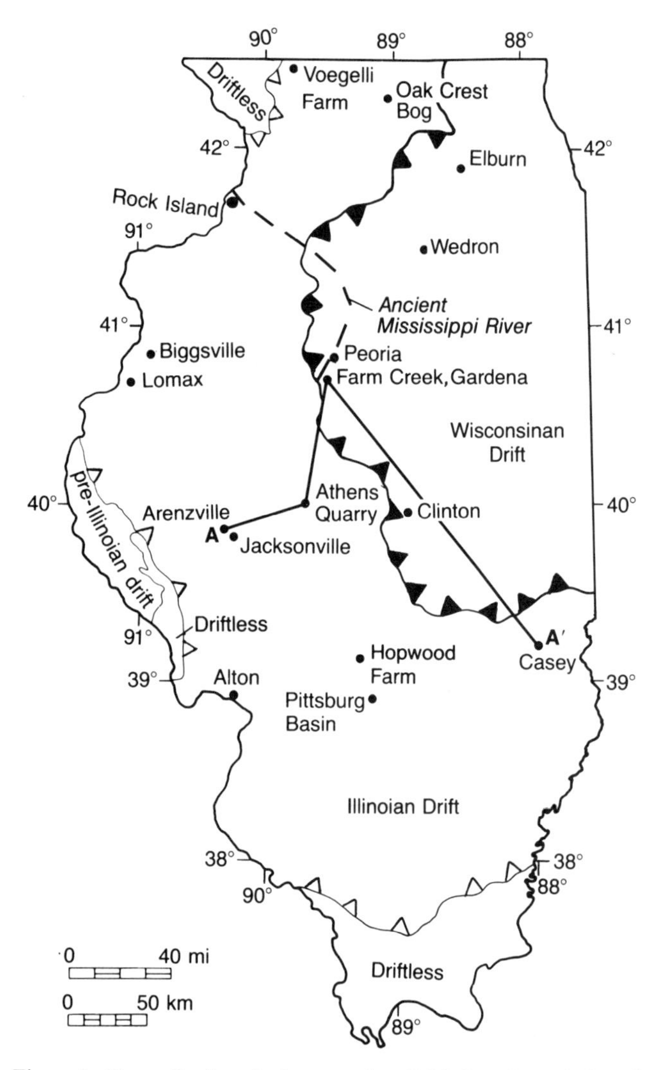

Figure 1. Generalized geologic map of surficial deposits and sites discussed in text. Solid sawtooth pattern denotes limit of Wisconsinan drift; open sawtooth pattern denotes limit of Illinoian drift. Section line A-A′ for Figure 3.

colluvium (Berry Clay Member of the Glasford Formation), Altonian loess (Roxana Silt), Farmdalian colluvium (Robein Silt), Woodfordian loess (Morton Loess, Peoria Loess, and Richland Loess) and Woodfordian till (Wedron Formation). The historical development of this stratigraphic sequence was discussed in McKay (1979a) and Follmer (1978, 1979).

Willman and Frye's (1970) stratigraphic framework has been modified (Fig. 2). The Winnebago Formation in northern Illinois, originally thought to be temporally correlative with Roxana Silt, is now considered Illinoian (Kempton and others, 1985). Plano Silt is no longer considered a viable lithostratigraphic unit, and is included in the lower part of the Robein Silt (Curry and Kempton, 1985).

LITHOSTRATIGRAPHY

The last interglacial-glacial transition is represented by lithologic units that commonly have pedogenic features. These units overlie Illinoian glacigenic diamicton (till) and stratified sediment of the Glasford and Winnebago formations (Fig. 2).

Berry Clay

Part of the Glasford Formation, the Berry Clay was deposited and pedogenically modified during the Sangamonian Age and Altonian Subage. The Berry has two facies: a soft, leached and gleyed, upward-fining loam to clay diamicton (accretion gley), and a less-common, more-oxidized facies composed of loam diamicton. Both facies are interpreted to be colluvial in nature and have been modified by soil formation under a nonglacial climate (Willman and others, 1963, 1966; Frye and others, 1974; Follmer, 1982, 1983). The base of the Berry Clay is locally demarcated by a stone line or, more commonly, a sandy, gravelly zone above pedogenically modified glacigenic sediment of the Glasford and Winnebago formations (Figs. 2 and 3).

Roxana Silt

We recognize two members, the lower Markham Silt Member and upper Meadow Loess Member, and an informal facies, the sandy-silt facies, in the Roxana Silt (Fig. 3). Willman and Frye (1970) also recognized the McDonough Loess Member and associated Pleasant Grove Soil, but these do not appear to be valid (McKay, 1979a), and are not recognized here.

The Markham Silt Member is composed of leached, pedogenically altered material that contains a moderate amount of sand and traces of gravel. It is distinguished from underlying material of the Glasford Formation by upward increases in the percentages of silt, vermiculite, and relatively unweathered minerals (Frye and others, 1974).

The Meadow Loess Member composes the uppermost 80% to 90% of the Roxana Silt (Fig. 3), and is composed of massive silt loam that is calcareous, oxidized, and contains weak pedogenic modifications such as few biopores and fine sesquioxide

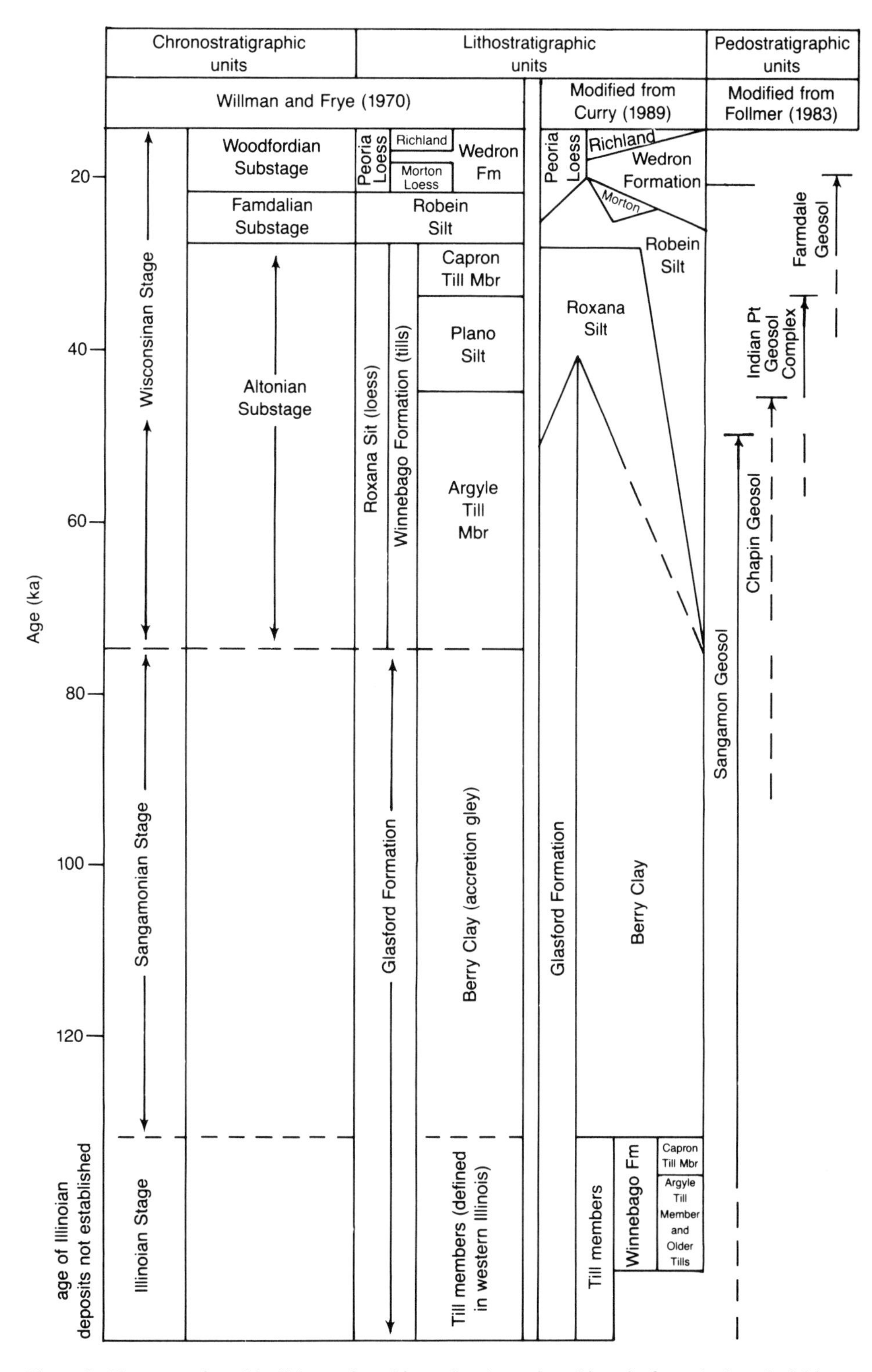

Figure 2. Chronostratigraphic, lithostratigraphic, and pedostratigraphic units from the last glacial-interglacial-glacial transition in Illinois.

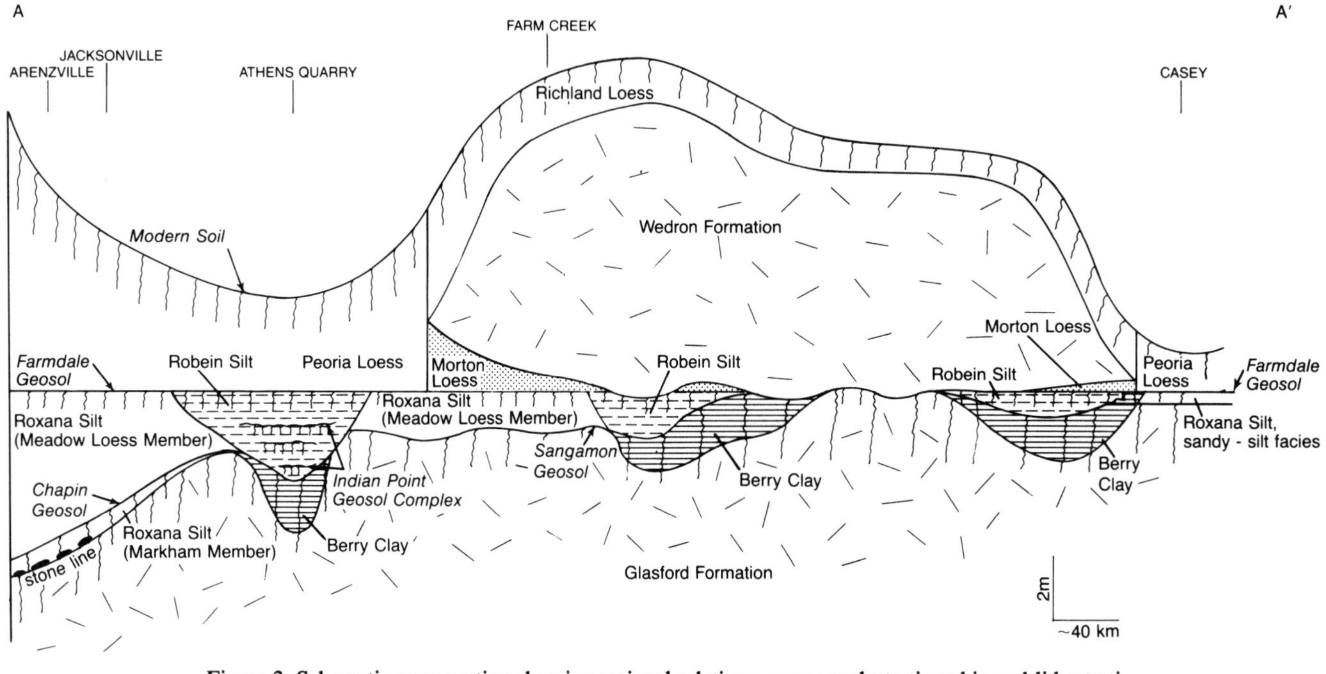

Figure 3. Schematic cross section showing regional relations among pedostratigraphic and lithostratigraphic units. Actual thicknesses of units are under labeled sites. Data from Follmer and others (1979, Arenzville, Farm Creek); Frye and others (1974, Jacksonville); Follmer (1983, Casey) and this chapter (Athens Quarry). Line of section shown in Figure 1.

concretions (Follmer and others, 1979). Where it is thick along the Illinois River and to the south along the Mississippi River, the Meadow Loess Member may be divided into three zones based on color, clay mineralogy, and carbonate mineralogy (McKay, 1979a). Differences in mineralogy are attributed to shifting compositions of glaciofluvial source areas and to varying degrees of pedogenic alteration.

The distal portion of the Roxana, informally referred to as the sandy-silt facies by Johnson and others (1972; Fig. 3), is characterized by lithologic and pedogenic features that are similar in character to the Markham Silt Member. However, they are not temporally correlative because the Markham is pedogenically altered silt bounded at the top by the Meadow Loess Member (Fig. 3), whereas in eastern Illinois the sandy-silt facies composes the entire Roxana silt.

Robein Silt

The Robein Silt is composed of stratified, leached, organic-rich silt loam, and in most cases appears to be reworked Roxana Silt. Evidence of colluviation, such as dark and light layers or laminae, distinguishes the Robein from the macroscopically massive Roxana that contains relatively little organic matter. In typical sections, Robein Silt has O, A, and Bg horizons of the Farmdale Geosol. Although Robein Silt is not a regionally continuous unit, locally it aids in differentiating gray to yellow-brown (10YR Munsell hue) Peoria Loess above from the reddish-gray (7.5YR) Roxana Silt below. The type section of the Robein Silt at Farm Creek (Willman and Frye, 1970) is inappropriate because the Robein is not stratified there. However, a suitable paratype section is at the Athens Quarry (Follmer and McKay, 1987).

Woodfordian lithostratigraphic units

The Woodfordian units that may overlie the Robein or Roxana silts are the Morton Loess, Peoria Loess, or Wedron Formation (Fig. 3). The Morton loess is overlain by the Wedron Formation, which is composed of glacigenic diamicton or associated sediment. For practical reasons, Peoria Loess is limited to the Woodfordian loess beyond the limit of the Wedron Formation; late Woodfordian loess that overlies the Wedron Formation is Richland Loess (Figs. 2 and 3). Below the modern soil, all loess units are massive, dolomitic silt loam. Near source valleys, the units locally include lenses of fine-grained to very fine-grained stratified sand deposited by eolian traction. Lacustrine sediment may occur below or is intercalated with loess, especially in tributary valleys along the Illinois and Mississippi river valleys (Hajic and others, 1991).

PEDOSTRATIGRAPHY

Background

The geosol is the fundamental unit of pedostratigraphy (North American Commission on Stratigraphic Nomenclature, 1983). As used here, a geosol is a mappable, stratiform body of material with soil characteristics formed on a previous land surface, with an upper surface that now constitutes both a material and stratigraphic boundary. A formalized geosol is defined at its type locality as a buried soil that contains catenal members. Where a geosol merges with the present land surface, the geosol loses much of its stratigraphic utility, but continues to be important for understanding the modern soil. Here we review the independent use of pedostratigraphy and lithostratigraphy in the glacial drift of Illinois, and briefly describe some of the formal and informal geosols present in sediments that were formed between about 140–22 ka.

Sangamon Geosol

The Sangamon Geosol is most commonly formed in Illinoian glacigenic sediments and Sangamonian colluvium and alluvium. The most distinctive aspect of the Sangamon Geosol is the B horizon, but evidence of pedogenesis, such as oxidation along joints, may extend far below the B horizon in the C horizon. In very poorly drained environments, the upper Sangamon Geosol is developed in Berry Clay (colluvium); the pedogenic character includes abundant, continuous cutans, biopores, krotovina (crayfish burrows) and gleyed colors. In environments where the Sangamon Geosol was better drained, the soil usually is developed in Illinoian deposits. The B horizons of the most oxidized profiles are often reddish (7.5YR Munsell hues) to brownish (10YR hues) and have medium-blocky structure and continuous cutans. There is an abundance of fine clay (<1 μm) in the B horizons of Sangamon profiles independent of drainage class.

Chapin Geosol

Thorp and others (1951) first suggested that the upper part of the Sangamon interglacial soil in the midwestern United States was siltier than the lower part because early-deposited loess had been weathered and mixed by pedoturbation into the upper solum. This relation is common in interglacial soils buried by loess (Ruhe, 1976), and was called the "basal mixed zone" of the loess by Schumacher and others (1988) and Miller and others (1988). Laboratory data reveal that in this stratigraphically complex zone, characteristics of the A horizon, such as biopores and organic matter, gradually are replaced upward by characteristics of loess, such as a more massive nature, greater silt content, and more abundant unweathered minerals (Frye and others, 1974). These zones are generally less than 1 m thick, and form as long as the rate of sedimentation does not exceed the rate of pedogenesis, especially as affected by pedoturbation. Where loess is thin, the basal mixed zone has an upward increase of vermiculite and medium to coarse silt ratio, indicating an addition of far-traveled loess to the soil (Follmer, 1983; Curry, 1989).

As a basis for defining the contact within the basal mixed zone between Sangamonian and Wisconsinan Stages, Frye and others (1974) gave priority to changes in lithology rather than pedological characteristics. They interpreted that the Sangamon Geosol had been truncated by erosion and then buried by thin colluvium (Markham Silt) that contained a portion of unweathered minerals from the initial deposition of Wisconsinan loess. The soil that formed in the Markham Silt was designated the Chapin Geosol, rather than the Sangamon Geosol (Willman and Frye, 1970), because priority was placed upon the lithology of the parent material rather than the evidence of continuity of the solum. Ruhe (1969) assumed a similar set of events in Iowa and Indiana, but gave priority to the pedogenic continuity by naming the younger soil the late Sangamon, and also recognized that much time passed before the Wisconsinan loess buried the Sangamon surface. Therefore, in terms of soils and the interrelated sediment record, the last interglacial-periglacial transition in the central midwestern United States is generally represented by an unconformity or localized weathered colluvium.

Indian Point Geosol Complex

At Athens Quarry (Fig. 1), the type locality of the Sangamon Geosol (Follmer and others, 1979), several minor soils, informally named the Indian Point geosol complex, occur in the lower part of the Robein Silt (redeposited Roxana Silt) between the Sangamon Geosol and Farmdale Geosol (Figs. 3 and 4). The geosol complex is a sequence of AO-Bg couplets that range from 3 to 15 cm thick. The layers are commonly broken and convoluted, suggesting that they were deformed by cryoturbation. At most locations in the quarry, the composite sequence is about 1 m thick and overlies about 1 m of massive Meadow Loess Member of the Roxana Silt with common biopores. A gleyed Sangamon Geosol profile underlies the Roxana (Fig. 4).

We chose not to include the Indian Point couplets with the overlying Farmdale Geosol because the poorly drained facies of the Farmdale at most localities is a simple AO/A/Bg profile usually more than 50 cm thick. More important, we believe the complex has significance for high-resolution stratigraphy because the couplets represent multiple soil-forming and depositional events. Elsewhere in Illinois where the Robein is less than about 1.5 m thick, the complex, if it had existed, has evolved into what appears to be a single profile. Multiple soil profiles with abundant organic matter are also rare within thick deposits of Roxana Silt, although upon close inspection, the Roxana contains abundant evidence of pedogenesis, such as biopores and small sesquioxide concretions.

Farmdale Geosol

The colluvial, organic-rich facies of the Farmdale Geosol is developed in the Robein Silt. The solum of a typical profile is generally less than 1 m thick. There is a distinctive AO horizon containing abundant wood fragments, a black A horizon, and Bg

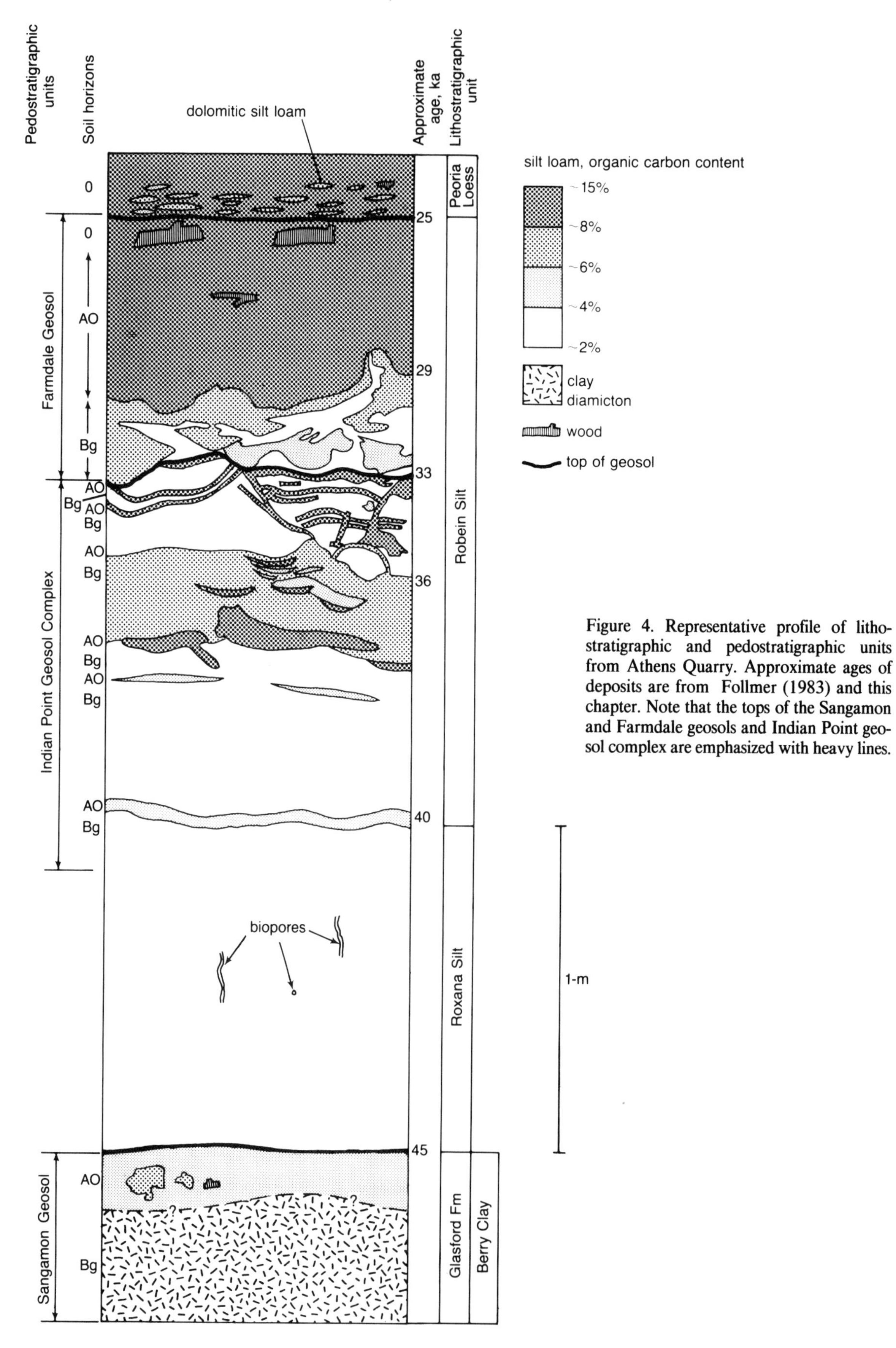

Figure 4. Representative profile of lithostratigraphic and pedostratigraphic units from Athens Quarry. Approximate ages of deposits are from Follmer (1983) and this chapter. Note that the tops of the Sangamon and Farmdale geosols and Indian Point geosol complex are emphasized with heavy lines.

horizon containing abundant biopores in layers of sediment with varying organic carbon content. The layers are locally convoluted, perhaps due to cryoturbation (Fig. 4). The in situ facies of the Farmdale Geosol are most common in the top of the Roxana Silt. These profiles are weakly developed but the depth of leaching may be greater than 2 m (Follmer, 1983).

Discussion of lithostratigraphy and pedostratigraphy

Most localities in Illinois that record the Sangamonian-Wisconsinan transition contain Berry Clay, Roxana Silt, Robein Silt, Peoria Loess, or Morton Loess. The loessial units are characterized by a pedoturbated basal zone, and all units are modified by varying degrees of pedogenesis. Laboratory analyses, particularly particle-size distribution and mineralogy of the <2 μm fraction, are useful in lithostratigraphic correlation and identification of materials within pedogenic profiles. An example of such a study is that of Curry (1989), who examined the evidence for the lack of Altonian (early Wisconsinan) glaciation in northeastern Illinois. He found consistent characteristics of particle-size and clay-mineral data, as well as similar radiocarbon ages, in pedogenically altered material above Illinoian glacigenic diamicton and below late Wisconsinan diamicton and loess across much of Illinois (Fig 5). The characteristics of the Berry Clay, Roxana Silt, Robein Silt, as well as the Sangamon Geosol and Farmdale Geosol, can be traced from their type sections in the Peoria area to northern Illinois. Such relations become less distinct and blurred where the loess is thinner. In sloping areas of thin loess in northern Illinois, sediments and soils that record the Sangamonian-Wisconsinan transition commonly have been truncated, and this has led to numerous stratigraphic interpretations (Follmer and Kempton, 1985; Curry and Kempton, 1985; Curry, 1989).

Mixed zones also are distinguishable in the Farmdale Geosol, where the basal parent material is composed of nonloessial sediment. For example, data from a Farmdale profile near the valley bottom along the bluffs of the Mississippi River at Lomax, Illinois, show two gradations in particle-size distribution (Curry, 1990; Fig. 6). At the base of the profile, there is a gradual change from well-sorted sand to clay that probably reflects a change from fluvial to paludal deposits, and may not necessarily be due to pedoturbation. The ratio of medium to coarse silt (16–32 μm/32–63 μm) increases upward to about 15 as organic carbon (not plotted) and clay content also increase. These data indicate sedimentation in a pond or lake, probably on a floodplain. Subsequently, in the upper 40 cm, silt becomes more abundant relative to clay, while the medium to coarse silt ratio decreases, probably indicating an influx of loess. Radiocarbon age estimates bounding this unusual mixing zone are 21, 250 $\pm$ 250 (ISGS-1730) and 24,720 $\pm$ 320 yr B.P. (ISGS-1720; Fig 6 and and Table 1), and indirectly record the advance of Woodfordian glaciers across northern Illinois (Kempton and Gross, 1971; Johnson and Hansel, 1987). The sediment, which has been locally reworked, possibly by cryoturbation and bioturbation, is correlated with the Robein Silt.

The clayey lacustrine sediment (Equality Formation) above the Robein Silt is laminated and has high medium to coarse silt ratios that indicate sediment sorting by deposition in a paludal or fluvial environment. Subsequently, the ratio decreases from >15 to <2 as the sediment becomes siltier, suggesting proximity to the loess source, probably in response to the diversion of the ancient Mississippi River southwest from the Rock Island area, which resulted from blockage of the former course north of Farm Creek by Woodfordian glaciers (Fig. 1; Glass and others, 1964). The age of this event is bracketed by the 21,250 $\pm$ 250 yr B.P. age from the top of the Robein and by an age of 20,320 $\pm$ 120 yr B.P. (ISGS-136; Coleman, 1974) from wood fragments found in sand that truncates the top of the Robein Silt at a site 250 m southwest of the profile illustrated in Figure 6. These ages agree with radiocarbon ages on wood and humic material in Peoria Loess in the Alton area (Fig. 1) that also suggest that the diversion of the Mississippi occurred at about 20,500 yr B.P. (McKay, 1979a).

CHRONOSTRATIGRAPHY

Chronostratigraphic boundaries are isochrons determined or estimated at type sections and are placed at the boundaries of lithostratigraphic or pedostratigraphic units. Where available, radiocarbon ages are used to establish the age of the isochron and time equivalency away from the type section. The intra-Wisconsinan Farmdalian-Woodfordian Substage contact is defined at the Farm Creek type locality; the age was originally estimated as 22 ka by Willman and Frye (1970) and later revised to 25 ka by Follmer and others (1979). At the type locality, the Farmdalian-Woodfordian contct coincides with the top of the Robein Silt and Farmdale Geosol. Away from the Illinois River Valley, the top of lithostratigraphic and pedostratigraphic units is coincident, but this contact does not necessarily coincide with the chronostratigraphic boundary. For example, the time represented by the Robein Silt and Farmdale Geosol extends into the Woodfordian at the Lomax section (Fig. 6).

We propose that the Athens Quarry sections, the type locality of the Sangamon Geosol (Follmer, 1983), be the paratype section of the Sangamonian Stage because the paratype sections designated by William and Frye (1970) are incomplete and inaccessible for detailed study. At the original paratype sections, William and Frye (1970) and Frye and others (1974) suggested that the age of the Sangamonian-Wisconsinan boundary is about 75 ka. However, radiocarbon-based sedimentation rates of Roxana Silt suggest an age of about 50 ka (McKay, 1979b; Leigh, 1991). At Athens Quarry, the age of the A horizon of the Sangamon Geosol is about 45 ka (Follmer, 1983). However, the age of the parent material is necessarily older than the age of the organic matter in the soil, and we concur with McKay's (1979b) estimate of 50 ka as the approximate age of the Sangamonian-Wisconsinan contact.

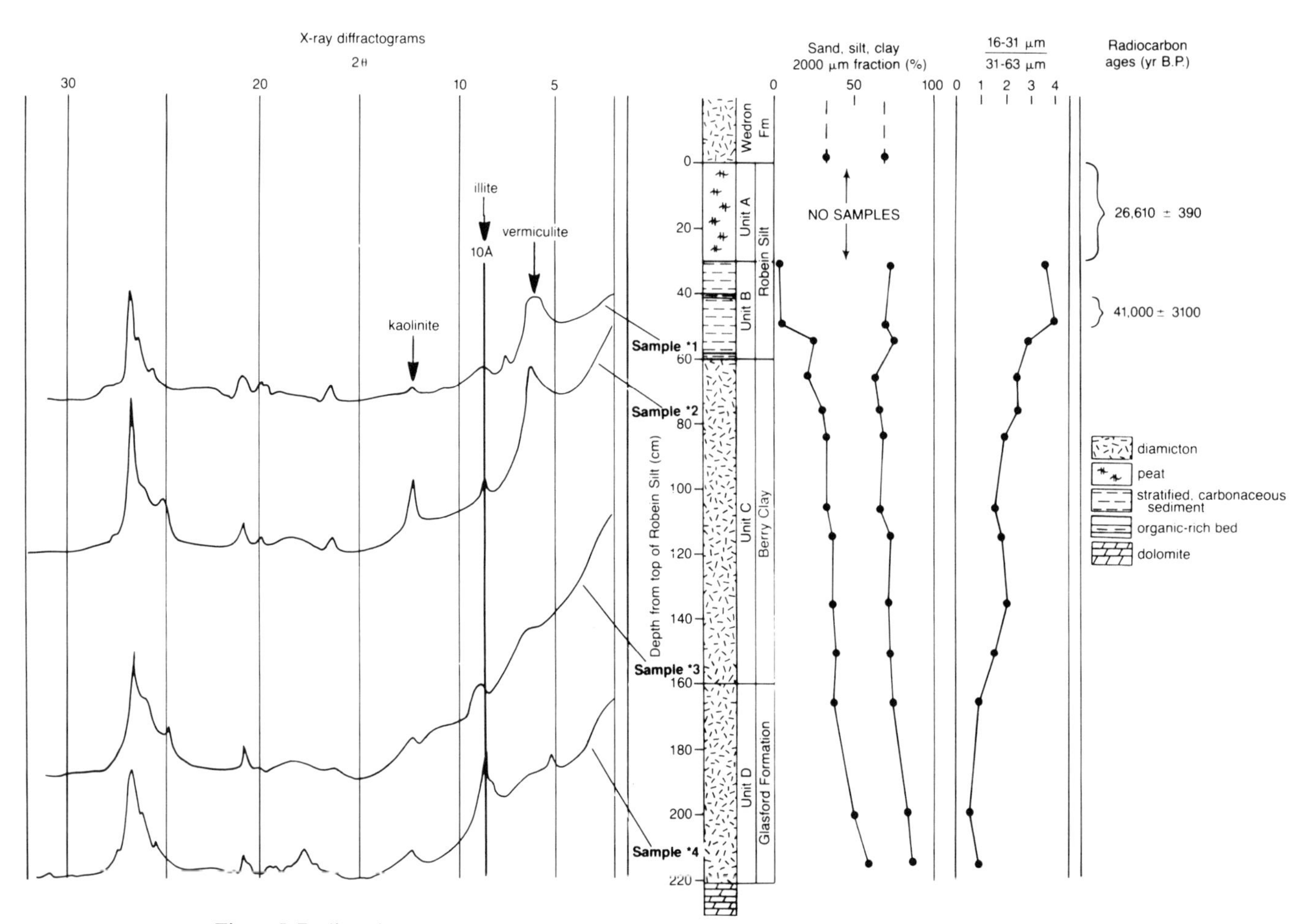

Figure 5. Radiocarbon ages, particle-size distribution, medium to coarse silt ratios, and X-ray diffractograms of subsamples of a core near Elburn, Illinois (modified from Curry, 1989).

We have different levels of confidence with respect to the chronology of the interval of interest. We are most confident with deposits associated with radiocarbon ages younger than about 30,000 yr B.P. because ages of wood, sediment, and extracted humic materials from the same stratigraphic interval are in good agreement. We are less confident with stratigraphic or pedogenic horizons that are >30,000 yr B.P. because ages of wood, soil, and extracted humic materials are dissimilar and commonly are not in chronological order, especially for samples that contain small amounts of organic carbon. We have the least confidence with age estimates >50,000 yr B.P. determined by any method. For example, the age of loess indicated by thermoluminescence methods (TL) in Illinois agrees with the radiocarbon chronology, but TL ages beyond the radiocarbon limit are questionable because of anomalous fading (Canfield, 1985).

Electron spin resonance (ESR) has provided age estimates of Sangamonian vertebrate teeth and gar scales from Hopwood Farm (Blackwell and others, 1990). The ages depend on many factors, with the most critical being uranium uptake in the fossil and the dose rate from the surrounding sediment. The age estimate of a *Mammut americanum* (Mastodon) molar from the top of the fossiliferous sequence ranged from 73 ± 9 ka (early uptake) to 137 ± 17 ka (recent uptake), while a *Casteroides ohioensis* (giant beaver) incisor dates at > 92 ± 2 ka to < 137 ± 35 ka. These ages indicate that the Sangamonian Age correlates with Oxygen Isotope Stage 5 of the deep-sea record.

In order to better delimit the age of the Sangamonian-Wisconsinan boundary, we obtained radiocarbon ages on organic-rich sediment from a new excavation pit at the Athens Quarry (Pit 10; Table 1). We wished to verify ages determined from previous studies (Follmer and others, 1979; Follmer, 1983) and to determine the average accumulation rate of Robein Silt in order to extrapolate the age of its lower contact, the Sangamonian-Wisconsinan Stage boundary. The results are inconclusive, and

→

Figure 6. Radiocarbon ages, lithology, particle-size distribution, and medium to coarse silt ratios from a profile at Lomax, Illinois. Note the change in scale above and below datum. Bars indicate data from sampled intervals; dots are data from spot samples.

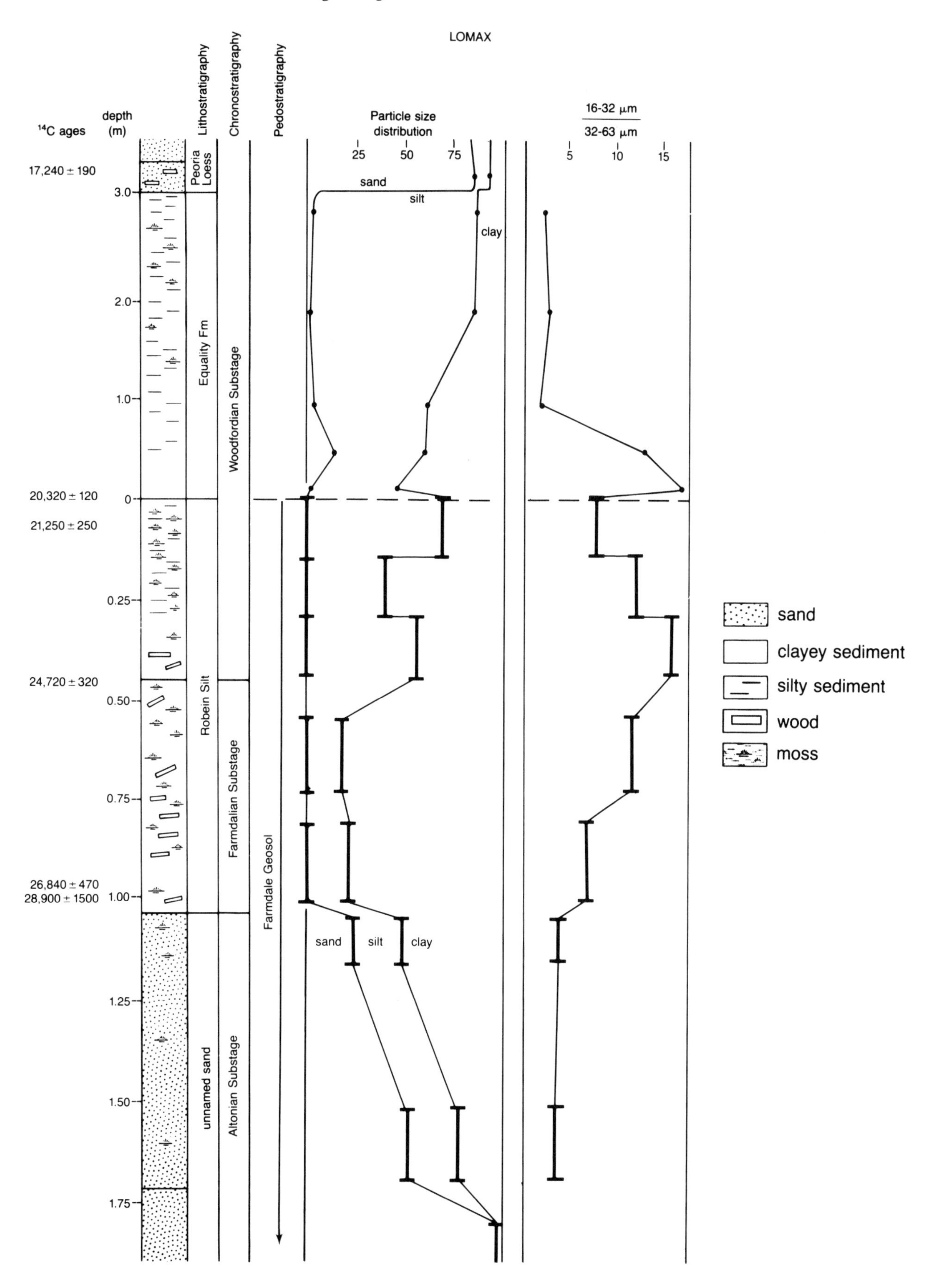
LOMAX
^{14}C ages
depth (m)
Lithostratigraphy
Chronostratigraphy
Pedostratigraphy
Particle size distribution
25
50
75
16-32 μm
32-63 μm
5
10
15
17,240 ± 190
3.0
2.0
1.0
0
20,320 ± 120
21,250 ± 250
0.25
24,720 ± 320
0.50
0.75
26,840 ± 470
28,900 ± 1500
1.00
1.25
1.50
1.75
Peoria Loess
Equality Fm
Robein Silt
unnamed sand
Woodfordian Substage
Farmdalian Substage
Altonian Substage
Farmdale Geosol
sand
silt
clay
sand
clayey sediment
silty sediment
wood
moss

TABLE 1. LIST OF RADIOCARBON AGES FROM ATHENS QUARRY, BIGGSVILLE, CLINTON, GARDENA, LOMAX, PITTSBURG BASIN, AND WEDRON QUARRY*

ISGS Number	Material	Stratigraphic Unit	Geosol	Depth (m)	Organic Carbon (%)	$\delta^{13}C$ (-‰)	Age† (^{14}C B.P.)
Pit 1			ATHENS QUARRY				
534	Wood	Peoria Loess	....	2.2 to 2.3	n.d.	n.d.	22,170 ± 450
536	Wood	Robein Silt	....	3.3 to 3.4	n.d.	n.d.	25,170 ± 200
Pit 3							
654	Whole soil	Robein Silt	Indian Pt.	6.0 to 6.2	1.2	n.d.	38,920 ± 1,100
684	Carbonized wood, seeds, organic clay	Berry Clay	Sangamon	6.4 to 6.5	n.d.	29.4	41,770 ± 1,100
688	Humic acid	Berry Clay	Sangamon	6.4 to 6.5	n.d.	28.8	35,560 ± 900
Pit 4							
870	Whole soil	Robein Silt	Indian Pt.	5.4 to 5.5	n.d.	28.8	35,750 ± 620
883	Whole soil	Robein Silt	Indian Pt.	5.7 to 5.8	n.d.	26.9	37,100 ± 1,200
Pit 10; profile 1							
1865	Wood	Peoria Loess	....	5.3	....	26.9	20,410 ± 280
1878	Whole soil	Robein Silt	Farmdale	5.6 to 5.7	13.9	27.0	27,770 ± 450
1879	Whole soil	Robein Silt	Farmdale	5.7 to 5.8	8.0	28.2	28,090 ± 470
1880	Whole soil	Robein Silt	Farmdale	5.8 to 5.9	7.8	28.1	28,240 ± 460
1861	Whole soil	Robein Silt	Farmdale	5.9 to 6.0	10.9	28.6	28,190 ± 460
1881	Whole soil	Robein Silt	Farmdale	6.0 to 6.1	8.6	28.6	29,620 ± 630
1897	Whole soil	Robein Silt	Farmdale	6.1 to 6.2	3.1	28.9	27,330 ± 290
1860	Whole soil	Robein Silt	Indian Pt.	6.2 to 6.3	2.9	28.9	32,800 ± 3,200
1889	Whole soil	Robein Silt	Indian Pt.	6.3 to 6.4	4.6	28.2	34,230 ± 710
1859	Whole soil	Robein Silt	Indian Pt.	6.4 to 6.5	8.8	28.6	43,800 ± 3,000
1890	Whole soil	Robein Silt	Indian Pt.	6.5 to 6.6	2.4	28.7	34,310 ± 640
Pit 10; profile 2							
1873	Whole soil	Robein Silt	Farmdale	5.7 to 5.8	13.2	28.1	29,220 ± 630
1883	Humic acid	Robein Silt	Farmdale	5.7 to 5.8	41.0	27.3	28,440 ± 310
1898	Whole soil	Robein Silt	Indian Pt.	5.9 to 6.0	6.2	27.6	36,160 ± 950
1905	Humic acid	Robein Silt	Indian Pt.	5.9 to 6.0	30.0	21.2	31,110 ± 450
1910	Whole soil	Robein Silt	Indian Pt.	6.3 to 6.4	3.4	28.2	36,930 ± 770
1904	Whole soil	Robein Silt	Indian Pt.	6.7 to 7.0	2.7	27.1	36,800 ± 1,200
			BIGGSVILLE				
1231	Soil	Robein Silt	Farmdale	4.42 to 4.54	....	29.4	21,410 ± 290
Beta-4129	Wood, soil	Robein Silt	Farmdale	6.52 to 6.47	....	....	27,870 ± 420
			CLINTON				
828	Moss	Morton Loess	....	....	....	30.2	20,670 ± 280
			GARDENA				
532	Moss	Morton Loess	....	....	....	n.d.	19,680 ± 460
531	Whole soil	Morton Loess	....	....	....	n.d.	25,370 ± 310
			LOMAX				
1832	Wood	Peoria Loess	....	3.0+	n.d.	28.6	17,240 ± 190
136	Wood	Equality Formation	....	0.0	n.d.	n.d.	20,320 ± 120
1730	Whole soil	Robein Silt	Farmdale	0 to 0.1	n.d.	29.0	21,250 ± 250
1720	Whole soil	Robein Silt	Farmdale	....	n.d.	28.6	24,720 ± 320
1635	Whole soil	Robein Silt	Farmdale	1.0 to 1.1	n.d.	28.9	26,840 ± 470
1637	Wood	Robein Silt	Farmdale	1.1 to 1.2	n.d.	25.0	28,900 ± 1,500
			PITTSBURG BASIN				
47	Gyttja	Lake Michigan	....	2.00	....	n.d.	21,370 ± 810
65	Gyttja	Lake Michigan	....	2.35	....	n.d.	24,200 ± 800
67	Gyttja	Lake Michigan	....	2.40	....	n.d.	34,000 ± 1,200
13	Gyttja	Lake Michigan	....	3.17 to 3.22	....	n.d.	>40,000
			WEDRON QUARRY				
1486	Wood fragments	Morton Loess	....	....	....	29.1	21,400 ± 470

*Athens Quarry (Follmer and others, 1979; Follmer, 1983; this report); Biggsville (Baker and others, 1989); Clinton, Gardena (Follmer and others, 1979); Lomaz, Pittsburg Basin (Grüger, 1972a, 1972b); and Wedron Quarry (Baker and Sullivan, 1986).
n.d. = not determined.
†Radiocarbon years before 1950.

illustrate some of the complexities in obtaining reliable radiocarbon ages on material >30,000 yr B.P.

We collected 12 continuous, 10 cm block samples along profile 10-1 from the base of the Peoria Loess down to the top of the Sangamon Geosol. These samples were pretreated by boiling in 2N HCl for 1.5 h to remove carbonate and acid-soluble fulvic acid. The age of the whole soil residue was determined. Along profile 10-2, about 10 m from profile 10-1, four whole soil samples were collected so that the humic substances could be extracted and dated (Table 1). The humic acid fraction was extracted from bulk samples by first treating with 0.1N HCl at a 1:10 ratio for about one week, washing in distilled water, and then adding a solution of 0.1N NaOH for 24 h (Stevenson, 1982). The dissolved humic acid was precipitated with HCl, centrifuged, and freeze dried.

Many of the radiocarbon ages of the bulk samples from both profiles in pit 10 are out of stratigraphic order. Moreover, they are inconsistent with previously obtained ages from similar horizons (Table 1, Fig. 7). Several studies have shown inconsistencies in radiocarbon ages from buried A horizons (e.g., Goh and others, 1977; Gilet-Blein and others, 1980; Kigoshi and others, 1980). Errors may occur during sample collection and laboratory preparation, and from root penetration from suprajacent horizons, pedoturbation, migration of humic fractions, or metabolism of humic substances by soil microbes (Mahaney and Boyer, 1986). A fundamental uncertainty is radiocarbon concentration of the paleoatmosphere, which is assumed to vary within about 20% of the modern, pre-atomic bomb atmosphere (Stuiver, 1978). However, the magnetic dipole moment of the Earth (Barbetti, 1980) may have been as much as 50% greater than that of 30,000 yr B.P. (Vogel, 1983).

Our samples include more reliable radiocarbon ages derived from material (samples) rich in organic carbon. For example, two samples at depths of 6.1–6.2 and 6.5–6.6 m from profile 10-1 are chronologically out of order; in both cases, the samples are organic carbon poor relative to suprajacent samples (Fig. 7A). In addition, the bulk soil and humic acid ages of the organic carbon-rich Farmdale Geosol are statistically indistinguishable, whereas the whole soil age of the top of the Indian Point Geosol complex is more than 3500 yr older than the humic-acid age. Follmer (1983) observed similar discrepancies in ages of a bulk sample and humic-acid extract from the top of the Sangamon Geosol (Table 1). The greater age of the whole soil sample suggests that humic acids, probably derived from the Farmdale Geosol, have migrated down through the profile and have contaminated the sub-Farmdale material. Therefore, we conclude that the mobile humic substances have affected the radiocarbon age of samples relatively poor in organic carbon.

In summary, we are suspicious of the accuracy of radiocarbon ages that exceed 30,000 yr B.P. from buried A-horizon material in colluviated loess. Inconsistencies in the ages of samples collected from adjacent horizons, as well as samples collected in stratigraphic order, make tenuous our goal of estimating the age of the Sangamonian-Wisconsinan boundary at pit 10-1 (Fig. 7).

FOSSILS

Introduction

The Pittsburg Basin (Fig. 1) contains sediment that probably spans the last two glacial/interglacial transitions. Although the sediment also contains molluscs, ostracodes, and diatoms, it has been studied only for pollen and plant macrofossils (Grüger, 1972a). Late Illinoian to Sangamonian fossil pollen, vertebrates (King and Saunders, 1986), ostracodes (Curry and Forester, 1991), diatoms, phytoliths, insects, molluscs, and vertebrates are preserved at Hopwood Farm (Fig. 1), but fossils are not preserved in sediment presumably deposited during the last interglacial-glacial transition. Pollen and plant macrofossil analyses from Oak Crest Bog (Meyers and King, 1985) and Voegelli Farm (Whittecar and Davis, 1982; Fig. 1) lack a Sangamonian (interglacial) component. All fossil-bearing sediments at these latter sites are of Altonian age and yield finite radiocarbon ages. Farmdalian and Woodfordian flora and fauna are preserved at a number of sites described below.

Invertebrates

Molluscan fauna, primarily pulmonate gastropods, are abundant and well represented in Sangamonian, Farmdalian and Woodfordian deposits (Leonard and Frye, 1960; Leonard and others, 1971; Willman and others, 1971). Staplin (1963a, 1963b) described freshwater ostracodes from Pleistocene sediment in Illinois. Studies are in progress on ostracodes and molluscs from long, lacustrine sections such as Hopwood Farm (Curry and Forester, 1991). Environmental interpretations based on paleohydrology are possible by characterizing biozones and chemistry of these fossils (DeDeckker and Forester, 1988).

Fossil insects with boreal affinities are described from Morton Loess at the Gardena and Clinton sections, from the lower Peoria Loess at the Athens Quarry (Carter, 1985; Morgan, 1987) and from Robein Silt at Biggsville (Carter, 1985; Schwert and Ashworth, 1988); the age of the sediment is ca. 28–23 ka (Table 1). The insect assemblage indicates a mean annual temperature of about 0 °C (Morgan, 1987). Carter (1985) found no evidence of climatic cooling at Biggsville or the Athens Quarry, although at Biggsville bryophyte macrofossils indicate a change from boreal to subarctic conditions at about 22.7 ka (Baker and others, 1989). At Wedron, Garry and others (1990) found plant and insect fossils with an age of 21,460 ± 470 yr B.P. (ISGS-1486) that are indicative of boreal and tundra environments. Other sites indicating stadial conditions include Gardena and Clinton, where Morton Loess is capped with moss that yielded ages of 19,680 ± 460 (ISGS-532) and 20,670 ± 280 yr B.P. (ISGS-828), respectively. The insects from this stratigraphic horizon suggest mean annual temperatures as low as –7 or –8 °C. (Morgan, 1987). Fossil insects are present also in the Robein Silt and Equality Formation at Lomax (Fig. 1) and in late Illinoian and Sangamonian lacustrine deposits at Hopwood Farm, but remain to be described.

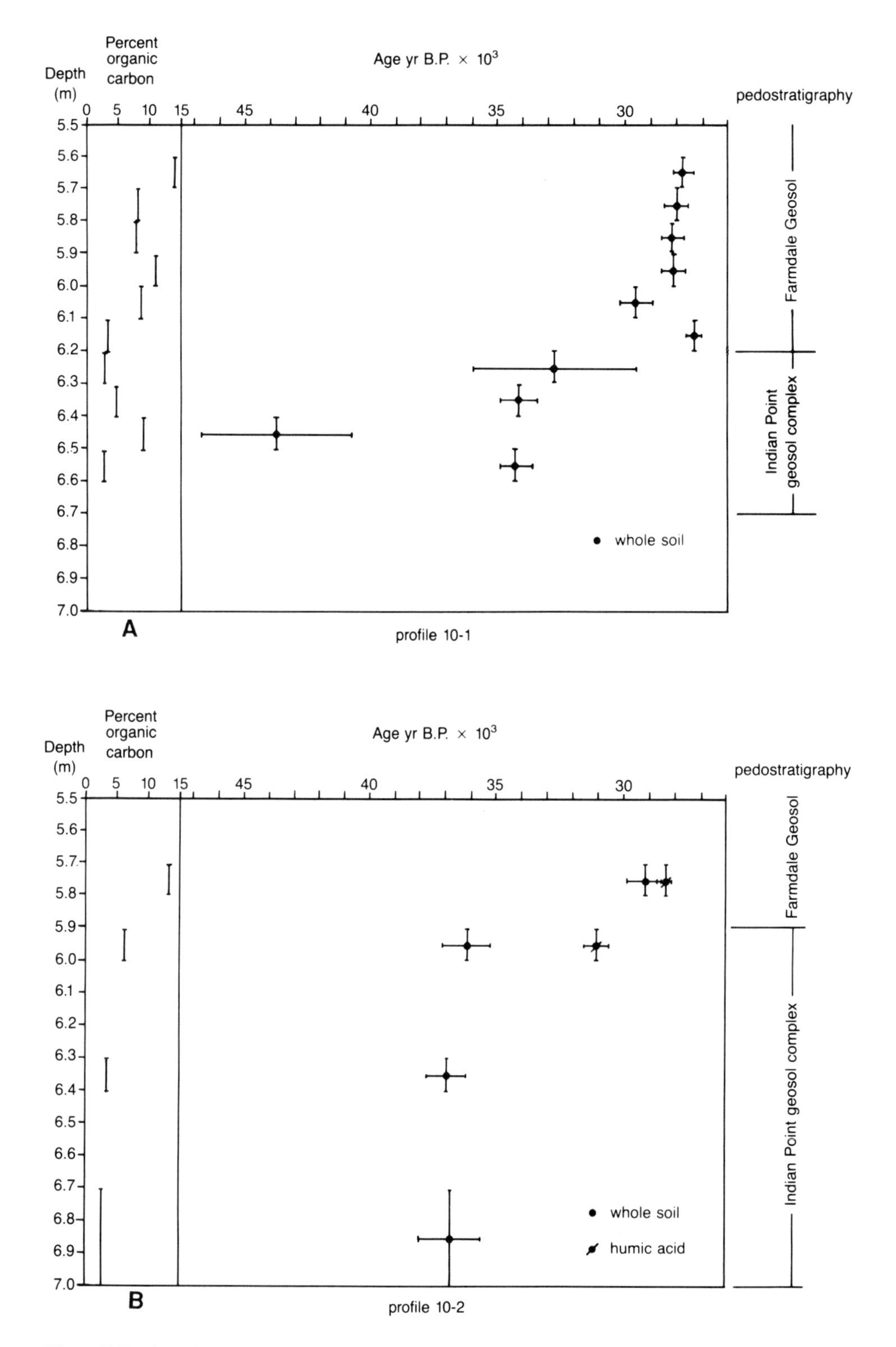

Figure 7. Radiocarbon age vs. depth along profiles 1 (A) and 2 (B) in Pit 10 at Athens Quarry, Illinois. The top of the Sangamon Geosol is a depth of 7 m along both profiles. Solid bars represent 68% confidence level for each age (see Table 1).

Vertebrates

Hopwood Farm contains a diverse late Illinoian-Sangamonian fauna, including giant tortoise (*Geochelone crassiscutata*), mastodon (*Mammut americanum*), giant beaver (*Castoroides ohioensis;* King and Saunders, 1986), short-nosed gar (*Lepisosteus platostomus*), and snapping turtle (*Chelydra serpetina;* Neal Woodman, 1988, personal commun.). *G. crassiscutata* did not burrow, and thus could not hibernate (Hibbard, 1960), suggesting that Sangamonian winters were much warmer than today. At Lomax, teeth of meadow vole (*Microtus pennsylvanicus*) are in Equality Formation strata deposited from 20,500 and 17,000 yr. B.P. (Russell Graham, 1989, personal commun.).

Pollen

Pollen from the Illinoian-Sangamonian transition is recorded in lacustrine and colluvial sediment at Hopwood Farm and the Pittsburg basin. In both cases, a basal, late Illinoian interval is interpreted from abundant spruce (*Picea;* as much as 50%) and pine (*Pinus;* about 50%; Fig. 8), indicating that the surrounding area was covered by a coniferous forest. A transition to interglacial conditions is indicated by pollen zones 2a, and H2a and H2b at the Pittsburg basin and Hopwood Farm, respectively. Pollen zone 2b at the Pittsburg basin and zones H3a and H3b at Hopwood Farm may be correlative because of the upward decline in the percentage of *Picea* and *Pinus* (Fig. 8). King and Saunders (1986) interpreted that the sediment of pollen zones H3a and H3b at Hopwood Farm was deposited during maximum interglacial conditions during the Sangamonian Age, inferred to have been hotter and dryer than anything evidenced in midwestern Holocene deposits. However, the nonarboreal pollen assemblages from pollen zones H3a and H3b are similar to upper Holocene assemblages from intermittent shallow ponds in northern Illinois and Iowa (Grimm, 1989). The assemblages are dominated by local taxa, and little regional information is gained by examining pollen from these environments. For example, Chenopodiaceae and Amnanthaceae prefer a substate of desiccated lacustrine or marshy sediments, and the predominance of these families in pollen zone H3a suggests unstable water levels at Hopwood Farm during the Sangamonian. *Cypridopsis vidua* is abundant in pollen zone H3a, suggesting an unstable lacustrine environment (Curry and Forester, 1991). Pollen, as well as other fossil remains, is absent above pollen zone H3a at Hopwood Farm; the sediment above this zone is composed largely of smectite-rich clay characteristic of Berry Clay (accretion gley; Frye and others, 1960; Willman and others, 1966).

The Sangamonian pollen record from the Pittsburg Basin contains two zones above transitional zone 2a. Pollen zone 2b is dominated by grass and *Ambrosia,* much like zone H3a at Hopwood Farm, but lacks abundant Chenopodiaceae and Amnanthaceae, suggesting stable lake levels. Pollen zone 2c is dominated by arboreal pollen, including taxa that do not occur together in modern deciduous forests; the species we have selectively used in Figure 8 (*Picea, Pinus, Betula* [birch], *Ambrosia,* Gramineae, Chenopodiaceae, and Amnanthaceae) are nearly absent (Grüger, 1972a).

At the Pittsburg Basin, Grüger (1972a) interpreted the transition from interglacial to glacial conditions in pollen zone 3. The contact between pollen zones 2c and 3a is interpreted as a change in vegetation from deciduous forests to prairie, including *Ambrosia,* Chenopodiaceae, Amnanthaceae, and Gramineae (Grüger, 1972a). The abruptness of the change in pollen spectra across zones 2 and 3 suggested to Grüger (1972a) that the contact is an unconformity. General cooling through zone 3 is suggested by the gradual increase of *Pinus* and decline of *Ambrosia,* whereas the abrupt rise of Chenopodiaceae and Amnanthaceae suggests locally exposed lake sediments.

In pollen zone 4, an increase in *Picea* suggests a cooling trend; more stable lake levels are indicated by the decline in *Ambrosia,* Chenopodiaceae, and Amnanthaceae. The age of pollen zone 4 is late Altonian, Farmdalian, and Woodfordian (Fig. 8; Grüger, 1972a, 1972b).

The pollen record in the lacustrine succession at the Pittsburg Basin is important because it is the only fossiliferous record unaffected by pedogenesis that apparently spans the Sangamonian, Wisconsinan, and Holocene. We cannot be certain, however, how Grüger's (1972a) pollen zones correspond to the chronostratigraphic boundaries. Grüger's core came from near the center of the Pittsburgh Basin, and further analysis of cores that transect the basin may reveal the pollen zones associated with the first Wisconsinan loess and any geosols. Webb and Webb (1988) believed, on the basis of statistical analyses, that the sediment accumulation rate in the Pittsburg Basin (just less than 10cm/1000 yr) suggests a nonconstant process of sediment accumulation. However, there is no physical evidence for desiccation, and the core contains abundant molluscs and ostracodes in pollen zones 2 and 3.

Voegelli Farm (Whittecar and Davis, 1982) and Oak Crest Bog (Meyers and King, 1985) are sites in northern Illinois that have yielded finite radiocarbon ages between ca. 47 and 24 ka. Without exception, the pollen spectra during this time are dominated by *Pinus* and *Picea,* with subordinate *Betula, Salix* (willow), and other taxa indicating forest or open woodland (Heusser and King, 1988). In contrast, pollen records from south of about 40° N, such as the Pittsburg basin (Grüger, 1972b), have lower percentages of *Picea,* and higher percentages of Gramineae, *Quercus* (oak), and other deciduous trees. Grüger (1972b) and Baher and others (1991) suggest that the prairie-forest boundary was within Illinois during the Altonian, but Heusser and King (1988) believed that only a coniferous forest was present and that the pollen of more thermophillous taxa are extraregional.

Farmdalian and lower Woodfordian sediments from Biggsville (Baker and others, 1989) and Lomax (Fig. 1) record the last periglacial-glacial transition. At both sites organic carbon–rich sediment is ~1.5 m thick, bounded by radiocarbon ages of 21,410 ± 290 (ISGS-1231) and 27,870 ± 420 (Beta-4129) and

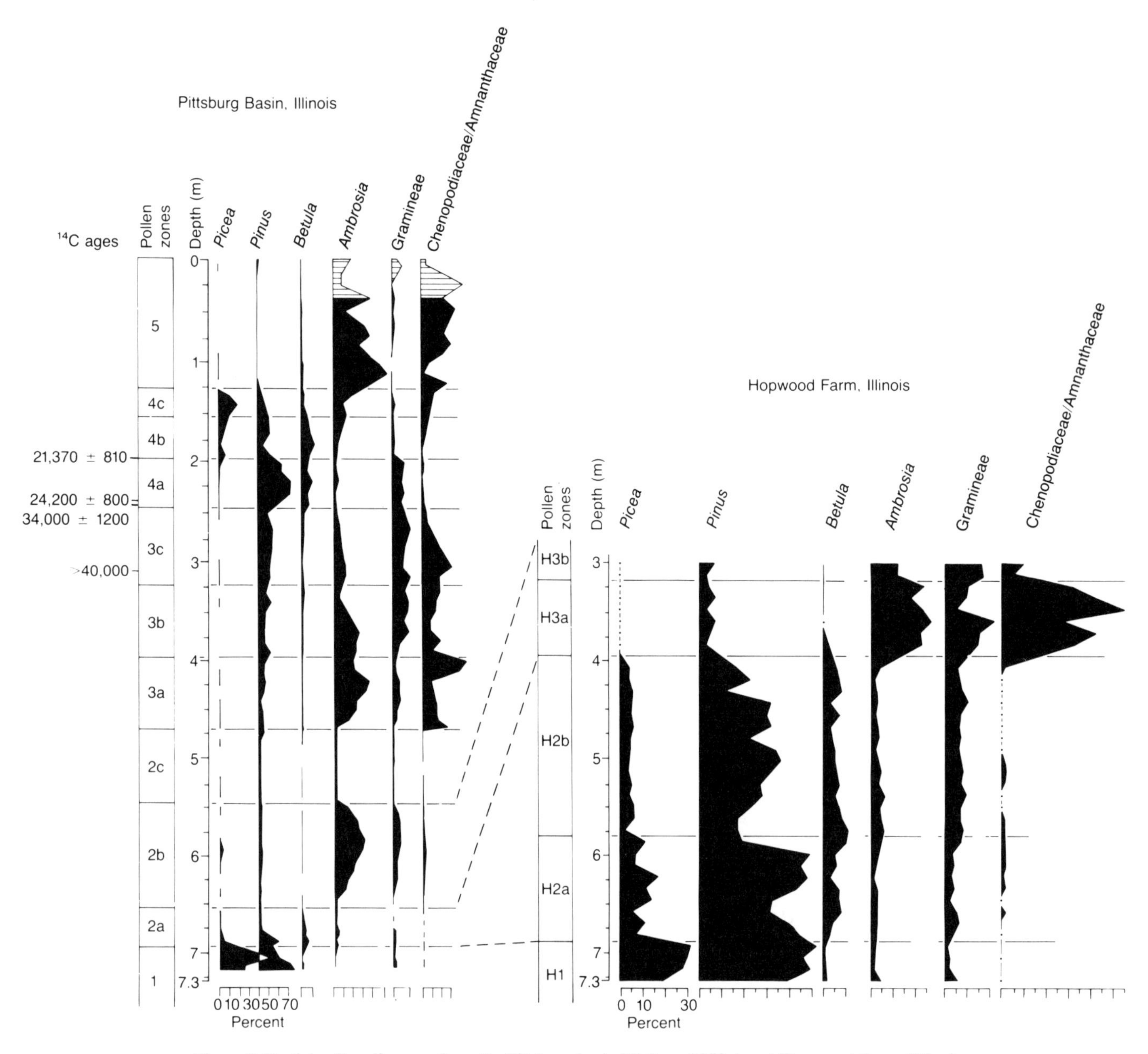

Figure 8. Partial pollen diagrams from the Pittsburg basin (Grüger, 1972a) and Hopwood Farm, Illinois (King and Saunders, 1986) comparing percentages of *Picea, Pinus, Betula, Ambrosia,* Graminae, and Chenopodiaceae/Amnanthaceae.

21,250 ± 250 (ISGS- 1730) and 28,900 ± 1500 yr B.P. (ISGS-1637; Table 1), respectively. Pollen in Robein silt from Biggsville indicates a cooling trend from cool temperate to subarctic conditions. From 27.9 to 26.6 ka, a *Picea-Pinus banksiana* (spruce-pine) forest surrounded the site, after which *Pinus* declined. Pollen influx indicates that *Picea* did not become more abundant regionally, even though it accounts for about 80% of the pollen spectra. At about 24.0 ka, the sediment at both sites became silty, suggesting incorporation of loess. Macrofossils and pollen from Biggsville indicate an open forest composed of *Picea mariana* (black spruce), *Larix laricina* (larch), and *Betula glandulosa* (shrub birch) analogous to modern boreal to subarctic sites. In sediment dated from about 22.7 to 21.4 ka, fossils of *Selaginella selaginoides* (spikemoss) indicate a subarctic environment similar to that extending north of the treeline in Canada (Baker and others, 1989).

At Wedron (Baker and Sullivan, 1986; Garry and others, 1990; Fig. 2), macrofossils of arctic plants *Dryas intergrifolia* (arctic avens) and *Vaccinium uliginosum* var. *alpinum* (arctic blueberry), and subarctic *Selaginella selaginoides* and *Betula glandulosa* are described from proglacial lacustrine sediment that yielded an age of 21,400 ± 470 yr B.P. (ISGS-1486). The arctic and subartic pollen composition and macroflora present from 22.7 to 21.4 ka at Biggsville and Wedron are rare or absent in all other post-Sangamonian pollen records from Illinois. This negative evidence indicates a nonglacial climate in Illinois during the Altonian.

SUMMARY AND DISCUSSION

We interpret climate on the basis of pollen, sediment, and soil (Table 2) because there are no isotopic records or other means to estimate average temperatures or moisture. An interglacial is defined here as a period of time when landscape stability prevailed, and colluviation and alluviation were the dominant sedimentological processes. Very poorly drained interglacial soils are accretion gley (cumulic aquic Mollisols), whereas better-drained soils are deeply leached Alfisols or Ultisols with significant mineral alteration in the solum (Willman and others, 1966; Follmer, 1983). Interglacial pollen assemblages include abundant deciduous trees, Gramineae, and *Ambrosia*.

Maximum interglacial conditions occurred at about 123 ka on the basis of correlation to the orbitally tuned chronostratigraphy of deep-sea sediments of Martinson and others (1987). Pollen assembages from the Pittsburg basin suggest general cooling from what are interpreted as a Sangamonian to Wisconsinan flora. It is not certain if the floristic change of interglacial vegetation (deciduous trees, grasses, *Ambrosia*) to periglacial vegetation (grasses and coniferous trees) at the Pittsburg Basin directly corresponds to a Sangamonian-Wisconsinan contact, based on the lithic criteria of initiation of loessial inputs to the lacustrine sediment. At the Athens Quarry, the type locality of the Sangamonian Stage and Sangamon Geosol, the age of the boundary is about 50 ka, which coincides with an age estimate based on the accumulation rate of Roxana Silt in the lower Illinois River Valley (McKay, 1979b) and in southwestern Wisconsinan (Leigh, 1991).

The late Altonian, Farmdalian, and early Woodfordian Subage of the Wisconsinan age were affected by periglacial environments from about 50 to 23–25 ka. The deposits of this age are characterized by loess and the vegetation was dominated by coniferous trees. The climate and sedimentation during this time were probably strongly influenced by stadial conditions, although the location of the glacier is unknown (Winters and others, 1988; Curry, 1989; Johnson and Follmer, 1989). Altonian and Farmdalian soils are relatively thin and contain well-preserved organic-rich horizons in poorly drained environments (hemic and sapric Histosols). In relatively well drained environments, the soils contain little pedogenic clay and the minerals have not been significantly altered, but the depth of leaching may be more than 2 m (Cryochrepts). Commonly, there is evidence for permafrost such as cryoturbated soil horizons and ice-wedge casts (Johnson, 1990).

A glacial is a period of time when landscapes are substantially modified by the direct or proximal influence of glacial ice. Sediments include loess, all types of till, and lacustrine and fluvial sediment deposited in ice-dammed lakes, supraglacial lakes, and along meltwater streams. Oxidized glacial soils are thin zones with root tubules; they commonly have platy structure, silt coatings on peds, and cryoturbation features (Cryorthents). Very poorly drained glacial soils are characterized by degraded humus

TABLE 2. SUMMARY OF THE CHRONOSTRATIGRAPHY, SOILS, AND FOSSILS OF THE LAST INTERGLACIAL TO GLACIAL TRANSITION IN ILLINOIS

Chronostratigraphy		Soils			Fossils		Climatic Interpretation
Stage	Substage	Well-Drained	Very Poorly Drained	Pedostratigraphy	Plant Microfossils, Pollen and Bryophytes	Vertebrates	
Wisconsinan	Woodfordian	Cryorthents	Cryaquents		*Selaginella selaginoides*, *Dryas intergrifolia*	....	Glacial
	Farmdalian	Ochrepts	Histosols	Farmdale Geosol	*Pinus, Picea*	....	Periglacial
	Altonian						
	upper	Cryorthents	Cryaquents	Indian Pt. geosol complex	*Pinus, Picea*	....	Periglacial
	lower	(superposed profiles)		Chapin Geosol	Graminaea(?)	....	Periglacial
Sangamonian		Alfisols and Mollisols		Sangamon Geosol	Deciduous trees, Graminaea, *Ambrosia*	*Geochelone crassiscutata*	Interglacial
Illinoian	Jubileean	....	....	....	*Pinus, Picea*	....	Periglacial

layers and weak expression of O/A/Bg horizon sequences. Glacial vegetation includes evidence of taxa that occur along or north of the treeline in modern Canada, including spores of spikemoss, and macrofossils of shrub birch, arctic avens, and arctic blueberry (Baker and Sullivan, 1986; Baker and others, 1989; Garry and others, 1990).

The transition from periglacial to glacial vegetation in Illinois is detailed only at Biggsville (Baker and others, 1989). Here earliest late Wisconsinan loess was deposited under periglacial conditons at about 24 ka, when black spruce, larch, and shrub birch formed an open forest. The transition occurred at 22.7 ka when spikemoss, an indicator of subarctic climate, became dominant. At the Athens Quarry, Peoria Loess was deposited from about 25 to 22 ka on the floor of a boreal forest, but younger plant or moss remains indicative of subarctic conditions are lacking (King, 1979).

In summary, the last interglacial-glacial transition in Illinois has two parts; the first is a transition from interglacial to periglacial conditions and then a transition from periglacial to glacial environments. Interglacial conditions are determined by mature soils and pollen records in lacustrine sediments; periglacial to glacial conditions are determined by immature, cryoturbated soils fossils, loess, and till associated with the maximum extent of Woodfordian glaciation (Table 2). The age of the last interglacial maximum in Illinois, about 123 ka, is correlated from the orbitally tuned chronostratigraphy of deep-sea sediments (Martinson and others, 1987). Electron spin resonance ages or Sangomonian vertebrate teeth and gar scales support this general age assignment. Between 75 and 50 ka, the conditions shifted from interglacial to periglacial when the first Wisconsinan loess, the Roxana Silt, was deposited on Berry Clay and other sediment affected by Sangamonian pedogenesis. The beginning and duration of this climatic shift is poorly known. No early Altonian (75 to 50 ka) sediment record has been found in Illinois that resolves the question of when the interglacial-periglacial transition occurred. Late Altonian and Farmdalian periglacial conditions were associated with loessial sedimentation and boreal forests. Glacial conditions first affected Illinois at 42°N about 25 ka and extended south to about 41°N by 23 ka. Pollen and plant and bryophyte macrofossils that today are found north of the treeline in Canada are the most diagnostic indicators of true glacial conditions in Illinois during the Wisconsinan other than obvious glacial deposits such as till. Loess deposited under glacial conditions contain little organic matter, and there is sparse evidence for pedogenesis except for irregular oxidized zones and few root channels.

ACKNOWLEDGMENTS

We thank W. H. Johnson, A. K. Hansel, P. U. Clark, and P. Lea for their helpful discussions and critical reviews of the mansucript. Humic-acid extractions were done under the direction of F. J. Stevenson of the Department of Agronomy at the University of Illinois at Urbana-Champaign. Radiocarbon ages were determined by C. L. Liu at the Radiocarbon Laboratory at the Illinois State Geological Survey.

REFERENCES CITED

Baker, R. G., and Sullivan, A. E., 1986, Paleobotanical records from Biggsville and Wedron; new information *in* Quaternary records of central and northern Illinois: Illinois State Geological Survey Guidebook 20, p. 68–69.

Baker, R. G., Sullivan, A. E., Hallberg, G. R., and Horton, D. G., 1989, Vegetational changes in western Illinois during the onset of late Wisconsinan glaciation: Ecology, v. 70, p. 1363–1376.

Baker, R. G., Schwert, D. P., Bettis, E. A. III, Kemmis, T. J., Horton, D. G., and Semken, H. A., 1991, Mid-Wisconsinan stratigraphy and paleoenvironments at the St. Charles site in south-central Iowa: Geological Society of America Bulletin, v. 103, p. 210–220.

Barbetti, M., 1980, Geomagnetic strength over the last 50,000 years and changes in atmospheric concentration: Emerging trends: Radiocarbon, v. 22, p. 192–199.

Blackwell, B. B., Schwarcz, H. P., Saunders, J. J., Woodman, N., and Curry, B. B., 1990, Dating the Sangamon: Electron Spin Resonance (ESR) dating mammal teeth and gar scales from Hopwood Farm, Montgomery County Illinois: Geological Society of America Abstracts with Programs, v. 22, no. 7, p. A85.

Canfield, H. E., 1985, Thermoluminescence dating and the chronology of loess deposition in the central United States [M.S. thesis]: Madison, University of Wisconsin, 159 p.

Carter, K. D., 1985, Middle and late Wisconsinan (Pleistocene) insect assemblages from Illinois [M.S. thesis]: Fargo, University of North Dakota, 124 p.

Clark, P. U., and Lea, P., 1986, Reappraisal of early Wisconsin glaciation in North America: Geological Society of America Abstracts with Programs, v. 18, no. 6, p. 565.

Coleman, D. D., 1974, Illinois State Geological Survey Radiocarbon Dates V: Radiocarbon, v. 16, p. 105–117.

Curry, B. B., 1989, Absence of Altonian glaciation in Illinois: Quaternary Research, v. 31, p. 1–13.

Curry, B. B., 1990, Fossiliferous Farmdalian and Woodfordian sediments at Lomax, Illinois, *in* Hammer, W., and Hess, D. F., eds., Geology field guidebook (Geological Society of America, North-Central section meeting): Macomb, Western Illinois University, p. F12–F17.

Curry, B. B. and Forester, R. M., 1991, Paleoenvironments and lithology of the Hopwood Farm site, Montgomery County, Illinois: Geological Society of America Abstracts with Programs, v. 23, no. 3, p. 9.

Curry, B. B., and Kempton, J. P., 1985, Reinterpretation of the Robein and Plano Silts: Geological Society of America Abstracts with Programs, v. 18, no. 6, p. 557.

DeDeckker, P., and Forester, R. M., 1988, The use of ostracodes to reconstruct continental paleoenvironmental records, *in* DeDeckker, P., Colin, J., and Peypouquet, J., eds., Ostracoda in the earth sciences: New York, Elsevier, p. 175–199.

Dreimanis, A., and Goldthwait, R. P., 1973, Wisconsin glaciation in the Huron, Erie and Ontario lobes, *in* Black, R. F., Goldthwait, R. P., and Willman, H. B., eds., The Wisconsinan stage: Geological Society of America Memoir 136, p. 71–106.

Follmer, L. R., 1978, The Sangamon Soil in its type area—A review, *in* Mahaney, W. C., ed., Quaternary soils: Norwich, England, Geo Abstracts, p. 125–165.

——, 1979, A historical review of the Sangamon Soil, *in* Wisconsinan, Sangamonian and Illinoian stratigraphy in central Illinois: Illinois State Geological Survey Guidebook 13, p. 79–91.

——, 1982, The geomorphology of the Sangamon surface: Its spatial and temporal attributes, *in* Thorn, C., ed., Space and time in geomorphology: Binghamton Symposia in Geomorphology, International Series No. 12: London, Allen & Unwin, p. 117–146.

——, 1983, Sangamonian and Wisconsinan pedogenesis in the midwestern United States, *in* Porter, S. C., ed., Late Quaternary environments of the United States, Volume 1, The late Pleistocene: Minneapolis, University of Minnesota Press, p. 138–144.

Follmer, L. R., and Kempton, J. P., 1985, A review of the Esmond Till Member, *in* Illinoian and Wisconsinan stratigraphy and environments in northern Illinois; the Altonian revised: Illinois State Geological Survey Guidebook 19, p. 139–155.

Follmer, L. R., and McKay, E. D., III, 1987, Farm Creek, central Illinois: A notable Pleistocene section, *in* Biggs, D. L., ed., North-Central Section of the Geological Society of America: Boulder, Colorado, Geological Society of America Centennial Field Guide Vol. 3, p. 231–236.

Follmer, L. R., McKay, E. D., Lineback, J. A., and Gross, D. L., 1979, Wisconsinan, Sangamonian and Illinoian stratigraphy in central Illinois: Illinois State Geological Survey Guidebook 13, p. 1–68.

Frye, J. C., Willman, H. B., and Glass, H. D., 1960, Gumbotil, accretion gley and the weathering profile: Illinois State Geological Survey Circular 295, p. 39.

Frye, J. C., Follmer, L. R., Glass, H. D., Masters, J. M., and Willman, H. B., 1974, Earliest Wisconsinan sediments and soils: Illinois State Geological Survey Circular 485, p. 12.

Fullerton, D. S., 1986, Stratigraphy and correlation of glacial deposits from Indiana to New York and New Jerrsey: Quaternary Science Reviews, v. 5, p. 23–38.

Garry, C. E., Schwert, D. P., Baker, R. G., Kemmis, T. J., Horton, D. G., and Sullivan, A. E., 1990, Plant and insect remains from the Wisconsinan interstadial/stadial transition at Wedron, north-central Illinois: Quaternary Research, v. 33, p. 387–399.

Gilet-Blein, N., Marien, G., and Evin, J., 1980, Unreliability of C-14 dates from organic matter of soils: Radiocarbon, v. 22, p. 919–929.

Glass, H. D., Frye, J. C., and Willman, H. B., 1964, Record of Mississippi river diversion in the Morton Loess of Illinois: Illinois Academy of Science Transactions, v. 57, p. 24–27.

Goh, K. M., Mollow, B.P.J., and Rafter, T. A., 1977, Radiocarbon dating of Quaternary loess deposits, Bank's Peninsula, Canterbury, New Zealand: Quaternary Research, v. 7, p. 177–196.

Gooding, A. M., 1975, The Sidney interstadial and late Wisconsinan history in Indiana and Ohio: American Journal of Science, v. 275, p. 993–1011.

Grimm, E. C., 1989, Palynological and plant-macrofossil studies of the Tonica thermokarst, La Salle County, Illinois: Geological Society of America Abstracts with Programs, v. 21, no. 4, p. 13.

Grüger, E., 1972a, Late Quaternary vegetation development in south-central Illinois: Quaternary Research, v. 2, p. 217–231.

—— , 1972b, Pollen and seed studies of Wisconsinan vegetation in Illinois: Geological Society of America Bulletin, v. 83, p. 2715–2734.

Hajic, E. R., Johnson, W. H., and Follmen, L. R., 1991, Quaternary deposits and landforms, confluence region of the Mississippi, Missouri and Illinois rivers, Missouri and Illinois: 38th Field Conference Midwest Friends of the Pleistocene, University of Illinois, Urbana, 106 p.

Hibbard, C. W., 1960, An interpretation of Pliocene and Pleistocene climates in North America: President's Address: Papers of the Michigan Academy of Science, Arts and Letters, 62nd Annual Report, p. 5–25.

Huesser, L. E., and King, J. E., 1988, North America, with special emphasis on the development of the Pacific coastal forest and prairie/forest boundary prior to the last glacial maximum, *in* Huntley, B., and Webb, T., III, eds., Vegetational history: Dordecht, The Netherlands, Kluwer Academic Publishers, p. 193–236.

Johnson, W. H., 1986, Stratigraphy and correlation of the glacial deposits of the Lake Michigan Lobe prior to 14 ka B.P.: Quaternary Science Reviews, v. 5, p. 17–22.

—— , 1990, Ice-wedge casts and relict patterned ground in central Illinois and their environmental significance: Quaternary Research, v. 33, p. 51–72.

Johnson, W. H., and Follmer, L. R., 1989, Source and origin of Roxana Silt and middle Wisconsinan glacial activity: Quaternary Research, v. 31, p. 319–331.

Johnson, W. H., and Hansel, A. K., 1987, Fluctuations of the late Wisconsinan (Woodfordian) Lake Michigan Lobe in Illinois, U.S.A.: Programme and Abstracts, International Union for Quaternary Research, XIIth International Congress, Ottawa, Canada, p. 195.

Johnson, W. H., Follmer, L. R., Gross, D. L., and Jacobs, A. M., 1972, Pleistocene stratigraphy of east-central Illinois: Illinois State Geological Survey Guidebook 9, p. 97.

Kempton, J. P., and Gross, D. L., 1971, Rate of advance of the Woodfordian (late Wisconsinan) glacial margin in Illinois: Stratigraphic and radiocarbon evidence: Geological Society of America Bulletin, v. 82, p. 3245–3250.

Kempton, J. P., Berg, R. C., and Follmer, L. R., 1985, Revision of stratigraphy and nomenclature of glacial deposits in central northern Illinois: Illinois State Geological Survey Guidebook 19, p. 1–19.

Kigoshi, K., Suzuki, N., and Shiraki, M., 1980, Soil dating by fractional extraction of humic acid: Radiocarbon, v. 22, p. 853–857.

King, J. E., 1979, Pollen analysis of some Farmdalian and Woodfordian deposits, central Illinois: Illinois State Geological Survey Guidebook 13, p. 109–113.

King, J. E., and Saunders, J. J., 1986, *Geochelone* in Illinois and the Illinois-Sangamonian vegetation of the type region: Quaternary Research, v. 25, p. 89–99.

Leigh, D. S., 1991, Origin and Paleoenvironment of upper Mississippi Valley Roxana Silt [Ph.D. dissertation] University of Wisconsin, Madison, 186 p.

Leonard, A. B., and Frye, J. C., 1960, Wisconsinan molluscan faunas of the Illinois Valley region: Illinois State Geological Survey Circular 304, p. 32.

Leonard, A. B., Frye, J. C., and Johnson, W. H., 1971, Illinoian and Kansan molluscan faunas in Illinois: Illinois State Geological Survey Circular 461, p. 24.

Mahaney, W. C., and Boyer, M. G., 1986, Microflora distribution in paleosols: A method for calculating the validity of radiocarbon-dated surfaces: Soil Science, v. 142, p. 100–107.

Martinson, D. G., Pisias, N. G., Hays, J. D., Imbrie, J., Moore, T. C., Jr., and Shackleton, N. J., 1987, Age dating and the orbital theory of the ice ages: Development of a high-resolution 0 to 300,000-year chronostratigraphy: Quaternary Research, v. 27, p. 1–29.

McKay, E. D., 1979a, Wisconsinan loess stratigraphy of Illinois, *in* Wisconsinan, Sangamonian and Illinois stratigraphy in central Illinois: Illinois State Geological Survey Guidebook 13, p. 95–108.

—— , 1979b, Stratigraphy of Wisconsinan and older loesses in southwestern Illinois, *in* 43rd Annual Tri-state Geological Field Conference: Illinois State Geological Survey Guidebook 14, p. 37–67.

Meyers, R. L., and King, J. E., 1985, Wisconsinan interstadial vegetation of northern Illinois, *in* Illinoian and Wisconsinan stratigraphy and environments in north Illinois; the Altonian revised: Illinois State Geological Survey Guidebook 19, p. 75–86.

Miller, B. J., Schumacher, B. A., Lewis, G. C., Rehage, J. A., and Spicer, B. E., 1988, Basal mixing zones in loesses of Louisiana and Idaho: II. Formation, spatial distribution and stratigraphic implications: Soil Science Society of America Journal, v. 52, p. 759–764.

Morgan, A. V., 1987, Late Wisconsin and early Holocene paleoenvironments of east-central North America, *in* Ruddiman, W. F., and Wright, H. E., eds., North America and adjacent oceans during the last deglaciation: Boulder, Colorado, Geological Society of America, The Geology of North America, v. K-3, p. 353–370.

North American Commision on Stratigraphic Nomenclature, 1983, North American stratigraphic code: The American Association of Petroleum Geologists Bulletin, v. 67, p. 841–875.

Ruhe, R. V., 1969, Quaternary Landscapes of Iowa: Ames, Iowa State University Press, 255 p.

—— , 1976, Stratigraphy of mid-continent loess, U.S.A., *in* Mahaney, W. C., ed., Quaternary stratigraphy of North America: Strousburg, Pennsylvania, Dowden, Hutchinson, and Ross, Inc., p. 197–211.

Schumacher, B. A., Lewis, G. C., Miller, B. J., and Day, W. J., 1988, Basal mixing zones in loesses of Louisiana and Idaho: I. Identification and characterization: Soil Science Society of America Journal, v. 52, p. 753–758.

Schwert, D. P., and Ashworth, A. C., 1988, Late Quaternary history of the northern beetle fauna of North America: A synthesis of fossil and distributional evidence: Entomological Society of Canada Memoir 141, p. 93–107.

Staplin, F. L., 1963a, Pleistocene Ostracoda of Illinois. Part I. Subfamilies Candoninae, Cyprinae, general ecology, morphology: Journal of Paleontology, v. 37, p. 758–797.

——, 1963b, Pleistocene Ostracoda of Illinois. Part II. Subfamilies Cycloprinae, Cypridopinae, Hyocyprinae; Familes Darwinulidae and Cytheridae. Stratigraphic ranges and assemblage patterns: Journal of Paleontology, v. 37, p. 1164–1203.

Stevenson, F. J., 1982, Humus Chemistry: New York, Wiley, 443 p.

Stuiver, M., 1978, Radiocarbon timescale tested against magnetic and other dating methods: Nature, v. 273, p. 271–274.

Thorp, J., Johnson, W. M., and Reed, E. C., 1951, Some post-Pliocene buried soils of the central United States: Journal of Soil Science, v. 2, p. 1–19.

Vogel, J. C., 1983, C-14 variation during the upper Pleistocene: Radiocarbon, v. 25, p. 213–218.

Webb, R. S., and Webb, T., 1988, Rates of sediment accumulation in pollen cores from small lakes and rivers of eastern North America: Quaternary Research, v. 30, p. 284–297.

Whittecar, G. R., and Davis, A. M., 1982, Sedimentology and palynology of middle Wisconsinan deposits in the Pecatonica River Valley, Wisconsin and Illinois: Quaternary Research, v. 17, p. 228–240.

Willman, H. B., and Frye, J. C., 1970, Pleistocene stratigraphy of Illinois: Illinois State Geological Survey Bulletin 94, p. 204.

Willman, H. B., Glass, H. D., and Frye, J. C., 1963, Mineralogy of glacial tills and their weathering profiles in Illinois, Part I: Glacial tills: Illinois State Geological Survey Circular 347, p. 55.

Willman, H. B., Glass, H. D., and Frye, J. C., 1966, Mineralogy of glacial tills and their weathering profiles in Illinois, Part II: Weathering profiles: Illinois State Geological Survey Bulletin 66, p. 76.

Willman, H. B., Leonard, A. B., and Frye, J. C., 1971, Farmdalian lake deposits and faunas in northern Illinois: Illinois State Geological Survey Circular 467, p. 12.

Winters, H. A., Alford, J. J., and Rieck, R. L., 1988, The anomalous Roxana Silt and mid-Wisconsinan events in and near southern Michigan: Quaternary Research, v. 29, p. 25–35.

MANUSCRIPT ACCEPTED BY THE SOCIETY SEPTEMBER 6, 1991

Printed in U.S.A.

Geological Society of America
Special Paper 270
1992

Ages of the Whitewater and Fairhaven tills in southwestern Ohio and southeastern Indiana

Barry B. Miller
Department of Geology, Kent State University, Kent, Ohio 44242
William D. McCoy
Department of Geology and Geography, University of Massachusetts, Amherst, Massachusetts 01003
William J. Wayne
Department of Geology, University of Nebraska, Lincoln, Nebraska 68588
C. Scott Brockman
Division of Geological Survey, Ohio Department of Natural Resources, Columbus, Ohio 43224

ABSTRACT

Alloisoleucine/isoleucine (aIle/Ile) ratios obtained from fossil mollusc shells collected at localities in southwestern Ohio and southeastern Indiana, where they occur in silt beds associated with the Whitewater and Fairhaven tills, indicate a pre-Wisconsinan age for these tills, which had previously been thought to be early or middle Wisconsinan.

The aIle/Ile ratios in shells from beneath the buried soil (Sidney soil) and till exposed near Sidney, Ohio, are most similar to values in shells obtained from Illinoian sediments at Clough Creek in Hamilton County, Ohio; Mechanicsburg southwest, Illinois; and Trousdale Mine in Vermillion Co., Indiana. The first well-developed weathering profile in the sequence above the implied Illinoian age silt at the Sidney cut, therefore, probably represents Sangamonian, early and middle Wisconsinan weathering. Molluscs from an organic silt, exposed near the base of the Bantas Fork cutbank section, also have aIle/Ile ratios that are similar to those measured in shell recovered from the silt at the Sidney cut and from the silt inclusion in inferred Illinoian till at Clough Creek. These data indicate that the organic silt is pre-Wisconsinan. Therefore, the Fairhaven Till, which overlies the silt at the Bantas Fork locality, could be pre-Wisconsinan and the weathering profile developed in the Fairhaven Till may be correlative with the Sangamon Soil of Illinois.

The New Paris Interstade silt overlies Whitewater Till at the American Aggregates quarry at Richmond, Indiana. Shells from the silt have aIle/Ile ratios that are intermediate between those obtained from inferred Illinoian age sediments at Bantas Fork, Sidney cut, and Clough Creek, and magnetically reversed sediments at Handley Farm, near Connersville, Fayette County, Indiana. These data suggest a pre-Illinoian age for the silt unit and the underlying Whitewater Till.

INTRODUCTION

The extent and indeed the very existence of an early Wisconsinan ice advance into southwestern Ohio and southeastern Indiana is still the subject of debate after decades of study (Clark and Lea, 1986). Recent reviews of late Quaternary till stratigraphy along the southern margin of the Laurentide Ice Sheet (Fullerton, 1986; Vincent and Prest, 1987) accept an early Wisconsinan age for the Whitewater and Fairhaven tills on the basis of the position of these tills above a soil or beds interpreted to be Sangamonian in age, and below nonglacial sediments containing wood that yields nonfinite radiocarbon ages (Fig. 1).

Miller, B. B., McCoy, W. D., Wayne, W. J., and Brockman, C. S., 1992, Ages of the Whitewater and Fairhaven tills in southwestern Ohio and southeastern Indiana, *in* Clark, P. U., and Lea, P. D., eds., The Last Interglacial-Glacial Transition in North America: Boulder, Colorado, Geological Society of America Special Paper 270.

REGIONAL CHRONOSTRATIGRAPHY	LOCAL PLEISTOCENE LITHOSTRATIGRAPHIC UNITS
WISCONSINAN STAGE	Knightstown Till Crawfordsville Till Shelbyville Till CONNERSVILLE INTERSTADE SILTS Fayette Till SIDNEY INTERSTADE SILTS Fairhaven Till NEW PARIS INTERSTADE SILTS Whitewater Till
SANGAMON STAGE	SANGAMON WEATHERING
ILLINOIAN STAGE	Richmond Till ABINGTON INTERSTADE SILTS Centerville Till

Figure 1. Relation of Pleistocene lithostratigraphic units in southwestern Ohio and southeastern Indiana to regional chronostratigraphic framework (modified from Goldthwait and others, 1981).

The sequence of sediments exposed at the Sidney cut, Bantas Fork, and American Aggregates quarry have played an important role in the development of the concept of early Wisconsinan ice deposits in this area. New data obtained from amino-acid epimerization studies of molluscan shell collected from these localities are more compatible with a pre-Wisconsinan placement for these deposits. In this chapter we compare amino-acid data from these sites with values obtained from late Wisconsinan, Illinoian, and pre-Illinoian age sediments within the area (Fig. 2) and discuss the significance of these findings with respect to the previous early-Wisconsinan age assigned to these deposits.

LITHOSTRATIGRAPHY

Whitewater Till (American Aggregates Quarry)

The Whitewater Till has been considered to be the basal unit of the Wisconsinan sequence in this area based upon its position (at some localities) above a weathering profile interpreted to be correlative with the Sangamon soil of Illinois (Gooding, 1963, 1975), and beneath nonglacial sediments that yield nonfinite radiocarbon ages (Goldthwait and others, 1981; Fullerton, 1986; Vincent and Prest, 1987). The Whitewater Till is correlated throughout the Whitewater basin on the basis of its blue-gray color, northwest-trending fabric, and the presence of pink to red-brown till inclusions. The type section for the Whitewater Till is the American Aggregates quarry in Richmond, Indiana, where it rests upon a till with oxidized joints which are interpreted to be the eroded remnant of a formerly more extensive Sangamonian weathering profile (Gooding, 1963). Gooding (1963) introduced the name "New Paris Interstade" for the nonglacial organic, calcareous, fossiliferous sediments above the Whitewater Till, and designated the American Aggregates quarry as the type section for the interstadial deposits. Wood samples from the New Paris Interstade unit at that locality have consistently yielded nonfinite ^{14}C ages >40,000 yr B.P. (Table 1; Fullerton, 1986). The molluscs used in this study are from the New Paris organic silt unit directly above the Whitewater Till at their type locality (Fig. 3).

Fairhaven Till (Bantas Fork)

The Fairhaven Till at its type section is exposed in a stream cutbank about 3.2 km (2 mi) north of Fairhaven, Preble County, Ohio, where it rests upon a discontinuous sand and the Whitewater Till, which has yielded a >45,160 yr B.P. (ISGS—590) age. The Sidney weathering profile has been identified at the top of the Fairhaven Till at this section (Goldthwait and others, 1981).

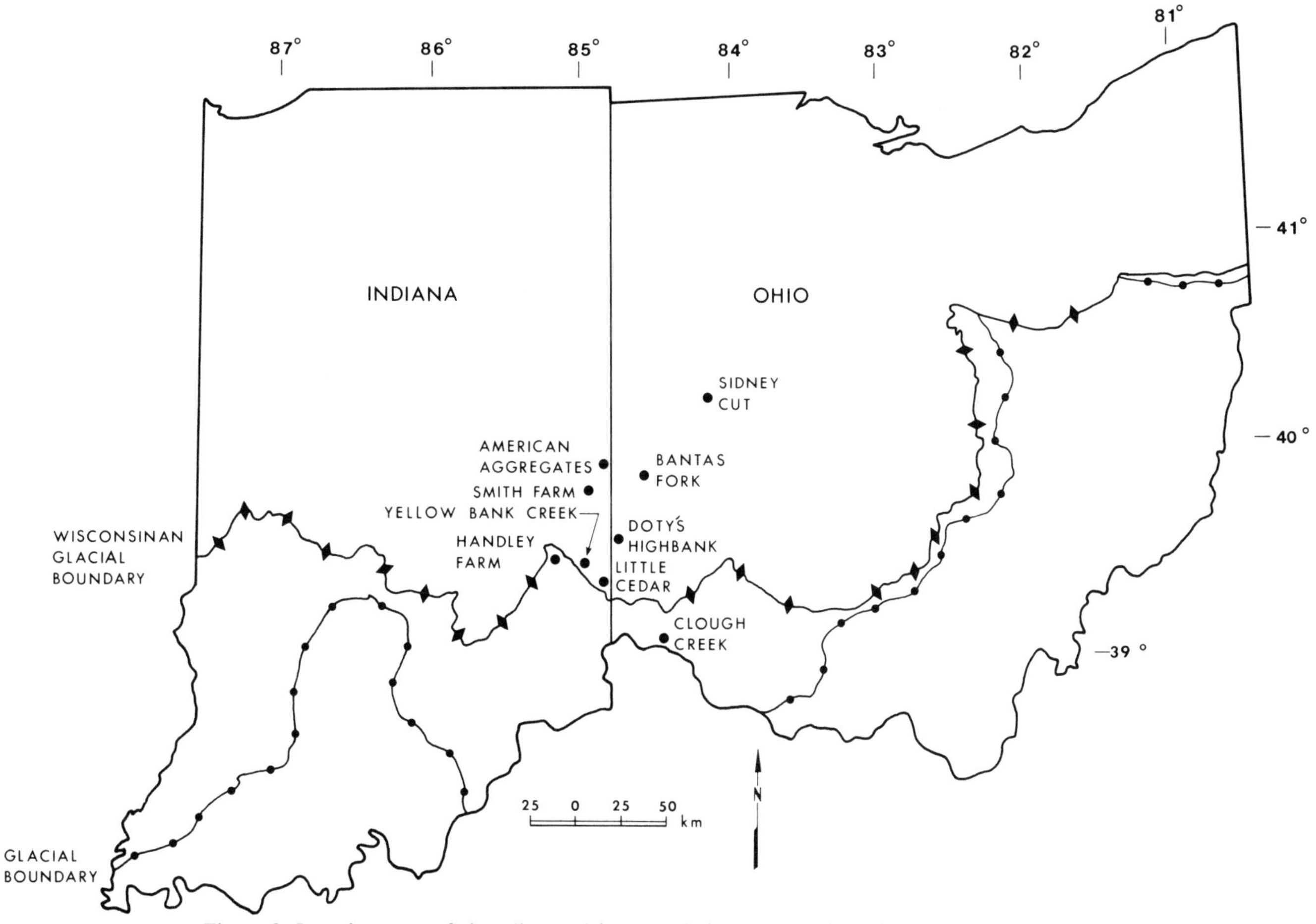

Figure 2. Location map of sites discussed in text relative to the Wisconsinan and pre-Wisconsinan maximum glacial boundaries.

Nonglacial organic sediments containing molluscs and plant remains separate the Whitewater Till from the overlying Fairhaven Till at several sties in this area of Ohio and Indiana (Gooding, 1963; Goldthwait and others, 1981; Fullerton, 1986). At a cutbank along Bantas Fork (Fig. 4), the Fairhaven Till rests upon a calcareous, organic-rich silt containing molluscs and wood (New Paris Interstade); the wood has yielded an age of 44,800±1700 yr B.P. (Table 1). The Fairhaven Till here is leached to a depth of up to 75 cm; Goldthwait and others (1981) attributed the leaching to weathering during the Sidney Interstade.

Sidney weathering interval (Sidney cut)

Goldthwait and others (1981) doubted the middle Wisconsinan age implied for the Fairhaven Till by the finite radiocarbon age from Bantas Fork. They favored a long, nonglacial middle Wisconsinan, the Sidney weathering interval, on the basis of the following: (1) the intensity of the weathering profile developed on the Whitewater Till at several sites in the area south of the southern margin of the Fairhaven Till; (2) the apparent absence of a well-developed weathering profile between the Whitewater and Fairhaven tills at sites where they do occur in superposition; and (3) the observation that Huron-Erie lobe tills of middle Wisconsinan age have not been identified to the north of the study area in Ohio or Indiana.

Goldthwait and others (1981) interpreted the buried soil exposed in a railroad cut south of Sidney, Ohio (Forsyth, 1965), as a consequence of a long middle Wisconsinan interstade, the Sidney weathering interval; the till in which the Sidney soil is developed should be correlative with one of the early Wisconsinan tills of southwestern Ohio and southeastern Indiana.

The few shells available for analysis from the Sidney cut are from a thin, but deeply buried silt (units 2 and 3; La Rocque and

TABLE 1. RADIOCARBON AGE DETERMINATIONS FOR WISCONSINAN SITES IN OHIO AND INDIANA

Locality	Laboratory Number	Date*	Source†
Yellow Bank Creek	DIC-2728	20,020 ± 200-300	1
	Beta-11,550	19,570 ± 220	2
Little Cedar Creek	I-1007	20,290 ± 800	3
Doty's Highbank	W-92	19,980 ± 500	4
	ISGS-761	20,210 ± 260	
	I-10184	20,500 ± 420	
Sidney Cut	W-188	23,000 ± 800	5
	W-356	22,480 ± 800	
Bantas Fork	ISGS-726	44,800 ± 1,700	4
American Aggregates	ISGS-1054	>50,000	2
	L-478B	>40,500	3
Clough Creek	PIT-0512	>45,000	6

*^{14}C yr B.P. (1950).
†1 = Miller (1985); 2 = Miller (unpublished); 3 = Gooding (1965); 4 = Goldthwait and others (1981); 5 = Forsyth (1965); 6 = T. Lowell, personal commun. (1989).

Forsyth, 1957), that was formerly exposed about 10 m below the buried soil (Fig. 5). Two radiocarbon dates from a log resting on the Sidney paleosol (unit 5, Fig. 5) place an upper limit on the age of the buried soil of about 23,000 yr B.P. (Table 1).

Calibration sites

Amino-acid data from additional sites in the region, where sample age is less equivocal, are included herein for comparison with the data from Bantas Fork, American Aggregates, and the Sidney cut, the three localities at which the interpreted age of the fossil-bearing units are suspect. These additional data include late Wisconsinan samples from Yellow Bank Creek, Little Cedar Creek and Doty's Highbank (Fig. 2). The age of about 20,000 yr B.P. for the fossil molluscs at these sites is based on multiple radiocarbon ages from wood (Table 1). The shells used in this study are from the same units as the radiocarbon-dated wood.

Another collection of shell was made from a cutbank exposure along a tributary of Clough Creek, in southeastern Hamilton County, Ohio. The analyzed molluscs were recovered from a silt inclusion in an unnamed till (Fig. 6). An Illinoian age for the till at this site is inferred from: (1) its location, 20 km south of the Hartwell Moraine (the mapped late Wisconsinan till boundary) and within an area of Ohio that has been mapped as Illinoian drift (Goldthwait and others, 1961); (2) the depth of carbonate leaching of this till unit at other localities in the area—the leaching is at least 40% deeper south of the Wisconsinan till boundary than on the surface till to the north; (3) lithologic differences between the surface tills north and south of the Hartwell Moraine (Brockman, unpublished data); and (4) *Fossaria* shells from silt included in the till that have aIle/Ile ratios similar to values obtained from shells of the same taxon from Illinoian sediments of the Glasford Formation and Petersburg Silt of Illinois and Indiana (see aIle/Ile values for shells from the Trousdale Coal Mine and Mechanicsburg southwest in Table 2).

A small collection of *Catinella* shell from Illinoian (Abington Interstade) sediments exposed at Smith Farm was analyzed (unit 2 in Gooding, 1963, Table 2).

Molluscs studied from the Handley Farm locality (Fig. 7) were collected from a clay unit exposed on the floor of the creek (unit 3 of Kapp and Gooding, 1974). Magnetic studies of the laminated clay overlying unit 3 at that locality suggest that it was deposited during the Matuyama Reversed Polarity Chronozone (Bleuer, 1976) and therefore is older than 740 ka (Izett and Wilcox, 1982).

AMINOSTRATIGRAPHY

Substantial literature documents the successful application of amino-acid data from fossil mollusc shell to the solution of Quaternary stratigraphic problems (Wehmiller, 1982; Bradley, 1985). Studies by McCoy (1987), Scott and others (1983), Miller and others (1987), and Clark and others (1989) have demonstrated the stratigraphic utility of the epimerization of isoleucine in nonmarine molluscan shells.

The rate of isoleucine epimerization is dependent on the molluscan taxon involved and is sensitive to temperature. The temperature history and age of most samples, however, are usually unknown. Aminostratigraphy permits direct use of amino-acid epimerization ratios as indicators of relative age, usually within a limited geographic region where close proximity permits the assumption that the sites in question have had similar temperature histories (Wehmiller, 1982).

Within the limits of the study area, highest and lowest mean annual temperatures differ by only 2.2 °C. (Armington, 1941; Fisher, 1941). For the amino acid data reported here, therefore, it is assumed that discordant aIle/Ile ratios between samples of the same taxon are due to differences in age.

Laboratory methods

Shells from four taxa of molluscs (*Fossaria, Stenotrema, Catinella,* and *Succinea*) from a total of nine sites in southwestern Ohio and southeastern Indiana (Fig. 2, Table 1) have been analyzed at the University of Massachusetts Amino Acid Geochronology Laboratory (AGL).

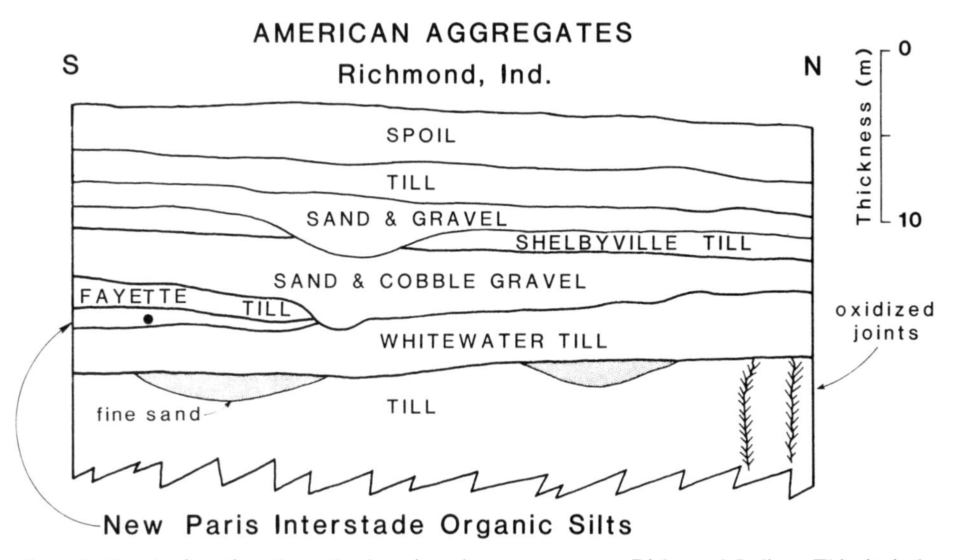

Figure 3. Sketch of stratigraphy at the American Aggregates quarry, Richmond, Indiana. This site is the type section for both the New Paris Interstade and the Whitewater Till. Molluscs were collected from the unit indicated by the black dot (modified from Gooding, 1965).

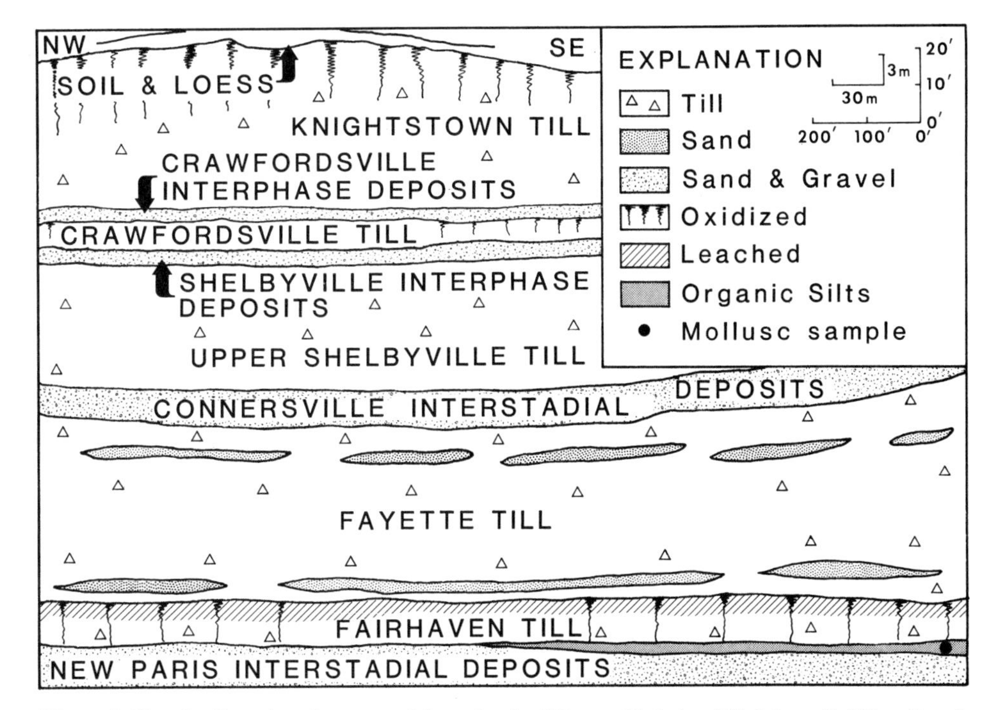

Figure 4. Sketch of stratigraphy exposed in cutbank of Bantas Fork (modified from Goldthwait and others, 1981).

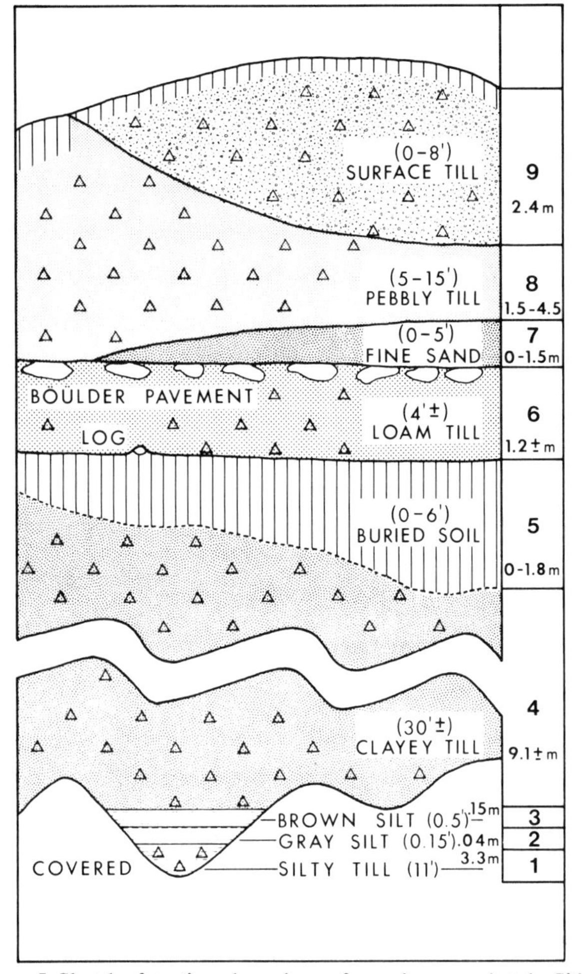

Figure 5. Sketch of stratigraphy as it was formerly exposed at the Sidney Cut. The molluscs used in this study are from units 2 and 3. The Sidney soil is unit 5 (modified from Forsyth, 1965).

Sample preparation involved repeated sonification and washing in purified water, followed by dissolution and hydrolysis in 6N HCl for 22 h at 110 °C under a nitrogen atmosphere. The aIle/Ile ratios in the total acid hydrolysate were determined by cation-exchange liquid chromatography. The results of our analyses of the Interlaboratory Comparison (ILC) samples (Wehmiller, 1984) were given in Miller and others (1987). Recent analyses of these standards have yielded comparable results and are not reproduced here.

When sample size permitted, three separate preparations and analyses were made for the total hydrolysate. The number of shells used for each preparation varied with the size of the individuals being analyzed. Smaller individual shell size usually results in greater variance between separate preparations from the same site. Replicate preparations and analyses of shells from a single locality are generally within 10% of their mean value. The results of these analyses are summarized in Table 2 and Figure 8.

Results

The epimerization ratios for the four genera used in this study (*Succinea, Catinella, Stenotrema,* and *Fossaria*) show significant differences for samples from the same lithostratigraphic units (Table 2, Fig. 8). The ratios suggest that *Succinea* and *Catinella* epimerize faster than *Stenotrema* and *Fossaria.* Direct comparison of ratios must consider the relative racemization rates.

The aIle/Ile hydrolysate ratios for the 22 samples analyzed in this study are plotted by locality and genus in Figure 8. The lowest aIle/Ile ratios (all less than 0.10) are found in shells collected from late Wisconsinan deposits at Dotys Highbank, Yellowbank Creek, and Little Cedar Creek (group 1, Fig. 8).

A second grouping of ratios includes samples of (1) *Fossaria* from inferred Illinoian silt at Clough Creek; (2) *Stenotrema, Ca-*

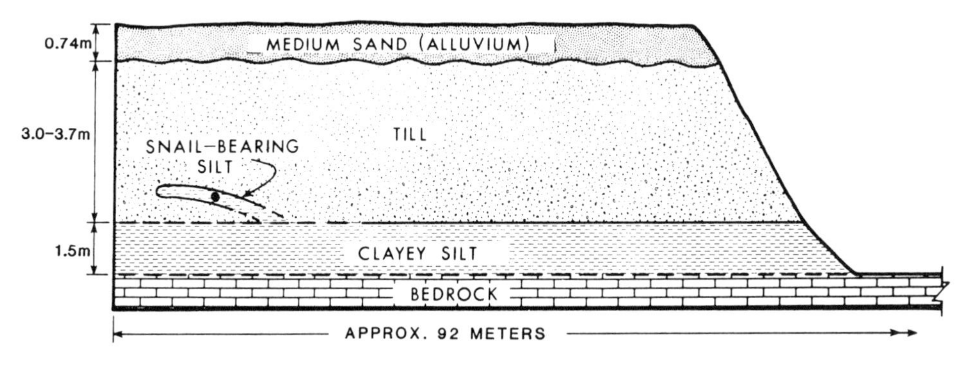

Figure 6. Sketch of stratigraphy exposed at tributary of Clough Creek, in southeastern Hamilton County, Ohio. Molluscs from this site were recovered from the silt inclusion (solid black circle).

TABLE 2. SUMMARY OF AILE/ILE TOTAL ACID HYDROLYSATE RATIOS

Locality	AGL* Number	Taxon	Number Analyzed†	Hydrol	Standard Deviation
Yellow Bank Creek	262	*Stenotrema*	3	0.060	± 0.008
	435	*Catinella*	2	0.071	± 0.020
	437	*Fossaria*	1	0.054	
	442	*Fossaria*	1	0.047	
Little Cedar Creek	383	*Stenotrema*	3	0.049	± 0.008
	311	*Catinella*	2	0.080	± 0.002
Doty's Highbank	364	*Catinella*	2	0.051	± 0.004
	444	*Catinella*	1	0.053	
Clough Creek	642	*Fossaria*	3	0.17	± 0.01
	714	*Fossaria*	3	0.16	± 0.02
Bantas Fork	123	*Catinella*	3	0.19	± 0.02
	713	*Catinella*	2	0.22	± 0.01
Sidney Cut	785	*Succinea*	1	0.22	
	786	*Stenotrema*	1	0.19	
	787	*Catinella*	1	0.23	
Trousdale Coal Mine (Vermillion County, Indiana)	639	*Fossaria*	3	0.20	± 0.02§
Mechanicsburg, S.W. (Sangamon County, Illinois)	580	*Fossaria*	2	0.20	± 0.01**
Smith Farm	1352	*Catinella*	2	0.33	± 0.01
American Aggregates	125	*Catinella*	2	0.50	± 0.01
	715	*Catinella*	2	0.50	± 0.02
	1016	*Catinella*	3	0.53	± 0.06
	1017	*Fossaria*	3	0.46	± 0.01
Handley Farm	1177	*Succinea*	2	0.78	± 0.01
	1178	*Catinella*	1	0.84	

*AGL = Amino Acid Geochronology Laboratory, University of Massachusetts.
†Number of separate preparations analyzed.
§Sample is from a slit between the Smithboro Till (Glasford Formation) and Hillery Till (Banner Formation).
**Sample is from silt in Glasford Formation.

tinella and *Succinea* from a silt beneath the weathering profile and till at the Sidney cut; and (3) *Catinella* from the organic silt near the base of the section at Bantas Fork (group II, Fig. 8). The range of aIle/Ile values within this group are partly due to differences between the faster epimerizers *Succinea* and *Catinella* versus the slower epimerizers *Fossaria* and *Stenotrema*. As a group, these aIle/Ile ratios are significantly higher than the values obtained from the known late Wisconsinan samples.

Succinea and *Catinella* from unit 3 at Handley Farm yield the highest total aIle/Ile ratios (group V, Fig. 8) for any of the study sites. These high values are consistent with the magnetically reversed signature of unit 3 at this site.

Catinella from the type section of the New Paris Interstade at the American Aggregates quarry yield aIle/Ile ratios (group IV, Fig. 8) that are intermediate between the values obtained for this taxon from unit 2 at Smith Farm (group III, Fig. 8) and unit 3 at Handley Farm.

DISCUSSION

The aIle/Ile ratios of shell from beneath the buried soil and till exposed in the Sidney cut (Fig. 5) are most similar to shell values obtained from sediments of interpreted Illinoian-age exposed at Mechanicsburg southwest, Trousdale Coal Mine, and Clough Creek. Molluscs from the Sidney cut are thus thought to be coeval with these fossil-bearing Illinoian silt units. The first well-developed weathering profile in the sequence above the Illinoian silts at the Sidney cut, therefore, may include the Sangamonian weathering interval.

Molluscs from the organic silt exposed near the base of the Bantas Fork cutbank section also have aIle/Ile ratios that are similar to those measured in shell recovered from (1) the silt at the Sidney Cut; (2) the silt inclusion in Illinoian till, at Clough Creek; (3) the Mechanicsburg southwest site; and (4) the Trousdale Coal Mine locality. The amino-acid data suggest that the organic silt at Bantas Fork is pre-Wisconsinan, and that the 44,800 ± 1700 yr B.P. age on wood collected from the same unit as the snails is too young. On the basis of this interpretation, the Fairhaven Till at this locality should be pre-Wisconsinan, and the buried soil developed in the Fairhaven Till (Fig. 4) probably represents weathering that took place during Sangamonian, early and middle Wisconsinan time.

AIle/Ile ratios for shell from the organic silt that overlies the Whitewater Till at the American Aggregates quarry suggest a much greater age for that unit than the sampled horizons at Bantas Fork, the Sidney cut, or the Smith Farm. Although the ratios are lower than those obtained for *Succinea* and *Catinella* from magnetically reversed sediments at Handley Farm, the aIle/Ile ratios are higher than shell values obtained from known Illinoian sediments at sites in Indiana and Illinois (Miller and others, 1988). On the basis of these aIle/Ile values, the calcareous organic silt at the American Aggregates quarry is interpreted to be pre-Illinoian.

CONCLUSIONS

The amino-acid data obtained during this study suggest the following interpretations that are summarized in Table 3.

1. If the Fairhaven and Whitewater tills are pre-Wisconsinan, then the basal Wisconsinan till in the area is the late Wisconsinan Fayette Till.

2. The Sidney weathering interval developed in the unit 4 till at the Sidney cut probably represents the last interglacial as well as early and middle Wisconsinan weathering, as does the Sangamon soil at many places in its type area of Sangamon County, Illinois.

3. The Fairhaven Till at Bantas Fork, the till at Clough

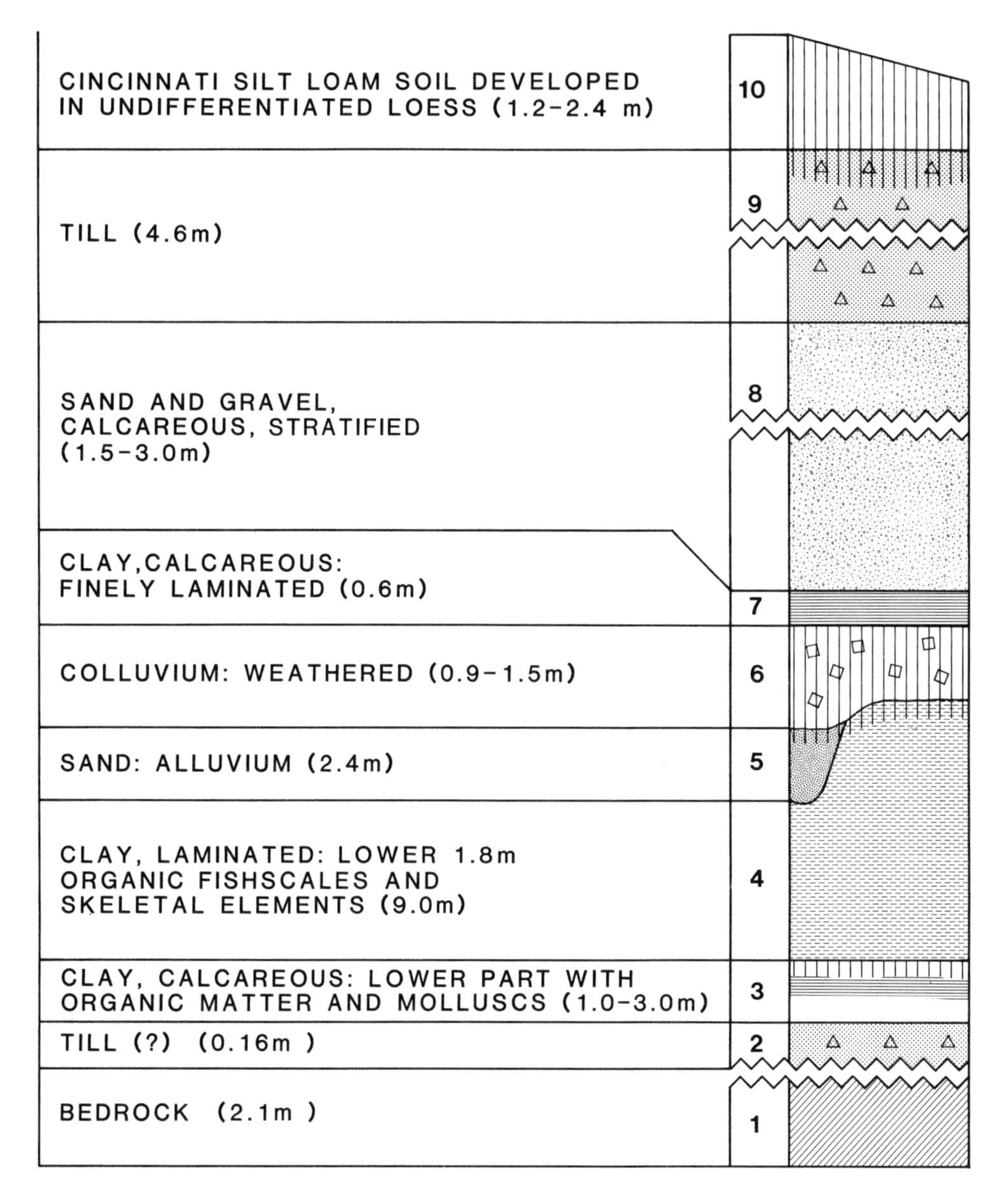

Figure 7. Sketch of stratigraphy at Handley Farm. The mollusc samples are from unit 3 (modified from Kapp and Gooding, 1974).

Creek, and the unit 4 till that overlies the mollusc-bearing silt at the Sidney cut probably were deposited by a late Illinoian ice advance into the area.

4. The hydrolysate ratios obtained from the Sidney cut and Bantas Fork molluscs suggest that they are approximately coeval with the fossil-bearing silt inclusion at Clough Creek, and are probably younger than the Illinoian Abington Interstade.

5. The Abington Interstade is represented by the shells from unit 2 at Smith Farm.

6. *Catinella* from the organic silt at the American Aggregate quarry yield aIle/Ile ratios that indicate that this unit is younger than the magnetically reversed sediments at Handley Farm but is probably pre-Illinoian.

7. The Whitewater Till at its type section, in this interpretation, probably represents an early-middle Pleistocene ice advance into the area.

8. The oldest Pleistocene fossils in the area are represented by material from unit 3 at the Handley Farm section.

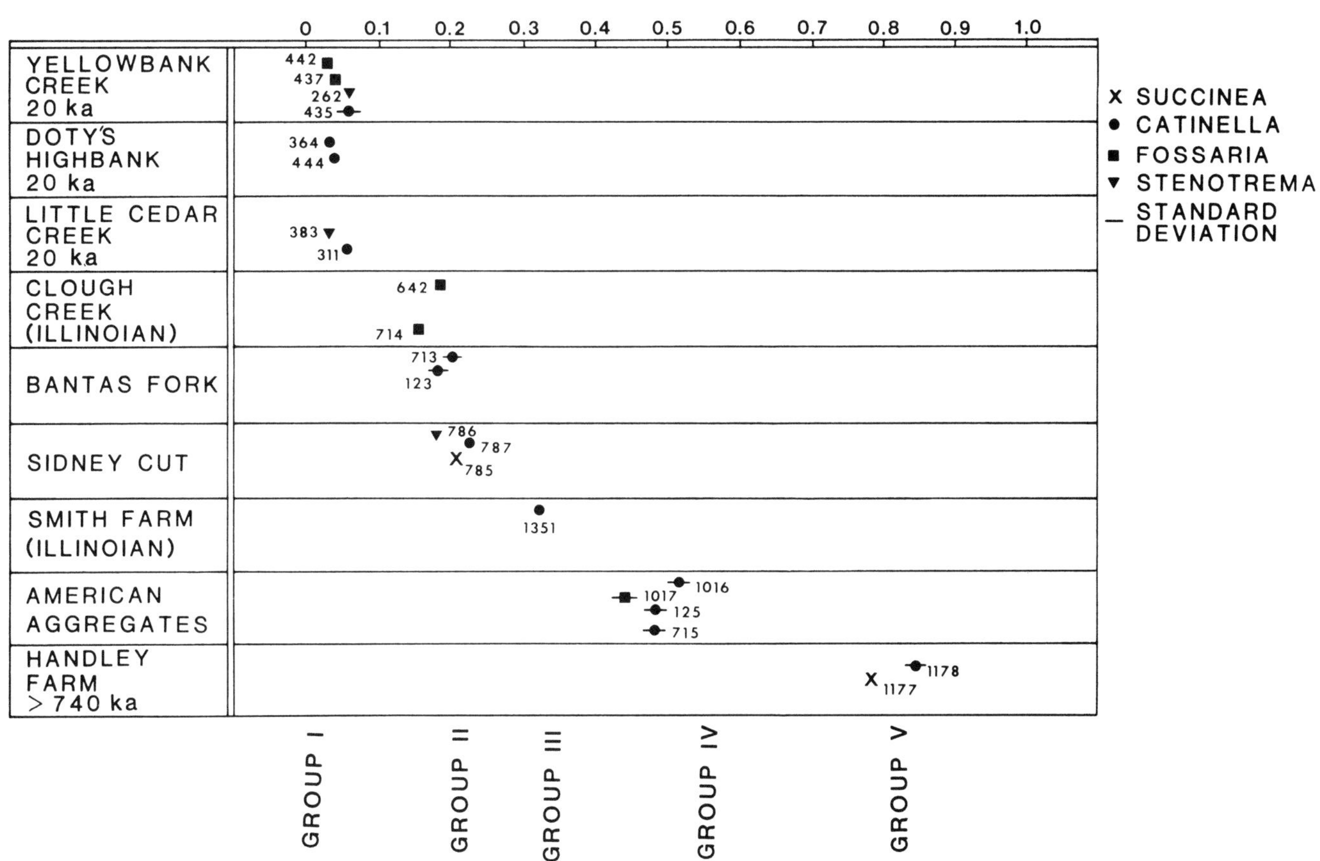

Figure 8. Plot of total Hydrolysate values for *Fossaria, Catinella, Succinea* and *Stenotrema* from nine localities in southwestern Ohio and southeastern Indiana.

TABLE 3. SUMMARY OF PLEISTOCENE STRATIGRAPHY IN STUDY AREA

Wisconsinian Stage	Knightstown Till Crawfordville Till Shelbyville Till Fayette Till
Wisconsinan-Sangamonian	Sidney weathering interval; unit 5 at the Sidney Cut
Illinoian Stage	Fairhaven Till at Bantas Fork; silt and till at Clough Creek; units 1, 2, 3, and 4 at Sidney cut; unit 2 at Smith Farm
Pre-Illinoian	New Paris organic silt at American Aggregates quarry Whitewater Till and pre-Whitewater till at American Aggregates quarry Pre-Whitewater till; unit 2 at Handley Farm

ACKNOWLEDGMENTS

We thank D. Perry Stewart, Miami University, for his generous help in collecting fossils from Bantas Fork and Doty's Highbank, to Floyd Nave, Wittenberg University, who provided the shell from the Smith Farm site, and to Michael W. Kotansky, Kent State University, who drafted the figures. John Szabo, University of Akron, provided useful comments on an earlier draft of this paper. We also thank Peter Lea, Bowdoin College, John Wehmiller, University of Delaware, and W. Hilt Johnson, University of Illinois, for their constructive criticism and editing. Some of the research reported here was supported by National Geographic Society Grant 2903-84 to Miller and National Science Foundation Grants EAR-86181228 to Miller and EAR-8618230 to McCoy.

REFERENCES CITED

Armington J. H., 1941, Climate of Indiana, *in* Hambidge, G., ed., Climate and Man: Yearbook of Agriculture: Washington, D.C., U.S. Government Printing Office, p. 852–861.

Bleuer, N. K., 1976, Remnant magnetism of Pleistocene sediments of Indiana: Indiana Academy of Science, v. 85, p. 277–294.

Bradley, R. S., 1985, Quaternary paleoclimatology: Winchester, Massachusetts, Allen and Unwin Inc., 472 p.

Clark, P. U., and Lea, P. D., 1986, Reappraisal of early Wisconsinan glaciation in North America: Geological Society of America Abstracts with Programs, v. 18, no. 6, p. 565.

Clark, P. U., Nelson, A. R., McCoy, W. D., Miller, B. B., and Barnes, D. K., 1989, Quaternary aminostratigraphy of Mississippi Valley loess: Geological Society of America Bulletin, v. 101, p. 918–926.

Fisher, J. C., 1941, Climate of Ohio, *in* Hambidge, G., ed. Climate and Man: Yearbook of Agriculture: Washington, D.C., U.S. Government Printing Office p. 1055–1076.

Forsyth, J. L., 1965, Age of the buried soil in the Sidney, Ohio, area: American Journal of Science, v. 263, p. 251–297.

Fullerton, D. S., 1986, Stratigraphy and correlation of glacial deposits from Indiana to New York and New Jersey: Quaternary Science Reviews, v. 5, p. 23–52.

Goldthwait, R. P., White, G. W., and Forsyth, J. L., 1961, Glacial map of Ohio: U.S. Geological Survey Miscellaneous Geologic Investigations Map I-316, scale 1:500000.

Goldthwait, R. P., Stewart, D. P., Franzi, D. A., and Quinn, M. J., 1981, Quaternary deposits of southwestern Ohio, *in* Roberts, T. G., ed., Geomorphology, hydrogeology, geoarchaeology, engineering geology (Geological Society of America annual meeting field trip guidebook): Falls Church, Virginia, American Geological Institute, p. 409–432.

Gooding, A. M., 1963, Illinoian and Wisconsinan glaciations in the Whitewater basin, southeastern Indiana, and adjacent areas: Journal of Geology, v. 71, p. 665–682.

—— , 1965, Southeastern Indiana: *in* Guidebook for Field Conference G, Great Lakes–Ohio Valley: Boulder, Colorado, International Association for Quaternary Research, p. 43–53.

—— , 1975, The Sidney Interstadial and late Wisconsinan history in Ohio and Indiana: American Journal of Science, v. 275, p. 993–1011.

Izett, G. A., and Wilcox, R. E., 1982, Map showing localities and inferred distributions of the Huckleberry Ridge, Mesa Falls, and Lava Creek ash beds (Pearlette family ash beds) of Pliocene and Pleistocene age in the western United States and southern Canada: U.S. Geological Survey Miscellaneous Geologic Investigations Series Map I-1325, scale 1:4000000.

Kapp, R. O., and Gooding, A. M., 1974, Stratigraphy and pollen analysis of Yarmouthian Interglacial deposits in southeastern Indiana: Ohio Journal of Science, v. 74, p. 226–238.

La Rocque, A., and Forsyth, J., 1957, Pleistocene molluscan faunules of the Sidney Cut, Shelby County, Ohio: Ohio Journal of Science, v. 57, p. 81–89.

McCoy, W. D., 1987, Quaternary aminostratigraphy of the Bonneville basin, western United States: Geological Society of America Bulletin, v. 98, p. 99–112.

Miller, B. B., 1985, Radiocarbon-dated molluscan assemblages from Ohio-Indiana: Their climatic significance: National Geographic Society Research Reports, v. 21, p. 305–312.

Miller, B. B., McCoy, W. D., and Bleuer, N. K., 1987, Stratigraphic potential of amino acid ratios in Pleistocene terrestrial gastropods: An example from west-central Indiana: Boreas, v. 16, p. 133–138.

Miller, B. B., McCoy, W. D., and Johnson, W. H., 1988, Aminostratigraphy of pre-Wisconsinan deposits in Illinois: Geological Society of America Abstracts with Programs, v. 20, no. 7, p. 345.

Scott, W. E., McCoy, W. D., Shroba, R. R., and Rubin, M., 1983, Reinterpretation of the exposed record of the last two cycles of Lake Bonneville, western United States: Quaternary Research, v. 20, p. 261–285.

Vincent, J. S., and Prest, V. K., 1987, The early Wisconsinan history of the Laurentide ice sheet: Geographic Physique et Quaternaire, v. 51, p. 199–213.

Wehmiller, J. F., 1982, A review of amino acid racemization studies in Quaternary molluscs: Stratigraphic and chronologic applications in coastal and interglacial sites, Pacific and Atlantic coasts, United States, United Kingdom, Baffin Island and tropical islands: Quaternary Science Reviews, v. 1, p. 83–120.

—— , 1984, Interlaboratory comparison of amino acid enantiomeric ratios in fossil Pleistocene mollusks: Quaternary Research, v. 22, p. 109–120.

Manuscript Accepted by the Society September 6, 1991

Printed in U.S.A.

Geological Society of America
Special Paper 270
1992

Reevaluation of early Wisconsinan stratigraphy of northern Ohio

John P. Szabo
Department of Geology, University of Akron, Akron, Ohio 44325

ABSTRACT

Three sections critical to the interpretation of upper Pleistocene deposits in northern Ohio are the Titusville, Pennsylvania, site and the Garfield Heights and Mt. Gilead sites in Ohio. At Titusville, Pennsylvania, the relation of peat dated at about 40,000 yr B.P. to the Titusville till is unclear. The interpretation of early Wisconsinan glaciation at Garfield Heights, Ohio, has been based on the Garfield Heights till and a subsequently derived accretion gley, both of which overlie a truncated paleosol (Sangamonian soil?) and two formerly exposed weathered tills. Current interpretation assigns the Garfield Heights till to the Illinoian. Previous interpretations for a middle or early Wisconsinan age of the Millbrook till of north-central Ohio have been placed in question by thermoluminescence ages of 146 ± 25 ka and 124 ± 16 ka on overlying loess at Mt. Gilead, Ohio. Much of what has been called Millbrook till has been traced to areas previously mapped as Illinoian. Thus, there is no evidence to support early to middle Wisconsinan glaciation in northern Ohio and adjacent northwestern Pennsylvania. During the early to middle Wisconsinan, landscapes in Ohio may have been geomorphically unstable, and deposits from this time may be in buried valleys or may have been eroded by late Wisconsinan glaciers.

INTRODUCTION

Interpretation of stratigraphic sequences and tentative age assignments of some units in northern Ohio have been critical to the inference of early or middle Wisconsinan glaciation along the southern margin of the Laurentide Ice Sheet. Evidence for glaciation in North America during this time frame, however, has been inferred from poorly delimited stratigraphic relations. Clark and Lea (1986) questioned the validity of many of the ages of deposits assigned to the early Wisconsinan (isotope stages 5d-4, 116–64 ka).

Correlation ideally should be based on the tracing of lithostratigraphic units over a geographic area in which datable materials are distributed. Correlations made among inferred early to middle Wisconsinan deposits in northern Ohio are less than ideal for several reasons. First, the ages of some units and subsequent correlatives have been assigned without any numerical control. For example, the ages of pre–late Wisconsinan deposits at Garfield Heights, Ohio, have been inferred from their stratigraphic relation to a paleosol, which has been assigned different ages by different workers (White, 1968, 1982; Szabo and Miller, 1986; Fullerton, 1986). Second, stratigraphic relations of some units are not reliably tied to units from which reasonably valid ages have been obtained. At Titusville, Pennsylvania, the age of a till has been inferred from radiocarbon ages of a peat that is topographically lower than the till, but the till does not overlie the peat in any exposure. Third, there is poor exposure of deposits possibly dating from the early to middle Wisconsinan. The climate in northern Ohio may not have been suitable for abundant plant growth during this time, and geomorphic processes operating during this time may have removed older organic deposits, loess, and/or a Sangamonian soil. Furthermore, much of the area was overridden by late Wisconsinan ice, possibly eroding older sediments. Holocene streams may not be exhuming older, buried valleys that may contain Wisconsinan deposits. Last, if deposits from the early to middle Wisconsinan do exist, they do not include material datable by most standard methods for this time frame.

The purpose of this paper is to reevaluate early Wisconsinan events in northern Ohio with respect to a few key sections, described below, where evidence for an early to middle Wisconsinan glaciation has been inferred from poorly delimited strati-

Szabo, J. P., 1992, Reevaluation of early Wisconsinan stratigraphy of northern Ohio, *in* Clark, P. U., and Lea, P. D., eds., The Last Interglacial-Glacial Transition in North America: Boulder, Colorado, Geological Society of America Special Paper 270.

graphic relations. Reference is also made to recent studies of early Wisconsinan events from adjacent areas in Michigan, Illinois, and Ontario.

KEY SECTIONS

Age assignments of Wisconsinan units in northern Ohio and northwestern Pennsylvania have relied largely upon sections exposed at Titusville, Pennsylvania, and at Garfield Heights, Ohio (Fig. 1). A section recently exposed on the north side of the fairgrounds at Mt. Gilead, Ohio, (Fig. 1) has required a revision of the age of units previously assigned to the early Wisconsinan (Totten and Szabo, 1987). The influence of these three sections on the interpretation of Wisconsinan events in north-central Ohio is discussed here.

Titusville section

The Titusville section is near the glacial boundary in northwestern Pennsylvania (Fig. 1). Leverett (1934) mapped this area as inner Illinoian; White and others (1969) revised this age based on radiocarbon ages of peat from this section. The section exposed in the Strawbridge pit (Fig. 2) often has been used as a reference to assign ages to Wisconsinan tills in northwestern Pennsylvania and northeastern and north-central Ohio (White and others, 1969; White, 1982). Prior to the discovery and dating of peat at this section, tils that underlie the Kent till were considered to be Illinoian (Leverett, 1934; Winslow and others, 1953; Smith and White, 1953; Droste and Tharin, 1958; Shepps and others, 1959).

The Strawbridge pit exposes up to 2 m of olive-brown, firm, leached Titusville till overlying 11 m of oxidized sandy gravel interpreted by White and others (1969) as outwash. A separate section in the wall of the present valley of Pine Creek, about 180 m away, exposes 7 m of sandy gravel overlying 0.3 m of peat having ^{14}C ages (White and others, 1969) of 31,400 ± 2,100 yr B.P. (I-1465); 35,000 +2,385/-2,835 yr B.P. (OWU-315); and 39,000 +4,900/-2,900 yr B.P. (I-1845). This peat is underlain by 0.9 m of intercalated sand, silt, and peat, which, in turn, overlie a thin, fibrous brown and black peat having ^{14}C ages (White and others, 1969) of 40,500 ± 1,000 yr B.P. (GrN-4996); >37,000 yr B.P. (OWU-316); and >42,000 yr B.P. (I-1771). Gray sand and gravel 0.7 m thick underlie the lower dated peat.

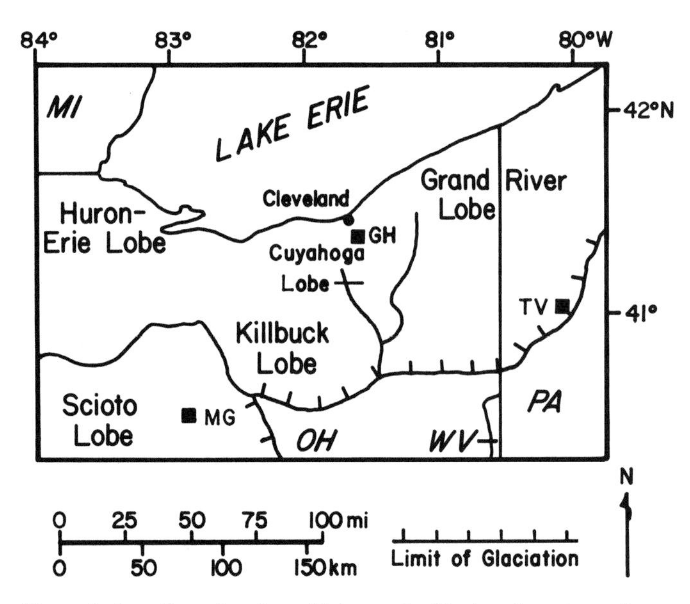

Figure 1. Location of major sublobes and critical sections in northern Ohio and northwestern Pennsylvania. TV, Titusville; GH, Garfield Heights; MG, Mt. Gilead. The limit of glaciation is the maximum extent of Pleistocene ice.

In the original interpretation of the section (Fig. 2a), White and Totten (1965) and White and others (1969) correlated gravel underlying the Titusville till in the Strawbridge pit to gravel exposed in the present valley. Thus, White and Totten (1965) and White and others (1969) interpreted the Titusville till as younger than 40 ka, but older than the late Wisconsinan Kent till, placing the ice advance that deposited Titusville till in the middle Wisconsinan. Presumably correlative units (Mogadore, Millbrook, and Jelloway tills) also were assigned to the middle Wisconsinan glaciation (White and others, 1969; Dreimanis and Goldthwait, 1973; White, 1982; Szabo, 1987).

Craft (1976) and Dreimanis (1982) questioned the validity of correlations at the Titusville section. Craft (1976) recognized that the degree of oxidation of the gravel is similar to that of the Illinoian Mapledale gravel (White and others, 1969). Dreimanis (1982) argued that the stratigraphic relation of the gravel in the Strawbridge pit to those exposed in the present valley is unclear because of the length of the covered section. Fullerton (1986) reviewed the numerous ^{14}C ages of the peat at the Titusville section and concluded that the peat may be part of an interstadial or postglacial alluvial sequence. Totten and Szabo (1987) also recognized that, because the part of the section containing the Titusville till is separated from the peat locality by about 180 m of covered section, the gravel underlying the till may not be the same gravel as that exposed at the peat site (Fig. 2b). Therefore, although the peat has a ^{14}C age of about 40,000 yr B.P., its stratigraphic relation to the Titusville till is uncertain, and the age of the Titusville till at its type section is unknown.

Garfield Heights section

The section exposed in a gravel pit along the valley of Mill Creek in Garfield Heights, Ohio, has been the subject of several investigations (Table 1). The section is southeast of Cleveland (Fig. 1), within the area glaciated by the Cuyahoga lobe of the Laurentide Ice Sheet. The deposits at this section are unique because no similar deposits have been found in northeastern Ohio. Although Mill Creek joins the Cuyahoga Valley 3 km to the west, the Pleistocene stratigraphy of the Cuyahoga Valley is different (Szabo and Miller, 1986; Szabo, 1987). The section in Mill Creek Valley (Fig. 3) exposes late Wisconsinan sediments that overlie a diamicton. This diamicton is interpreted as early Wisconsinan accretion gley and Garfield Heights till because of

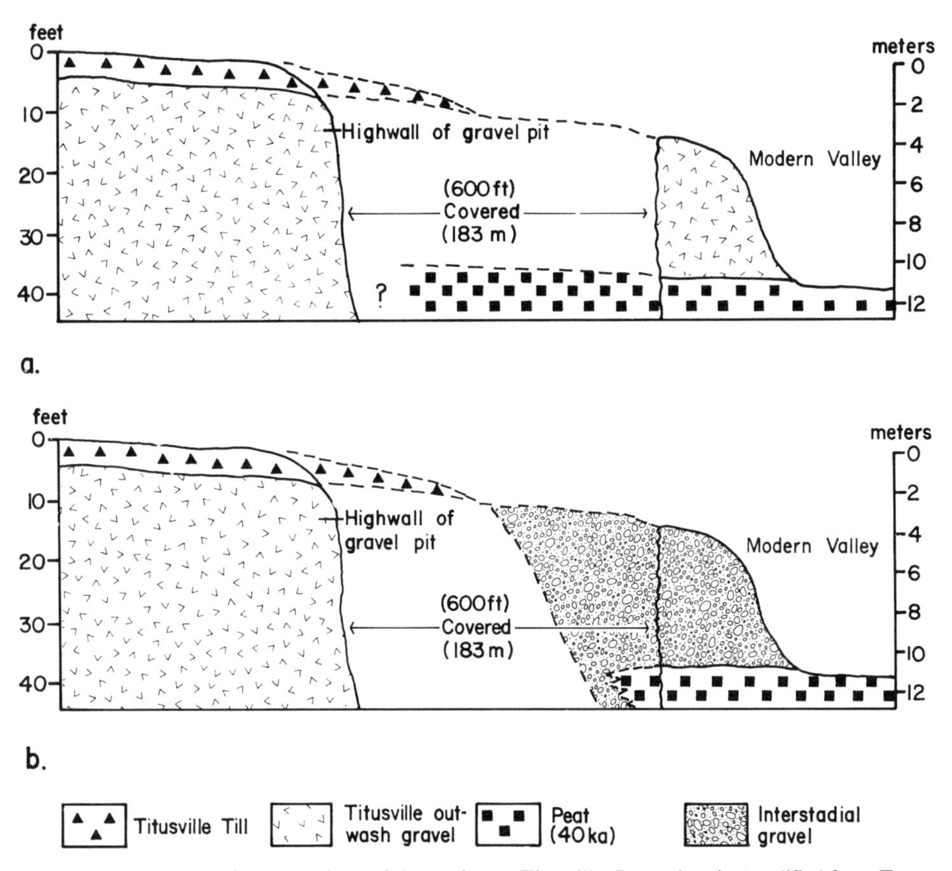

Figure 2. Two possible interpretations of the section at Titusville, Pennsylvania (modified from Totten and Szabo, 1987). a: Interpretation of White and others (1969). b: Interpretation of gravel exposed in present valley as post-Titusville till alluvium.

its stratigraphic placement above a possible Sangamonian soil developed in gravel. The age of a red till formerly exposed above the gravel in which the Sangamonian soil is developed is also unknown.

Most workers (White, 1953, 1968; Berti, 1971; Fullerton, 1986; Miller and Szabo, 1987) agree that at least two and possibly three late Wisconsinan tills (Fig. 3, unit 7) overlie lacustrine sediments (unit 6) containing logs dated at 24,520 +695/−760 yr B.P. (DIC-38). These lacustrine sediments overlie fossiliferous loess (unit 5) that comprises two units, referred to here as the upper and lower loesses. Wood in the upper loess yielded a ^{14}C age (Fullerton, 1986) of 28,195 ± 535 yr B.P. (K-361-3). Unweathered samples of the lower loess contain clay-mineral and carbonate-mineral suites different from those of the dated upper loess (Szabo and Miller, 1986). The upper part of the lower loess contains involutions suggestive of a cold climate (Berti, 1975); Fullerton (1986) attributed the cold climate to the earliest late Wisconsinan advance of the Ontario-Erie lobe. Investigators have found loess in the upper part of the lacustrine deposits (unit 6) and in a colluvial zone (Miller and Wittine, 1972; Miller and Szabo, 1987) that spans the lower part of the lacustrine deposits (unit 6) and the top part of the upper loess (unit 5).

A yellowish-brown to greenish-gray diamicton (unit 4) underlies the lower loess. Its structure ranges from platy to blocky to massive, and its matrix texture (<2 mm) varies among and within measured sections (Szabo and Miller, 1986). Pebble bands are common, and joints are coated with iron and manganese oxides. Clay minerals consist of vermiculite, montmorillonite, illite-montmorillonite, and heterogeneous swelling material (Szabo and Miller, 1986).

Dreimanis (1971), Berti (1975), and Szabo and Miller (1986) interpreted this diamicton as an accretion gley and/or till (Table 1). Berti (1975) and Dreimanis (1989, personal commun.) interpreted the unweathered part of the diamicton as a subglacial till on the basis of its fissility, high density, pebble lithology, and heavy-mineral composition, and named it the Garfield Heights till.

A gravel (units 1–3), which has not been exposed recently, underlies the diamicton interpreted as Garfield Heights till. White (1968, 1982) and Szabo and Miller (1986) interpreted the remnants of a paleosol found in the upper part of the gravel (unit 3) as the lower part of the B horizon of a Sangamonian soil developed in Illinoian gravel.

Part of the stratigraphic sequence (Table 1) described by

TABLE 1. VARIOUS INTERPRETATIONS OF THE SECTION EXPOSED AT GARFIELD HEIGHTS, OHIO

Time Division	White (1968)	Berti (1971) Dreimanis (1971)	Berti (1975)	Szabo and Miller (1986) Miller and Szabo (1987)	Fullerton (1986)	Age (ka)
Late Wisconsinan	Hiram Till Kent Till Lake sediments Upper loess	Hiram Till Kent Till Lake sediments Upper loess	Hiram Till Kent Till Lake sediments Upper loess	Hiram and/or Lavery tills Loess Two Kent Tills Lake sediments Colluvium Upper loess	Hiram Till Loess Lavery Till Two Kent Tills Lake sediments Upper loess	
						32
Middle Wisconsinan	Lower loess	Lower loess	Lower loess Accretion gley	Lower loess Accretion gley	Lower loess	
						65
Early Wisconsinan		Garfield Heights Till "red" till	Garfield Heights Till	Garfield Heights Till		
						80
Sangamonian	Gley Paleosol on gravels			Paleosol on gravel	Accretion gley	
						130
Illinioan	Gravel	Gravel	Gravel	Gravel	Garfield Heights Till "red" till gravel	
Pre-Illinoian					Weathered tills Paleosol on gravels	

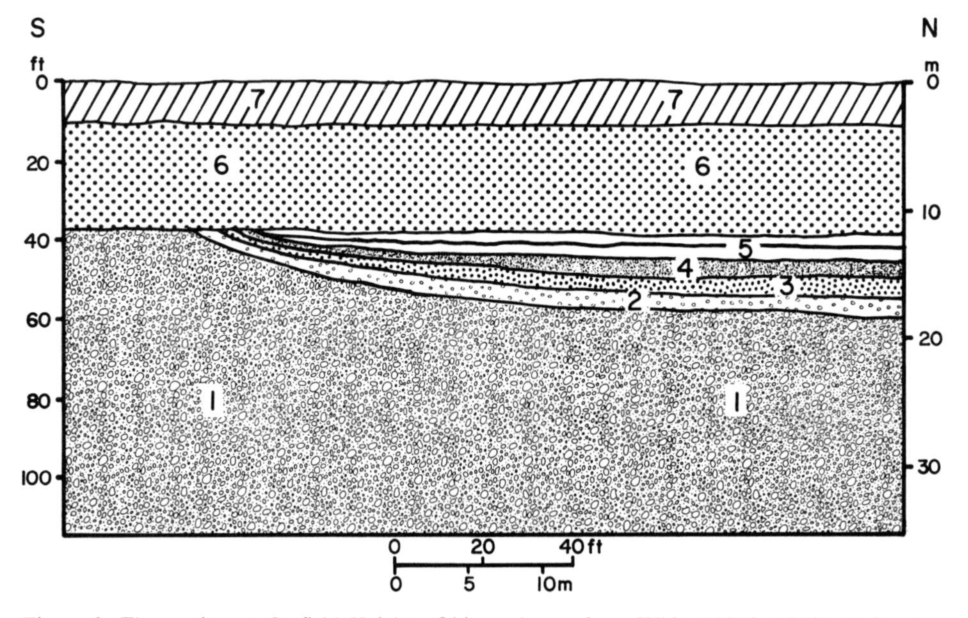

Figure 3. The section at Garfield Heights, Ohio, redrawn from White (1968, 1982). Units: 1—unoxidized and unleached gravel; 2—oxidized and unleached gravel; 3—paleosol developed in leached gravel; 4—diamicton; 5—upper and lower loess; weathered zone is in top of lower loess; 6—laminated lacustrine silt and clay (varves?); 7—Hiram and/or Lavery till.

Fullerton (1986) is based on an exposure in a gravel pit to the south of the section described above, where additional gravel units and weathered tills once were exposed (Fullerton and Groenwold, 1974). Fullerton observed a red till lying in a sequence above the Sangamonian soil but below the Garfield Heights till. The red till at Garfield Heights may correlate to the red Keefus till found elsewhere in northeastern Ohio beneath Titusville till (White and Totten, 1979; White, 1982). However, analyses of red till inclusions in late Wisconsinan till sheets (Bruno, 1988) reveal a range of clay mineralogies, indicating that not all red tills can be correlated by color alone. The stratigraphic sequence described by Fullerton (1986) cannot be evaluated further here, however, because the southern gravel pit has been filled. Only part of the late Wisconsinan stratigraphic sequence is currently exposed in the western wall of this pit.

Age assignments of deposits beneath the upper loess are speculative. The primary bound on the interpretation of the ages of the Garfield Heights till and the red till has been the age assigned to the red paleosol (unit 3) developed in the underlying gravel. White (1968, 1982) assigned a Sangamonian age to the red paleosol because it resembled some exposures of the Sangamon soil in Illinois. He may have assigned the accretion gley (unit 4) this same age because other Sangamonian soils in Illinois are accretion gleys. White (1968, 1982) regarded the lower loess as Altonian (middle Wisconsinan). As a result of these age assignments, the Garfield Heights till was assigned to the early Wisconsinan (Dreimanis, 1971; Berti, 1975). Szabo and Miller (1986) suggested that an early Wisconsinan age is equally likely for the lower loess, as also suggested by Dreimanis (this volume). I collected a sample from an organic zone 10 cm below the contact between the upper and lower loesses that was dated at 27,390 ± 350 yr B.P. (ISGS-1949), and thus establishes a minimum age for the older loess. If the lower loess is assigned an early Wisconsinan age and the accretion gley developed in the upper part of the diamicton (unit 4) represents an interval of weathering, the Garfield Heights till may be older than early Wisconsinan.

Fullerton (1986) assigned the Garfield Heights till to the Illinoian on the basis of reinterpretation of the age of the paleosol in the gravel (unit 3): he assigned the paleosol to the early Pleistocene because of the degree of weathering, and because it lay beneath a well-developed paleosol in two intensely weathered tills of interpreted pre-Illinoian age. Dreimanis (this volume) correlated the Garfield Heights till to the Bradtville till, which he now assigns to the Illinoian on the basis of reinterpretation of the stratigraphy on the Canadian side of Lake Erie.

In summary, the chronostratigraphy of pre–late Wisconsin deposits at Garfield Heights hinges on two units: (1) the lower, unweathered part of the diamicton named the Garfield Heights till; and (2) the underlying gravel and the paleosol developed in its upper part, which plunge below a group of units no longer exposed in a gravel pit south of the currently exposed section. However, there are no numerical ages to delimit the timing of deposition of units below the top of the lower loess. Age assignments of these units are therefore speculative and subject to revision.

Mt. Gilead Fairgrounds section

Another section that has been critical to the interpretation of the ages of pre–late Wisconsinan tills in northern Ohio is along the edge of the Allegheny escarpment at Mt. Gilead, Ohio (Fig. 1). The tills at this section are interpreted to be the Millbrook till. The Millbrook till (White, 1961) underlies late Wisconsinan Navarre, Hayesville, and Hiram tills deposited by the Killbuck lobe. White (1961) and Dreimanis and Goldthwait (1973) interpreted the Millbrook till as Illinoian on the basis of the greater depth of weathering observed in it compared to that of the overlying late Wisconsinan tills. Totten (1973) and White (1982), however, correlated the Millbrook till with the Titusville till, thus suggesting a middle to early Wisconsinan age.

Several tills are exposed in a series of stream cuts on the north side of Whetstone Creek on the north side of the fairgrounds on the southwestern edge of Mt. Gilead. Two separate outcrops about 37 m apart present a composite stratigraphic sequence from the top of the bluff to stream level (Fig. 4). The oldest (unit 1, Fig. 4) is dark gray, firm, calcareous, stoney till that crops out in the western section. Gravel (unit 2) separates this till from another dark gray, firm, calcareous till (unit 3); sand and gravel (unit 4) of varying thickness overlie this till and separate it from a dark gray-brown, firm, calcareous, pebbly third till (unit 5). This third till can be traced eastward to an outcrop (Fig. 4) where it is overlain by olive-brown to dark gray, firm, massive, calcareous till (unit 6). Brownish-yellow, massive to weakly laminated, friable, calcareous silt (unit 7), which overlies the fourth till, is interpreted as a loess (Totten and Szabo, 1987). Yellowish-brown, calcareous Navarre till (unit 8) overlies the loess.

The four lower tills (units 1, 3, 5, and 6) have field characteristics similar to those of the Millbrook till (White, 1961), and are jointly referred to informally as Millbrook till B. These four tills can be distinguished on the basis of carbonate contents in the <0.074 mm fraction (Totten and Szabo, 1987). The average carbonate contents of these tills, from oldest to youngest, are: Millbrook till BIV, 2% calcite and 9% dolomite; Millbrook till BIII, trace of calcite and 11% dolomite; Millbrook till BII, 6% calcite and 18% dolomite; and Millbrook till BI, no calcite and 5% dolomite. Complementary laboratory data make it possible to trace these subunits of the Millbrook till across several counties (Totten and Szabo, 1987; Viano, 1986). These data suggest, therefore, that the Millbrook till originally defined by White (1961) consists of several subunits, each having a characteristic lithology reflecting a different source area.

Two samples of Mt. Gilead loess (unit 7, Fig. 4) were collected for thermoluminescence (TL) dating. A TL age from the middle of this loess is 125 ± 16 ka (Alpha 3018), and a TL age from the bottom of the loess had a partial age of 145 ± 25 ka (Alpha 2870). Alpha Analytic Inc. used the combination of the regen, residual, and R-beta methods to produce TL ages. The TL

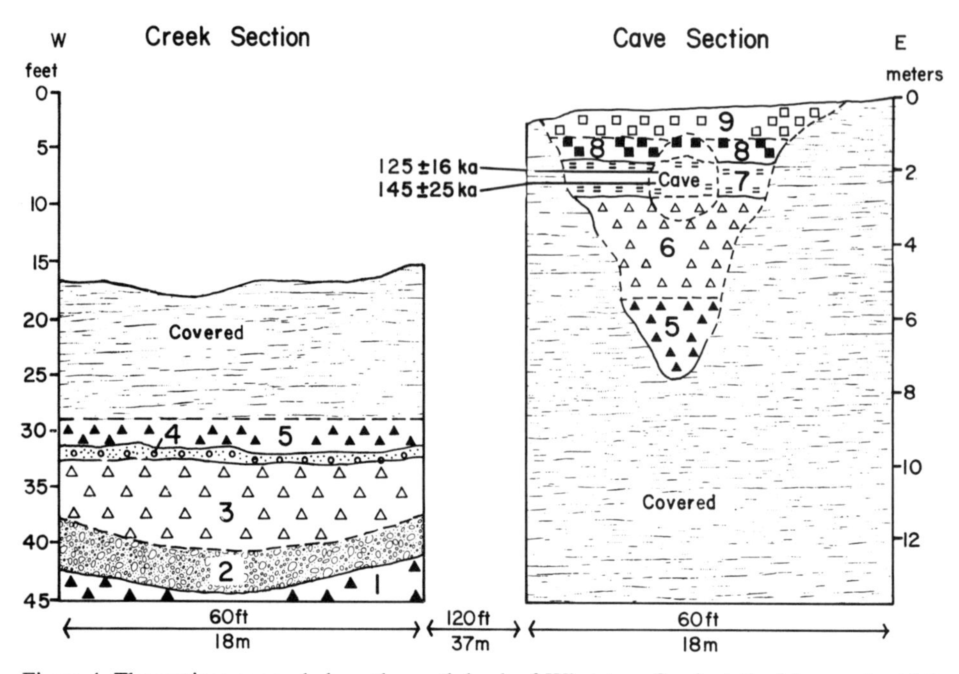

Figure 4. The sections exposed along the north bank of Whetstone Creek at the fairgrounds at Mt. Gilead, Ohio (modified from Szabo and Totten, 1987). Units: 1—Millbrook till BIV; 2—gravel; 3—Millbrook till BIII; 4—sand and gravel; 5—Millbrook till BII; 6—Millbrook till BI; 7—silt (loess?); 8—Navarre till; 9—covered Navarre till?

age on the bottom of the loess is the average of the regen and residual ages; the R-beta method gave unexplainable behavior and could not be used (J. Stipp, 1986, personal commun.). The ages of the loess are not statistically different, but have not been tested against data from other laboratories. The TL ages oppose the assignment of this till to either middle or early Wisconsinan time. Thus the Millbrook subunits may be Illinoian, as originally suggested by White (1961) and Dreimanis and Goldthwait (1973).

Two additional tills, which have characteristics of the Millbrook till but overlie the subunits of Millbrook till B at Mt. Gilead, are designated informally as Millbrook till U and Millbrook till A. These two tills are also found in stratigraphic succession over the Millbrook B subunits north and west of Mt. Gilead (Totten and Szabo, 1987). The stratigraphic relation of these tills to the loess dated by the TL method at Mt. Gilead is unknown. Late Wisconsinan Navarre till overlies Millbrook tills U and A north of Mt. Gilead. The age of these two tills is therefore delimited only by late Wisconsinan Navarre till overlying them and the Illinoian Millbrook B subunits underlying them.

Three hypotheses for the age of Millbrook tills U and A are considered (Table 2). First, the tills may be Illinoian: on the basis of the lithology of the Millbrook units, this would require a major shift in source areas for successive Millbrook subunits (Totten and Szabo, 1987). Millbrook till A contains two-thirds as much calcite as dolomite, whereas the underlying Millbrook till BI has no calcite and 5% dolomite. This change may represent shifting of adjacent sublobes during the Illinoian, whereby one sublobe advanced from the north (no calcite and low dolomite of Millbrook till BI), and the other subsequently advanced from the west-northwest (higher carbonate of Millbrook till A) (Totten and Szabo, 1987).

The second hypothesis suggests the placement of Millbrook tills U and A in early Wisconsinan time (Table 2), as suggested by Fullerton (1986). Lithologic characteristics of Millbrook till A may represent an advance of early Wisconsinan ice which deposited the till from a different source area than pre-Wisconsinan Millbrook till BI.

A third hypothesis (Table 2) suggests that Millbrook tills U and A were deposited in the early part of the late Wisconsinan. If the correlation of the Navarre till of the Killbuck lobe to the Kent till of the Cuyahoga and Grand River lobes is correct, and the ages of lacustrine sediments beneath the Kent till at Garfield Heights are acceptable, there is no evidence for significant ice advances southward into north-central Ohio during the late Wisconsinan prior to the event that deposited the Navarre and Kent tills. Fullerton (1986) suggested an ice advance to the Allegheny escarpment at about 28 ka, on the basis of radiocarbon ages in the basal part of the Catfish Creek till on the northern side of Lake Erie and the ^{14}C age on wood in the upper loess at Garfield Heights. No sites in north-central Ohio have datable materials that support any of the three hypotheses.

I favor the first hypothesis (Illinoian age) because Stanley Totten (1988, personal commun.) and I have not found any weathering profiles developed in the Millbrook B tills where Millbrook tills U and A overlie them. In addition, the moraines of

TABLE 2. POSSIBLE AGE ASSIGNMENTS FOR MILLBROOK TILLS U AND A IN NORTH-CENTRAL OHIO

Time	Hypothesis 1	Hypothesis 2	Hypothesis 3
Late Wisconsinan	Hiram Till Hayesville Till Navarre Till	Hiram Till Hayesville Till Navarre Till	Hiram Till Hayesville Till Navarre Till Milbrook Till U Millbrook Till A
Middle Wisconsinan			
Early Wisconsinan		Millbrook Till U Millbrook Till A	
Sangamonian			
Illinoian	Millbrook Till U Millbrook Till A Millbrook Till BI Millbrook Till BII Millbrook Till BIII Millbrook Till BIV	Millbrook Till BI Millbrook Till BII Millbrook Till BIII Millbrook Till BIV	Millbrook Till BI Millbrook Till BII Millbrook Till BIII Millbrook Till BIV

north-central Ohio consist of cores of Millbrook till overlain by late Wisconsinan tills (Totten and Szabo, 1985). The cores of the most distal moraines are composed of Millbrook tills BIII and BIV. The cores of the moraines contain progressively younger Millbrook subunits northward toward the Defiance moraine, which has a core formed of Millbrook till U (Totten and Szabo, 1985). Thus, a sequence of probable Illinoian moraines was overridden by late Wisconsinan ice.

Within north-central Ohio, surface tills in areas mapped as Illinoian along the northeastern margin of the Scioto lobe (Goldthwait and others, 1961) were thought to be older than the Millbrook till. Viani (1986) and Totten and Szabo (1987) have demonstrated, however, that surface till in the Illinoian area in Knox County east of Mt. Gilead is composed of Millbrook tills BI, BIII, and BIV. Stratigraphic and geomorphic evidence and the TL ages thus support the age assignment of till mapped as Illinoian in north-central Ohio.

DISCUSSION

The sections at Titusville, Garfield Heights, and Mt. Gilead show no evidence to support an early or middle Wisconsinan expansion of the Laurentide Ice Sheet into north-central Ohio or northwestern Pennsylvania. This agrees with studies of early and middle Wisconsinan deposits in Illinois (Kempton and others, 1985; Curry, 1989; Curry and Follmer, this volume), which also negate an incursion of early or middle Wisconsinan ice south of the latitude of the Illinois-Wisconsin state line. Furthermore, Dreimanis (this volume) restricts early Wisconsinan ice to the northern end of the Erie basin.

Discrimination of subunits of the Millbrook till and the tentative age assignment of them to the Illinoian affects the age assignment of other pre-Wisconsinan tills of northeastern Ohio. Tentative correlations of lithologic units in north-central and northeastern Ohio are suggested in Table 3. The Navarre, Hayesville, and Hiram tills and their correlatives are assigned to the late Wisconsinan (White, 1982). White (1982) correlated the Millbrook, Mogadore, and Titusville tills. I correlate the Mogadore and Titusville tills to the Millbrook BI subunit (Table 3) on the basis of physical characteristics and carbonate content. These tills generally are massive, overconsolidated lodgment tills that weather olive brown and contain no calcite and a small amount of dolomite. Millbrook till A has been traced into the area glaciated by the eastern part of the Killbuck lobe (Totten and Szabo, 1985), and correlates to the Northampton till (Szabo, 1987) of the Cuyahoga lobe (Table 3), on the basis of similar textural, lithologic, and mineralogic characteristics. The Northampton till has not yet been traced into the area glaciated by the Grand River lobe. Correlatives of the older Millbrook subunits have not been found on the Appalachian Plateau. Placement of the Garfield Heights till is problematic. It contains less dolomite than the Mogadore till in the adjacent Cuyahoga Valley (Szabo and Miller, 1986), and may be older than the Mogadore till (Fullerton, 1986). An exposure of probable Mogadore till is found at a section stratigraphically higher than the type section of the Garfield Heights till (Szabo and Miller, 1986). Dreimanis (this volume) correlates the Bradtville till on the north shore of Lake Erie to the Garfield Heights till.

The lack of evidence for glaciation during the early to middle Wisconsinan in northeastern and north-central Ohio is comparable to other studies in the midcontinent. Kempton and others (1985) investigated the type area of the Altonian or middle Wisconsinan tills in Illinois and discovered a Sangamon soil developed on many of these tills, thus suggesting that they are Illinoian. Curry (1989) and Curry and Follmer (this volume) showed that pedogenesis in Illinois, that began during the last interglaciation,

TABLE 3. CORRELATIONS OF LITHOLOGIC UNITS IN NORTHEASTERN AND NORTH-CENTRAL OHIO

Western Scioto Lobe (Central Lowlands)	Eastern Scioto Lobe and Killbuck Lobe (Central Lowlands and Plateau)	Cuyahoga Lobe (Plateau)	Grand River Lobe (Plateau)
n.f.*	n.f.	n.f.	Ashtabula Till
Hiram Till	Hiram Till	Hiram Till	Hiram Till
Hayesville Till	Hayesville Till	Lavery Till	Lavery Till
Navarre Till	Navarre Till	Kent Till	Kent Till
Millbrook Till U	Millbrook Till U	n.f.	n.f.
Millbrook Till A	Millbrook Till A	Northampton Till	n.f.
Millbrook Till BI	Millbrook Till BI	Mogadore Till	Titusville Till
Millbrook Till BII	Millbrook Till BII	n.f.	n.f.
n.f.	Millbrook Till BIII	n.f.	Keefus Till (?)
n.f.	Millbrook Till BIV	n.f.	n.f.
n.f.	Butler Till	n.f.	Mapledale Till (?)

Note: Age of units increases downward.
*n.f. = Not found.
(?) = Very tentative correlation.

continued up to the late Wisconsinan, thus precluding any glaciation during the early to middle Wisconsinan. Winters and others (1988) examined well logs and dated orgnaic matter from wells in western Michigan. The organic matter was deposited during the middle Wisconsinan, indicating that the southern part of Michigan must have been free of ice at that time. Johnson and Follmer (1989) suggested that the upper Great Lakes area may have been affected by middle Wisconsinan glaciation because a glacial source is needed for the Roxana silt.

Recent work on the north sides of the Great Lakes basins also indicates restricted glaciation during the early to middle Wisconsinan. Eyles and Westgate (1987) and Eyles and Williams (this volume) argued that the stratigraphic sequence at the Scarborough Bluffs near Toronto, Ontario, precludes glaciation of the southern Great Lakes region during the middle to early Wisconsinan. Diamictons previously interpreted as tills were reinterpreted as part of a glaciolacustrine complex (Eyles and Eyles, 1983). Further examination of the Sunnybrook drift, however, argues for grounded ice in the Ontario basin during the early Wisconsinan (Hicock and Dreimanis, 1989; this volume). Dreimanis (this volume) proposed that the Bradtville drift in the Lake Erie basin is Illinoian and that unit B of the overlying Tyrconnell Formation was deposited in an early Wisconsinan proglacial lake. Thus, the southern limit of the Laurentide Ice Sheet during the early Wisconsinan may have been restricted to the northern sides of Lakes Erie and Ontario.

One of the puzzling problems in the Pleistocene stratigraphy of northern Ohio is the apparent paucity of sediments or soil formed during the Sangamonian. Outcrops of Sangamonian soil are found in southwestern Ohio (Thomas Lowell and Leon Follmer, 1988, personal communs.). Other than a possible truncated Sangamonian soil and/or gley at Garfield Heights, however, no other sites in northern Ohio have yielded firm evidence of Sangamonian weathering. One possible indication of Sangamonian weathering is the presence of occasional oxidized zones in Millbrook till underlying younger tills, or the presence of oxidized and cemented gravel between tills, but some of these zones may have been produced by ground water. Possible explanations for the scarcity of evidence of Sangamonian weathering in northern Ohio are that the Sangamonian soil was stripped off during periods of geomorphic instability (Follmer, 1983) in early or middle Wisconsinan time, or that it was eroded during successive glaciations. Sangamonian deposits may be preserved in buried valleys not currently being exhumed by present streams.

ACKNOWLEDGMENTS

I thank Stanley Totten and Charles Carter for their comments on this manuscript and for fruitful discussions on the Pleistocene stratigraphy of northern Ohio. The manuscript was also improved by the comments of Aleksis Dreimanis, Ernie Muller, and Ned Bleuer. The work of numerous graduate students in both the field and the laboratory over the years is gratefully appreciated. The Ohio Geological Survey graciously provided samples and laboratory supplies.

REFERENCES CITED

Berti, A. A., 1971, Palynology and stratigraphy of the mid-Wisconsin in the eastern Great Lakes region, North America [Ph.D. thesis]: London, Ontario, University of Western Ontario, 160 p.

—— , 1975, Paleobotany of Wisconsin interstadials, eastern Great Lakes region, North America: Quaternary Research, v. 5, p. 591–619.

Bruno, P. W., 1988, Lithofacies and depositional environments of the Ashtabula Till, Lake and Ashtabula counties, Ohio [M.S. thesis]: Akron, Ohio, University of Akron, 207 p.

Clark, P. U., and Lea, P. D., 1986, Reappraisal of early Wisconsin glaciation in North America: Geological Society of America Abstracts with Programs, v. 18, p. 565.

Craft, J. L., 1976, Stop III. White City sand and gravel pit, Titusville, PA, *in* Ward, A. N., Jr., Chapman, W. F., Lukert, M. T., and Craft, J. L., eds., Bedrock and glacial geology of northwestern Pennsylvania in Crawford, Forest, and Venango counties: Pennsylvania Geologists, 41st Annual Field Conference, Guidebook, p. 28–33.

Curry, B. B., 1989, Absence of Altonian glaciation in Illinois: Quaternary Research, v. 31, p. 1–13.

Dreimanis, A., 1971, The last ice age in the eastern Great Lakes region, North America, *in* Ters, M., ed., Etudes sur le Quaternaire dans le Monde: International Union for Quaternary Research, VIIIth Congress, v. 1, p. 69–75.

—— , 1982, Middle Wisconsin substage in its type region, the eastern Great Lakes and Ohio River basin: Quaternary Studies in Poland, v. 3, part 2, p. 21–28.

Dreimanis, A., and Goldthwait, R. P., 1973, Wisconsin glaciation in the Huron, Erie, and Ontario lobes, *in* Black, R. F., Goldthwait, R. P., and Willman, H. B., eds., The Wisconsinan Stage: Geological Society of America Memoir 136, p. 71–106.

Droste, J. B., and Tharin, J. C., 1958, Alteration of clay minerals in Illinoian till by weathering: Geological Society of America Bulletin, v. 69, p. 61–68.

Eyles, C. H., and Eyles, N., 1983, Sedimentation in a large lake: A reinterpretation of the late Pleistocene stratigraphy of Scarborough Bluffs, Ontario, Canada: Geology, v. 11, p. 146–152.

Eyles, N., and Westgate, J. A., 1987, Restricted regional extent of the Laurentide ice sheets in the Great Lakes basins during early Wisconsin glaciation: Geology, v. 17, p. 537–540.

Follmer, L. R., 1983, Sangamon soil and Wisconsinan pedogenesis in the midwestern United States, *in* Porter, S. C., ed., Late Quaternary environments of the United States: Volume 1, The late Pleistocene: Minneapolis, Minnesota, University of Minnesota Press, p. 138–144.

Fullerton, D. S., 1986, Stratigraphy and correlation of glacial deposits from Indiana to New York and New Jersey, *in* Sibrava, V., Bowen, D. Q., and Richmond, G. M., eds., Quaternary glaciations in the Northern Hemisphere: Quaternary Science Reviews, v. 5, p. 23–37.

Fullerton, D. S., and Groenwold, G. H., 1974, Quaternary stratigraphy at Garfield Heights (Cleveland), Ohio—Additional observations: Geological Society of America Abstracts with Programs, v. 6, p. 509–510.

Goldthwait, R. P., White, G. W., and Forsyth, J. L., 1961, Glacial map of Ohio: U.S. Geological Survey Miscellaneous Geologic Investigations Map I-316, scale 1:500,000.

Hicock, S. R., and Dreimanis, A., 1989, Sunnybrook drift indicates a grounded early Wisconsin glacier in the Lake Ontario basin: Geology, v. 17, p. 169–172.

Johnson, W. H., and Follmer, L. R., 1989, Source and origin of Roxana Silt and middle Wisconsinan midcontinent glacial activity: Quaternary Research, v. 31, p. 319–331.

Kempton, J. P., Berg, R. C., and Follmer, L. R., 1985, Revisions of the stratigraphy and nomenclature of glacial deposits in central Illinois, *in* Berg, R. C., Kempton, J. P., Follmer, L. R., and McKenna, D. P., eds., Illinoian and Wisconsinan stratigraphy and environments in northern Illinois: Midwest Friends of the Pleistocene, 32nd Field Conference, Guidebook, p. 1–20.

Leverett, F., 1934, Glacial deposits outside the terminal moraine in Pennsylvania: Pennsylvania Topographic and Geologic Survey, 4, Bulletin G7, 123 p.

Miller, B. B., and Szabo, J. P., 1987, Garfield Heights: Quaternary stratigraphy of northeastern Ohio, *in* Biggs, D. L., ed., North-Central Section of the Geological Society of America: Boulder, Colorado, Geological Society of America, Centennial Field Guide, v. 3, p. 399–402.

Miller, B. B., and Wittine, A. H., 1972, The origin of late Pleistocene deposits at Garfield Heights, Cuyahoga County, Ohio: Ohio Journal of Science, v. 76, p. 305–313.

Shepps, V. C., White, G. W., Droste, J. B., and Sitler, R. F., 1959, Glacial geology of northwestern Pennsylvania: Pennsylvania Geological Survey General Geology Report 32, 59 p.

Smith, R. C., and White, G. W., 1953, The groundwater resources of Summit County, Ohio: Ohio Division of Water Bulletin 27, 130 p.

Szabo, J. P., 1987, Wisconsinan stratigraphy of the Cuyahoga Valley in the Erie basin, northeastern Ohio: Canadian Journal of Earth Sciences, v. 24, p. 279–290.

Szabo, J. P., and Miller, B. B., 1986, Pleistocene stratigraphy of the lower Cuyahoga valley and adjacent Garfield Heights, Ohio (Geological Society of America North-Central Section meeting guidebook, field trip 2): Kent, Ohio, Kent State University Geology Department, 62 p.

Totten, S. M., 1973, Glacial geology of Richland County, Ohio: Ohio Geological Survey Report of Investigation 88, 55 p.

Totten, S. M., and Szabo, J. P., 1985, Pre-Woodfordian till and the age of moraines in north-central Ohio: Geological Society of America Abstracts with Programs, v. 17, p. 329.

—— , 1987, Pre-Woodfordian stratigraphy of north-central Ohio: Midwest Friends of the Pleistocene, 34th Field Conference, Guidebook, 92 p.

Viani, C. W., 1986, Stratigraphy and mineralogy of tills in Knox County, Ohio [M.S. thesis]: Akron, Ohio, University of Akron, 98 p.

White, G. W., 1953, Sangamon soil and early Wisconsin loess at Cleveland, Ohio: American Journal of Science, v. 251, p. 369–376.

—— , 1961, Classification of Wisconsin glacial deposits in the Killbuck lobe, northeast-central Ohio: U.S. Geological Survey Professional Paper 424-C, p. C71–C73.

—— , 1968, Age and correlation of Pleistocene deposits at Garfield Heights (Cleveland), Ohio: Geological Society of America Bulletin, v. 79, p. 749–752.

—— , 1982, Glacial geology of northeastern Ohio: Ohio Geological Survey Bulletin 68, 75 p.

White, G. W., and Totten, S. M., 1965, Wisconsin age of the Titusville Till (formerly called "inner Illinoian"), northwestern Pennsylvania: Science, v. 148, p. 234–235.

—— , 1979, Glacial geology of Ashtabula County, Ohio: Ohio Geological Survey Report of Investigations 112, 48 p.

White, G. W., Totten, S. M., and Gross, D. L., 1969, Pleistocene stratigraphy of northwestern Pennsylvania: Pennsylvania Geological Survey, ser. 4, Bulletin G55, 88 p.

Winslow, J. D., White, G. W., and Webber, E. E., 1953, Water resources of Cuyahoga County, Ohio: Ohio Division of Water Bulletin 26, 123 p.

Winters, H. A., Alford, J. J., and Rieck, R. L., 1988, The anomalous Roxana silt and mid-Wisconsinan events in and near southern Michigan: Quaternary Research, v. 29, p. 25–35.

MANUSCRIPT RECEIVED BY THE SOCIETY SEPTEMBER 6, 1991

Printed in U.S.A.

Geological Society of America
Special Paper 270
1992

Early Wisconsinan in the north-central part of the Lake Erie basin: A new interpretation

Aleksis Dreimanis
Department of Geology, University of Western Ontario, London, Ontario N6A 5B7, Canada

ABSTRACT

The Bradtville drift, the Canning till, and their correlatives in southwestern Ontario have been previously thought to be early Wisconsinan in age. Here another alternative is offered, whereby the Bradtville drift is assigned to the Illinoian stage, the lowermost Member A of the overlying Tyrconnell Formation to the late Sangamonian; Eowisconsinan, or the earliest part of early Wisconsinan, and its Member B to the early Wisconsinan substage. The age of the Canning till is still unknown.

Member A of the Tyrconnell Formation is an accretion gley that formed about 20 m below the present level of Lake Erie, thus requiring a low outlet for the Erie basin. At that time, the Erie basin was drained probably by the buried Erigan channel, which extends about 50 m below the present level of Lake Erie.

Member B of the Tyrconnell Formation is varved glaciolacustrine silt and clay, the deposition of which required a rise of lake level above the present one. This rise could have been caused by the Ontario lobe overriding the Niagara peninsula, possibly as far as Gowanda, New York; however, the ice margin remained in the eastern part of Lake Erie.

The above hypothesis is supported by available lithologic and paleoecologic data from the region adjoining the north-central and eastern part of Lake Erie, but supporting numerical age determinations beyond the range of radiocarbon dating are still lacking.

INTRODUCTION

An upper Pleistocene stratigraphic unit can be assigned to the early Wisconsinan with certainty only if it is overlain by a unit of unquestionable middle Wisconsinan age and underlain by a unit of equally certain Sangamonian interglacial age. In the Lake Erie basin, no stratigraphic sequence completely satisfies this requirement because of the lack of unequivocal evidence that would permit correlation of soils or sediments with the fully developed Sangamon paleosol in its type area in Illinois (Follmer, 1983, and references therein).

However, sections are known (Fig. 1) wherein radiocarbon-dated middle Wisconsinan sediments containing cool-climate flora and fauna are underlain by sediments or soils that could belong to older episodes of middle Wisconsinan, early Wisconsinan, or pre-Wisconsinan time. The most thoroughly investigated composite Pleistocene section (Fig. 2, Table 1) is in the Plum Point–Bradtville–Port Talbot area along the north shore of Lake Erie. Dreimanis (1957) first proposed the threefold subdivision of the last glaciation into late, middle, and early Wisconsin substages for southwestern Ontario on the basis of evidence from this section.

The evidence and reasoning for the prevailing middle and early Wisconsinan assignments and terminology (Table 2) are discussed in this chapter as background for the new alternative interpretation discussed later and also presented in Table 2. On the basis of numerical ages, the oldest being 47.6 ± 0.4 ka (GrN-2601; Vogel and Waterbolk, 1972), Member C of the Tyrconnell Formation is assigned a middle Wisconsinan age in both alternative interpretations (Table 2). The chronostratigraphic assignments of the underlying lithostratigraphic units (Member B of the Tyrconnell Formation and sediments below it) have been and are

Dreimanis, A., 1992, Early Wisconsinan in the north-central part of the Lake Erie basin: A new interpretation, *in* Clark, P. U., and Lea, P. D., eds., The Last Interglacial-Glacial Transition in North America: Boulder, Colorado, Geological Society of America Special Paper 270.

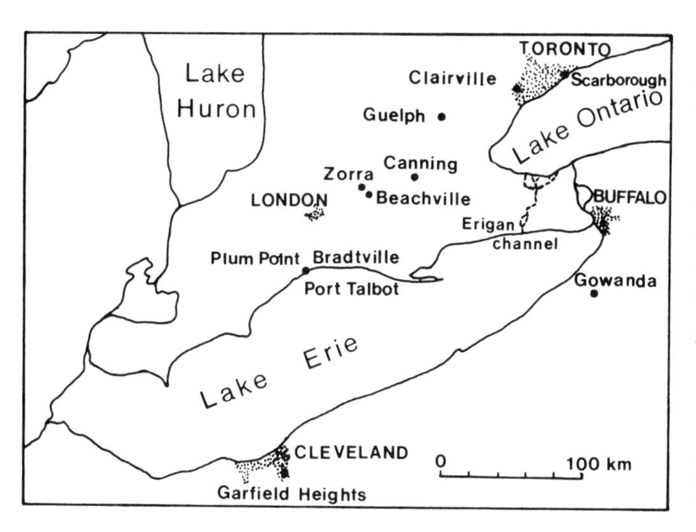

Figure 1. Location map of sites mentioned in text.

now based mainly upon sedimentologic, paleoenvironmental, and paleoclimatologic information from test drillings and their core samples. Because this information is mainly of an interpretive nature, more alternative interpretations are always possible.

PLUM POINT-BRADTVILLE-PORT TALBOT AREA

First informal stratigraphic assignments 1957–1958

The first assignments of early Wisconsinan age for sediments in the north-central area of the Lake Erie basin were based upon field examination of lake cliffs at Bradtville (Fig. 2), preliminary investigations of microfossils and macrofossils and lithology of the sediments, and seven finite and nonfinite radiocarbon ages (24.6 to >39.0 ka) on wood and gyttja from overlying sediments (Dreimanis, 1957, 1958). Consequently, an early Wisconsinan age was assigned to the dolomite-rich rhythmite unit now known as Member B of the Tyrconnell Formation (Table 1, Fig. 2) and also, by lithologic similarity, to a dolomite-rich sandy till exposed in an isolated section 0.2 km northeast of the radiocarbon-dated gyttja of the middle Wisconsinan Port Talbot interstadial (Dreimanis, 1958). This till, later called the Dunwich till, remained an enigma for many years, but, after having been identified at several more sites, it has been included in the late Wisconsinan Catfish Creek drift (Dreimanis, 1987), and is no longer related to Member B of the Tyrconnell Formation.

Early Wisconsinan reassignments 1963–1972

Because the Lake Erie cliff sections did not provide enough stratigraphic information, five test holes were drilled down to bedrock (Fig. 2, sites 1–5) in the Plum Point–Bradtville area during 1962–1963. Two new stratigraphic units were discovered below the rhythmites or varved clay: a green, carbonate-poor clay named Port Talbot I, and the gray to reddish, calcareous Bradtville drift (Fig. 2) between the green clay and the Devonian bedrock (Dreimanis, 1963, 1964). As a result, the middle Wisconsinan/early Wisconsinan boundary was lowered to the boundary between the Port Talbot I and Bradtville drift.

McKenzie (1964) completed a detailed investigation of the

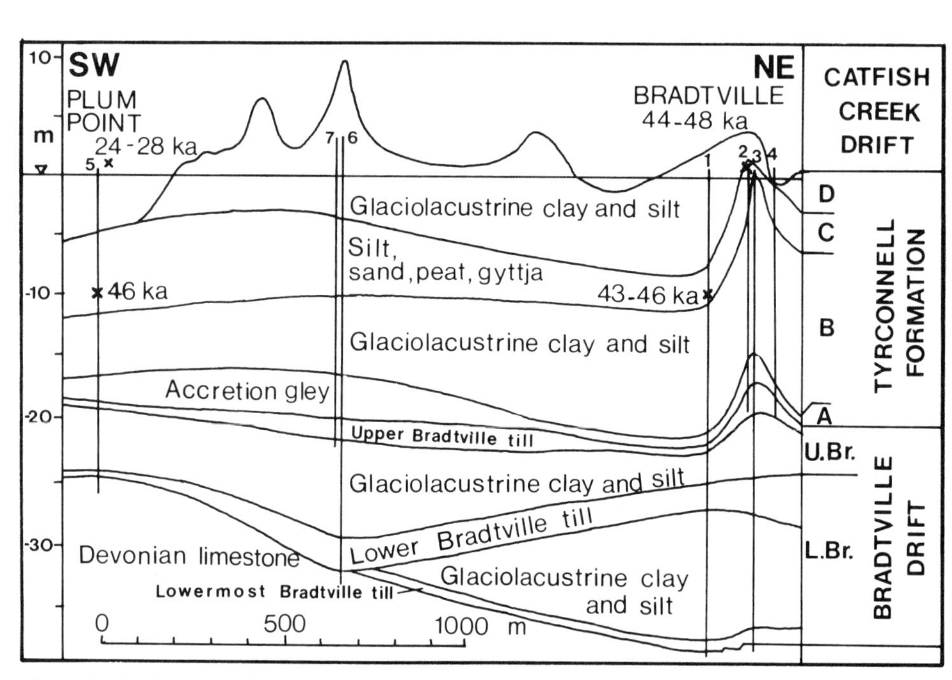

Figure 2. Geological profile section through the lower 10 m of Lake Erie bluffs and the underwater section between Plum Point and Bradtville. 1–7 = test drillings. A–D are members of Tyrconnell Formation. Members of the Bradtville drift: U.Br. = upper Bradtville, L. Br. = lower Bradtville. Crosses: sites of radiocarbon-dated wood and gyttja. Triangle = Lake Erie level.

TABLE 1. LITHOSTRATIGRAPHIC TERMINOLOGY FOR THE PLUM POINT-BRADTVILLE SECTION (FIG. 2) IN THREE REFERENCES OF 1966–1987

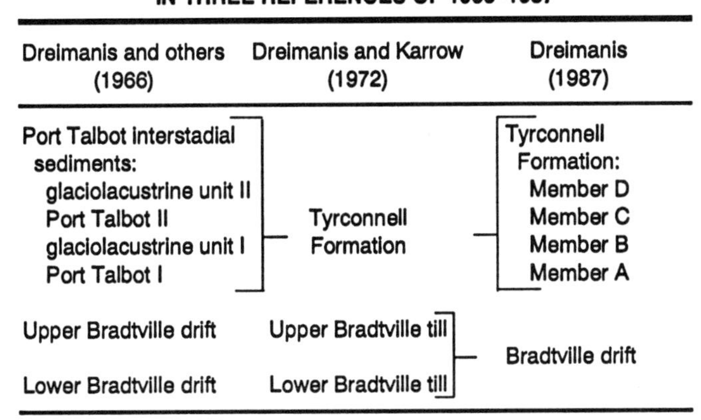

Dreimanis and others (1966)	Dreimanis and Karrow (1972)	Dreimanis (1987)
Port Talbot interstadial sediments: glaciolacustrine unit II, Port Talbot II, glaciolacustrine unit I, Port Talbot I	Tyrconnell Formation	Tyrconnell Formation: Member D, Member C, Member B, Member A
Upper Bradtville drift	Upper Bradtville till	Bradtville drift
Lower Bradtville drift	Lower Bradtville till	

TABLE 2. TWO ALTERNATIVES FOR CHRONOSTRATIGRAPHIC ASSIGNMENTS TO THE LITHOSTRATIGRAPHIC UNITS IN THE PLUM POINT-BRADTVILLE AREA DISCUSSED IN THIS CHAPTER

Lithostratigraphy		Chronostratigraphy
1. Dominant assignment 1963-1988:		
Tyrconnell Formation		Middle Wisconsinan
Bradtville drift		Early Wisconsinan
2. Alternative proposed in this chapter:		
Tyrconnell Formation	Members C and D	Middle Wisconsinan
	Member B	Early Wisconsinan
	Member A	Early Wisconsinan, Eowisconsinan, or Late Sangamonian
Bradtville drift		Illinoian

above five test drillings. His data, as well as palynologic, macrofossil, and lithologic investigations, and additional radiocarbon ages as old as 47.6 ka, were summarized in Dreimanis and others (1966) (Table 1). Several larger-diameter (7 cm) test holes that were continuously sampled at several intervals were drilled in the area in 1965 (Fig. 2, sites 6 and 7, and others outside the Fig. 2 section). Subsequently, pollen and plant macrofossils were studied by Berti (1971, 1975), and clay minerals and other lithologic components were studied by Quigley and Dreimanis (1972). The last three reports dealt mainly with the Tyrconnell Formation (name proposed by Dreimanis and Karrow, 1972; Table 1), but the paleoecologic interpretations of the Tyrconnell Formation were relevant to repeated cross-checking on the middle/early Wisconsinan boundary. Particular attention was paid to the origin of the green clay (Port Talbot I), which Quigley and Dreimanis (1972) finally interpreted as accretion gley, and to its boundary relation with the underlying upper Bradtville till and the overlying glaciolacustrine unit I (Table 1, Fig. 3, Member B).

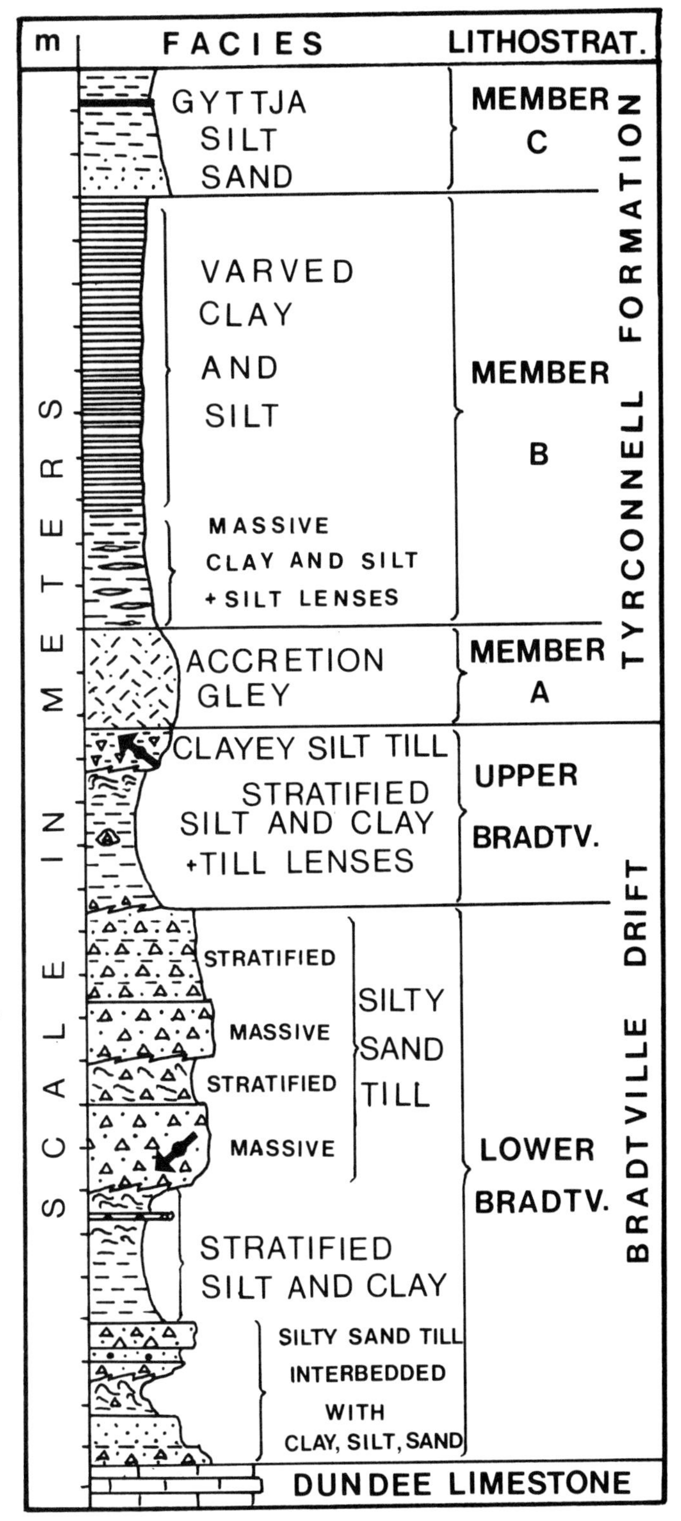

Figure 3. Facies of the Bradtville drift and Members A, B, and C of the Tyrconnell Formation at Bradtville, a combined vertical column based mainly upon test drillings 1 to 4. Arrows represent paleoglacial flow direction derived from fabric measurements of granules in oriented cores, with N. at the top (McKenzie, 1964, Fig. 9).

Both boundaries appeared to be transitional, suggesting a sedimentologic continuum without any major break (McKenzie, 1964). This continuity, however, was concluded mainly from lithologic characteristics.

The palynologic investigations by Berti (1971, 1975) confirmed previous conclusions, based on fossil remains, that the Tyrconnell Formation sediments overlying the green clay were deposited in a moist and cool forest-tundra environment. However, the Port Talbot I green clay was formed under warmer and dryer conditions (Berti, 1975), as concluded from the relatively high abundance of oak (*Quercus*) pollen associated with herb pollen (Fig. 4).

Because the accretion gley (Port Talbot I) occurred 12–22 m below the present lake level in a tectonically stable area, the level of the ancestral Lake Erie of that time must have been at least 22 m lower than at present. Such a paleoenvironmental situation may have resembled early Lake Erie at the end of the late Wisconsinan (Calkin and Feenstra, 1985; Coakley and Lewis, 1985, and references therein), when the outlet area in the Niagara peninsula was glacioisostatically depressed after the retreat of the Erie-Ontario lobe. By analogy, it was reasoned that the formation of accretion gley could have occurred at the end of any pre–middle Wisconsinan glacial retreat from the Erie basin. Because numerical-age determinations were not available for this event, the end of the early Wisconsinan and the end of Illinoian glaciation were considered as two possible alternatives. The former appeared to be more probable because of the apparent continuity in the sedimentation cycle from the Port Talbot I to the radiocarbon-dated Port Talbot II (Table 1).

The following sequence of events (Dreimanis, 1988, p. A3)

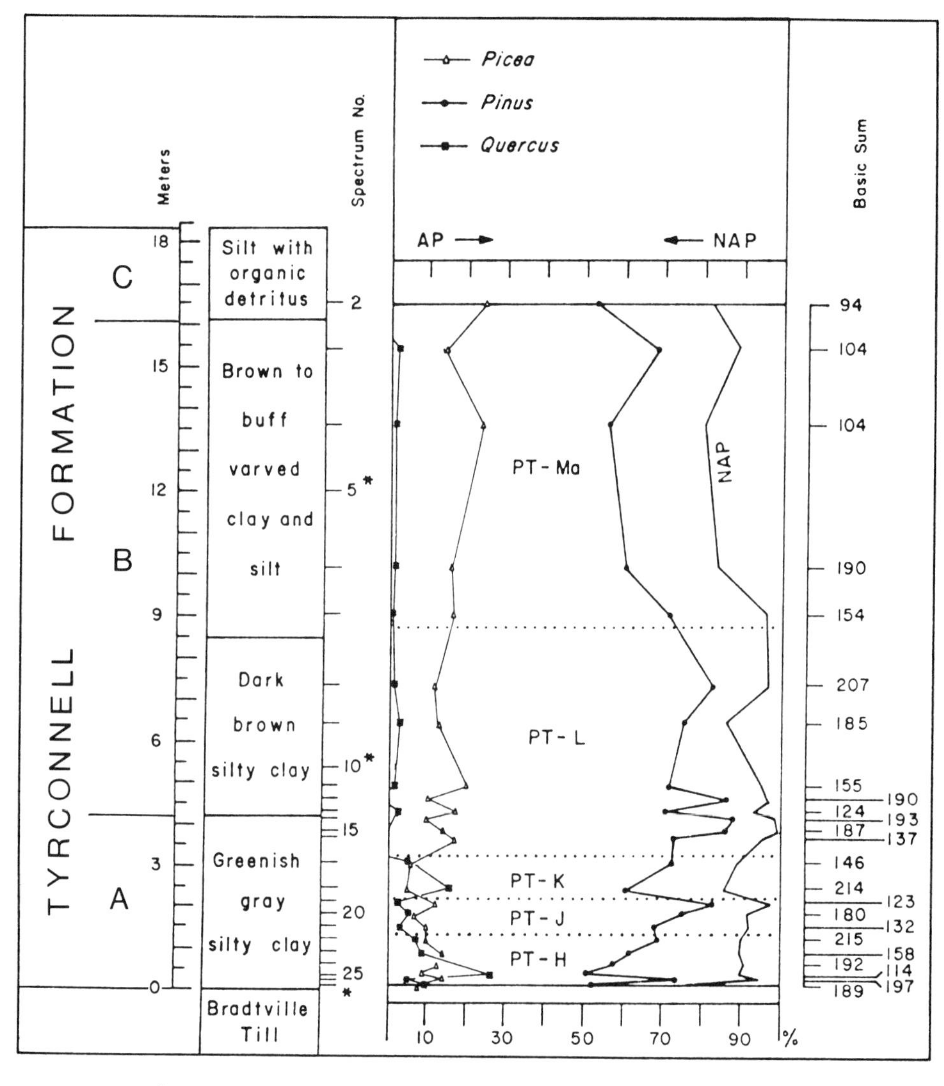

Figure 4. Summary pollen diagram of the Tyrconnell Formation, Members A and B, test drillings 2 and 3 (after Berti, 1975, Fig. 5).

was concluded from the lithostratigraphic units and their sedimentologic and paleoenvironmental interpretations.[1]

A vigorous glacial advance from the N.E. eroded the pre-Wisconsinan sediments and deposited the lowermost till of the Lower Bradtville Drift. The following glacial retreat and a re-advance of the same lobe resulted in the deposition of the Lower Bradtville glacio-lacustrine clayey silt and a second layer of till. The lobe withdrew again, and the Upper Bradtville glaciolacustrine silty clay was laid down. Finally the Ontario lobe overran the area, depositing the Upper Bradtville till. After retreat of the Ontario lobe, the glacio-isostatically depressed Niagara Penninsula kept the ancestral Lake Erie level at least 20m lower than the present one for a time interval sufficiently long to allow the deposition of up to 3m of decalcified accretion gley soil of the Middle Wisconsinan Tyrconnell Formation. Subsequent rise of lake level produced 5–10m of rhythmites interpreted as varves representing about 100 years. They are overlain by organic-rich sediments radiocarbon dated 48-42ka BP. The above Middle Wisconsinan events are mentioned here because the rhythmites could be interpreted as an Early *Wisconsinan glaciolacustrine sediment overlying the* Sangamonian Interglacial soil. However, this alternative is rejected because it cannot explain how the accretion gley with a cool-climate pollen assemblage could have been deposited more than 20m below the present lake level at the end of the Sangamonian Interglacial.

However, while preparing visual materials for the oral presentation of the above paper, I began to reevaluate the current interpretation following the suggestion of Clark and Lea (1986) for a reappraisal of the existing stratigraphic concepts of the early Wisconsinan. The result is presented in the next section, and a brief summary in Dreimanis (1991, p. 283–284).

NEW ALTERNATIVE TIME-STRATIGRAPHIC CORRELATION

Summary of the new proposal

I propose that the Bradtville drift is Illinoian, Member A of the Tyrconnell Formation is late Sangamonian or Eowisconsinan, and Member B of the Tyrconnell Formation is early Wisconsinan.

Rechecking the original data

Two sets of conditions originally supported the inclusion of the entire sequence of the Bradtville drift and the Tyrconnell Formation (Table 1) in the lower half of Wisconsinan glaciation: (1) the apparent lithologic continuity along every boundary between the formations or their members, and (2) the ease of accommodating the deposition of accretion gley 20 m below the present lake level.

When reexamining the boundary contacts in the field at Bradtville and in the 1965 drill cores 6 and 7 (Fig. 2), and the analytical data from the contact zones, I found now that most of them with two exceptions, were transitional. The boundary between Members C and B in the Tyrconnell Formation that is exposed at lake level at bore hole 3 (Fig. 2) is marked by a discontinuity between the faintly stratified, fossiliferous gray sand and/or silt of Member C and the underlying rhythmically bedded, buff, clayey silt of Member B, although their carbonate and pollen content is nearly identical (Dreimanis and others, 1966, Figs. 3 and 4; Berti, 1975, Fig. 4). The sediment source of both members apparently was the same, but the depositional environment was different: a deep proglacial lake for Member B and a shallower lake with no association to glacial ice for Member C.

The other, more-pronounced discontinuity (Fig. 3) was noted in the drill cores of bore holes 6 and 7 between the unweathered, reddish, calcite-rich silty clay upper Bradtville till, and the overlying greenish-gray, partly to completely leached, sandier accretion gley (Port Talbot I in Fig. 4 of Dreimanis and others, 1966). The clay-mineral data from these units (Quigley and Dreimanis, 1972) suggest that chlorite of the Bradtville till was first changed to smectite by oxidation weathering and subsequently redeposited in the reducing environment as an abundant component of the accretion gley or a shallow-pond deposit. The time interval involving oxidation, redeposition, and reduction could be of considerable length. It is unlikely that pollen would be preserved in the interval of oxidation, and, therefore, the pollen association would represent the time of the deposition of accretion gley in a reducing environment. The abundance of pollen in the accretion gley is very high and they are well preserved (Berti, 1971, Appendices D, E, and F). The very high percentage of oak (*Quercus*) pollen (zone PT-H) is conspicuous in the lower part of accretion gley: 15%–40% (Dreimanis and others, 1966, Fig. 7; Berti, 1971, Figs. 16 and 17; Berti, 1975, Figs. 4 and 5), whereas pine (*Pinus*) dominates at its top; spruce (*Picea*) is of secondary importance. Another small oak (*Quercus*) maximum occurs 1–2 m below the upper contact (zone PT-K, Fig. 4). Such a sequence would be more probable at the end of a nonglacial interval than at its beginning. However, the latter alternative was preferred by Dreimanis and others (1966) and by Berti (1971, 1975), more on paleogeographic than palynologic grounds. Berti (1975, p. 616), however, expressed some reservations:

The Port Talbot I Interval, with mean July temperatures approaching those of today, might be questioned as being interglacial by some workers. Correlation of this interval (which is beyond the range of carbon-14) with others poses serious problems with respect either to climate or to age.

Bradtville Drift

Because this report deals mainly with the early Wisconsinan, and the Bradtville drift is now proposed to be Illinoian in age, its characteristics and depositional history will not be discussed here in any detail. The presence of Illinoian glacial deposits in the Erie basin is quite probable because Illinoian tills, although without

[1]This conclusion is quoted here from the abstract of Dreimanis (1988). Words in italics were omitted from the original by a word-processor's printing error.

any numerical-age determinations, have been widely reported from the downglacier area, south of the eastern Great Lakes (Fullerton, 1986, and references therein; Szabo, this volume). In the upglacier area, the nearest till assigned with some certainty to the Illinoian glaciation is the York till at Toronto, which underlies Sangamonian interglacial sediments (Terasmae, 1960; Karrow, 1984).

Member A of the Tyrconnell Formation

I propose that Member A of the Tyrconnell Formation (Table 2), formerly called Port Talbot I, was deposited during the oxygen-isotope substages 5a, 5b, and 5c, which would correspond to the latest part of Sangamonian stage (Fulton, 1984; Fulton and Prest, 1987), called also Eowisconsin (Richmond and Fullerton, 1986). If Karrow's (1984, Fig. 7) chronostratigraphic assignments for the Great Lakes–St. Lawrence region are applied, considering the Sangamonian equivalent to the warmest part of the last interglacial (125 ka, or isotope substage 5e), then Member A of the Tyrconnell Formation was deposited during the earliest part of the early Wisconsinan. Berti's (1971, 1975) pollen zones PT-K and PT-H (Fig. 4), with the oak maxima (15% in PT-K and up to 40% in PT-H) and sporadic presence of other deciduous tree pollen, probably corresponded to the relatively warm oxygen-isotopic substages 5a and 5c. The very beginning of the sedimentation of accretion gley, below the lower oak maximum (Fig. 4), could have been triggered by the cold climatic episode corresponding to the oxygen-isotopic substage 5d. The upper part of the accretion gley is already in the pollen zone PT-L; it has a boreal pollen assemblage (Fig. 4), indicating a cooler climate.

No lacustrine sediment was noted between the accretion gley (Member A of the Tyrconnell Formation) and the upper Bradtville till in all seven test drillings (Fig. 2). Because the upper Bradtville drift contains waterlain till, some lacustrine sediment must have been deposited over it during the glacial retreat from the Plum Point–Port Talbot area. This sediment was probably eroded prior to the deposition of accretion gley. The Plum Point–Port Talbot area is on a southward-inclined regional slope. Erosion would have been common along such a slope, if the base level was more then 22 m underneath the present lake level. Such a situation would be possible if the Erie basin was drained via the now-buried Erigan channel (Fig. 2) during the Sangamonian stage, because this channel was about 50 m lower than the present level of Lake Erie (Flint and Lolcama, 1986, and references therein). The limited retreat of the waterfall in the Erigan channel (Flint and Lolcama, 1986) suggests that only the Erie watershed was drained at that time, and the very low elevation of the channel precluded the existence of a lake in the Erie basin. The ancestral upper Great Lakes of the Sangamonian interglacial probably drained through the buried Laurentian valley that entered Lake Ontario at Toronto (Spencer, 1890; White and Karrow, 1971; Rieck and Winters, 1982; Eyles and others, 1985).

The length of time of the deposition of the accretion gley of Member A may be estimated from the pollen diagrams discussed above. If the pollen zones PT-H and PT-K, with the oak maxima (Fig. 4), correspond to the oxygen-isotope substages 5c and 5a, and the intervening PT-J, with boreal pollen assemblages, corresponds to 5b, then about 30 ka would have been required (Shackleton, 1987, and references therein) to deposit the 1–4 m of accretion gley. A lengthy time of sedimentation of the accretion gley is supported also by its mineralogical composition, structure, and thickness. The presence of smectite requires first an oxidation of iron chlorite, but the absence of the brown oxidation color suggests a slow redeposition in a reducing environment, where the brown is converted to green (Quigley and Dreimanis, 1972). The near absence of carbonate throughout the main part of the accretion gley, except along its top and bottom (Dreimanis and others, 1966, Fig. 4; Quigley and Dreimanis, 1972, p. 993), suggests either leaching in place or redeposition of the leached A and B horizons of soil from the upslope areas, without inclusion of the C horizon. Either is a slow process, considering that the underlying Bradtville drift contains up to 45% carbonate in its silt and clay fraction (Quigley and Dreimanis, 1972, p. 993). The massive structure and considerable thickness, up to 4 m, of the accretion gley also supports slow accumulation, accompanied probably by in-place weathering.

Member B of the Tyrconnell Formation

Member B is a 6– 12-m-thick glaciolacustrine silt and clay that is rhythmically laminated throughout its upper two-thirds, but is nearly massive with occasional silt lenses in the lower one-third. The dominant color is buff to brown, except for the greenish-gray basal part. The rhythmically laminated upper two-thirds are interpreted to be varved, containing ~100 varves, each of them several centimetres thick (Dreimanis and others, 1966, p. 312). The lower massive part contains <20% carbonate, whereas the varved upper part is rich in carbonate (30%–50%), particularly dolomite. The greenish color and downward-decreasing carbonate content, in addition to the absence of any sharp boundary with underlying Member A, suggest a gradual transition from Member A to B by reworking of local accretion-gley material. The rise of water level in the proglacial lake was probably gradual, and the preexisting local sediments became reworked by subaquatic gravity flows and currents, as suggested by scattered silt and fine-sand lenses and some folding in the nearly massive clay. By analogy with the composition of the late Wisconsinan Catfish Creek drift, the high dolomite content in the upper two-thirds of Member B suggests that the main influx of sediments into the lake was provided by glacial meltwater from the north.

Pollen assemblages of Member B (pollen zones PT-L and PT-Ma in Berti, 1975) are dominated by pine; spruce and non-arboreal pollen (NAP) are secondary. However, Berti (1975,- p. 600) noted:

> a decrease in *Pinus*, and increase in *Betula, Alnus*, Gramineae, and *Sphagnum*, consistent with the glacial conditions implied by the varved sediment. Ice apparently advanced into a pine-spruce-birch-alder forest. The vegetation had a much more boreal character than during deposition of the preceding zone.

Member B probably represents a brief interval, a couple of centuries long, when the Ontario lobe, which deposited the Sunnybrook drift in the Toronto area (Hicock and Dreimanis, 1989; this volume), overrode the Niagara escarpment and blocked the eastward outlet of the contemporaneous lake in the Erie basin (Fig. 5). A proglacial lake with a high lake level developed, probably similar to some of the proglacial lakes that preceded early Lake Erie at the end of the late Wisconsinan (Calkin and Feenstra, 1985, Fig. 2). The main meltwater input into this lake was from the north, and the ice margin was relatively close, as suggested by the thick varves.

End of Early Wisconsinan glacigenic sedimentation

The proglacial varved silty clay of Member B is abruptly terminated at its upper boundary (Fig. 3), where the lower part of the overlying Member C consists of coarser, sandy and silty lacustrine sediments containing macroscopic plant and animal remains. These sediments represent a lowering of lake level to at least 10 m below the present one, as suggested by the presence of peat at the depth of 10 m below the lake level at Plum Point-(Fig. 2; test drilling no. 5). A woody peat layer offshore, below the present lake level, must exist also at Bradtville, where peat balls and slabs of peat of the same age have been washed ashore (Dreimanis and others, 1966).

The lowering of lake level was caused by the retreat of the Ontario lobe from the Niagara peninsula, thus opening the isostatically depressed eastward low outlet. This event terminated the influence of the early Wisconsinan Ontario lobe events in the Lake Erie basin.

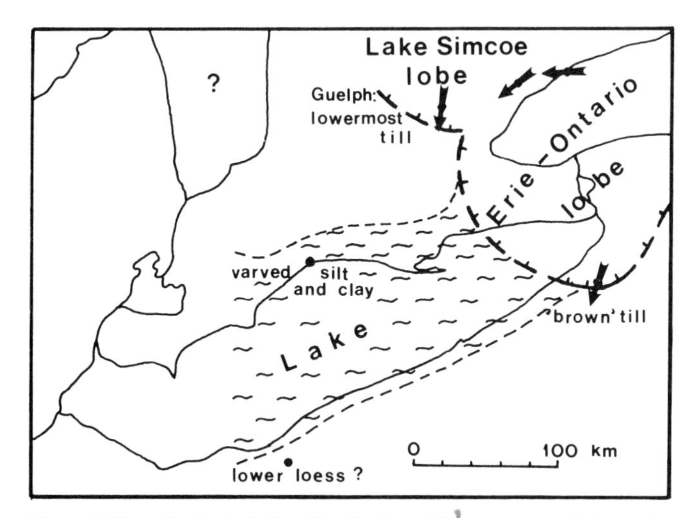

Figure 5. Hypothetical relative distribution of lake waters and glacier ice during the deposition of glaciolacustrine varved clay (Member B of the Tyrconnell Formation) in the Plum Point–Bradtville area, the "brown" till at Gowanda, the lowermost till at Guelph, and the Sunnybrook drift at Toronto. For paleo-ice-flow direction at Toronto, see Hicock and Dreimanis (this volume).

OTHER POSSIBLE EARLY WISCONSINAN DEPOSITS IN THE LAKE ERIE BASIN

Canning, Ontario, and surrounding area

Karrow (1963) suggested that a nearly pebble-free silty clay till exposed along Nith River at Canning, Ontario, named the Canning till, is of early Wisconsinan age. In texture and heavy-mineral content (Karrow, 1987, Table 4), it resembles the Sunnybrook and upper Bradtville tills. This lithologic similarity is merely due to the lobal affinity of all three tills: they were deposited by the Ontario or Ontario-Erie lobe, but not necessarily during the same glacial episode. Karrow (1987, p. 39) wrote:

> The Canning Till is believed to extend northwest along the Nith River valley as far as new Hamburg. Throughout its extent it underlies the sandy Catfish Creek Till (Cooper, 1975). The age of the Canning Till is still unknown and its relationship to dated organic interstadial deposits in nearby areas (Innerkip, Glen Allan, Guelph, Toronto) has not been established. The author (Karrow, 1963) favoured an Early to Middle Wisconsinan age in 1963, and Cowan (1975) classified correlative till in the Woodstock area under Early Wisconsinan. Cooper (1975), however, suggested that it was younger than the Port Talbot Interstade, which would make it Middle Wisconsinan or early Late Wisconsinan in age.

Karrow (1987, p. 35) tentatively considered the "lower beds" underlying the Canning till along the Nith River to be of early Wisconsinan age. However, the same age uncertainties for the Canning till apply to the "lower beds": both are older than the Catfish Creek drift, but their age is still conjectural.

The possible early Wisconsinan tills in the Woodstock area, mentioned in the above quotation, are the correlatives of the Canning till in Domtar quarry at Beachville, Ontario (Cowan, 1975, p. 17–19), and tills A and B in the Canada Cement quarry near Zorra (Westgate and Dreimanis, 1967) that contain reworked plant remains, mainly *Pinus* pollen.

Because of the lack of numerical ages beyond the range of radiocarbon dating, the stratigraphic assignments of all of the above tills are based on their provenance, their relative stratigraphic position, and a comparison with the Sunnybrook drift and/or Bradtville drift. Therefore, the present reinterpretation of the Bradtville drift as being Illinoian influences the stratigraphic assignments between the Toronto and the Plum Point–Bradtville areas, permitting two alternative correlations. Selected areas between them are therefore discussed briefly below.

Guelph, Ontario

Karrow and others (1982, p. 1870) described a section containing a paleosol developed in gravel exposed along Victoria Road in Guelph, and concluded the following.

> A glacial advance deposited the lowest till and probably during the retreat the overlying gravel was laid down as glaciofluvial outwash. The age of the advance is unknown but it is at least as old as early Wisconsinan and could be of Illinoian age.

The intensity of soil development in this gravel is interpreted as "intermediate between that of the Sangamon and Sidney soils." The heavy-mineral content of the "lowest till" suggests a western Grenville source (Karrow and others, 1982, p. 1859), and it was probably deposited by the Georgian Bay–Lake Simcoe lobe (Fig. 5).

The Karrow and others (1982) alternative of early Wisconsinan age of the lowest till and gravel is accepted here because the high dolomite content in the rhythmically bedded Member B of the Tyrconnell Formation at Bradtville (Dreimanis and others, 1966, Figs. 3 and 4) suggests a glacial-meltwater source to the northeast. Consequently, the ice margin of the Georgian Bay–Lake Simcoe lobe may have been in the Guelph area (Fig. 5).

Gowanda Hospital, New York

Calkin and others (1982, p. 1110) described "a well developed soil, characterized by deep leaching, strong heavy mineral etching and depletion, and chlorite and illite alteration . . . developed in red 'Collins Till' . . . . Overlying woody peat with a terminal date of 52,000 yr. b.p. bears a spruce-rich, forest-tundra pollen assemblage." The peat is overlain by an unnamed brown till that has north-northeast–dipping shear planes at its base. The next younger till is the Thatcher till, which "Is typical of surface tills in western New York, south of Buffalo" (Calkins and others, 1982, p. 1120), and is therefore of late Wisconsinan age.

The overall interpretation of the stratigraphy of the Gowanda Hospital sections permitted Calkin and others (1982) to tentatively assign the "brown" till to the middle Wisconsinan Cherrytree stadial, and therefore to correlate it "at least in part, with the Titusville Till in Pennsylvania" (Calkins and others, 1982, p. 1139). Fullerton (1986, p. 30–31) agreed with this lithostratigraphic correlation, but considered the Titusville till and, consequently, the "brown" till of Gowanda, to be early Wisconsinan in age. All these stratigraphic assignments are based upon long-distance correlations of isolated sections, and therefore alternative interpretations are always possible.

I tentatively agree with Fullerton's age reassignment of the "brown" till to the early Wisconsinian because of the strongly developed soil on the next-older Collins till; however, more numerical ages are needed. The two currently available radiocarbon ages, 51.6 +1.9 −1.5 ka (GrN-133) and 47.2 ± 0.8 ka (QL-134), are on wood from carbonaceous silt that is laterally connected with a ground-water–bearing channel gravel (Calkin and others, 1982, Fig. 2), and therefore could have been contaminated by younger carbon. The two ages on wood from the base of the "brown" till are nonfinite: >38 ka (W-866) and >48.4 ka (GrH-5486). All of the above dates are listed in Calkin and others (1982, p. 1120).

If the "brown" till is indeed early Wisconsinan, then it could have been deposited by the advance of the Ontario lobe, which blocked the Lake Erie outlet during deposition of Member B of the Tyrconnell Formation at Plum Point–Bradtville, Ontario (Fig. 5). The north-northeast–dipping shear planes at the base of this till (Calkins and others, 1982, p. 1116) suggest that the paleoflow of ice was from north-northeast, and the northwest fabrics (Calkins and others, 1982, p. 1117, Table 2) are probably transverse fabrics. According to Calkin and others (1982, p. 1119), the "brown" till is composed mainly of a mixture of weathered and unweathered Collins till. Therefore, its lithologic composition cannot be used for lithostratigraphic correlations, as has been done by Fullerton (1986, p. 30) in support of the correlation of the brown till with the Titusville till. In addition, Calkin and other's (1982, p. 1120) negative statement, that "The brown till cannot yet be correlated with other stratigraphic sections beyond this site" indirectly supports my hypothesis (Fig. 5) that the Gowanda Hospital site is at the terminal position of the early Wisconsinan advance of the Ontario lobe, and that the Titusville till is older.

Garfield Heights, Ohio

The possible early Wisconsinan sites of Ohio are not discussed here because Szabo discusses them in this volume; still, I refer briefly to the Garfield Heights site of White (1953, 1968). I (Dreimanis, 1971) proposed the name "Garfield Heights Till" to a weathered and leached diamicton separating White's (1953, 1968) Sangamonian humic gley from the underlying weathered Illinoian gravel (these age assignments are from White [1968]). I assigned the till to the early Wisconsinan and correlated it with the upper Bradtville till (Dreimanis and Karrow, 1972) by its heavy-mineral composition and stratigraphic position, but postponed the publication of the description of the section pending further detailed investigation.

Miller and Wittine (1972) and Szabo and Miller (1986) have investigated and described adjoining sections, but they did not recognize the Garfield Heights till in them, nor did they make any stratigraphic changes. Fullerton (1986) considered the Garfield Heights till to be Illinoian and White's (1953) humic gley and the underlying gravel to be pre-Illinoian.

Dreimanis's (1971) and Berti's (1971, 1975) section probably adjoins to Szabo and Miller's (1986) Section B. Szabo and Miller's (1986) 2.5-m-thick accretion gley probably corresponds to our combined accretion gley (0.6–1.2 m thick) and the underlying weathered Garfield Heights till, which was measured as 1–4 m thick along a 12-m-long section in 1969. The till is dense and fissile, and has a consistent north-northeast to south-southwest pebble fabric and striated clasts; its pebble and heavy-mineral composition was completely different from that of underlying weathered gravel.

If my correlation of the Garfield Heights and Bradtville tills is correct, then to be consistent with the assignment of Illinoian age to the Bradtville drift proposed herein, the Garfield Heights till is also Illinoian, as suggested by Fullerton (1986). The "lower loess" of White (1968), then, may be early Wisconsinan (Fig. 5).

CONCLUSIONS

For the past 25 years, the Bradtville drift, Canning till, and their lithostratigraphic correlatives in southwestern Ontario have been thought to be early Wisconsinan. This age assignment would imply that southwestern Ontario, at least as far as the central part of Lake Erie, was glaciated at that time.

Another, still hypothetical, alternative is offered in this chapter by assigning the Bradtville drift to the Illinoian glaciation, the lower Member A of the overlying Tyrconnell Formation to the late Sangamonian, Eowisconsinan, or the earliest part of early Wisconsinan and Member B of the Tyrconnell Formation to the early Wisconsinan.

Member A is an accretion gley containing a pollen assemblage that suggests an upward cooling trend representing an interglacial-glacial transition. It may be equivalent to oxygen-isotope substages 5a-5c. Because the soil was formed about 20 m below the present lake level, the Erie basin was probably drained by the buried Erigan channel (Fig. 1).

Member B of the Tyrconnell Formation is a glaciolacustrine varved clay and silt deposited in a proglacial lake during ~200 years. Because lake level must have been much higher than before, the Ontario lobe had overrun the Niagara escarpment and blocked the eastern outlet of the ancestral Lake Erie. The glacier advanced probably as far as Gowanda, New York (Fig. 5), and deposited the "brown" till of Calkin and others (1982). This "brown" till is correlated to the Sunnybrook drift at Toronto.

The lithologic composition of the lowermost till at Guelph (Karrow and others, 1982) suggests that it was deposited by the Georgian Bay–Lake Simcoe lobe; its age is probably early Wisconsinan (Fig. 5), but it could be also older. The merging of the Lake Simcoe and the Ontario lobes northwest of Toronto is implied by the glaciotectonically deformed lower parts of the Sunnybrook till at the Clairville dam (Hicock and Dreimanis, 1985).

Events involved in both the above alternative interpretations of the Bradtville drift, Canning till, and their correlatives, as well as overlying and underlying nonglacial sediments, need to be dated by applicable numerical-age determinations, such as the thermoluminescence method.

ACKNOWLEDGMENTS

The research for this project has been funded by a grant from the Natural Sciences and Engineering Research Council of Canada. I am grateful to Peter Barnett, Stephen R. Hicock, John P. Szabo, an anonymous reviewer, and both editors for improving the manuscript by critical reading.

REFERENCES CITED

Berti, A. A., 1971, Palynology and stratigraphy of the mid-Wisconsin in the eastern Great Lakes region, North America [Ph.D. thesis]: London, Canada, University of Western Ontario, 178 p.

——, 1975, Paleobotany of Wisconsinan interstadials, eastern Great Lakes region, North America: Quaternary Research, v. 5, p. 591–619.

Calkin, P. E., and Feenstra, B. H., 1985, Evolution of the Erie-basin Great Lakes, *in* Karrow, P. F. and Calkin, P. E., eds., Quaternary evolution of the Great Lakes: Geological Association of Canada Special Paper 30, p. 149–170.

Calkin, P. E., Muller, E H., and Barnes, J. J., 1982, The Gowanda Hospital interstadial site, New York: American Journal of Science, v. 282, p. 1110–1142.

Clark, P. U., and Lea, P. D., 1986, Reappraisal of early Wisconsin glaciation in North America: Geological Society of America Abstracts with Programs, v. 18, p. 565.

Coakley, J. P., and Lewis, C.F.M., 1985, Postglacial lake levels in the Erie basin, *in* Karrow, P. F., Calkin, P. E., eds. Quaternary evolution of the Great Lakes: Geological Association of Canada Special Paper 30, p. 195–212.

Cooper, A. J., 1975, Pre-Catfish Creek tills of the Waterloo area: [M.S. thesis]: Waterloo, Ontario, University of Waterloo, 178 p.

Cowan, W. R., 1975, Quaternary geology of the Woodstock area, Ontario: Ontario Division of Mines Geological Report 119, 91 p.

Dreimanis, A., 1957, Stratigraphy of the Wisconsin glacial stage along the northwestern shore of Lake Erie: Science, v. 126, p. 166–168.

——, 1958, Wisconsin stratigraphy at Port Talbot on the north shore of Lake Erie, Ontario: Ohio Journal of Science, v. 58, p. 65–84.

——, 1963, New test drillings in lower Wisconsin deposits at Port Talbot, Ontario: Geological Society of America Special Paper 76, 51 p.

——, 1964, Notes on the Pleistocene time-scale in Canada, *in* Osborne, F. F., ed., Geochronology in Canada: Royal Society of Canada Special Publication 8, p. 139–156.

——, 1971, The last ice age in the eastern Great Lakes region, North America, *in* Ters, M., ed., Etudes sur le Quaternaire dans le Monde (VIII e Congress INQUA Paris 1969): Centre National de la Recherche Scientifique, v. 1, p. 69–75.

——, 1987, The Port Talbot interstadial site, southwestern Ontario, *in* Roy, D. C., ed., Northeastern section of the Geological Society of America: Boulder, Colorado, Geological Society of America, Centennial Field Guide, v. 5, p. 345–348.

——, 1988, The early Wisconsinan in the Erie basin: Geological Society of America Abstracts with Programs, v. 20, no. 7, p. A3.

——, 1991, The Laurentide Ice Sheet during the Last Glaciation: A review and some current reinterpretations along its southern margin, *in* Frenzel, B., ed., Klimageschichtliche Probleme der letzten 130000 Jahre: Mainz, Akademie der Wissenschaften und der Literatur; Stuttgart, Gustav Fischer Verlag, p. 267–291.

Dreimanis, A., and Karrow, P. F., 1972, Glacial history of the Great Lakes–St. Lawrence region, the classification of the Wisconsin(an) Stage, and its correlatives: Proceedings, International Geological Congress, 24th Montreal, section 12, p. 5–15.

Dreimanis, A., Terasmae, J., and McKenzie, G. D., 1966, The Port Talbot interstade of the Wisconsin glaciation: Canadian Journal of Earth Sciences, v. 3, p. 305–325.

Eyles, N., Clark, B. M., Kaye, B. G., Howard, K.W.F., and Eyles, C. H., 1985, The application of basin analysis techniques to glacial terrains: An example from the Lake Ontario basin, Canada: Geoscience Canada, v. 12, p. 22–32.

Flint, J. J., and Lolcama, J., 1986, Buried ancestral drainage between Lakes Erie and Ontario: Geological Society of America Bulletin, v. 97, p. 75–84.

Follmer, L. R., 1983, Sangamon and Wisconsinan pedogenesis in the midwestern United States, *in* Porter, S. C., ed., Quaternary environments of the United States, Volume 1, The late Pleistocene: Minneapolis, University of Minnesota Press, p. 138–144.

Fullerton, D. S., 1986, Stratigraphy and correlation of glacial deposits from

Indiana to New York and New Jersey, *in* Šibrava, V., Bowen, D. Q., and Richmond, G. M., eds., Quaternary glaciations in the Northern Hemisphere: Quaternary Science Reviews, v. 5, p. 23–36.

Fulton, R. J., 1984, Summary: Quaternary stratigraphy of Canada, *in* Fulton, R. J., ed., Quaternary stratigraphy of Canada—A Canadian contribution of IGCP Project 24: Geological Survey of Canada Paper 84-10, p. 1–5.

Fulton, R. J., and Prest, V. K., 1987, Introduction, The Laurentide ice sheet and its significance: Geographie Physique et Quaternaire, v. 41, p. 181–186.

Hicock, S. R., and Dreimanis, A., 1985, Glaciotectonic structures as useful ice-movement indicators in glacial deposits: Four Canadian case studies: Canadian Journal of Earth Sciences, v. 22, p. 339–346.

—— , 1989, Sunnybrook drift indicates a grounded early Wisconsinan glacier in the Lake Ontario basin: Geology, v. 17, p. 169–172.

Karrow, P. F., 1963, Pleistocene geology of the Hamilton-Galt area: Ontario Department of Mines, Geological Report 16, 68 p.

—— , 1984, Quaternary stratigraphy and history, Great Lakes–St. Lawrence region, *in* Fulton, R. J., ed., Quaternary stratigraphy of Canada—A Canadian contribution to IGCP Project 24: Geological Survey of Canada Paper 84-10, p. 137–153.

—— , 1987, Quaternary geology of the Hamilton-Cambridge area, southern Ontario: Ontario Geological Survey Report 255, 94 p.

Karrow, P. F., Hebda, R. J., Presant, E. W., and Ross, G. J., 1982, Late Quaternary inter-till paleosol and biota at Guelph, Ontario: Canadian Journal of Earth Sciences, v. 19, p. 1857–1872.

McKenzie, G. D., 1964, The type section of the Port Talbot interstadial [M.S. thesis]: London, Ontario, University of Western Ontario, 144 p.

Miller, B. B., and Wittine, A. H., 1972, The origin of late Pleistocene deposits at Garfield Heights, Cuyahoga County, Ohio: Ohio Journal of Science, v. 72, p. 305–313.

Quigley, R. M., and Dreimanis, A., 1972, Weathered interstadial green clay at Port Talbot, Ontario: Canadian Journal of Earth Sciences, v. 9, p. 991–1000.

Richmond, G. M., and Fullerton, D. S., 1986, Introduction to Quaternary glaciations in the United States of America, *in* Šibrava, V., Bowen, D. Q., and Richmond, G. M., eds., Quaternary glaciations in the Northern Hemisphere: Quaternary Science Reviews, v. 5, p. 3–10.

Rieck, R. L., and Winters, H. A., 1982, Low-altitude organic deposits in Michigan: Evidence for pre-Woodfordian Great Lakes and paleosurfaces: Geological Society of America Bulletin, v. 93, p. 726–734.

Shackleton, N. J., 1987, Oxygen isotopes, ice volume and sea level: Quaternary Science Reviews, v. 6, p. 183–190.

Spencer, J.W.W., 1890, Origin of the basins of the Great Lakes of America: American Geologist, v. 7, p. 86–97.

Szabo, J. P., and Miller, B. B., 1986, Pleistocene stratigraphy of the lower Cuyahoga valley and adjacent Garfield Heights, Ohio trip guidebook, (Geological Society of America North-Central Section meeting guidebook, field trip number two): Kent, Ohio, Kent State University, 61 p.

Terasmae, J., 1960, A palynological study of Pleistocene interglacial beds at Toronto, Ontario: Geological Survey of Canada Bulletin 56, p. 24–40.

Vogel, J. C., and Waterbolk, H. T., 1972, Groningen radiocarbon dates X: Radiocarbon, v. 14, p. 6–110.

Westgate, J. A., and Dreimanis, A., 1967, The Pleistocene sequence at Zorra, southwestern Ontario: Canadian Journal of Earth Sciences, v. 4, p. 1127–1143.

White, G. W., 1953, Sangamon soil and early Wisconsin loess at Cleveland, Ohio: American Journal of Science, v. 251, p. 362–368.

—— , 1968, Age and correlation of Pleistocene deposits at Garfield Heights (Cleveland), Ohio: Geological Society of America Bulletin, v. 79, p. 749–752.

White, O. L., and Karrow, P. F., 1971, New evidence for Spencer's Laurentian River (Great Lakes Research Conference, 14th, Proceedings): International Association of Great Lakes Research, p. 394–400.

Manuscript Accepted by the Society September 6, 1991

Printed in U.S.A.

Geological Society of America
Special Paper 270
1992

The sedimentary and biological record of the last interglacial-glacial transition at Toronto, Canada

Nicholas Eyles
Glaciated Basin Research Group, Department of Geology, University of Toronto, Scarborough Campus, Scarborough, Ontario M1C 1A4, Canada
Nancy E. Williams
Division of Life Sciences, University of Toronto, Scarborough Campus, Scarborough, Ontario M1C 1A4, Canada

ABSTRACT

A substantial Sangamon interglacial (isotope stage 5) and subsequent Wisconsin glacial sedimentary record is preserved at Toronto, Canada. The age of individual stratigraphic units is poorly constrained, however. An interglacial sequence records climatic deterioration from warm temperate to subarctic. The interglacial *Don Beds,* resting on a presumed Illinoian till, were deposited on a storm-influenced shoreface of an ancestral Lake Ontario in water depths that increased over the recorded time interval from about 2 to 20 m. Pollen and faunal analyses identify a climatic deterioration in the upper Don from warm-temperate conditions, with mean annual temperatures some 2 °C warmer than at present, to cool temperate, with temperatures lowered by about 3 °C. Continued cooling is recorded in overlying deeper-water deltaic sediments of the *Scarborough Formation,* but later climatic amelioration and the return of mixed forest are suggested by the pollen record and by caddisfly fauna. The youngest deltaic sediments, lying immediately below a Wisconsin glacial complex, were deposited in a subarctic setting in an ice-dammed lake, with local mean annual temperatures depressed by at least 7 °C.

This chapter identifies the likely continuity of the Don Beds and Scarborough Formation and places them in an expanded "Sangamon" interglacial possibly equivalent to the whole of stage 5 (i.e., 130–75 ka). The biological record at Toronto for this stage indicates one or more phases of cooling when continental-glacier ice may have developed in North America; this record can be favorably compared with the marine oxygen-isotope record with its evidence of increased global ice volumes during stage 5. The subsequent Wisconsin record at Toronto is not well constrained by radiometric age dating but indicates that maximum regional expansion of the Laurentide Ice Sheet in the eastern Great Lakes area occurred after 25 ka.

INTRODUCTION

Almost a century has elapsed since A. P. Coleman (1894) first identified organic-rich interglacial sediments below glacial sediments in Taylors Brickyard, Toronto. The site, now better known as the Don Valley Brickyard (Figs. 1 and 2), was first exposed in 1882 and is now preserved by the Province of Ontario as a site of outstanding geologic significance.

Coleman's recognition of the Don Beds as interglacial was based on the diversity of included flora and fauna and the stratigraphic position of the beds between till sheets. Coleman (1932), summarizing more than 40 years of research, argued that the Don Beds belonged to a pre-Illinoian interglacial but subsequent workers are agreed that the sequence in the Don Valley most likely dates from the last (Sangamon) interglacial. Overlying del-

Eyles, N., and Williams, N. E., 1992, The sedimentary and biological record of the last interglacial-glacial transition at Toronto, Canada, *in* Clark, P. U., and Lea, P. D., eds., The Last Interglacial-Glacial Transition in North America: Boulder, Colorado, Geological Society of America Special Paper 270.

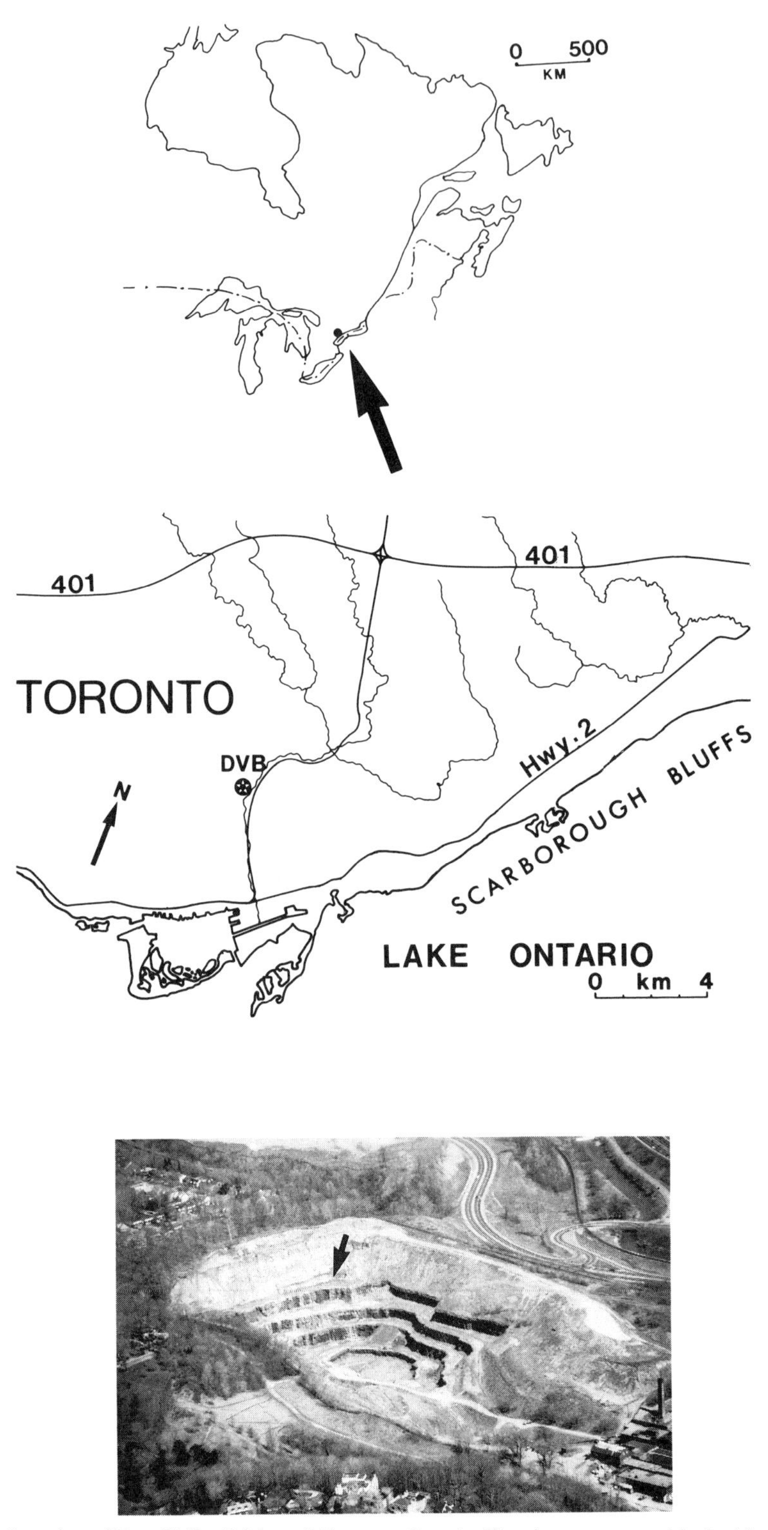

Figure 1. Location of Don Valley Brickyard, Toronto, Canada. The site exposes a critical mid-continent record (arrow in photograph) of climate change during the last interglacial-glacial transition.

Figure 2. A: Don Valley Brickyard in 1913 showing, between arrows, last interglacial Don Beds (DB) resting on bouldery Illinoian York Till (YT; Figs. 3 and 4). The wagon is located on a bedrock surface of low relief forming part of a regional "Tertiary" valley system draining from Georgian Bay to the Lake Ontario basin (the Laurentian Channel; Fig. 11). Photograph by A. P. Coleman. B: The north slope of the Don Valley Brickyard at the present day. Arrows demarcate the interglacial stratigraphy logged in Figure 4. DB, Don Beds; SF, Scarborough Formation. The dashed line is an erosion surface overlain by the Pottery Road gravels (see text).

taic beds (Scarborough Formation) are only partially exposed in the brickyard, but crop out extensively along the Lake Ontario coastline. These contain a record of the climatic deterioration accompanying the initial expansion of the Wisconsin Laurentide Ice Sheet (Fig. 3).

The purpose of this chapter is to present the results of combined sedimentologic and biologic studies of the interglacial record at Toronto. This chapter stresses (1) the lack of an adequate age-dating framework for most of the record, and (2) the relation between the local physical environment and the faunal and floral record. Ongoing work emphasizes the important controls on species preservation by local factors such as water depth, chemistry, and incidence of storm waves; integrated biologic and sedimentologic work is needed to isolate these local environmental controls before regional climatic signals are apparent. In this chapter we critically evaluate the existing evidence for or against glaciation during oxygen-isotope stage 5 (ca. 130–75 ka) and comment on the relative value of different biological groups in interpreting climate. This chapter should be regarded as a progress report; our ideas on the Quaternary stratigraphy of Toronto are evolving rapidly as more and more subsurface data become available as a result of accelerated urbanization. By way of emphasis, Sangamon interglacial beds correlative with those of the Don Valley Brickyard have been recently discovered along the Rouge River Valley some 35 km to the east of the Don Valley (Eyles and Boyce, 1992).

BRIEF HISTORY OF INVESTIGATIONS

Large-scale excavations in connection with brickmaking began systematically at the Don Valley Brickyard site in 1891; Upper Ordovician shales (the Georgian Bay Formation) were used as raw material; overlying Quaternary sediments were used for variation in brick color. Substantial numbers of fossil remains molluscs and wood) were exposed; Coleman (1894, 1895) identified these as interglacial and correlated them with earlier sporadic finds of leaves, wood, and shells along the nearby Don River reported by Chapman (1861), Bell (1861), and Dawson and Penhallow (1890). Early accounts of fossils from the nearby Scarborough Bluffs include Hinde's (1877) identifications of diatoms, algae, moss, spores, wood, seeds, ostracodes and molluscs, as well as Scudder's (1895, 1900) work on beetles. Tentative climatic reconstructions of the interglacial climate were made by Upham (1895a, 1895b).

A long series of investigations, beginning with those sup-

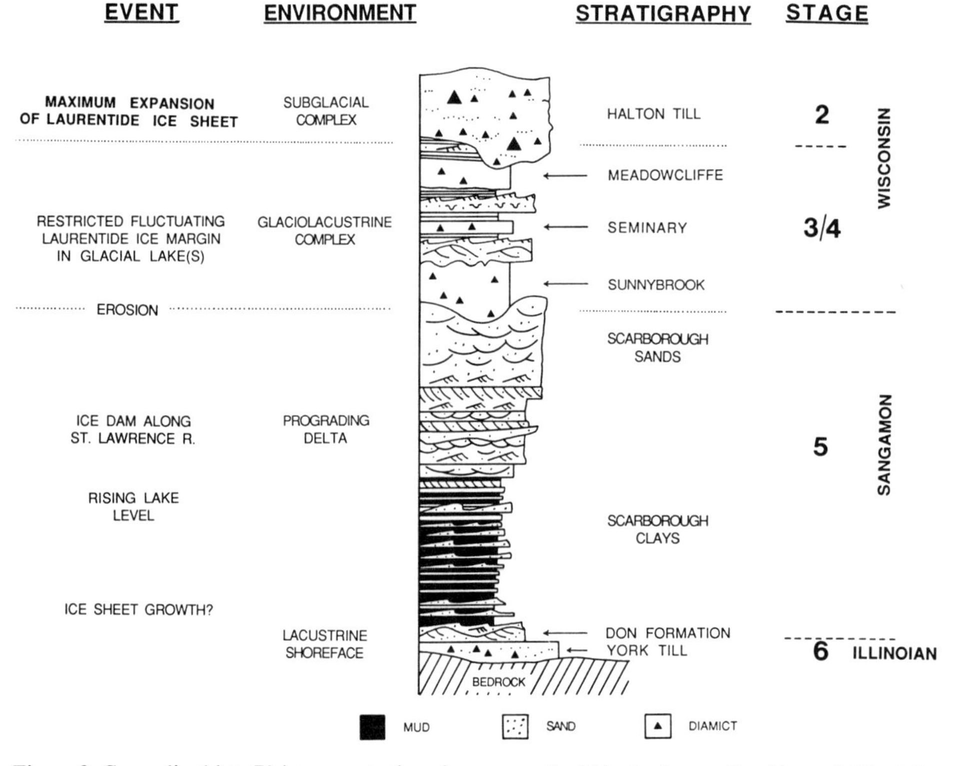

Figure 3. Generalized late Pleistocene stratigraphy preserved within the Laurentian Channel (Fig. 11). Thin sands and silty-clay rhythmites that occur between the Sunnybrook, Seminary, and Meadowcliffe diamicts are referred to as the Thorncliffe Formation. The climate record of the last interglacial-glacial transition is preserved within shoreface and deltaic sediments of the Don Beds and Scarborough Formation (Figs. 3, 4, 6, 7). Note that the correlation with marine oxygen-isotope stages is *tentative* and does not imply a continuous, complete sedimentary succession (see Fig. 10 and text).

ported by the British Association for the Advancement of Science (1900) through Wright (1914), Watt (1958), Dreimanis and Terasmae (1958) and Karrow (1967), has assigned the lowermost till at the Don Valley Brickyard (the York Till of Terasmae, 1960) to the Illinoian, the overlying warm-climate interglacial Don Beds to the Sangamon, and the uppermost glacial deposits to the Wisconsin (Fig. 3). It should be noted that the cool-climate deltaic deposits (the Scarborough Formation) overlying the Don Beds (Figs. 3 and 4) were initially assigned to the Sangamon, and collectively with the Don Beds, were identified as the Toronto Formation (Chamberlin, 1895; Coleman, 1932). Subsequently the two units were separated on the basis of a presumed unconformable relation, the Scarborough sediments being placed into the Wisconsin (see Karrow, 1984). Recent sedimentological work (Eyles, 1987; Eyles and Clark, 1988, 1989), together with palynological evidence (see below), however, suggest that the Don Beds and the Scarborough Formation are a continuous sequence recording overall climatic deterioration and rising lake levels. The Sangamon interglacial record at Toronto is argued herein to start at the base of the Don Beds and terminate at the top of the Scarborough Formation below a regionally widespread early Wisconsin glacial diamict (the Sunnybrook till of Karrow, 1967) and associated fan-delta gravels (the Pottery Road Formation; Karrow, 1974) (see below; Figs. 3 and 4).

Sedimentological studies of the Don Beds and Scarborough Formation were reported by Kelly and Martini (1986) and Eyles and Clark (1988). Modern paleoecological studies have included work on pollen and plant macrofossils (Terasmae, 1960), diatoms (Duthie and Mannada Rani, 1967), insects (Morgan, 1972; Morgan and Morgan, 1980; Williams and Morgan, 1977; Williams and others, 1981), molluscs (Hui and others, 1969; Kalas, 1975; Kerr-Lawson, 1985), ostracodes (Poplawski and Karrow, 1981), and cladocerans (Hann and Karrow, 1984). Fossils from some of these plant and animal groups (Fig. 5) have been recovered in continuous sequences throughout the Don Beds and overlying Scarborough Formation outcrops or in boreholes. However, there remain significant gaps in interpretation with regard to any one fossil group because of variable preservation. In addition, groups are of different value in interpreting local climate. An integrated approach, emphasizing the biological along with the sedimentological framework, is therefore essential.

SEDIMENTOLOGY OF THE LAST INTERGLACIAL DEPOSITS

The Don Beds (Fig. 4) record shallow-water sedimentation on a muddy lacustrine shoreface environment along the storm-influenced margin of an ancestral Lake Ontario. The overlying Scarborough Formation records the commencement of prodelta sedimentation by a large southward-flowing river associated with rising interglacial lake levels, cooling climate, and subsequently the presence of an ice dam along the St. Lawrence River Valley. Almost the full thickness of the Scarborough Formation (about 50 m) is exposed above modern lake level along the Scarborough Bluffs (Fig. 6). At the Brickyard, only the lower 5 m are preserved below an erosion surface, at which rest poorly-sorted early Wisconsin gravels (the Pottery Road Formation; Karrow, 1974). These gravels record the cutting and filling of erosional valleys into the interglacial stratigraphy as the result of the progradation of a large meltwater-fed fan delta (Eyles, 1987; see below). The Pottery Road sediments contain reworked faunal remains that include deer, bison, bear, and muskox, and pollen indicative of a subarctic forest-tundra setting (Churcher and Karrow, 1977).

York Till, Don Beds

The base of the Pleistocene outcrop in the Don Valley Brickyard is represented by a bouldery diamict about 1 m thick, overlain by a blue sandy clay containing well-preserved and articulated fresh-water molluscs, some of which today are restricted to the Mississippi drainage basin (Kalas, 1975). The unit can be more clearly seen on photographs taken when the site was being actively quarried (Fig. 2A). The diamict, as currently exposed, appears to be a wave-washed remnant of a till (the Illinoian York Till of Terasmae, 1960; Karrow, 1967, 1974) that is widespread in the Toronto area and crops out at Woodbridge about 30 km to the north. The blue clay forms the base of the Don Beds and is overlain by about 6 m of sand and gravelly sands, in beds up to 1 m thick, interbedded with mud strata. The thickness of muddy strata increases upward, composing a well-defined fining-upward sequence (Fig. 4). Sands display well-developed hummocky and swaley cross-stratification, truncated by wave ripples and draped with mud (Fig. 7C), a succession interpreted to record storm waves and subsequent periods of fair weather on a shallow (2 m) lacustrine shoreface. Detailed facies descriptions were presented in Eyles and Clark (1987, 1988). Reconstruction of water depths (see below) indicates that depths increased to at least 20 m toward the top of the Don Formation. The current elevation of the site (10 m above the modern lake) suggests a water plane about 30 m above Lake Ontario toward the top of the Don Beds prior to the start of prodelta sedimentation by a large river. Rising interglacial water levels can perhaps be rationalized by continued isostatic rebound of the lake outlet east of present-day Kingston. Present water levels at Toronto are rising by about 0.23 m per 100 years (Clark and Persoage, 1970; Kite, 1972) as a result of differential postglacial uplift of the eastern lake basin. Elevated interglacial lake levels at Toronto are consistent with a last interglacial sea-level highstand, of 7 m above the modern level, between 120 and 130 ka (Bard and others, 1990).

In the upper part of the Don Beds, hummocky and swaley cross-stratified sand facies become thinner and show well-developed lenticular bedding with "linsen" sand lenses composed of wave ripples enclosed in mud. Large isolated clasts are present throughout the Don Beds and are interpreted as the product of rafting by seasonal lake or river ice. Mud strata are extensively bioturbated, indicating an oxygenated lake floor.

Reconstruction of wave heights using linear wave theory (Eyles and Clark, 1988) identifies a relatively high energy shore-

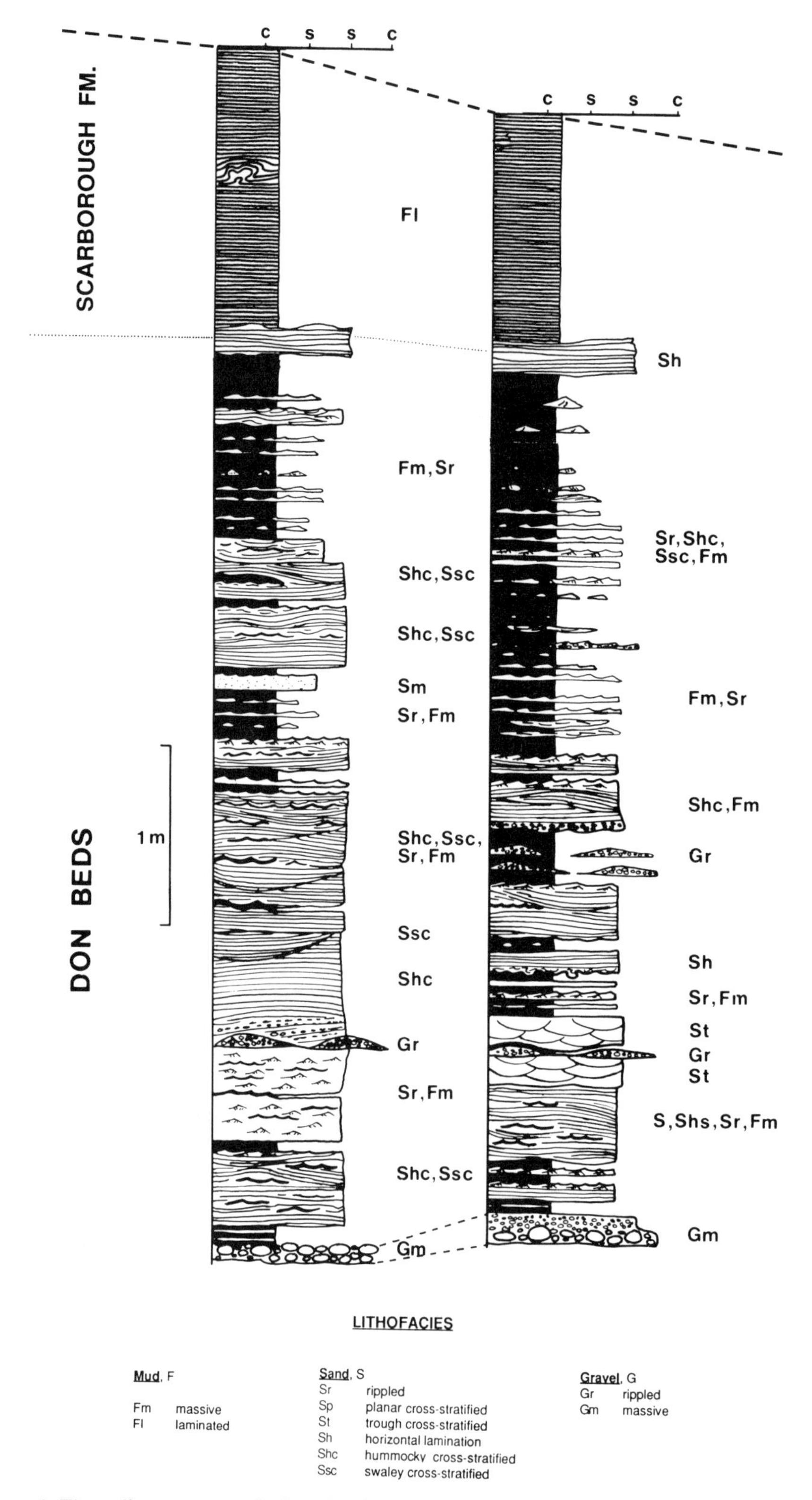

Figure 4. The sedimentary record of the last interglacial-glacial transition exposed in the Don Valley Brickyard (Fig. 2B). The Don Beds and overlying Scarborough Formation clays contain a record of climatic deterioration from warm temperate to southern boreal conditions; the record at this site is truncated by an erosion surface (heavy dashed line) below early Wisconsin outwash (Pottery Road gravels; see text). Uppermost Scarborough Formation sediments are exposed along Scarborough Bluffs (Fig. 6).

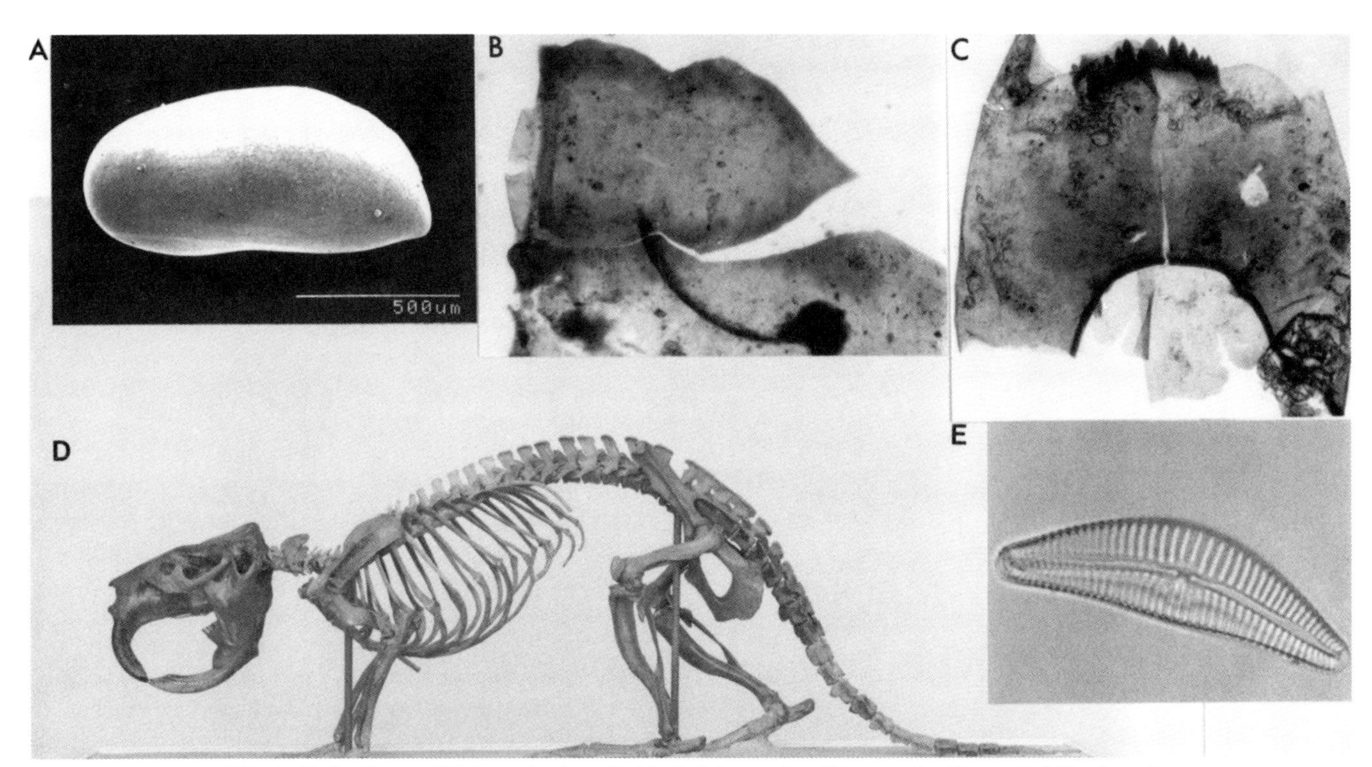

Figure 5. Fossil material from the interglacial-glacial record at Toronto. A—Ostracod: left valve of *Candona caudata*. B—Caddisfly: frontoclypeus and parietal of *Rhyacophila* sp. Same scale as in A. C—Chironomid: head capsule of *Chironomus* sp. Same scale as in A. D—Giant Beaver (*Castoroides ohioensis*); length of skeleton is 1.6 m. E—Diatom: *Cymbella* sp.

line with waves up to 6.5 m high with periods of about 5 s; similar short, steep waves are developed in Lake Ontario during sustained easterly winds (maximum wave height = 4 m, period = 9 s). Sand and gravel wave ripples within the Don Beds are consistently oriented north-south, suggesting that dominant winds were also from the east during the last interglacial. Littoral influences decrease upward in the Don Beds and are reflected in the type and quality of preservation of fossil material (see next section).

Scarborough Formation (Scarborough clays, sands)

Karrow (1967) defined the top of the Don Beds at a weathered carbonate-poor sand bed showing strong oxidation colors at a height of about 6 m above the base of the sequence (Figs. 4 and 7A). We follow Coleman (1932), who argued that there is no marked sedimentological discontinuity between the Don Beds and overlying Scarborough Formation, the two being a single lacustrine sequence (the Toronto Formation of Coleman, 1932; see above). The pollen record (Fig. 8) is consistent with a gradual transition from the Don Beds into the fine-grained prodelta facies of the Scarborough Formation (see below). The latter are composed of organic-rich silty-clay rhythmites up to 2 cm thick; Coleman (1932) reported more than 650 rhythmites at this section and suggested an annual control (e.g., Fielding, 1984; Hazeldine, 1984). Trace fossils, thin detrital peats, and siderite concretions are common; organic carbon contents range from about 0.01% to 1.35%.

At the Don Valley Brickyard, prodelta rhythmites of the Scarborough Formation are truncated by overlying Pottery Road Formation gravels, but a complete delta sequence nearly 50 m thick is exposed along the lakeshore to the east (Fig. 6). The delta, capped by a Wisconsin glaciolacustrine complex (Fig. 3), contains cool temperate, boreal, and subarctic insects consistent with climatic deterioration prior to the direct entry of ice into the Ontario basin (see below).

The basal portions of the exposed delta, up to about 20 m above present lake level, show a range of rhythmically laminated and bioturbated silts and clays deposited by underflows (quasi-continuous turbidity currents) down the delta front. The presence of smectite clay suggests advanced Sangamon pedogenesis in the drainage basin. Overlying delta-front facies, almost 30 m thick, show an upward increase in the number of rippled and parallel-laminated sand beds interbedded with rhythmites. Within the delta-front sequence, massive silty sand beds up to about 4 m thick, containing "rip-up" rafts, are broadly channelized (widths up to 400 m) and record downslope failure of the delta front as slump sheets. Sediment gravity-flow facies (debris flow) occur at the top of crudely developed thickening-upward cycles, suggesting that delta-front instability was initiated by increasing

Figure 6. Coarsening upward Scarborough Formation delta from prodelta muds at the lake level (Scarborough clays; Fig. 3) to uppermost Scarborough sands deposited under subarctic conditions (section 2581 of Eyles and Eyles, 1983). The delta is overlain by a thick glaciolacustrine complex; arrow shows the contact (Fig. 3).

fluvial discharge and sediment loading (Eyles, 1987, p. 1015; see below).

Above the 30 m level, facies are sand dominated (the Scarborough sands; Figs. 3, 6) and rest abruptly on underlying fine-grained facies. Sands show considerable variability, but planar and trough cross-bedded, horizontally laminated and rippled facies dominate and indicate a sandy braided-river system; cross-bedded facies record the migration of large in-channel dunes and are capped by rippled and parallel-laminated sands (e.g., Miall, 1978; Walker and Cant, 1984). Large crosscutting channels several tens of metres wide and up to 2 m deep contain detrital peats on the foresets of individual cross-beds. The suggested depositional environment is one of a large sandy braid plain.

The precise genetic relation between the upper Scarborough Formation, showing broad channels floored by Pottery Road Formation gravels, and the overlying Wisconsin glaciogenic sequence is not clear. The elevation of the uppermost deltaic beds above modern lake level (+45 m) and inferred subarctic conditions suggests but does not prove, an indirect glacial influence on sedimentation as a result of an ice dam along the eastern lake outlet. High lake levels could also reflect, however, enhanced isostatic rebound of the basin outlet particularly as the tectonics of this area (the Frontenac arch) are not well known. Debris flows within the uppermost Scarborough Formation, argued to be glaciogenic in origin by Kelly and Martini (1986), are slump sheets formed on the unstable delta front (see above). Similar uncertainty surrounds interpretation of the origin of the broad channels cut on the delta top; these may indicate a phase of low lake level, or alternatively could also result from subaqueous downcutting through noncohesive sands under conditions of enhanced sediment and runoff to the delta in a subarctic environment. Given this level of uncertainty regarding the relation between the Scarborough Formation and overlying glaciogenic beds, and the suggested continuity between the Don Beds and Scarborough Formation, the latter is assigned to oxygen-isotope stage 5 as part of an extended Sangamon "interglacial." This recognizes, however, the likely presence of glaciers in the region.

The contact between the basal part of the Sunnybrook diamict and underlying Scarborough Formation is variable. In places, the two units are interbedded, but locally the contact is disconformable and at a few locations shows deformation structures such as small-scale folding and shearing. The latter were interpreted as "glacitectonic" by Hicock and Dreimanis (1989) and ascribed to subglacial deformation below an early Wisconsin ice sheet that reached the southern margins of the Great Lakes. The Sunnybrook diamict was correspondingly interpreted as a subglacial deposit. The structures described by these workers are not persistent as rapid (50 cm/yr) cliff erosion demonstrates. The structures described by these workers are no longer present at that outcrop. We regard local deformation structures at the base of the Sunnybrook as scours made by floating masses of ice impacting on the floor of a large proglacial lake (see below).

BIOLOGICAL RECORD

The biological record provides specific information regarding both the local environment and the regional climate (Fig. 9); this information further supports many of the conclusions drawn from analysis of sedimentary facies.

Palynological investigations of the Don Beds and Scarborough Formation (Terasmae, 1960; McAndrews, unpublished) are of particular significance in demonstrating that the two successions are part of a single sedimentary sequence. Figure 9 is a composite diagram based on the exposures in the Don Brickyard (across the Don/Scarborough contact) and at Scarborough Bluffs (through correlative and overlying beds of the Scarborough). Data from the Don Brickyard show an unbroken trend of increasing cool-climate indicators such as spruce, pine, and birch that is matched by a decrease in such species as oak, ash, and elm across the Don/Scarborough contact. These data are inconsistent with the presence of a major erosional disconformity between the Don

Figure 7. The Don Beds at the Don Valley Brickyard. A: Site shown in Figure 4. Arrow identifies contact with overlying Scarborough Formation. B: Same site as above, showing overall fining-up character of Don Beds. C: Gravel-wave ripples (outlined), overlain by swaley cross-stratified sands (A) and wave-rippled sands with mud drapes (B).

and Scarborough, but do not rule out the existence of a hiatus between the two (see above).

Pollen (Terasmae, 1960; McAndrews, unpublished) and plant macrofossils (Dawson and Penhallow, 1890) from the lowermost 3 m of the Don Formation show that the area supported a diverse deciduous forest, including oak(*Quercus*), elm (*Ulmus*), birch (*Betula*), ironwood (*Ostrya/Carpinus*), ash (*Fraxinus*), maple (*Acer*), basswood (*Tilia*), cedar (*Thuja*), beech (*Fagus*), hickory (*Carya*), poplar (*Populus*), sweet gum (*Liquidambar*), and willow (*Salix*). Plant macrofossils identified by Kerr-Lawson (1985) from the Don Beds were mainly aquatics and herbs, but also included specimens of balsam (*Populus balsamifera*), spruce (*Picea*), pine (*Pinus*), birch (*Betula*), alder (*Alnus*), and willow (*Salix*) in the upper Don Beds; spruce, pine, and ash (*Fraxinus*) at a slightly lower level; and ash, eastern white cedar (*Thuja occidentalis*), and willow (*Salix*) in the lower part of the sequence. The remainder of the Don Beds and the lower Scarborough Formation show increasing spruce, pine, alder, and birch, and a corresponding fall in the representation of deciduous species. Small quantities of deciduous pollen recur in the uppermost part of the Scarborough Formation clays before disappearing again in the overlying sands.

Much of the early work on fossils from the Don Beds and Scarborough Formation described some species as extinct forms, e.g., Dawson and Penhallow (1890) for plant macrofossils from the Don Beds, and Scudder (1895, 1900) for beetles from the Scarborough Formation. More recently, corrections to some nomenclature and identifications have been made by others, e.g., Baker (1931) for molluscs, Brown (1942) and Warner (1986) for plants, or have been suggested to be needed (e.g., Morgan, 1975). Most plant and invertebrate fossils are now identified as extant forms. Nevertheless, a high standard of identification and interpretation appears to have been maintained by early fossil-plant collectors, as shown by reexamination of old collections and new samples from some of the same localities (Warner, 1986). Recent work, for the most part, has confirmed early interpretations.

Diatom, ostracode, aquatic mollusc, and caddisfly assem-

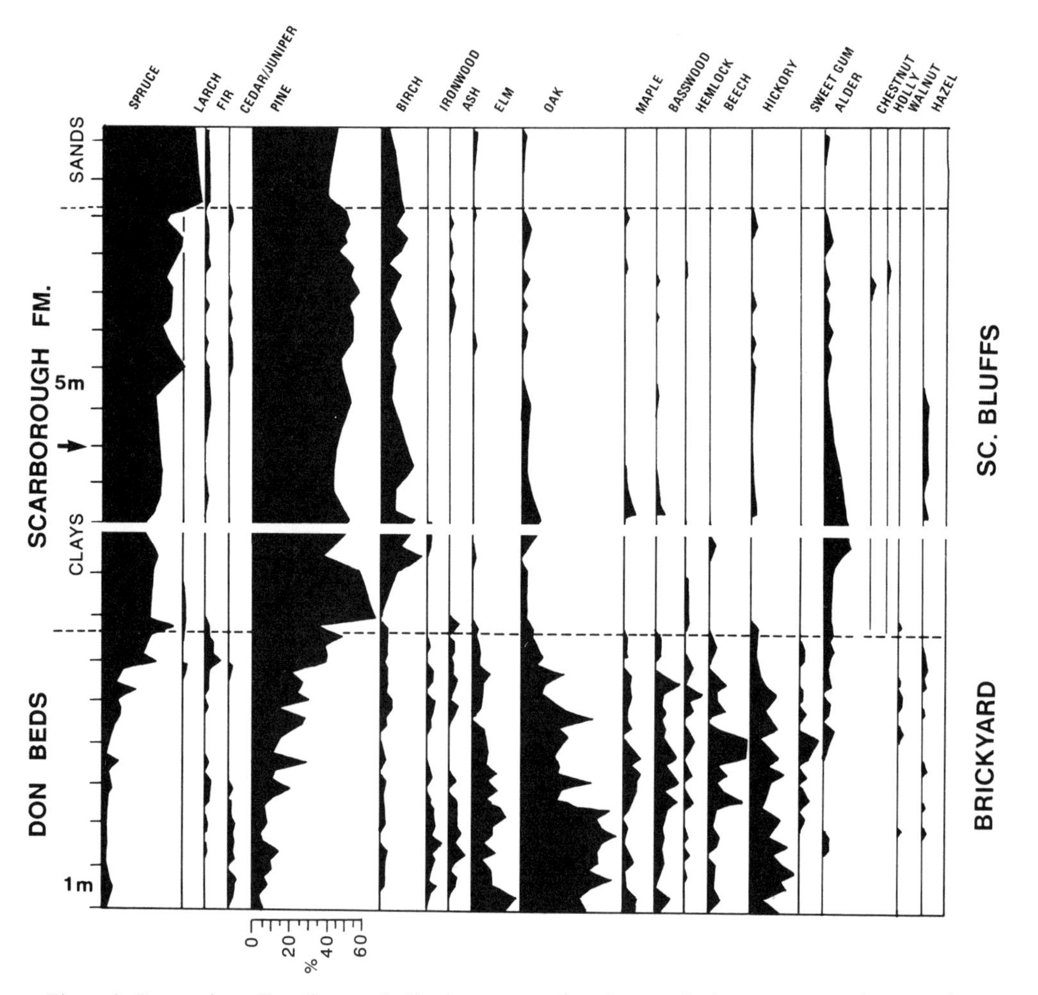

Figure 8. Composite pollen diagram (indicating percent of total tree pollen) across contact between the Don Beds and Scarborough Formation at the Don Valley Brickyard. Note continuity of pollen record across contact. Uppermost pollen data from Scarborough Formation are from correlative sections along Scarborough Bluffs. Redrawn from Terasmae (1960) by J. McAndrews, N. Eyles, and N. Williams. Note scale change at arrow; 1 m intervals below arrow and 5 m intervals above.

blages from the Don Beds are all mixtures of lotic and lentic species indicating deposition in shallow water (Duthie and Mannada Rani, 1967; Kalas, 1975; Kerr-Lawson, 1985; Poplawski and Karrow, 1981; Williams and Morgan, 1977; Williams and Eyles, unpublished). The composition and diversity of the mollusc and caddisfly assemblages suggest an important fluvial influence during deposition of the Don Beds.

Cladocera are particularly useful in recording changes in water depths. Hann and Karrow (1984) identified 30 species of chydorid cladocerans and three genera of nonchydorid Cladocera in a Don/Scarborough core from the brickyard, and calculated the percentage representation of littoral (shallow water) and planktonic (open water) species at 38 levels. Their data identified four major littoral peaks and show an overall increase in planktonic species upward in the section, confirming the deepening trend shown by calculations of water depths using linear wave theory applied to wave ripples (see above). Predominant chydorid species throughout the Don Beds were *Monospilus dispar, Eurycercus longirostris, Alona quadrangularis,* and *Alona affinis,* although the bosminid, *Bosmina longirostris,* was the dominant cladoceran throughout much of the lower half of the Don Beds. Low numbers of cladoceran remains were found in the upper 1 m of the Don Beds and in all of the Scarborough Formation exposed in the Don Brickyard core.

Diatom distributions are more closely linked with water chemistry than with climate (Hustedt, 1954; Duthie and Mannada Rani, 1967), and fossil assemblages are therefore good indicators of pH and trophic status. Duthie and Mannada Rani (1967) identified about 200 species of diatoms from the Don Beds and Scarborough Formation, and estimated that the Don

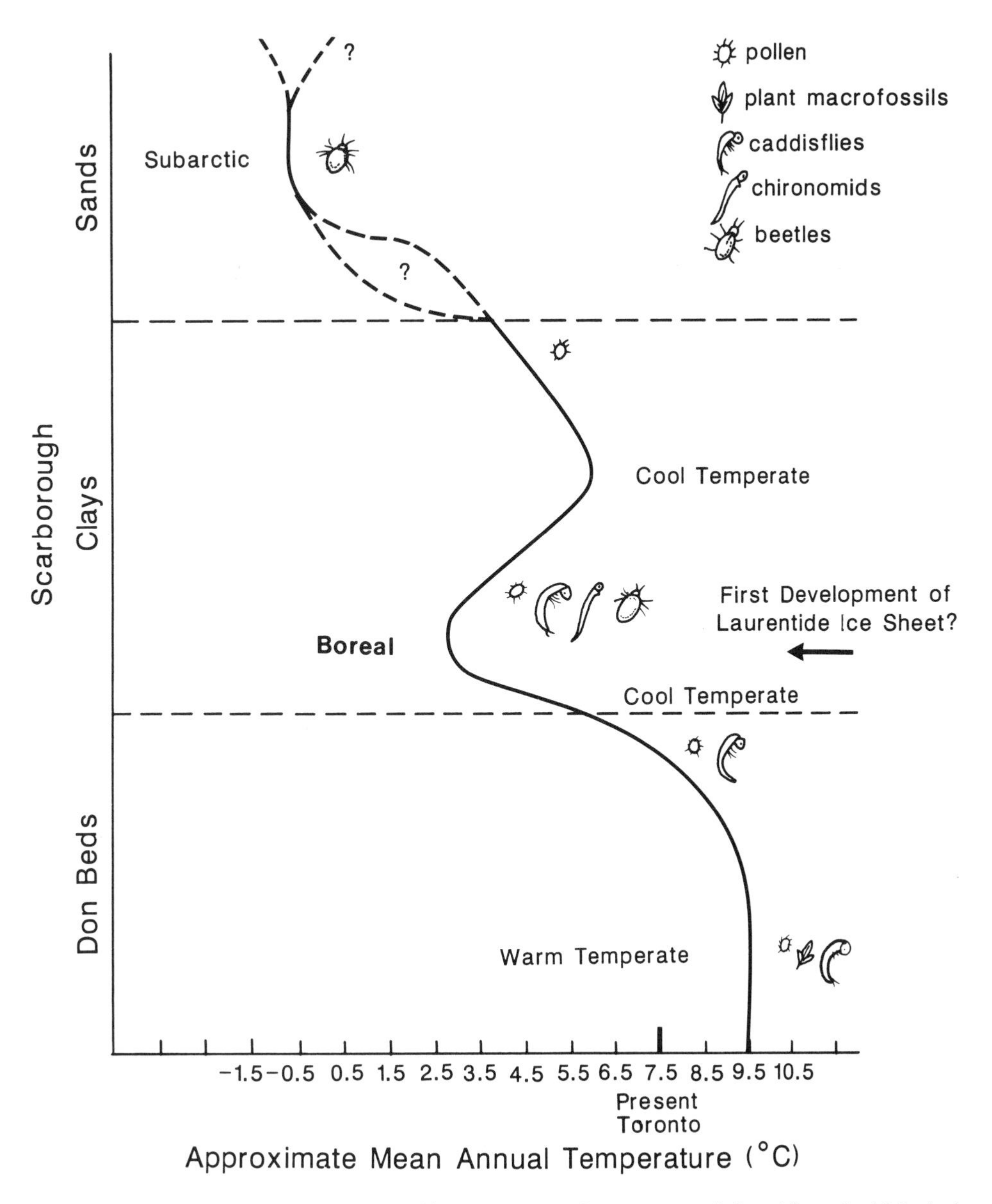

Figure 9. The last interglacial-glacial transition; mean annual temperatures inferred from the biological record (groups indicated by symbols) in the Don Beds and Scarborough Formation.

Beds in the brickyard were deposited in neutral to moderately alkaline (pH 7.5–9.0), mesotrophic to eutrophic water. Many of the most-abundant species occur today in lakes and slowly-flowing rivers, the overall assemblage suggesting deposition in a fresh-water estuary. The assemblage is comparable to the extant diatom flora in the vicinity of the lower Great Lakes. Diatoms were erratically distributed in the sequence, with only 11 of 32 samples containing frustrules. The upper 2 m of the Don Beds were barren, and only one sample from the Scarborough Formation contained diatoms. In the Scarborough sample, the absence of several mesotrophic-eutrophic indicators found in the Don samples suggests that the trophic level may have decreased.

Cladocerans (Hann and Karrow, 1984), diatoms (Duthie and Mannada Rani, 1967), and plant macrofossils (Kerr-Lawson, 1985) indicate the presence of abundant shallow-water macrophytes at various times during accumulation of the Don Beds, suggesting the presence of relatively quiet water episodes between storms along the lakeshore.

Ostracodes present in the Don Beds (Poplawski and Karrow, 1981) have an erratic distribution; there are few valves in the high-energy littoral sediments in the lower part of the formation, and none in the upper 1 m of sediments as a result of poor preservation. The dominant ostracode in the Don Formation is *Candona rawsoni,* a temperate species that lives in shallow (<20 m) lakes. *Candona caudata,* which occurs in streams and lakes with some current action, and *Candona* cf. *subtriangulata,* a cold, deep-water species (Delorme, 1969), are the dominant forms in the lower Scarborough Formation. Consistently thinner and more fragile shells within the Scarborough Formation, in comparison with those within the Don Beds, suggest slight lowering of pH, consistent with cooler waters (Poplawski and Karrow, 1981).

Caddisfly, beetle, and chironomid (midge) faunas provide physical and biological details about the rivers feeding the lake and the nature of vegetation within the drainage basin. A few of the 60 caddisfly taxa, in 16 families, recovered from the Don Beds and Scarborough Formation occur at the present day along wave-washed shores of the Great Lakes (*Hydropsyche confusa, Hydropsyche alterans, Cheumatopsyche* sp.), but the majority are river inhabitants whose habitat preferences indicate that the interglacial rivers were characterized by a variety of substrate grain sizes (silt, sand, gravel), shallow riffles and deeper pools, epilithic algae, and abundant suspended organic matter, probably derived largely from an influx of autumn-shed leaves (Williams and Morgan, 1977; Williams and others, 1981; Williams and Eyles, unpublished). A high proportion (64%) of the caddisfly species from the Don Beds, and a lesser but still-important number (36%) of those from the Scarborough Formation, are filter feeders dependent upon suspended matter for food.

Morgan and Morgan (1976) described deciduous-environment beetles from the Don Beds in the brickyard, and Morgan (1975) identified about 100, mainly boreal, beetle taxa from the Scarborough Formation at the Scarborough Bluffs. Insects recovered from the Scarborough Formation, at a site near Highland Creek, include at least 42 caddisfly species, 22 chironomid genera, with members of the Chironomini predominating, and about 35 beetle species in 12 families (Williams and others, 1981). These insects were predominantly boreal.

Kalas (1975) identified 55 species of molluscs in the Don Beds including prosobranchs and unionid and sphaeriid clams, all of temperate character, and Kerr-Lawson (1985) identified an additional 18 species. Vertebrate remains recovered from the Don Beds include fish such as pike (*Esox* sp.) and catfish (*Ictalurus punctatus*), mammals such as woodchuck (*Marmota arctos*) mammoth or mastodon (Proboscidea), white-tailed deer (*Odocoileus virginianus*), bison (*Bison* sp.), and giant beaver (*Castoroides ohioensis*) (Fig. 5).

PALEOCLIMATIC RECONSTRUCTIONS

Terrestrial and aquatic insects and plants provide the best clues to past climatic conditions, because these organisms have relatively long life cycles and are directly influenced by macroclimate. Small aquatic organisms with short life cycles spent entirely in the water, on the other hand, are buffered by the aquatic environment, and only indirectly influenced by climate (Williams, 1989). Warner and Hann (1987) discussed the utility of aquatic organisms as paleoclimatic indicators, but unfortunately did not distinguish between insects with terrestrial adults, as opposed to wholly aquatic organisms.

Figure 9 shows the most probable trend of mean annual temperature during deposition of the Don Beds and Scarborough Formation based on currently available insect, pollen, and plant macrofossil data. Ostracod, diatom, mollusc, and cladoceran data appear to be consistent with this interpretation. It should be stressed, however, that detailed insect, pollen, and plant macrofossil data are lacking for much of the Scarborough Formation, and a more complex pattern of climate change may ultimately be determined.

Terasmae (1960) identified a hiatus between the Illinoian sediments and the Don Beds, and this is supported by the present study (see below). A high diversity of temperate caddisfly and tree species, including a number whose ranges no longer extend as far north as Toronto, are found throughout most of the Don Beds. Coleman (1932) reported leaves of trees such as pawpaw (*Asiminor triloba*), osage orange (*Maclura pomifera*), black locust (*Robinia pseudacacia*), iron oak (*Quercus stellata*), blue ash (*Fraxinus quadrangulata*), and southern white cedar (*Chamaecyparis thyoides*) whose current northern distribution limits occur well south of Toronto. Whereas the presence of some of these exotic trees requires confirmation, they are not the only southern species reported. Williams and Eyles (unpublished) found southern caddisflies such as *Potamyia flava* and several members of the *Hydropsyche scalaris* group. A climate similar to central Pennsylvania is indicated by these indicators, with a mean annual temperature of about 2 °C above present Toronto conditions.

In the upper half of the Don Beds and lowermost silty clays of the Scarborough Formation, the gradual increase in importance of spruce and pine and decline in deciduous species (Fig. 9) indicate a cooling climate. At this time the plant and insect assemblages (beetles, chironomids, and caddisflies) are consistent with a cool-temperate climate such as perhaps currently occurs on the northern margins of Lake Superior. Members of the *Hydropsyche morosa* group of caddisflies, found in Ontario and farther north today, occur in the upper Don Beds and predominate in the lower Scarborough Formation, where they occur along with boreal beetles, including *Helophorus sempervarians, Phloeotribus piceae* and *Polygraphus rufipennis* (Williams and others, 1981). The data indicate a boreal climate at the time of deposition with mean annual temperatures about 2.5 °C, some 5 °C below their present level (Fig. 9). It is particularly significant to note that recent marine data show that initial development of the Laurentide Ice Sheet occurred while terrestrial climate and sea surface temperatures were "close to interglacial optima" (Miller and Vernal, 1992, p. 244). High latitude (65–80°N) ice sheet growth during the last interglacial is suggested by the marine data and is confirmed by the terrestrial record of the Don and Scarborough formations at Toronto.

During deposition of the remaining Scarborough clays, climate may have ameliorated sufficiently to allow the return of several deciduous tree species in a mixed-forest environment, although differences in the depositional environment and/or pollen preservation also could determine the pollen pattern. The overlying sand, in contrast, contains subarctic beetles, indicating renewed cooling and a mean annual temperature about 0 °C during the final stages of deltaic sedimentation.

Some general comments are in order with regard to the climate reconstruction shown in Figure 9. Certain fossil groups such as diatoms, cladocerans, and ostracodes are more dependent upon variables such as water depth and chemistry than macroclimate, and are therefore relatively poor climatic indicators. In addition, local sedimentary controls on faunal preservation dictate the completeness of the fossil record, and extreme caution must therefore be used in interpreting biological data. Emphasis should be placed on indicator species, rather than on species diversity. Reliance on species diversity, particularly in interpreting assemblages of small, aquatic organisms can be misleading. For example, Hann and Karrow (1984) suggested that the basal blue clay at the Don Valley Brickyard contained a low-diversity cladoceran fauna typical of boreal-subarctic habitats. They suggested that the section therefore recorded a gradual transition from the Illinoian glaciation to the Sangamon interglaciation, in contrast to pollen studies that indicated a marked hiatus (Terasmae, 1960). However, apparent increases or decreases in species richness can be highly dependent upon sample preservation, and we believe this is likely to be the case for the cladoceran data in question. Williams and Eyles (unpublished) have shown that the basal Don Beds contain a low-diversity assemblage of robust fossils, which they have interpreted to result from reworking in shallow storm-influenced waters, rather than indicating cool climate. Small numbers of southern caddisflies were recovered from the basal Don Bed, thus confirming Terasmae's results.

Local environmental effects are clearly apparent also in the upper Don Beds. At the Don Valley Brickyard, cladoceran species richness declines at the top of the Don Beds and Cladocera in the lower Scarborough Formation are sparsely represented by boreal and subarctic species. In contrast, a core from another site in Toronto (Leaside) shows an apparent *increase* in species richness over the same interval (Hann and Karrow, 1984). Only when large volumes of sediment are processed are specimens of delicate, and therefore poorly preserved, species recovered.

DISCUSSION

Coleman (1932) argued that the Don Beds and the Scarborough Formation formed a continuous sedimentary sequence that he defined as the Toronto Formation. Later workers, starting with Gray (1950), separated the Don Beds and the Scarborough on the basis of a presumed hiatus between the two. Sedimentological (Eyles and Clark, 1988) and paleoecologic work (Figs. 4, 6, 8, 9), however, strongly support Coleman's notion of a single depositional sequence containing a record of climatic fluctuations prior to the entry of Wisconsin glaciers into the area. The continuity between the Don and Scarborough is of particular importance given the great uncertainty surrounding the age of the overlying Sunnybrook diamict.

No buried soils of periglacial structures are reported from the uppermost part of the Scarborough Formation though these sediments were deposited in a severely cold climate (Fig. 9). Nonetheless, the top of the formation could represent a major unconformity since the age of the overlying Sunnybrook diamict is not accurately known. It is possible that the entire succession from Sunnybrook to Halton (Fig. 3) is late Wisconsin in age. The only constraints on the age of this succession is provided by radiocarbon age dates from detrital peat, in the range 28–32 ka, found just below the Halton Till (see below). Thus the glaciolacustrine diamicts and sands below the Halton could record ice damming of an ancestral Lake Ontario by an expanding Laurentide ice sheet during the earlier phases of the late Wisconsin. If correct, then where are deposits of early and middle Wisconsin age? Does the Scarborough Formation span not only the end of the last interglacial but also extend well into the Wisconsin?

The dating of the Toronto Pleistocene succession is clearly poorly delimited and a major effort is underway to strengthen the dating control. At the moment, a tentative chronologic scheme can be suggested by comparing the climatic and stratigraphic record at Toronto with the deep-ocean oxygen-isotope record; the single biggest influence on the latter is the Laurentide Ice Sheet, and thus there should be at least some parallelism between the continental record at Toronto and the marine isotope record (see below). In this light, the following tentative reconstruction of paleoenvironments in the Ontario basin is suggested.

Fulton and others (1984) argued that both the Don Beds and Scarborough Formation should be placed within oxygen-isotope stage 5. This is supported herein because of the argued continuity of the two deposits and the record of climatic fluctuations from warm-temperate to boreal to subarctic contained in the sedimentary sequence (Fig. 9). These fluctuations, possibly corresponding to changes in glacier ice cover in North America, are best placed within oxygen-isotope stage 5 with its evidence of global cooling and increased ice volume at 110 ka (stage 5.4), and 90 ka (stage 5.2; Fig. 10). There is agreement that the glaciolacustrine diamict that caps the interglacial sequence (Fig. 3) records the earliest phases of the Wisconsin glaciation at about 74 ka (stage 5/4 boundary). A thermoluminescence age for this diamict (the Sunnybrook "Till") of 66 ± 7 ka is reported by Berger (1984), but this is subject to revision (Berger, 1989, personal commun.). The high-resolution marine oxygen-isotope record (Martinson and others, 1987; Fig. 10) suggests that the close of stage 6 ("Illinoian"), perhaps represented at Toronto by the York Till, occurred 135 ka, stage 5 lasting until 74 ka (see below). This chronology may provide some age limitations on the Don Beds–Scarborough record at Toronto (Fig. 10). The high lake levels inferred from the Don Formation are consistent with a high interglacial sea-level stand (see above).

The Don Beds and overlying Scarborough Formation have been selectively preserved within a major bedrock low that connects the bedrock basins of the present Lake Ontario and Georgian Bay (Fig. 11). Defined to the west by the Niagara Escarpment, this 30-km-wide depression in the bedrock surface was the former route taken by the "Laurentian River," which is

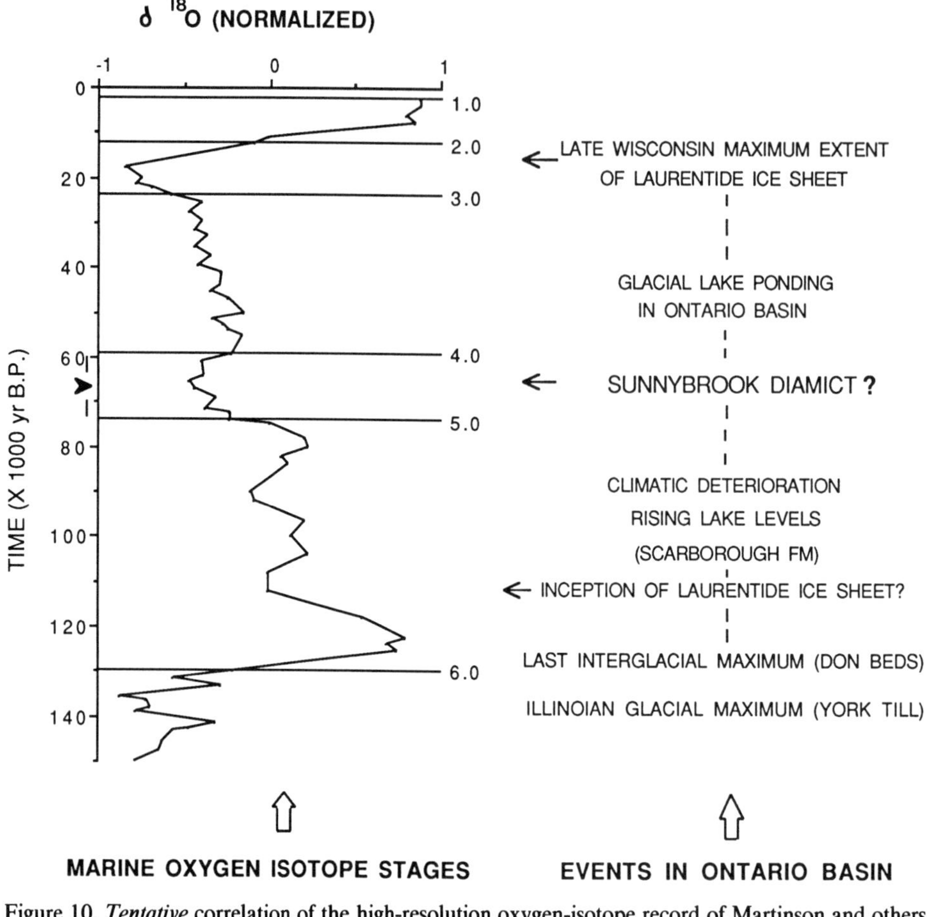

Figure 10. *Tentative* correlation of the high-resolution oxygen-isotope record of Martinson and others (1987) with events in the Ontario basin. The Laurentide Ice Sheet may have been initiated as early as 110 ka, only extending beyond the Toronto area during stage 2 after ca. 25 ka.

thought to have been the principal drainage outlet of the preglacial Great Lakes prior to the Quaternary glacial diversion through the Niagara River (Spencer, 1890). The Scarborough Formation records a substantial river within the Laurentian Channel; a subsurface program of drilling and geophypsical logging shows the delta to be present across a wide area of southern Ontario (at least 1,000 km^2) and several distributary lobes have been identified (see Eyles, 1987).

The stratigraphic record at Toronto is of considerable significance for ice-sheet modeling. It can be suggested, based on data shown in Figure 9, that the Laurentide Ice Sheet was initiated over Keewatin and Quebec-Labrador during deposition of the upper Don Beds and lower Scarborough clays when mean annual temperatures at Toronto were about 3 °C lower than at present. Comparison with the marine oxygen isotope record suggests that this initial cooling phase may be equivalent to the stage 5e/5d boundary when ice first started to form in high latitudes (see Miller and Vernal, 1992). During deposition of the upper sands of the Scarborough Formation, the drainage basin may have been largely free of boreal forest cover with an extensive tundra braid plain along the Laurentian Channel fed by meltstreams from an expanding ice sheet (Fig. 12B). The channels cut into the delta top and filled by the Pottery Road Formation record an erosive interval of unknown duration; marked fluctuations in water depths are typical of ice-dammed lakes. Alternatively, these channels may have been cut subaqueously (see also Eyles and Clark, 1988, p. 1119).

Eyles (1987) and Eyles and Westgate (1987) suggested that the early and middle phases of the Wisconsin glaciation are recorded at Toronto by a glaciolacustrine complex recording deposition in extensive glacial lakes on the margin of a thin and highly dynamic Laurentide Ice Sheet (Figs. 3 and 12). Schwarcz and Eyles (1991) have reported the stable isotopic compositions (O, C) of two species of benthic ostracodes, *Candona caudata* and *C. subtriangulata,* present as well-preserved juveniles and adults in the lower Scarborough Formation and overlying Wisconsin glaciolacustrine diamicts (Westgate and others, 1987; Rutka and Eyles, 1989) (Fig. 13). Marked secular trends in $\delta^{18}O$ and $\delta^{13}C$ for both species suggest the progressive deepening of the basin and the increased addition of isotopically-light glacial

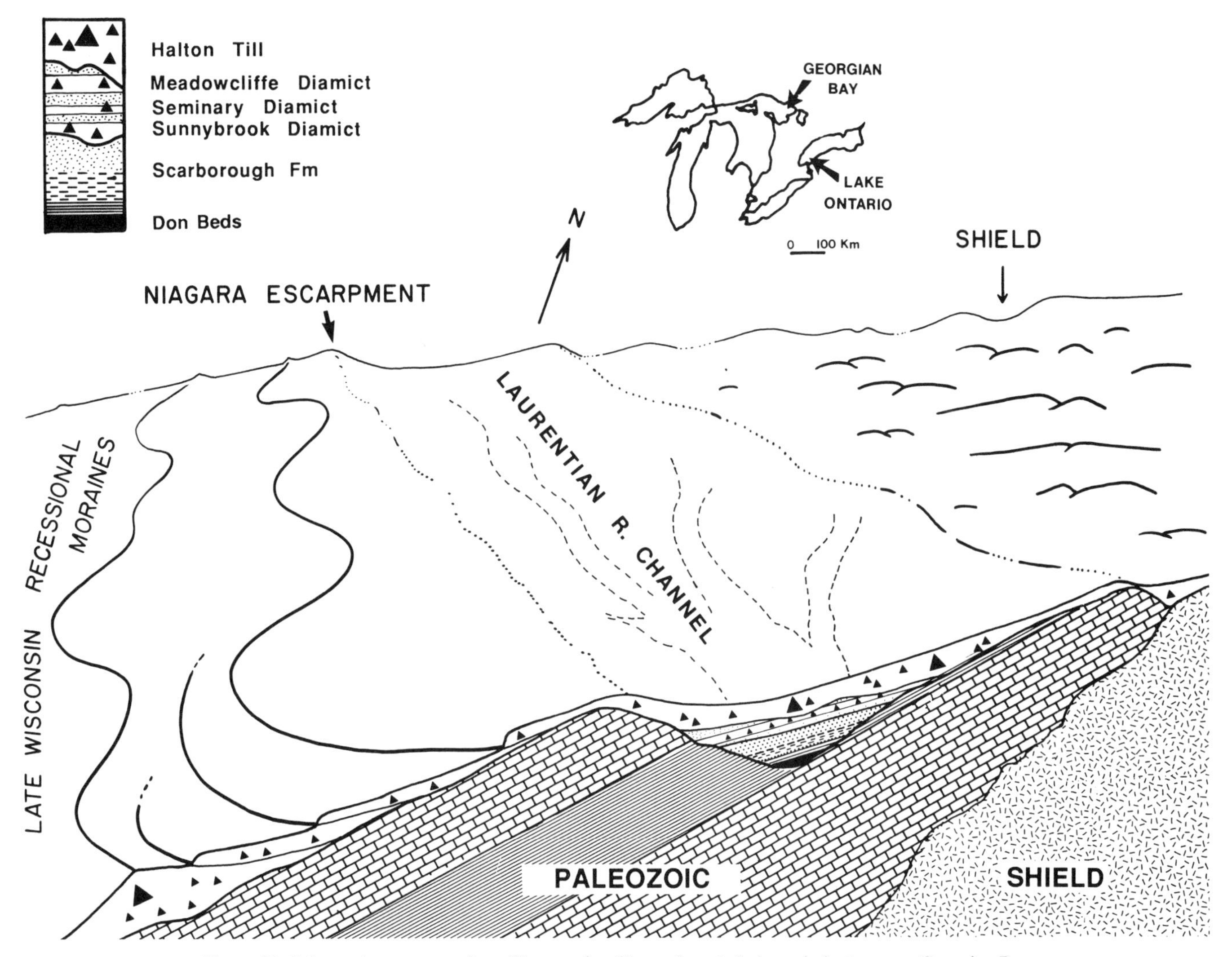

Figure 11. Schematic representation of Laurentian Channel eroded along shales between Georgian Bay and Lake Ontario. Late Pleistocene sediments, containing a record of the last interglacial-glacial transition (Don Beds, Scarborough Formation), are selectively preserved along the channel. Early and middle Wisconsin environments are recorded by a glaciolacustrine complex comprising the Sunnybrook, Seminary, and Meadowcliffe diamicts. The channel is capped by a late Wisconsin till (Halton Till) deposited when the Laurentide Ice Sheet attained its maximum extent in the mid-continent (see Figs. 3 and 10). Possible Illinoian York Till not shown.

meltwaters. These data are not consistent with a subglacial origin for the Sunnybrook diamict as a "deformation till," such as proposed by Hicock and Dreimanis (1989, this volume). The presence of delicate ostracode valves also indicates that the "glacially crushed" quartz grains found in the Sunnybrook by Mahaney (1990) were emplaced by other means, most likely as ice-rafted debris. Isotope and sedimentological data suggest that diamicts accumulated principally from suspended sediment plumes and ice rafting; local scouring by floating ice and sediment gravity flows on the lake floor were also significant. It is argued that the alternation between fine-grained, clay-rich diamicts (Sunnybrook, Seminary, Meadowcliffe) and intervening rhythmites and sands of the Thorncliffe Formation (Fig. 13) is the result, primarily, of changing water depths in the basin and displacement of the lake shoreline. In contrast to the fine-grained Sunnybrook and associated diamict units, the Halton Till is a coarse-grained bouldery diamict that in places is more than 50 m thick and rests on a marked erosional surface that truncates and in many places glacitectonizes the older stratigraphy (e.g., Boyce and Eyles, 1991). The till has a drumlinized upper surface with morainic ridges and outwash deposits.

The picture of restricted ice extent during the early and middle phases of the Wisconsin glaciation in the Ontario basin is in general agreement with the recognition of similar ice-free epi-

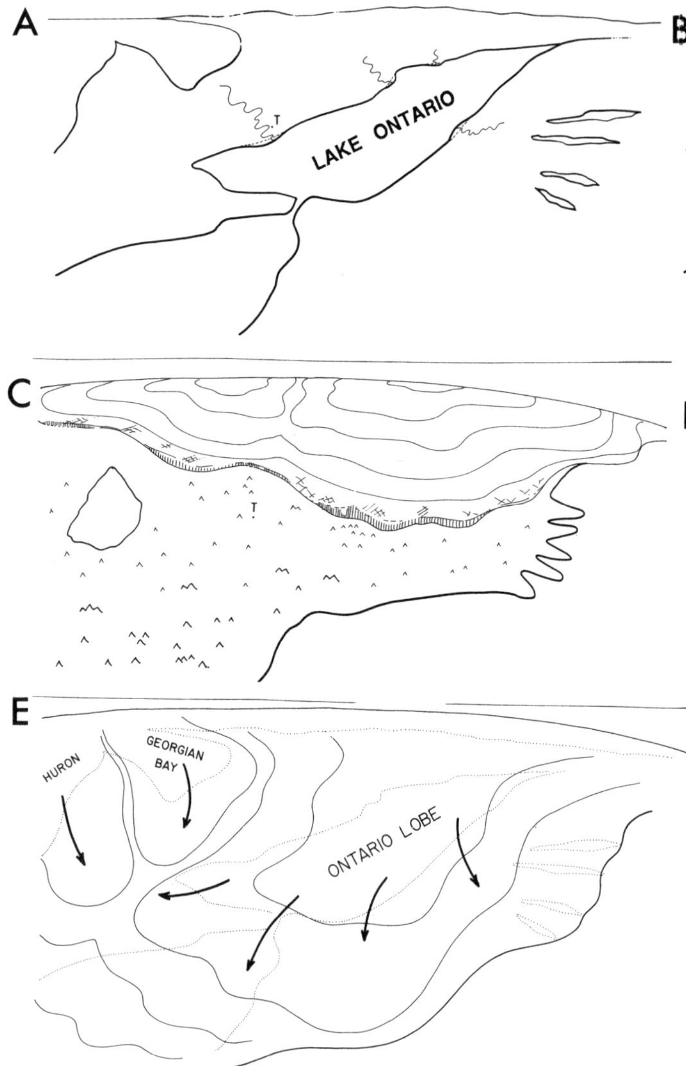

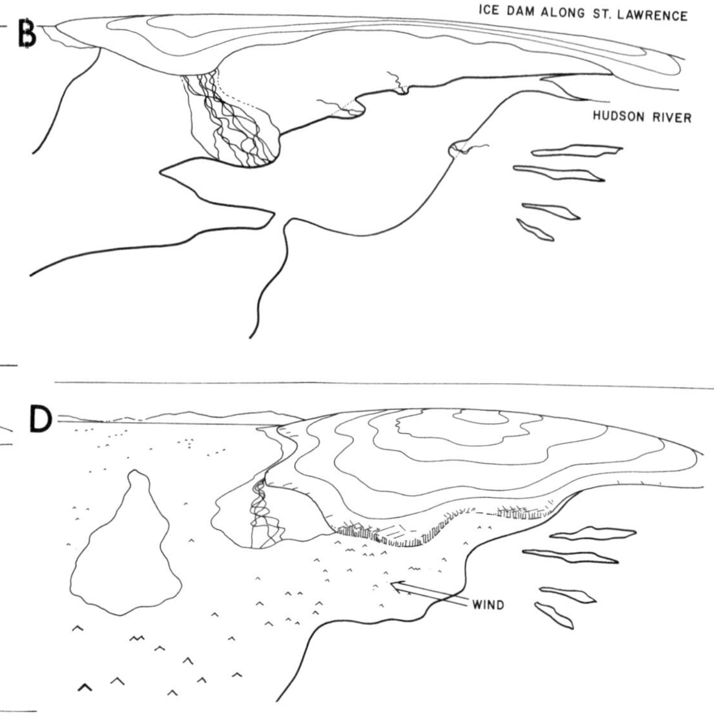

Figure 12. Schematic cartoons illustrating *suggested* paleogeographic evolution of the Ontario basin from the last glaciation ca. 125 ka to the late Wisconsin maximum of the Laurentide Ice Sheet. T: Toronto. A: Don Beds; B: Scarborough sands; C: Early-middle Wisconsin glaciolacustrine complex (Fig. 13) showing high-level ice-dammed lake in which muds and ice-rafted debris (diamict facies) accumulate at times of high lake level and sandy facies (Thorncliffe Formation; Fig. 3) accumulate at times of lowered levels (D). D: Fluctuating ice margin, changes in lake depth, and growth of storm-influenced deltas. E: Late Wisconsin maximum; deposition of coarse-grained Halton Till and erosion of underlying sediments.

sodes in the Hudson Bay area (see Andrews and others, 1983) and at an increasing number of sites in the mid-continent (see discussion in Schwarcz and Eyles, 1991). Age correlative units in the southern Michigan basin are represented not by glacial sediments but by loess sequences (Follmer, 1983; Curry, 1989; Johnson and Follmer, 1989; Curry and Follmer, this volume). Perhaps only in the late Wisconsin (stage 2; <22 ka) did the Laurentide Ice Sheet advance across the Great Lakes as a series of surges (e.g., Attig and others, 1985). The youngest radiocarbon age on wood below late Wisconsin till in southern Ontario is 23 ka from St. David's Gorge near Niagara Falls (Hobson and Terasmae, 1969). Ages in the range 28 to 32 ka were reported by Morner (1973) and Berti (1975) from silts below the late Wisconsin Halton Till at Toronto, but these ages are from reworked plant detritus and are probably too old.

CONCLUSIONS

The sedimentary and biological record at Toronto represents as long a record of Sangamon to late Wisconsin environments (isotope stages 5, 4, 3, 2) as has been found anywhere to date in eastern North America. The possibility of major unconformities in this record is real, however (e.g., upper Scarborough Formation and Pottery Road Formation), and it is emphasized that the currently available numerical dating control is very poor. The biological record shows a complex pattern of climate deterioration following deposition of the Don Beds, which may record the initiation of the Laurentide Ice Sheet. Mean annual temperatures derived from pollen, plant macrofossils, caddisflies, beetles, and chironomids offer excellent prospects for climate modeling. The late Pleistocene stratigraphy at Toronto comprises transgressive

Figure 13. Early to middle(?) Wisconsin glaciolacustrine complex exposed along Scarborough Bluffs (section 1481 of Eyles and Eyles, 1983). The figure (arrow) stands at the contact with the underlying Scarborough Formation delta (Fig. 3). Note tabular layer-cake geometry of diamict and interbedded deltaic sands and silty-clay rhythmites. A, Sunnybrook; B, Seminary; C, Meadowcliffe diamict units (Fig. 3).

shoreface sediments (Don Formation), deltaic sediments (Scarborough Formation), a glaciolacustrine complex (Sunnybrook, Seminary, Meadowcliffe diamicts and associated facies), and subglacially deposited till (Halton Till). While we again emphasize the poor numerical dating control and the likelihood of disconformities in the succession, the stratigraphy has an inherent continuity since it forms a *progradational succession* such as would be deposited at the margin of an ice sheet advancing into, halting, and eventually crossing a very large and deep lake basin.

ACKNOWLEDGMENTS

We are very grateful to Carolyn Eyles, Mike Kerr, Marg Rutka, John Westgate, and Dudley Williams for their comments and help during the course of our investigations in the Don Valley Brickyard. We thank Jock McAndrews for permission to use Figure 9 and, together with Jim Ritchie and John Birks, for discussions of the pollen record. These investigations were financed by operating grants from the Natural Sciences and Engineering Research Council of Canada to Eyles and by the award of an NSERC Post-Doctoral fellowship to Williams. Frank Kershaw and Clive Roberts of Metro Toronto Parks and Property Department are thanked for their logistical support. The manuscript was reviewed by an anonymous reviewer, Ardith K. Hansel, Peter Clark, and Peter Lea; we thank them for their constructive and very helpful comments.

REFERENCES CITED

Andrews, J. T., Shilts, W. W., and Miller, G. H., 1983, Multiple glaciations of the Hudson Bay lowlands, Canada, since deposition of the Missinaibi (last interglacial?) Formation: Quaternary Research, v. 19, p. 18–37.

Attig, J. W., Clayton, L., and Mickelson, D. M., 1985, Correlation of late Wisconsin glacial phases in the western Great Lakes area: Geological Society of America Bulletin, v. 96, p. 1585–1593.

Baker, F. C., 1931, A restudy of the interglacial molluscan fauna of Toronto, Canada: Illinois Academy of Science Transactions, v. 23, p. 358–366.

Bard, E., Hamelin, B., and Fairbanks, R. G., 1990, U-Th ages obtained by mass spectrometry in corals from Barbados: Sea level during the past 130,000 years: Nature, v. 346, p. 456–458.

Bell, R., 1861, On the occurrence of freshwater shells of our post-Tertiary deposits: Canadian Naturalist and Geologist, v. 6, p. 42–51.

Berger, G. W., 1984, Thermoluminescence dating studies of glacial silts from Ontario: Canadian Journal of Earth Sciences, v. 21, p. 1393–1399.

Berti, A. A., 1975, Paleobotany of Wisconsin interstadials, eastern Great Lakes region, North America: Quaternary Research, v. 5, p. 591–619.

Boyce, J., and Eyles, N., 1991, Drumlins carved by deforming till streams below the Laurentide ice sheet: Geology, v. 19, p. 787–790.

British Association for the Advancement of Science, 1900, Canadian Pleistocene flora and fauna: v. 70, p. 328–339.

Brown, R. W., 1942, Supposed extinct Maples: Science, v. 96, p. 15.

Chamberlin, T. C., 1895, The classification of American glacial deposits: Journal of Geology, v. 3, p. 270–277.

Chapman, E. J., 1861, Some notes on drift deposits of Western Canada and on the ancient extension of the lake area of that region: Canadian Journal, new series, v. 6, p. 221–229.

Churcher, C. S., and Karrow, P. F., 1977, Late Pleistocene muskox (Ovibos) from the Early Wisconsin at Scarborough Bluffs, Ontario, Canada: Canadian Journal of Earth Sciences, v. 14, p. 326–331.

Clark, R. H., and Persoage, N. P., 1970, Some implications of crustal movements in engineering planning: Canadian Journal of Earth Sciences, v. 7, p. 628–633.

Coleman, A. P., 1894, Interglacial fossils from the Don Valley: American Geologist, v. 13, p. 85–95.

——, 1895, Glacial and interglacial deposits near Toronto (Ontario): Journal of Geology, v. 3, p. 622–645.

——, 1932, The Pleistocene of the Toronto region: Ontario Department of Mines, v. 41, no. 7, entire issue.

Curry, B. B., 1989, Absence of Altonian glaciation in Illinois: Quaternary Research, v. 31, p. 1–13.

Dawson, W., and Penhallow, D. P., 1980, On the Pleistocene flora of Canada: Geological Society of America Bulletin, v. 1, p. 311–334.

Delorme, C. D., 1969, Ostracodes as Quaternary paleoecological indicators: Canadian Journal of Earth Science, v. 6, p. 1471–1476.

Dreimanis, A., and Terasmae, J., 1958, Stratigraphy of Wisconsinglacial deposits of Toronto area, Ontario: Geological Association of Canada Proceedings, v. 10, p. 119–136.

Duthie, H. C., and Mannada Rani, R. G., 1967, Diatom assemblages from Pleistocene interglacial beds at Toronto, Ontario: Canadian Journal of Botany, v. 45, p. 2249–2261.

Eyles, C. H. and Eyles, N., 1983, Sedimentation in a large lake: A reinterpretation of the late Pleistocene stratigraphy at Scarborough Bluffs, Ontario, Canada: Geology, v. 11, p. 146–152.

Eyles, N., 1987, Late Pleistocene depositional systems of Metropolitan Toronto and their engineering and glacial geological significance: Canadian Journal of Earth Sciences, v. 24, p. 1009–1021.

Eyles, N., and Clark, B. M., 1987, Storm-dominated deltas and ice-scouring in a late Pleistocene ancient glacial lake: Geological Society of America Bulletin, v. 100, p. 793–809.

——, 1988, Last interglacial sediments of the Don Valley Brickyard, Toronto, Canada, and their paleoenvironmental significance: Canadian Journal of Earth Sciences, v. 25, p. 1108–1122.

——, 1989, "Reply *to* Comments *on* "Last interglacial sediments of the Don Valley Brickyard, Toronto, Canada and their paleoenvironmental significance": Canadian Journal of Earth Sciences, v. 26, p. 1083–1086.

Eyles, N., and Boyce, J., 1992, Earth Science Survey of the Rouge Valley Park, Ontario Ministry of Natural Resources, Central Region, Aurora, Ontario: Open-file Geological Report 9113, 82 p. and 4 folded maps.

Eyles, N., and Schwarcz, H. P., 1991, Stable isotope record of the last glacial cycle from lacustrine ostracodes: Geology, v. 19, p. 257–260.

Eyles, N., and Westgate, J. A., 1987, Restricted regional extent of the Laurentide ice sheet in the Great Lakes basins during early Wisconsin glaciation: Geology, v. 15, p. 537–540.

Fielding, C. R., 1984, Upper delta plain lacustrine and fluviolacustrine facies from the Westphalian of the Durham coalfield, NE England: Sedimentology, v. 31, p. 547–567.

Follmer, L. R., 1983, Sangamon and Wisconsinan pedogenesis in the midwestern United States, *in* Porter, S. C., ed., The late Pleistocene: Minneapolis, University of Minnesota Press, p. 138–144.

Fulton, R. J., Karrow, P. F., Lasalle, P., and Grant, D. R., 1984, Summary of Quaternary stratigraphy and history, eastern Canada, *in* Fulton, R. J., ed., Quaternary stratigraphy of Canada: A Canadian contribution to IGCP Project 20: Geological Survey of Canada Paper 84-10, p. 193–210.

Gray, A. B., 1950, Sedimentary facies of the Don Member (Toronto Formation) [M.A. thesis]: Toronto, Canada, University of Toronto.

Hann, B. J., and Karrow, P. F., 1984, Pleistocene paleoecology of the Don and Scarborough formations, Toronto, Canada; based on Cladoceran microfossils at the Don Valley Brickyard; Boreas, v. 13, p. 377–391.

Hazeldine, R. S., 1984, Muddy deltas in freshwater lakes and tectonism in the Upper Carboniferous Coalfield in NE England: Sedimentology, v. 31, p. 811–822.

Hicock, S. R., and Dreimanis, A., 1989, Sunnybrook drift indicates a grounded early Wisconsin glacier in the Lake Ontario basin: Geology, v. 17, p. 169–172.

Hinde, G. J., 1877, The glacial and interglacial strata of Scarborough Heights and other localities near Toronto, Ontario: Canadian Journal, new series, v. 15, no. 94, p. 388–413.

Hobson, G. D., and Terasmae, J., 1969, Pleistocene geology of the buried St. David's Gorge, Niagara Falls, Ontario: Geophysical and palynological studies: Geological Survey of Canada Paper 68-67, 16 p.

Hui, H. T., Fernando, C. H., and Karrow, P. F., 1969, Mollusca of the Toronto Interglacial: American Zoologist, v. 9, p. 615–616.

Hustedt, F., 1954, Die Diatomeenflora des Interglazials von Oberohe in der Luneburger Heide. Abhandlungen Naturwissenschaftliche Verein Brumen, v. 33, p. 431–455.

Johnson, W. H., and Follmer, L.R., 1989, Source and origin of Roxana Silt and Middle Wisconsinan mid-continent glacial activity: Quaternary Research, v. 31, p. 319–331.

Kalas, L. L., 1975, Malacological evidence of interglacial environments at Toronto, Ontario, Canada: A quantitative approach: Quaternary Non-Marine Paleoecology Conference, Program with Abstracts, Waterloo, Ontario.

Karrow, P. F., 1967, Pleistocene geology of the Scarborough area: Ontario Department of Mines Geological Report 46, 108 p.

——, 1974, Till stratigraphy in parts of southwestern Ontario: Geological Society of America Bulletin, v. 85, p. 761–768.

——, 1984, Quaternary stratigraphy and history, Great Lakes–St. Lawrence Region, *in* Fulton, R. J., ed., Quaternary stratigraphy of Canada: A Canadian contribution to IGCP Project 24: Geological Survey of Canada Paper 84-10, p. 137–153.

Kelly, R. I., and Martini, I. P., 1986, Pleistocene glaciolacustrine deltaic deposits of the Scarborough Formation, Ontario, Canada: Sedimentary Geology, v. 47, p. 27–52.

Kerr-Lawson, L., 1985, Gastropods and plant macrofossils from the Quaternary

Don Formation (Sangamonian Interglacial), Toronto, Ontario [M.S. thesis]: Waterloo, Canada, University of Waterloo, 135 p.

Kite, G. W., 1972, An engineering study of crustal movement around the Great Lakes: Department of the Environment, Inland Waters Branch, Technical Bulletin no. 63, 56 p.

Mahaney, W. H., 1990, Macrofabrics and quartz microstructures confirm glacial origin of Sunnybrook drift in the Lake Ontario basin: Geology, v. 18, p. 145–148.

Martinson, D. G., Pisias, N. G., Hays, J. D., Imbrie, J., Moore, T. C., and Shackleton, N. J., 1987, Age dating and the orbital theory of the Ice Ages: Development of a high resolution 0 to 300,000-year chronostratigraphy: Quaternary Research, v. 27, p. 1–29.

Miall, A. D., 1978, Lithofacies types and vertical profile models in braided river deposits: A summary, *in* Miall, A. D., ed., Fluvial sedimentology: Canadian Society of Petroleum Geologists Memoir 5, p. 597–604.

Miller, C. H., and Vernal, A., 1992, Will greenhouse warming lead to Northern Hemisphere ice-sheet growth? Nature, v. 355, p. 244–246.

Morgan, A., 1972, The fossil occurrence of *Helophorus arcticus* Brown (Coleoptera: Hydrophilidae) in Pleistocene deposits of the Scarborough Bluffs, Ontario: Canadian Journal of Zoology, v. 50, p. 555–558.

——, 1975, Fossil beetle assemblages from the Early Wisconsin Scarborough Formation, Toronto, Canada: Quaternary Non-marine Paleoecology Conference, Program with Abstracts, University of Waterloo, Ontario.

Morgan, A., and Morgan, A. V., 1976, Climatic interpretations from the fossil insect faunas of the Don and Scarborough formations, Toronto, Ontario, Canada: Geological Society of America Abstracts with Programs, v. 8, p. 1020.

Morgan, A. V., and Morgan, A., 1980, Faunal assemblages and distributional shifts of Coleoptera during the late Pleistocene in Canada and the northern United States: Canadian Entomologist, v. 112, p. 1105–1128.

Morner, N. A., 1973, The Plum Point Interstadial: Age, climate and subdivision: Canadian Journal of Earth Sciences, v. 8, p. 1423–1431.

Poplawski, S., and Karrow, P. F., 1981, Ostracods and paleoenvironments of the late Quaternary Don and Scarborough formations, Toronto, Ontario: Canadian Journal of Earth Sciences, v. 18, p. 1497–1505.

Rutka, M., and Eyles, N., 1989, Ostracod faunas in a large late Pleistocene ice-dammed lake, southern Ontario, Canada: Palaeogeography, Palaeoclimatology, Palaeoecology, v. 73, p. 61–76.

Schwarcz, H., and Eyles, N., 1991, Laurentide Ice Sheet extent inferred from stable isotopic composition (O, C) of ostracodes at Toronto, Canada: Quaternary Research, v. 35, p. 305–320.

Scudder, S. H., 1985, The Coleoptera hitherto found fossil in Canada: Geological Survey of Canada, Contributions to Canadian Paleontology, v. 2, p. 27–56.

——, 1900, Additions to the coleopteran fauna of the interglacial clays of the Toronto district: Geological Survey of Canada, Contributions to Canadian Paleontology, v. 2, p. 67–92.

Spencer, J. W., 1890, Origin of the basins of the Great Lakes of America: American Geologist, v. 7, p. 86–97.

Terasmae, J., 1960, A palynological study of Pleistocene interglacial beds at Toronto, Ontario: Geological Survey of Canada Bulletin, v. 56, p. 24–40.

Upham, W., 1895a, The climatic conditions shown by North American interglacial deposits: American Geologist, v. 15, no. 5, p. 273–295.

——, 1895b, The climatic conditions shown by North American interglacial deposits: American Geologist, v. 16, no. 2, p. 105–106.

Walker, R. G., and Cant, D. J., 1984, Sandy fluvial systems, *in* Walker, R. G., ed., Facies models: Geological Association of Canada Reprint Series I, p. 71–89.

Warner, B. G., 1986, Early work in Quaternary botany in Canada: Geoscience Canada, v. 13, p. 39–44.

Warner, B. G., and Hahn, B. J., 1987, Aquatic invertebrates as paleoclimatic indicators?: Quaternary Research, v. 28, p. 427–430.

Watt, A. K., 1954, Correlation of the Pleistocene geology as seen in the subway with that of the Toronto region, Canada: Geological Association of Canada Proceedings, v. 6, p. 69–81.

Westgate, J. A., Chen, F. J., and Delorme, L. D., 1987, Lacustrine ostracodes in the late Pleistocene Sunnybrook diamict of southern Ontario, Canada: Canadian Journal of Earth Sciences, v. 24, p. 2330–2335.

Williams, N. E., 1989, Factors affecting the interpretation of caddisfly assemblages from Quaternary sediments: Journal of Paleolimnology, v. 1, p. 241–248.

Williams, N. E., and Morgan, A. V., 1977, Fossil caddisflies (Insecta: Trichoptera) from the Don Formation, Toronto, Ontario, and their use in paleoecology: Canadian Journal of Zoology, v. 55, p. 519–527.

Williams, N. E., Westgate, J. A., Williams, D. D., Morgan, A., and Morgan, A. V., 1981, Invertebrate fossils (Insecta: Trichoptera, Diptera, Coleoptera) from the Pleistocene Scarborough Formation at Toronto, Ontario, and their paleoenvironmental significance: Quaternary Research, v. 16, p. 146–166.

Wright, G. F., 1914, Age of the glacial deposits in the Don Valley, Toronto, Ontario: Geological Society of America Bulletin, v. 25, p. 205–214.

Manuscript Submitted to the Society October 1988
Manuscript Accepted by the Society September 6, 1991

Geological Society of America
Special Paper 270
1992

Sunnybrook drift in the Toronto area, Canada: Reinvestigation and reinterpretation

Stephen R. Hicock and Aleksis Dreimanis
Department of Geology, The University of Western Ontario, London, Ontario N6A 5B7, Canada

ABSTRACT

Our reinvestigation indicates that Sunnybrook drift represents a glaciolacustrine-subglacial-glaciolacustrine cycle, the Ontario lobe being the main glacial agent. Glacier ice overrode the present Toronto area during Sunnybrook deposition, deforming the substrate and depositing subglacial till. We propose that Sunnybrook drift comprises three or four members: a diamictic Sunny Point member (mainly till) sandwiched by two glaciolacustrine members (Sylvan Park and Bloor), and possibly a subglacial fluvial member (Pottery Road) associated with till deposition.

The Sylvan Park member represents glaciolacustrine fine sediment that was overridden, deformed, and incorporated by ice that deposited the Sunny Point member. Parallel striated stone pavements, parallel and transverse stone and magnetic fabrics, stone facet and stoss-lee relations, and shear planes within the Sunny Point member indicate that it was formed by a combination of lodgement and subglacial deformation. Deposition under active grounded ice is further supported by the sharply erosive contact of the Sunny Point member with the top of the buried Scarborough delta, as well as glaciotectonic fractures and folds in substrata. Sunnybrook drift deposition concluded with the glaciolacustrine Bloor Member. Regional correlation suggests that the glaciolacustrine members were not formed far inland from the Scarborough Bluffs.

INTRODUCTION: A REVIEW OF PREVIOUS WORK

Overview

Reconstructing the early Wisconsinan history of the eastern Great Lakes hinges on the genetic interpretation of the Sunnybrook drift (assuming that it represents the early Wisconsinan substage in the Ontario basin). This is because glacial events in the Ontario basin must have influenced areas beyond it, which, taken together, bear on the climatic history of the Great Lakes. The Sunnybrook drift, also known by other names to 1969, has been considered a key early Wisconsinan stratigraphic marker in the eastern Great Lakes region. Most authors recognized both till and glaciolacustrine sediments as its constituents, but, until 1983, the till was emphasized. An exclusively glaciolacustrine origin was then proposed as an alternative, initiating a lively and continuous debate.

The various arguments of this controversy are discussed as background for our reinvestigation. This study is based upon the application of multiple criteria, with an emphasis on measurement of deformation structures, fabric, and erosional features, particularly along the base of the Sunnybrook drift where it caps the buried Scarborough delta in the Scarborough Bluffs.

Early work

The Sunnybrook drift (Table 1) is a key stratigraphic marker along the north shore of Lake Ontario and is crucial to interpreting the extent and activity of early Wisconsinan glacial ice in the eastern Great Lakes region. It overlies the Scarborough Formation and underlies the Thorncliffe Formation (Terasmae, 1960; Karrow, 1967, 1969, 1984b). The Sunnybrook formation was named "Sunnybrook till" by Terasmae (1960) and the second term was capitalized by Karrow (1967). Historically, others have applied the same descriptive usage of the term "till" in the Ontario basin (Hinde, 1877; Coleman, 1895, 1913, 1933, 1941;

Hicock, S. R., and Dreimanis, A., 1992, Sunnybrook drift in the Toronto area, Canada: Reinvestigation and reinterpretation, *in* Clark, P. U., and Lea, P. D., eds., The Last Interglacial-Glacial Transition in North America: Boulder, Colorado, Geological Society of America Special Paper 270.

Wilson, 1901; Watt, 1954, 1957, 1968; Dreimanis and Terasmae, 1958; Ostry, 1962; Lajtai, 1969; White, 1971; Stupavsky and others, 1974; Sharpe, 1980).

Karrow (1967) also named the glaciolacustrine layered mud overlying the "Sunnybrook Till" as the Bloor Member of the Thorncliffe Formation. However, he later assigned it to the Sunnybrook, concluding that "the more inclusive term 'Sunnybrook Drift' can more appropriately be applied where till and stratified sediments occur together" (Karrow, 1969, p. 13; see also Lajtai, 1969). Karrow (1984b, p. 143) defined the term "Sunnybrook Till" as the "lowest member of Sunnybrook Drift"; however, it is not always the lowest subunit (see Table 2).

Sedimentologic studies

Early work emphasized the lithostratigraphic role of the "Sunnybrook Till" in the Toronto area. The first detailed sedimentologic analysis of the Sunnybrook drift was by Eyles and Eyles (1983) and Eyles and others (1983). Their conclusion that it was exclusively a large lake deposit initiated a lively debate over the status of early Wisconsinan glaciation in the Eastern Great Lakes. They claimed a conspicuous absence of (1) major erosive contacts, (2) glaciotectonic structures within and beneath Sunnybrook diamicton, and (3) buried landforms associated with grounded glacier ice. Instead, they reported and illustrated by photographs various soft-sediment structures and contact relations common in subaquatic sediments. The term "Sunnybrook drift" was changed to "Sunnybrook unit," or "Sunnybrook diamict complex," and "till" to "diamict."

In the ensuing discussion, Karrow (1984a, p. 185) pointed out that "evidence of grounded ice action was seen at the Seminary section during field work in the early 1960's" and that a definite preferred fabric was measured in other sections by Ostry (1962) and Karrow (1967). In their reply, Eyles and Eyles (1984, p. 189) disregarded these pebble- and micro-fabric data, as well as the magnetic fabric measured by Stupavsky and others (1974). They also dismissed the lithologic data of Dreimanis and Terasmae (1958), Karrow (1967), and Lajtai (1969), when they disputed Sharpe (1984) by stating that "a westward direction of ice flow is not known."

Sharpe (1984, p. 186) and Gravenor and Wong (1987, p. 2040) pointed out that Eyles and Eyles (1983) and Eyles and others (1983) had included stratified and flow-deformed nondiamictic muds in their "Sunnybrook diamict facies" and that most of their evidence for lacustrine origin was probably from these glaciolacustrine sediments. Most of the detailed evidence of Eyles and Eyles (1983), including photographs, is from the Dutch Church and Sylvan Park areas, where the Sunnybrook drift fills erosional depressions on the surface of the Scarborough delta (Fig. 1, southwest and northeast ends). Such depressions are the most likely places for accumulation of aquatic sediments, either in a lake or under a glacial lobe terminating in a proglacial lake.

The late Wisconsinan Halton drift was deposited by such a glacial lobe, and therefore provided a model for comparison with the Sunnybrook drift (Sharpe and Barnett, 1985; Sharpe, 1987, p. 342–343 and 1988, p. 25). Sedimentary structures and contact relations similar to those listed by Eyles and Eyles (1983) as proof of nonglacial lacustrine origin for the Sunnybrook drift were found in the Halton drift in association with till.

Gravenor (1984, p. 187–188) also criticized the ambiguity of several of the sedimentologic criteria of Eyles and Eyles (1983).

> Neither load structures nor paleomagnetism can be used to determine the origin of clastic sediment, because both are postdepositional phenomena. . . . Although the stringers of sand and silt in the diamicton and other associated stratified sediments suggest that deposition of the diamicton took place in a subaquatic environment, their presence does not prove or disprove the hypothesis that the diamicton accumulated by a "rain-out" of debris from floating ice.

Kelly and Martini (1986, p. 51) investigated the Scarborough Formation and its contact with the Sunnybrook drift in many sections along the Scarborough Bluffs, and concluded that

TABLE 1. MAJOR SANGAMONIAN TO MIDDLE WISCONSINAN LITHOSTRATIGRAPHIC UNITS IN THE TORONTO AREA*

Chronostratigraphy	Lithostratigraphy
middle Wisconsinan	Thorncliffe Formation
early Wisconsinan	Sunnybrook drift Pottery Road Formation (local valley fills) Scarborough Formation
Sangamonian	Don Formation

*After Karrow, 1984b.

TABLE 2. EARLY WISCONSINAN LITHOSTRATIGRAPHIC UNITS AND THEIR DOMINANT FACIES IN THE SCARBOROUGH BLUFFS OF FIGURE 1

Formation	Member	Facies Descriptions
Sunnybrook drift	Bloor Member	Stratified to massive glaciolacustrine mud
	Sunny Point member	Mainly silty to clayey massive matrix-supported diamicton with a prominent stone pavement along its base
	Sylvan Park member	Stratified and deformed glaciolacustrine mud and sand
Scarborough Formation	Scarborough sand	Deltaic sand
	Scarborough silt and clay	Stratified to massive lacustrine clay, silt, and silty sand

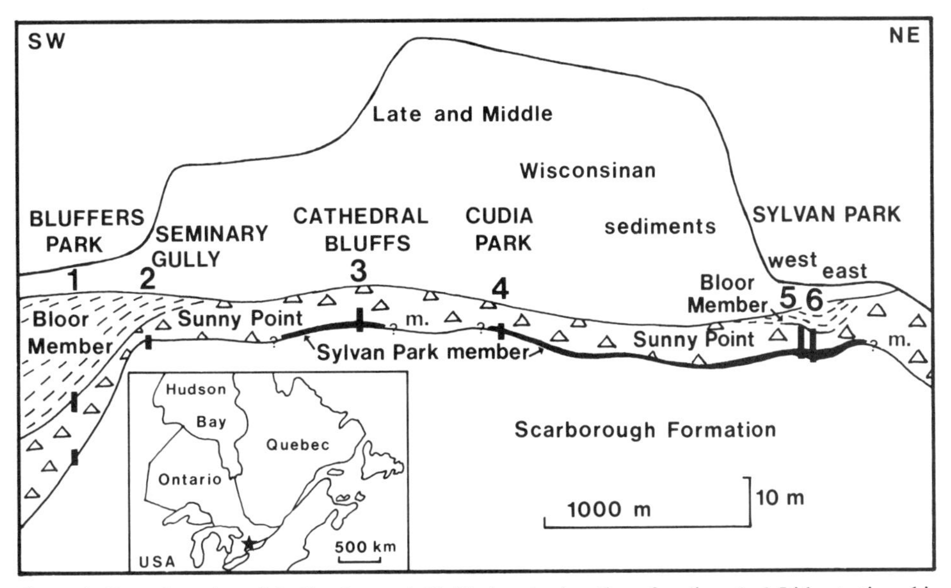

Figure 1. Central portion of the Scarborough Bluffs showing location of sections 1–6. Lithostratigraphic boundaries are approximate (modified from Karrow, 1967, Fig. 15; Eyles and Eyles, 1983, Fig. 1). Inset map shows location of the Scarborough Bluffs.

the glacier "readvanced into the area and eventually deposited the overlying Sunnybrook Formation and locally deformed the underlying sediments."

Magnetic and pebble fabrics

Lawson (1979, p. 644), from a very thorough fabric investigation of sediment gravity flows and melt-out till at Matanuska Glacier in south-central Alaska, concluded that "pebble fabric cannot be used by itself to define the genesis of a particular deposit." An exception may be a subaquatic "rain-out" sediment if it has not undergone any lateral resedimentation (Domack and Lawson, 1985). A magnetic fabric of fine particles, being very sensitive to any obstructions such as clasts (Stupavsky and Gravenor, 1975), would show great deviation from the theoretically expected patterns. Nevertheless, fabric provides information about the orientation of the local strain ellipsoid and the implied stress field at each sampling site. Only in association with other criteria (see below) can it be useful for deciphering sediment genesis (Boulton, 1971; Rappol, 1985; Dowdeswell and Sharp, 1986).

On the basis of new detailed data of magnetic and pebble fabric in the Sunnybrook diamicton facies in two sections at Sylvan Park, Gravenor and Wong (1987, p. 2038) concluded, in agreement with the previous magnetic and pebble-fabric data (Ostry, 1962; Karow, 1967; Stupavsky and others, 1974), that the Sunnybrook diamicton was subglacially deposited as till.

However, the fabrics are not typical of a lodgement till and to some extent resemble fabrics found in subaquatic debris flow deposits. It is suggested that the glacier that deposited the Sunnybrook diamicton was buoyed up by high pore water pressure in the accumulating sediment at the sole of the glacier. Under these conditions, the till either flowed away from the zone of release at the base of the glacier and(or) was being lightly smeared by a partly grounded glacier.

Eyles and others (1983, 1987) also investigated the magnetic fabric (anisotropic magnetic susceptibility, or AMS) in the same part of the Scarborough Bluffs as Gravenor and Wong (1987). However, they included in their section the upper third of the Sunnybrook drift that belongs to the Bloor Member (called mud, silt, and clay in Gravenor and Wong, 1987, Fig. 2), which they labeled "diamict." Eyles and others (1987) maintained that the massive Sunnybrook diamicton facies is a "subaqueous rain-out deposit," even though the stereonet characteristics of a "rain-out" fabric (Eyles and others, 1987, Fig. 10) appear in only 15% of the samples. They explained that the AMS fabric characteristics of the remaining majority of the Sunnybrook "diamict" samples that produced their B- and C-type fabrics with strong maximum (K_{max}) and minimum (K_{min}) axial orientations, were a result of resedimentation by gravity flow (Eyles and others, 1987, p. 2450–2455). Although the directions of K_{max} and K_{min} vary among sampling sites, they are not as inconsistent as claimed by Eyles and others (1987, p. 2450). In fact, most directions of K_{max} are similar to those measured by Gravenor and Wong (1987) and the directional measurements of clast fabric. Some of the B-type fabrics of Eyles and others (1987, Figs. 8–10) are very similar to Boulton's (1971, Fig. 12) magnetic fabrics for lodgement till studied beneath a Svalbard glacier. We agree with Eyles and others (1987, p. 2452–2453) that most Sunnybrook "diamict" AMS fabrics have resulted from resedimentation and deformation, but instead of forming in a proglacial environment (Eyles and others, 1987, Fig. 10 A–C), we invoke a subglacial setting (see below).

Microfossils

The presence of microfossils has been evidence for the lacustrine origin of the Sunnybrook drift. Eyles and Eyles (1987, Fig. 17), Rutka and Eyles (1989), and very similar papers by Eyles and Schwarcz (1991) and Schwarcz and Eyles (1991) reported ostracodes in the entire section of the Sunnybrook drift at Sylvan Park, in the same area where the magnetic fabric was investigated (see above). Westgate and others (1987) and Rutka and Eyles (1989) also identified ostracodes from another section of the Sunnybrook drift in the Rouge River valley, about 10 km north of the Scarborough Bluffs. The above papers considered the presence of lacustrine ostracodes in the Sunnybrook diamicton as unambiguous evidence for its lacustrine origin (see also Eyles and Westgate, 1987). If supplemented by field observations of Gravenor and Wong (1987) at the Sylvan Park sections, however, the granulometric data of Eyles and Eyles (1987, Fig. 17) and Westgate and others (1987, Fig. 4) show that the upper parts of the sections where ostracodes have been found are not all "diamicts," as labeled by Eyles and Eyles (1987), Westgate and others (1987), Rutka and Eyles (1989), Eyles and Schwarcz (1991), and Schwarcz and Eyles (1991). The sediments are lacustrine mud of dominantly clayey silt and, therefore, belong to the glaciolacustrine Bloor Member; it is not surprising that they contain ostracodes.

Nevertheless, ostracodes and other microfossils have also been found in many bona fide tills. In subglacial till in Spitsbergen, Boulton (1970b, p. 238) reported that "abundant well-preserved shells occur, some of which, even the most delicate, are still articulated although ligaments have rotted away." In Siberia, the presence of marine microfossils in tills caused heated arguments for about 15 years among the so-called "marinists" and "glacialists." This controversy gradually subsided only after multicriterial regional investigations had proven that the sediments in question were unequivocally glacial (Troitskiy, 1975; Kaplianskaya and Tarnogradskiy, 1984, and references therein). In Scandinavia, friable microfossils are used for stratigraphic differentiation of tills (Kronborg and Knudsen, 1985; Ringberg and others, 1984; Houmark-Nielsen, 1987). In northern Manitoba, where the Laurentide Ice Sheet overrode marine sediments, well-preserved microfossils (foraminifera and ostracodes) have been found in several tills, in which "even quite fragile species are well preserved" (Nielsen and others, 1986, p. 1647). Nielsen (1988, p. 1718), in discussing Sunnybrook ostracodes reported by Westgate and others (1987), concluded that "The presence of delicate unbroken microfossils in a diamicton should not be used solely as a criterion for differentiating glaciolacustrine or glaciomarine sediments from till deposits."

In the Sunnybrook diamicton from Rouge River (Westgate and others, 1987, Fig. 2, samples b and c), only one-third of the ostracodes are complete valves, whereas two-thirds are damaged (ranging from small fragments to valves with minor damage), suggesting that they have been reworked. The uppermost diamicton sample contains very few ostracodes, most of them damaged. Eyles and Eyles (1987, Fig. 17), Rutka and Eyles (1989), Eyles and Schwarcz (1991), and Schwarcz and Eyles (1991) offered no quantitative data on the preservation of ostracode valves at Sylvan Park. The mixture of large and small ostracode valves in the Sunnybrook diamicton at Rouge River, emphasized by Westgate and others (1987) as supportive evidence for in situ occurrence, may be equally supportive for the diamicton being a till, because till is commonly a mixed (nonsorted) sediment.

Deep and cold water ostracode shells, *Candona subtriangulata,* are mixed with the shallower and warmer water species, *C. candata* (Rutka and Eyles, 1989) in Sunnybrook diamicton at Sylvan Park. Rutka and Eyles (1989, p. 72) suggested "mixing of species by resedimentation processes." Eyles and Schwarcz (1991) and Schwarcz and Eyles (1991) published their oxygen and carbon isotope data from the same ostracodes. The great variability of $\delta^{18}O$ and $\delta^{13}C$ values (ranging from −8 to −17‰ PDB for $\delta^{18}O$ and from −4 to −10‰ PDB for $\delta^{13}C$; identical Figs. 3 in Eyles and Schwarcz, 1991, and Schwarcz and Eyles, 1991, which they claim are clear, consistent data) are most simply explained by subglacial mixing and viscous shearing (Rutka and Eyles's resedimentation?) of proglacially deposited lacustrine mud (the Sylvan Park member, discussed later). Such a heterogeneous mixture would not be expected in undisturbed glaciolacustrine sediments as postulated by Eyles and Schwarcz (1991) and Schwarcz and Eyles (1991), in spite of their claims to the contrary. Instead, they invoke a complicated explanation involving changing water depths, algae photosynthesis, changing inputs from organic matter, and isotopic and thermal stratification of a large lake. Subglacial mixing would also account for the hiatus (undated with question mark) of the ostracode and sediment records in their identical Figs. 1 (Eyles and Schwarcz, 1991, and Schwarcz and Eyles, 1991). The identical figures also show an apparently erosive contact of Sunnybrook drift on underlying Scarborough sediments, which Eyles and Eyles (1983, p. 150) originally claimed was "typically interbedded, conformable, and in many places loaded."

Chronostratigraphy

The Sunnybrook drift was thought to be early Wisconsinan even when it was called "lower till" in the Toronto area by Dreimanis and Terasmae (1958). This age assignment has been retained by most subsequent workers, mainly on the basis of its stratigraphic position. The Sunnybrook drift is overlain by the middle Wisconsinan Thorncliffe Formation, which contains plant remains ranging in radiocarbon age from >50,000 yr B.P. in its lower member to 28,000 yr B.P. in its upper member (Lowdon and others, 1971, p. 272–274). Three samples collected by A. A. Berti contained friable leaves of *Dryas* and other plants, and thus he discounted contamination from the fossiliferous Scarborough or Don formations (Berti, 1971, p. 106). Thermoluminescence dates of 35.9 ± 5.4 ka (Berger, 1984, p. 1396) and 32.7 ± 5.5 ka (Lamothe and others, 1984, p. 166) on silt grains in the upper Thorncliffe member agree with the radiocarbon age.

The Sunnybrook drift is underlain in the Toronto area by

the Scarborough Formation, which overlies the Don Formation, generally accepted to be of Sangamonian age (Karrow, 1984b, p. 141). The Scarborough Formation contains deltaic sand that was deposited in a lake at least 45 m higher than modern Lake Ontario, thus requiring the present lake outlet to have been blocked. Plant remains in the Scarborough Formation suggest "a climate 6° C cooler than present (Terasmae, 1960), which is compatible with the idea of ice existing in the St. Lawrence valley" (Karrow, 1984b, p. 142). Thus the Scarborough Formation marks the beginning of the expansion of the Laurentide Ice Sheet following the Sangamon interglaciation.

Massive silt from the Sunnybrook drift at Woodbridge was dated at 66.5 ± 6.8 ka (Berger, 1984, p. 1396). However, because one date is normally unsatisfactory for a stratigraphic unit, more age determinations are required, especially from the Scarborough Bluffs. Nevertheless, the 66 ka age is compatible with the stratigraphic assignment of the Sunnybrook drift to the early Wisconsinan substage.

Figure 2. Section 6 at Sylvan Park, looking east. Sunny Point member is 5 m thick.

THIS REINVESTIGATION

Criteria

The above brief review of the Sunnybrook drift controversy shows that several sedimentologic criteria have been applied to support either a lacustrine or a glacial origin for the diamicton, particularly when the criteria were used singularly or selectively. However, a single criterion, or even a few interrelated criteria are insufficient (because of the polygenetic character of till) to conclude unequivocally whether or not a diamicton is till (Dreimanis, 1989, p. 19). This applies especially to those genetic varieties of till such as deformation till and flow till, in which shear and flow deformation of soft-sediment inclusions have occurred. Some of their structures, if casually observed, may resemble those of nonglacial sediment-flow deposits. Differences exist, but they can be recognized only be detailed structural investigations, in combination with other criteria over a large area.

A particularly discriminating approach is the qualitative and quantitative investigation of deformation structures, fabric, and erosional features (e.g., Åmark, 1985), all of which reflect the dynamic activity of associated processes. The dynamic behavior of glacier ice differs from that of flowing water, sediment flows, or slumps. Therefore, each of these agents leaves distinctive imprints in the sediments formed by them. Investigations of such combined characteristics have not been done on the Sunnybrook diamicton facies; therefore, we concentrated on these formerly overlooked features in six sections along a 3 km portion of the Scarborough Bluffs, and in gullies that dissect the bluffs (Fig. 1). Field work was done during the fall of 1987 and spring of 1988, and observations were added from notes and photographs taken during our yearly visits to geologic sections in the Toronto area.

Sites investigated

We began our reinvestigation with the Sylvan Park sections (Fig. 1, sections 5 and 6; Figs. 2 and 3) because several sedimentologic and lithologic aspects of the Sunnybrook drift had been investigated before: by Karrow (1967, section A1016), Eyles and Eyles (1983, section 2581; 1987, stop 1), Eyles and others (1983, section 2581; 1987, sections 2581, 2985, 3085), Kelly and Martini (1986, near sections GIWA, GIWB, CAX), and Gravenor and Wong (1987). We then reinvestigated Cudia Park (section 4, close to Karrow's [1967] section 1012; Eyles and Eyles's [1983], section 2081; between Kelly and Martini's [1986] sections COU and CAT), Cathedral Bluffs gully (section 3, near Karrow's [1967] section 1011; at Eyles and Eyles's [1983] sections 0681 and 1281; at Sharpe and Barnett's [1985] section, p. 262–263; and near Sharpe's [1987] site 1 and the Bluffers Park and Seminary gully sections (sections 1 and 2, near Ostry's [1962] section, p. 14; at Karrow's [1967] section A1009; at Eyles and Eyles's [1983] sections 1581 and 1681; and near Kelly and Martini's [1986] section BRI). All of these sections (except for 1) are on top of the Scarborough delta, believed by Coleman (1933, p. 30) to have "presented a solid obstruction" to glacier flow.

Methodology

Sections were carefully cleaned and documented. The orientation of sets of parallel striae, stone long axes, and planar structures were measured with a Silva compass. Grain-size analyses were done by a combined hydrometer-sieve method with sand-silt and silt-clay boundaries of 0.063 and 0.002 mm. (American

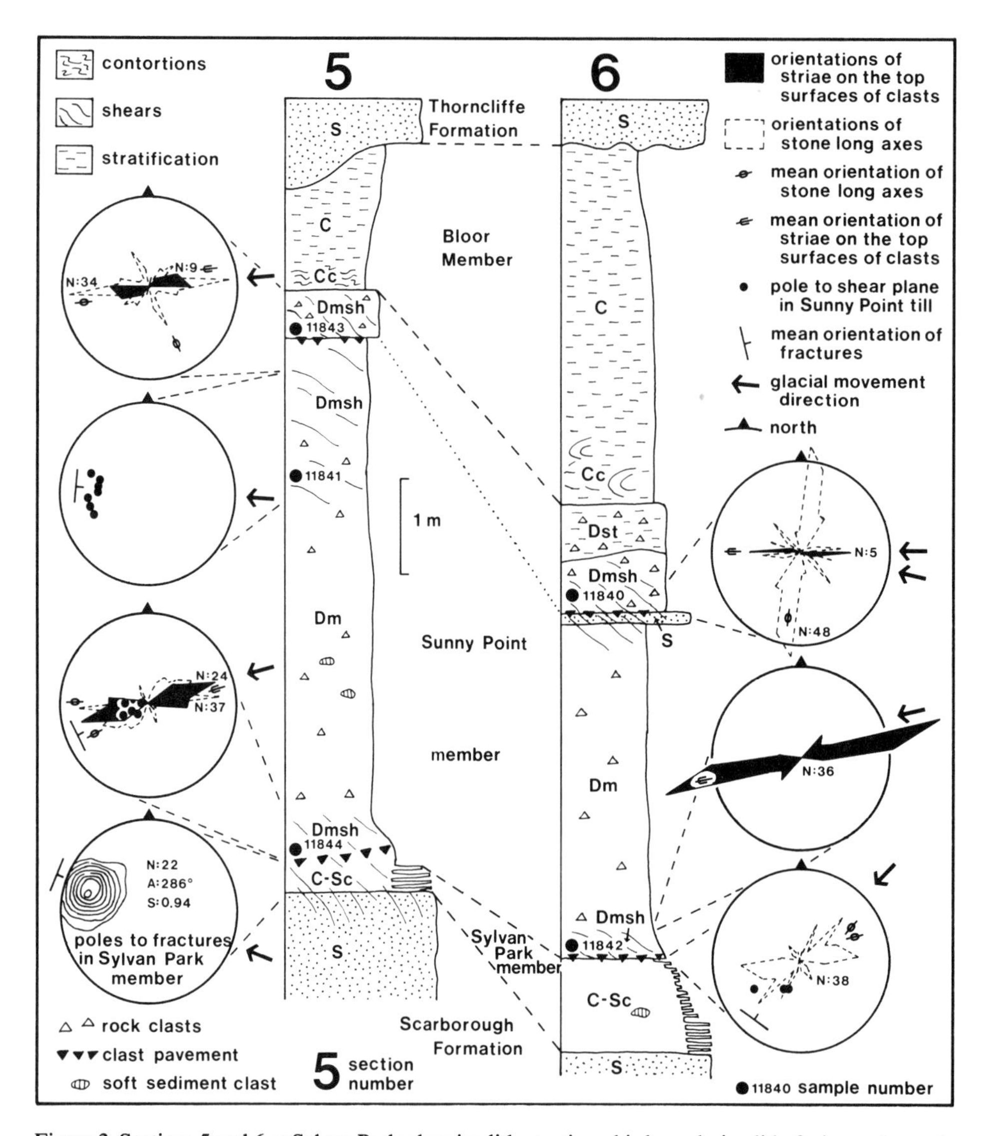

Figure 3. Sections 5 and 6 at Sylvan Park, showing lithostratigraphic boundaries, lithofacies, and sample localities. Structural data include lower-hemisphere equal-area spherical projection of Sylvan Park member structures (N = number of measurements; A = azimuth of eigenvector; S = strength of eigenvector, maximum 1.0; contoured at 1, 3, 5, 7, 9, 11, 13, and 15 points per 5% area), and rose diagrams of stone fabric and striae. Lithofacies abbreviations: C = clay and silty clay; D = diamicton; S = sand. Structural abbreviations: c = contorted; m = massive; sh = sheared; st = stratified.

Society for Testing and Materials, 1972). Carbonate analyses were performed using Chittick apparatus (on 0.85 g mud samples; Dreimanis, 1962). Stones were collected in the field (many of them marked with north arrows on upper surfaces) and their glacial surface markings were further studied in the laboratory. Clast lithologies were identified in the laboratory under a binocular microscope. Heavy minerals were recovered from the 0.250- to 0.125 mm fraction with liquid sodium polytungstate (specific gravity 2.9), and garnets were counted under a binocular microscope.

Overview of results and conclusions

Our reinvestigation provided strong evidence that the diamicton facies of the Sunnybrook drift is subglacial till (Hicock and Dreimanis, 1988, 1989; Dreimanis, 1991). We (Hicock and Dreimanis, 1988) summarized the main evidence in the following statement:

Clast pavements with consistently oriented parallel striae (E-W ±20°) occur 0.3–1 m above the base of Sunnybrook drift and in its middle part.

Clast alignments within and between pavements are mainly parallel or transverse to striae. Shear planes above and below pavements within diamicton, and drag folds and tension fractures in underlying sediment strike normal to striae. Stoss and lee relations on clasts, dips of shear planes, axial planes, and tension fractures indicate glacial advance over the area towards W (±20°).

We (Hicock and Dreimanis, 1988) concluded that "most of the till is reworked glaciolacustrine mud mainly transported by subsole drag with easterly-derived glacial debris."

Lower-hemisphere, equal-area, spherical projections and rose diagrams of the orientation of deformational structures and fabric were presented from four sections (Hicock and Dreimanis, 1989, Fig. 1); the granulometric composition of tills investigated were presented as Figure 3 (Hicock and Dreimanis, 1989); and six selected photographs of some of the field evidence as Figure 2 (Hicock and Dreimanis, 1989). The space restrictions for that paper and for our abstract (Hicock and Dreimanis, 1988), did not allow us to present all the results of our reinvestigation, so we take this opportunity to expand on the above. We suggest that the "Sunnybrook Till," proposed by Karrow (1984b, p. 143) as the lowest member of the Sunnybrook drift, be redefined as the middle member and renamed the Sunny Point member. Furthermore, we suggest that the glaciolacustrine unit a of Hicock and Dreimanis (1988, 1989) be considered as the lowest member, here named the Sylvan Park member after Sylvan Park, where it is best exposed. In addition, we propose that the Pottery Road Formation of Karrow (1974) be included as a glaciofluvial member in the Sunnybrook drift, if reinvestigation warrants such a change; it may be part of the glaciofluvial member proposed by Lajtai (1969) and inferred by Sharpe and Barnett (1985).

SUNNYBROOK DRIFT

Subdivisions

The Sunnybrook Formation is a regionally recognized lithostratigraphic marker in the greater Toronto area. In the Scarborough Bluffs, it consists of three members (Table 2, Fig. 2), informally labeled units, a, b, and c in Hicock and Dreimanis (1989), and herein called the Sylvan Park member, the Sunny Point member (till), and the Bloor Member, respectively. Karrow (1984b) listed two: the "Sunnybrook Till" and the Bloor Member. Lajtai (1969) considered three informal members: glaciofluvial sediment, till, and varved clay. Formalization of Sunnybrook drift and its members, with stratotypes assigned, will be done after a more thorough regional investigation in the greater Toronto area.

The next two sections of this chapter are on the sedimentologic and lithologic character of the lower two members of the Sunnybrook drift. Subglacial deformational and erosional features encountered within them, and along their contacts with the underlying Scarborough sand, will be discussed later. The uppermost Bloor Member is generally accepted as glaciolacustrine mud, and was not studied by us.

Sylvan Park member

General Character. The Sylvan Park member consists of interlayered mud, clayey to silty diamicton, and sand up to 1.5 m thick (Figs. 3 and 4, a–e). Probably because of the thinness, Karrow (1967, 1969, 1984a) did not separate it from overlying "Sunnybrook Till" (our Sunny Point member).

Sharpe and Barnett (1985, Fig. 9a and p. 262) referred to our Sylvan Park member as "interbedded diamictons and sand" in the lower part of the Sunnybrook diamicton sequence. Kelly and Martini's (1986, Figs. 18, 19, and p. 42) part b of their facies association 7 of the Scarborough Formation probably corresponds to our Sylvan Park member: it consists of "parallel-laminated sand with scattered mud pebbles overlain by interlayers of massive to cross-laminated sand (Sr) and silty clay (Cl)." Sharpe (1987, p. 341), in discussing Kelly and Martini's (1986) uppermost facies association, also suggested that it "may belong to the overlying Sunnybrook." Gravenor and Wong (1987, Fig. 2 and p. 2039) called our Sylvan Park member a "transition zone" from the Scarborough sand to the Sunnybrook, and included it within the Scarborough Formation. Eyles and Eyles (1983, Fig. 1 and p. 150) and Eyles and others (1983, Figs. 1 and 5, p. 13–15) labeled it merely as an interbedded basal contact zone of Sunnybrook "diamict" with sand of the Scarborough Formation.

Contacts and structures. In those sections where the Sylvan Park member is absent, a sharp erosional contact separates the Sunny Point member from the underlying Scarborough sand (Figs. 4f and 5, a, c, and d). At section 4 (Fig. 4f), an ~1-m-long, partly sheared block of stratified Sylvan Park member sediments is visible as an inclusion within the bottom part of the Sunny Point member. Apparently, the Sylvan Park member has been eroded at these places.

The Sylvan park member is commonly deformed by small faults (Fig. 4, b and c), dewatering structures, and loading structures (Fig. 4e). Where undeformed, it is subhorizontally layered (Figs. 4 a–c), including wavy and lenticular bedding that was probably formed by alternating current and suspension deposition in standing water (Reineck and Singh, 1973, p. 100–101). Its lower contact with the Scarborough sand is sharp and is commonly marked by a dark mud lamina (Fig. 4, b and c; Kelly and Martini, 1986, Fig. 19; Gravenor and Wong, 1987, Fig. 3). The upper contact with the Sunny Point diamicton (part c of facies association 7 of Kelly and Martini, 1986) is either sharp (Fig. 4d; Kelly and Martini, 1986, Fig. 19; Gravenor and Wong, 1987, Fig. 3; Hicock and Dreimanis, 1989, Fig. 2d, middle), or transitional, parts of the Sylvan park member being sheared up into the Sunny Point member (Fig 4, d and e; Hicock and Dreimanis, 1989, Fig. 2d, left side). The proportion of mud or diamicton interbeds increases upward (Fig. 4a; Hicock and Dreimanis, 1989, Fig. 2a). Some mud layers are finely laminated, but most sand interbeds appear to be structureless to crudely bedded, unlike the uppermost Scarborough sand, which is clearly rippled and cross-stratifiled (Fig. 4, b and c). In some places mud layers

ST
SP
Sc
a
ST
ST
SP
Sc
d
x
ST
SP
Sc
b
x
ST
ST
SP
Sc
e
ST
x
SP
Sc
c
ST
SP
ST
ST
Sc
f

are draped over small sand mounds that resemble incomplete or starved ripples (Fig. 4, a–c). Mud clasts occur within sand beds and vice versa. Rock clasts in the diamicton layers are pebble size to granule size, and they are lithologically similar to those of the overlying diamicton.

Stratigraphic assignment. Kelly and Martini (1986) considered their entire facies association 7 to be the uppermost unit of the Scarborough Formation, and pointed out that (p. 48) "the upper parts (b, c) of association 7 indicate lacustrine conditions increasingly affected by an approaching glacier." We prefer to include parts b and c of their association 7 in the Sunnybrook drift because of the lithologic similarity of the diamictic and muddy interbeds of their part b to the Sunnybrook. The gradual upward-increasing dominance of silt and clay over sand and the presence of dolostone and sandstone clasts in their part b, are more consistent with the character of the Sunny Point member (Karrow, 1967; Hicock and Dreimanis, 1989) than with the underlying Scarborough Formation. Thus we consider their b to be equivalent to our Sylvan Park member, and their part c to the basal portion of the Sunny Point member.

Sunny Point member

Definition, texture, and subunits. In anticipation of Sunnybrook formalization, and in order to satisfy the requirement of the North American Commission on Stratigraphic Nomenclature (1983) that members have different names than formations, the "Sunnybrook Till" is here renamed the Sunny Point member after Sunny Point Crescent, a Toronto (Scarborough) street that parallels the west bank of the Cathedral Bluffs ravine (site 3 in Fig. 1). The Sunny Point member in most sections of the Scarborough Bluffs consists of 7 to 10 m of dominantly massive, silty to clayey diamicton, labeled Dmm (matrix-supported massive "diamict") by Eyles and Eyles (1983). Karrow (1967, p. 32) described it as "fine grained till, varying from a silt till to a silty clay till with low pebble content." The basal 0.5 to 1 m is commonly more sandy or silty than above (Table 3), because of incorporation of the underlying Sylvan Park member and/or Scarborough sand. For further granulometric data from the Scarborough area, see Ostry (1962, Fig. 10), Karrow (1967, Appendix D), Eyles and Eyles (1987, Fig. 17), Gravenor and Wong (1987, p. 2039), Westgate and others (1987, samples b–d), and Hicock and Dreimanis (1989, Fig. 3); however, note that the silt and clay percentages vary among the above reports, because 0.002 and 0.004 mm were variously adopted as the silt/clay boundary.

In section 6, we noticed that the lighter, buff-colored, upper 1.2 m of the Sunny Point member (Figs. 2 and 3) is also more sandy (Table 3, sample 11840) than the lower 3.5 m, probably due to incorporation of underlying sand (Fig. 3, facies S, 1.2–1.3 m below the top of the Sunny Point diamicton). The westward continuation of this uppermost Sunny Point subunit, which is marked by a clast pavement at its base (Fig. 3), is considerably less sandy than (Table 3, section 5, sample 11843), and as clayey as, the underlying till. We do not know whether the uppermost subunit of the Sunny Point is a local feature restricted to the western part of Sylvan Park (sites 5 and 6), or if it is laterally more extensive, because the upper part of the Sunny Point member was inaccessible for study at sections 1 to 4.

Structures. In sections 5 and 6, the uppermost subunit of Sunny Point diamicton appears to be faintly stratified. It may correspond to the stratified Dms facies of Eyles and others (1983, Fig. 5), 4–5 m above the base of the Sunnybrook. However, the conspicuous flow noses shown in Figure 5 of Eyles and others (1983) were not seen in our section; they were probably eroded during the intervening period. We found only a small recumbent fold with its upper limb attenuated westward; the entire subunit has been so strongly sheared that the fold was barely visible.

Recumbent fold and flow noses are common in sediment flows (Boulton, 1968; Lawson, 1979). However, similar structures may also be formed by subsole glacial drag, as observed in Iceland (Boulton, 1987, p. 41). Both origins are possible in section 6, but because the flow noses, recumbent fold, attenuation of sheared sediment lenses, and striae on clasts (Fig. 3) consistently point westward, subglacial drag is invoked as their final formational process.

Where the Sunny Point member is described as massive diamicton, weakly developed laminations are noticeable in X-ray images (Eyles and others, 1987, Fig. 11), or "faint discontinuous stringers of silt 1–2 mm thick" have been seen at various elevations (Gravenor and Wong, 1987, p. 2039). We have noted such "stratification," particularly in those parts of diamicton that have been sheared; the "stratification" is commonly parallel to shear planes (Fig. 5b), and therefore is probably related to shear deformation.

Figure 4. Basal part of Sunnybrook drift in sections 3–6. Abbreviations: Sc = Scarborough sand; SP = Sylvan Park member; ST = Sunny Point till. Knife is 20 cm long. a: Section 6, Scarborough sand overlain by 1.0 m of Sylvan Park member, separated from the Sunny Point member by a clast pavement (arrow). b: Section 5, cross-laminated Scarborough sand overlain by 0.3 m of Sylvan Park member and the Sunny Point member. Note the lighter-colored bottom 15 cm of the Sunny Point member (up to the top of the square boulder x) that consists mainly of homogenized silt of the Sylvan Park member with a few clasts, which we interpret as deformation till. c: Close-up of right side of b, showing the transitional boundary between the normal-faulted Sylvan Park member and the base of the Sunny Point member near x. d: Section 5, 2–3.5 m to the right of b. Pocket knife (15 cm long) is at the base of the contorted Sylvan Park member. The Sunny Point till is massive, with holes 10–20 cm above its base where clasts were collected. e: Section 3, showing in ascending order Scarborough sand, the strongly deformed Sylvan Park member, sheared bottom 20–30 cm of the Sunny Point till, and the massive Sunny Point till above x. f: Section 4, showing a sharply eroded surface separating the Scarborough sand from the overlying Sunny Point member. Clast pavement (arrow) separates the 0.5-m-thick deformation till from overlying, massive fissile lodgement till. A 1-m-long sheared block of the Sylvan Park member is included within the deformation till. Centimeter card for scale (under ST).

Figure 5. Sections showing sharp erosional contacts at the base of, or within, the Sunny Point till. Abbreviations as in Figure 4. a: Section 4 about 3 m east of Figure 4f, with steeply dipping extensional faults in the Scarborough sand. Some of the faults have undergone displacement of their tops downglacier by subsole drag. b: Section 5, showing overhanging base of the uppermost meter of till at a clast pavement. Clast positions are marked x and shallow sole marks by arrows. c: Section 2, showing east-west–trending sole marks at the contact of the Sunny Point till with the Scarborough sand. Note a few light-colored silt clasts and lenses (arrowed) within the lower 5 cm of till. Knife is 20 cm long. d: Section 1, showing sharp contact of the Sunny Point till (dark area below lighter slopewash) with several steeply dipping faults within the underlying Scarborough sand below scale card.

Most shear planes observed in the Sunny Point diamicton were planar and rose gently in the inferred downglacier direction (Fig. 6, uppermost diagram in section 4). Commonly they are coated with silt or sand entrained from underlying sediment which was attenuated downglacier (Fig. 4 e and f).

Fissility was observed at several levels in the Sunny Point diamicton, but mainly above clast pavements (e.g., in site 4 for more than 1 m above the clast pavement, and in site 5 for about 0.5 m above both clast pavements).

Clast pavements. All reports on the Sunny Point diamicton mention that its clast content is low. However, a concentration of clasts along the base of this member was noticed by Eyles and Eyles (1983, Fig. 1), Sharpe and Barnett (1985, Fig. 12, b and c), and particularly by Kelly and Martini (1986, Figs. 18 and 19), who called it a "stone line" and used it "as an arbitrary convenient contact between facies association 7 and the overlying glacigenic Sunnybrook Formation" (Kelly and Martini, 1986, p. 45). In some places, the clast concentration is at the basal contact of Sunny Point diamicton (Fig. 4, a and b), whereas elsewhere it rises to 0.5 m above its base (Fig. 4f). Where the clast concentration is within the diamicton, part c of Kelly and Martini's (1986) facies association 7 probably belongs to the Sunny Point member rather than to the Scarborough Formation (as they arbitrarily assigned it), because the diamicton above and

TABLE 3. GRANULOMETRIC AND CARBONATE DATA OF SUNNYBROOK DRIFT, SCARBOROUGH BLUFFS SECTIONS 1-6*

Sample Number	Section Number	Material	Height Above Base of Sunnybrook (m)	Sand† (%)	Silt† (%)	Clay† (%)	Carbonate† (%)	Calcite/ Dolomite
	1	*Sunny Point member:*						
01582		Massive diamicton	7	....	....	....	25	2.8
01583		Massive diamicton	6	....	....	....	18	2.2
	3	*Sunny Point member:*						
11851		Massive diamicton	1.7	17	55	28	32	4.2
11849		Massive diamicton	1.0	18	52	30	9	1.4
11848		Sheared diamicton	0.3	42	31	27	10	1.3
	3	*Sylvan Park member:*						
11847		Mud interbed	0.05	36	40	24	11	1.8
	4	*Sunny Point member:*						
11845		Massive fissile diamicton	1	10	42	48	9	0.9
11846		Sheared diamicton	0.4	23	35	42	12	3.0
	5	*Sunny Point member:*						
11843		Massive fissile diamicton	6	11	42	47	9	1.2
11841		Massive sheared diamicton	4	11	40	49	9	1.0
11844		Massive sheared diamicton	0.1	17	55	28	9	1.3
	6	*Sunny Point member:*						
11840		Sheared diamicton on sand lens	4	28	49	23	24	1.0
11842		Massive diamicton	0.1	14	51	35	9	1.2

*For locations see Figure 1.
†Sand = 2 to 0.06 mm; Silt = 0.06 to 0.002 mm; Clay = <0.002 mm; Carbonates of <0.06 mm fraction.

below this basal clast pavement is lithologically similar (Table 3, sites 3 to 5).

There is another clast pavement, although less conspicuous than the basal one, 1–2 m below the top of the Sunny Point diamicton at Sylvan Park sections 5 and 6 (Fig. 3). In 1987–1988, we saw it at the base of an overhanging shelf of hard compact diamicton in section 5 (Fig. 5b) and over a 5–10-cm-thick subhorizontal sand lens in section 6.

Clast provenance. The provenance of clasts in the Sunny Point diamicton is dominantly distant (50%–70%). Most rock fragments were derived from the area northeast of Lake Ontario: Beekmantown dolostone, Potsdam or Nepean sandstone, igneous and metamorphic rocks from the Precambrian shield (Table 4). In Sylvan Park sections 5 and 6, the abundance of distantly-derived rocks increases from 50% in its basal part to 70% in its upper 1–2 m. A pebble sample from the middle of the Sunny Point diamicton in the Sylvan Park area, investigated by Karrow (1967, Fig. 15, section 1017), was also dominated (90%) by clasts of distant provenance. However, local Ontario basin rocks (limestone and shale) dominate in the western part of the Scarborough Bluffs, southwest of the Seminary section. Limestone content is also high at the eastern end of the Scarborough Bluffs (Karow, 1967, Fig. 15).

Small pebble- to granule-sized, rounded or lens-shaped, soft-sediment clasts of silt or clay were noted in the middle part of the Sunny Point diamicton in sections 4 and 5, as well as in the basal part of sections 2 to 5.

Matrix provenance. The carbonate content of the silt and clay fraction of the Sunny Point diamicton is low (9%–12%) in most samples, proportions of calcite and dolomite being about equal (Table 3); this is in agreement with most carbonate data of Karrow (1967, Fig. 15). Because the gasometric method used for carbonate determination tends to exaggerate calcite proportions in fine-grained samples (Locat and Bérubé, 1986), the actual percentage of dolomite may be even higher. The main source of dolomite is the St. Lawrence lowland. In some samples, mainly from the upper part of the Sunny Point diamicton, the carbonate content exceeds 20% (Table 3, samples 1582, 11840, and 11851; Karrow, 1967, Fig. 15, sites 1023 and 1035). Westgate and others (1987, p. 233) also reported an upward increase (at Cathedral Bluffs, our section 3) caused mainly by calcite enrichment, with total carbonate content approaching 25%.

Heavy minerals were investigated in only three samples from sections 1 and 4. They were rich in garnets (20%, 21%, and 36%) that had purple/red ratios of 0.7, 1.1, and 1.1, respectively. The garnet-color ratios agree well with those obtained by Ostry (1962; 11 analyses, 0.4–1.5, mean 1.0), and suggest the eastern Grenville Province as the distant source area (Gwyn and Dreimanis, 1979).

The above lithologic data support the conclusions of Dreimanis and Terasmae (1958), Ostry (1962), and Karrow (1967, 1969, 1984a) that the Sunny Point diamicton was deposited by the Ontario glacial lobe.

EVIDENCE FOR SUBGLACIAL EROSION AND DEPOSITION

We paid particular attention to the dynamic effects of depositional agents when investigating the two lower members of the Sunnybrook drift and their basal contacts, considering both glacial and glaciolacustrine alternatives. We found consistent interrelations among (1) the orientation of parallel sets of striae on stones, (2) the sets of fractures, faults, and recumbent folds in the sediments, (3) clast fabrics, and (4) basal sole marks along a lateral distance of more than 3 km. This consistency means that the geologic agent responsible for them acted on a large scale and moved in the same direction throughout. By considering other sedimentologic and lithologic evidence, we concluded that a grounded or partially grounded glacier deposited the Sunny Point diamicton as subglacial till (Hicock and Dreimanis, 1989). Most of the deformation structures are small scale (appearing in vertical intervals of 0.1–1 m), and therefore could have been overlooked by Eyles and Eyles (1983) and Eyles and others (1983), who claimed that they are absent. In most sections deformation structures are covered by slopewash, so considerable cleaning of sections is required to find them and to trace them laterally.

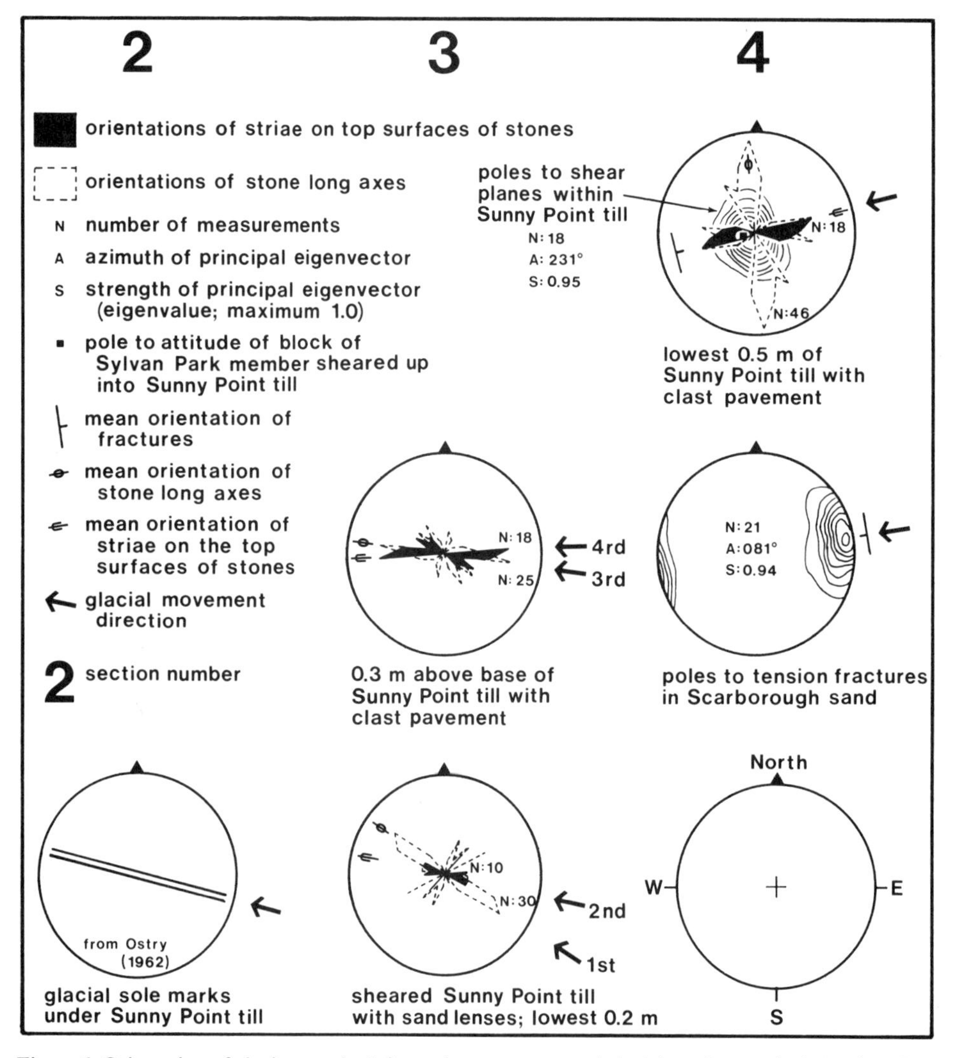

Figure 6. Orientation of glaciotectonic-deformation structures and glacial erosion marks in sections 2–4, plotted according to sequence of deformation events. Lower-hemisphere, equal-area, spherical projections contoured at 1, 3, 5, 7, 9, 11, 13, and 15 points per 5% area.

TABLE 4. CLAST TYPE PERCENTAGES IN SUNNY POINT MEMBER, SECTIONS 3 THROUGH 6

Provenance and Lithologic Varieties	Sections 3 and 4 Base of Till	Sections 5 and 6 Base of Till	Sections 5 and 6 1 to 2 m Below Top of Till
Clasts from area northeast of Lake Ontario	54	50	70
-Igneous and metamorphic	10	14	17
-Dolostone	30	24	21
-Sandstone	14	12	32
Clasts from Lake Ontario basin	46	50	30
-Limestone	45	42	29
-Shale	<1	08	01
-Calcite concretion from Scarborough Formation?	<1	0	0
Number of clasts counted	144	280	155

Stone pavements

The most telling erosional phenomenon is the consistent east-west orientation of striae on stone pavements at the base of the massive diamicton of the Sunny Point member (Figs. 3 and 6; for pavements, see Fig. 4, a, b, and f). In addition, stone long axes in this pavement (and above it) are consistently either parallel or transverse to striae on stones (Figs. 3 and 6; Hicock and Dreimanis, 1989, Fig. 1), which agrees with the results of Ostry (1962), Karrow (1967, p. 32) and Gravenor and Wong (1987, p. 2041). The dominance of transverse stones in some places (e.g., Fig. 3, section 6, above upper pavement) suggests compressive subglacial flow during transport prior to their emplacement (cf. Boulton, 1970, p. 219; 1971, p. 46–48; Stanford and Mickelson, 1985, p. 222) so that the long axis of the strain ellipsoid would have aligned with the direction of maximum extension and transverse to the flow direction. Striated facets superimposed on stones of various shapes and orientations (Fig. 7, a and c; Hicock and Dreimanis, 1989, Fig. 2c) also indicate strong abrasion of some stones after their lodgement.

Lower contact of diamicton; intradiamicton and subdiamicton deformations

Where the Sunny Point diamicton is in direct contact with Scarborough sediments, the contact is sharp (Figs. 4f and 5, a, c, and d). Tension faults (Figs. 4c, 5a) occur in the underlying sand, striking generally transverse to stone-striae orientations (Figs. 3 and 6; cf. Åmark, 1985). Sand-filled shear (thrust) planes (Figs. 4, e and f, and 5a) within the Sunny Point diamicton, and axial planes of folds in the deformed Sylvan Park member (Fig. 4d), also strike generally normal to stone striae (Fig. 3) and to the direction of maximum compression in the inferred strain ellipsoid. At site 4, a 1-m-long sediment lens (Fig. 4f) within the base of diamicton strikes normal to striae on the stone pavement above it (Fig. 6). The lens comprises interbedded and folded sand and mud that can be traced eastward along a sand-coated thrust plane to the underlying Scarborough sand (Hicock and Dreimanis, 1989, Fig. 2d); it thins out within the diamicton to the west. We interpret this lens to be part of the Sylvan Park member that was sheared up and deformed by overriding glacial ice during deposition of the Sunny Point diamicton.

Many of the above data correspond to information in Hicock and Dreimanis (1989) from sites 3–6. In 1978, at site 2 (Fig. 1) on the east side of the Seminary gully, we observed elongated and parallel straight grooves in the base of the Sunny Point diamicton (Fig. 5c), close to the place where Karrow had seen them during the early 1960s (Karrow, 1984b, p. 185). These grooves were oriented east-west and match the 285° orientation of similar structures at the base of the Sunny Point diamicton measured by Ostry (1962, p. 14) on the opposite (western) side of the wide Seminary gully, and photographed at another section where the base of the diamicton was overhanging (Ostry, 1962, Plate 5A). This consistent orientation over several hundreds of meters suggests that the grooves represent sole marks formed by glacial fluting of the substrate, similar to those described by Westgate (1968) from Alberta, as opposed to less-consistent scours formed by mudflows that are more influenced by local topography. The orientation of the sole marks at Seminary gully is also consistent with the magnetic fabric of Sunny Point diamicton in this area (maximum susceptibility at 277°–297°, as measured by Stupavsky and others [1974] at sites S3–S6), and with shear planes in underlying Scarborough sand observed by W. R. Church (1989, oral commun.). At site 1, a sharp erosional contact could be seen between the Sunny Point diamicton and underlying Scarborough sand (Fig. 5d), as well as steeply dipping normal faults in the Scarborough sand (tops bent as in a vertical deckof cards), which we interpret as having been caused by drag of overriding ice.

Other authors have also reported evidence from the Sunnybrook diamicton suggesting grounded ice and have interpreted it as till. Hinde (1877, Fig. 1), Coleman (1895, p. 644; 1901,

Figure 7. Clasts from the lower stone pavement, oriented with north at the top of each photo. Black bars are 2 cm long. a and b are the same limestone clast from section 6. a shows the faceted and striated top and white impact marks (arrowed) on the upglacier east side. b is the underside, with most parallel striae concentrated at its downglacier west side. c: Upper side of an angular limestone clast from section 6 with a sharply planed, striated facet on its highest part, left side. d: Upper surface of subrounded limestone cobble from section 5 with two crossing sets of parallel striae on its lower surface. The various sides have five differently oriented sets of striae, suggesting that the stone had been rotated during its abrasive transport.

p. 305; 1933, p. 19), and Wilson (1901, p. 166) recognized that folded and faulted sediments were directly overlain by the diamicton at the eastern end of the bluffs. Wilson (1901, p. 166–167) attributed the deformation to glacial overriding of the Scarborough delta and postulated that most glacial disturbance would have been on the upglacier side of this obstruction.

In summary, the above evidence indicates that the Ontario glacial lobe was grounded on top of the Scarborough delta during deposition of the Sunny Point member. The consistently striated stone pavement at the base of the massive and fissile unit, upsheared Scarborough sand within the diamicton, and deformation in underlying sediment also indicate that the glacier was moving over its substrate. Therefore, we conclude that Sunny Point diamicton is a subglacial till. Similar conclusions, based on comparable criteria, were reached by Åmark (1985) in an investigation of till near Skåne, Sweden.

GENESIS OF TILL IN THE SECTIONS INVESTIGATED

Several varieties of subglacial or basal till may form under a moving glacier: lodgement, deformation, and, locally, subglacial melt-out and subglacial flow till (Dreimanis, 1989, Table 13 and p. 40–54; see also Boulton, 1982, Figs. 1 and 2). Our proposed genesis for the Sunny Point member invokes a dominance of deformation and lodgement processes, with local occurrences of subglacial flow till.

Lodgement

The following characteristics suggest lodgement as the mechanism for emplacement of clasts in the basal clast pavement, particularly at sections 3, 4, and 6: (1) small deformation structures at the inferred downglacier[1] contact of several clasts, suggesting ploughing of the substratum (Boulton, 1982, Fig. 2b); (2) a "double stoss-lee form" (as in Krüger, 1984) seen on many clasts, with their upper surfaces usually more abraded and striated than their lower surfaces (Fig. 7, a and b; Sharp, 1982, p. 478); (3) faceted upper surfaces on many clasts, with numerous striae parallel to the inferred direction of glacial movement, even where the long axis of a clast does not coincide with that direction; and (4) dominantly parallel or transverse orientation of the long axes of clasts in relation to the inferred direction of glacial movement.[1] A laterally traceable planar erosional contact is apparent between the surface facets on clasts in sections 3, 4, and 6 (Fig. 4, a and f; Hicock and Dreimanis, 1989, Fig. 2, a, b and d), probably representing the position of the sole of a moving grounded glacier at the time of facet formation.

Deformation

In the basal clast pavement of section 5, the above lodgement features were not as pronounced as in sections 3, 4, and 6. For example, (1) one-third of the stones with stoss-lee features had their stoss ends on the westward side, in reverse relation to the inferred ice flow direction from east to west; (2) several elongated clasts were oriented vertically; (3) facets were less abundant and occurred on various sides of some stones; and (4) some stones were striated in various directions on several sides (Fig. 7d). All of these observations suggest that the clasts had been rotated prior to their final emplacement in the pavement (Hicock, 1991). In section 5, the surface of the pavement was quite irregular; the tops of clasts varied in elevation by up to 10 cm over short distances. Apparently the basal clast pavement of section 5, and the till surrounding the clasts, was transported by subsole drag and deposited as deformation till (e.g., Elson, 1961, 1989; Boulton, 1987), although some large clasts were probably lodged (Fig. 4b, square boulder on left side; Hicock, 1991, Fig. 11). Similar pavements at sections 3, 4, and 6, and associated till at least up to 0.5 m above the pavements, were deposited mainly by lodgement (criteria in preceding paragraph). At sections 3 and 4, the lowermost 0.4–0.5 m of till beneath the basal pavement contains many sheared lenses of the glaciolacustrine sediment of the Sylvan Park member (Fig. 4f), typical of deformation till.

Combination of lodgement and deformation

Judging from the above descriptions of the basal clast pavement and associated tills above and beneath them, the lowermost zone of the Sunny Point till was formed and deposited by two alternating processes: subsole deformation and lodgement (Hicock, 1991, Fig. 12). These processes also participated in the formation and deposition of the upper clast pavements and overlying tills at sections 5 and 6 (Fig. 3). The upper pavement and its overlying till at section 6 display most of the characteristics previously listed for a deformation till pavement of section 5 (see Hicock, 1991, Table 1). In addition, the clast fabric is transverse to striae, and the till displays shearing and local attenuated recumbent folding. However, the upper pavement of section 5 has the characteristics of a lodgement pavement (see Hicock, 1991, Table 2), although facets on the upper surfaces of stones are not as strongly developed as in the lower pavement of section 6.

At most sections where the till between the upper and lower pavements was accessible for close examination, we saw occasional small soft-sediment (silt or clay) clasts and thin sheared silt or silty sand lenses, either subhorizontal or gently rising to the west. The soft-sediment clasts were also noticed by Eyles and Eyles (1983) and Eyles and others (1983, 1987), who used them as a supporting criterion for lacustrine sedimentation. Indeed, soft-sediment clasts would not survive the lodgement process in the strict sense. However, they would not be destroyed while being transported in a viscous or plastic clay- and silt-rich sediment moved by subsole drag and subsequently deposited as deformation till (Elson, 1961, 1989; Boulton, 1987). A strong dilating effect, enhanced by high pore-water pressure, would also permit the preservation of fragile ostracode shells that were transported together with enclosing glaciolacustrine mud by subsole drag. This would cause ostracodes to be mixed and would readily explain the erratic isotope trends in identical Figures 3 of Eyles and Schwarcz (1991) and Schwarcz and Eyles (1991).

In such a mobile mass, the magnetic fabric would be quite variable from place to place, being influenced by local obstructions such as stones (Stupavsky and Gravenor, 1975) or areas of easier flow. Such variability was displayed by several of the magnetic-fabric diagrams in Gravenor and Wong (1987, Figs. 5 and 6) and Eyles and others (1987, Figs. 8 and 9) from Sylvan Park sections 5 and 6. Gravenor and Wong (1987, p. 2038) concluded that "the fabrics are not typical of a lodgement till and to some extent resemble fabrics found in subaquatic debris flow deposits." Because deformation till is formed at least partly by squeeze flow of water-saturated mud, its magnetic fabric should be similar to that of subaquatic debris flows (Dreimanis, 1989). This conclusion was supported by Stanford and Mickelson (1985, p. 225), who suggested that deforming till may behave like a debris flow, and by Dowdeswell and Sharp (1986, p. 707), who concluded that pebble fabrics in the upper deformed horizon of lodgement till (Boulton's [1987] A horizon, or deformation till) resemble sediment flows.

In conclusion, our investigation of the structural, fabric, and

---

[1]The direction of glacial movement was concluded from a consistent agreement among the orientation of four features: (1) sets of parallel striae on clasts, (2) stoss-lee forms of clasts, (3) deformation structures within and beneath till, and (4) clast fabric in till.

erosional features in six sections along the Scarborough Bluffs, together with relatd data published by many other authors, suggest that the Sunny Point member was deposited mainly by subglacial deformation and lodgement, deformation dominating.

As in many subglacial tills described from other areas (e.g., Eyles and others, 1982; Dreimanis and others, 1987; Dreimanis, 1989), subglacial channel deposits, filled either with stratified sorted sediment or diamicton of various origins, may be present as lenses within the Sunny Point member. One lens of silty sand was observed beneath the upper clast pavement in our section 5. If the flow lobes shown in Eyles and others (1983, Fig. 5a) were formed by gravity flows, then a lens or layer of subglacial subaquatic sediment flows of unknown dimensions occurred above the upper clast pavement of section 6. Other lenses are shown by Eyles and Eyles (1983, Fig. 1) in several sections, particularly at Bluffers Park, where the Sunnybrook drift fills an erosional valley. These channel fills, discussed below in a regional context in association with lodgement and deformation till, offer further supporting evidence for a subglacial origin for the Sunny Point member.

J. Shaw (1989, written commun.) pointed out the similarities between the characteristics of the Sunny Point member and tills that he studied near Edmonton, Alberta (Shaw, 1982, 1987). Shaw concluded that his tills were formed by a combination of lodgement, melt out, and subglacial debris flow. It is possible that subglacial melt out of lodged, debris-rich glacier ice was also involved in forming parts of the Sunny Point till where subhorizontal thin lenses of silt occur. However, subglacial melt-out features indicating stone settling, bending of layered strata around stones, or convex-up sorted sediment lenses were not found in the Sunny Point till.

EARLY WISCONSINAN GLACIAL FLOW: SCARBOROUGH BLUFFS AND ADJOINING AREA

Directional data from the Sunnybrook drift in the Scarborough Bluffs (Fig. 8) suggest a gradual shift in the direction of ice movement over the top of the Scarborough delta. Initial deformation of the Sylvan Park member and underlying Scarborough sand was by ice advancing west-northwest out of the Ontario basin (the record of this initial deformation was not preserved at section 4, where the Sylvan Park member has been eroded). Subsequent lodgement of the basal stone pavement and deposition of the lower part of Sunny Point till was by ice moving west and west-southwest. During formation of the upper part of the till at section 5 by viscous deformation, ice flow was toward the west-northwest. Similar fabrics ranging from west-southwest to west-northwest were also recorded by Ostry (1962, Fig. 19) from eight sites over an area extending from the Don River to Duffin Creek.

The early northwestward flow direction may represent initial radial flow of the Ontario lobe out of its basin. However, as lobes and other components of the growing Laurentide Ice Sheet coalesced, they probably influenced local flow directions. Such coalescence could have occurred under the influence of the southwestward-flowing Simcoe-Kawartha ice mass and the north side of the Ontario lobe, forcing ice movement along the north shore of Lake Ontario to the southwest (Fig. 9). During the ensuing glacial retreat, radial flow out of the lake basin was restored within the Ontario lobe, as evidenced by clast fabrics from the upper part of the Sunny Point till, including the uppermost transverse fabric over the upper clast pavement at site 6 (Fig. 3) and Ostry's (1962) fabrics 1, 6, and 25 (Fig. 10).

SUNNYBROOK DRIFT IN THE TORONTO AREA: LATERAL AND VERTICAL CHANGES OF LITHOFACIES

Table 2 lists the dominant facies associations in each of three members of the Sunnybrook drift along the central portion of the Scarborough Bluffs. Judging from their lithologic composition and directional evidence of glacial movement (Tables 3 and 4; Figs. 3 and 6), the three members are related to the Ontario glacial lobe that entered the Lake Ontario basin from the northeast.

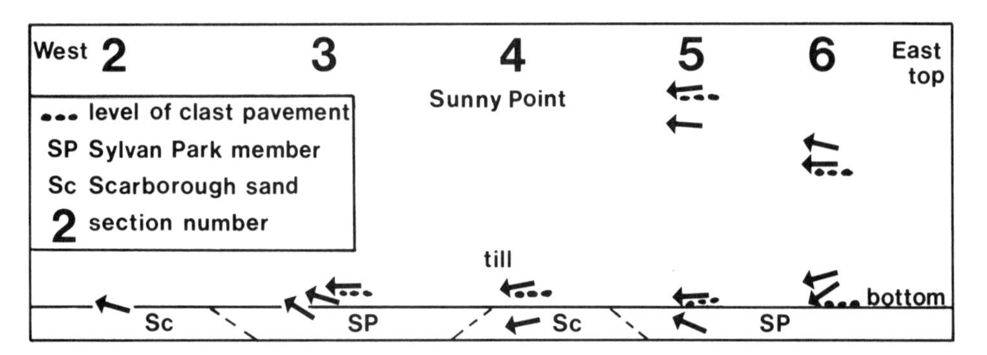

Figure 8. Summary diagram of vectors of glacial movement during deformation of the Scarborough sand and Sylvan Park member, and subsequent deposition of the Sunny Point member. Based on glaciotectonic-deformation structures, till fabric, and glacial erosion marks at sites 2–6. North at top, not drawn to scale.

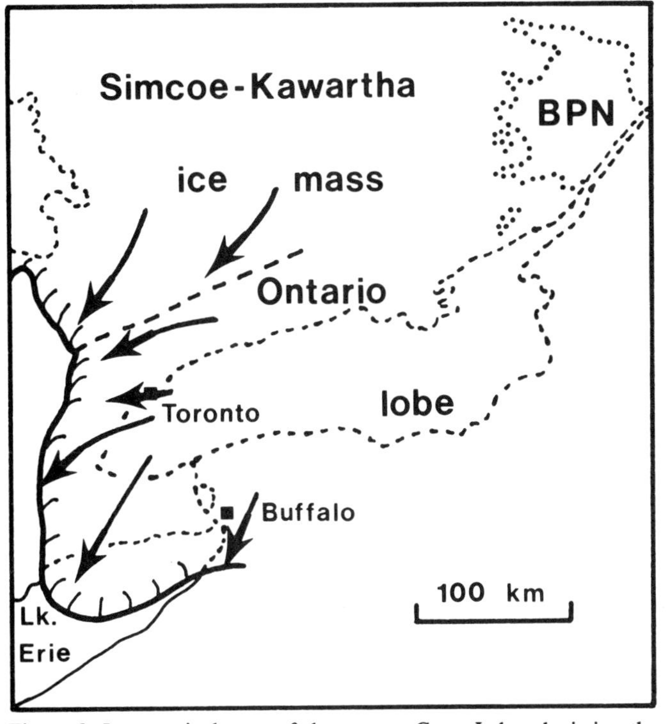

Figure 9. Ice-marginal map of the eastern Great Lakes depicting the merging Ontario and Simcoe-Kawartha ice masses, which caused southwestward flow over Toronto during Sunny Point till deposition. Short dashes outline the present margins of the eastern Great Lakes, and long dashes, the ice lobe boundary. BPN is the main source area for Beekmantown dolostone and Potsdam-Nepean sandstone clasts found in the Sunnybrook drift. BPN is partly surrounded by the eastern Grenville subprovince of the Precambrian shield.

The central section of the Scarborough Bluffs records only the events along the north shore of Lake Ontario over the highest part of the Scarborough delta. To obtain a regional picture, it would be appropriate to compare our results with inland localities and with the southern shore of Lake Ontario. Unfortunately, no adequately defined or described correlatives of the Sunnybrook drift are available for comparison from the south shore. However, Sunnybrook drift has been identified in many places in the Toronto area (see Introduction).

We present but a few sites in Figure 10 (localities) and in Figure 11 that show elevations of the base of the Sunnybrook drift and the distribution of its three members. Sites were chosen where the Sunnybrook drift has been identified with some certainty by its lithology and/or stratigraphic position, and where sufficient evidence (published and unpublished) was available on the occurrence of facies associations typical of any of the three members (Table 2). Differences in facies descriptions and interpretations by different observers are quite possible, and are readily apparent when comparing the facies assignments given to the Sunnybrook drift in our section 6 (Eyles and Eyles, 1983, section 2581; Gravenor and Wong, 1987, section 1), as well as at Rouge River (Westgate and others, 1987; see our reinterpretation of the section in the discussion of microfossils). All three members of the Sunnybrook drift (Table 2) are present along most of our sections in the Scarborough Bluffs. In the greater Toronto area, the base of the Sunnybrook drift occurs at elevations of 70–120 m near Lake Ontario, but higher (140–160 m) inland (Fig. 11).

Toronto subway sections

In the University Avenue subway sections (Lajtai, 1969, Fig. 1), between Yonge Street and Rosedale Valley, rhythmically laminated clay at the base of the Sunnybrook drift may correspond to our Sylvan Park member. Such clay is absent along the basal contact of the Sunny Point member along University Avenue south of College Street. From College Street to Wellington Street, the Sunny Point till truncates several underlying formations with a southward-dipping erosional contact (cf. Sharpe, 1980, section N-S along Yonge Street subway). Sunny Point till in the University subway sections contains more sand than at other sites in Toronto (Table 5). This is probably due to incorporation of Scarborough sand, which is absent beneath the Sunny Point member in the southern part of downtown Toronto (Lajtai, 1969, Fig. 1, section B-B; Sharpe, 1980, section N-S).

Lajtai (1969) and Sharpe (1980) encountered the Bloor Member along the Toronto subway excavations in a large area around the intersections of the Bloor Street, Avenue Road, and Yonge Street subways. The member also occurs at the Don Valley Brickyard (Terasmae, 1960, Danforth beds). According to Karrow (1967, p. 36), the Bloor Member "appears to occupy depressions in the surface of Sunnybrook Till, with which it is closely associated," and this association also applies to our sections (Fig. 1). This close association of the Bloor Member to the Sunnybrook (Karrow, 1969) was disregarded when Eyles and Clark (1988) included the Bloor Member sediments in the lower Thorncliffe Formation.

Buried channel fills: A fourth member of Sunnybrook drift?

Because glaciolacustrine sediments tend to accumulate in preexisting depressions, we would expect both the Sylvan Park and Bloor members to occur over the buried channels incised into the Scarborough delta at Dutch Church and Don Valley Brickyard. However, the Sylvan Park member is not encountered in these channels. Therefore, we begin to question various previous interpretations that consider erosion of these channels, and their infilling by the Pottery Road Formation, as predating the Sunnybrook drift (Karrow, 1967, 1969, 1974, 1984b; Sharpe, 1987, p. 343; Eyles and Clark, 1988, p. 1119). These interpretations are different, and it is not surprising that Karrow (1984b, p. 143) concluded his description of the Pottery Road Formation with the statement: "The limited exposure of the Pottery Road Formation has left it as a poorly understood facet of the stratigraphy of Toronto."

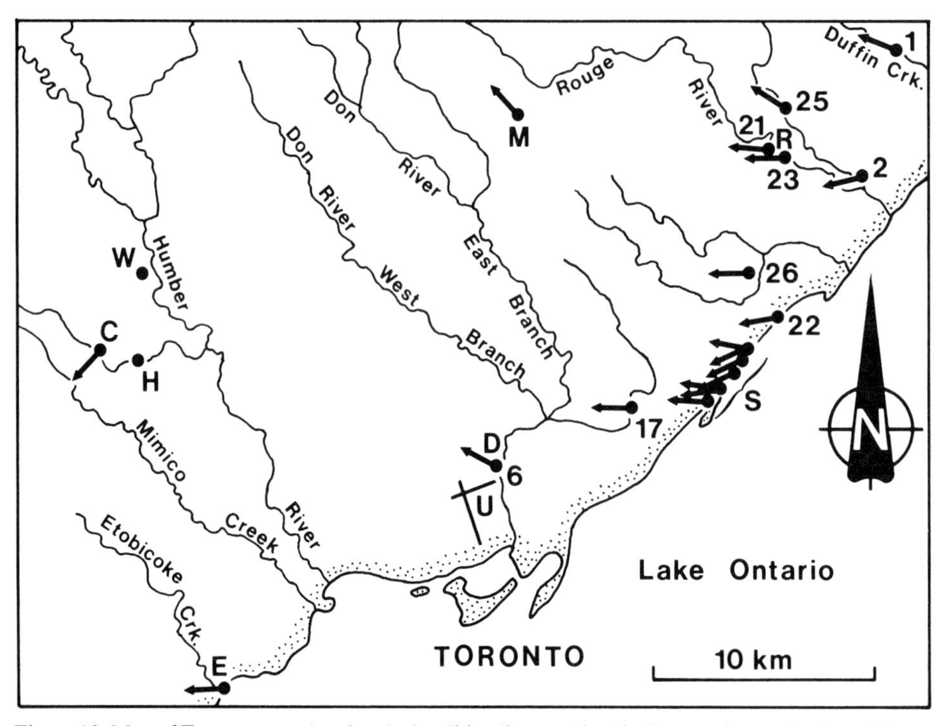

Figure 10. Map of Toronto area showing the localities discussed in this chapter. Arrows indicate inferred directions of glacial movement during deposition of the Sunny Point member as suggested by our reinvestigation along the Scarborough Bluffs (S), Ostry's (1962) pebble- and micro-fabric data (sites for his samples 1–26), Dreimanis and Terasmae's (1958) till-fabric measurement at M; Hicock and Dreimanis's (1985, Fig. 3) glaciotectonic deformations and striae at C, and Watt's data on boulder striae at E (personal communication, *in* Dreimanis and Terasmae, 1958, p. 124). Abbreviations of localities: C = Clairville dam, D = Don Valley Brickyard, E = Etobicoke Creek, H = Humber River west, M = Markham gravel pit, R = Rouge River, U = University subway cross-sections, W = Woodbridge railway cut and borrow pit.

We would promote a hypothesis that the buried channels cut into the Scarborough delta, and filled or partially filled by the Pottery Road sediments, were formed at least in part by subglacial meltwater streams (Sharpe and Barnett, 1985, p. 267), penecontemporaneous with the deposition of the Sunny Point member. The Sylvan Park member of the Sunnybrook drift was probably eroded by such streams in the area of Don Valley Brickyard. This hypothesis is supported by Boulton and Hindmarsh's (1987) theoretical discussion on the association of tunnel valleys with deformable beds in areas where ice sheets flowed over soft sediments. If our interpretation of the genesis of the Sunny Point member by subglacial deformation (deformable bed) is correct, then hydraulic conditions under the Ontario glacial lobe would also be favorable for erosion of large channels into the Scarborough delta.

Published paleocurrent directions for the Pottery Road Formation vary. Karrow (1969, p. 12; 1989, p. 1079) and Eyles and Clark (1988, p. 1113; 1989, p. 1086) have concluded from their observations that paleocurrents flowed southward (southwestward to southeastward) in the channel at the Don Valley Brickyard. However, in another similar channel fill at the end of Livingston Avenue, "bedding of the gravel indicates a northward current" (Karrow, 1969, p. 11). We would expect westward to northward currents under the margin of an advancing Ontario lobe. However, if the Ontario lobe had eventually merged with the Simcoe-Kawartha ice mass (Fig. 9), then subglacial meltwater streams could have also flowed southward, especially if calving bays formed along the edge of the Ontario lobe in deeper water.

No provenance data are available from the Pottery Road Formation to test the above hypothesis. Although reworked sediments of the Scarborough and Don formations probably dominate the channel fills (Karrow, 1969, p. 11–12), some lithologic components derived from distant sources may be present. Clasts typical of either the Ontario lobe or the Simcoe-Kawartha ice mass would assist the interpretation of origin and stratigraphic assignment of the sand and gravel now classified as the Pottery Road Formation.

There are some indications that sorted sediments of the Pottery Road Formation are interbedded with Sunnybrook diamictons. Sharpe and Barnett (1985, p. 269) observed at Don Valley Brickyard that gravel and sand "become interbedded with the top of the channel fill." Eyles and Clark (1988, p. 1113) stated that

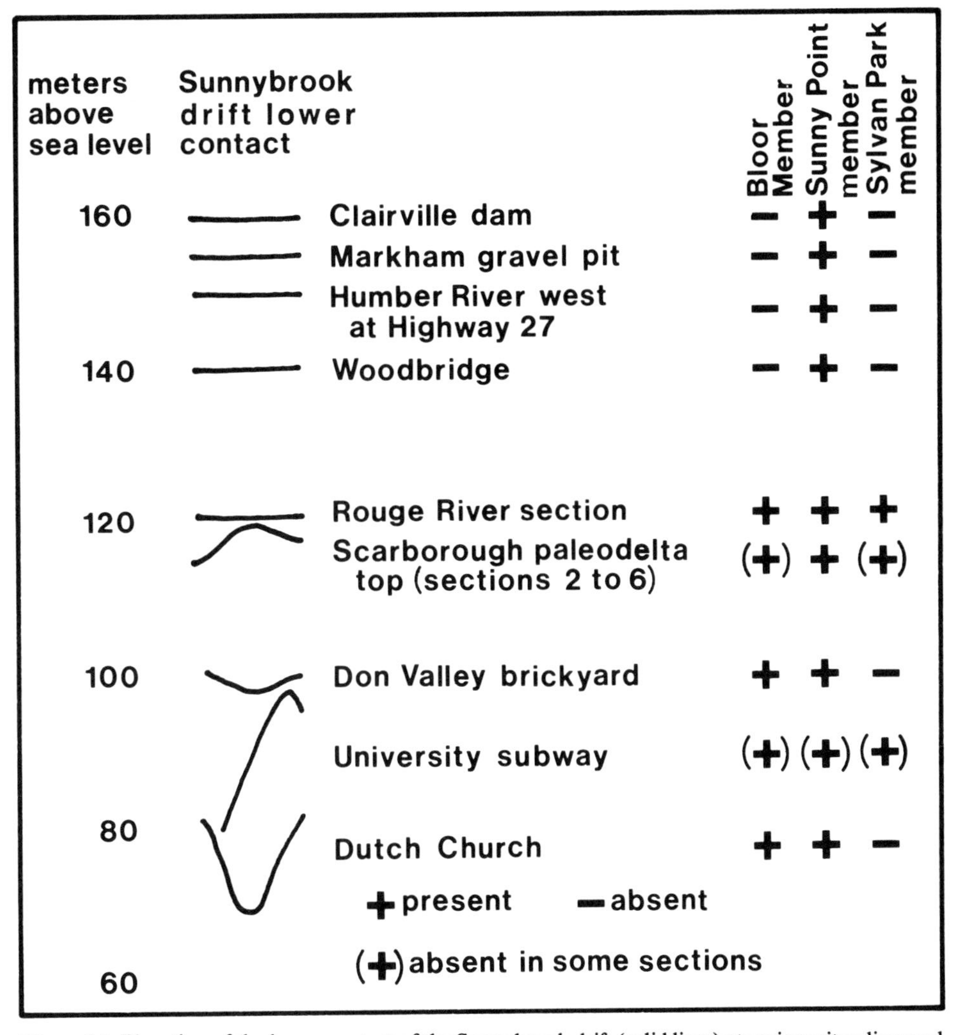

Figure 11. Elevation of the lower contact of the Sunnybrook drift (solid lines) at various sites discussed in text. The presence or absence of each of the three Sunnybrook drift members are indicated in columns to the right.

"the contact between the Pottery Road gravels and the overlying diamict complex is not well exposed." Apparently, even the contact relations between the Pottery Road sediments and the Sunnybrook drift are not yet sufficiently clear. It is possible that at least some parts of the glaciofluvial member of the Sunnybrook drift described by Lajtai (1969, p. 11–12) from the University Avenue subway sections in Toronto are correlative with the Pottery Road sediments, as interpreted by us. Lithologically, Lajtai's glaciofluvial member is similar to the Sunny Point member, in pebble, heavy mineral, and carbonate composition (Lajtai, 1969, Table IV). Karrow (1969, p. 11) also considered that some association may exist between the channel sand (later named by him the Pottery Road Formation) and Lajtai's (1969) "glaciofluvial members." However, Karrow (1969, p. 11) stated that "these relationships remain unexplained."

The stratigraphic position of Lajtai's (1969) glaciofluvial member is in the lower part of the Sunnybrook drift, in contact with the underlying Scarborough Formation, between the Sunny Point member and the lower glaciolacustrine member (our Sylvan Park member), or even as channel fills in the Sunny Point member (Lajtai, 1969, Fig. 1). The elevation (90–110 m) of the glaciofluvial member of the Sunnybrook drift is similar to that of the Pottery Road Formation. This correspondence should be reinvestigated.

If our suggestion that the Pottery Road Formation and Lajtai's (1969) glaciofluvial member of the Sunnybrook drift are equivalent is valid, then together they should be added as a fourth lithostratigraphic member of the Sunnybrook drift.

The part of the Sunny Point member that fills the upper portions of the channels at Dutch Church and the Don Valley Brickyard differs from the dominantly massive Sunny Point till over the Scarborough delta in the Scarborough Bluffs (Eyles and Eyles, 1983, Fig. 1; Eyles and Clark, 1988, Fig. 4). The channels contain a higher proportion of stratified diamicton and sorted

TABLE 5. LITHOLOGIC DATA OF PROBABLE SUNNY POINT MEMBER AT SITES C, D, M, R, U, AND W*†

Site	Locality (Reference)	Matrix Percentage				Calcite/ Dolomite	Percent Garnets	Purple/Red Garnets	Pebble Percentage§				
		Sand	Silt	Clay	Carbo-nates				IM	SS	DS	LS	SH
C	Clairville dam**	17	59	24	9 to 12	1.2 to 21	14 to 27	0.7 to 1.0	...	...	...	...	...
D	Don Valley brickyard**	22	57	21	28	2.2	36	0.7	...	...	...	...	...
M	Markham gravel pit**	23 to 50	26 to 38	22 to 42	13 to 28	0.3 to 1.0	11 to 19	1.0 to 1.8	10	07	02	78	03
R	Rouge River (Westgate and others, 1987)	2 to 7	28 to 48	45 to 70	11 to 24	...	...	...	12	03	02	81	02
U	University subway (Lajtai, 1969)	26 to 31	31 to 36	32 to 37	11 to 16	1.2 to 1.9	...	1.4 to 2.1	...	...	...	...	...
W	Woodbridge railwaycut**	8 to 18	31 to 63	24 to 61	10	0.8	29	0.7	...	...	...	...	...
	Woodbridge railway cut (White, 1971)	8 to 20	34 to 59	28 to 57	12 to 25	1.4 to 6?	...	...	11	13	25	44	07

*Single number for single analysis; range for 3 analyses, except Lajtai's (1969) data which represent the middle 50% range of 44 analyses.
†Locations on Figure 9.
§IM = igneous and metamorphic rocks; SS = sandstone; LS = limestone; DS = dolostone; SH = shale.
**From the University of Western Ontario Quaternary Datafile.

interbeds or lenses. Such facies associations are compatible with our hypothesis of a subglacial channel fill comprising both glacial and glaciofluvial sediments.

Sunny Point member correlatives

In their Sunnybrook "diamict assemblage" at the Don Valley Brickyard, Eyles and Clark (1988, p. 1114) described "deformed diamict and sand facies with deformed flow noses and complex folds, indicating postdepositional movement to the south," and presented convincing photographs of these features in their Figure 7. However, on two earlier visits in 1972 and 1980, we observed and photographed westward-sheared sand at the base of the lowest diamicton unit, and westward-sheared sand lenses in a diamicton layer at an intermediate level within the diamicton complex. These structures suggest that the deforming stress came from the east, and the planar, consistently north-south–oriented shear surfaces indicate that the stress was probably applied by the sole of the Ontario glacial lobe. The flow structures observed by Eyles and Clark (1988) may represent gravity deposits in subglacial channels. Further detailed structural, fabric, and lithologic investigations are still required at this site.

Sediments of both the lower (Sylvan Park) and upper (Bloor) glaciolacustrine members have not been reported by others or observed by us at the high-elevation sections shown in Figure 11, which are 20–40 m higher than the top of the Scarborough delta. It appears that the Sunny Point till was deposited here on land, above the level of proglacial lakes, and, therefore, that the glaciolacustrine members of the Sunnybrook drift are absent here. In the area southwest of Woodbridge (Fig. 10, sites C, H, W), the Sunny Point member consists, as it does at the Scarborough Bluffs, of low-carbonate clay and silt (Table 5) with sand lenses. The pebble composition at site W also resembles that of the Scarborough Bluffs. However, at site C, the orientation of glaciotectonic folding and striae on boulders in the lower part of Sunny Point till (Hicock and Dreimanis, 1985, Fig. 3) suggest that glacial movement was from the northeast (Fig. 10, site C). Our interpretation is that the till at site C was first deposited at the northwest margin of the Ontario lobe, but subsequently dragged and deformed by the Simcoe-Kawartha ice mass advancing from the northeast (Fig. 9). Further structural and fabric investigations are still required at other sites of Sunny Point till in the adjoining area to test this hypothesis.

The Markham gravel pit (Fig. 10, site M; Fig. 11) was examined in 1954 (Dreimanis, 1956), and its brief description published in Dreimanis and Terasmae (1958), prior to Terasmae (1960) naming it "Sunnybrook till." The till we call the Sunny Point member, and its correlatives, were then called "lower till," and were assigned an early Wisconsinan age. The "lower till" at the Markham pit is coarser than Sunny Point till at Toronto and contains more limestone clasts (about 80%; Dreimanis and Terasmae, 1958, Fig. 3; lithologic data in Table 2). Its heavy mineral and matrix carbonate compositions are similar to those of Sunny Point tills of the Toronto area, and pebble fabric is from southeast to northwest (Fig. 10, site M); most bullet-shaped stones point southeast. The last three parameters suggest that the "lower till" was deposited by the Ontario lobe. Although the high percentage of limestone pebbles differs between Markham and most Sunny Point diamictons of the Toronto area, the most easterly of the Sunny Point samples from the Scarborough Bluffs also contains about 70% limestone (Karrow, 1967, Fig. 15). Such a high limestone content is compatible with a limestone source under the north side of the Ontario lobe (north shore of eastern Lake Ontario). No glaciolacustrine sediment was found associated with the "lower till" at the Markham pit. However, the silty sand matrix of this till is more similar to the terrestrial tills deposited on high ground north of Lake Ontario (Dreimanis, 1961, Table IId) than to the Sunny Point till at the Scarborough Bluffs, which was derived from fine glaciolacustrine sediment.

Matrix carbonate and glacial lobe affiliation

The carbonate and calcite contents of the mud fraction of the Sunnybrook drift tend to increase upward, as previously recorded by Lajtai (1969, Table IV) and Westgate and others (1987). Some of our samples, taken from the upper part of the Sunny Point member (Table 3, samples 11851, 11840, 1582, 1583), contain 18%–32% carbonates. Late Wisconsinan tills north of Toronto that were deposited by the Simcoe-Kawartha ice mass, or its easterly correlative, which had become part of the Ontario lobe, contain higher carbonate percentages (Dreimanis and Terasmae, 1958; Dreimanis, 1961). Thus, it is possible that the more northerly Kawartha ice also supplied glacial debris to the Ontario lobe during the latest phases of deposition of the Sunnybrook drift. More investigations are required to test this hypothesis.

CONCLUSIONS

In view of the debate over a glaciolacustrine vs. a combined subglacial-glaciolacustrine origin of the Sunnybrook drift, we concentrated our reinvestigation on the lower part and basal contact of this important stratigraphic marker over the top of the Scarborough delta (Fig. 1). Our results strongly suggest that the Sunnybrook drift represents a glaciolacustrine-subglacial-glaciolacustrine depositional cycle. The deposition of its lowermost member, named the Sylvan Park, began in a proglacial lake during advance of the Ontario lobe from the St. Lawrence lowlands. The Ontario lobe subsequently advanced over this muddy glaciolacustrine sediment, deforming much of it by subsole drag.

The Sunny Point diamicton, the second and normally the thickest member of the Sunnybrook drift, consists mainly of this deformed sediment as well as debris added from the wet-based glacier. We recognize other subglacially deposited components in this member: lodgement till, soft sediment blocks, and lenses and layers of sorted sediments. Previously published facies descriptions and our field observations also suggest the presence of subglacial waterlain flow till and undermelt diamictons. The latter three appear to be prevalent over buried channels. These and similar erosional channels incised into the Scarborough delta, together with their fills (particularly the Pottery Road Formation), probably do not predate the Sunnybrook drift, as concluded by most previous authors. We suggest that these channel fills are part of the Sunnybrook drift, and that they are similar, and at least partly equivalent, to Lajtai's (1969) glaciofluvial member of the Sunnybrook drift.

The Ontario lobe probably advanced as far as the eastern Lake Erie basin. Dreimanis (this volume and 1991) has proposed correlation of the "brown till" at Gowanda Hospital, New York (Calkin and others, 1982), with the Sunnybrook drift at Toronto.

When the early Wisconsinan Ontario lobe retreated from the Toronto area, the uppermost unit of the Sunnybrook drift (Bloor Member) was deposited in a glacial lake. Because both the Bloor and the Sylvan Park members are absent in the northern part of metropolitan Toronto, that area must have been above the level of proglacial lake waters, both at the beginning and conclusion of deposition of the Sunnybrook drift.

ACKNOWLEDGMENTS

Thorough and helpful reviews of the manuscript were completed by J. Shaw, D. R. Sharpe, P. D. Lea, P. U. Clark, and P. F. Karrow, and editing, by S. Upson, to whom we are most grateful. We thank C. H. Eyles for helping us locate her Sylvan Park sections. W. Harley prepared stone thin sections and sand grain mounts for pebble and mineral identification. G. Shields provided photographic services. Research was supported by Natural Sciences and Engineering Research Council of Canada grants to the authors.

REFERENCES CITED

Åmark, M., 1985, Subglacial deposition and deformation of stratified drift at the formation of tills beneath an active glacier—An example from Skåne, Sweden, Geological Society of Denmark Bulletin, v. 34, p. 75–81.

American Society for Testing and Materials, 1972, Standard method for particle-size analysis of soils, *in* Annual book of standards D422-63: Philadelphia, Pennsylvania, p. 112–122.

Berger, G. W., 1984, Thermoluminescence dating studies of glacial silts from Ontario: Canadian Journal of Earth Sciences, v. 21, p. 1393–1399.

Berti, A. A., 1971, Palynology and stratigraphy of the Mid-Wisconsinan in the Great Lakes region, North America [Ph.D. thesis]: London, Canada, The University of Western Ontario, 178 p.

Boulton, G. S., 1968, Flow tills and related deposits on some Vestspitsbergen glaciers: Journal of Glaciology, v. 7, p. 391–412.

——, 1970a, On the origin and transport of englacial debris in Svalbard glaciers: Journal of Glaciology, v. 9, p. 213–229.

——, 1970b, On the deposition of subglacial and melt-out tills at the margins of certain Svalbard glaciers: Journal of Glaciology, v. 9, p. 231–245.

——, 1971, Till genesis and fabric in Svalbard, Spitsbergen, *in* Goldthwait, R. P., ed., Till, A symposium: Columbus, Ohio State University Press, p. 41–72.

——, 1982, Subglacial processes and the development of glacial bedforms, *in* Davidson-Arnott, R., Nickling, W., and Fahey, B. D., eds., Research in glacial, glaciofluvial, and glaciolacustrine systems: Norwich, England, Geobooks, p. 1–31.

——, 1987, A theory of drumlin formation by subglacial sediment deformation, *in* Menzies, J., and Rose, J., eds., Drumlin symposium: Rotterdam, Netherlands, A.A. Balkema, p. 25–80.

Boulton, G. S., and Hindmarsh, R.C.A., 1987, Sediment deformation beneath glaciers: Rheology and geological consequences: Journal of Geophysical Research, v. 92, no. B9, p. 9059–9082.

Calkin, P. E., Muller, E. H., and Barnes, J. H., 1982, The Gowanda Hospital interstadial site, New York: American Journal of Science, v. 282, p. 1110–1142.

Coleman, A. P., 1895, Glacial and inter-glacial deposits near Toronto: Journal of

Geology, v. 3, p. 622–645.
——, 1901, Glacial and interglacial beds near Toronto: Journal of Geology, v. 9, p. 285–310.
——, 1913, Geology of the Toronto region, *in* Faull, J. H., ed., The natural history of the Toronto region: Toronto, Canada, The Canadian Institute, p. 51–81.
——, 1933, The Pleistocene of the Toronto region: Ontario Department of Mines Annual Report, v. 41, Part VII, 61 p.
——, 1941, The last million years: Toronto, Canada, University of Toronto Press, 216 p.
Domack, E. W., and Lawson, D. E., 1985, Pebble fabric in an ice-rafted diamicton: Journal of Geology, v. 93, p. 577–591.
Dowdeswell, J. A., and Sharp, M. J., 1986, Characterization of pebble fabrics in modern terrestrial glacigenic sediments: Sedimentology, v. 33, p. 699–710.
Dreimanis, A., 1956, Pleistocene stratigraphy of the Markham gravel pit, north of Toronto, Ontario: Ottawa, Geological Survey of Canada, 24 p. (unpublished).
——, 1961, Tills of southern Ontario, *in* Legget, R. F., ed., Soils in Canada: Toronto, Canada, University of Toronto Press, p. 80–96.
——, 1962, Quantitative gasometric determination of calcite and dolomite by using Chittick apparatus: Journal of Sedimentary Petrology, v. 32, p. 520–529.
——, 1989, Tills: Their genetic terminology and classification, *in* Goldthwait, R. P., and Matsch, C. L., eds., Genetic classification of glacigenic deposits: Rotterdam, Netherlands, A.A. Balkema, p. 17–83.
——, 1991, The Laurentide Ice Sheet during the Last Glaciation: A review and some current reinterpretations along its southern margin, *in* Frenzel, B., ed., Klimageschichtliche Probleme der letzten 130000 Jahre: Mainz, Akademie der Wissenschaften und der Literatur, Stuttgart, Gustav Fischer Verlag, p. 267–291.
Dreimanis, A., and Terasmae, J., 1958, Stratigraphy of Wisconsin glacial deposits of Toronto area, Ontario: Geological Association of Canada Proceedings, v. 10, p. 119–135.
Dreimanis, A., Hamilton, J. P., and Kelly, P. E., 1987, Complex subglacial sedimentation of Catfish Creek till at Bradtville, Ontario, Canada, *in* Van der Meer, J.J.M., ed., Tills and glaciotectonics: Rotterdam, Netherlands, A.A. Balkema, p. 73–87.
Elson, J. A., 1961, The geology of tills, *in* Penner, E., and Butler, J., eds., Proceedings, 14th Canadian Soil Mechanics Conference: Ottawa, Canada, National Research Council of Canada, Associate Committee on Soil and Snow Mechanics, Technical Memorandum 69, p. 5–17.
——, 1989, Comment on glacitectonite, deformation till, and comminution till, *in* Goldthwait, R. P., and Matsch, C. L., eds., Genetic classification of glacigenic deposits: Rotterdam, Netherlands, A.A. Balkema, p. 85–88.
Eyles, C. H., and Eyles, N., 1983, Sedimentation in a large lake: A reinterpretation of the late Pleistocene stratigraphy at Scarborough Bluffs, Ontario, Canada: Geology, v. 11, p. 146–152.
——, 1984, Reply *to* Comment *on* "Sedimentation in a large lake: A reinterpretation of the late Pleistocene stratigraphy at Scarborough Bluffs, Ontario, Canada": Geology, v. 12, p. 188–190.
Eyles, N., and Clark, B. M., 1988, Last interglacial sediments of the Don Valley Brickyard, Toronto, Canada, and their paleoenvironmental significance: Canadian Journal of Earth Sciences, v. 25, p. 1108–1122.
——, 1989, Reply *to* Comment *on* "Last interglacial sediments of the Don Valley Brickyard, Toronto, Canada, and their paleoenvironmental significance": Canadian Journal of Earth Sciences, v. 26, p. 1083–1086.
Eyles, N., and Eyles, C. H., 1987, Glaciolacustrine and subglacial facies of the Ontario basin: Geological Society of America, Penrose Conference field guide.
Eyles, N., Day, T. E., and Gavican, A., 1987, Depositional controls on the magnetic characteristics of lodgement tills and other glacial diamict facies: Canadian Journal of Earth Sciences, v. 24, p. 2436–2458.
Eyles, N., and Westgate, J. A., 1987, Restricted regional extent of the Laurentide Ice Sheet in the Great Lakes basins during early Wisconsin glaciation: Geology, v. 15, p. 537–540.
Eyles, N., Sladen, J. A., and Gilroy, S., 1982, A depositional model for stratigraphic complexes and facies superimposition in lodgement tills: Boreas, v. 11, p. 317–333.
Eyles, N., Eyles, C. H., and Day, T. E., 1983, Sedimentologic and palaeomagnetic characteristics of glaciolacustrine diamict assemblages at Scarborough Bluffs, Ontario, Canada, *in* Evenson, E. B., Schlüchter, C., and Rabassa, J., eds., Tills and related deposits: Rotterdam, Netherlands, A.A. Balkema, p. 23–45.
Eyles, N., and Schwarcz, H. P., 1991, Stable isotope record of the last glacial cycle from lacustrine ostracodes: Geology, v. 19, p. 257–260.
Gravenor, C. P., 1984, Comment *on* "Sedimentation in a large lake: A reinterpretation of the late Pleistocene stratigraphy at Scarborough Bluffs, Ontario, Canada": Geology, v. 12, p. 187–188.
Gravenor, C. P., and Wong, T., 1987, Magnetic and pebble fabrics and origin of the Sunnybrook Till, Scarborough, Ontario, Canada: Canadian Journal of Earth Sciences, v. 24, p. 2038–2046.
Gwyn, Q.H.J., and Dreimanis, A., 1979, Heavy mineral assemblages in tills and their use in distinguishing glacial lobes in the Great Lakes region: Canadian Journal of Earth Sciences, v. 16, p. 2219–2235.
Hicock, S. R., 1991, On subglacial stone pavements in till: Journal of Geology, v. 99, p. 607–619.
Hicock, S. R., and Dreimanis, A., 1985, Glaciotectonic structures as useful ice-movement indicators in glacial deposits: Four Canadian case studies: Canadian Journal of Earth Sciences, v. 22, p. 339–346.
——, 1988, Sunnybrook Drift re-examined at Scarborough Bluffs, Lake Ontario, Canada: Geological Society of America Abstracts with Programs, v. 20, no. 7, p. A246.
——, 1989, Sunnybrook Drift indicates a grounded early Wisconsinan glacier in the Lake Ontario basin: Geology, v. 17, p. 169–172.
Hinde, G. J., 1877, The glacial and interglacial strata of Scarboro Heights, and other localities near Toronto, Ontario: Canadian Journal, new series, v. 15, no. 5, p. 388–413.
Houmark-Nielsen, M., 1987, Pleistocene stratigraphy and glacial history of the central part of Denmark: Geological Society of Denmark Bulletin, v. 36, p. 1–189.
Kaplianskaya, F. A., and Tarnodgradskiy, V. D., 1984, Zapadno-Sibirskaya ravnina, *in* Krasvov, I. I., ed., Stratigrafiya SSSR: Moscow, Nedra, Chetvertichnaya Sistema, v. 2, p. 227–270.
Karrow, P. F., 1967, Pleistocene geology of the Scarborough area: Ontario Department of Mines, Geological Report 46, 108 p.
——, 1969, Stratigraphic studies in the Toronto Pleistocene: Geological Association of Canada Proceedings, v. 20, p. 4–16.
——, 1974, Till stratigraphy in parts of southwestern Ontario: Geological Society of America Bulletin, v. 85, p. 761–768.
——, 1984a, Comment *on* "Sedimentation in a large lake: A reinterpretation of the late Pleistocene stratigraphy at Scarborough Bluffs, Ontario, Canada": Geology, v. 12, p. 185.
——, 1984b, Quaternary stratigraphy and history, Great Lakes–St. Lawrence region, *in* Fulton, R. J., ed., Quaternary stratigraphy of Canada: Geological Survey of Canada Paper 84-10, p. 137–153.
——, 1989, Last interglacial sediments of the Don Valley Brickyard, Toronto, Canada, and their paleoenvironmental significance: Discussion: Canadian Journal of Earth Sciences, v. 26, p. 1078–1082.
Kelly, R. I., and Martini, I. P., 1986, Pleistocene glacio-lacustrine deltaic deposits of the Scarborough Formation, Ontario, Canada: Sedimentary Geology, v. 47, p. 27–52.
Kronborg, C., and Knudsen, K. L., 1985, Om Kvartaeret ved Rugard: en forelobig unndersogelse: Dansk geologisk Foreningens, Årskrift for 1984, p. 37–48 (in Danish with English summary).
Krüger, J., 1984, Clasts with stoss-lee form in lodgement tills: A discussion: Journal of Glaciology, v. 30, p. 241–243.
Lajtai, E. Z., 1969, Stratigraphy of the University subway, Toronto, Canada: Geological Association of Canada Proceedings, v. 20, p. 17–23.
Lamothe, M., Dreimanis, A., Morency, M., and Raukas, A., 1984, Thermo-

luminescence dating of Quaternary sediments, *in* Mahaney, W. C., ed., Quaternary dating methods: Amsterdam, Netherlands, Elsevier Science Publishers, p. 153–170.

Lawson, D. E., 1979, Sedimentological analysis of the western terminus region of the Matanuska Glacier, Alaska: U.S. Army Corps of Engineers, Cold Regions Research and Engineering Laboratory, Report 79-9, 96 p.

Locat, J., and Bérubé, M. A., 1986, L'influence de la granulometrie sur la mesure des carbonates par la methode du Chittick: Géographie physique et Quaternaire, v. 40, p. 331–336.

Lowdon, J. A., Robertson, I. M., and Blake, W., Jr., 1971, Geological Survey of Canada radiocarbon dates XI: Radiocarbon, v. 13, p. 255–324.

Nielsen, E., 1988, Lacustrine ostracodes in the late Pleistocene Sunnybrook diamicton of southern Ontario, Canada: Discussion: Canadian Journal of Earth Sciences, v. 25, p. 1717–1720.

Nielsen, E., Morgan, A. V., Morgan, A., Mott, R. J., Rutter, N. W., and Causse, C., 1986, Stratigraphy, paleoecology, and glacial history of the Gillam area, Manitoba: Canadian Journal of Earth Sciences, v. 23, p. 1641–1661.

North American Commission on Stratigraphic Nomenclature, 1983, North American stratigraphic code: American Association of Petroleum Geologists Bulletin, v. 67, p. 841–875.

Ostry, R. C., 1962, An analysis of some tills in Scarborough Township and vicinity [M.A. thesis]: Toronto, Canada, University of Toronto, 67 p.

Rappol, M., 1985, Clast-fabric strength in tills and debris flows compared for different environments: Geologie en Mijnbouw, v. 64, p. 327–332.

Reineck, H. E., and Singh, I. B., 1973, Depositional sedimentary environments: Berlin, Springer-Verlag, 439 p.

Ringberg, B., Holland, B., and Miller, U., 1984, Till stratigraphy and provenance of the glacial chalk rafts at Kvarnby and Angdala, southern Sweden: Striae, v. 20, p. 79–90 (in Swedish with English summary).

Rutka, M. A., and Eyles, N., 1989, Ostracod faunas in a large late Pleistocene ice-dammed lake, southern Ontario, Canada: Palaeogeography, Palaeoclimatology, Palaeoecology, v. 73, p. 61–76.

Schwarcz, H. P., and Eyles, N., 1991, Laurentide Ice Sheet extent inferred from stable isotopic composition (O, C) of ostracodes at Toronto, Canada: Quaternary Research, v. 35, p. 305–320.

Sharp, M., 1982, Modification of clasts in lodgement tills by glacial erosion: Journal of Glaciology, v. 28, p. 475–481.

Sharpe, D. R., 1980, Quaternary geology of Toronto and surrounding area: Ontario Geological Survey Preliminary Map P2204, scale 1:100 000.

——, 1984, Comment *on* "Sedimentation in a large lake: A reinterpretation of the late Pleistocene stratigraphy at Scarborough Bluffs, Ontario, Canada": Geology, v. 12, p. 186–187.

——, 1987, Quaternary geology of the Toronto area, Ontario, *in* Roy, D. C., ed., Northeastern section of the Geological Society of America: Boulder, Colorado, Geological Society of America, Centennial Field Guide, v. 5, p. 339–344.

——, 1988, The internal structure of glacial landforms: An example from the Halton till plain, Scarborough Bluffs, Ontario: Boreas, v. 17, p. 15–26.

Sharpe, D. R., and Barnett, P. J., 1985, Significance of sedimentological studies on the Wisconsinan stratigraphy of southern Ontario: Géographie physique et Quaternaire, v. 34, p. 255–273.

Shaw, J., 1982, Melt-out till in the Edmonton area, Alberta, Canada: Canadian Journal of Earth Sciences, v. 19, p. 1548–1569.

——, 1987, Glacial sedimentary processes and environmental reconstruction based on lithofacies: Sedimentology, v. 34, p. 103–116.

Stanford, S. D., and Mickelson, D. M., 1985, Till fabric and deformational structures in drumlins near Waukesha, Wisconsin, U.S.A.: Journal of Glaciology, v. 31, p. 220–228.

Stupavsky, M., and Gravenor, C. P., 1975, Magnetic fabric around boulders in till: Geological Society of America Bulletin, v. 86, p. 1534–1536.

Stupavsky, M., Gravenor, C. P., and Symons, D.T.A., 1974, Paleomagnetism and magnetic fabric of the Leaside and Sunnybrook tills near Toronto, Ontario: Geological Society of America Bulletin, v. 91, p. 593–598.

Terasmae, J., 1960, A palynological study of Pleistocene interglacial beds at Toronto, Ontario: Geological Survey of Canada Bulletin 56, Part II, p. 23–41.

Troitskiy, S. L., 1975, Modern anti-glacialism, critical essay: Moscow, Nauka (in Russian).

Watt, A. K., 1954, Correlation of the Pleistocene geology as seen in the subway, with that of the Toronto region, Canada: Geological Association of Canada Proceedings, v. 6, p. 69–82.

——, 1957, Pleistocene geology and ground-water resources of the Township of North York, York County: Ontario Department of Mines, 6th Annual Report, Part 7, 64 p.

——, 1968, Pleistocene geology and ground-water resources, township of Etobicoke: Ontario Department of Mines Geological Report 59, 50 p.

Westgate, J. A., 1968, Linear sole markings in Pleistocene till: Geological Magazine, v. 105, p. 501–505.

Westgate, J. A., Chen, F. J., and Delorme, L. D., 1987, Lacustrine ostracodes in the late Pleistocene Sunnybrook diamicton of southern Ontario, Canada: Canadian Journal of Earth Sciences, v. 24, p. 2330–2335.

White, O. L., 1971, Pleistocene geology of the Bolton area (30M-13), southern Ontario: Ontario Department of Mines and Northern Affairs, Geological Branch Open-File Report 5067, 249 p.

Wilson, A.W.G., 1901, Physical geology of central Ontario: Transactions of the Canadian Institute, v. 7, p. 139–186.

Manuscript Accepted by the Society September 6, 1991

Printed in U.S.A.

Geological Society of America
Special Paper 270
1992

On the age of the penultimate full glaciation of New England

R. N. Oldale and S. M. Colman
U.S. Geological Survey, Woods Hole, Massachusetts 02543

ABSTRACT

Tills that discontinuously underlie the late Wisconsinan till throughout New England represent the penultimate full glaciation of the region. In southern New England, the late Wisconsinan till and the tills that locally underlie it are informally referred to as upper and lower tills, respectively. For the most part, the ages of the lower tills are not firmly established, and regional correlations between occurrences of lower till, including those on Long Island, New York, are tenuous. Where a lower till underlies deposits having limiting middle Wisconsinan radiocarbon ages (e.g., the Montauk till member of the Manhassett Formation on Long Island at Port Washington, New York, and the lower till at New Sharon, Maine), many workers have assigned the till an early Wisconsinan age. However, lower tills throughout much of New England may be Illinoian or older in age and may correlate with a lower till exposed at Sankaty Head, Nantucket Island, Massachusetts, that is pre-Sangamonian in age. The till at Sankaty Head lies below marine beds containing marine faunas indicative of sea-water temperatures both warmer and slightly cooler than those off Nantucket today and that have uranium-thorium and amino-acid racemization (AAR) age estimates suggesting a Sangamonian age (marine oxygen-isotope stage 5).

The lower till at Sankaty Head and the Montauk till member on Long Island were deposited during a full glaciation of New England that was at least as extensive as the late Wisconsinan advance of the Laurentide ice. Global ice-volume data from the marine oxygen-isotope record and the late Pleistocene eustatic sea-level record inferred from raised coral terraces support an advance of this magnitude during marine oxygen-isotope stage 6, but not during stage 4.

An early Wisconsinan age of the southern New England lower tills and, hence, of the penultimate glaciation there is problematic in terms of the pre-Sangamonian age of the lower till on Nantucket, and in terms of the late Pleistocene global ice-volume and sea-level records. An Illinoian age for the tills and for the penultimate full glaciation of New England is compatible with all the available evidence except some equivocal radiocarbon ages and AAR age estimates.

INTRODUCTION

Two superposed tills, usually called the upper and lower till, occur throughout New England. The upper till is part of a ubiquitous drift sheet that occurs subaerially and offshore. A late Wisconsinan age is firmly established by middle Wisconsinan limiting finite radiocarbon dates that range from about 40,800 to 21,500 yr B.P. from Long Island, New York (Sirkin and Stuckenrath, 1980; Sirkin, 1982) and by correlation to radiocarbon-dated glaciomarine deposits that occur along the coast from Boston, Massachusetts, to the Canadian border (Stone and Borns, 1986), and offshore within the Gulf of Maine (Fig. 1) (Tucholke and Hollister, 1973; Schnitker, 1988).

Lower tills occur discontinuously beneath the late Wisconsinan drift sheet; some or all of these tills represent the penultimate full glaciation of New England. Unfortunately, not enough information is available on the pre–late Wisconsinan deposits in New England and Long Island, New York, to establish the age of

Oldale, R. N., and Colman, S. M., 1992, On the age of the penultimate full glaciation of New England, *in* Clark, P. U., and Lea, P. D., eds., The Last Interglacial-Glacial Transition in North America: Boulder, Colorado, Geological Society of America Special Paper 270.

these tills unequivocally. This lack of data has allowed much speculation over many years on the age of the glacial and interglacial deposits that stratigraphically underlie the ubiquitous late Wisconsinan drift. We address the questions of the age of the lower tills and the age of the penultimate full glaciation of New England by examining pre–late Wisconsinan deposits on Nantucket Island, Massachusetts, and Long Island, New York, as well as the lower tills throughout the New England mainland (Fig. 1). We also examine indirect evidence, including global ice-volume and eustatic sea-level records for the late Quaternary.

The Illinoian and Sangamonian stages and the Wisconsinan substages (early, middle, and late), as used in this chapter, are considered to be equivalent to the corresponding marine oxygen-isotope stages (hereafter called isotope stages) 6, 5, and 4 through 2, respectively (Fig. 2).

LOWER TILL AT SANKATY HEAD, NANTUCKET ISLAND

On Nantucket Island, Massachusetts, a lower drift is exposed below interglacial or interstadial marine deposits in the

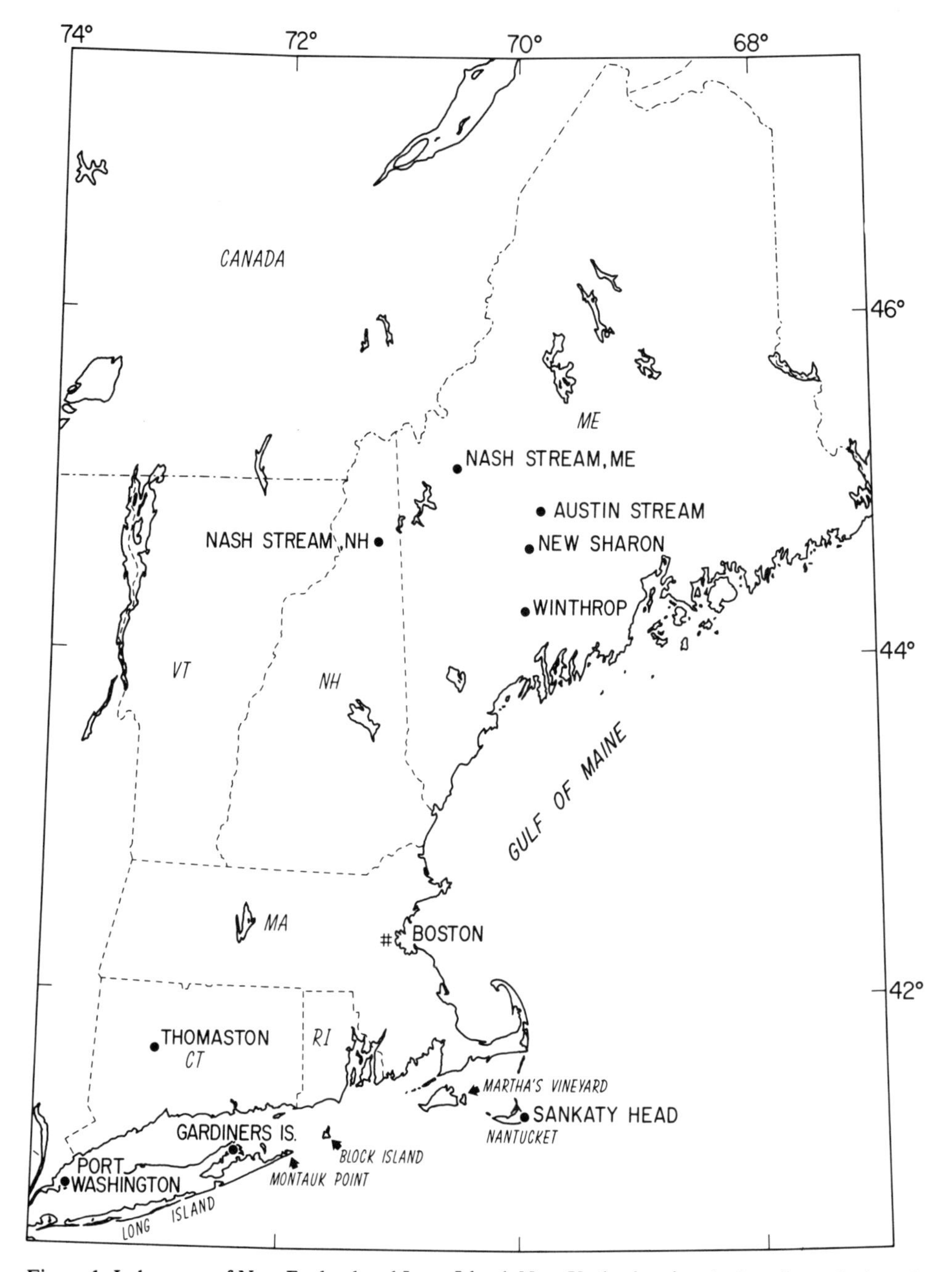

Figure 1. Index map of New England and Long Island, New York, showing the locations of selected lower till exposures and of Gardiners Island, the type location for the Gardiners Clay.

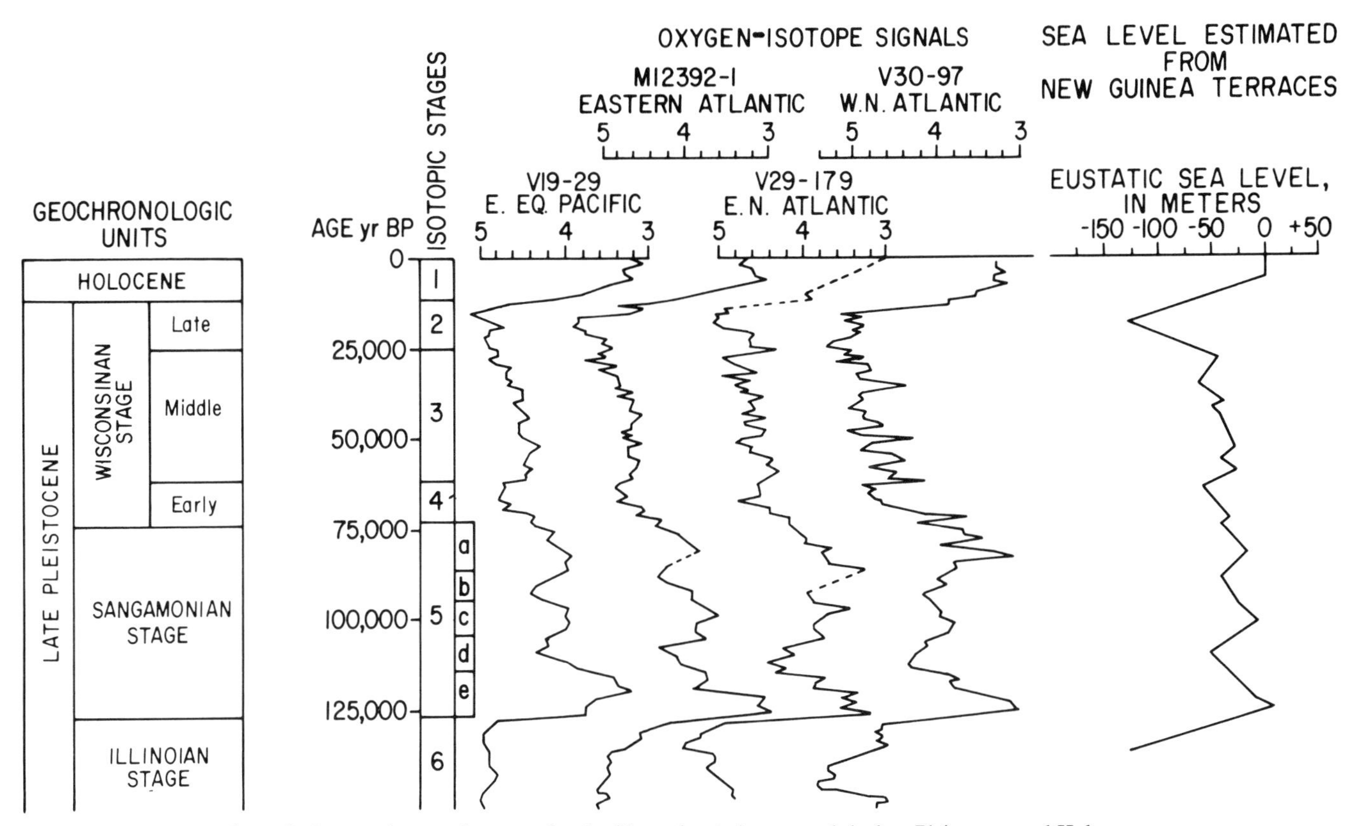

Figure 2. Oxygen-isotope data correlated with sea-level changes and the late Pleistocene and Holocene geochronologic units used in this chapter. Marine oxygen-isotope curves used to provide an estimate of ice volume are modified from Ruddiman and McIntyre (1981); scales are in $\delta^{18}O$ (Peedee belemnite). Curve representing eustatic sea level relative to present sea level is modified from Shackleton (1987).

Sankaty Head Cliff (Fig. 3) and was also encountered in a deep borehole (Oldale and others, 1982). The drift is composed of stratified sand and gravel of fluvial and deltaic origin, rhythmically laminated lacustrine silt and clay, and a layered to massive diamicton. The diamicton contains striated stones and is interpreted as till (Oldale and Eskenasy, 1983). It is oxidized to a depth of 2 to 3 m below its upper contact. Detrital wood fragments from the stratified sand and gravel were radiocarbon dated at 26,200 ± 500 yr B.P. (W-4615).

The overlying marine deposits include the abundantly fossiliferous Sankaty sand (Oldale and others, 1982). The Sankaty sand is made up of two units that contain distinct molluscan and ostracode faunas. The fauna from the lower part of the Sankaty Sand indicates sea-water temperatures off Nantucket up to 4°C warmer than present, whereas the fauna from the upper part indicates sea-water temperatures from 3 to 7°C colder than present (Gustavson, 1976; Oldale and others, 1982). A shell from the lower unit yielded a finite radiocarbon age of 31,900 ± 730 yr B.P. (W-4205). This age is considered unreliable because the shell has been recrystallized and was probably contaminated by ground-water bicarbonate. The wood from the underlying stratified drift was probably also contaminated by younger organic matter (Oldale and others, 1982). A uranium-thorium age of 133,000 ± 7,000 yr B.P. on detrital coral from the upper unit is supported by amino-acid racemization (AAR) age estimates of 120–140 ka for shells from both the upper and lower units of the Sankaty sand (Oldale and others, 1982). Thus, the Sankaty sand appears to be Sangamonian in age, and the lower till at Sankaty Head is consequently Illinoian or older.

MONTAUK TILL

Till and outwash below the late Wisconsinan drift has been recognized on Long Island, New York (Sirkin, 1982; Rampino and Sanders, 1981). The till is the Montauk till member of the Manhassett Formation (Fuller, 1914), commonly called the Montauk till; it is compact, distinctly banded or laminated, locally contains clay beds and consists of a sandy to silty matrix containing abundant stones (Mills and Wells, 1974; Nieter and others, 1975; Sirkin, 1982). The Montauk till in all places is faulted and folded, the result of galciotectonism, which complicates its stratigraphic position relative to the deposits above and below it (Mills and Wells, 1974; Nieter and others, 1975; Sirkin and Stuckenrath, 1980). Although Sirkin (1982) reported that the Montauk till is unoxidized, Mills and Wells (1974) reported that outwash that grades into the till is oxidized to depths as great

Figure 3. Photograph of the Sankaty Head cliff showing the lower till (T) overlain by interstratified very fine sand, silt, and till (SST); fossiliferous Sankaty sand (S); and drift of Wisconsinan age (W). This wide-angle photograph distorts horizontal and vertical distances. Distance represented by the width of the photograph is about 24 m, cliff height at center of photograph is about 33 m.

as 4.5 m. Uncertainty as to the island-wide correlations among occurrences of lower till caused Sirkin (1982) to apply new names to lower tills found outside of the type locality of the Montauk till on the Montauk Peninsula. Other occurrences of tills lithologically and stratigraphically similar to the Montauk till are found on Block Island, Rhode Island, and on Martha's Vineyard (Fuller, 1914; Woodworth and Wigglesworth, 1934; Sirkin, 1982). In a geologically less-complicated setting, south of the thrust moraines on Long Island and on the adjacent continental shelf, drill holes penetrated an upper late Wisconsinan drift, a fossiliferous marine deposit (the Wantagh Formation), and a lower drift composed of outwash (Rampino and Sanders, 1981). The lower drift was named the Merrick Formation and was considered equivalent to the Montauk till (Rampino and Sanders, 1981). These lower drifts on Long Island, Block Island, and Martha's Vineyard are undated.

The Montauk till is stratigraphically underlain by the Gardiners clay (Fuller, 1914; Stone and Borns, 1986). Gardiners clay is here restricted to marine clay containing fossil assemblages that indicate sea-water temperatures somewhat warmer to somewhat cooler than present sea-water temperatures off Long Island (Gustavson, 1976). The Gardiners clay lacks radiometric ages, but Stone and Borns (1986) reported AAR age estimates of 80 to 130 ka—made by Belknap (1979)—from shells in the Gardiners clay. However, Belknap (1980) reported AAR age estimates that indicated both isotope stage 5 and isotope stage 7 to 9 ages for the shells in the Gardiners clay.

Late Wisconsinan drift ubiquitously overlies the Montauk till, and in at least two localities, at Port Washington and Old Bethpage, New York, about 20 km to the east-southeast, the Montauk till is overlain by marine clay, oyster reef beds, salt- and fresh-water peat, or clay containing wood and pollen (Sirkin, 1982). Radiocarbon ages from these deposits range from >43,800 to 21,750 yr B.P. (Sirkin and Stuckenrath, 1980; Sirkin, 1982). In the subsurface of the inner shelf off southern Long Island, peat and shell material from a deposit considered to be equivalent to the Wantagh Formation provided a radiocarbon age of 28,150 yr B.P. (Rampino and Sanders, 1981).

LOWER TILLS OF THE NEW ENGLAND MAINLAND

Throughout most of mainland New England, tills occur locally below the widespread late Wisconsinan drift sheet. These tills are informally referred to as the lower till or, in some cases, the drumlin till, because of their common occurrence within drumlins. The age and origin of these lower tills have been disputed for many years. It is not our intention to review completely the previous work related to these deposits or to detail the numerous arguments about the age of these lower tills. Koteff and Pessl (1985) and Stone and Borns (1986) provided good reviews. Some of the better-studied sections containing a lower till (Fig. 1) include Thomaston, Connecticut (Newton, 1978), the Boston Harbor drumlins (Kaye, 1982; Newman and others, 1987; Newman and others, 1990), and Nash Stream, New Hampshire (Koteff and Pessl, 1985). Multiple till sections in Maine include New Sharon (Caldwell, 1959, 1960; Borns and Calkin, 1977; Weddle, 1988; Weddle and Retelle, 1988; Weddle and others, 1989), Austin Stream (Borns and Calkin, 1977; Weddle and Retelle, 1988), Nash Stream (Borns and Calkin, 1977), and Winthrop (Thompson, 1982).

Lower tills throughout mainland New England are generally similar in color, grain size, structure, and oxidation or weathering (Weddle and others, 1989). They are olive to olive-gray or olive-brown in their upper parts and mostly dark gray in their lower parts. The matrix in the lower tills is generally sandy silt and clay or clayey to silty sand. Coarse clasts are sparse and large boulders are mostly absent. These tills are locally stratified or layered and are commonly jointed. A maximum of about 10 m of weathering of the lower tills is suggested by color changes that occur in many exposures (Koteff and Pessl, 1985). This depth of weathering is also indicated by mineralogical changes within the upper part of the till. For example, an upward decrease in abundance of chlorite and a corresponding increase in vermiculite was observed in the lower tills exposed in Boston Harbor drumlins (Newman and others, 1990). A similar change in mineralogy, a downward increase in illite associated with an upward increase in mixed-layer clay minerals, was seen in a deeply oxidized till in Maine (Weddle and others, 1989). Weathering is also suggested by the complete leaching of shells in the upper part of the lower till in the Boston Harbor drumlins (Newman and others, 1990). However, the depth or degree of weathering of other exposures of lower tills is commonly less, and has been attributed to erosion of the weathered zone by late Wisconsinan ice.

Shells that indicate warm sea-water temperatures are reworked into the lower till of the Boston Harbor drumlins. *Mercenaria* shells incorporated into the lower till produced AAR age estimates within the time of isotope stage 5 (Belknap, 1980), suggesting an early Wisconsinan age for the till. However, previous AAR age estimates on these shells correspond to isotope stage 7 (Belknap, 1979), which would allow the lower till to be the Illinoian (isotope stage 6) in age. No radiometric age determinations are available from the lower tills.

At New Sharon, Maine, a lower till is overlain by an organic-bearing layer that was inferred to be a paleosol (Caldwell, 1960). Wood fragments from the organic-bearing layer produced radiocarbon ages of >35,000 and >52,000 yr B.P. (Caldwell, 1960; Borns and Calkin, 1977). Because of these limiting radiocarbon ages, the lower till was considered to be pre–late Wisconsinan in age. However, the lower till and organic-bearing layer at New Sharon described by Caldwell (1960) and by Borns and Calkin (1977) were reexamined by Weddle. He found that the organic-bearing horizon that produced the >52,000 yr B.P. age was likely allochthonous (Weddle, 1988, Weddle and others, 1989), and that the lower till lacked the degree of weathering characteristic of other lower tills of mainland New England (Weddle and others, 1989). For these reasons, both deposits are now considered to represent a single late Wisconsinan glaciation (Weddle and others, 1989).

In most other places, the mainland lower tills are directly overlain by drift of late Wisconsinan age and rest directly on bedrock or, rarely, on unconsolidated deposits that are undated. Thus, the actual age of the lower tills cannot be established.

DISCUSSION

A uranium-thorium age of 133,000 ± 7,000 yr from detrital coral, AAR age estimates in the same range, and a fauna that indicates a warmer-than-present seawater temperature show that the marine beds at Sankaty Head, Nantucket, are Sangamonian (isotope stage 5) in age. Thus, the underlying drift, including till, is older than Sangamonian. We infer that the lower drift on Nantucket is most likely Illinoian (isotope stage 6) in age.

At present, unequivocal numerical-age determinations for the drift that contains the Montauk till and units that lie stratigraphically above and below it do not exist. Arguments for assigning an early Wisconsinan age to the Montauk till may be justified on the basis of its stratigraphic position (1) below units radiocarbon-dated as a middle Wisconsinan, and (2) above the Gardiners clay, inferred to be Sangamonian on the basis of AAR age estimates. Even if the middle Wisconsin radiocarbon ages are correct, however, they provide only an upper limit for the age of the underlying till. However, the middle Wisconsinan ages may be incorrect, and thus misleading. For example, middle Wisconsinan radiocarbon ages used to establish a higher-than-present middle Wisconsinan sea level have long been suspect (Broecker, 1965; Thom, 1973; Bloom, 1983; Colman and others, 1989). These ages have been questioned because they are inconsistent with ice-volume estimates from the oxygen-isotope record and with many sea-level records (Fig. 2) that indicate a middle Wisconsinan eustatic sea level well below present sea level (Bloom, 1983). AAR age estimates used to infer a Sangamonian age (isotope stage 5) for the Gardiners clay should be viewed for what they are: estimates that depend on an inferred temperature history and a kinetic model that may or may not be correct. The arguments of Fuller (1914, 1937) for an older-than-Sangamonian age for the Gardiners clay and for a pre-Sangamonian (probably Illinoian) age for the Montauk till and its possible equivalents on Block Island, Rhode Island, and Martha's Vineyard, Massachusetts, are still valid.

Estimates of the ages of the lower tills on the mainland have ranged from Illinoian or older (Stone, 1975; Oldale and Eskenasy, 1983), to early Wisconsinan (Kaye, 1961; Koteff and Pessl, 1985; Stone and Borns, 1986), to late Wisconsinan (Denny, 1958; Chute, 1966). Arguments for a late Wisconsinan age were mostly based on an interpretation that the lower and upper tills are the subglacial lodgement facies and the ablation facies, respectively, of a single till sheet. According to this interpretation, the differences between the lower and upper tills are the result of factors other than weathering (Denny, 1958; Chute, 1966). Clasts of weathered till incorporated into overlying upper till (Pessl, 1966; Pease, 1970) tend to refute this argument. In addition, mineralogical and chemical studies of lower till (Stone, 1975; Newton,

1978; Newman and others, 1990) have reinforced the interpretation that the lower tills locally are deeply weathered and thus are older than late Wisconsinan.

A pre-Wisconsinan age for the lower tills has been proposed based on their degree and depth of weathering. Stone (1975) argued that the lower till was as weathered as tills of Illinoian age in the midwestern United States. Newton (1978) assigned an early Wisconsinan age to the New England lower tills because the degree and depth of weathering were similar to tills assigned that age in Wisconsin state (Mickelson and others, 1984). However, the Wisconsin tills are now considered to be most likely Illinoian in age (Matsch and Schneider, 1986). Similarly, tills in Illinois, previously assigned an early Wisconsinan age (Altonian), are now considered to be Illinoian in age (Curry, 1989). Most recently, a detailed study of the weathering of the lower till exposed in the Boston Harbor drumlins has shown that the alteration of minerals in the weathered till is similar to that found in weathered till of Illinoian age in the midwestern United States and requires considerably longer subaerial exposure than was available during the middle Wisconsinan substage (Newman and others, 1990). However, the argument that the color and mineralogical changes in the upper part of lower tills represent everywhere a soil that developed subaerially during an interglacial stage or interstade was questioned during a 1989 New England Intercollegiate Geologic Conference field trip (see Weddle and others, 1989). It is unlikely that the A horizon of the solum was "surgically" removed from the lower tills throughout New England by late Wisconsinan glacial erosion. In addition, it was observed that, if the color change in the upper part of the till represents a B horizon, it is much thicker than the B horizon in older tills elsewhere. Thus, the arguments that some lower tills are a facies of the late Wisconsinan till and that changes in the upper part of some lower tills are a product of processes other than subaerial weathering (e.g., ground-water alteration) have yet to be put completely to rest.

Arguments for an early Wisconsinan age for the lower tills of the mainland are most often based on correlations with tills inferred to be early Wisconsinan in age on the basis of middle Wisconsinan radiocarbon age determinations in overlying deposits. Other arguments correlate the mainland lower tills with the lower till in Boston Harbor drumlins, which has been inferred to be early Wisconsinan in age because it contains shells thought to be Sangamonian in age. For example, Stone and Borns (1986) correlated the lower tills in New England with the Montauk till on Long Island and with the lower till in the Boston area. Koteff and Pessl (1985) correlated the lower till in New Hampshire with the lower till at New Sharon, Maine. In both papers, the authors were careful to point out the tenuous nature of these regional correlations and the uncertainty of the age of the lower tills of New England and Long Island.

We propose correlation of the lower till at Sankaty Head with most or all of the mainland lower tills. Arguments for this correlation are presented by Oldale and Eskenasy (1983). Such a correlation is reasonable, if only because the ice that deposited the lower till at Sankaty Head advanced across New England. However, because the correlation is long range and because the Sangamonian (isotope stage 5) age of the Sankaty sand only limits the age of the underlying lower till, this proposed correlation falls well short of establishing unequivocally the age of the New England lower tills.

The inability to fix the age of the lower tills requires that we look elsewhere for evidence of the age of the penultimate full glaciation of New England. Late Pleistocene ice-volume estimates from the marine oxygen-isotope record, and the eustatic sea-level record from sets of coral terraces, such as those in New Guinea, provide such evidence (Fig. 2).

The Laurentide Ice Sheet, including the New England sector, was the largest ice sheet of the last glacial stage. (In the manner of Fulton and Prest [1987] we have restricted the Laurentide Ice Sheet to the Wisconsinan glacial stage.) Its growth and subsequent decay, which may have represented 60% to 70% of the decrease in global ice volume at the end of the Wisconsinan glaciation (Fulton and Prest, 1987), were dominant influences on the marine oxygen-isotope record (Andrews, 1987) and on eustatic sea levels. Individually, the lower till at Sankaty Head and the Montauk till of Long Island represent a southward glacial advance that was approximately the same magnitude as the maximum advance of the Laurentide ice. Assuming the New England sector was representative of the ice sheet as a whole, an ice advance of this magnitude should be represented in the marine oxygen-isotope record and in the eustatic sea-level record by indications of ice volume roughly equal to or greater than that of isotope stage 2. Marine oxygen-isotope curves (Ruddiman and McIntyre, 1981; Martinson and others, 1987; Shackleton, 1987) and the eustatic sea-level curve, based on marine terraces in New Guinea (Shackleton, 1987), indicate that maximum ice volumes during isotope stage 6 were roughly equivalent to those of isotope stage 2 (Fig. 2). On the other hand, they indicate that ice volumes during isotope stages 3 and 4 were considerably less than those during isotope stage 2.

Estimates of ice volume based on the marine oxygen-isotope record and on the history of eustatic sea-level change suggest that the penultimate full glaciation of New England probably occurred during Illinoian time (isotope stage 6). They also suggest that during early and middle Wisconsinan time (isotope stages 3 and 4), the Laurentide Ice Sheet was probably characterized by significant amounts of ice. This ice may have covered parts of Canada and, occasionally, parts of northernmost New England, but was probably insufficient to advance into southern New England. These conclusions are supported by Eyles and Westgate (1987), and Eyles and Williams (this volume), who believe the lower and middle Wisconsinan deposits, including diamictons, along the north shore of Lake Ontario to be lacustrine sediments. They believe that these lower and middle Wisconsinan lacustrine sediments preclude an extensive advance of the Laurentide ice during these times, and require a limited ice advance that only dammed the St. Lawrence River Valley. Even if interpretations of more extensive ice in the Ontario basin during early Wisconsinan time (Hicock and Dreminis, 1989, this volume; Mahaney, 1990) are correct, ice may still not have advanced in much of northern New England until after 46 ka (Hillaire-Marcel and Causse, 1989).

SUMMARY AND CONCLUSIONS

The lower tills in New England and Long Island, New York, have been assigned ages that range from late Wisconsinan to Illinoian or older, and regional correlations between occurrences of lower till have covered the region from Long Island to Canada. In truth, the ages of most of the older tills are unknown and regional correlations between lower tills are unproven. However, the lower till at Sankaty Head on Nantucket Island, Massachusetts, and the Montauk till of Long Island, New York, individually represent a full glaciation at least as extensive a the late Wisconsinan glaciation of New England. Because the age of these southernmost lower tills is not known, they cannot establish the age of this penultimate glaciation, and evidence must be looked for elsewhere. We think that the answer lies in the record of the marine oxygen-isotope ratios and the record of eustatic sea level (Fig. 2). Both records indicate that the last time global ice-volumes were equal to or somewhat greater than those that occurred during late Wisconsinan time was during the Illinoian stage (isotope stage 6). Because of this, we believe that the penultimate full glaciation of New England, represented by the lower till at Sankaty Head and by the Montauk till of Long Island, is probably Illinoian in age. This conclusion is supported by the Sangamonian age of the Sankaty sand, which overlies the lower till on Nantucket. The lower tills on the mainland are reasonably, but equivocally, correlated with this penultimate full glaciation.

Greater-than-present ice volumes and lower-than-present eustatic sea levels in isotope stage 4 and 3 probably indicate that the transition from interglacial (Sangamonian) to full glacial (late Wisconsinan) was characterized by a persistent Laurentide Ice Sheet of significant but varying volume. Thus, early and middle Wisconsinan advances of the Laurentide Ice Sheet may have reached parts of northern New England, but no evidence suggests that they reached southern New England. It appears that only during the late Wisconsinan did the Laurentide ice advance to southern New England and Long Island before its final and complete retreat.

REFERENCES CITED

Andrews, J. T., 1987, The late Wisconsin glaciation and deglaciation of the Laurentide Ice Sheet, *in* Ruddiman, W. F., and Wright, H. E., Jr., eds., North America and adjacent oceans during the last deglaciation: Boulder, Colorado, Geological Society of America, The Geology of North America, v. K-3, p. 13–37.

Belknap, D. F., 1979, Application of amino acid geochronology to stratigraphy of late Cenozoic marine units of the Atlantic Coastal Plain [Ph.D. thesis]: Newark, University of Delaware, 550 p.

—— , 1980, Amino acid geochronology and the Quaternary of New England and Long Island: American Quaternary Association, 6th Biennial Meeting, Orono, Maine, August 18–20, 1980, Abtracts and Program, p. 16–17.

Bloom, A. L., 1983, Sea level and coastal morphology of the United States through the late Wisconsin glacial maximum, *in* Wright, H. E., Jr., ed., Late Quaternary environments of the United States, Volume 1, The late Pleistocene: Minneapolis, Minnesota, University of Minnesota Press, p. 215–229.

Borns, H. W., and Calkin, P. C., 1977, Quaternary glaciation, west-central Maine: Geological Society of America Bulletin, v. 88, p. 1773–1784.

Broecker, W. S., 1965, Isotope geochemistry and the Pleistocene climate record, *in* Wright, H. E., Jr., and Frey, D. C., eds., The Quaternary of the United States: Princeton, New Jersey, Princeton University Press, p. 737–753.

Caldwell, D. W., 1959, Glacial lake and glacial marine clays of the Farmington area, Maine: Maine Geological Survey Special Geological Studies, ser. 3, 48 p.

—— , 1960, The surficial geology of the Sandy River Valley from Farmington to Norridgewock, Maine, Trip C, (New England Intercollegiate Geological Conference, 52nd Annual Meeting, Rumford, Maine, October 8–9, 1960, Guidebook for field trips in west-central Maine): Rumford, Maine, p. 19–23.

Chute, N. E., 1966, Geology of the Norwood Quadrangle, Norfolk and Suffolk counties, Massachusetts: U.S. Geological Survey Bulletin 1163-B, p. B1–B78.

Colman, S. M., Mixon, R. B., Rubin, M., Bloom, A. L., and Johnson, G. H., 1989, Comment *on* "Late Pleistocene barrier island sequence along the southern Delmarva Peninsula: Implications for middle Wisconsinan sea levels": Geology, v. 17, p. 84–85.

Curry, B. B., 1989, Absence of Altonian Glaciation in Illinois: Quaternary Research, v. 31, p. 1–13.

Denny, C. S., 1958, Surficial geology of the Canaan area, New Hampshire: U.S. Geological Survey Bulletin 1061-C, p. 73–100.

Eyles, N., and Westgate, J. A., 1987, Restricted regional extent of the Laurentide Ice Sheet in the Great Lakes basins during early Wisconsin glaciation: Geology, v. 15, p. 537–540.

Fuller, M. L., 1914, The geology of Long Island: U.S. Geological Survey Professional Paper 82, 231 p.

—— , 1937, Comment *on* "Correlation of late Pleistocene marine and glacial deposits of New Jersey and New York": Geological Society of America Bulletin, v. 47, p. 1982–1994.

Fulton, R. J., and Prest, V. K., 1987, The Laurentide Ice Sheet and its significance: Géographie Physique et Quaternaire, v. 41, p. 181–186.

Gustavson, T. C., 1976, Paleotemperature analysis of the marine Pleistocene of Long Island, New York, and Nantucket Island, Massachusetts: Geological Society of America Bulletin, v. 87, p. 1–8.

Hicock, S. R., and Dreimanis, A., 1989, Sunnybrook drift indicates a grounded early Wisconsin glacier in the Lake Ontario basin: Geology, v. 17, p. 169–172.

Hillaire-Marcel, C., and Causse, C., 1989, The late Pleistocene Laurentide glacier: Th/U dating of its major fluctuations and $\delta^{18}O$ range of the ice: Quaternary Research, v. 32, p. 125–138.

Kaye, C. A., 1961, Pleistocene stratigraphy of Boston, Massachusetts: U.S. Geological Survey Professional Paper 424-B, p. 73–76.

—— , 1982, Bedrock and Quaternary geology of the Boston area, Massachusetts: Geological Society of America Reviews in Engineering Geology, v. 5, p. 25–39.

Koteff, C., and Pessl, F., Jr., 1985, Till stratigraphy in New Hampshire: Correlations with adjacent New England and Quebec, *in* Borns, H. W., Jr., LaSalle, P., and Thompson, W. B., eds., Late Pleistocene history of northeastern New England and adjacent Quebec: Geological Society of America Special Paper 197, p. 1–12.

Mahaney, W. C., 1990, Macrofabrics and quartz microstructures confirm glacial origin of Sunnybrook drift in the Lake Ontario basin: Geology, v. 18, p. 145–148.

Martinson, D. G., Pisias, N. G., Hays, J. D., Imbrie, J., Moore, T. C., Jr., and Shackleton, N. J., 1987, Age dating and the orbital theory of the ice ages: Development of a high-resolution 0 to 300,000-year chronostratigraphy: Quaternary Research, v. 27, p. 1–29.

Matsch, C. L., and Schneider, A. F., 1986, Stratigraphy and correlation of the glacial deposits of the glacial lobe complex in Minnesota and northwestern Wisconsin, *in* Sibrava, V., Bowen, D. Q., and Richmond, G. M., eds., Quaternary glaciations in the Northern Hemisphere: Oxford, Pergamon Press, p. 59–64.

Mickelson, D. M., Clayton, L., Baker, R. W., Mode, W. N., and Schneider, A. F., 1984, Pleistocene stratigraphic units of Wisconsinan: Wisconsin Geological and Natural History Survey Miscellaneous Paper 84-1.

Mills, H. C., and Wells, D. W., 1974, Ice-shove deformation and glacial stratigraphy of Port Washington, Long Island, New York: Geological Society of America Bulletin, v. 85, p. 357–364.

Newman, W. A., Berg, R. C., Rosen, P. S., and Glass, H. D., 1987, Pleistocene stratigraphy of the Boston Harbor drumlins: Geological Society of America Abstracts with Programs, v. 19, no. 1., p. 32.

—— , 1990, Pleistocene stratigraphy of the Boston Harbor drumlins, Massachusetts: Quaternary Research, v. 34, p. 148–159.

Newton, R. M., 1978, Stratigraphy and structure of some New England tills [Ph.D. thesis]: Amherst, University of Massachusetts, 241 p.

Nieter, W., Nemickas, B., Koszalka, E. J., and Newman, W. S., 1975, The late Quaternary geology of the Montauk Peninsula: Montauk Point to Southampton, Long Island, New York, *in* Wolff, M. P., ed., Guidebook to field excursions, 47th meeting New York State Geological Association: p. 129–156.

Oldale, R. N., and Eskenasy, D. M., 1983, Regional significance of pre-Wisconsinan till from Nantucket Island, Massachusetts: Quaternary Research, v. 19, p. 302–311.

Oldale, R. N., and 7 others, 1982, Stratigraphy, structure, absolute age, and paleontology of the upper Pleistocene deposits at Sankaty Head, Nantucket Island, Massachusetts: Geology, v. 10, p. 246–252.

Pease, M. H., Jr., 1970, Pleistocene stratigraphy observed in a pipeline trench in east-central Connecticut and its bearing on the two-till problem: U.S. Geological Survey Professional Paper 700-D, p. D36–D48.

Pessl, F., Jr., 1966, A two-till locality in northeastern Connecticut: U.S. Geological Survey Professional Paper 550-D, p. D89–D93.

Rampino, M. R., and Sanders, J. E., 1981, Upper Quaternary stratigraphy of southern Long Island, New York: Northeastern Geology, v. 3, p. 116–128.

Ruddiman, W. F., and McIntyre, A., 1981, Oceanic mechanisms for amplification of the 23,000-year ice-volume cycle: Science, v. 212, p. 617–627.

Schnitker, D., 1988, Timing and environment of deposition of the offshore equivalent to the Presumpscot Formation: Geological Society of America Abstracts with Programs, v. 20, no. 1, p. 68.

Shackleton, N. J., 1987, Oxygen isotopes, ice volume and sea level: Quaternary Science Reviews, v. 6, p. 183–190.

Sirkin, L., 1982, Wisconsinan glaciation of Long Island, Ne York, to Block Island, Rhode Island, *in* Larson, G. J., and Stone, B. D., eds., Late Wisconsinan glaciation of New England: Dubuque, Iowa, Kendall/Hunt, p. 35–59.

Sirkin, L., and Stuckenrath, R., 1980, The Portwashingtonian warm interval in the northern Atlantic coastal plain: Geological Society of America Bulletin, v. 91, p. 332–336.

Stone, B. D., 1975, The Quaternary geology of the Plainfield and Jewett City quadrangles, central-eastern Connecticut [Ph.D. thesis]: Baltimore, Johns Hopkins University, 217 p.

Stone, B. D., and Borns, H. W., Jr., 1986, Pleistocene glacial and interglacial stratigraphy of New England, Long Island, and adjacent Georges Bank and Gulf of Maine, *in* Sibrava, V., Bowen, D. Q., and Richmond, G. M., eds., Quaternary glaciations in the Northern Hemisphere: Oxford, Pergamon Press, p. 39–52.

Thom, B. G., 1973, Dilemma of high interstadial sea levels during the last glaciation: Progress in Geography, v. 5, p. 167–246.

Thompson, W. B., 1982, Recession of the late Wisconsinan ice sheet in coastal Maine, *in* Larson, G. J., and Stone, B. D., eds., Late Wisconsinan glaciation of New England: Dubuque, Iowa, Kendall/Hunt, p. 211–228.

Tucholke, B. E., and Hollister, C. D., 1973, Late Wisconsin glaciation of the southwestern Gulf of Maine: New evidence from the marine environment: Geological Society of America Bulletin, v. 84, p. 3279–3296.

Weddle, T., 1988, On the lack of early and middle Wisconsinan deposits at New Sharon, Maine—Implications for middle Wisconsinan ice centers: American Quaternary Association, 10th Biennial Meeting, Amherst, Massachusetts, June 6–8, 1988, Program and Abstracts, p. 160.

Weddle, T., and Retelle, M., 1988, Till stratigraphy in Maine: Existing problems and alternative models: Geological Society of America Abstracts with Programs, v. 20, no. 1, p. 78.

Weddle, T. K., Stone, B. D., Thompson, W. B., Retelle, M. J., Caldwell, D. W., and Clinch, J. M., 1989, Illinoian and late Wisconsinan tills in eastern New England: A transect from northeastern Massachusetts to west-central Maine, *in* Berry, A. W., Jr., ed., New England Intercollegiate Geological Conference, 81st annual meeting, guidebook for field trips in southern and west-central Maine: p. 25–49.

Woodworth, J. B., and Wigglesworth, E., 1934, Geography and geology of the region including Cape Cod, the Elizabeth Islands, Nantucket, Martha's Vineyard, No Mans Land, and Block Island: Harvard College Museum of Comparative Zoology Memoir 52, 322 p.

Manuscript Accepted by the Society September 6, 1991

Printed in U.S.A.

Geological Society of America
Special Paper 270
1992

Sangamonian and early Wisconsinan events in the St. Lawrence Lowland and Appalachians of southern Quebec, Canada

Michel Lamothe
Département des sciences de la terre, Université du Québec à Montréal, C.P. 8888, Succ. "A", Montréal, Québec H3C 3P8, Canada
Michel Parent
Centre géoscientifique de Québec, Geological Survey of Canada, 2700 Einstein, C.P. 7500, Sainte-Foy, Québec G1V 4C7, Canada
William W. Shilts
Terrain Sciences Division, Geological Survey of Canada, 601 Booth, Ottawa, Ontario K1A 0E8, Canada

ABSTRACT

In the St. Lawrence Lowland of southern Quebec, an early Wisconsinan glacial advance deposited the Levrard Till. This glacial event, known as the Nicolet Stade, is tentatively correlated with marine oxygen-isotope stage 4. Radiocarbon and thermoluminescence ages bracket the Nicolet Stade between 90 and 70 ka. This advance was preceded and followed by periods of free drainage during which were deposited the Lotbiniere Sand and St. Pierre Sediments, two nonglacial units dated at or beyond the limit of the radiocarbon method. Available evidence suggests that the Deschaillons Varves were deposited ca. 80 ka in a large glacial lake that was impounded in front of the Laurentide Ice Sheet as it advanced up the St. Lawrence Valley.

In the Appalachian Uplands, fluvial and lacustrine sediments of the Massawippi Formation were probably deposited at the end of the Sangamonian Interglacial. These sediments underlie the Chaudiere Till, a unit in which the occurrence of distinctive lithological indicators is taken as evidence that a regional episode of westward to southwestward ice flow prevailed at the onset of the last glaciation.

The proposed paleogeographic reconstruction suggests that the development of an independent ice cap in the northeastern Appalachians played a key role during the early part of the Wisconsinan Glaciation in southern Quebec. This independent ice mass flowed southwestward across the Appalachian Uplands of southern Quebec and eventually coalesced with the Laurentide Ice Sheet, which was advancing up the St. Lawrence Valley.

INTRODUCTION

In southern Quebec, as in many other glaciated regions of North America, the timing and paleoclimatic significance of geologic events that occurred during the early phases of the last glaciation are inferred from (1) rarely preserved organic-bearing sediments and their paleoecological record (e.g., St. Pierre Sediments); (2) regional lithological and sedimentological analysis of till (e.g., the Chaudiere Till) and related glacial sediments; and (3) the results of not yet fully reliable geochronological methods, such as thermoluminescence (TL) and uranium-thorium (U-Th) dating. In spite of some chronological and stratigraphical uncertainties, the complex sequence of events that characterized the beginning of the last glaciation are of interest not only to Pleistocene stratigraphers, but also to other scientists interested in paleoclimatic reconstructions and modeling.

In this chapter we describe briefly the stratigraphic units that record Sangamonian and early Wisconsinan events in southern Quebec (Lamothe, 1987) and discuss their regional significance. The chronostratigraphic term "Sangamonian" is herein utilized according to the definition proposed by Fulton (1984), and it thus refers to the time interval between 130 and 80 ka. This chapter

Lamothe, M., Parent, M., and Shilts, W. W., 1992, Sangamonian and early Wisconsinan events in the St. Lawrence Lowland and Appalachians of southern Quebec, Canada, *in* Clark, P. U., and Lea, P. D., eds., The Last Interglacial-Glacial Transition in North America: Boulder, Colorado, Geological Society of America Special Paper 270.

also introduces a tentative paleogeographic framework that depicts the configuration of water bodies, glacial margins, and ice-flow lines that may have prevailed at that time in this part of North America.

PHYSIOGRAPHY AND BEDROCK GEOLOGY

Southern Quebec can be divided into three main physiographic regions (Bostock, 1970; Fig. 1). The Laurentian Highlands occupy the northwestern part of the area and are underlain by Precambrian crystalline metamorphic (e.g., paragneiss) and intrusive (e.g., anorthosite and granite) rocks. This shield terrain is in fault contact with relatively undisturbed Cambrian to Ordovician shelf rocks of the Central St. Lawrence Lowland. The complexly folded Paleozoic strata and intrusive complexes of the Appalachian Mountains extend across the southeastern part of the area. The Appalachians are subdivided into the Eastern Quebec Uplands, the Sutton and Notre-Dame mountains, and, along the international border, the Megantic Hills. This area is underlain by metasedimentary and metavolcanic rocks of the northeast-southwest–trending Appalachian orogen.

Bedrock lithology and structure together with preglacial landscape partly controlled ice-flow patterns and the development of ice-marginal water bodies during the Pleistocene. The St. Lawrence Lowland was inundated periodically by large bodies of water. These episodes of either marine or glaciolacustrine inundation reflect a delicate balance between ice-front fluctuations across the St. Lawrence Valley in the vicinity of Quebec City, regional glacio-isostatic subsidence or uplift, and global sea-level changes. The resulting Pleistocene depositional record of the St. Lawrence Lowland is dominated by waterlaid units.

Glacial ice in the St. Lawrence Lowland also caused periodic impounding of drainage in northward-flowing tributaries within the Appalachian region, forming glacial lakes. Although areally the most widespread Pleistocene sediment in this area is till, significant volumes of glaciolacustrine sediments separate till sheets below 430 m (above sea level) in major valleys. The lithological and geochemical composition of till provide significant limitations on the interpretation of the Appalachian regional lithostratigraphic record (Shilts, 1973, 1981; Parent, 1987b). Till composition has been used as a major criterion for correlation and for deciphering till provenance and directions of glacial

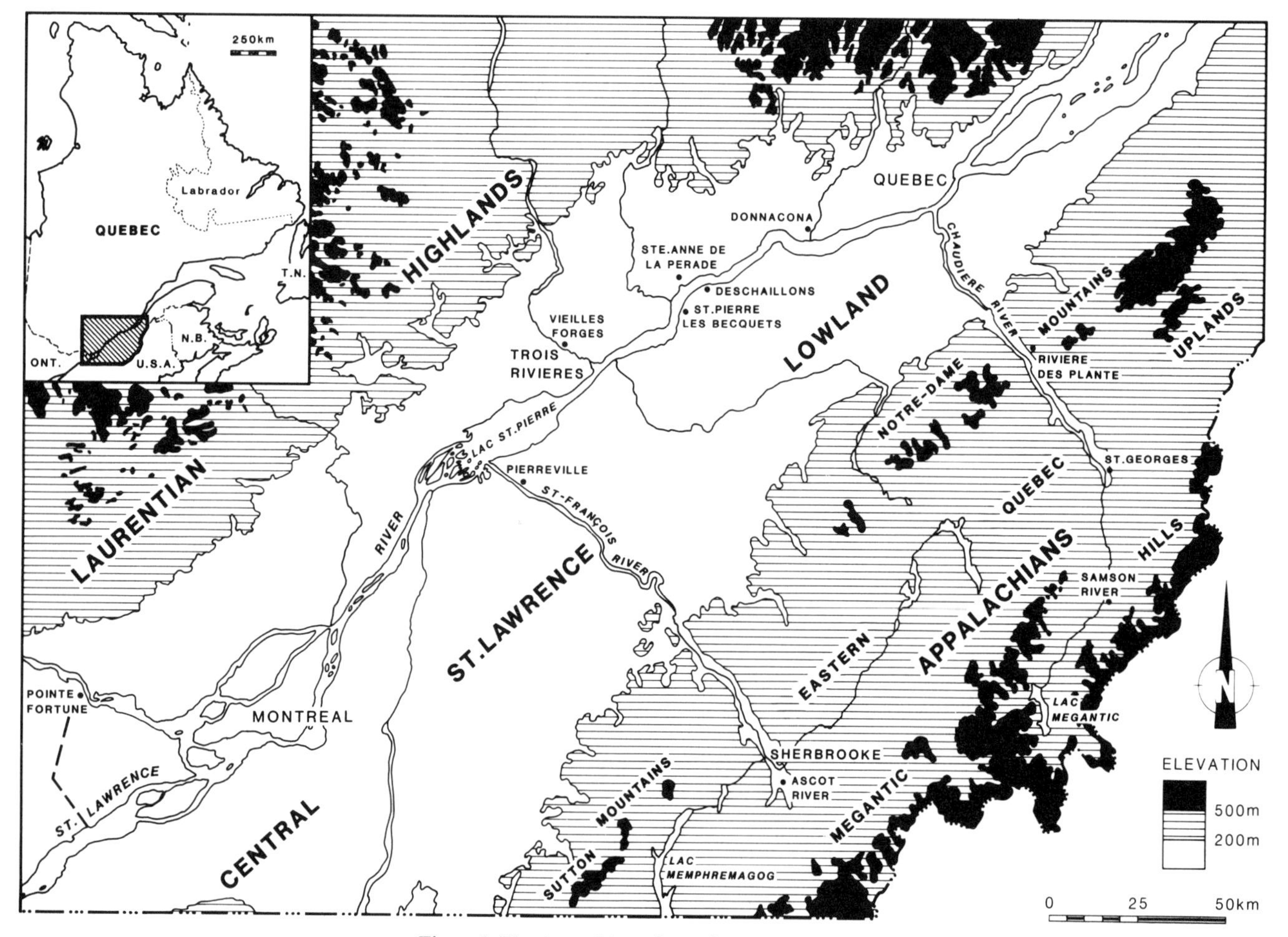

Figure 1. Physiographic regions of southern Quebec.

transport. These methods have been applied successfully for several years in the Appalachian region, where any occurrence of shield-type erratics, such as Precambrian gneiss, is used as evidence for deposition from a Laurentide ice mass or for its nearby presence (McDonald, 1967). Distinctive local indicators are provided by ultramafic and mafic rocks of the ophiolite complex which extends over 200 km along the Appalachian fold belt, from the international border to the Chaudiere River, and slightly beyond. Areas adjacent to the ophiolite belt have been the subject of extensive glacial-transport modeling, particularly for mineral exploration purposes (Shilts, 1973), and the dispersal of ultramafic debris in the surface till defines Wisconsinan ice-flow patterns. Ice overriding intrusive Devonian granitic rocks, which occur as isolated stocks in the southeastern part of the Quebec Appalachians, produced prominent southeastward-trending dispersal trains as well as numerous erratics displaced southwestward and westward of their source during earlier glacial advances.

SANGAMONIAN AND EARLY WISCONSINAN RECORD OF SOUTHERN QUEBEC

St. Lawrence Lowland

For three decades, the Pleistocene stratigraphic framework of the Central St. Lawrence Lowland revolved around the St. Pierre Sediments (Gadd, 1955, 1971; Occhietti, 1982), a nonglacial fluvial and lacustrine lithostratigraphic unit that contains peat layers and organic-rich sand characterized by boreal-type pollen assemblages (Terasmae, 1958). The St. Pierre beds were thought to have been deposited during a single nonglacial event postdating and predating Laurentide ice advances that deposited the Becancour and Gentilly tills (Gadd, 1971; Occhietti, 1982). The two ice advances were thought to be Wisconsinan, mainly because of the absence of warm-climate indicators in the organic beds (Terasmae, 1958).

Recent investigations by Lamothe (1989) provided evidence that the St. Pierre organic-bearing sediments, which had long been assigned to a single nonglacial event, instead represent two distinct nonglacial sequences. These two units, formally defined by Lamothe (1989) as the Lotbiniere Sand and the St. Pierre Sediments, are separated by the Levrard Till, a unit deposited during a glacial advance that had not been recognized during previous investigations. Thus, the Pleistocene depositional sequence of the lowland (Figs. 2 and 3) is now thought to record three glacial advances (St. Lawrence, Nicolet, and Trois-Rivieres stades) separated by two nonglacial events (Grondines and Les Becquets interstades), each represented by organic-bearing units (the Lotbiniere Sand and the St. Pierre Sediments; Lamothe, 1989).

Until 1984, the geochronology of pre–Champlain Sea sediments in the St. Lawrence Lowland was based solely on radiocarbon dating of wood and peat fragments found in the St. Pierre Sediments. Several ages were obtained from these beds, but only three were reported as finite, and they are at the limit of the method. Wood fragments collected from the St. Pierre Sediments at the type section were dated by De Vries at 65,300 ± 1400 yr B.P. (GrN-1799; Vogel and Waterbolk, 1972). Ages of 66,500 ± 1600 yr B.P. (GrN-1711; Vogel and Waterbolk, 1972) and $74,700^{+2700}_{-2000}$ ± 2700 yr B.P. (QL-198; Stuiver and others, 1978) were also measured on wood obtained from a St. Pierre correlative at Pierreville (Fig. 1). The oldest age was obtained through thermal diffusion isotopic enrichment. These dates are not considered as geologically different because of the poor reliability of ^{14}C dating in this time range.

Several calcareous concretions of the Deschaillons Varves have been dated ca. 35 ka by conventional radiocarbon dating (Hillaire-Marcel and Page, 1981; Lamothe and others, 1983; Lamothe, 1985). Lamothe (1985) suggested that this minute radiocarbon activity is the result of complex carbonate diagenesis. These ages are therefore considered unreliable for the dating of this glaciolacustrine unit and are not used in this chapter as chronostratigraphic tools. Hillaire-Marcel and Causse (1989) have published U-Th ages obtained on concretions of the Deschaillons Varves. These ages suggest that the concretions were first precipitated ca. 80 ka. These results support an age assignment of 90–70 ka for Lake Deschaillons, as earlier proposed by Lamothe (1985, 1989) on the basis of TL dating (see below).

Lamothe (1985) and Lamothe and Huntley (1988) reported a series of TL ages that were obtained on the waterlaid units of the stratigraphic sequence. Apparent TL ages were obtained on the fine-grained fraction (4–11 μm) using a partial bleach technique in which the TL from light-sensitive traps was separated from total TL (Wintle and Huntley, 1980). Ages ranging from >60 ka to 86.3 ± 17 ka were measured on samples of the Deschaillons Varves and their correlatives, all collected in the St. Pierre les Becquets and Pierreville areas. An apparent TL age of 135 ka that had been obtained on a sample of the Pierreville Varves (Nicolet Stade) is considered invalid, based on a discussion presented in Lamothe and Huntley (1988). Finite TL ages of 61.2 ± 11 and 61.1 ± 9.2 ka were also calculated for the St. Pierre Sediments exposed at Pierreville (Lamothe, 1984). These apparent TL ages compare favorably with the ^{14}C ages obtained on pieces of wood from the same site, considering the large uncertainties associated with both methods. A TL age of 27.7 ± 4.3 ka was also obtained on a sample collected at the top of the St. Pierre Sediments at the type section, but this young TL age is thought to be the result of sedimentary recycling and does not reflect the true age of the sediment.

Correlation of Pleistocene events in the St. Lawrence Lowland with the deep-sea oxygen-isotope record (Shackleton, 1969; Shackleton and Opdyke, 1973) is based on available numerical ages and on the following paleogeographic assumption (Fig. 3): any significant buildup of glacier ice on the continent will probably affect normal drainage by blocking the lower St. Lawrence Valley, hence causing large ice-dammed lacustrine basins to form in the central part of the lowland (Gadd, 1971).

Although no "warm" interglacial sediments have yet been

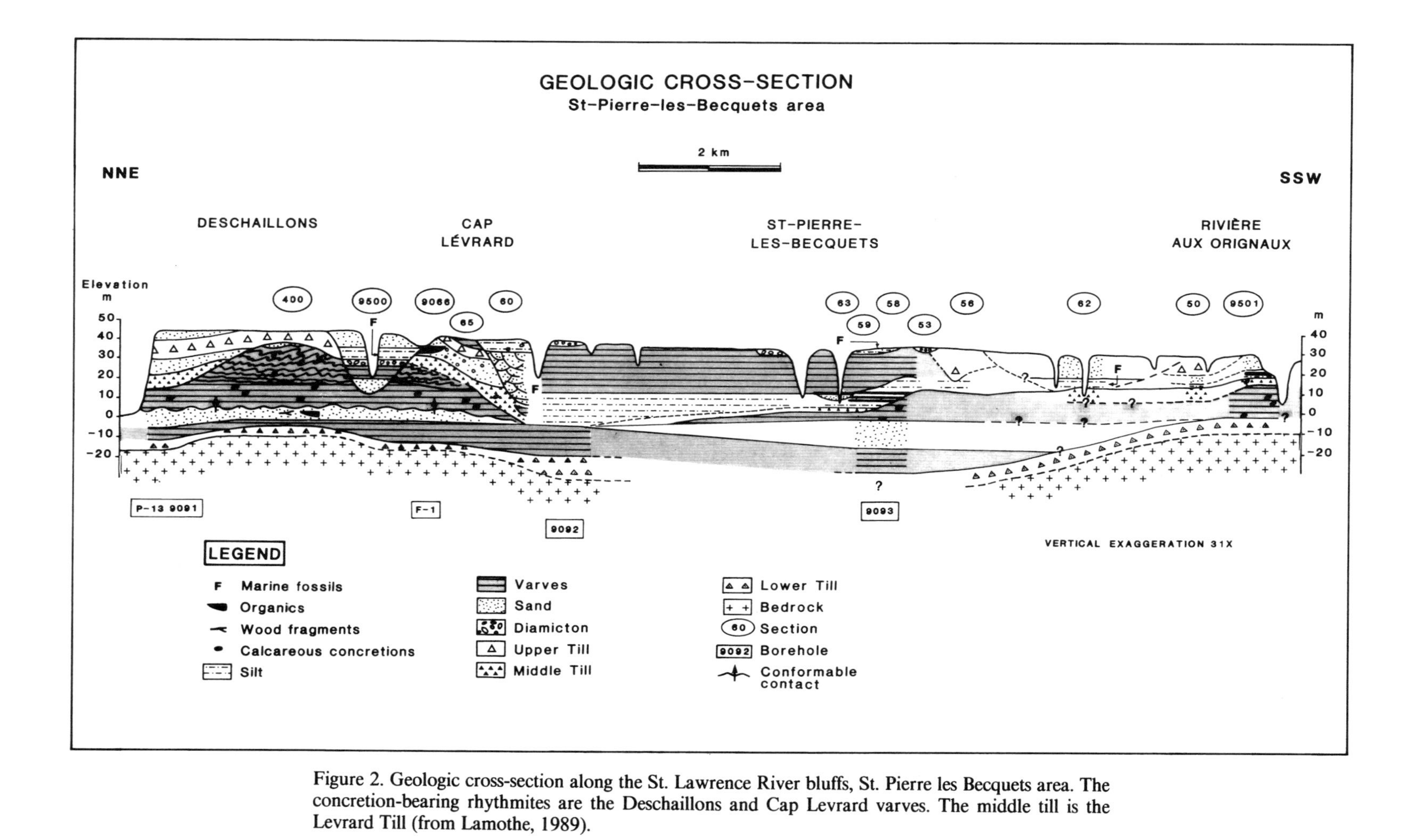

Figure 2. Geologic cross-section along the St. Lawrence River bluffs, St. Pierre les Becquets area. The concretion-bearing rhythmites are the Deschaillons and Cap Levrard varves. The middle till is the Levrard Till (from Lamothe, 1989).

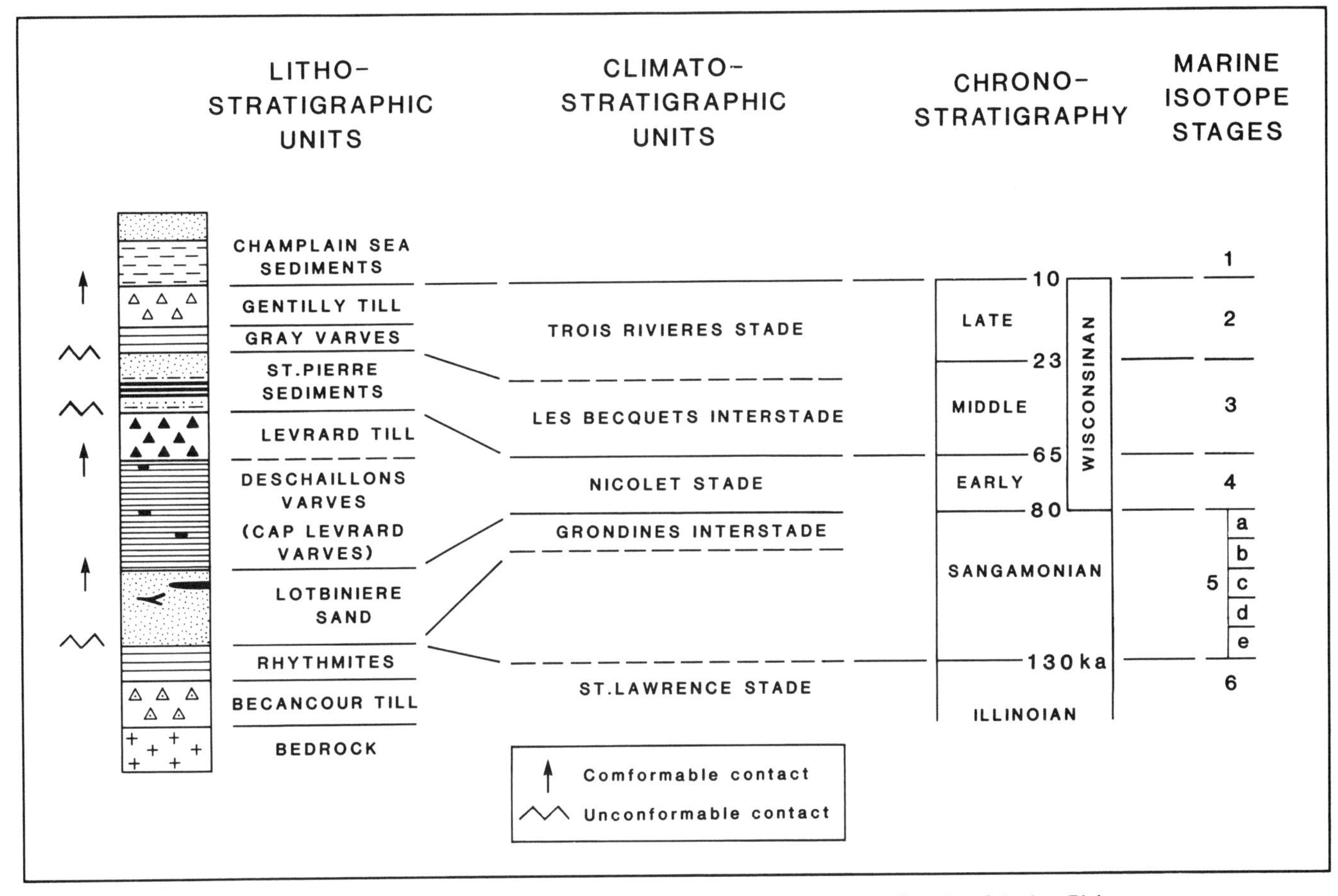

Figure 3. Pleistocene lithostratigraphy, climatostratigraphy, and chronostratigraphy of the late Pleistocene sediments of the St. Lawrence Lowland (from Lamothe, 1989), and correlation with the marine isotope record (Shackleton, 1969).

reported in the central part of the lowland, the St. Lawrence Stade (the Becancour Till and associated rhythmites) is thought to be Illinoian, particularly in light of the recent discovery of true interglacial sediments at Pointe-Fortune, 80 km west of Montreal (Fig. 1; Anderson and others, 1990). Till underlying the interglacial sediments at this site (the Rigaud Till of Anderson and others, 1990) has been correlated with the Becancour Till of the central part of the St. Lawrence Lowland by Veillette and Nixon (1984). The St. Lawrence Stade is most likely pre-Sangamonian, and it is tentatively correlated with isotopic stage 6 of Shackleton and Opdyke (1973).

The Grondines and Les Becquets interstades represent periods of free drainage during which nonglacial organic-bearing sediments were deposited. In view of the conformable relation between the Lotbiniere Sand and the Deschaillons Varves (see below), the age of the lower interstadial unit is probably ca. 90–80 ka (Lamothe, 1989). The Grondines Interstade is thus correlated with the latest part of marine isotope stage 5 (substage 5a of Shackleton, 1969).

The Lotbiniere Sand is conformably overlain by rhythmites of the Deschaillons Varves (= Cap Levrard Varves; Fig. 3), indicating that free drainage in the St. Lawrence Valley was blocked by an advancing glacier. This glacial event, the Nicolet Stade, is the earliest Wisconsinan stade in the eastern Great Lakes–St. Lawrence region (Dreimanis and Karrow, 1972). Geochronological data (TL and U-Th) and varve counting (Hillaire-Marcel and Page, 1981; Lamothe, 1985) suggest that by 80 ka, a large glacial lake (Lake Deschaillons of Karrow, 1957) was impounded in front of the ice margin, and that this lake may have lasted as much as 2500 years. Fine-grained bottom sediments of this glacial lake were overridden by ice that ultimately deposited the Levrard Till. This glacial advance may have occurred between 90 and 70 ka, because the nonglacial St. Pierre Sediments overlying the Levrard Till are dated at or beyond the limit of the radiocarbon method, and because the TL ages are ca. 60–65 ka (Lamothe, 1984).

On the basis of ^{14}C and TL ages of the St. Pierre Sediments, the apparent TL ages obtained on the Deschaillons Varves, and other correlatives, and uranium-thorium dating on calcareous concretions of the Deschaillons Varves (ca. 80 ka; Hillaire-Marcel and Causse, 1989), the Lotbiniere Sand, Deschaillons Varves, Levrard Till, and St. Pierre Sediments are collectively and tentatively correlated with the latest part of marine oxygen-isotope stage 5 (5a), stage 4, and the earliest part of stage 3 of Shackleton (1969) and Shackleton and Opdyke (1973).

Pre-Gentilly glaciomarine sediments were recently discovered by Ferland and Occhietti (1990) at Ste. Anne de la Perade (Fig. 1), demonstrating that marine conditions prevailed in the St. Lawrence Lowland some time prior to Les Becquets Interstade (Fig. 3). A ^{14}C age of >35,590 yr B.P. (AMS; BETA-28404) was obtained on *Nucula* sp. shells from the marine unit. Ferland and Occhietti (1990) believe that this unit, formally defined by them as "Argile de la Pérade," is a correlative of the St. Pierre Sediments. This correlation is based on the presence of marine Demosponge spicules in a unit overlain by the St. Pierre Sediments at the type section. However, the possibility that these spicules may have been reworked from some older unit cannot be ruled out at the moment, because other marine fossils typical of the "Argile de la Pérade" are absent at this site. This marine unit is not included in the lithostratigraphic sequence of the St. Lawrence Lowland in Figure 3 of this chapter, because its correct stratigraphic position was not established at the time of writing.

The upper part of the lithostratigraphic sequence (the Gray Varves, Gentilly Till, and Champlain Sea Sediments) is thought to be middle to late Wisconsinan in age.

Appalachians of southern Quebec

The regional stratigraphic framework records three advances of the Laurentide Ice Sheet across the southeastern Quebec Appalachians and into northern New England (McDonald, 1967; McDonald and Shilts, 1971; Shilts, 1970). These glacial episodes are represented, from oldest to youngest, by the Johnville, Chaudiere, and Lennoxville tills (Fig. 4).

Fabric measurements and the occurrence of Precambrian erratics in the Johnville Till demonstrate that this till is a product of Laurentide ice. Following deposition of the Johnville Till, free drainage conditions toward the St. Lawrence River were established in the region. This event is recorded by the nonglacial fluvial sediment facies of the Massawippi Formation. The fluvial episode was followed by the formation of lakes, probably arms of larger lakes dammed by ice entering the St. Lawrence Lowland. Pollen spectra from the lake sediments indicate the presence of a boreal forest in the region during the later time of deposition of the Massawippi Formation (McDonald and Shilts, 1971). Radiocarbon ages obtained from disseminated organic debris concentrated from lacustrine sediments of the Massawippi Formation are beyond the limit of the method: (1) >54,000 yr B.P. (Y-1683; McDonald, 1967) at the Ascot River type section; (2) >40,000 yr B.P. (GSC-1084; Shilts, 1981) at the Riviere Grande Coulee section; and (3) >53,000 yr B.P. (GSC-4728 HP) at the Riviere des Plante.

The most complete facies record of the Massawippi Formation yet known from the Appalachians is exposed at several sections along the Riviere des Plante (Figs. 1 and 4; Shilts and Blais, 1989). In 1988–1990, the Massawippi sequence was best exposed and least disturbed in section B of Shilts and Smith (1987). There, compressed peat and organic silt rest in fluvial channels and within flood-plain sediments of the ancestral Riviere des Plante, which had cut down through the Johnville Till and into the underlying glacially deformed sediments. Macrofossils isolated from the peat collected in 1988 were examined by Lynn Ovenden, who reported that "the lack of diversity and the interwoven, highly organic nature of the peat indicates [that it is] an in situ peat deposit." This peat is significant in that it is the first in situ peat deposit found in the Appalachian Uplands. During preliminary examination of macrofossils from samples collected in 1989, seeds of plants currently growing in or south of the vicinity of the Riviere des Plante were found (J. V. Matthews, Jr., 1989, personal commun.). This suggests that the lower part of the Massawippi Formation may have been deposited when local climate was at least as warm as today. Examination of sparse, disseminated organic detritus from an overlying stony diamicton led Matthews and others (1987) to state that during that time, this part of Quebec was covered by a spruce forest that contained some larch and a species of *Alnus* typical of northern boreal forests. Organic remains in the glaciolacustrine facies that overlies the peat-bearing colluvium contain evidence of arctic flora and fauna (Matthews and others, 1987). Although much further paleoecological work remains to be done, these observations suggest that the Massawippi Formation may be truly interglacial, and that the tundra and boreal fauna and flora usually associated with the Massawippi interval reflect conditions that prevailed just prior to the onset of the Chaudiere glaciation.

The Chaudiere glaciation is recognized as a complex event. During its early phase, referred to as the "Maritime Ice Cap" phase by Shilts (1981), local ice advanced from the east or northeast and deposited a lower till member. In many exposures of the Chaudiere Till, this lower member is overlain by an upper member that records displacement of this "local" ice by Laurentide ice (McDonald and Shilts, 1971) or a complex history of shifting ice-flow directions within an Appalachian-based ice cap (Parent, 1987a, 1987b). Directional and compositional data collected from the Chaudiere Till at the Riviere des Plante and Ascot River sections, as well as observations made at surface localities (presented below), are taken as evidence that, at the onset of the Chaudiere glacial phase, an independent ice cap developed in the northeastern Appalachians, in Maine and/or New Brunswick.

The Riviere des Plante sections are fortuitously located along the southeast edge of a northeast-striking ophiolite complex. Tills deposited from glaciers flowing from the northwest, such as the Johnville Till, contain strong geochemical and mineralogical "fingerprints" (principally Ni, Cr, Co, magnetite, and chromite) derived from the ultramafic rocks of the complex. For example, the clay fraction (<2 μm) of the Johnville Till commonly contains more than 1000 ppm Ni, and 40% of the clasts consist of ultramafic rocks. By comparison, tills deposited by glaciers flowing from the east or northeast are depleted in these components, except where they include material reworked from the underlying units. At the Riviere des Plante, the Chaudiere Till is impover-

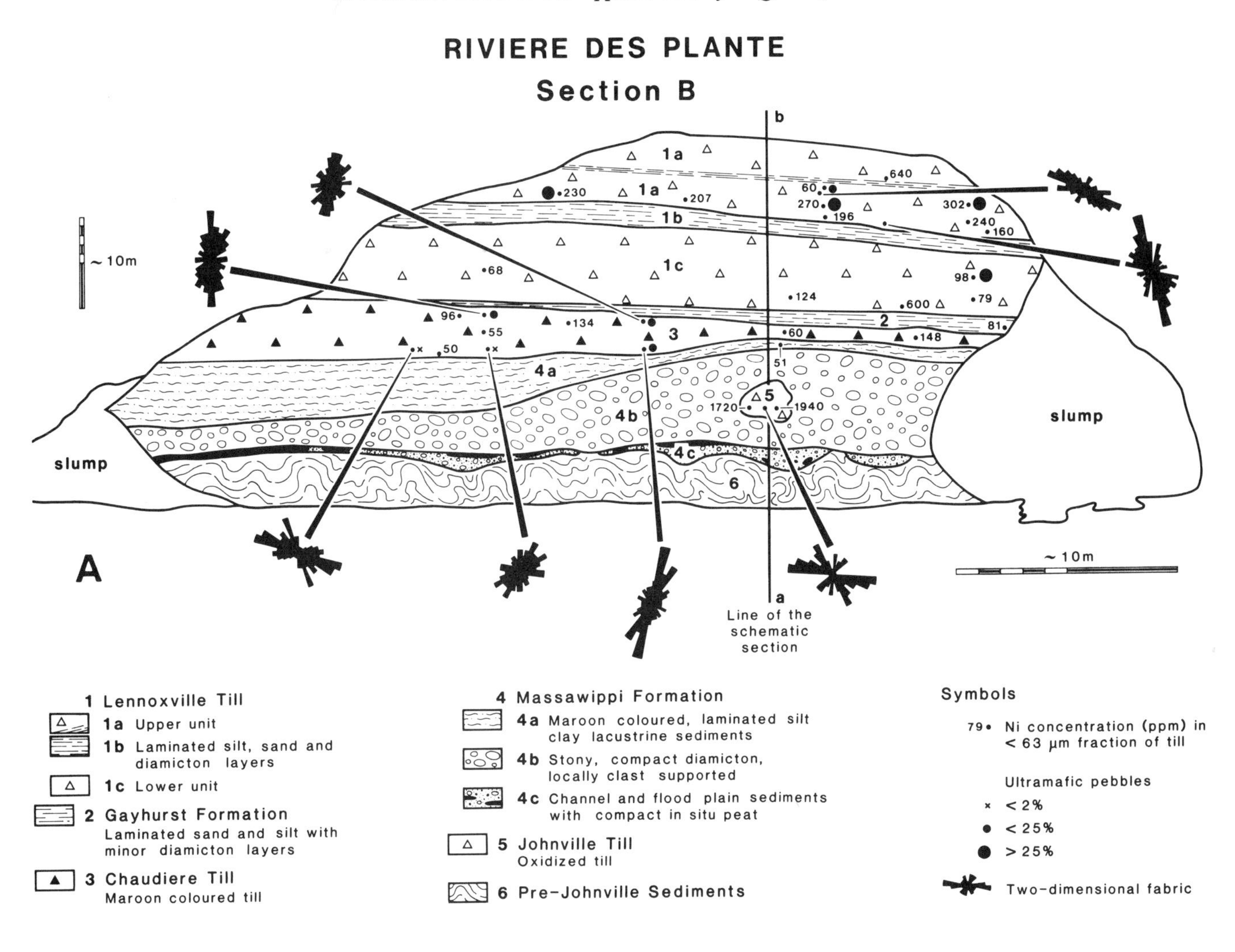

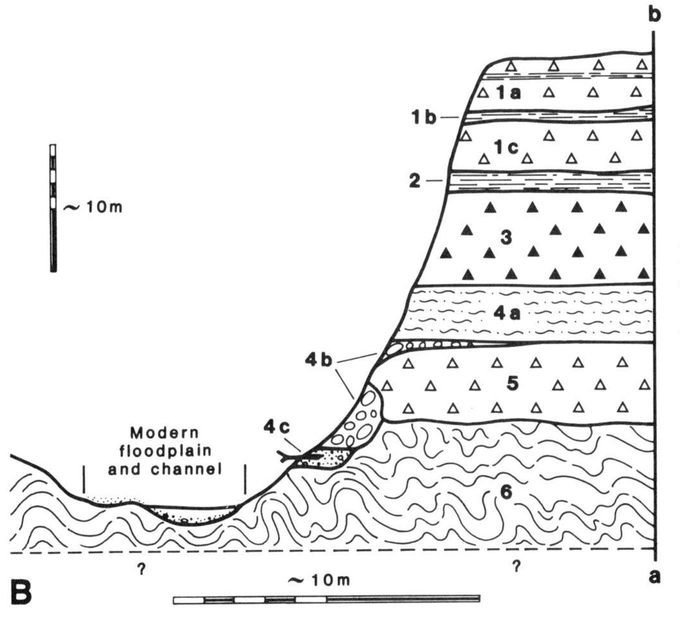

Figure 4. A: Pleistocene lithostratigraphic succession, geochemistry, and clast lithology at Riviere des Plante section B (from Shilts and Blais, 1989). B: Schematic section at right angle to exposed section along vertical line a-b.

ished in ultramafic clasts compared to the underlying Johnville Till and the overlying Lennoxville Till. It is equally impoverished in trace elements typical of ultramafic rocks, its <2 μm fraction having typically less than 100 ppm Ni (Shilts and Blais, 1989). Ice must have been flowing at any azimuth more easterly than the 040°–050° strike of the ophiolite belt, i.e., 060°–090°, to have produced drift with no ultramafic components (other than material reworked from earlier Johnville-Massawippi deposits). In a nearby section (section D of Shilts and Smith, 1987), the Chaudiere Till is intercalated with the Johnville Till and deformed Massawippi sediments in a series of thrust plates. The attitude of the thrusted beds suggests that the stacking was produced by basal drag or by freezing on and subsequent upshearing of subglacial deposits at a time when the Chaudiere glacier (Laurentide phase) was flowing from 005°, a direction represented by several fabrics measured in the Chaudiere Till at section B (Fig. 4).

In the Saint-Francois River watershed, directional and compositional data recorded by McDonald (1967) and Parent (1987a, 1987b) suggest that the Chaudiere Till records a regional episode of southwestward to westward ice flow. At the Ascot River (Fig. 5), field observations of clast lithology have revealed that distinctive granodiorite and granite clasts, presumably derived from Devonian stocks located northeast of the Ascot River section (Fig. 6) occur in low abundance throughout the Chaudiere Till, but are absent in the Lennoxville Till (McDonald, 1967; Parent, 1987a, 1987b). The Chaudiere Till contains very rare Precambrian clasts, and the few ultramafic clasts (two) that were found near the top of the unit may well be derived from small ultramafic bodies located east of Weedon (Fig. 6), hence northeast of the section. By comparison, the Lennoxville Till consistently contains ultramafic and Precambrian clasts, though in low abundance (1%). Striations measured on the top surface of boulders occurring in the basal bouldery zone of the Chaudiere Till, clast fabrics and subtill thrust faulting also indicate that ice that deposited the lower member of the Chaudiere Till was flowing westward to southwestward (Parent, 1987a, 1987b).

Evidence for southwestward ice flow some time prior to the subsequent (Lennoxville) ice advance comes also from the presence, at the surface, of Devonian granite erratics displaced southwest of their known outcrop (McDonald, 1967; McDonald and Shilts, 1971). They occur as scattered surface boulders at a number of localities outside of prominent late Wisconsinan southeast-trending Lennoxville dispersal trains, documented by

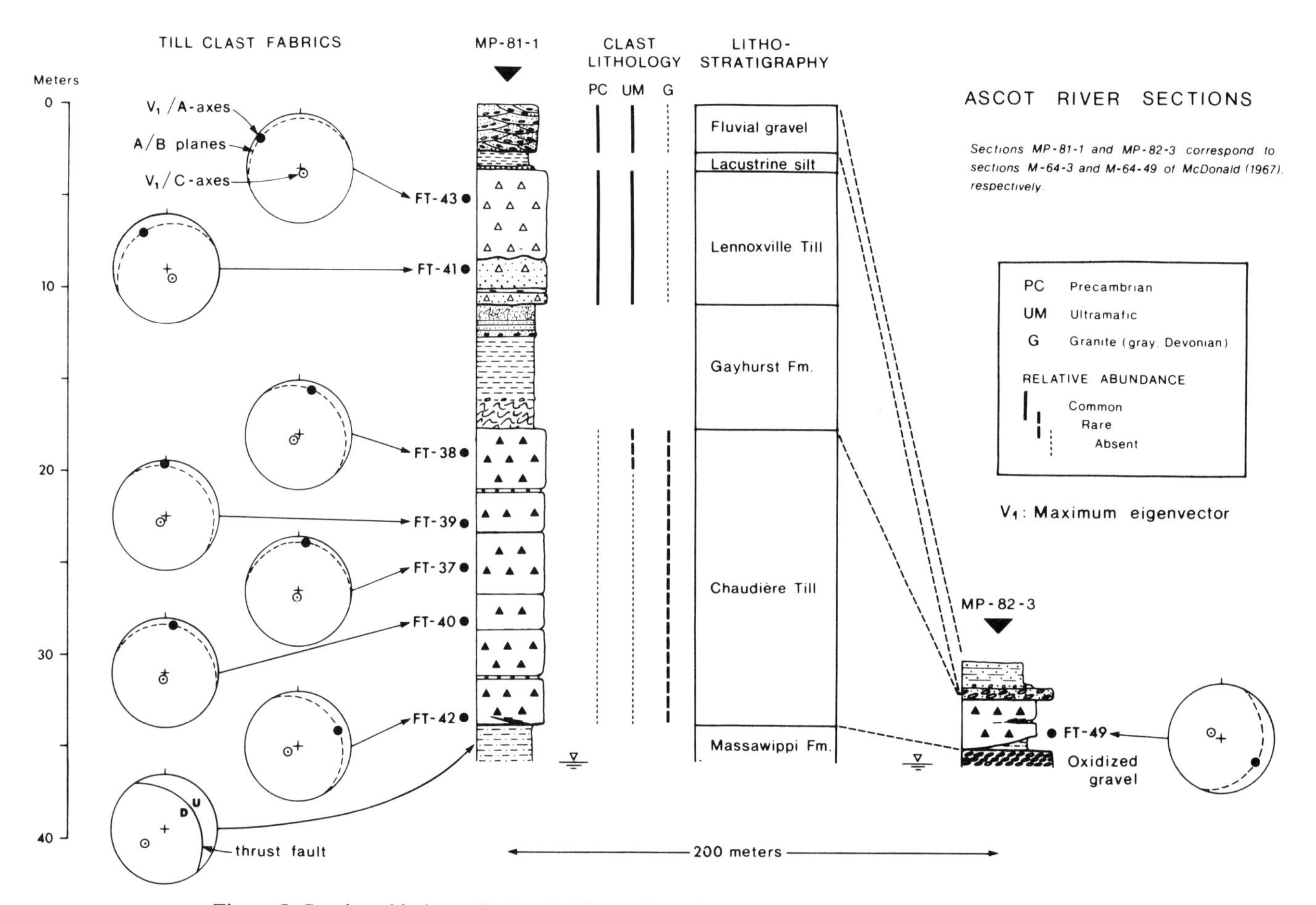

Figure 5. Stratigraphic logs, till clast fabrics, and glaciotectonic deformations, Ascot River sections (modified from Parent, 1987a, 1987b).

Shilts (1973) and Parent (1987b). Additional localities north and northwest of the Stoke Mountains were found by Parent (1987b). These along with localities previously reported by McDonald (1967) are plotted in Figure 6. Also shown is the area of occurrence of erratics of pillow-lava that were presumably transported southwest of their known sources (Elson, 1987). The areal distribution of these occurrences discloses a fairly distinct, although residual, dispersal pattern extending southwestward from the Devonian granitic stocks. This provides further evidence of southwestward glacial transport over distances exceeding 40 km during the Chaudiere episode. These residual trains also indicate that the subsequent advance of Laurentide ice that deposited the Lennoxville till failed to fully reentrain glacial debris that had been transported westward during the Chaudiere glaciation. The above observations thus provide strong support for a westward to southwestward sequence of shifting ice flows during at least part of the Chaudiere glacial phase.

The end of the Chaudiere glaciation was marked by a short-lived retreat of Appalachian and Laurentide ice to the Appalachian front. Glacier ice in the St. Lawrence Valley impounded

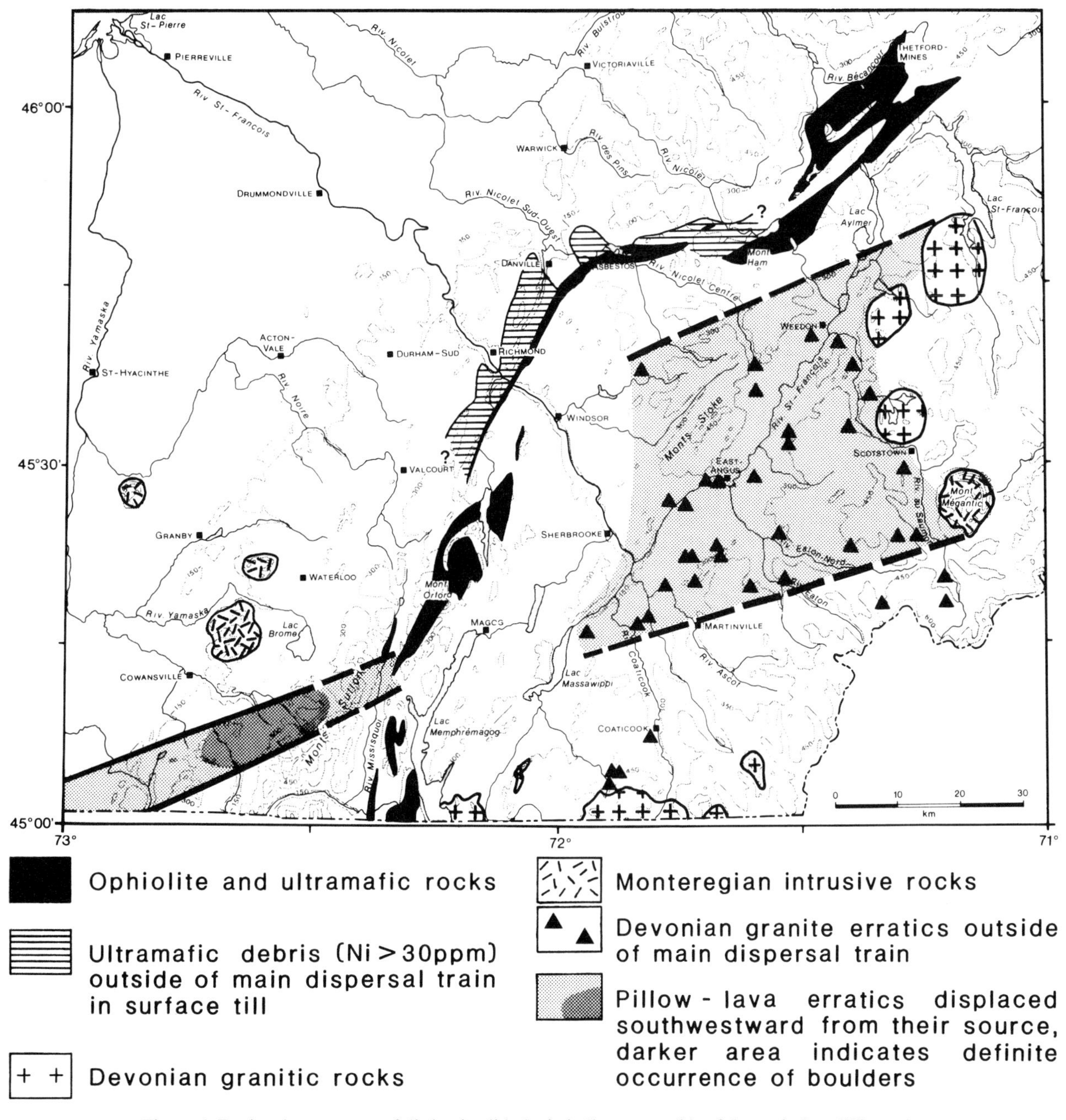

Figure 6. Regional occurrence of distinctive lithologic indicators outside of the main Late Wisconsinan southeastward dispersal trains in the Appalachians of southern Quebec (modified from Parent, 1987a).

glaciolacustrine waters in the northward-flowing Chaudiere and St. Francois valleys, resulting in deposition of locally thick glaciolacustrine and deltaic sediments of the Gayhurst Formation. Only a rather unsatisfactory radiocarbon age of >20,000 yr B.P. (GSC-1137; McDonald and Shilts, 1971) has been obtained thus far from disseminated organic debris in the generaly unfossiliferous sediments of the Gayhurst Formation, which is thought to be Middle Wisconsinan in age. The last major glacial advance resulted in the deposition of the Lennoxville Till, the surface till of the region.

DISCUSSION

The Pleistocene lithostratigraphic frameworks of both the St. Lawrence Lowland and the Appalachians of southern Quebec comprise three tills. In the lowlands, fluvial sediment facies (the Lotbiniere Sand and St. Pierre Sediments) represent unimpeded drainage in the St. Lawrence Valley and separate three till units. In the Appalachians, only sediment of the Massawippi Formation was deposited under normal drainage conditions. The Gayhurst Formation, represented by clay-rich laminated sediments more than 100 m thick in places, records a short-lived glaciolacustrine event during which ice must have been standing in the St. Lawrence Lowland in order to dam high-level glacial Lake Gayhurst. Problems in correlating the stratigraphic sequences of the two areas were discussed by McDonald (1971), Gadd and others (1972), Gadd (1976), and LaSalle (1984).

The Becancour and Johnville tills have long been correlated on the basis of similar stratigraphic position, and recently collected data provide further support for this correlation. It now seems likely that the Johnville and Becancour tills are remnants of more extensive Illinoian till sheets, particularly because interglacial sediments overlie till at Pointe Fortune, 80 km west of Montreal (Fig. 1; Anderson and others, 1990). These glacial sediments were probably deposited some time during stage 6 of the oxygen-isotope chronology (Shackleton and Opdyke, 1973).

The St. Pierre Sediments were correlated with the Massawippi Formation on the basis of their lithofacies and their similar fossil fauna and floral contents (McDonald and Shilts, 1971; Gadd and others, 1972; Gadd, 1976; Shilts, 1981; Matthews, 1987). Evidence presented by Lamothe (1989) that organic sediments formerly assigned to the "St. Pierre Interval" actually represent two distinct nonglacial units, the Lotbiniere Sand and St. Pierre Sediments, together with the possibility that the Massawippi Formation may be partly interglacial, calls for a reevaluation of the St. Pierre–Massawippi correlation. Two of several possible correlation schemes are discussed below. They are illustrated in Figure 7 as hypotheses A and B.

Hypothesis A

On the basis of ^{14}C and TL ages, and in the absence of warm climate indicators in the central part of the lowland, it can be assumed the Lotbiniere Sand and St. Pierre Sediments were deposited during isotope stage 5a and stage 3, respectively (Fig. 3). Because part of the Massawippi Formation may represent the end of the last interglacial, correlation between the cool facies of the Massawippi Formation and the Lotbiniere Sand is possible. The Lotbiniere Sand is overlain by sediments of the Nicolet Stade (the Deschaillons Varves and Levrard Till). In the Appalachians, the Massawippi Formation is overlain by the Chaudiere Till, a glacial unit deposited in part from local ("Appalachian") ice masses that displaced distinctive erratics over a distance of at least 40 km toward southwest. In this hypothesis, the Levrard and Chaudiere tills are correlated (Fig. 7).

Available geochronological data for sediments underlying (Deschaillons Varves) and overlying (St. Pierre Sediments) the Levrard Till, the volume of continental ice required to explain the presence of glacier ice in the Central St. Lawrence Lowland, and assignment of part of the Massawippi Formation to the last interglacial, indicate that the Levrard Till was deposited during marine oxygen-isotope stage 4. The glacial advance of the Nicolet Stade and the early Wisconsinan growth of Appalachian ice along the continental margin of northeastern North America are in accord with open sea-surface conditions in the North Atlantic, storm-track patterns, and continental ice volumes estimated at 75% of the stage 2 glacial maximum at the 5/4 boundary (Ruddiman and others, 1980). This large increase of global ice volume recorded at the stages 5/4 transition in the deep-sea oxygen-isotope stratigraphy from the North Atlantic (Ruddiman and others, 1980) further supports the hypothesis that sediments of the Nicolet Stade were deposited during stage 4 rather than during an earlier substage of isotope stage 5. In this correlation scheme, the beginning of the last glaciation in early Wisconsinan time is characterized by the coalescence of Appalachian-based ice flowing toward southwest and Laurentide ice channelled up the St. Lawrence River while damming a proglacial lake in the lowland (Fig. 8).

This paleogeographic reconstruction implies that there was complete deglaciation of the Appalachians some time between the deposition of the Chaudiere Till and the glacial advance that deposited the Gayhurst Formation and the Lennoxville Till. However, there is as yet little evidence to suggest such a significant deglaciation has occurred between the onset of the Chaudiere phase and the end of the Lennoxville phase (Shilts, 1981). There is no apparent break in deposition (i.e., no oxidation of Gayhurst or Chaudiere sediments, no fluvial sediments, no organic sediments) or any other evidence that the Chaudiere or St. Francois valleys were open to the St. Lawrence from the onset of the Wisconsinan until late glacial time. Therefore, another scheme of correlation is presented as an alternative to hypothesis A.

Hypothesis B

In hypothesis B (Fig. 7), the Chaudiere-Gayhurst-Lennoxville sequence is correlated with the Gentilly Till, and the Lotbiniere–Deschaillons–Levrard–St. Pierre units are correlated with the Massawippi Formation. The nonglacial Massawippi Formation

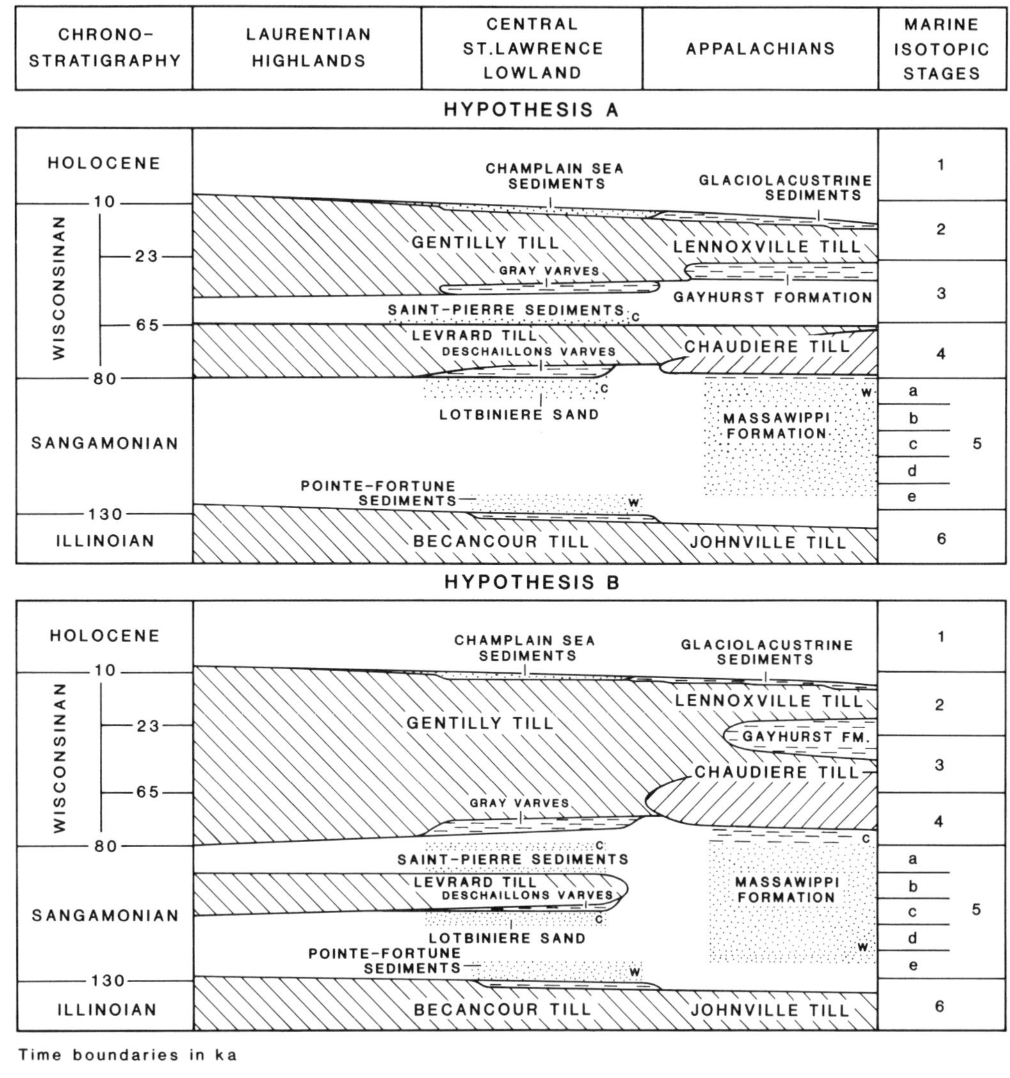

Time boundaries in ka

Till deposited by Laurentide ice

Till deposited by Appalachian ice

Waterlaid sediments deposited during periods of free drainage in the St.Lawrence Lowland
c: cool climate indicators
w: warm climate indicators

Glaciolacustrine sediments deposited during periods of blocked drainage in the St.Lawrence Lowland or the Appalachians

Figure 7. Two hypothetical time-space correlations between the Late Pleistocene depositional records of the St. Lawrence Lowland and the Appalachians of southern Quebec and the marine isotope record (Shackleton, 1969).

may therefore represent the transition from the Sangamonian (i.e., stage 5 of Shackleton and Opdyke, 1973) to the beginning of the early Wisconsinan (stage 4 of Shackleton and Opdyke, 1973). The glacial advance during which the Deschaillons Varves and the Levrard Till were deposited in the Central St. Lawrence Lowland would then be recorded in the Appalachians by the arctic floral and faunal remains that were discovered by Matthews and others (1987) in the upper part of the Massawippi Formation. In this hypothesis, the Nicolet Stade is correlated with marine isotope stages 5b and 5c of Shackleton (1969). According to this correlation, the ^{14}C, U-Th, and TL ages would be too young. Indeed, underestimation of TL ages has been documented by Debenham (1985) for >50 ka sediments of northwestern Europe. This TL misbehavior is being investigated, but it may be the result of some technical misconceptions (Balescu and Lamothe, 1992).

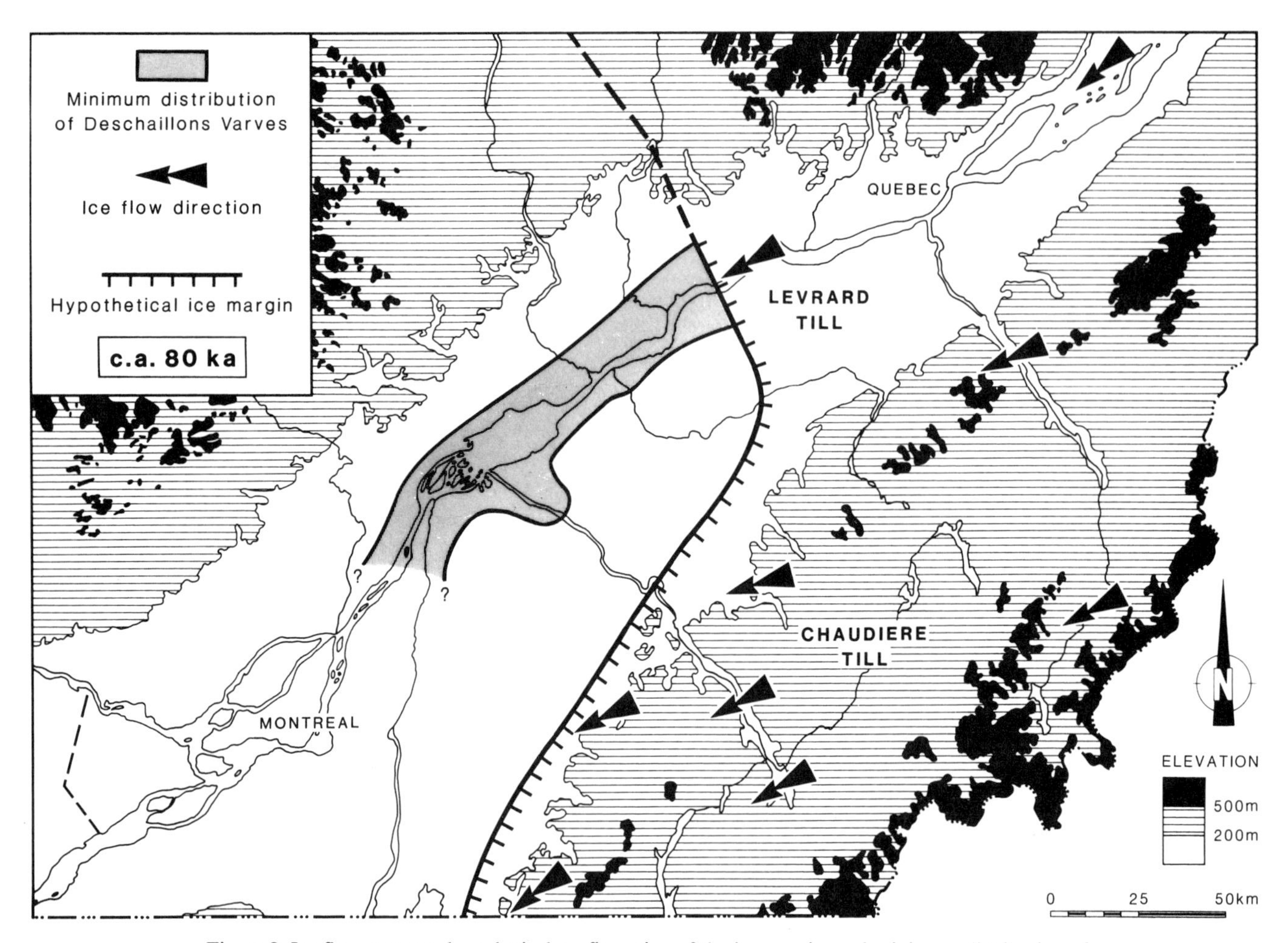

Figure 8. Ice-flow patterns, hypothetical configuration of the ice margin, and minimum distribution of early Wisconsinan glaciolacustrine sediments (Deschaillons Varves) in southern Quebec during an early phase of the last glaciation, ca. 80 ka, for hypothesis A. Ice-flow directions for the Appalachians are deduced from till fabric and composition. Those for the St. Lawrence Lowland are inferred from the necessity of blocking drainage in the St. Lawrence Valley.

CONCLUSION

The growth of Appalachian ice along the North Atlantic continental margin and its subsequent coalescence with Laurentide ice at the onset of the last glaciation (stage 5/4 boundary) represents a unique event documented in the regional terrestrial stratigraphy. In the earliest part of the last glaciation, Maritime ice would have reached southern Quebec prior to coalescence with the main continental ice sheet, as suggested by the regional depositional and erosional record.

In the St. Lawrence Lowland, a large ice-dammed lake, Lake Deschaillons, formed in early Wisconsinan time. The existence and duration of glacial Lake Deschaillons is associated with the Laurentide Ice Sheet, which was channelled up the St. Lawrence River. In the Appalachian Uplands, assuming that the Massawippi Formation was deposited at the end of the Sangamonian Interglacial, available compositional and fabric data for the overlying Chaudiere Till support the concept that the growth of an Appalachian-based ice cap took place at the onset of the last glaciation.

In conclusion, the timing of events that took place in southern Quebec during the last interglaciation-glaciation transition remains largely unresolved. These events will be clarified only by further geochronological investigation and more detailed paleoecological studies focused on the Late Sangamonian–early Wisconsinan time period. We hope that our confidence in the various dating techniques that extend beyond the range of ^{14}C dating will grow so that we may secure reliable numerical ages for units predating the last glacial maximum.

ACKNOWLEDGMENTS

This review paper is the result of many years of investigation carried out by the authors in the course of their doctoral studies and later, as research scientists with the Geological Survey of

Canada. The authors thank A. Dreimanis (University of Western Ontario), B. C. McDonald, and E. H. Muller (Syracuse University) for their moral, academic, and financial support. Parent and Lamothe were supported by NSERC Grant A4215; research conducted by Shilts was funded under Geological Survey of Canada Projects 660055 and 690095. We also thank M. Laithier who drafted the figures. The original manuscript was greatly improved by R. J. Fulton and an anonymous reviewer.

REFERENCES CITED

Anderson, T. W., Matthews, J. V., Jr., Mott, R. J., and Richard, S. H., 1990, The Sangamonian Pointe-Fortune site, Ontario-Quebec border: Géographie physique et Quaternaire, v. 44, p. 271–287.

Balescu, S., and Lamothe, M., 1992, The blue emission of K-feldspar coarse grains and its potential for overcoming TL age underestimation: Quaternary Science Reviews, v. 11, p. 45–51.

Bostock, H. S., 1970, Physiographic regions of Canada: Geological Survey of Canada Map 1254A, scale 1:5 000 000.

Debenham, N. C., 1985, Use of UV emissions in TL dating of sediments: Nuclear Tracks and Radiation Measurements, v. 10, p. 717–724.

Dreimanis, A., and Karrow, P. F., 1972, Glacial history of Great Lakes–St. Lawrence region, the classification of the Wisconsin(an) stage and its correlatives: International Geological Congress, 24th, Montreal, Report Section 12, p. 5–15.

Elson, J. A., 1987, West-southwest glacial dispersal of pillow-lava boulders, Phillipsburg-Sutton region, Eastern Townships, Quebec: Canadian Journal of Earth Sciences, v. 24, p. 985–991.

Ferland, P., and Occhietti, S., 1990, L'Argile de la Pérade: Nouvelle unité marine antérieure au Wisconsinien supérieur, vallée du Saint-Laurent, Québec: Géographie Physique et Quaternaire, v. 44, p. 159–172.

Fulton, R. J., 1984, Summary: Quaternary stratigraphy in Canada, *in* Fulton, R. J., ed., Quaternary stratigraphy of Canada—A Canadian contribution to IGCP Project 24: Geological Survey of Canada Paper 84-10, p. 1–5.

Gadd, N. R., 1955, Pleistocene geology of the Becancour map-area, Quebec [Ph.D. thesis]: Urbana, University of Illinois, 191 p.

——, 1971, Pleistocene geology of the St. Lawrence Lowland with selected passage from an unpublished manuscript: *The St. Lawrence Lowland*, by J. W. Goldthwaith: Geological Survey of Canada Memoir 359, 153 p.

——, 1976, Quaternary stratigraphy in southern Quebec, *in* Mahaney, W. C., ed., Quaternary stratigraphy of North America: Stroudsburg, Dowden, Hutchinson and Ross, p. 37–50.

Gadd, N. R., Lasalle, P., Dionne, J. C., Shilts, W. W., and McDonald, B. C., 1972, Quaternary geology and geomorphology of southern Quebec: International Geological Congress, 24th Montreal Guidebook, Field Excursion A44-C44, 70 p.

Hillaire-Marcel, C., and Causse, C., 1989, Chronologie Th/U des concrétions calcaires des varves du lac glaciaire de Deschaillons (Wisconsinien Inférieur): Canadian Journal of Earth Sciences, v. 26, p. 1041–1052.

Hillaire-Marcel, C., and Pagé, P., 1981, Paléotemperatures isotopiques du Lac glaciaire de Deschaillons, *in* Mahaney, W. C., ed., Quaternary paleoclimate: Norwich, GeoAbstracts, p. 273–298.

Karrow, P. F., 1957, Pleistocene geology of Grondines map-area, Quebec [Ph.D. thesis]: Urbana, University of Illinois, 97 p.

Lamothe, M., 1984, Apparent thermoluminescence age of St. Pierre sediments at Pierreville, Quebec, and the problem of anomalous fading: Canadian Journal of Earth Sciences, v. 21, p. 1406–1409.

——, 1985, Lithostratigraphy and geochronology of the Quaternary deposits of the Pierreville and St. Pierre les Becquets areas, Quebec [Ph.D. thesis]: London, University of Western Ontario, 227 p.

——, ed., 1987, Pleistocene stratigraphy in the St. Lawrence Lowland and the Appalachians of southern Quebec; a field guide, *in* Delisle, C. E., and Bouchard, M. A., eds., Collection Environnement et Géologie, Volume 4: Montréal, Université de Montréal, 280 p.

——, 1989, A new framework for the Pleistocene stratigraphy of the Central St. Lawrence Lowland, southern Quebec: Géographie Physique et Quaternaire, v. 43, p. 119–129.

Lamothe, M., and Huntley, D. J., 1988, Thermoluminescence dating of late Pleistocene sediments, St. Lawrence Lowland, Quebec: Géographie Physique et Quaternaire, v. 42, p. 33–44.

Lamothe, M., Hillaire-Marcel, C., and Pagé, P., 1983, Découverte de concrétions calcaires striées dans le Till de Gentilly, basses terres du Saint-Laurent, Québec: Canadian Journal of Earth Sciences, v. 20, p. 500–505.

Lasalle, P., 1984, Quaternary stratigraphy of Quebec: A review, *in* Fulton, R. J., ed., Quaternary stratigraphy of Canada—A Canadian contribution to IGCP Project 24: Geological Survey of Canada Paper 84-10, p. 155–171.

Matthews, J. V., Jr., 1987, Macrofossils of insects and plants from southern Quebec, *in* Lamothe, M., ed., Pleistocene stratigraphy in the St. Lawrence lowland and the Appalachians of southern Quebec: A field guide: Montréal Université de Montréal, p. 166–181.

Matthews, J. V., Jr., Smith, S. L., and Mott, R. J., 1987, Plant macrofossils, pollen and insect fossils of arctic affinity from Wisconsinan sediments in Chaudiere Valley, southern Quebec: Geological Survey of Canada Paper 87-1A, p. 165–175.

McDonald, B. C., 1967, Pleistocene events and chronology in the Appalachian region of southeastern Quebec, Canada [Ph.D. thesis]: New Haven, Connecticut, Yale University, 161 p.

——, 1971, Late Quaternary stratigraphy and deglaciation in eastern Canada, *in* Turekian, K. K., ed., The late Cenozoic glacial ages: New Haven, Connecticut, Yale University Press, p. 331–353.

McDonald, B. C., and Shilts, W. W., 1971, Quaternary stratigraphy and events in southeastern Quebec: Geological Society of America Bulletin, v. 82, p. 683–698.

Occhietti, S., 1982, Synthèse lithostratigraphique et paléoenvironments du Quaternaire au Québec méridional: Géographie Physique et Quaternaire, v. 36, p. 15–49.

Parent, M., 1987a, Pleistocene stratigraphy in the Appalachians of southern Quebec: The Asbestos-Valcourt and Sherbrooke area, *in* Lamothe, M., ed., Pleistocene stratigraphy in the St. Lawrence Lowland and the Appalachians of southern Quebec: A field guide: Montréal, Université de Montréal, p. 102–139.

——, 1987b, Late Pleistocene stratigraphy and events in the Asbestos-Valcourt region, southeastern Quebec [Ph.D. thesis]: London, University of Western Ontario, 320 p.

Ruddiman, W. F., McIntyre, A., Niebler-Hunt, V., and Durazzi, J. T., 1980, Oceanic evidence for the mechanism of rapid Northern Hemisphere glaciation: Quaternary Research, v. 13, p. 33–64.

Shackleton, N. J., 1969, The last interglacial in the marine and terrestrial record: Royal Society of London Proceedings, v. B174, p. 135–154.

Shackleton, N. J., and Opdyke, N. D., 1975, Oxygen isotope and paleomagnetic stratigraphy of equatorial Pacific core V28-238: Oxygen isotope temperatures and ice volumes on a 10^5 and 10^6 year scale: Quaternary Research, v. 3, p. 39–55.

Shilts, W. W., 1970, Pleistocene geology of the Lac Megantic region, southeastern Quebec, Canada [Ph.D. thesis]: New York, Syracuse University, 154 p.

——, 1973, Glacial dispersal of rocks, minerals, and trace elements in Wisconsinan till, southeastern Quebec, Canada, *in* Black, R. F., Goldthwait, R. P., and Willman, H. B., eds., The Wisconsinan Stage: Geological Society of America Memoir 136, p. 189–219.

——, 1981, Surficial geology of the Lac Megantic area, Quebec: Geological Survey of Canada Memoir 397, 102 p.

Shilts, W. W., and Blais, A., 1989, Field guide–St. Joseph de Beauce region, *in* LaSalle, P., ed., 52nd Annual reunion, Friends of the Pleistocene, , Québec: Ministère des Transports du Québec, p. 3–26.

Shilts, W. W., and Smith, S. L., 1987, Pleistocene stratigraphy in the Appalachians of southern Quebec: The Chaudiere Valley, *in* Lamothe, M., ed., Pleistocene stratigraphy in the St. Lawrence lowland and the Appalachians of southern Quebec: A field guide: Montréal, Université de Montréal, p. 72–101.

Stuiver, M., Heusser, C. J., and Yan, I. C., 1978, North American glacial history extended to 75000 years ago: Science, v. 200, p. 16–21.

Terasmae, J., 1958, Contributions to Canadian palynology. Part I: The use of palynological studies in Pleistocene stratigraphy. Part II: Non-glacial deposits in the St-Lawrence lowlands, Quebec: Geological Survey of Canada Bulletin 46, 28 p.

Veillette, J. J., and Nixon, F. M., 1984, Sequence of Quaternary sediments in the Belanger sand pit, Pointe-Fortune, Quebec-Ontario: Géographie Physique et Quaternaire, v. 38, p. 59–68.

Vogel, J. C., and Waterbolk, H. T., 1972, Groningen radiocarbon dates XL: Radiocarbon, v. 14, p. 6–110.

Wintle, A. G., and Huntley, D. J., 1980, Thermoluminescence dating of ocean sediments: Canadian Journal of Earth Sciences, v. 17, p. 348–360.

Manuscript Accepted by the Society September 6, 1991

Printed in U.S.A.

Geological Society of America
Special Paper 270
1992

The pre–late Wisconsinan chronology of Nova Scotia, Canada

R. R. Stea
Centre for Marine Geology, Dalhousie University, Department of Natural Resources, Halifax, Nova Scotia B3H 3J5, Canada
R. J. Mott
Geological Survey of Canada, 601 Booth Street, Ottawa, Ontario, K1A 0E8 Canada
D. F. Belknap
Institute of Quaternary Studies, University of Maine, Orono, Maine 04469
U. Radtke
Geography Institute, University of Dusseldorf, Dusseldorf, Federal Republic of Germany

ABSTRACT

The dating of events older than 50 ka, the limit of the radiocarbon method, has been a major drawback in assessing the chronology of the Quaternary. Several new methods have been applied to the dating of pre–late Wisconsinan organic beds in Nova Scotia. These methods include U/Th disequilibrium dating of wood and shells, amino-acid racemization dating of shells, and wood and electron spin resonance dating of shells. These methods are not without problems, and must be assessed together, and in concert with geologic evidence, in establishing a chronology.

Evidence of the penultimate interglacial (marine isotopic stage 7) has been found in southern Nova Scotia. A raised marine platform, forest beds beneath till, and glacially-resedimented marine deposits were all formed during the last interglacial or Sangamonian stage (stage 5). Middle Wisconsinan U/Th and radiocarbon dates are questionable, so the chronology of post-Sangamonian events is not well constrained. Post-Sangamonian erosional and depositional stratigraphy indicates that at least four phases of ice flow have affected the Nova Scotia region. The earliest of these flows was a major advance that crossed the Gulf of St. Lawrence and Bay of Fundy (ice-flow phase 1). Later, separate ice caps and divides formed in areas adjacent to the province and the province itself (ice-flow phases 2–4). There is evidence for ice retreat between phases 1 and 2 in offshore areas and locally on land. Nova Scotia was probably covered with ice throughout the Wisconsinan stage (marine isotopic stages 4 to 2).

INTRODUCTION

Many sections with pre–late Wisconsinan organic sediments have been found in Nova Scotia. Dates obtained from wood at these sites are generally beyond the range of the radiocarbon method. The lack of a suitable dating method is a major drawback to the establishment of a chronology beyond 40,000 yr B.P.

In Nova Scotia a relative glacial chronology has been developed through the integration of detailed mapping on the surface and facies and provenance analyses of till sheets in stratigraphic sections. Till sheets can be linked to mapped ice flows through erratic-dispersal studies, fabric, and their relations to underlying striated bedrock and boulder pavements (Stea, 1984; Stea and others, 1985, 1987). In Nova Scotia the close juxtaposition of varied tectonic terranes and detailed bedrock mapping allows for accurate assessment of ice-flow directions from erratic dispersal. Difficulties with this approach, however, arise in lithostratigraphic correlation across the terranes and physiographic regions of Nova Scotia (Fig. 1). A till deposited during one glacial flow will show compositional facies changes across bedrock-contact zones, as well as depositional variations due to ice dynamics. Striation trends associated with one flow may display variations due to topographic deflection. Several flows of differing ages may have the same trends. The relative glacial chronology is interpreted by

Stea, R. R., Mott, R. J., Belknap, D. F., and Radtke, U., 1992, The pre-late Wisconsinan chronology of Nova Scotia, Canada, *in* Clark, P. U., and Lea, P. D., eds., The Last Interglacial-Glacial Transition in North America: Boulder, Colorado, Geological Society of America Special Paper 270.

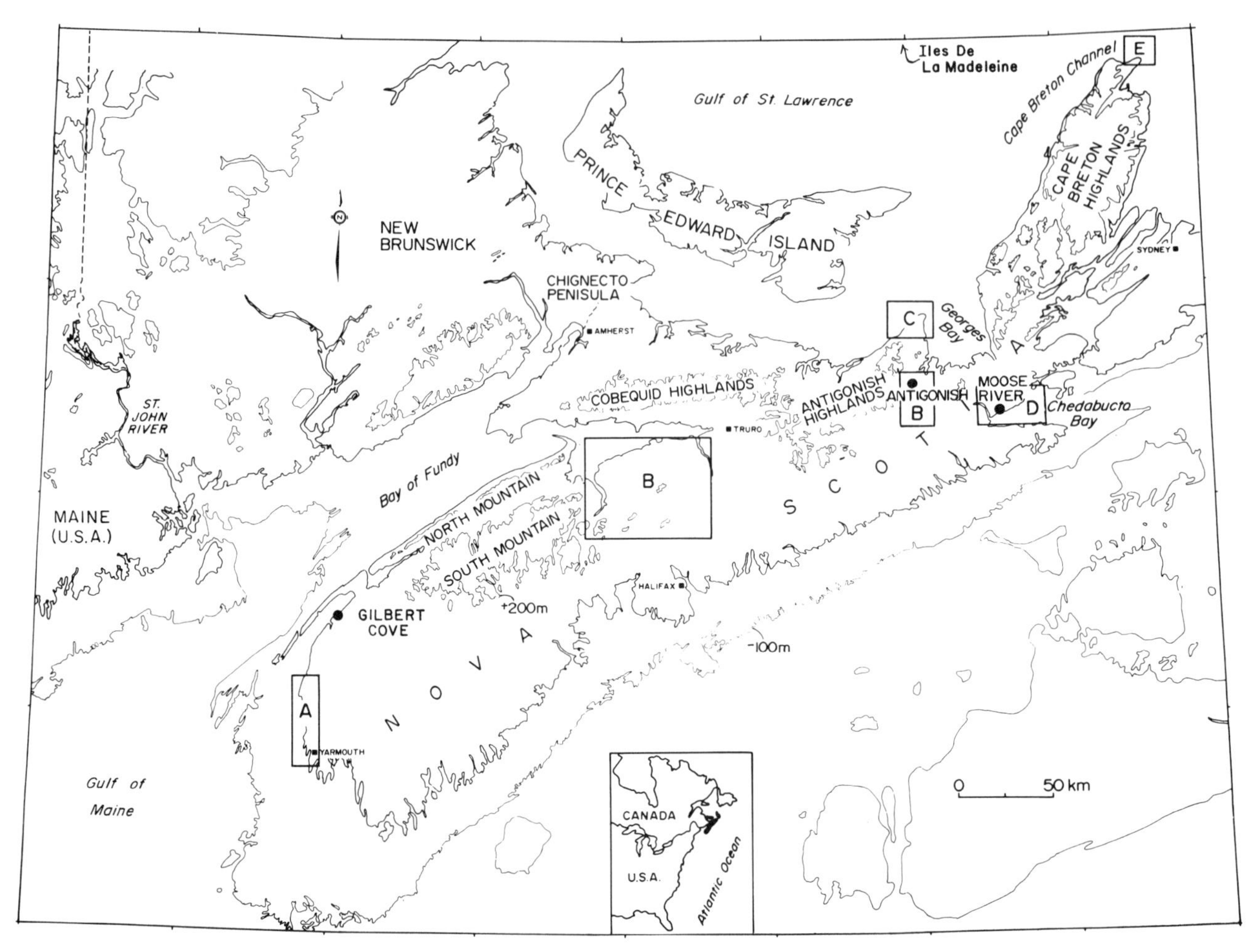

Figure 1. Nova Scotia physiography and the locations of the type stratigraphic areas and sites in the province. A = Southwestern Nova Scotia, B = Central Nova Scotia, C = Cape George sections, D = Chedabucto Bay sections, E = Bay St. Lawrence.

the detailed analysis of the till sheets at many of the type localities in the province.

Recently, several new dating methods have been tested in Nova Scotia. Vernal and others (1986) have presented a chronology based on U-series disequilibrium using wood. Grant and King (1984) summarized the results of amino-acid racemization (AAR) measurements on wood. Wehmiller and others (1988) analyzed amino acids of pre–late Wisconsinan shells and correlated these with calibrated sites along the eastern U.S. seaboard. We present additional radiocarbon and amino-acid dating results, the results of electron spin resonance (ESR), and U/Th dating of molluscs.

The purposes of this chapter are to review and to reevaluate the established relative chronology, to assess the validity of some of the regional correlations, and to present the new age data in the context of the established relative chronologies.

METHODS

Amino-acid racemization dating (AAR)

Wehmiller and others (1988) described the AAR technique used for this study, gas chromatography of total amino acids, which has been developed from previous work such as Belknap (1979) and Wehmiller and Belknap (1982). Shells of two species of mollusc and a barnacle were collected from the Cape George and Bay St. Lawrence sections (Table 1). Standard laboratory preparation procedures (Belknap, 1979; Wehmiller and Belknap, 1982) were used to clean, prepare, and analyze the samples. Using amino-acid dextro to levro stereoisomer (D/L) ratios as relative stratigraphic indicators within a limited geographic area (aminostratigraphy) assumes that nearby localities will have similar Quaternary temperature histories. In order to correlate

TABLE 1. AMINO-ACID RACEMIZATION DATA FROM SHELLS IN MAINLAND NOVA SCOTIA

Sample	Location	Lat/Long	Species	D/L LEU	Allo/ Iso	Asp/ Leu	Kinetic Model Age Assignments EQT - 0.2 °C
DFB-88-1	Cape George, N.S. Unit 4, +24 m amsl	45°53.5'N 67°54.0'W	*Mercenaria mercenaria*	0.192 ± 0.005	0.212 ± 0.010	6.7 ±	80 to 110 ka Oxygen Isotope Stage 5
DFB-88-2	Cape George, N.S. Unit 4, +24 m amsl	45°53.5'N 67°54.0'W	*Mercenaria mercenaria*	0.173 ± 0.001	0.177 ± 0.008	7.7 ± 0.2	
				0.182 ± 0.013 mean			
DFB-88-3	Bay St. Lawrence Unit 4, +18 m amsl	47°00.7'N 60°26.9'W	*Megayoldia thraciaeformis*	0.124 ± 0.004	0.131	3.9 ± 0.2	
DFB-88-4	Bay St. Lawrence Unit 4, +18 m amsl	47°00.7'N 60°26.9'W	*Megayoldia thraciaeformis*	0.171 ± 0.001	0.182 ± 0.016	2.7 ± 0.4	50 to 100 ka?
DFB-88-5	Bay St. Lawrence Unit 4, +18 m amsl	47°00.7'N 60°26.9'W	*Balanus balanus*	0.174 ± 0.009	0.219 ± 0.008	21.6 ± 6	
				0.148 ± 0.003 mean			

amino-acid ratios over a larger distance, it is necessary to formulate models of variation of D/L ratios with temperature as well as age. This approach, known as kinetic-age modeling, requires a model of effective Quaternary temperature (EQT) based on current mean annual temperature (CMAT) and expected variations over time. This model is uncertain, but can be delimited by two sets of data: the CLIMAP Project Members (1976) model of sea-surface temperatures 18,000 yr B.P. and the Shackleton and Opdyke (1973) oxygen-isotope curve. The latter supplies the shape of the climatic-change curve, and the former supplies the minimum temperature at glacial maximum. Because temperature effects increase exponentially at about 16.5%/°C, the resultant integrated effective temperature is not a simple average. In order to determine the age of a fossil, the D/L ratio must be compared to a calibrated kinetic model using the range of EQT models. Wehmiller and others (1988) presented the latest refinement to the kinetic age model. This *Mercenaria*-leucine model is based on U-series and radiocarbon calibration at more than 23 locations along the U.S. east coast; other localities and different genera are also used.

Electron spin resonance dating (ESR)

The principles of ESR dating were described in several publications (Henning and Grun, 1983; Ikeya, 1985). The methodology was described in Grün (1989). The ESR "signal" is a manifestation of the numbers of unpaired, free electrons in the sample created by the input of energy from radioactive decay. The intensity of the signal increases with age. The age is determined by the formula:

$$\text{Age (year)} = \frac{\text{accumulated dose (gy)}}{\text{annual dose (mgy/yr)}} .$$

The annual dose, produced by the radioactive elements in the sample and its surroundings, was determined by analyses of U, Th, and K in the sample and the surroundings. Problems with ESR dating of molluscs have been outlined by Katzenberger and others (1988). The main problem is in the identification of the dating "signal" from other interferences. Dates older than 150 ka should be considered minimum ages. Mollusc ESR dates are used in this study in concert with other independent dating techniques and with geologic inference.

EROSIONAL STRATIGRAPHY

Ice-flow phases

A relative ice-flow chronology has emerged in Nova Scotia through the mapping of striations and other glacial landforms (Stea and Fowler, 1979; Stea and others, 1985; Stea and Finck, 1988; Grant, 1988; Stea and Myers, 1990). Ice-flow trends can be traced over broad regions of Nova Scotia using striations (Stea and Finck, 1984). In addition, crosscutting relations and the preservation of older, weathered striations on lee-side surfaces has allowed the development of an erosional stratigraphy. Events defined by discrete, regionally mappable trends of striations are called ice-flow phases. The patterns of ice flow mapped by striations are verified by till fabric, dispersal studies, and the orientation of glacial landforms such as eskers and drumlins (Stea, 1984; Stea and others, 1989). The sequence of ice-flow phases has been discerned from superimposed striation sites and through correlation with stacked till sheets of known provenance (Stea, 1984).

Each of the ice-flow phases produced at least one recognizable till sheet, sometimes with lodgement and melt-out facies.

Phase 1. Striation patterns, distinctive erratics, till fabric, and striated boulder pavements suggest that the earliest and most extensive ice flows in Nova Scotia were eastward and southeastward (Fig. 2). These have been designated phases 1a and 1b. Several widely spaced striation sites reveal evidence of a distinct eastward flow, later overrun by southeastward-trending ice. In fact, the eastward ice flow may represent a separate, older phase of glaciation. Erratic trains of igneous rocks from the Cobequid highlands and trains of basaltic rocks from North Mountain are oriented southeastward and can be traced to the Atlantic Coast,

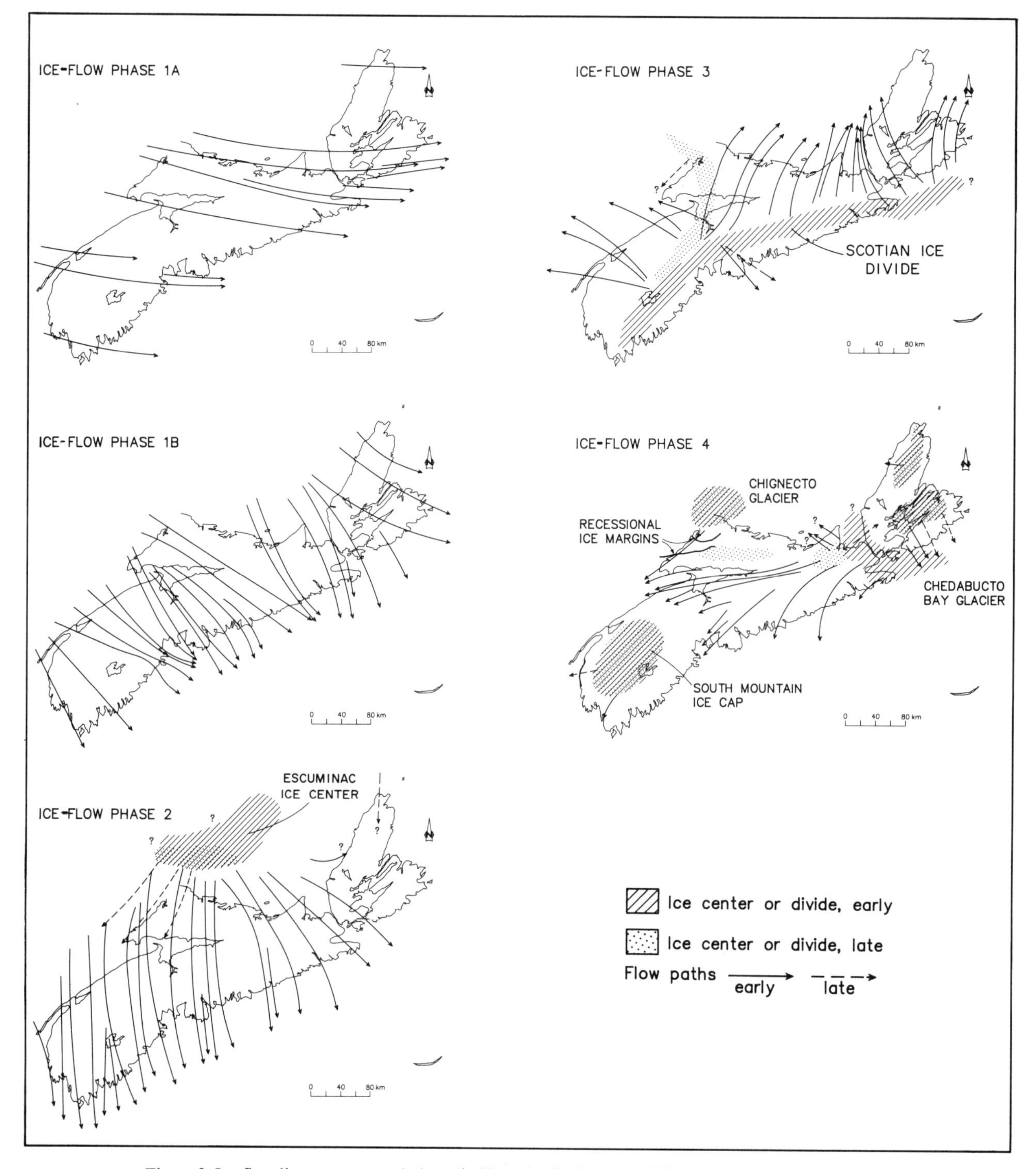

Figure 2. Ice-flow lines, centers, and phases in Nova Scotia during the Wisconsinan stage (Cape Breton phases after Grant, 1988).

up to 120 km down ice (Fig. 3; Nielsen, 1976). This phase may represent the main Laurentide ice flow. Evidence of its passage across New Brunswick, however, is equivocal (Rampton and others, 1984; Rappol, 1989; Pronk and others, 1989). An anorthosite boulder in western Prince Edward Island, derived from the Canadian Shield, suggests that Laurentide ice did cross the region at some time (Prest and Nielsen, 1987).

Phase 2. The second major ice-flow event was southward and southwestward from the Escuminac ice center in the Prince Edward Island region (Prest and others, 1972; Rampton and others, 1984; Fig. 2). This flow phase is analogous to the Acadian Bay lobe of Goldthwait (1924) and the "Fundian" glacier of Shepard (1930). Goldthwait (1924) envisioned southward flow from a Laurentide source across the Gulf of St. Lawrence. Ice-flow trends in Prince Edward Island (Prest, 1973) and adjacent New Brunswick (Rampton and others, 1984) do not reflect a pervasive southward flow, but suggest radial flow from a local center. This event is recorded by southward striae crossing earlier southeastward-trending striae at many localities on the upland regions of Nova Scotia and New Brunswick. Red-beds from northern mainland Nova Scotia and the Carboniferous basins in the Prince Edward Island region were transported southward onto the metamorphic and igneous Cobequid and Meguma terranes of mainland Nova Scotia. Dispersal of distinctive Cobequid highland erratics was also southward (Grant, 1963). Schnikter (1987) reported evidence of southward dispersal of red clastic material from the Bay of Fundy into the Gulf of Maine. Ice flow from the Escuminac ice center appears to swing to the southeast in the eastern region of Nova Scotia and Cape Breton Island. Here, the evidence for this ice flow becomes difficult to distinguish from earlier or later ice flows. M. Parent and H. Josenhans (1990, personal commun.) report northward flow from the Escuminac ice center on the Iles de la Madeleine (Magdalen Islands) and in the Gulf of St. Lawrence. Vigorous northward flow during the ensuing phase 3 appears to have eliminated the depositional evidence for phase 2 in the coastal lowlands. This ice sheet may not have overridden the Cape Breton highlands.

Phase 3. During this event ice flow was northward and southward from a divide (Scotian ice divide) across the axis of the Nova Scotia peninsula. Granitic debris from the South Mountain batholith was transported northward onto the North Mountain basalt cuesta (Figs. 1 and 2; MacNeill, 1951; Hickox, 1962). Erratics from the Cobequid highlands can be found throughout the Carboniferous lowlands to the north (Fig. 4; Stea and Finck, 1984). Northward-trending striations can be traced across the northern mainland of Nova Scotia (Fig. 2). This well-documented northward ice flow was clearly in response to the development of an ice divide in southern Nova Scotia (Fig. 2). This divide may have formed as a result of marine incursion into the Bay of Fundy (Prest and Grant, 1969) or increased precipitation after recession of the Escuminac ice cap (MacNeill and Purdy, 1951; Hickox, 1962). Northward ice flow from the Scotian ice divide was probably synchronous with the ice dome off Cape Breton Island proposed by Grant (1977). The divide can be

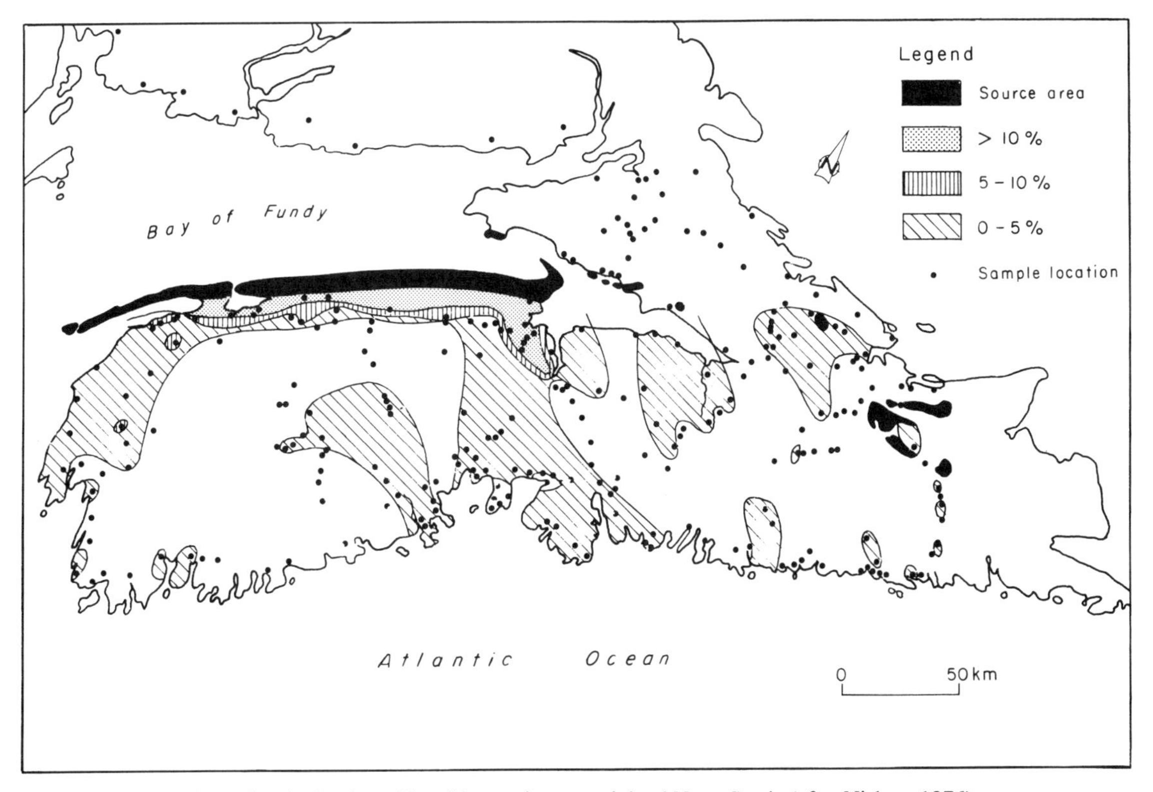

Figure 3. Distribution of basaltic erratics on mainland Nova Scotia (after Nielsen, 1976).

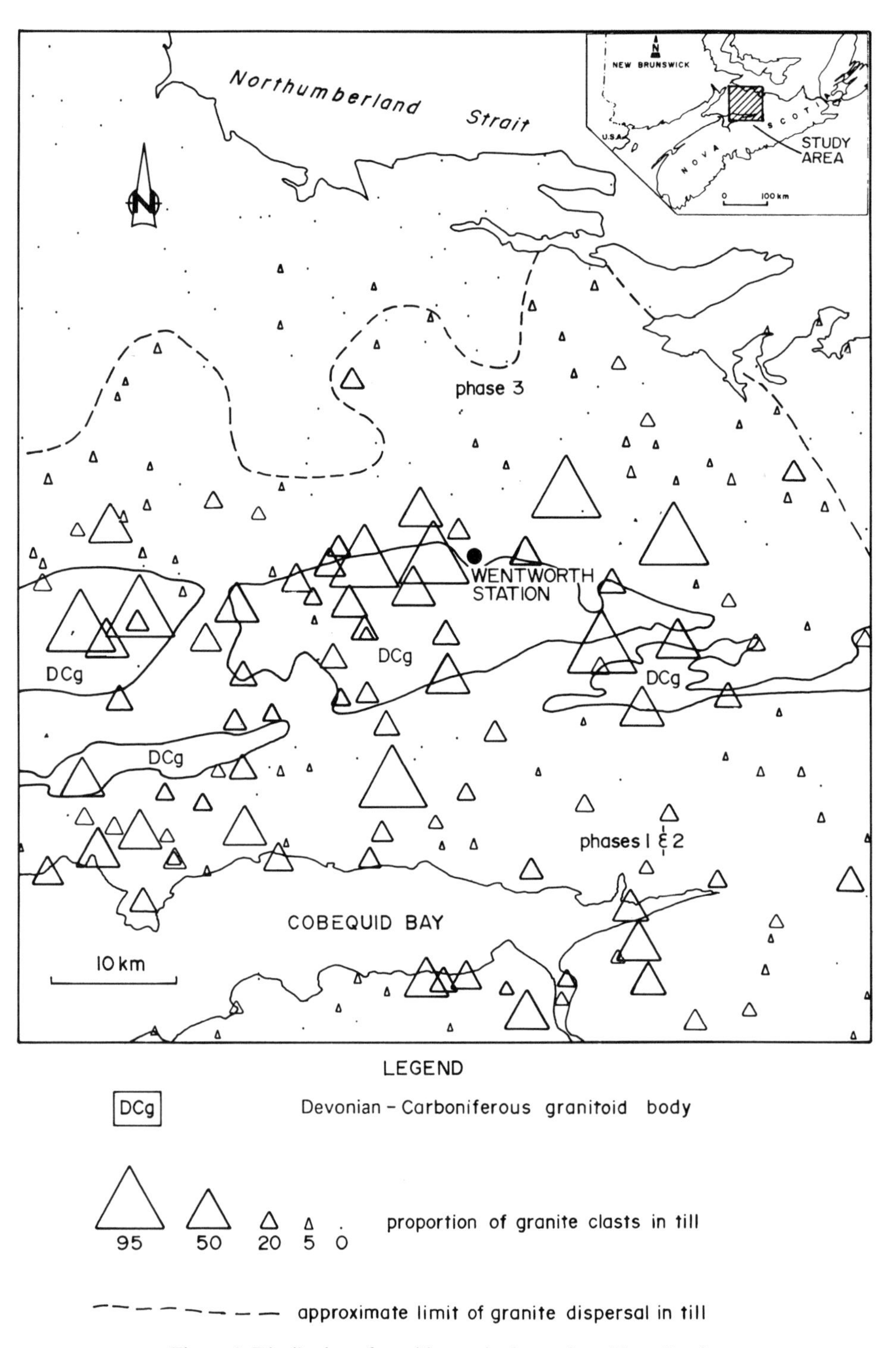

Figure 4. Distribution of granitic erratics in northern Nova Scotia.

traced from areas south of the Antigonish highlands (Myers and Stea, 1986) offshore into Chedabucto Bay. Here, ice flow was funneled out of Georges Bay through the Cape Breton channel and into the Cabot Strait. Northward ice flow was less intense in the western Cobequid highlands. As marine drawdown began to take effect, the ice may have been increasingly directed into the Bay of Fundy merging with southwestward ice streams from New Brunswick.

Phase 4. During this final phase, remnant ice caps developed from the Scotian ice divide (Fig. 2). Eskers and striations cut across features formed by earlier ice flows. Ice caps or glaciers that formed over the Chignecto Peninsula and southern Nova

Scotia had recessional margins on land marked by hummocky moraine, glaciofluvial deposits, and the pinch-out of till sheets. Ice flow during this last phase was strongly funneled westward into the Bay of Fundy and eastward into the Atlantic Ocean. Erosional features and deposits relating to these late-glacial ice caps are restricted to low-lying areas.

DEPOSITIONAL STRATIGRAPHY AND CHRONOLOGY

In this section we describe five stratigraphic type areas in mainland Nova Scotia that feature till sheets interbedded with nonglacial deposits with dateable organic beds. In some areas studies of pebble lithology and fabric have allowed correlation of till units with the erosional stratigraphy discussed above.

Southwestern Nova Scotia

Much of the stratigraphic information in this section is from Grant (1976, 1980a) and Stea and Grant (1982). Figure 5 is a compilation of the stratigraphy of the Yarmouth-Digby coast, where the best sections are exposed.

A prominent Quaternary feature in southwestern Nova Scotia is an abrasion platform, 4–6 m above sea level, developed across different rock types in the area. Grant (1980b) interpreted this feature as an emerged wave-cut platform representing equilibrium between crustal rebound and sea-level rise during the last interglacial. It is the "rock" on which much of the stratigraphy of Maritime Canada is based.

Overlying the rock bench are nine major lithostratigraphic units (Fig. 5). The lowest is an oxidized bouldery gravel (BG—Unit 1). Above this is a stony, massive diamicton (Dm—Unit 2) composed mainly of locally-derived clasts, interpreted as a till formed by a Nova Scotia–based ice cap (Grant, 1980a). Unit 3 is an indurated, grayish-red, matrix-supported diamicton that contains a significant percentage of New Brunswick–derived erratics such as polymictic conglomerate, gneiss, and syenogranite. Fabric, structure, and underlying bedrock deformation strongly suggest that this unit, named the Red Head till (Grant, 1980a) is a lodgement till. Grant (1980a) described a glacio-tectonically deformed and striated phyllite rock surface underlying this unit in some localities which indicates an ice flow between 120° and 150°. Till fabric and striated, imbricate, bullet-nose boulders within the unit at Salmon River also indicate a southeastward ice flow, suggesting a correlation with ice-flow phase 1 of the erosional stratigraphy (Fig. 2). This unit contains shell fragments derived from the Bay of Fundy dated at >38,000 yr B.P. (GSC-695).

Above the Red Head till at the Salmon River section and at other localities is a bed of gray sand and intercalated diamictons which contain a rich, warm-water molluscan fauna (Wagner, 1977). The bed ranges from 2 to 8 m above mean sea level. Above the sand at many sections are three distinct, massive diamicton units (units 5, 6, and 8) interpreted as tills, each with a

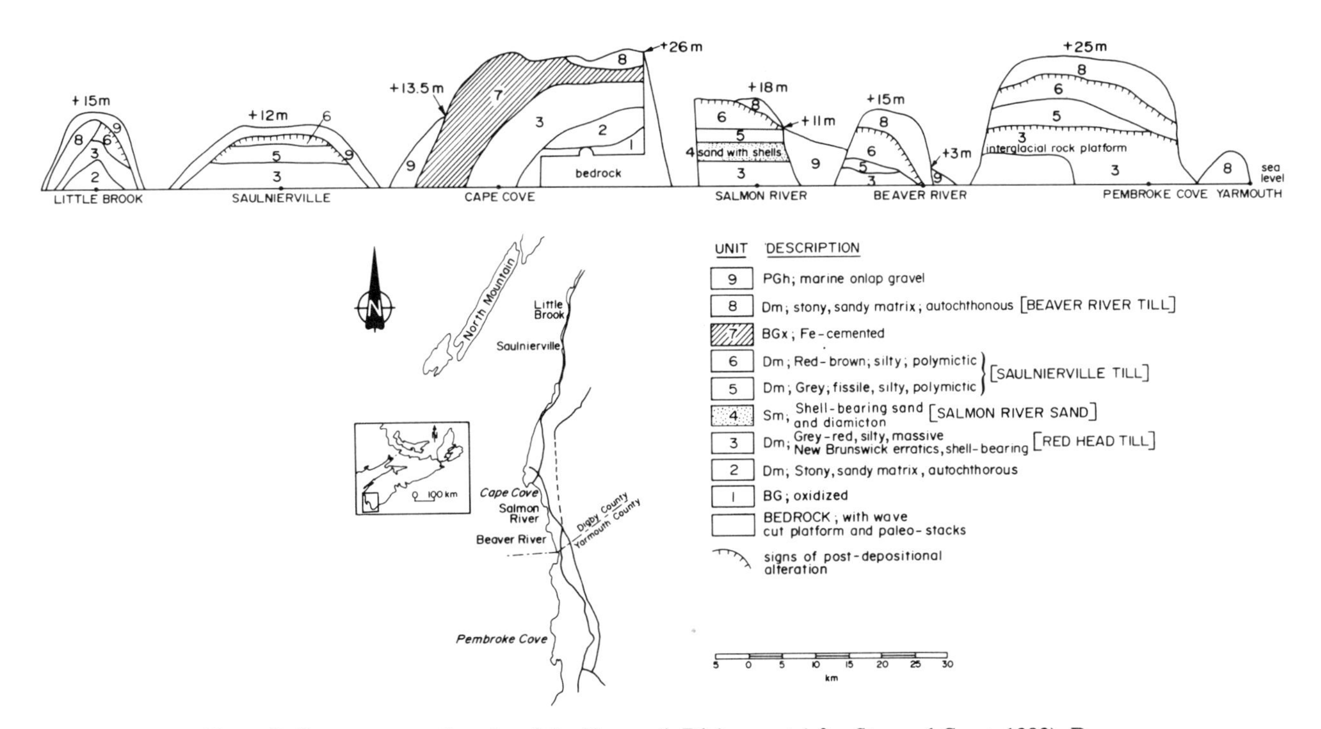

Figure 5. Quaternary stratigraphy of the Yarmouth-Digby coast (after Stea and Grant, 1982). Dm = matrix-supported diamicton; PGh = parallel-laminated, pebble-grade gravel; BGx = cross-laminated, boulder-grade gravel; SM = massive sand; BG = cobble to boulder gravel.

distinct color and pebble lithology. Units 5 and 6 contain North Mountain basaltic erratics and may correlate with regional southward ice flow during ice-flow phase 2 (Fig. 2). Unit 8 is a stony, "immature" till of local provenance, believed to have formed during ice-flow phases 3 and 4 or during local ice-divide development in Nova Scotia in the late Wisconsinan.

Clarke and others (1972) interpreted the Salmon River sand to be Sangamonian on the basis of the warm-water (temperate) fauna and the extinct index fossil *Atractodon stonei.* Shells from the Salmon River sand produced middle Wisconsinan finite radiocarbon and U/Th ages (Grant, 1980a). Nielsen (1974) reinterpreted the Salmon River sand as a middle Wisconsinan glaciomarine deposit, based on the intimate association of the lower part of the sand with diamicton units and the middle Wisconsinan dates. Paleodepth analysis of the fauna casts doubt on the in situ origin of the sand. Wagner (1977) used the paleodepth of the molluscan fauna in the sand and the present elevation of the deposit to infer that sea level would have been 16–31 m higher than present. This is 8–23 m higher than estimates of interglacial sea levels from oxygen-isotope data and the New Guinea coral-terrace record (Shackelton, 1987). Grant and King (1984) suggested two possible solutions to this paradox: (1) the sand is part of a thrust sheet; or (2) the sand represents a warm-water, glaciomarine deposit of middle Wisconsinan age, and the high inferred sea levels are due to isostatic depression of the crust (Nielsen, 1974).

Wehmiller and others (1988) provided new amino-acid age estimates of type sections in the region. *Mercenaria* fragments within the Red Head till and the Salmon River sand at the type sections at Salmon River and Saulnierville have D/L leucine ratios averaging 0.24. These are labeled A in Figure 6. Wehmiller and others (1988) inferred a Sangamonian age for the Salmon River sand based on similar amino-acid ratios in the Sankaty sand in Massachusetts, where a solitary coral was dated by U/Th as 133,000 yr B.P. (Oldale and others, 1982). On the basis of the analysis by Wehmiller and others (1988), the source beds for the shell fragments in the Red Head till are also inferred to be Sangamonian. However, shells from a basal diamicton at Gilbert Cove, north of Saulnierville, gave D/L ratios suggesting source beds >250 ka, perhaps equivalent to oxygen-isotope stage 7 (Fig. 6).

Central Nova Scotia sections

Carboniferous evaporite basins in central Nova Scotia provide some of the best exposures of buried pre–late Wisconsinan forest beds (Fig. 7). At the gypsum quarries of East Milford and Miller Creek, Stea (1982) observed organic beds (unit 3) with wood fragments sandwiched between two till units (units 2 and 4). In each case the forest bed lay above the gypsum surface that was truncated by a till unit termed the Miller Creek till (unit 2; Fig. 7), which is oxidized on top. Associated with the organic bed at both sections are bedded silt and sand. A grayish-red, calcareous, massive, silty diamicton with few stones (East Milford till) overlies unit 3. Fabric and lithology of the East Milford till imply a southeastward ice flow correlative with ice-flow phase 1 (Fig. 2) of the erosional stratigraphy (Stea, 1982). The East Milford till is a distinctive marker unit in the region. The section is topped by a reddish, sandy diamicton (Hants till).

The organic horizon (unit 3) at East Milford consists of two facies based on pollen analysis (Mott and others, 1982). The lower facies (section B; Fig. 8) consists of four zones; zone 1, dominated by hardwoods (beech and elm); zone 2, birch; zone 3, fir and spruce; and zone 4, alder and fir. The upper facies (section C; Fig. 8) is dominated by spruce and pine. At Miller Creek, buried organic horizons were also found to contain two distinct pollen facies (Stea and Mott, 1990), although the stratigraphic relations between the two are uncertain. One facies is dominated by jack pine with some hardwood genera, and the other is dominated by spruce.

Radiocarbon ages obtained from wood fragments in unit 3 are from section B at East Milford (Fig. 8; >50,000 yr B.P.; GSC-1642). U/Th age dates of 84,900 (UQT-185) and 84,200 yr B.P. (UQT-186) were also obtained from section B. At Miller Creek the pine facies of unit 3 was dated at >52,000 yr B.P. (GSC-2694). Wood from a unit of unknown pollen facies affiliation produced a middle Wisconsinan finite date of 33,000 ± 2000 yr B.P. (I-3237) (MacNeill, 1969).

Addington Forks is located southeast of Antigonish, Nova Scotia, in a Carboniferous basin, a geologic setting similar to the quarry sections (Fig. 1). About 80 cm of silt with organic seams, organic silt, and highly compacted peat make up a nonglacial package of sediments sandwiched between a basal red-brown till and an overlying red till (Fig. 9).

Wood from the site collected by MacNeill (1969) produced a finite date of 33,700 ± 2300 yr B.P. (I-3236). However, wood (juniper) gave an age of >42,000 yr B.P. (GSC-1598; Prest, in Lowdon and Blake, 1973). A third age determination on spruce-tamarack wood gave an age of 36,000 ± 520 yr B.P. (GSC-3848). Roots from trees on the densely treed slope covering the deposit may be a source of contamination that produced the finite dates.

The abbreviated pollen diagram (Fig. 9) has been divided into five pollen assemblage zones (Mott and Grant, 1985). The basal zone (1) is dominated by Polypodiaceae and Osmunda (fern) spores and tree pollen of *Quercus* (oak) and *Betula* (birch). *Quercus* dominates in zone 2 along with significant amounts of *Carpinus*/*Ostrya* (blue beech/ironwood) pollen. *Quercus* decreases and *Pinus* (pine), *Carya* (hickory), and other thermophilous hardwood taxa such as *Fagus* (beech), *Tilia* (basswood), and *Ulmus* (elm) increase in zone 3. *Cephalanthus* (buttonbush) type and Gramineae along with other herbaceous taxa probably reflect the local flora. Osmunda spores attain a second maximum in this zone as well. Decline of *Pinus* pollen and thermophilous hardwood taxa in zone 4 accompanies increase in *Abies* (fir), *Picea* (spruce), and *Alnus* (alder). In zone 5 extremely large *Alnus* values supplant most other taxa.

Despite the two finite dates obtained, the pollen spectra

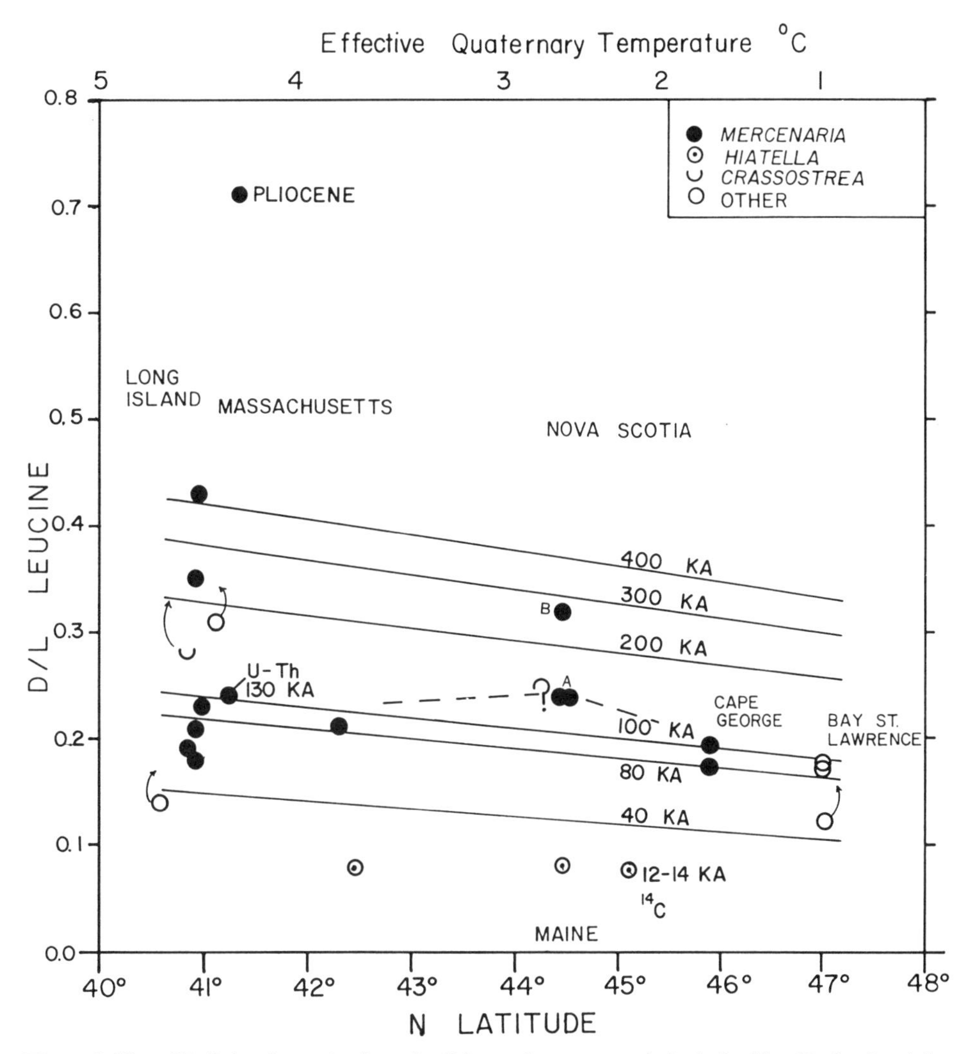

Figure 6. Plot of D/L leucine ratios from fossil invertebrates versus latitude for New England and the Maritimes. Kinetic age model isochrons after Wehmiller and others (1988). The Gilbert Cove determinations, labeled B, and the Saulnierville and Salmon River determinations, labeled A, are from Wehmiller and others (1988). Cape George and Bay St. Lawrence determinations are from this study.

indicate that the organic sediments were deposited under interglacial conditions. Therefore, the dates are considered spurious. The changing pollen spectra indicate that the waning phase of a warm climatic interval is represented. Forests with birch, oak, hickory, and other hardwoods were replaced by coniferous forests of balsam fir and spruce. Alder domination may indicate further cooling to subarctic conditions.

This deposit probably related to the cooling phase of substage 5e; i.e., conditions that were warmer than present declining to cool boreal to subarctic conditions. If this interpretation is correct, the till at the base of the section would be Illinoian. The overlying till probably relates to ice-flow phase 3, based on regional mapping (Stea and Myers, 1990).

In their synthesis of paleoecological work in Atlantic Canada, Mott and Grant (1985) and Vernal and others (1986) differentiated three periods of organic accumulation, termed palynostratigraphic units 1–3, that span a lengthy interglacial period. Each unit is believed to record climate change from cool to warmer to cool conditions, only unit 1 attaining true "interglacial" warmth in the palynological sense. Section B at East Milford and the pine facies at Miller Creek are correlated with palynostratigraphic unit 2, which is characterized by the presence of balsam fir with beech and oak. The section at Addington Forks (Fig. 9) is correlated with unit 1; its pollen assemblage suggests warmer than present climatic conditions or the climatic optimum of the last interglacial (oxygen-isotope stage 5e). Section C at East Milford and the spruce-dominated facies at Miller Creek correlate with palynostratigraphic unit 3, which is defined by a boreal assemblage of spruce and pine with little or no fir. The Miller Creek and East Milford pollen spectra do not indicate conditions

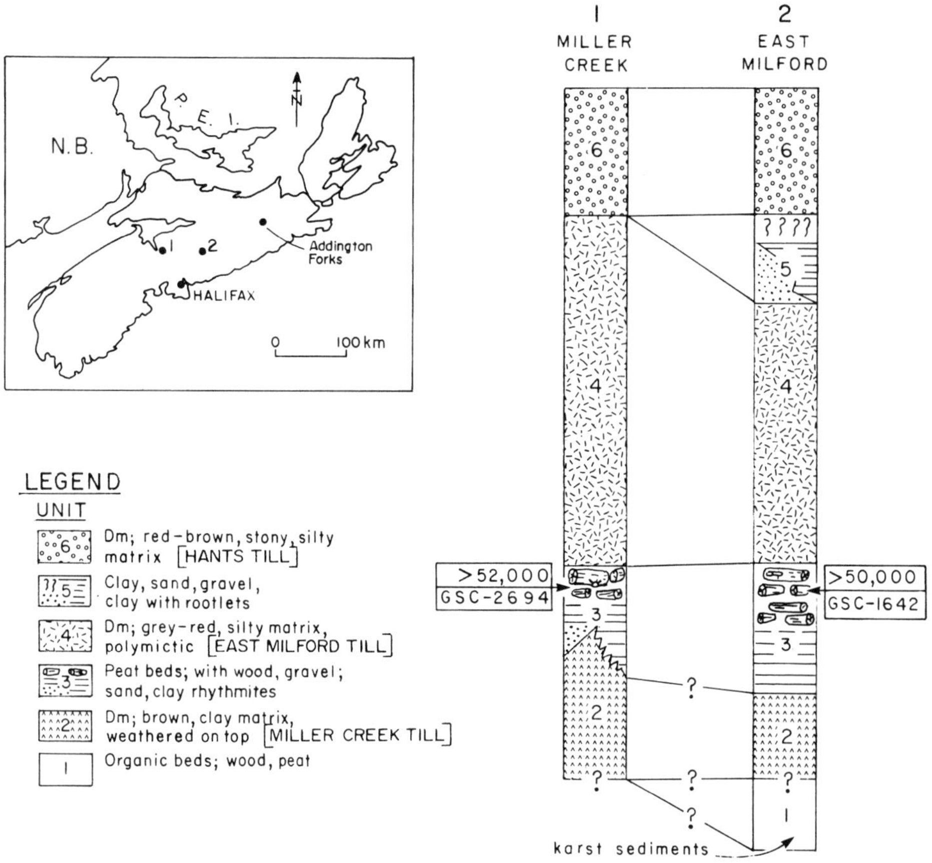

Figure 7. Quaternary stratigraphy of central mainland Nova Scotia. Dm = Matrix-supported diamicton.

comparable to the climatic optimum, although the waning phase of 5e is a possibility. It is more likely that these deposits relate to younger, cooler intervals, perhaps stages 5c and 5a. The organic units had been previously placed in the Sangamonian interglacial as defined by Fulton (1984) (75,000–125,000 yr B.P.) (Grant and King, 1984).

Cape George coastal sections

A number of coastal sections were measured from Arisaig on the Northumberland Strait to MacIssacs Point on the coast of Georges Bay (Fig. 10). Common to most of these sections is a sequence of stratified, well-sorted, sands and gravels (unit 2) resting on an emerged abrasion platform believed to be a remnant of a Sangamonian sea-level highstand (Grant, 1980a). The height of this rock bench varies from 4 to 6 m above mean sea level. At Cape George Point, this bench is developed on mafic rocks. The surface of the bench, when exhumed from under sand and till, is scalloped and smoothed.

Lying above the stratified sands in most of the sections is a distinctive, bouldery diamicton (unit 3), which varies in thickness from 4 to 20 m. The clasts are invariably angular and locally derived. This material may be a product of mass wasting from adjacent rock slopes during periglacial conditions, or a flow till, deposited by advancing glaciers that deposited the overlying till. This deposit may be analogous to the "head" deposits of Ireland and Wales (Warren, 1991).

Stratigraphically above this diamicton are two till units distinguished primarily by their pebble content. The lower till (unit 4) is grayish-red, has few stones, and a silty matrix. Robust fragments of *Mercenaria* (sp.) are commonly found near the base of this till unit. Thick exposures of this till rest on bedrock surfaces with striations trending 90°–130°. Large bullet-nose boulders at the base of some sections with stoss and lee form have long axes parallel to the bedrock striations and have parallel surface striations. This till sheet was formed by an ice sheet moving to the east and southeast and is therefore correlated with phase 1 of the erosional stratigraphy (Fig. 2).

The upper till (unit 5) is distinguished from the lower, shelly till by its abundance of stones and boulders. At some sections it is slightly sandier than the lower till. Bullet-nose boulders, with striations parallel to their long axes, generally trend northeastward. Striations on rock surfaces underneath this till unit along the coast trend from 335°–045°, suggesting that it was formed by a northward ice flow during phase 3 (Fig. 2).

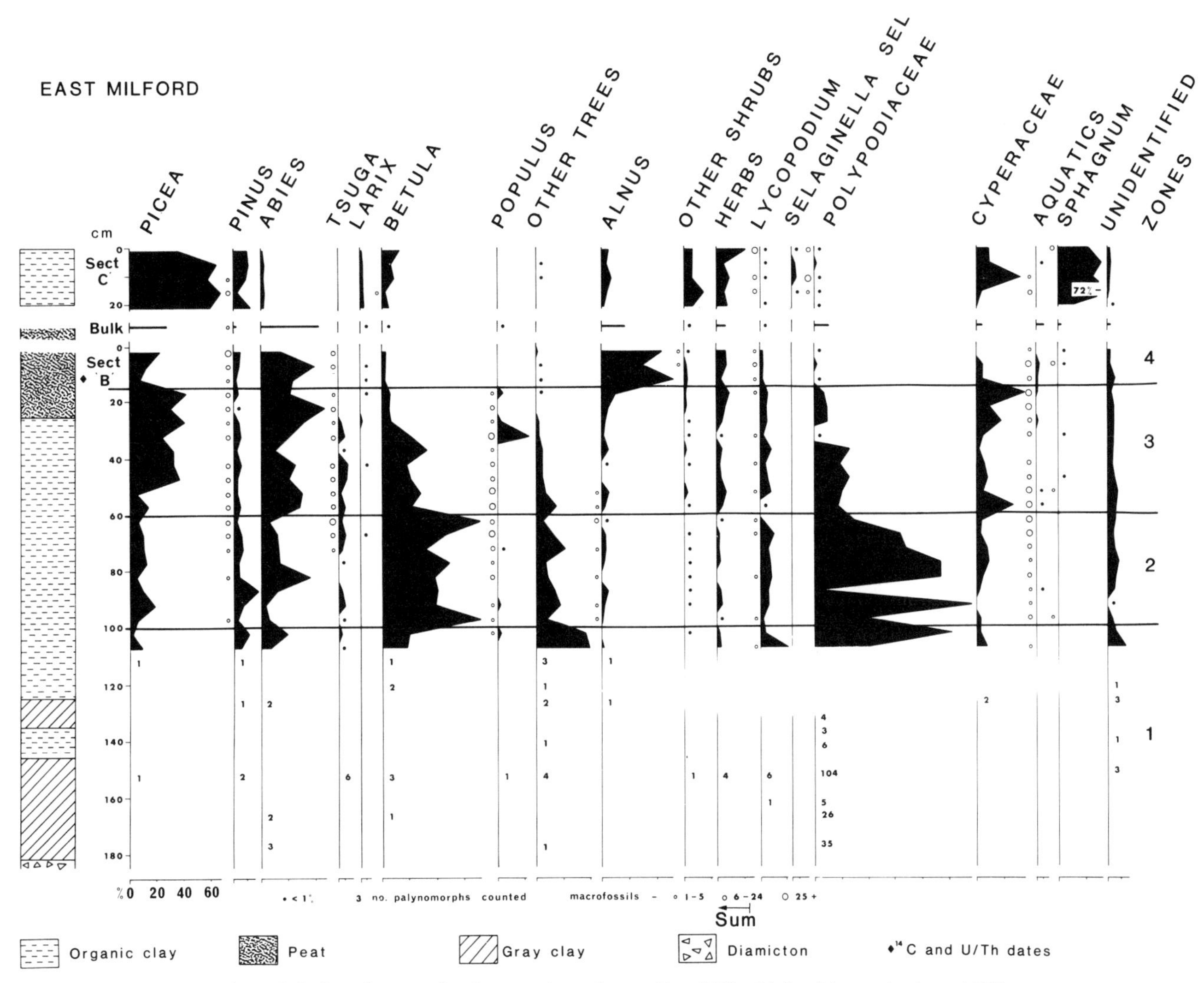

Figure 8. Pollen diagrams for the organic sections at East Milford (after Mott and others, 1982).

A radiocarbon age of >37,000 yr B.P. (GSC-4048) was obtained on one of the *Mercenaria* fragments in unit 4 at Cape George (D. R. Grant, 1984, personal commun.).

Mercenaria fragments were also collected from unit 4 from Cape George for AAR dating. *Mercenaria* is used as the standard calibration and correlation genus along much of the east coast from Nova Scotia to New England (Wehmiller and others, 1988) because of its moderate racemization rate, integrity of its massive shell, and ubiquity. The Cape George samples (DFB-88-1, OFB-88-2) have D/L ratios of 0.19 and 0.17 (Table 1). The kinetic model age assignments are 80 to 110 ka or oxygen-isotope stage 5.

Twelve shell fragments (*Mercenaria* sp.) from unit 4 at Cape George were dated using the ESR method (Table 2). The dates range from 99,700 to 184,000 yr B.P.; the mean value is 141,500 yr B.P. These dates were calculated assuming direct or early uranium accumulation. The geologic evidence summarized earlier and the amino-acid results suggest a stage 5 age (80,000–120,000 yr B.P.) for the shells in unit 4. The ESR ages, however, form two clusters, those in the expected range and those whose ages are much older, spanning marine isotope stage 6. The stage 6 age is unlikely because ice probably covered the region at that time. Although age underestimation is not common in ESR studies, this may represent a special case because the shells have been resedimented. A possible scenario for the changing environments of the shells would be as follows. The shells formed in interglacial seas 120,000–100,000 yr B.P. They were buried in a reducing marine environment, rich in organic matter with elevated U content. This was accompanied by early uptake of uranium in the shells. An ice advance 40–80 ka later removed the shells from this reducing offshore environment to a terrestrial, oxidizing environment relatively depleted in uranium. The annual dose was calculated on the till enclosing the shells, so the total dosage was underestimated.

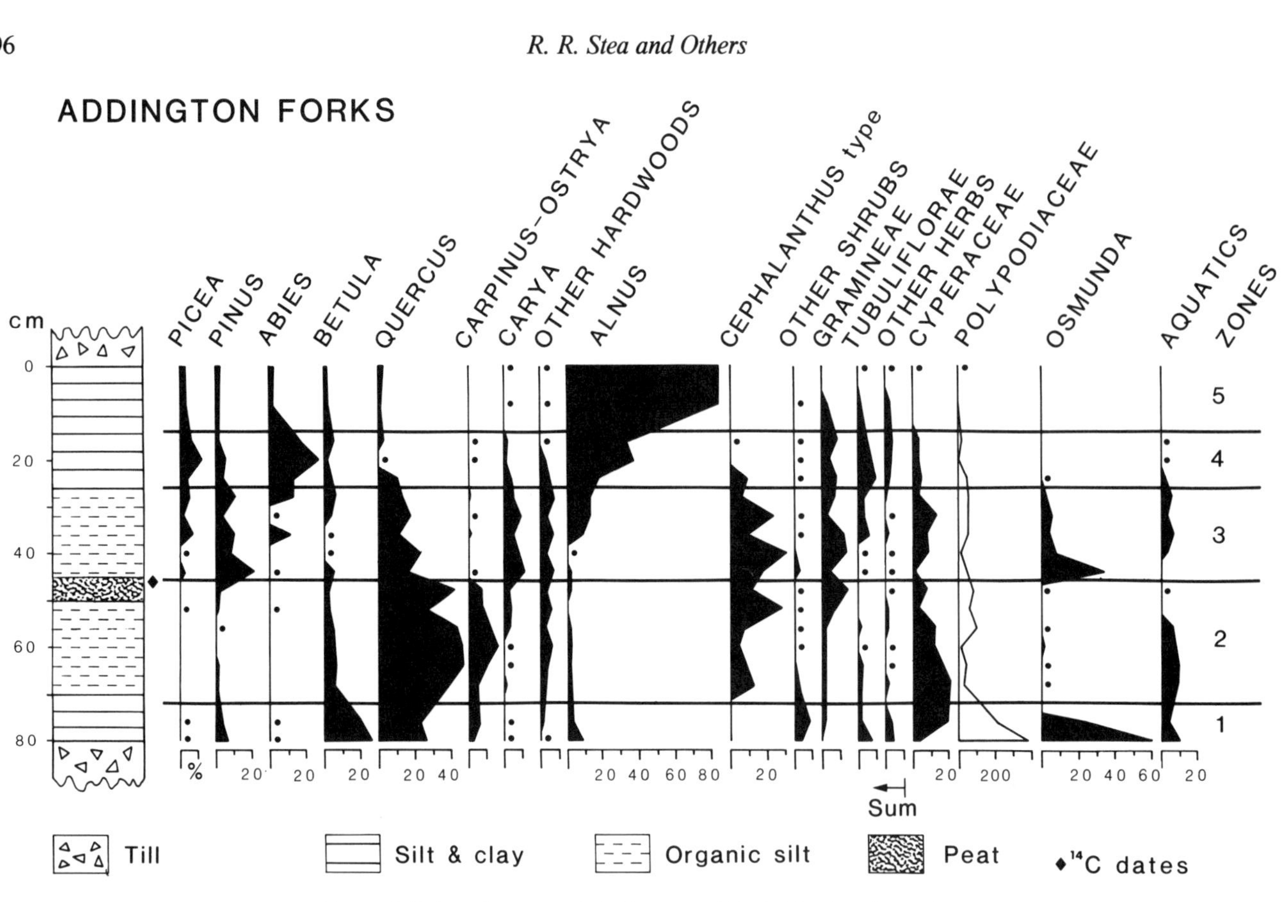

Figure 9. Pollen diagram and stratigraphy of the Addington Forks site.

Several of the ESR shell samples were dated using the U/Th disequilibrium method (Table 3). Only two samples provided enough of the Th daughter isotope to be effectively dated. The dated samples gave U/Th ages of 50 and 30 ka. These anomalously young ages suggest a postdepositional flux of uranium, and an open system. Mollusc samples do not generally provide reliable U/Th dates (Schwarcz and Blackwell, 1985).

Chedabucto Bay sections

Sections along the north shore of Chedabucto Bay reveal seven lithostratigraphic units lying above the emerged rock bench (Fig. 11). Along this shore there is a pre–late Wisconsinan peat bed with wood fragments that lies directly on the bench and predates the deposition of tills. The exposures along this coast are generally in drumlins or drumlinoids whose stoss and lee sides are not well developed.

The oldest surficial unit (unit 1) is a silt containing organics and compressed peat layers with large wood fragments, which lies directly on regolith from an abrasion platform 4–6 m above tide level.

Unit 2, a silty diamicton, is the thickest unit in the area. It forms the core of many of the drumlins, which are the sites of cliff erosion. The unit is interpreted as a till because of its massive and indurated nature, erratic content, and the abundance of bullet-nose boulders whose long axes parallel underlying bedrock striations (Fig. 11). The flow associated with this unit is eastward and southeastward, parallel to ice-flow phase 1 (Fig. 2). The presence of distinctive appinite (mafic pegmatite) boulders in the unit implies an ice flow either to the east or southeast from two of the only known stocks (J. B. Murphy, 1990, personal commun.). Ice flow across the stock north of Antigonish is consistent with the modal trend of the long axes of the till-embedded boulders (115°–125°). It appears from underlying bedrock striations that the initial flow was almost due east (phase 1a; Fig. 2). Shell fragments in the till are presumably derived from Georges Bay.

Units 3 and 4 are waterlaid units, which are relatively rare in these drumlin sections. Section 153, however, reveals fairly continuous, rhythmically-bedded, gravelly-sand and silty-sand beds that separate two massive diamicton units.

Unit 5 is a diamicton that is distinguished from unit 2 by greater stone content. This unit contains bullet-nose boulders that indicate a northwestward ice flow, parallel to the ice flow inferred by regional striation mapping (ice-flow phase 3; Fig. 2).

Spruce wood from the compressed peat produced a radiocarbon date of >49,000 yr B.P. (GSC-4419 HP). Preliminary pollen studies from the organic horizon (unit 1) show that it is spruce dominated and correlative with palynostratigraphic unit 3, representing a boreal forest environment. The organic interval's

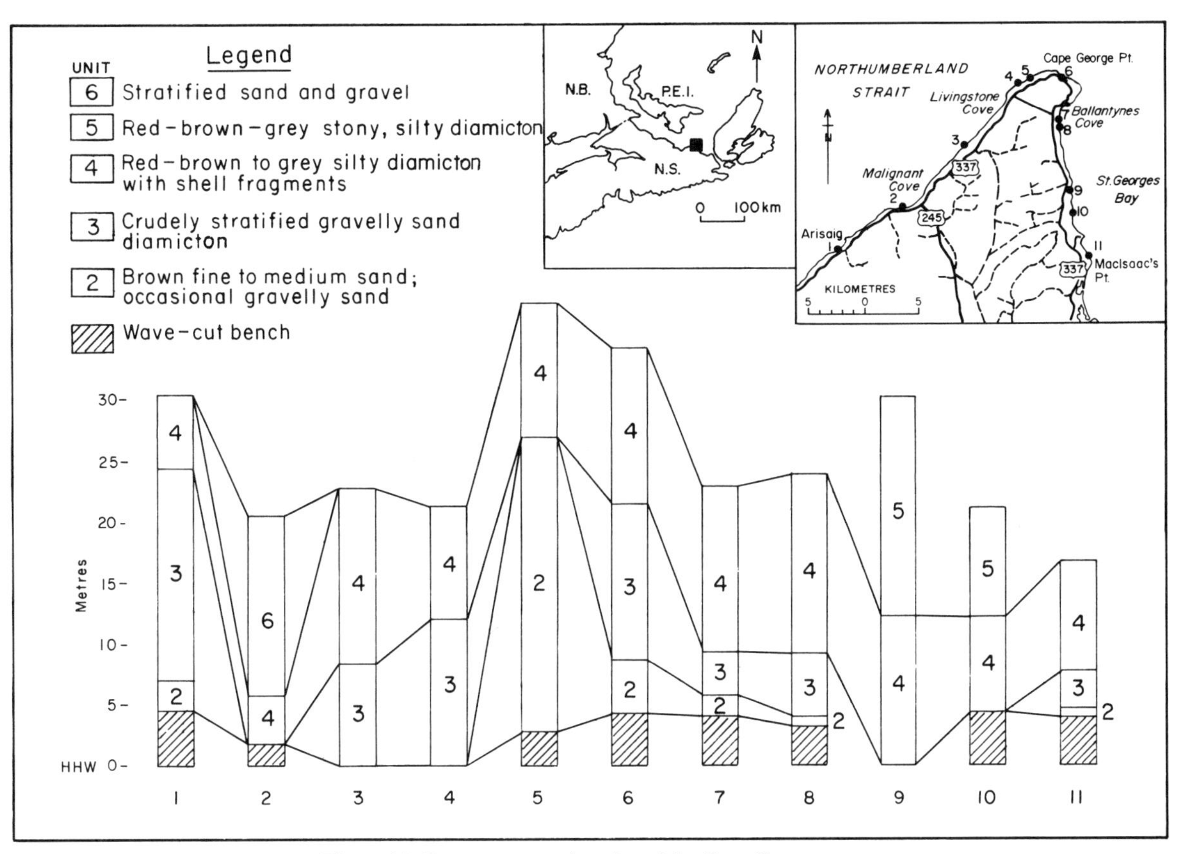

Figure 10. Quaternary stratigraphy of the Cape George area.

position on the raised marine bench suggests that its time of deposition postdates the climatic optimum of the last interglacial.

Bay St. Lawrence section

This section, described by various authors (Prest, 1957; Mott and Prest, 1967; Newman, 1971; Grant, 1975; Occhietti and Rutter, 1982; Vernal, 1983), is the only one in the region to feature the ~5 m platform, overlain by a pre-late Wisconsinan organic bed and a bed with marine shells similar in aspect to the Salmon River sand (Fig. 12). The section, however, lacks the till stratigraphy found in most of the mainland sections overlying these organic sediments. The Quaternary section is located over Carboniferous rocks that flank the Precambrian Cape Breton highlands. The highlands rise precipitously to an altitude of 300 m along the St. Lawrence fault a few hundred meters east of the section (Macdonald and Smith, 1980).

At the base of the section (Fig. 12) is an abrasion platform cut into metasiltstone bedrock 2–7 m above mean sea level. Above this platform is a bouldery diamicton (unit 1) interpreted as a till by Mott and Prest (1957) and Newman (1971). Unit 2 consists of stratified sand and well-rounded, cobble-gravel with sand beds, overlain by a laterally continuous bed of compressed peat and wood 0.2–1 m thick.

Unit 3, which overlies the peat-bearing unit, is a massive, clast- and matrix-supported, boulder-cobble diamicton, about 10 m thick. The clasts are subangular and angular and the matrix is generally a coarse sand. Newman (1971) described some crude stratification. A pebble count was made on 300 pebble-sized clasts of a bulk sample of the unit. Of the clasts, 22% are pelitic schists that crop out on the highlands 2–3 km east of the section (Macdonald and Smith, 1980). Gneiss and pegmatite derived from the steep highland slope 500 m east of the section only represent 15% of the total clast count. Vernal (1983) interpreted the unit as colluvium, but pebble lithologies suggest that mass wasting was not the only transport process involved. The high percentage of schist erratics suggests that the diamicton could be of glacial origin, or derived from an underlying glacial unit (unit 1?).

Unit 4 consists of three facies. At the base of the unit in some parts of the section is a thin, compact, massive to indistinctly stratified, matrix-supported silty diamicton. It contains many small, disarticulated shell fragments. Above this, separated by a sharp contact, is a grayish, sandy diamicton, less compact than the underlying diamicton, but also containing shell fragments. The upper diamicton is crudely bedded, with numerous silt stringers. At the top of unit 4 in some parts of the section is a brownish, laminated silt devoid of fossils. Unit 4 is, at maximum,

TABLE 2. ELECTRON SPIN RESONANCE DATA FROM CAPE GEORGE, NOVA SCOTIA

Sample Number	AD*	U Content Shell (ppm)	U Content (ppm)	Th Sediment (ppm)	K (%)	Thickness†	Internal Dose-Rate (m Gy/a)	External	ESR Age (a)
D-1391	108.4	0.6	2.0	7.0	0.24	2.2 (0.6)	0.1463	0.6893	130,000
D-1392	128.8	0.5	2.0	7.0	0.23	5.0 (0.9)	0.1415	0.6199	168,000
D-1393	200.4	6.0	2.0	7.0	0.30	7.2 (2.9)	0.1441	0.6074	99,000
D-1394	168.0	0.55	3.0	9.0	0.31	3.7 (2.5)	0.1568	0.8253	171,000
The following samples are shell samples only from the same site:									
D-1486	116.4	1.29	2.0	7.0	0.27	6.0 (1.0)	0.3295	0.6206	123,000
D-1487	131.5	1.0	2.0	7.0	0.27	5.0 (80.8)	0.3747	0.7309	119,000
D-1488	141.4	1.00	2.0	7.0	0.27	5.8 (0.8)	0.2777	0.6249	157,000
D-1489	209.1	1.71	2.0	7.0	0.27	4.6 (0.6)	0.4954	0.6397	184,000
D-1490	101.2	(1.00)	2.0	7.0	0.27	5.4 (0.8)	0.2484	0.6280	116,000
D-1491	111.2	1.70	2.0	7.0	0.27	5.4 (1.2)	0.4135	0.6209	108,000
D-1492	149.4	1.30	2.0	7.0	0.27	5.2 (1.0)	0.3564	0.6255	152,000
D-1493	179.4	1.50	2.0	7.0	0.27	5.2 (0.8)	0.4265	0.6294	170,000

*Accumulated dose Gy.
†Removed thickness mm.
Note: Alpha-Efficiency = 0.1; Water content of sediment = 15 percent (assumed); Values in parentheses = assumed; Beta-dose correction factors are considered for shells with removed surface (for details see Grün, 1989); U-content of shells determined by alpha-spectrometry (R. Hausmann, Geological Institute, University of Cologne); U and Th content of sediment (Gamma-spectrometry, H. Pietzner, Geological Landesamt, Krefeld); K-content, W. Sauer, Atomic Absorbtion Spectr. (AAS), Max-Planck-Institute, Düsseldorf; Calculation of ESR-ages with computer program DATA-IV, R. Grün, Subdepartment of Quaternary Research, University of Cambridge.

TABLE 3. U/Th DISEQUILIBRIUM DATING OF SHELL SAMPLES FROM CAPE GEORGE, NOVA SCOTIA*

Sample	U (ppm)	Th (ppm)	$^{234}U/^{238}U$	$^{230}Th/^{232}U$	$^{230}Th/^{234}U$	U/Th age (x 1,000 a)	Detrital Th-corrected Age (ka)
D-1486	1.29 ± 0.043		1.388 ± 0.049	...	...	...	
D-1487	5.167 ± 0.138	1.914 ± 0.336	1.371 ± 0.034	4.882 ± 0.852	0.433 ± 0.033	59.9 ± 6.2 5.9	50.2 ± 5.8 5.6
D-1489	1.707 ± 0.058		1.529 ± 0.058	...	...	...	
D-1491	1.704 ± 0.062	0.563 ± 0.189	1.362 ± 0.056	3.825 ± 1.371	0.305 ± 0.043	38.8 ± 6.7 6.3	30.0 ± 6.2 6.0

*R. Hausmann, personal communication, 1989.

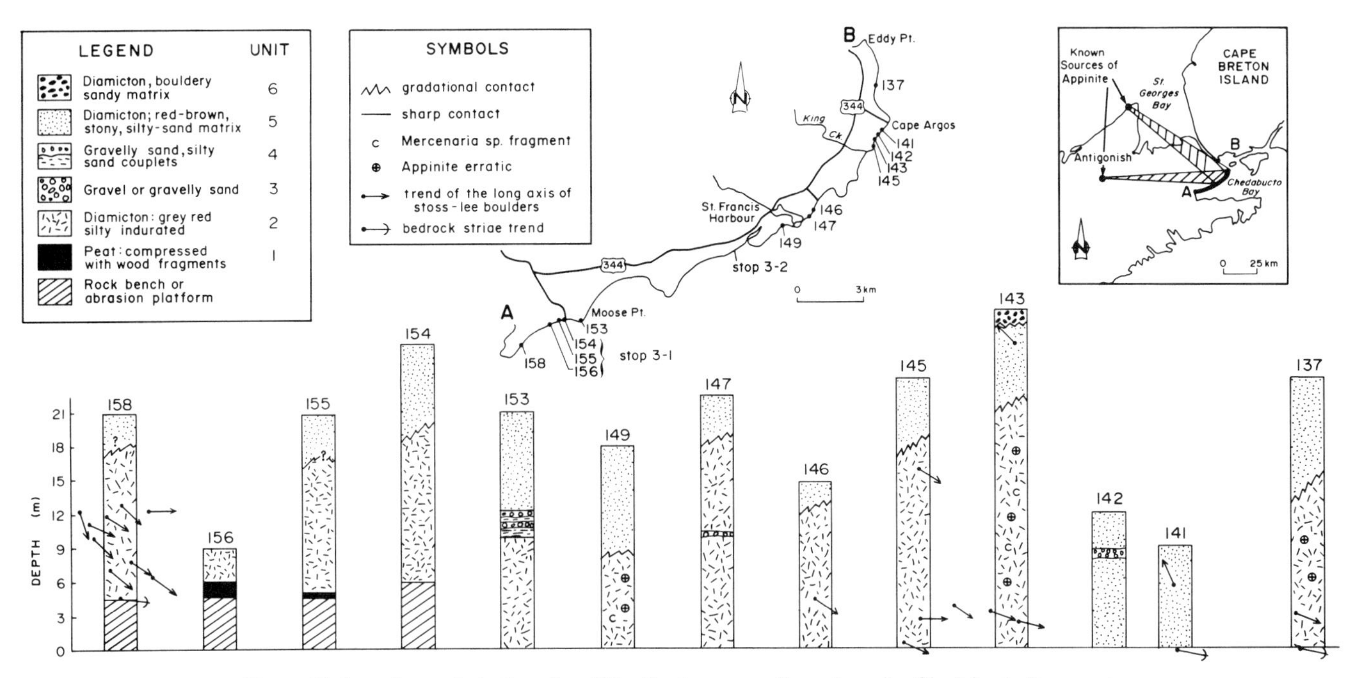

Figure 11. Location and stratigraphy of the Quaternary sections along the Chedabucto Bay coast.

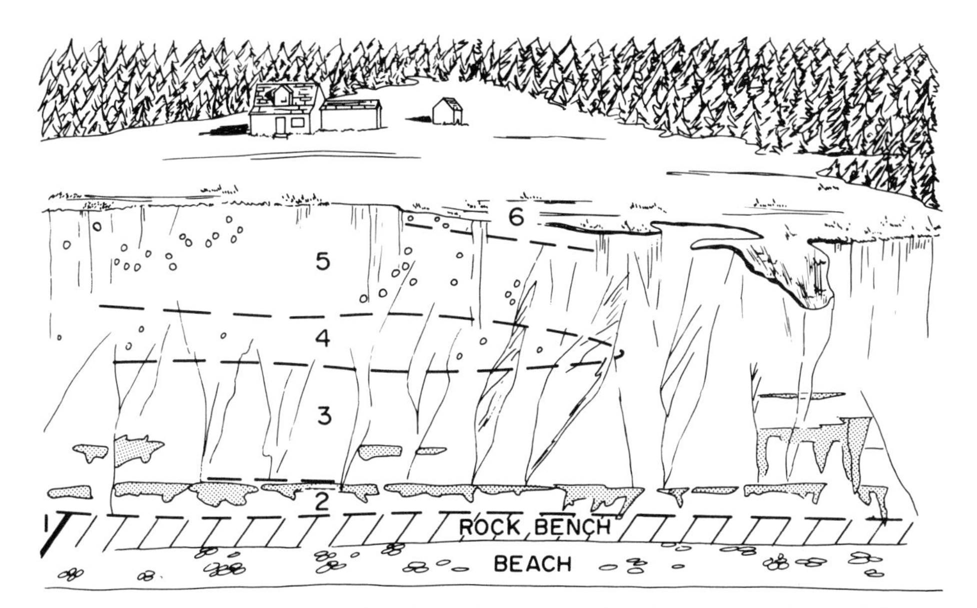

Figure 12. Quaternary stratigraphy of the Bay St. Lawrence section (drawn from a photograph in Grant and King, 1984; stratigraphy in part compiled from Newman, 1971, and Grant and King, 1984). Units 1 through 6 are described in the text.

3 m thick and extends across most of the exposure (Fig. 12). Guilbault (1982) interpreted the diamicton facies of unit 4 as a till because it contains an allochthonous assemblage of foraminifera with widely differing bathymetric ranges, some as great as 100 m. Guilbault (1982) argued against an in situ origin because the inferred paleodepths of the foraminifera and the present elevation of the unit, which imply an unlikely former RSL at +115 m above mean sea level. Newman (1971) also interpreted the diamicton facies of unit 4 as a till, on the basis of consistent southward-trending pebble fabrics that are parallel to bedrock striations found in the region (Grant, 1988). The sedimentology of the basal diamicton facies of unit 4 is consistent with a glacial origin. The top part of the unit, however, is water deposited, perhaps subglacially.

An unfossiliferous, bouldery diamicton (unit 5) overlies the shelly diamicton. This unit is similar in texture and structure to the lower, bouldery diamicton (unit 4), but has larger and more-abundant boulders. Clast lithologies are 50% granitic rocks, 17% sandstone (Carboniferous), 13% gneiss and pegmatite, and 7% schist. The remaining 13% is largely vein material. The granitic clasts that constitute much of the unit may be derived from granitoid injection veins, which are found throughout the shear rock faces of the highland slope. The abundance of schist suggests an alternative to a simple mass-wasting hypothesis. The lack of colluvium on the adjacent slopes of the highlands south of the section also suggests that this unit is not simply part of a talus cone. Newman (1971) suggested that the fabrics and lithologies of surface tills along the section imply a late northwestward ice flow off the highlands. The section is capped by a clast-supported, boulder-cobble layer, possibly water deposited (unit 6).

Pollen from the peat-bearing layer (unit 3) shows a complete cycle of climate change from tundra to boreal forest and back to tundra again (Vernal, 1983). Unit 3 correlated with palynostratigraphic unit 3 of Mott and Grant (1985). It had been assigned to an interstadial deposit of early Wisconsinan age (Mott and Prest, 1967; Newman, 1971) or middle Wisconsinan (Mott and Grant, 1985; Vernal and others, 1986) on the basis of finite radiocarbon and U/Th dates.

Radiocarbon ages on wood from this horizon are >38,300 (GSC-283), 44,200 ± 820 (GSC-3636) and >46,000 (GSC-3864) and >49,000 (GSC-4487) yr B.P. A U/Th disequilibrium age of 47,000 ± 4700 yr B.P. (UQT-178) was obtained from the peat bed (Mott and Grant, 1985; Vernal and others, 1986). The sample has a high $^{230}Th/^{232}Th$ ratio, implying little postdepositional influx of Th, so the date was considered to be valid (Vernal and others, 1986). Wood from which the U/Th determination was made, however, has a uranium content of 4 ppm. This is anomalous when compared with other analyses of wood from Nova Scotia (Mott and Grant, 1985), and suggests some form of postdepositional influx. The dating method (daughter-deficient method) depends on the separation of the geochemically mobile uranium parent isotope from its relatively inert thorium daughters. The dating "clock" is assumed to start after uptake of soluble uranium by the wood before or soon after burial. The major drawback is the closed-system assumption. Postdepositional influx of relatively insoluble thorium is unlikely, but soluble uranium may interact with the wood long after burial, especially in a peat bed interbedded with coarse gravel rich in pegmatite debris. Vernal and others (1986) acknowledged this and cautioned that the U/Th ages are to be considered as minimum ages only.

The shell-bearing diamicton (unit 4) contains a molluscan fauna (Wagner, 1977) indicative of marine conditions similar to modern or even warmer. Guilbault (1982) found thermophilous species of foraminifera together with species with ice-shelf affinity. Vernal (1983) interpreted the pollen in this unit as reworked from offshore beds of a former warm-climate interval and later redeposited in a tundra environment. The geologic evidence therefore suggests that unit 4 at Bay St. Lawrence is a glacially-resedimented deposit. Amino-acid ratios on shell fragments from the site imply a Wisconsinan interstadial age for the source beds (Occhietti and Rutter, 1982). Guilbault (1982) argued that, because the intertidal rock platform is of Sangamonian age, the overlying bed studied has to be Wisconsinan.

Shell (*Megayoldia thraciaeformis* and *Balanus*) fragments from unit 4 at Bay St. Lawrence were sampled for AAR age dating (Table 1). *Mercenaria* were not found in the unit. Other genera, however, can be compared to *Mercenaria* because slower racemizers typically have high Asp/Leu ratios (>8), whereas faster racemizers have low Asp/Leu ratios (<2) (Belknap, 1979). Thus, *Megayoldia* is expected to have racemization rates greater than *Mercenaria,* while *Balanus* should be much slower. This would imply an older age for the *Balanus* fragment than the *Mercenaria* fragments at Cape George (80–110 ka; Table 1) and younger ages for the *Megayoldia* (Fig. 13). Age estimates for these samples range from 50 to 120 ka (Table 1). More samples from Bay St. Lawrence are needed to clarify these trends.

DISCUSSION

The earliest known Quaternary deposits in Nova Scotia are >250 ka based on AAR ages of shells from a diamicton at Gilbert Cove, southwestern Nova Scotia (Wehmiller and others, 1988). A till unit (Miller Creek till) in central Nova Scotia underlying wood and peat beds of last interglacial age (stage 5) formed during an Illinoian glaciation or marine oxygen-isotope stage 6 (Fig. 9).

The rock bench or abrasion platform 4–6 m above mean sea level, which underlies many of the coastal sections, has not been dated directly but it is assumed to relate to the last time of minimum ice volume (substage 5e; Grant, 1980b). Geologic and dating evidence presented in this chapter do not refute this hypothesis. Two localities in the province show a forest bed overlying the bench. At both sites (Bay St. Lawrence, Moose Point), palynological analysis has demonstrated that the climate was cooler than today. If these beds were formed during the cooler phases of the last interglacial (stage 5), then the bench probably formed during the altithermal period of the last interglacial (stage 5e). It is possible, however, that the bench was formed during earlier interglacials and remained as a relict feature.

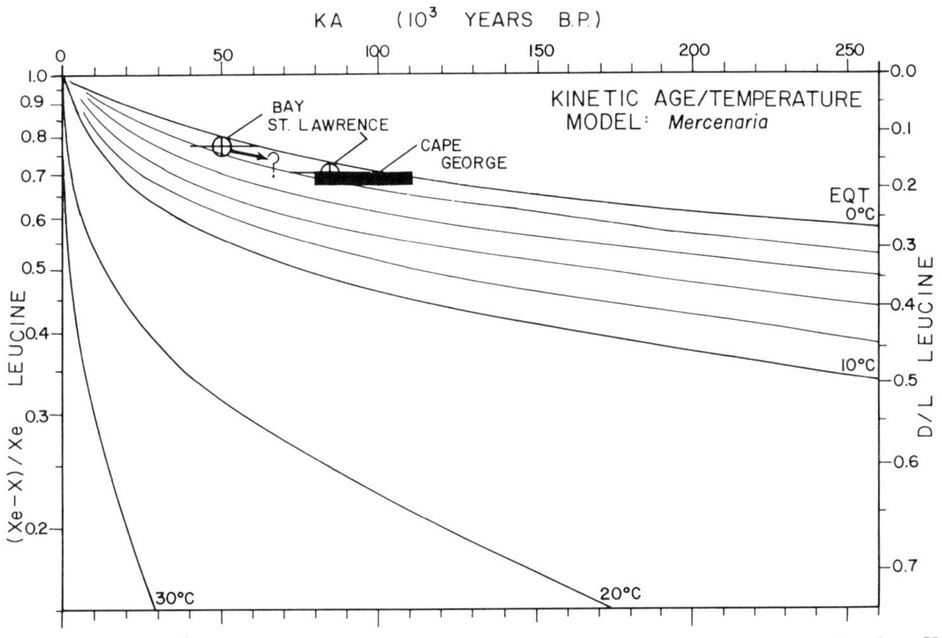

Figure 13. Kinetic age/temperature model after Wehmiller and others (1988). $Y = A + B\,(\ln[t])$, where $Y = (Xe - X)/Xe$, $Xe = 0.5$, X = D/L ratio; B = constant: 0.135; A = variable dependent on temperature ($A = 0.3999 - 0.0242(\mathrm{EQT})$; and t = time in 10^6 yr. EQT: Effective Quaternary temperature.

Shell-bearing diamictons in mainland Nova Scotia (Red Head till, unit 4, Cape George, Bay St. Lawrence) are correlative, using lithostratigraphic criteria. They overlie the raised marine platform and are overlain by several till sheets. The marine source beds for these shelly diamictons contain a temperate fauna, and may be temporally correlative with interglacial buried forest beds elsewhere in the province (Mott and Grant, 1985). Ages for these organics were determined by AAR, U/Th disequilibrium, and ESR. They appear to span the last interglacial (120–75 ka; as defined by Fulton, 1984). Palynological evidence implies that several distinct climatic half cycles are represented (Mott and Grant, 1985), rather than one warm interval. Vernal and others (1986), however, proposed an alternative chronology. Most of their U/Th ages confirm a Sangamonian age for the beds, except for the youngest unit (palynostratigraphic unit 3), which gave middle Wisconsinan ages. They suggested that these beds were part of an extended interglacial, characterized by three intervals both warmer and colder than the present that lasted until the middle Wisconsinan, with no glacial interruptions. Several factors mitigate against this conclusion. Wood from Bay St. Lawrence providing the youngest U/Th age (47 ka) has an anomalous U content, and was sampled from an environment where postdepositional U-migration is a strong possibility. Middle Wisconsinan dates on molluscs at Salmon River and Cape George (Table 2) are suspect for the same reason. The same wood beds that have produced finite radiocarbon dates have yielded infinite radiocarbon dates (Fig. 14) and U/Th dates as old as 98 ka.

With the questionable reliability of middle Wisconsinan U/Th and radiocarbon dates, the chronology of post-Sangamonian events is not well constrained. It is clear from post-Sangamonian erosional and depositional stratigraphy, however, that at least four phases of ice flow have affected the Nova Scotia region since the last interglacial (Fig. 2). The earliest of these flows was a major advance that crossed the Gulf of St. Lawrence and Bay of Fundy (ice-flow phase 1; Fig. 2). The Red Head till, East Milford till, and the shelly diamictons (unit 4) at Cape George and Bay St. Lawrence were formed during this phase (Fig. 15). Later, separate ice caps and divides flowed out of areas adjacent to the province and the province itself (ice-flow phases 2–4; Fig. 2). It is unlikely that four separate phases of ice flow from differing centers from outside and within the province itself could have occurred in the late Wisconsinan, especially with the possibility of intervening recessional intervals. It is more likely that the earliest advance (phase 1) occurred during marine isotope stage 4.

Evidence for major recessional intervals is found in the offshore, where a continuous, conformable sequence of glaciomarine sediments overlie and are intercalated with diamictons believed to represent the last grounded ice on the continental shelf (King and Fader, 1986). Recently published AMS dates (Gipp and Piper, 1989) place the base of the glaciomarine sediments on the central shelf at 18 ka. Ice did not retreat permanently from the central continental shelf until 15 ka (Gipp and Piper, 1989). The underlying till, called the Scotian Shelf drift, therefore is coeval with, or predates the late Wisconsinan maximum. The Scotian

Shelf drift, which contains basaltic erratics (King and Fader, 1986), has been correlated with ice-flow phase 1 of the erosional stratigraphy on land (Fundy Stade) (Fig. 2), a flow that crossed the Bay of Fundy and eroded basaltic rocks. Illinoian glaciations, however, also crossed the Bay of Fundy and may have been of greater intensity (Alam and Piper, 1977). The Scotian Shelf drift may encompass the correlatives of several till units on land. Stea and others (1991) correlated an acoustic facies within major moraines on the inner shelf (King and others, 1972) with the Beaver River till (Fig. 5) formed under the Scotian ice divide (phase 3, Fig. 2). This correlation would imply that flow from the Scotian ice divide was more extensive than generally depicted in the "minimum model" (Piper and others, 1986; Grant and King, 1984; Grant, 1989). This interpretation is consistent with striation data that imply extensive flow from the Scotian ice divide across highland regions in northern Nova Scotia above an elevation of

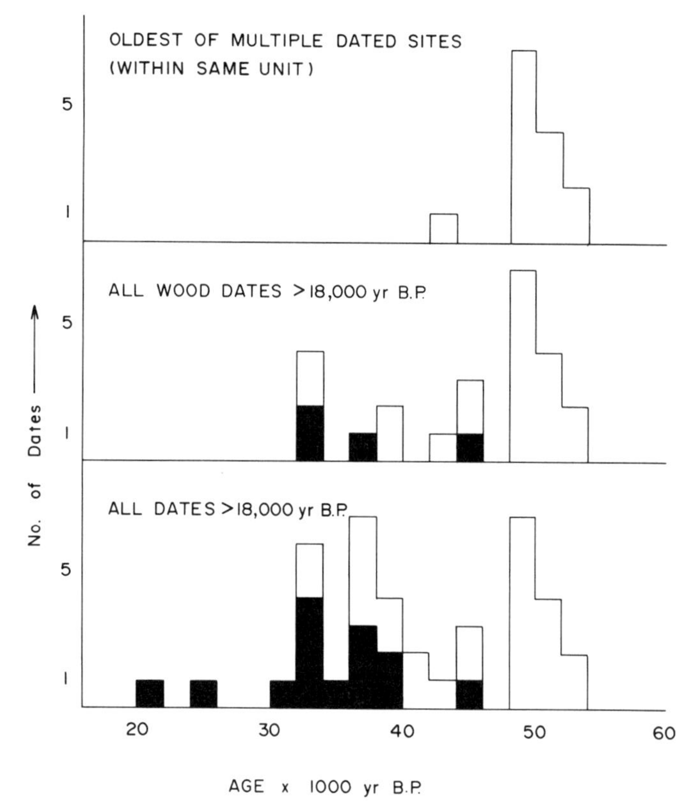

Figure 14. Histogram comparing all radiocarbon ages in Nova Scotia >18 ka. Solid bar represents a finite date, open bar represents a nonfinite date. The evidence for a widespread, terrestrial middle Wisconsinan nonglacial interval becomes less impressive when wood dates are segregated. The remaining sites with finite dates have all been redated and have produced infinite dates.

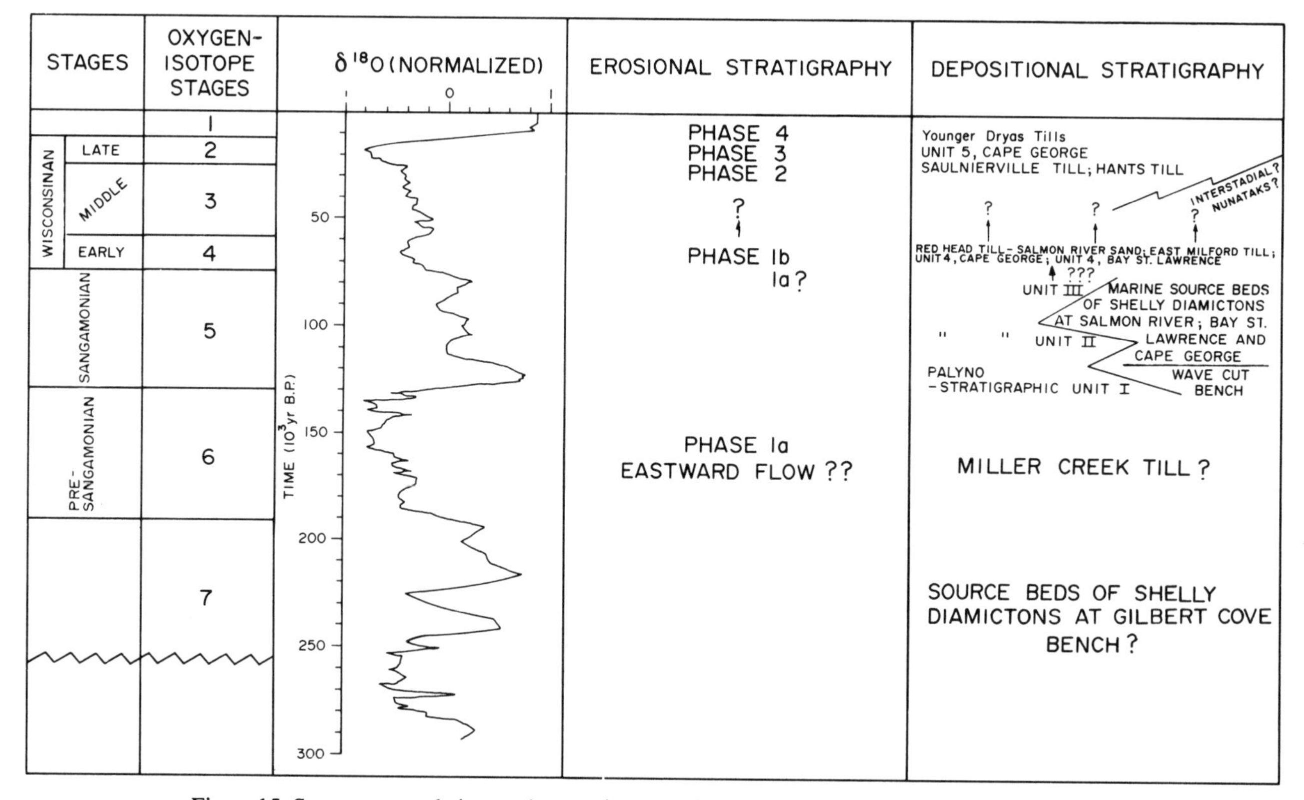

Figure 15. Summary correlations and age assignments for the stratigraphic units and ice flow phases described in the text. Dated oxygen isotope curve after Martinson and others (1987).

200–300 m (Stea and Finck, 1984). Offshore seismic evidence links the ice divide with ice streams adjacent to the Cape Breton highlands (Heiner Josenhans 1990, personal commun.). Piper and Fehr (1991) proposed that mid-Wisconsinan sediments were preserved on the inner shelf region, based on a shell date of 28 ka. Amos and Miller (1990) also reported mid-Wisconsinan shell dates on the outer shelf. A 37 ka date was obtained on marine sediments infilling tunnel valleys on the outer shelf (Boyd and others, 1988). If mid-Wisconsinan ice recession exposed the inner shelf, then late Wisconsinan ice can be bracketed between 28 and 15 ka. It is suggested that phases 2 and 3 were formed during this interval. Soil formation has been noted on the surface of phase 1 tills (Grant, 1980a; MacEachern and Stea, 1985). Phase 1 till units are often separated from later tills by widespread clay and sand layers (Grant and King, 1984) assigned to a mid-Wisconsinan interstadial (Grant, 1989). Newman (1971) obtained two dates on peat blocks within a till relating to the last ice-flow phases in northern Cape Breton Island. These dates imply that ice-free conditions prevailed, perhaps locally, between 20 and 24 ka. A "snapshot" of possible mid-Wisconsinan ice distribution is given in Figure 16.

The Bay St. Lawrence section differs from all others in the mainland because of lack of recognizable till units relating to regional ice-flow phases. This may be due to its unique topographic setting, adjacent to the highest plateau in Nova Scotia, which may have harbored local ice caps. Alternatively, the Cape Breton Channel (Fig. 1) and the Cabot Strait, deep marine channels, may have funneled ice away from the section or allowed ice-free conditions to have been maintained longer. These channels could also have propagated ice retreat into the region during the middle Wisconsinan (Fig. 16).

Ice-flow phase 1 appears to be the most extensive Wisconsi-

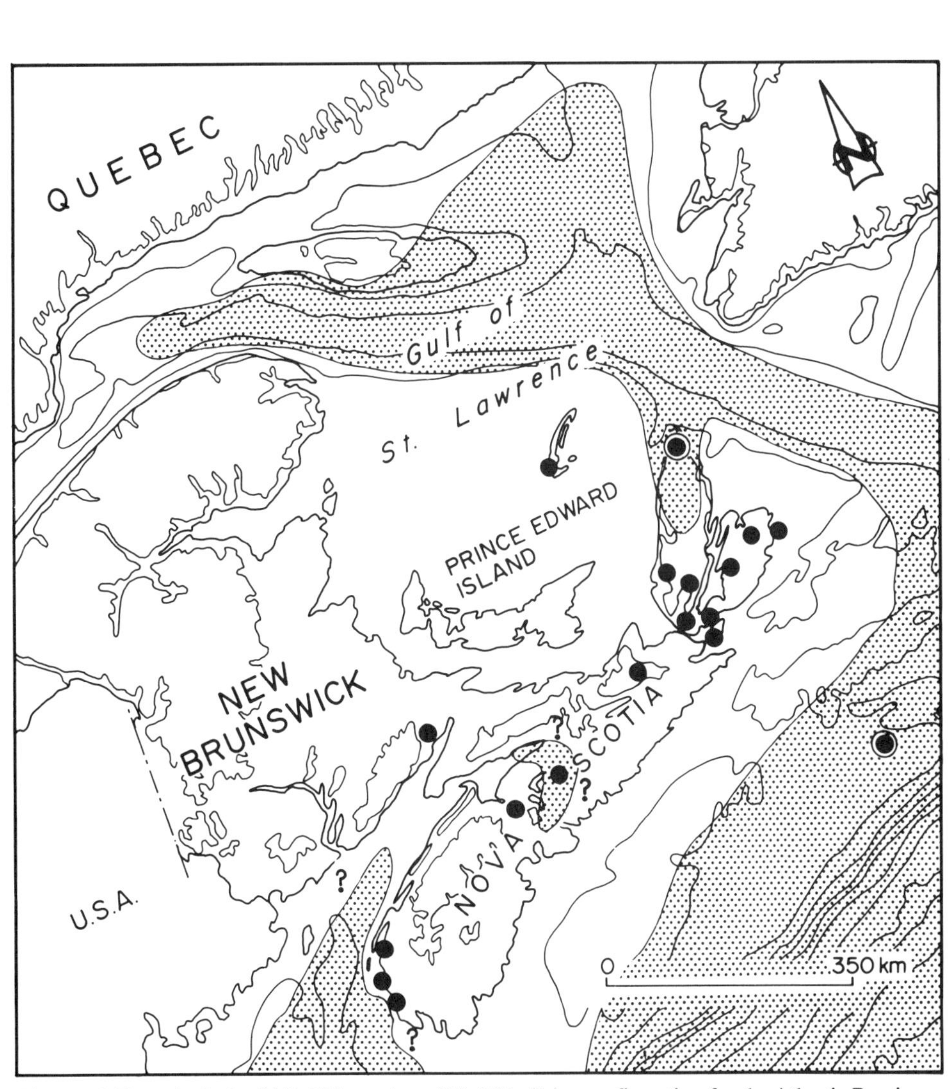

Figure 16. Hypothetical middle Wisconsinan (40–25 ka?) ice configuration for the Atlantic Provinces region. Dotted areas are ice free. Solid large dots represent sites with nonfinite radiocarbon dates; circled dots represent sites with finite radiocarbon dates in suitable stratigraphic context and not subsequently redated as nonfinite.

nan ice flow, and can be correlated across Nova Scotia and may have extended out to the edge of the continental shelf. Grant and King (1984) termed this event the Fundy stade. If the Fundy stade is related to the Laurentide center, then it would represent the culmination of a major continental ice buildup. There is no clear evidence, however, linking the Fundy stade with the Laurentide Ice Sheet. Furthermore, Rampton and others (1984), Pronk and Parkhill (1989), Pronk and others (1989), and Rappol (1986, 1989) did not define a pervasive Laurentide southeastward ice flow across New Brunswick; rather, they link the evidence of early eastward and southeastward ice flow in New Brunswick with Appalachian ice divides, an idea that Chalmers (1894) first espoused. In adjacent Maine, however, late southeastward ice flow has been linked to the Laurentide ice sheet by distinctive erratics (Genes and others, 1981).

To reconcile this apparently contradictory evidence, Prest and Grant (1969) proposed that Appalachian ice was a barrier to the Laurentide Ice Sheet in eastern Canada, or restricted it to defined channels such as the St. John Valley. Further evidence for separation between Laurentide-dominated Maine and Appalachian-dominated eastern Canada is in the pattern of strandlines across the region, an indication of the magnitude of former ice loads. If late Wisconsinan Laurentide ice flow was responsible for all southeastward flow trends in the region, then there should be little change in the elevations of strandlines along the Bay of Fundy coast, which is nearly perpendicular to the presumed Laurentide maximum flow (Denton and Hughes, 1981). Instead, the raised marine features decline from >60 m above mean sea level in Maine to <20 m above mean sea level at the head of the bay (Stea and others, 1987). It is our contention, therefore, that southeastward ice flow in eastern New Brunswick and Nova Scotia (phase 1) is not correlative with similar trends of late Wisconsinan age in Maine.

Ruddiman and others (1980) inferred that a significant part of the global ice accumulation at the stage 5/4 boundary occurred over relatively low latitudes in eastern North America adjacent to the relatively warm Atlantic Ocean. Shackelton (1987) calculated that the maximum ice volume during stage 4 was only 68% of stage 2. Ice-flow phase 1 in Nova Scotia and the Caledonia phase in southern New Brunswick (Rampton and others, 1988) may therefore represent the early buildup of the Appalachian ice (Grant, 1977) rather than the extension of the continental Laurentide Ice Sheet over the region.

ACKNOWLEDGMENTS

This chapter was conceived during the field trip in conjunction with NATO advanced Study Institutes Program course on Late Quaternary Sea-Level Correlation and Applications, July 19–30, 1987, Halifax, Nova Scotia, Canada. This conference was run in cooperation with IGCP Project 200. We thank D. B. Scott and Jane Barrett, Centre of Marine Geology, Dalhousie University, for organizing the field trip, and Jane Barrett for assistance with manuscript preparation.

REFERENCES CITED

Alam, M., and Piper, D.J.W., 1977, Pre-Wisconsin stratigraphy and paleoclimates of Atlantic Canada and its bearing on glaciation in Quebec: Geographie Physique et Quaternaire, v. 31, p. 15–22.

Amos, C. L., and Miller, A. L., 1990, The Quaternary stratigraphy of southwest Sable Island Bank, eastern Canada: Geological Society of America Bulletin, v. 102, p. 915–934.

Belknap, D. F., 1979, Application of amino acid geochronology to stratigraphy of the late Cenozoic marine units of the Atlantic coastal plain [Ph.D. thesis]: University of Delaware, Newark, 550 p.

Boyd, R., Scott, D. B., and Douma, M., 1988, Glacial tunnel valleys and Quaternary history of the outer Scotian Shelf: Nature, v. 333, no. 6168, p. 61–64.

Chalmers, R., 1894, Summary report on investigations of surface geology of southeastern New Brunswick and adjacent parts of Nova Scotia and Prince Edward Island: Geological Survey of Canada, Summary Report 1893, Annual Report 1893, no. 6, part a, p. 52–57.

Clarke, A. H., Grant, D. R., and MacPherson, E., 1972, The relationship of Atractodon stonei (Pilsbry) (Mollusca, Buccinidae) to Pleistocene stratigraphy and paleoecology of Southwestern Nova Scotia: Canadian Journal of Earth Sciences, v. 9, p. 1030–1038.

CLIMAP Project Members, 1976, The surface of the ice-age Earth: Science, v. 191, p. 1131–1137.

Denton, G. H., and Hughes, T. J., 1981, The last great ice sheets: Toronto, Ontario, John Wiley and Sons, Inc., 484 p.

Fulton, R. J., 1984, Summary: Quaternary stratigraphy of Canada, *in* Fulton, R. J., ed., Quaternary stratigraphy of Canada—A Canadian contribution to IGCP Project 24: Geological Survey of Canada Paper 84-10, p. 1–5.

Genes, A. N., Newman, W. A., and Brewer, T. B., 1981, Late Wisconsinan glaciation models of northern Maine and adjacent Canada: Quaternary Research, v. 16, p. 48–65.

Gipp, M. R., and Piper, D.J.W., 1989, Chronology of late Wisconsinan glaciation, Emerald Basin, Scotian Shelf: Canadian Journal of Earth Sciences, v. 26, p. 333–335.

Goldthwait, J. W., 1924, Physiography of Nova Scotia: Geological Survey of Canada Memoir 140, 103 p.

Grant, D. R., 1963, Pebble lithology of the tills of southeast Nova Scotia [M.S. thesis]: Halifax, Nova Scotia, Dalhousie University, 235 p.

——, 1975, Glacial style and the Quaternary stratigraphic record in the Atlantic Provinces, Canada, *in* Report of activities, Part B: Geological Survey of Canada Paper 75-1B, p. 109–110.

——, 1976, Reconnaissance of early and middle Wisconsinan deposits along the Yarmouth-Digby coast of Nova Scotia, *in* Report of activities, Part B: Geological Survey of Canada Paper 76-1B, p. 363–369.

——, 1977, Glacial style and ice limits, the Quaternary stratigraphic record, and changes of land and ocean level in the Atlantic Provinces, Canada: Geographie Physique et Quaternaire, v. 31, no. 3-4, p. 247–260.

——, 1980a, Quaternary stratigraphy of southwestern Nova Scotia: Glacial events and sea level changes: Geological Association of Canada and Mineralogical Association of Canada Guidebook, 63 p.

——, 1980b, Quaternary sea-level change in Atlantic Canada as an indication of crustal delevelling, *in* Morner, N. A., ed., Earth rheology, isostasy and eustasy: London, John Wiley and Sons, p. 201–214.

——, 1988, Surficial geology, Cape Breton Island, Nova Scotia: Geological Survey of Canada Map 1631A, scale 1:125,000.

——, 1989, Quaternary geology of the Atlantic Appalachian region of Canada: Chapter 5, *in* Fulton, R. J., ed., Quaternary geology of Canada and Greenland: Geological Survey of Canada, Geology of Canada, no. 1 (also Geological Society of America, The Geology of North America, v. K-1), p. 393–440.

Grant, D. R., and King, L. H., 1984, A stratigraphic framework for the Quaternary history of the Atlantic Provinces, *in* Fulton, R. J., ed., Quaternary stratigraphy of Canada—A Canadian contribution to IGCP Project 24: Geo-

logical Survey of Canada Paper 84-10, p. 173–191.
Grün, R., 1989, Die ESR-Altersbestimmungsmethode: Berlin, Springer-Verlag, 130 p.
Guilbault, J. P., 1982, The pre-late Wisconsinan foraminiferal assemblage of St. Lawrence Bay, Cape Breton Island, Nova Scotia, *in* Current research, Part C: Geological Survey of Canada Paper 82-1C, p. 39–43.
Henning, G. J., and Grun, R., 1983, ESR dating in Quaternary geology: Quaternary Science Reviews, v. 2, p. 157–238.
Hickox, C. F., Jr., 1962, Pleistocene geology of the central Annapolis Valley, Nova Scotia: Nova Scotia Department of Mines Memoir 5, 36 p.
Ikeya, M., 1985, Dating methods of Pleistocene deposits and their problems: IX. Electron spin resonance, *in* Rutter, N. W., ed., Dating methods of Pleistocene deposits and their problems: Geoscience Canada, Reprint Series 2, p. 73–87.
Katzenberger, O., Debuyst, R., DeCannière, P., Dejehet, F., Apers, D., and Barabas, M., 1988, Temperature experiments on mollusc samples: An approach to ESR signal identification, *in* 2nd International Symposium on ESR Dosimetry and Application: Munich, Neuherberg.
King, L. H., and Fader, G. B., 1986, Wisconsinan glaciation of the Continental Shelf of southeast Atlantic Canada: Geological Survey of Canada Bulletin 363, 72 p.
King, L. H., MacLean, B., and Drapeau, G., 1972, The Scotian Shelf submarine end-moraine complex, *in* Section 8; Marine Geology and Geophysics: International Geological Congress, 24th, Montreal, p. 237–249.
Lowdon, J. A., and Blake, W., Jr., 1973, Geological Survey of Canada radiocarbon dates XIII: Geological Survey of Canada Paper 73-7, p. 10–11.
Macdonald, A. S., and Smith, P. K., 1980, Geology of Cape North area, northern Cape Breton Island, Nova Scotia: Nova Scotia Department of Mines and Energy Paper 80-1, 60 p.
MacEachern, I. J., and Stea, R. R., 1985, The dispersal of gold and related elements in tills and soils at the Forest Hill Gold District, Guysborough County, Nova Scotia: Geological Survey of Canada Paper 85-18, 49 p.
MacNeill, R. H., 1951, Pleistocene geology of the Wolfville area, Nova Scotia [M.S. thesis]: Wolfville, Nova Scotia, Acadia University, 59 p.
——, 1969, Some dates relating to the dating of the last major ice sheet in Nova Scotia: Maritime Sediments, v. 5, no. 1, p. 3.
MacNeill, R. H., and Purdy, C. A., 1951, A local glacier in the Annapolis-Cornwallis Valley [abs.]: Nova Scotian Institute of Science Proceedings, v. 23, part 1, p. 111.
Martinson, D. G., Pisias, N. G., Hays, J. D., Imbrie, J., Moore, T. C., Jr., and Shackleton, N. J., 1987, Age dating and the orbital theory of the ice ages: Development of a high resolution 0 to 300,000 year chronostratigraphy: Quaternary Research, v. 27, p. 1–29.
Mott, R. J., and Grant, D.R.G., 1985, Pre-late Wisconsinan paleoenvironments in Atlantic Canada: Geographie Physique et Quaternaire, v. 39, p. 239–254.
Mott, R. J., and Prest, V. K., 1967, Stratigraphy and palynology of buried organic deposits from Cape Breton Island, Nova Scotia: Canadian Journal of Earth Sciences, v. 4, p. 709–724.
Mott, R. J., Anderson, T. W., and Matthews, J. V., Jr., 1982, Pollen and macrofossil study of an interglacial deposit in Nova Scotia, *in* Richard, P., ed., Special edition devoted to papers from the Eleventh INQUA Congress: Geographie Physique et Quaternaire, v. 36, no. 1-2, p. 197–208.
Myers, R. A., and Stea, R. R., 1986, Surficial mapping results: Pictou, Guysborough, and Antigonish counties, Nova Scotia, *in* Bates, J. L., ed., Mines and Minerals Branch, Report of activities 1985: Nova Scotia Department of Mines and Energy Report 86-1, p. 189–194.
Newman, W. A., 1971, Wisconsin glaciation of northern Cape Breton Island, Nova Scotia, Canada [Ph.D. thesis]: Syracuse, New York, Syracuse University, 117 p.
Nielsen, E., 1974, A mid-Wisconsinan glacio-marine deposit from Nova Scotia, *in* Abstracts Volume, International Symposium on Quaternary Environments: Toronto, York University,
——, 1976, The composition and origin of Wisconsinan tills in mainland Nova Scotia [Ph.D. thesis]: Halifax, Nova Scotia, Dalhousie University, 256 p.
Occhietti, S., and Rutter, N., 1982, Amino acids and Wisconsinan interstades in the Saint Lawrence Valley and Cape Breton Island, *in* Current research, Part B: Geological Survey of Canada Paper 82-1B, p. 301–305.
Oldale, R. N., Valentine, P. C., Cronin, T. M., Spiker, E. C., Blackwelder, B. W., Belknap, D. F., Wehmiller, J. F., and Szabo, B. J., 1982, Stratigraphy, structure, absolute age, and paleontology of the upper Pleistocene deposits at Sankaty Head, Nantucket Island, Massachusetts: Geology, v. 10, p. 246–252.
Piper, D.J.W., and Fehr, S. D., 1991, Radiocarbon chronology of late Quaternary sections on the inner and middle Scotian Shelf, *in* Current research, Part E: Geological Survey of Canada Paper 91-1E, p. 321–325.
Piper, D.J.W., Mudie, P. J., Letson, J.R.J., Barnes, N. E., and Iuliucci, R. J., 1986, The marine geology of the inner Scotian Shelf off the south shore, Nova Scotia: Geological Survey of Canada Paper 85-19, 65 p.
Prest, V. K., 1957, Pleistocene geology and surficial deposits: Geological Survey of Canada Economic Geology Series, no. 1, p. 443–495.
——, 1973, Surficial deposits of Prince Edward Island: Geological Survey of Canada Map 1366A, scale 1:126,720.
Prest, V. K., and Grant, D. R., 1969, Retreat of the last ice sheet from the Maritime Provinces–Gulf of St. Lawrence region: Geological Survey of Canada Paper 69-33, 15 p.
Prest, V. K., and Nielsen, E., 1987, The Laurentide ice sheet and long distance transport, *in* Kujansuu, R., and Saarnisto, M., eds., INQUA Till Symposium, Finland 1985: Geological Survey of Finland Special Paper 3, p. 91–102.
Prest, V. K., and nine others, 1972, Quaternary geology, geomorphology and hydrogeology of the Atlantic Provinces: International Geological Congress, 24th, Excursion Guidebook A61-C61, 79 p.
Pronk, A. G., and Parkhill, M. A., 1989, Why Laurentide ice indicators are absent in the Bay Des Chaleurs region, northern New Brunswick, *in* Abstract volume, Current Research in the Atlantic Provinces: Atlantic Geoscience Society Colloqium, p. 3.
Pronk, A. G., Bobrowsky, P. T., and Parkhill, M. A., 1989, An interpretation of the late Quaternary glacial flow indicators in the Baie des Chaleurs region, northern New Brunswick: Geographie Physique et Quaternaire, v. 43, no. 2, p. 179–190.
Rampton, V. N., Gauthier, R. C., Thibault, J., and Seaman, A. A., 1984, Quaternary geology of New Brunswick: Geological Survey of Canada Memoir 416, 77 p.
Rappol, M., 1986, Aspects of ice-flow patterns, glacial sediments, and stratigraphy in northwest New Brunswick, *in* Current research, Part B: Geological Survey of Canada Paper 86-1B, p. 223–237.
——, 1989, Glacial history of northwestern New Brunswick: Geographie Physique et Quaternaire, v. 43, no. 2, p. 191–206.
Ruddiman, W. F., McIntyre, A., Niebler-Hunt, V., and Durazzi, J. T., 1980, Oceanic evidence for the mechanism of rapid Northern Hemisphere glaciation: Quaternary Research, v. 13, p. 33–64.
Schnikter, D., 1987, Late glacial paleoceanography of the Gulf of Maine: International Union of Quaternary Research, 12th International Congress, Programme and Abstracts, p. 260.
Schwarcz, H. P., and Blackwell, B., 1985, Dating methods of Pleistocene deposits and their problems: II Uranium-Series disequilibrium dating, *in* Rutter, N. W., ed., Dating methods of the Pleistocene and their problems: Geoscience Canada Reprint Series 2, p. 9–19.
Shackelton, N. J., 1987, Oxygen isotopes, ice volume and sea level: Quaternary Science Reviews, v. 6, p. 183–190.
Shackleton, N. J., and Opdyke, N. D., 1973, Oxygen isotope and paleomagnetic stratigraphy of equatorial Pacific core V28-238: Oxygen isotope temperatures and ice volumes on a 10^5 year and 10^6 year scale: Quaternary; Research, v. 3, p. 39–55.
Shepard, F. P., 1930, Fundian faults or Fundian glaciers: Geological Society of America Bulletin, v. 41, p. 659–674.
Stea, R. R., 1982, The properties, correlation and interpretation of Pleistocene sediments in central Nova Scotia [M.S. thesis]: Halifax, Nova Scotia, Dalhousie University, 215 p.

——, 1984, The sequence of glacier movements in northern mainland Nova Scotia determined through mapping and till provenance studies, *in* Mahaney, W. C., ed., Correlation of Quaternary chronologies: Norwich, Geo-Books, p. 279–297.

Stea, R. R., and Finck, P. W., 1984, Patterns of glacier movement in Cumberland, Colchester, Hants, and Pictou counties, northern Nova Scotia, *in* Current research, Part A: Geological Survey of Canada Paper 84-1A, p. 477–484.

——, 1988, Quaternary geology of northern mainland Nova Scotia, sheets 10 and 11: Nova Scotia Department of Mines and Energy, maps 88-13 and 88-14, scale 1:100,000.

Stea, R. R., and Fowler, J. H., 1979, Minor and trace-element variations in Wisconsinan tills, eastern shore region, Nova Scotia: Nova Scotia Department of Mines and Energy Paper 79-4, 30 p.

Stea, R. R., and Grant, D. R., 1982, Pleistocene geology and till geochemistry of southwestern Nova Scotia (sheets 7 and 8) Nova Scotia Department of Mines and Energy Map 82-10, scale 1:100 000.

Stea, R. R., and Mott, R. J., 1990, Quaternary geology of Nova Scotia: Guidebook for 53rd Annual Friends of the Pleistocene Field Excursion: Nova Scotia Department of Mines and Energy Open-File Report 90-008, 85 p.

Stea, R. R., and Myers, R. A., 1990, Surficial geology of parts of Halifax, Pictou, Antigonish and Guysborough counties: Nova Scotia Department of Mines and Energy Map 90-6, scale 1:100,000.

Stea, R. R., Finck, P. W., and Wightman, D. M., 1985, Quaternary geology and till geochemistry of the western part of Cumberland County, Nova Scotia (sheet 9): Geological Survey of Canada Paper 85-17, 57 p.

Stea, R. R., and 10 others, 1987, Quaternary glaciations, geomorphology and sea level changes: Bay of Fundy region: NATO Advanced Course on Sea Level Correlations and Applications, field trip guidebook: Halifax, Nova Scotia, NATO, 79 p.

Stea, R. R., Turner, R. G., Finck, P. W., and Graves, R. M., 1989, Glacial dispersal in Nova Scotia: a zonal concept, *in* Dilabio, R.N.W., and Coker, W. B., eds., Drift Prospecting: Geological Survey of Canada Paper 89-20, p. 155–169.

Stea, R. R., Fader, G.B.J., and Boyd, R., 1991, Quaternary seismic stratigraphy of the inner shelf region, Eastern Shore, Nova Scotia, *in* Current research, part A: Geological Survey of Canada Paper 91-1A.

Vernal, A., de, 1983, Paleoenvironments du Wisconsinien par la palynologie dans la region du Bay St. Laurent, Ile du Cap Breton [these]: Montreal, Quebec, Universite de Montreal, 97 p.

Vernal, A., de Causse, C., Hillaire-Marcel, C., Mott, R. J., and Occhietti, S., 1986, Palynostratigraphy and Th/U ages of upper Pleistocene interglacial and interstadial deposits on Cape Breton Island, eastern Canada: Geology, v. 14, p. 554–557.

Wagner, F., J. E., 1977, Paleoecology of marine Pleistocene Mollusca, Nova Scotia: Canadian Journal of Earth Sciences, v. 14, p. 1305–1323.

Warren, W. P., 1991, Ireland 1991: Field guide for excursion: INQUA Commission on Formation and Properties of Glacial Deposits, 50 p.

Wehmiller, J. F., and Belknap, D. F., 1982, Amino acid age estimates, Quaternary Atlantic coastal plain; comparison with U-series dates, biostratigraphy, and paleomagnetic control: Quaternary Research, v. 18, p. 311–336.

Wehmiller, J. F., Belknap, D. F., Boutin, B. S., Mirecki, J. E., and Rahaim, S. D., and York, L. L., 1988, A review of the aminostratigraphy of Quaternary mollusks from United States Atlantic Coastal Plain sites, *in* Easterbrook, D. J., ed., Dating Quaternary sediments: Geological Society of America Special Paper 227, p. 69–110.

Geological Survey of Canada Contribution No. 47691 (Mott)

Manuscript Accepted by the Society September 6, 1991

Printed in U.S.A.

Geological Society of America
Special Paper 270
1992

Hudson Bay lowland Quaternary stratigraphy: Evidence for early Wisconsinan glaciation centered in Quebec

L. H. Thorleifson, P. H. Wyatt,* and W. W. Shilts
Geological Survey of Canada, 601 Booth Street, Ottawa, Ontario K1A 0E8, Canada
E. Nielsen
Manitoba Energy and Mines, 330 Graham Avenue, Winnipeg, Manitoba R3C 4E3, Canada

ABSTRACT

Information from river sections in the Hudson Bay lowland indicates that two pre-Holocene nonglacial episodes separated by glacial advances postdate the oldest recognized glaciation. Amino-acid data from *in situ* and transported marine shell fragments provide relative ages for glacial and nonglacial intervals. Absolute ages for nonglacial sediments as recent as mid-stage 3 were obtained from thermoluminescence (TL) data, although no finite radiocarbon ages have been obtained from wood. Deglaciation and deposition of the Bell Sea marine sediments are correlated to substage 5e by extrapolation from TL data. Ensuing stage 5 glaciation was dominated in Ontario by west-northwestward ice flow emanating from Quebec, and in Manitoba by southwestward ice flow. Deglaciation dated by TL at about 75 ka was followed by isostatic recovery and subaerial exposure in a climate which could have been warmer, but was no more than slightly colder than present. Extensive glaciolacustrine sediments deposited at the close of this interstade were TL dated at about 40 ka in Manitoba. If the TL method has systematically underestimated age, glaciolacustrine sedimentation may date to very late stage 5 or stage 4, or the two nonglacial episodes could be reassigned to substage 5e and stage 7. A resurgence of Quebec-derived ice that culminated as late Wisconsinan glaciation first flowed westward across the entire lowland, but was displaced in the north by southward ice flow. Southwestward and, locally, southward ice flow occurred during final ice retreat along a saddle extending across Hudson Bay and linking domes in Keewatin and Quebec.

INTRODUCTION

A complex Quaternary stratigraphic sequence is well exposed at numerous sites in the Hudson Bay lowland (Fig. 1). This record is located at the geographic center of Laurentide Ice Sheet glaciation. The surrounding shield terrane generally lacks a depositional record of glaciation and there are few exposures where thick drift occurs. In contrast, the Hudson Bay lowland is mantled by thick sediments produced by glacial erosion of Paleozoic and Proterozoic rocks. Rivers have cut into the nearly flat surface, exposing a stratigraphic record that is unlikely to differ from sequences underlying broad interfluves. This record provides information regarding the nature of past nonglacial episodes, the nature of ensuing glaciation, and therefore, the nature of the transition from interglacial to glacial conditions near the site of glacial inception.

Topography in the Hudson Bay lowland is subdued. Elevations rise gently from sea level to about 150 m at the contact with the shield. The Sutton Ridge reaches an elevation of 250 m. Broad topographic highs elsewhere in the lowland reach 200 m in

*Present address: Northwood Geoscience, 357 Preston Street, Suite 201, Ottawa, Ontario K1S 4M8, Canada.

Thorleifson, L. H., Wyatt, P. H., Shilts, W. W., and Nielsen, E., 1992, Hudson Bay lowland Quaternary stratigraphy: Evidence for early Wisconsinan glaciation centered in Quebec, *in* Clark, P. U., and Lea, P. D., eds., The Last Interglacial-Glacial Transition in North America: Boulder, Colorado, Geological Society of America Special Paper 270.

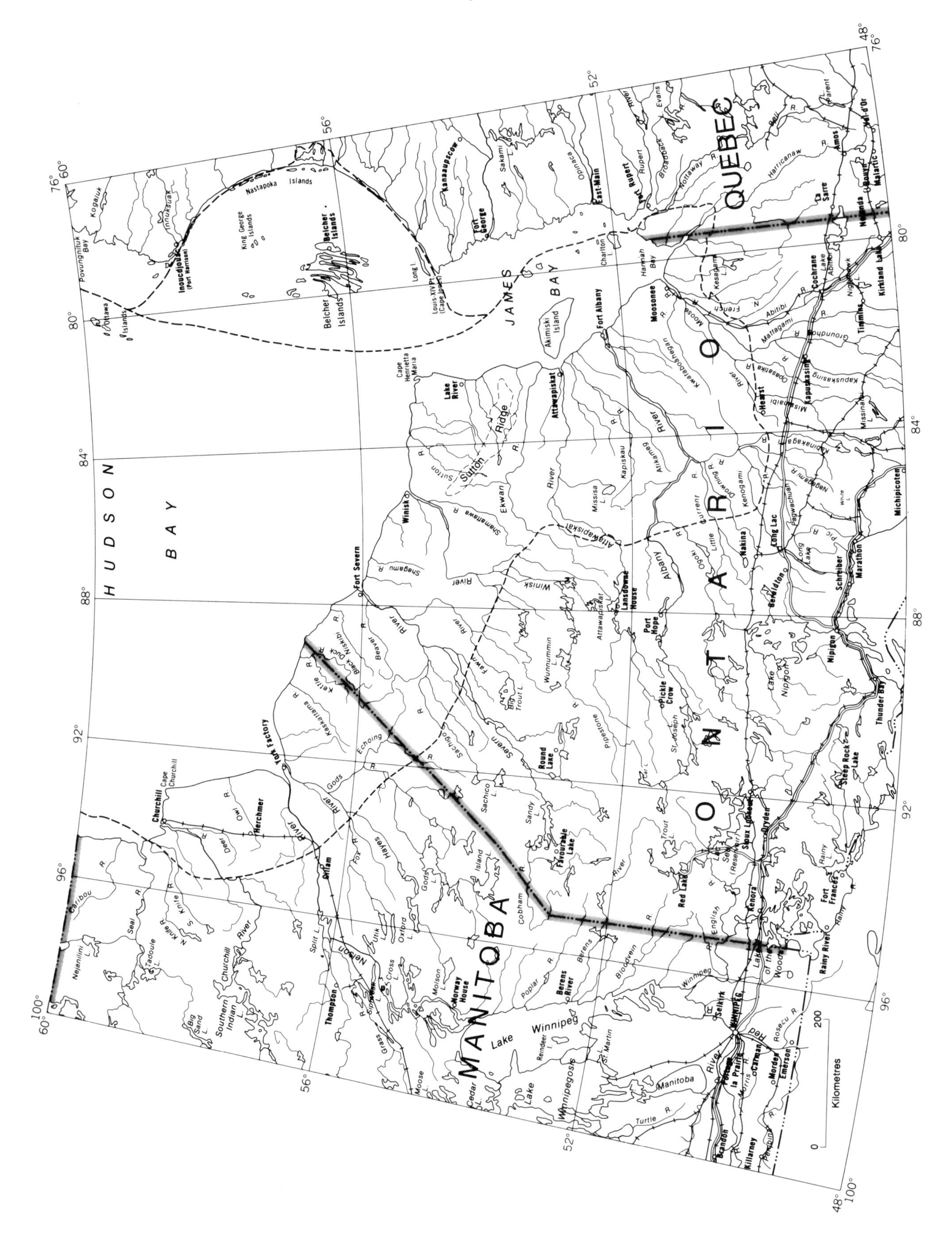

elevation. The surface of this immense gently sloping plain is poorly drained and covered to a large extent by peat. Mean annual temperatures ranges from 0 °C in the south to –7 °C in the north.

PREVIOUS RESEARCH

For more than a century, the Hudson Bay lowland has been a source of information regarding local glacial history and the functioning of the Laurentide Ice Sheet as a whole. Bell (e.g., 1887) recognized Paleozoic and Proterozoic erratics transported west to Manitoba, southwest across northern Ontario, and south to the Moose River basin, the southernmost part of the lowland. Tyrrell (1898) correlated multiple tills in the Hudson Bay lowland to multiple ice sheets separated in space and time and subsequently reported evidence for northwestward, "Patrician" ice flow in northern Ontario (Tyrrell, 1914). Terasmae (1958) and Terasmae and Hughes (1960) examined the Missinaibi beds, a sequence of marine sediments, peat, forest litter, and glaciolacustrine sediments that underlies till in the Moose River basin. McDonald (1969, 1971) conducted a reconnaissance examination of lowland stratigraphy and found nonglacial deposits that he correlated to the Missinaibi beds. Skinner (1973) examined Moose River basin stratigraphy in detail and assigned the marine member of the Missinaibi Formation, which he named the Bell Sea sediments, to isostatic depression during the onset of the last interglaciation. Stuiver and others (1978) used isotope enrichment techniques to obtain a radiocarbon age of >73,500 yr B.P. for wood from the Missinaibi Formation. Two or three tills overlying subaerial deposits, sometimes separated by waterlaid sediments, were reported for Ontario by McDonald (1969), Skinner (1973), and Thorleifson and others (1992) and for Manitoba by Netterville (1974), Klassen (1986), Nielsen and others (1986), and Dredge and others (1986). A late glacial readvance on the shield south of James Bay named the Cochrane (Hughes, 1965) was correlated to the upper till in parts of the lowland by Prest (1969, 1970). The tills are overlain by a glaciolacustrine sequence and marine sediments deposited in the early Holocene Tyrrell Sea (Lee, 1960), which inundated isostatically depressed terrain around Hudson Bay to altitudes of 150 m ca. 8 ka (Craig, 1969; Walcott, 1972; Dredge and Cowan, 1989).

It has not been conclusively determined whether a single center of Laurentide ice was established over Hudson Bay (Paterson, 1972), or whether multiple dispersal centers indicated by erratic transport remained active throughout the last glaciation (Shilts and others, 1979; Shilts, 1980; Andrews and Miller, 1979; Dyke and others, 1982; Dyke and Prest, 1987). According to Budd and Smith (1987), negligible basal velocities at the center of an ice sheet dictate that erratic dispersal patterns may not be sufficient evidence to preclude a single-centered ice sheet.

The concept of a stable core of ice over Hudson Bay during tens of thousands of years of Wisconsinan glaciation (Andrews, 1976; Sugden, 1977) was challenged by Shilts (1982) and Andrews and others (1983), who cited amino-acid evidence for repeated marine inundation of the Hudson Bay lowland during the Wisconsinan. They reported that isoleucine epimerization ratios from in situ and transported shells cluster into at least four distinct groups, the youngest being of Holocene age and the oldest derived from Bell Sea sediments. Great fluctuations in ice extent were inferred, due to climatic change or perhaps due to the inherent instability of a marine ice sheet (e.g., Thomas, 1979). Sea-level control of marine ice-sheet volume (e.g., Denton and others, 1986) would involve drawdown or deglaciation by calving of marine ice from sites such as the Hudson Bay lowland. The alternative of direct climatic control of ice volume (e.g., Budd and Smith, 1981) or the possible lack of ice-sheet components, such as ice streams that enable sea-level control of ice volume (Miller and others, 1988), raises the importance of field tests of theories relating to the behavior of the ice sheet through time.

GLACIAL STRATIGRAPHY

The Quaternary sediments of the Hudson Bay lowland are dominated by thick, massive matrix-supported diamicts (Fig. 2). The diamicts are consistently very compact and blocky, and are made up of about one-tenth by weight >2 mm gravel clasts and a matrix in which silt is typically slightly more abundant than nearly equal proportions of sand and clay. Massive diamicts up to 25 m in thickness include sharply defined deformed masses and lenses of sorted sediments, and have abrupt basal contacts associated with erosive features such as striated boulder pavements and shearing of underlying sediments (Warman, 1987). Where not obscured by marine sediments, the surface of the diamict sequences has the form of elongate flutes (Fig. 3). Prolate pebbles in the diamicts have a preferred orientation parallel to underlying boulder pavement striations and, in the uppermost unit, to the trend of the fluted surface. The features of most sediments throughout the lowland indicate subglacial sedimentation beneath a wet-based glacier (Eyles and others, 1982, 1983), hence the massive diamicts are basal tills.

Descriptions of tills are available for the Manitoba lowland (Netterville, 1974; Klassen, 1986; Nielsen and others, 1986; Dredge and others, 1986), the Severn-Winisk basin in northern Ontario (Thorleifson and others, 1992), and for the Moose River basin (Skinner, 1973). Five major till units can be determined in the entire lowland on the basis of (1) stratigraphic position, (2) provenance and amino-acid data, and (3) ice-flow direction inferred from flutes, fabric, and boulder-pavement striations.

Figure 1. Location map. The Hudson Bay lowland extends from the southern coast of Hudson Bay and the western coast of James Bay to the limit of Paleozoic terrane, shown by a dashed line. Unmetamorphosed Proterozoic sedimentary and volcanic rocks underlie the Belcher Island region of eastern Hudson Bay and the Sutton Ridge. Dominantly granitic and metamorphosed rocks of the Canadian Shield surround these areas.

Figure 2. Henday section, Nelson River, Manitoba. At this site, Tyrrell Sea sediments, the Sky Pilot till, and Long Spruce till are separated from the underlying Amery till by Nelson River sediments. This site has recently been inundated by hydroelectric development.

Winisk till

The Winisk till (Thorleifson and others, 1992) is the uppermost unit and represents a discrete ice flow. This relatively thin massive diamict underlies a belt of flutes that extends from the Fawn and Winisk rivers in the north to the Albany River in the south (Figs. 3 and 4). Fabric and striations on boulder pavements at the lower contact of the Winisk till parallel the flutes. The converging northern portion of the fluted belt is confined between a topographic high in the upper Fawn River area and the Sutton Ridge. Dredge and Cowan (1989) and Dyke and Prest (1987) attributed these flutes to many small ice-marginal lobations, but their model cannot be reconciled with the presence of a single till under a simple pattern of generally parallel flutes. Hence the Winisk till is considered the deposit of a single, simple event. Prest (1969) interpreted these flutes as the western flank of the Cochrane lobe. This scenario is doubted, however, on the basis of (1) the improbability of ice flow parallel to the postulated Cochrane ice margin, (2) the substantial rise in elevation along the western margin of this belt, and (3) the presence of an esker that trends southward within the belt before shifting to a southwest trend parallel to flutes west of the belt (Prest and others, 1968). These factors can, however, be accommodated by southward ice flow in a belt flanked by comparatively stagnant ice; i.e., an ice stream (Flint, 1971). This ice flow is thought to be a late feature of the rapidly diminishing late Wisconsinan ice sheet.

Boulton and Clark (1990) analyzed past ice-flow patterns of the Laurentide Ice Sheet using interpretation of lineations in satellite imagery. The flutes on the surface of the Winisk till were labeled drift lineation set 17, their most recent event.

Sky Pilot–Severn–Kipling tills

On the Fawn and Winisk rivers, the Severn till (Thorleifson and others, 1992) underlies the Winisk till. Southwest fabrics in this till parallel flutes that occur throughout the lowland, beyond the limits of the Winisk till (Figs. 3 and 4). The Sky Pilot till of Manitoba (Nielsen and others, 1986), the Gods River tills C and D (Netterville, 1974; Klassen, 1986), and the Kipling till of the Moose River basin (Skinner, 1973) all yield flute, fabric, striation, and compositional data indicating southwestward ice flow. Correlation of the Kipling till to units throughout the lowland rules out correlation of this unit to the Cochrane readvance. Instead, the Sky Pilot, Severn, and Kipling tills are correlated to late Wisconsinan flow along the flank of an ice divide in the form of an elongate ridge over Hudson Bay. Lack of ice flow into Keewatin or Quebec (Shilts, 1980) indicates that this ridge must have been concave along its axis; i.e., a saddle connecting two domes.

Drift lineation sets recognized by Boulton and Clark (1990) that relate to this episode of late Wisconsinan ice flow are their set 38 in Manitoba, set 20 in Ontario, and set 21 in Quebec. Although set 21 was assigned to the late Wisconsinan, immediately preceding the Winisk till, the other two sets were assigned to two different episodes in the early Wisconsinan. Extensive southwestward ice flow predating late glacial patterns was also reported on the basis of striations by Veillette (1986).

Glaciofluvial sediments lacking organic material underly the equivalent of the Sky Pilot till on Gods River (the Twin Creeks sediments; Netterville, 1974; Klassen, 1986) and the Kipling till in the Moose River basin (the Friday Creek sediments; Skinner, 1973). Netterville (1974), McDonald (1971) and Skinner (1973) suggested that deglaciation was required to explain these sediments, but a subglacial origin was proposed by Dredge and Nielsen (1985).

Long Spruce–Sachigo–Adam tills

Till yielding evidence for west-southwestward to westward ice flow underlies the tills described above throughout the lowland (Fig. 4). The Long Spruce till in Manitoba in places contains high concentrations of Proterozoic graywacke erratics derived from the east (Nielsen and others, 1986). Striated boulders at the base of the Adam till indicate a west-southwest orientation (Skinner, 1973). The Sachigo till, which underlies the Severn till at two sites on the Severn River (Thorleifson and others, 1992), produced sparse fabric and striation evidence for westward ice flow near the lower contact, but red carbonate erratics in the upper part of this till are derived from the north (Thorleifson, 1989). Similarly, Netterville (1974) and Klassen (1986) reported southward ice flow for till at this stratigraphic position. Therefore, the Long Spruce, Sachigo, and Adam tills were deposited by a major westward advance, which was followed by localized southward ice flow in the north. These tills overlie subaerial deposits throughout the lowland, and so they are attributed to the growth of the ice sheet that later culminated in the late Wisconsinan maximum.

Boulton and Clark (1990) recognized this ice-flow pattern as set 16 in northern Ontario and assigned it an early Wisconsinan age.

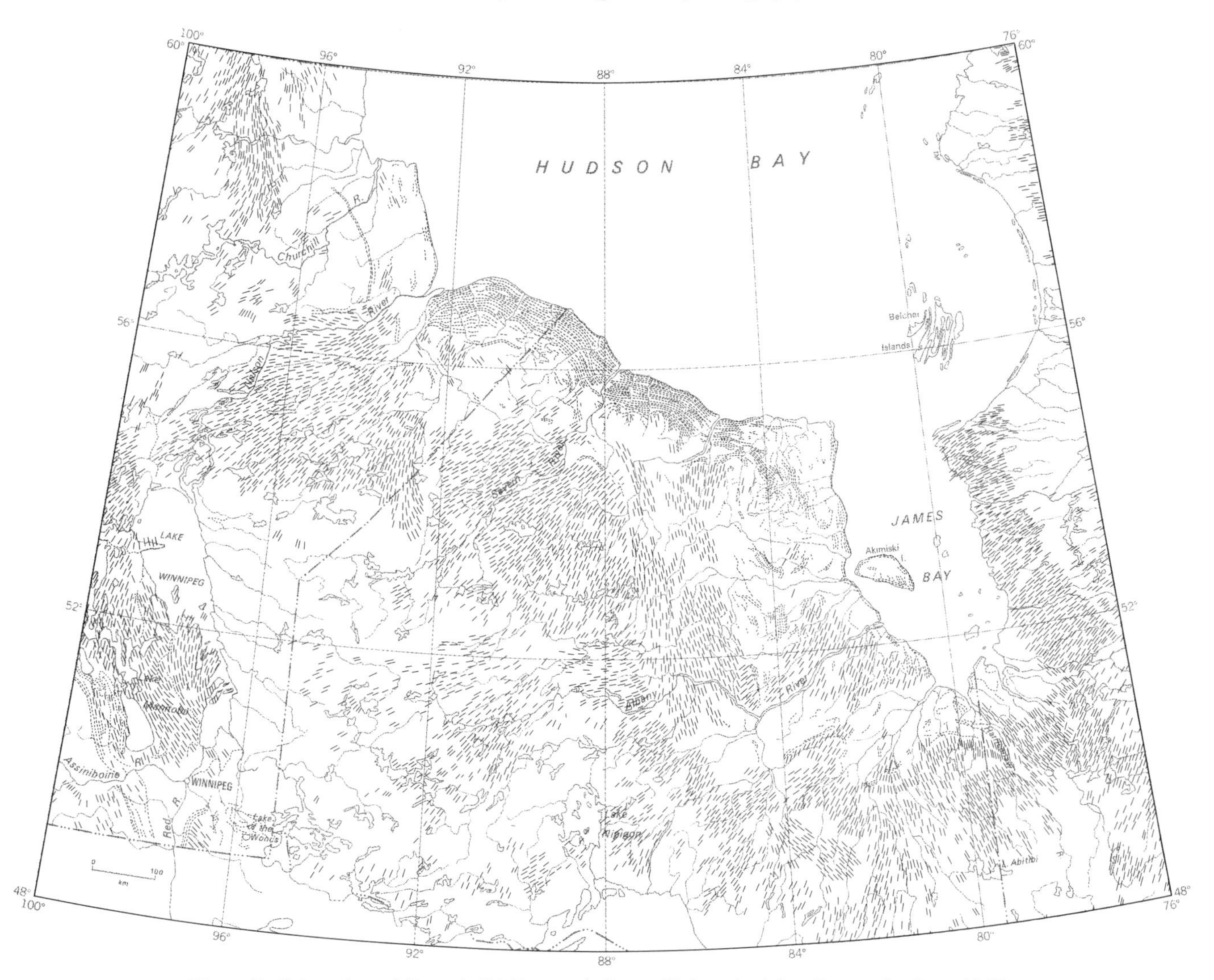

Figure 3. Orientation of flutes (solid line symbol) on till deposits (after Prest and others, 1968). Shoreline features, shown as dashed lines, obscure glacial morphology near Hudson Bay.

Amery-Rocksand tills

The Rocksand till underlies the Severn and Sachigo tills at four sections on the upper Severn River (Thorleifson and others, 1992). This till is slightly enriched in erratics derived from the east and yielded fabric results oriented northwest-southeast, parallel to striations indicating northwestward ice flow at Big Trout Lake (Tyrrell, 1914). Tyrrell combined this orientation with what he thought was north-northwestward ice flow for the Winisk till on the Fawn River to infer that an ice sheet had been centered in the District of Patricia of northern Ontario. This model is rejected, however, due to the reinterpretation of the Winisk till as a deposit of southward flow and because additional striation sites oriented west-northwest (Prest, 1963; Prest and others, 1968; Skinner, 1973; Veillette and Pomares, 1991) imply an ice source in Quebec (Fig. 4). Correlation to Manitoba on the basis of stratigraphic position implies equivalence of the Rocksand till and Amery tills (Nielsen and others, 1986; Klassen, 1986). Fabrics obtained by Nielsen and others (1986) from this till at the Henday section are oriented northwest, but other sites reported by these authors, as well as by Klassen (1986), indicate southwestward ice flow. In the Moose River basin, no till is obviously correlative, but Skinner (1973) obtained evidence for west-northwest ice flow from striated boulders in till that he correlated to the Kipling till, due to the lack of overlying till. Therefore, an additional till which was not recognized by Skinner (1973) must be present, or the uppermost of the pre-Adam tills, which occur on the upper Missinaibi River, may correlate to this episode of dominantly west-northwestward ice flow.

Boulton and Clark (1990) labeled lineations oriented west-

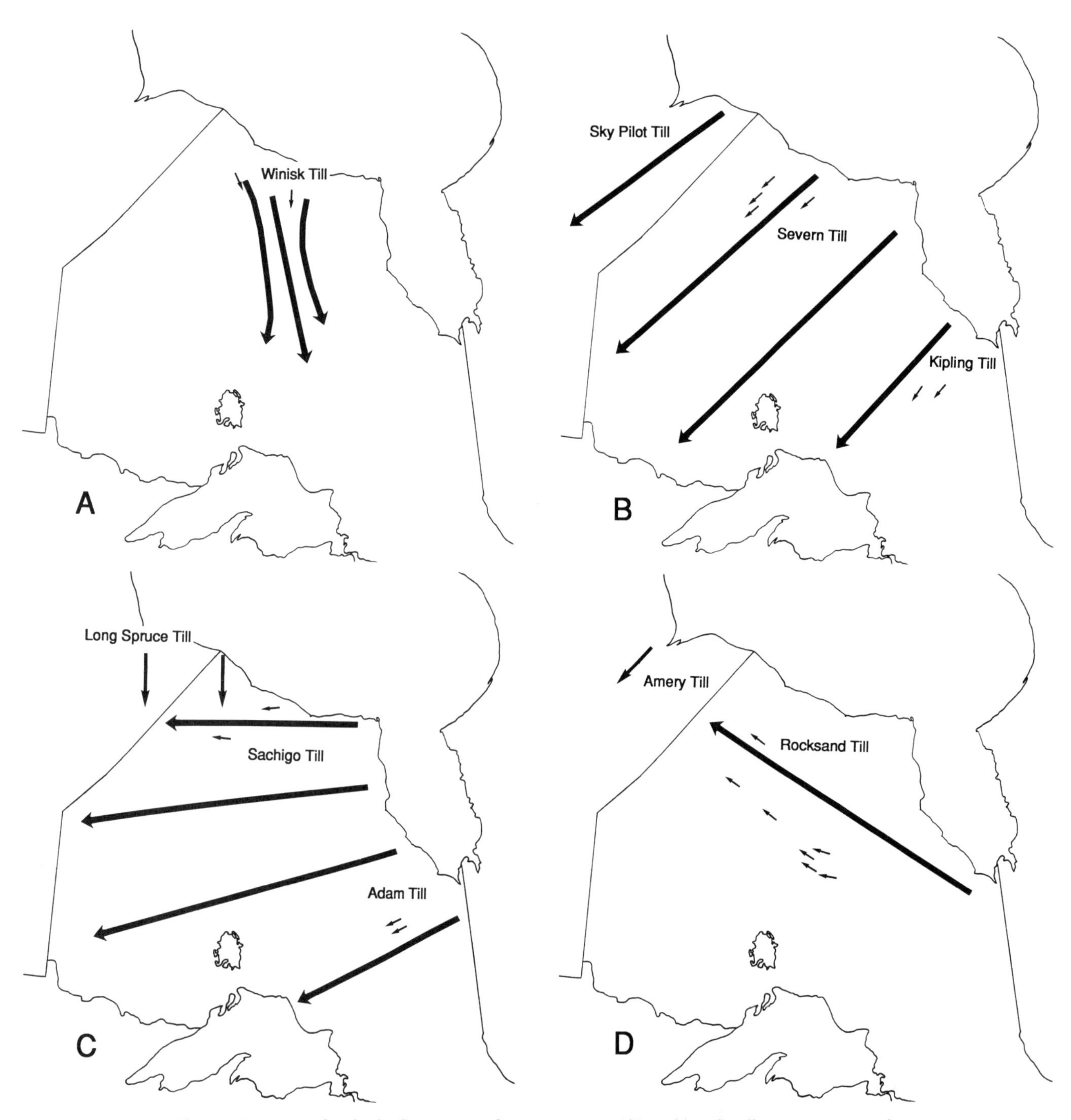

Figure 4. Sequence of major ice flow patterns from most recent (A) to oldest. Small arrows are examples of data from fabric and striations. Arrows of intermediate size indicate ice flow late in this episode (C) or contrasting with the trend in Ontario (D). Evidence for an older, dominantly southeast pattern is not shown.

northwest in northern Ontario as set 18, although the features were interpreted as the product of east-southeastward ice flow. An earliest Wisconsinan age was proposed.

Sundance-Shagamu tills

Underlying the Amery till in Manitoba is the Sundance till (Nielsen and others, 1986), deposited by Keewatin-derived southeastward ice flow. The Shagamu till, in the Severn-Winisk area (Thorleifson and others, 1992), is at this stratigraphic position and also is derived from the northwest. Southeastward ice flow was also determined for the second of three pre-Adam tills of the Moose River basin (Skinner, 1973). An old southeastward ice flow has also been reported for Quebec by Bouchard and Martineau (1985) and the shield south of James Bay (Veillette and others, 1989).

Boulton and Clark (1990) labeled old southeast drift lineations in northern Manitoba as set 34.

PRE-HOLOCENE NONGLACIAL STRATIGRAPHY

Pre-Holocene nonglacial and proglacial sediments (Fig. 5) occur at several sites across the Hudson Bay lowland (McDonald, 1969). The best known occurrences are the Missinaibi Formation of the Moose River basin (Skinner, 1973), and, in northern Manitoba, the Nelson River sediments (Nielsen and others, 1986) and the Gods River sediments (Netterville, 1974; Klassen, 1986). The Kabinakagami sediments, located west of the Moose River basin, were introduced by Shilts (1982) and Andrews and others (1983). Deposits in the central lowland, including the Fawn River gravel (Shilts, 1982; Andrews and others, 1983), were named the Fawn River sediments by Thorleifson and others (1992).

Figure 5. Adam Creek, Moose River basin, Ontario. At this site, the Adam till overlies glaciolacustrine rhythmites of the Missinaibi Formation lacustrine member (from Skinner, 1973; GSC Photograph 154890).

Nelson River sediments

The Nelson River sediments occur on Nelson River in northeastern Manitoba (Nielsen and others, 1986; Dredge and others, 1990; Berger and Nielsen, 1990). Total thickness ranges from 2 to 3 m, and sediments include gravel, thinly bedded sand, silt, and clay with dropstones at some exposures. Organic material 20 cm thick at two sections is composed of woody compressed peat. The Nelson River sediments underlie the Long Spruce till and overlie the Amery till.

Beetle remains from the Henday section indicate that the Nelson River sediments were deposited north of the tree line under conditions more severe than the present. Several taxa currently occupying tundra were encountered, whereas taxa characteristic of the boreal forest, such as bark beetles (Scolytidae), were absent. Analogous conditions were suggested to be the present Eskimo Point, District of Keewatin (Nielsen and others, 1986) or Churchill, Manitoba (Dredge and others, 1990). Beetles from the Limestone section may indicate an open-ground assemblage marginally south of the treeline (Dredge and others, 1990). In contrast, pollen from the sediments yielded dominant spruce (*Picea*) and pine (*Pinus*), suggesting conditions not substantially different from today. A woolly mammoth molar found on a bar in the Limestone River may have been derived from the Nelson River sediments (Nielsen and others, 1988). A tundra pollen assemblage was recovered from a pre-Amery till soil at the Sundance section.

Gods River sediments

The Gods River sediments (Netterville, 1974; Klassen, 1986; Dredge and others, 1990) consist of silt over wood, peat, sand, and gravel. Apparently correlative sediments, which possibly include a basal marine unit, occur on the Hayes (McDonald, 1969), Echoing, and Stupart rivers (Dredge and others, 1990).

Unlike the Nelson River sediments, bark beetles are present in the Gods River sediments (Dredge and others, 1990). Netterville (1974) reported pollen data from a 4-m-thick gyttja sequence within the Gods River sediments (Klassen, 1986; Dredge and others, 1990). The data were divisible into three zones based on shifts in tree pollen, primarily changes in spruce. The lower zone was characterized by low but increasing values for spruce and strong representation of Sphagnum and Cyperaceae, thus indicating a climate similar to the present tundra regions of northeastern Manitoba and southern Keewatin. The assemblage differed from these tundra sites in their higher content of birch

(*Betula*), alder (*Alnus*), and fern spores. The overlying middle zone contains higher values for tree pollen, including a marked inflection in the curves for spruce. This was interpreted by Netterville as the northward passage of the tree line and establishment of a forest-tundra environment similar to present. The upper pollen zone resembles the lower zone, but shows an upward decrease in spruce and other tree pollen. Hence a cooling climate and southward migration of the tree line were inferred.

Fawn River sediments

The Fawn River sediments include marine, fluvial, subaerial, and glaciolacustrine sediments (Thorleifson and others, 1992). Deglaciation of isostatically depressed terrain and opening of Hudson Bay are indicated by marine sediments on the Severn River (Wyatt, 1989). Isostatic recovery is indicated by the Fawn River gravel (Shilts, 1982; Andrews and others, 1983) and similar deposits on Beaver River. Subaerial exposure is indicated by 0.35-m-thick peat that overlies this gravel on Beaver River. Glaciolacustrine sediments overly subaerial deposits at sites on Severn River.

Spruce, larch (*Larix*), and bark beetle macrofossils in the Beaver River peat indicate trees near the site (Wyatt, 1989). Because this occurrence is presently near the tree line, a climate no colder than present is implied. The inferred paleoenvironment has modern analogues well to the south, so the possibility of climate warmer than present cannot be ruled out. Pollen in the peat (Wyatt, 1989; Mott and DiLabio, 1990) shows little vertical variation. Spruce is the dominant pollen taxon at values between 40% and 55%. Pine is second in abundance at about 20%. Birch, Ericaceae (heath), and Gramineae (grass) pollen make up 11% or less. *Sphagnum* spores are abundant. Low values for alder and willow (*Salix*) attest to the presence of these shrubs in suitable nearby locations. Small amounts of sage (*Artemisia*) and other herbaceous pollen taxa indicate that open, drier upland areas were not locally abundant. According to Wyatt (1989), northern boreal to forest tundra conditions, with climate as warm or warmer than present, are suggested by the data.

Kabinakagami sediments

The Kabinakagami sediments, named by Shilts (1982), occur at three sections on the Kabinakagami River, west of the Moose River basin. These sediments consist of contorted laminated silt and crossbedded sand, indicating southeastward paleocurrents. The sediments lack indications of subaerial exposure and are probably glaciofluvial in origin. A shell from this unit analyzed by Andrews and others (1983) was probably reworked. These sediments therefore are correlated to the Friday Creek sediments of the Moose River basin (Skinner, 1973).

Missinaibi Formation

Deposits correlated as the Missinaibi Formation by Skinner (1973) occur on the upper Missinaibi River and its tributaries, Adam Creek, Abitibi River, Moose River Crossing, and Kwataboahegan River. The upper Missinaibi River sites expose glaciolacustrine sediments, peat, wood and a weathering horizon. At Adam Creek, silty clay with wood fragments, gastropod shells, *Hiatella arctica* shells and fragments, marine foraminifera, and ostracoda overlies oxidized gravel, silt, and clay and is overlain by 4 m of glaciolacustrine rhythmites (Skinner, 1973; Fig. 5). Fossiliferous marine sediments underlying till on the Abitibi River were first reported by Prest (1966). The Moose River Crossing section exposes a subtill sequence of sand, silt, and clay containing marine *Hiatella arctica* shell fragments overlain by up to 1.5 m of silt-clay rhythmites. Peat and wood fragments are found at the contact between these two units. Sections on the Kwataboahegan River, just downstream from the confluence with the Mistuskwia River, expose marine sediments underlying till (McDonald, 1969; Skinner, 1973).

Foraminifera from the Abitibi River marine sediment dominated by *Elphidium clavatum* indicate deposition in cold, low-salinity bottom water (Skinner, 1973). Pollen analysis of the Missinaibi Formation on the Missinaibi, Pivabiska, Soweska and Opasatika rivers and from the Moose River Crossing section all yielded assemblages dominated by spruce and pine, similar to present assemblages (Skinner, 1973; Mott and DiLabio, 1990).

Correlation

Stratigraphic context with respect to tills, depositional environments, and pollen data suggest correlation of at least most occurrences of the Nelson River, Gods River, Fawn River, and Missinaibi sequences. Beetle evidence from the Nelson River sediments, the northernmost of the occurrences, for a paleoenvironment north of but near tree line contrasts with other sites.

These deposits are in every case, however, incomplete and fragmentary, so conclusive comparisons of aspects of the sequences cannot be made. Lack of differences in pollen data may be due to the lack of sensitivity of the northern boreal forest to shifts in climate. Two ages of nonglacial deposits were recognized for the Nelson River on the basis of lithostratigraphy (Nielsen and others, 1986), but subdivision is not possible elsewhere without reference to geochronological data.

GEOCHRONOLOGY

Radiocarbon

Radiocarbon dates have been obtained for organic material beneath till at several sites across the Hudson Bay lowland (Table 1). Out of 30 dates, 26 are indeterminate. Two accelerator dates on marine shells from Abitibi River yielded finite dates (Andrews, 1987), but shell material tends to acquire young carbon (e.g., Miller, 1985). Skinner (1973) obtained an age of 38,200 yr B.P. (GSC-1475) from an unleached shell sample, but a nonfinite date was obtained following acid treatment to remove the outer portion of the shells. A finite date of 37,400 yr B.P. was obtained by Wyatt (1989) from peat from Beaver River, but

TABLE 1. RADIOCARBON AGES FROM HUDSON BAY LOWLAND PRE-HOLOCENE ORGANIC MATERIAL

^{14}C (yr BP)	Laboratory number	Material	River	Reference
33,600	Arizona	Shell	Abitibi	Andrews, 1987
37,040	WAT-1378	Peat	Beaver	Wyatt, 1989
38,200	GSC-1475	Outer shell	Kwataboahegan	Skinner, 1973
>37,000	GSC-1475	Inner shell	Kwataboahegan	Skinner, 1973
40,040	TO-125	Shell	Abitibi	Andrews, 1987
>19,000	GSC-1535	Shell	Abitibi	Skinner, 1973
>29,630	Y-269		Missinaibi	McDonald, 1971
>30,840	Y-270		Missinaibi	McDonald, 1971
>32,000	GSC-3074	Wood	Churchill	Dredge and others, 1990
>35,800	GSC-83	Wood	Attawapiskat	Dyck and Fyles, 1963
>37,000	GSC-892	Wood	Echoing	McDonald, 1969
>37,000	GSC-2481	Wood	Stupart	Dredge and others, 1990
>38,000	GSC-4146	Peat	Beaver	Wyatt, 1989
>38,000	W-241		Missinaibi	McDonald, 1971
>38,000	W-242		Missinaibi	McDonald, 1971
>41,000	GSC-1736	Wood	Gods	Klassen, 1986
>41,000	GSC-1011	Peat	Severn	McDonald, 1969
>42,000	Gro-1921	Peat		McDonald, 1971
>42,000	Y-1165	Peat	Harricana	Stuiver and others, 1963
>42,600	L-396B		Missinaibi	McDonald, 1971
>43,000	GSC-4154	Peat	Beaver	Wyatt, 1989
>43,600	GSC-435	Wood	Little Abitibi	Lowdon and others, 1967
>49,000	GSC-4420HP	Wood	Nelson	Nielsen and others, 1988
>49,000	GSC-4471HP	Wood	Gods	Dredge and others, 1990
>50,000	GSC-5071HP	Wood	Missinaibi	This chapter
>51,000	GSC-4423HP	Peat	Beaver	Wyatt, 1989
>51,000	GSC-4444HP	Wood	Echoing	Dredge and others, 1990
>53,000	Gro-1435	Wood	Missinaibi	McDonald, 1971
>54,000	GSC-1185	Peat		McDonald, 1971
>72,500	QL-197	Wood	Missinaibi	Stuiver and others, 1978

subsequent analysis of the same material yielded a nonfinite age. There is, therefore, no convincing radiocarbon evidence indicating that any pre-Holocene organic material in the Hudson Bay lowland has an age less than 50 ka.

Beukens (1990) has indicated, however, that Geological Survey of Canada (GSC) background reference materials have ^{14}C concentrations equivalent to apparent ages of 42 and 45 ka, without correction for background. It could be argued that an uncontaminated wood sample with an age of 45 ka would be dated at >50 ka at the GSC lab. It seems unlikely, however, that a field sample could be less contaminated than the reference materials. Nevertheless, these uncertainties indicate that caution is required in the use of high finite or nonfinite radiocarbon dates.

Thermoluminescence

A TL age of 73 ± 10 ka was obtained for the marine unit of the Fawn River sediments by Forman and others (1987). Berger and Nielsen (1990) subsequently obtained TL ages of 32 to 46 ka from fresh-water silt and silty clay from the Nelson River sediments in northern Manitoba. Marine sediments were deposited early in a nonglacial episode prior to isostatic recovery, whereas glaciolacustrine sediments overlying subaerial deposits and directly underlying till are associated with the blockage of drainage by the advancing ice sheet, at the end of the interstadial or interglacial episode. An age difference sufficient to allow for isostatic recovery and deposition of peat is required, so the two ages are not contradictory.

A contradiction remains, however, with the radiocarbon record, which does not support subaerial exposure between 40 and 50 ka. A possible 40% underestimation of age by the TL method, as discussed by Berger and Nielsen (1990), would permit assignment of the glaciolacustrine sediments to an age greater than 50 ka, such as the end of substage 5a or stage 4 (Shackleton and Opdyke, 1973).

Rejection of GSC nonfinite radiocarbon dates and acceptance of TL age determinations by Berger and Nielsen (1990), however, would indicate the delay of glacial advance after late stage 5 deglaciation until mid-stage 3. This would imply a lack of ice cover during oxygen-isotope stage 4.

Amino acid

***Data from* in situ *shells*.** Shilts (1982) and Andrews and others (1983) obtained total alloisoleucine to isoleucine (aIle/Ile) ratios for *Hiatella arctica* shells collected from two marine units. Early Holocene Tyrrell Sea deposits produced values of ~0.03, whereas shells from the Bell Sea sediments produced ratios of ~0.22. Their data were confirmed by the analyses (Fig. 6) carried out by Wyatt (1989).

Shilts and Wyatt (1988) and Wyatt (1990) subsequently obtained ratios of 0.12 to 0.14 from two additional in situ marine deposits on the Severn and Abitibi rivers (Fig. 6). The name "Prest Sea" was proposed for the latter site by Shilts and Wyatt (1988) and the additional name "Abitibi River sediments" is proposed here. These values are equivalent to the older of the two intermediate clusters obtained from transported shells by Andrew and others (1983). For the Prest Sea and the Bell Sea to be the same age, e.g., 125 ka, a temperature history difference of 3 °C would be required. This is regarded as unlikely, given the proximity of the two Moose River basin sites.

The possibility that Prest Sea marine sediments were deposited in front of advancing ice at the end of the Bell Sea episode can be ruled out because Skinner (1973) observed glaciolacustrine sediments overlying subaerial deposits. Furthermore, in the Severn River basin, Prest Sea ratios were obtained from reworked *Hiatella arctica* fragments collected in gravel underlying peat.

Skinner (1973) correlated all Missinaibi Formation sites, including both marine units, to one episode. Amino-acid data indicate, however, two ages for marine sediments in the Moose River basin. Some of the organic deposits from which paleoecological data have been obtained may therefore date from the Prest Sea episode. Confident correlation of the Missinaibi Formation type locality at section 24M on the upper Missinaibi River to one of the marine units will not be possible until additional data, such as amino-acid ratios from reworked shells, are obtained.

***Thermal history*.** Andrews and others (1983) calculated an effective diagenetic temperature (EDT) of +0.6 °C for the assignment of Bell Sea ratios (0.22) to substage 5e (130 ka). A higher value, +2.2 °C, results from the higher mean ratio (0.24) obtained by Wyatt (1989), an age of 125 ka for the onset of substage 5e (Martinson and others, 1987), and updated Arrhenius parameters (Miller, 1985). This EDT is higher than values considered reasonable by Andrews and others (1983). Assignment of Bell Sea ratios (0.24) to stage 7 would imply a more reasonable EDT of –0.6 °C.

Comparison to the European record also favors assignment of the Prest Sea to substage 5e. Total aIle/Ile ratios of about 0.18 have been obtained from *Hiatella arctica* shells in deposits of the last interglacial in Belgium, the Netherlands, and Germany (Miller and Mangerud, 1985). The present mean annual temperature in northern Europe, about +8 °C, is well above values of 0 to –7 °C in the Hudson Bay lowland. A value of 0.14 would therefore seem reasonable for the last interglacial of Hudson Bay.

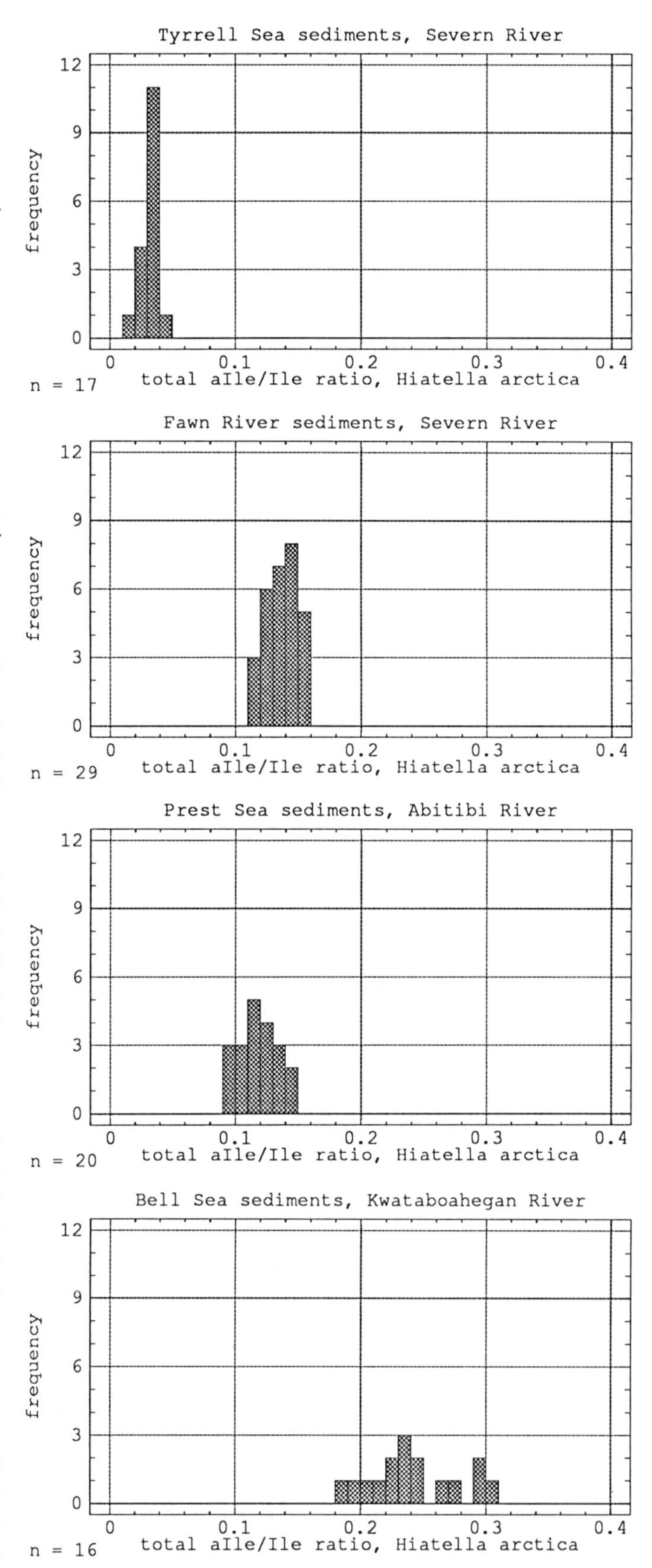

Figure 6. Amino-acid ratios from marine shells collected from raised marine sediments (Wyatt, 1989).

These considerations are not, however, considered sufficient evidence for the rejection of TL data that date marine sediments bearing shells with 0.14 ratios to about 75 ka. The various temperatures obtained from thermal history reconstructions may be within the range of reasonable values, given their sensitivity to the several inputs required. Regarding comparison to Europe, it is suggested that basal temperatures beneath the Laurentide Ice Sheet were probably warmer than the subaerial exposure experienced by northern Europe during much of the last glaciation.

The lack of a convincing conclusion based on thermal history modeling of amino-acid ratios leaves TL dating of Prest Sea sediments as the only opportunity to favor one chronology over another. Therefore, a substage 5a age is assigned to the Prest Sea. By extrapolation, the Bell Sea is assigned to substage 5e.

Data from glacially transported shells. Amino-acid data from tills in the Severn-Winisk area (Wyatt, 1989) allow ice-flow events to be fixed in time relative to marine incursions.

The Rocksand till yielded sparse data that span a range of values similar to and slightly lower than those derived from Bell Sea sediments (Fig. 7). The lack of Prest Sea values and the requirement for an ice mass to produce isostatic depression associated with the Prest Sea indicates correlation of the Rocksand till to post–Bell Sea, pre–Prest Sea time; i.e., mid-stage 5. The lack of Prest Sea ratios in the Rocksand till offers further evidence against the deposition of marine sediments in front of advancing Wisconsinan ice.

The Sachigo till yielded both Bell Sea and Prest Sea ratios (Fig. 7). These ratios suggest that this ice advance postdates the Prest Sea, but that the older Bell Sea shells were reworked.

The Severn till yielded many more shell fragments than the underlying till (Fig. 7). On the Winisk River, this till contains numerous Prest Sea values as well as what might be called a late Bell Sea cluster ca. 0.18. On the Severn River, Prest Sea and Bell Sea ratios are accompanied by an additional cluster at ~0.07, equivalent to the lower of the two intermediate clusters reported by Andrews and others (1983).

The Winisk till, which is primarily derived from offshore Hudson Bay, yielded values dominated by the 0.07 cluster (Fig. 7).

Ratios from tills thus include two clusters of values, at 0.07 and 0.18, that were not obtained from raised marine sediments (Fig. 8). These clusters are considered to be the offshore equivalents of the Prest Sea and the Bell Sea, respectively, as originally discussed by Andrews and others (1984). Whereas raised marine sediments were deposited at the onset of an interglacial or interstadial episode prior to isostatic recovery, offshore shells may be as much as tens of thousands of years younger. In addition to lesser age, offshore shells may have had a cooler thermal history, further contributing to lower ratios. Mean annual air temperature in the lowland ranges from 0 to –7 °C, but subsurface temperature is several degrees higher (Brown, 1973; Harris, 1981). Shells with radiocarbon ages of 7 ka derived from Tyrrell Sea sediments give total aIle/Ile ratios for *Hiatella arctica* of ~0.033, indicating an effective diagenetic temperature of 5.5 °C. Water at the bot-

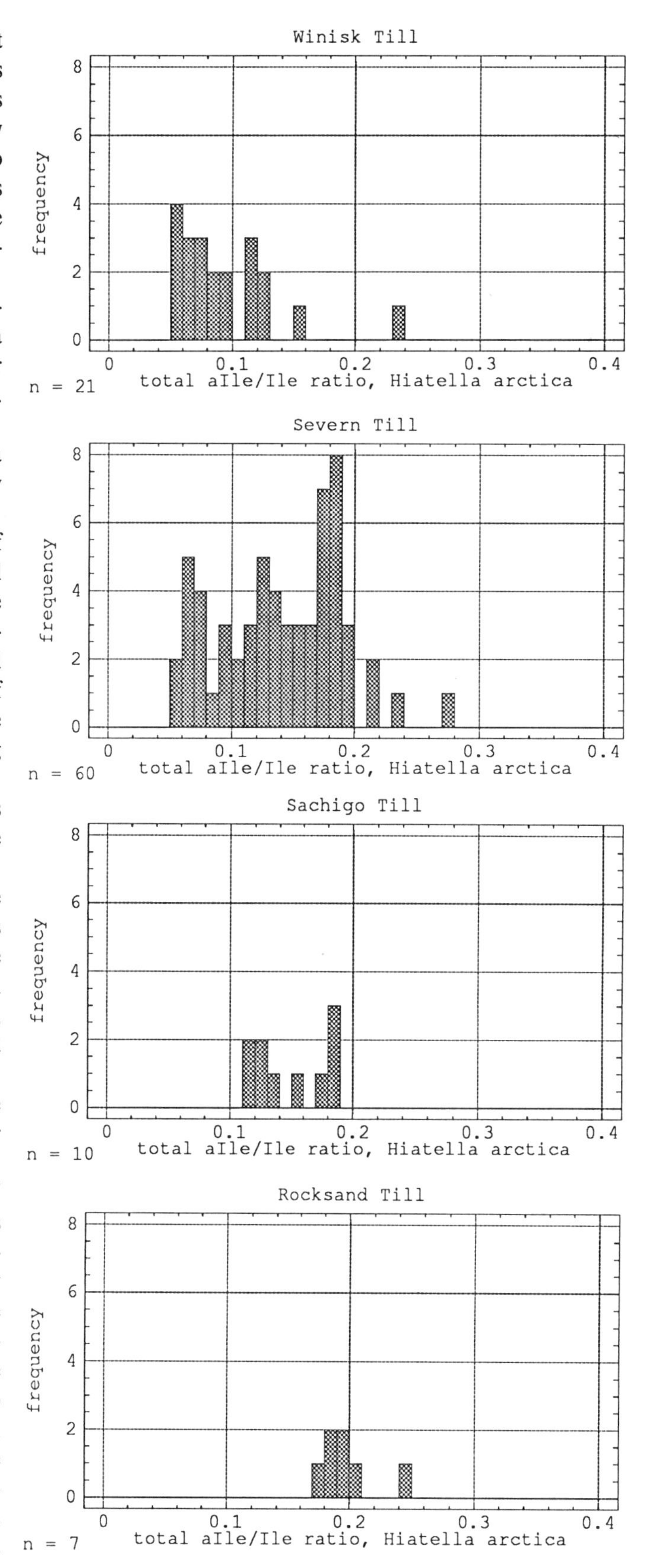

Figure 7. Amino-acid ratios from marine shell fragments collected from tills in the Severn and Winisk drainage basins (Wyatt, 1989).

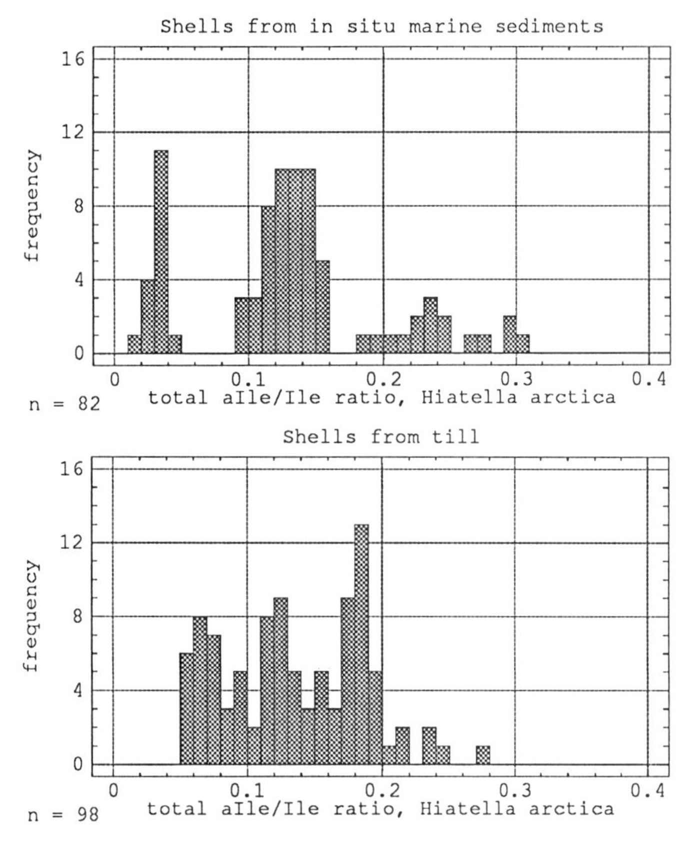

Figure 8. Comparison of amino-acid ratios from in situ and glacially transported shells (Wyatt, 1989).

tom of Hudson Bay currently has a temperature between 0 and −2 °C (Prinsenberg, 1986). Therefore, onshore shells were exposed to a warm episode in their thermal histories that was not experienced by their offshore counterparts. After glacial transport to a site above present sea level, deglaciation, and retreat of sea level, the offshore shells have only been exposed to a few thousand years of warm subaerial conditions.

The combination of age and thermal history offers an explanation for the additional clusters that is more compatible with known lithostratigraphy than the inference of additional glacial/deglacial events, whether by deglaciation (Andrews and others, 1983) or a calving bay in Hudson Bay (Dredge and Thorleifson, 1987).

An age of less than 40 ka for the end of the Prest Sea (Berger and Nielsen, 1990) would help to account for aIle/Ile ratios in till as low as 0.05, only slightly higher than Tyrrell Sea ratios of 0.03.

GLACIAL INCEPTION

The Hudson Bay lowland was glaciated early in any major glacial episode. Glacial-inception theories based on highland origin of ice sheets (Flint, 1943) or more extensive initial snow accumulation (Ives, 1957; Williams, 1978, 1979) involve the advance of ice sheets from surrounding land and coalescence in Hudson Bay. In contrast, the ice shelf freeze-down mechanism of Denton and Hughes (1981) suggests ice-sheet nucleation within Hudson Bay, possibly coincident with ice-sheet growth on land (Denton and Hughes, 1983).

Glaciation involving generally westward ice flow derived from Quebec dates to mid-stage 5 in both of the chronologies proposed here. Glacial-inception models such as those presented by Flint (1943) and Ives (1957) involving growth of an ice sheet in Quebec are therefore favored.

The alternative of inception in Hudson Bay (Denton and Hughes, 1981) would require the conclusion that early Wisconsinan ice over Hudson Bay was cold based. Sedimentation by approximately westward wet-based ice flow would then postdate the growth of an ice sheet frozen to its bed. Freeze-down of an ice shelf would have to be reconciled with paleoecological indications of climate similar to present obtained from organic material underlying glaciolacustrine sediments marking ice-sheet growth.

CONCLUSIONS

Till correlation and ice-flow patterns

The tills of the Hudson Bay lowland can be correlated on the basis of (1) stratigraphic position, (2) provenance and amino-acid data, and (3) ice-flow direction inferred from flutes, fabric, and striations on boulder pavements (Table 2).

As discussed by Boulton and Clark (1990), several coherent ice-flow patterns are recognizable in the geomorphic as well as the stratigraphic record. Great shifts in ice-dome position are implied by the various ice-flow directions.

No evidence for a Patrician center of glaciation in northern Ontario (Tyrrell, 1914) was obtained. Instead, a crucial component of Tyrrell's model, northward ice flow for the Winisk till, has been refuted. Additional evidence has, however, been obtained for west-northwestward ice flow across northern Ontario. This ice flow presumably was centered in Quebec.

In addition to this west-northwestward ice flow, an earlier southeastward trend and later westward, southward, southwestward, and local south-southeastward trends may be traced over long distances.

Pre-Holocene nonglacial stratigraphy

Skinner (1973) proposed a model for the Missinaibi Formation that involved marine inundation of isostatically depressed terrain, retreat of sea level, fluvial sedimentation, subaerial exposure, and glaciolacustrine sedimentation in front of the advancing ice. In each of the areas that have been studied, this model is compatible with available data (Table 2).

In the Moose River basin, however, two episodes of marine inundation are now recognized. Additional field and laboratory analyses will be required in order to correlate other sites to the two episodes.

TABLE 2. STRATIGRAPHIC CORRELATION

Nelson River 1*	Gods River 2*	Severn-Winisk 3*	Moose River Basin 4*	O^{18} Record TL	Alternative
POSTGLACIAL	POSTGLACIAL	POSTGLACIAL	POSTGLACIAL	1	1
Subaerial	Subaerial	Subaerial	Subaerial		
Fluvial	Fluvial	Fluvial	Fluvial		
Marine	Marine	Marine	Marine		
Glaciolacustrine	Glaciolacustrine	Glaciolacustrine	Glaciolacustrine		
GLACIATION	GLACIATION	GLACIATION	GLACIATION		
		Winisk Till		2	2
Sky Pilot Till	Tills C and D	Severn Till	Kipling Till	2	2
	Twin Creeks sediments		Friday Creek sediments		
Long Spruce Till	Till B	Sachigo Till	Adam Till	3	5
NELSON RIVER SEDIMENTS	GODS RIVER SEDIMENTS	FAWN RIVER SEDIMENTS	ABITIBI RIVER SEDIMENTS		
Glaciolacustrine	Glaciolacustrine	Glaciolacustrine		3	5d
Subaerial	Subaerial	Subaerial			
Fluvial	Fluvial	Fluvial	Fluvial		
		Marine	Marine (Prest Sea)	5a	5e
GLACIATION	GLACIATION	GLACIATION	GLACIATION		
Amery Till	Till A	Rocksand Till	Till?	5	6
			MISSINAIBI FORMATION		
			Glaciolacustrine		
Sundance Soil			Subaerial		
			Fluvial		
			Marine (Bell Sea)	5e	7
GLACIATION		GLACIATION	GLACIATION		
Sundance Till		Shagamu Till	Tills	6	8

*References: 1 = Nielsen and others, 1986; 2 = Netterville, 1974; Klassen, 1986; 3 = Thorleifson and others, 1992; 4 = Skinner, 1973; Shilts and Wyatt, 1988.

In the northern lowland, stratigraphic position with respect to tills and the lack of contradictory geochronological data imply the equivalence of the Nelson River, Gods River, and Fawn River sediments. TL data indicate that this prolonged episode, which included isostatic recovery and peat accumulation, extended from substage 5a to a time possibly as late as mid-stage 3.

Ice-sheet growth

The TL chronology indicates that glaciation of the Hudson Bay lowland during stage 5 was initiated by a west-northwestward advance of ice derived from Quebec. In Manitoba, equivalent till also shows evidence for southwestward ice flow.

If TL data are rejected, west-northwestward ice flow would be reassigned in an alternative chronology to stage 6. Stage 5 glacial inception in this alternative chronology would involve the westward ice flow that is recorded by the Long Spruce, Sachigo, and Adam tills. This implies an ice dome in Quebec well to the north of its earlier position. Fabrics on the Gods River and red carbonate erratics on the Severn River indicate that, in the north, ice flow shifted to southward late in this initial episode of continuous till sedimentation.

Wisconsinan glaciation

The Nelson River, Gods River, and Fawn River sediments, as well as marine sediment on the Abitibi River, were deposited during the final pre-Holocene nonglacial episode. Subsequently, two or three tills, in places separated by sorted sediments, were deposited. This sequence has been attributed to an early ice advance, intra-Wisconsinan deglaciation that fell just short of marine incursion, and a readvance to the late Wisconsinan maximum (e.g., McDonald, 1971).

A simpler scenario may be constructed if it is accepted that till sedimentation is concentrated in settings within a few hundreds of kilometers of the margin and that rippled sand and laminated silt can be deposited in a subglacial environment. The Long Spruce, Sachigo, and Adam tills could be attributed to the growth of the ice sheet. Sorted glaciofluvial sediments lacking organic material, such as the Twin Creeks sediments, may be a subglacial meltwater deposit formed when the ice sheet was extensive. The upper tills, Sky Pilot, Severn, and Kipling, would then be attributed to near-marginal flow from a saddle over Hudson Bay during retreat of the ice sheet. Local, late glacial perturbations of this pattern included a major ice stream in the Winisk River area.

ACKNOWLEDGMENTS

The manuscript was improved on the basis of reviews by A. S. Dyke and J. T. Andrews.

REFERENCES CITED

Andrews, J. T., 1976, The present ice age: The Cenozoic, *in* John, B., ed., The winters of the World: New York, John Wiley and Sons, p. 173–218.

—— , 1987, The late Wisconsin glaciation and deglaciation of the Laurentide Ice Sheet, *in* Ruddiman, W. F., and Wright, H. E., Jr., eds., North America and adjacent oceans during the last deglaciation: Boulder, Colorado, Geological Society of America, The Geology of North America, v. K-3, p. 13–37.

Andrews, J. T., and Miller, G. H., 1979, Glacial erosion and ice sheet divides, northeastern Laurentide Ice Sheet, on the basis of the distribution of limestone erratics: Geology, v. 7, p. 592–596.

Andrews, T. J., Shilts, W. W., and Miller, G. H., 1983, Multiple deglaciations of the Hudson Bay Lowlands, Canada, since deposition of the Missinaibi (last-interglacial?) Formation: Quaternary Research, v. 19, p. 18–;37.

—— , 1984, Reply *to* Discussion *of* "Multiple deglaciations of the Hudson Bay Lowlands, Canada, since deposition of the Missinaibi (last-interglacial?) Formation": Quaternary Research, v. 22, p. 253–258.

Bell, R., 1887, Exploration of portions of the Attawapiskat and Albany rivers: Geological Survey of Canada Annual Report 1886, part G, 38 p.

Berger, G. W., and Nielsen, E., 1990, Evidence from thermoluminescence dating for middle Wisconsinan deglaciation in the Hudson Bay Lowland of Manitoba: Canadian Journal of Earth Sciences, v. 28, p. 240–249.

Beukens, R. P., 1990, High-precision intercomparison at Isotrace: Radiocarbon, v. 32, p. 335–339.

Bouchard, M. A., and Martineau, G., 1985, Southeastward ice flow in central Quebec and its paleographic significance: Canadian Journal of Earth Sciences, v. 22, p. 1536–1541.

Boulton, G. S., and Clark, C. D., 1990, The Laurentide ice sheet through the last glacial cycle: The topology of drift lineations as a key to the dynamic behaviour of former ice sheets: Royal Society of Edinburgh Transactions: Earth Sciences, v. 81, p. 327–347.

Brown, R. J., 1973, Influence of climatic and terrain factors on ground temperatures at three locations in the permafrost region of Canada, *in* Permafrost, The North American Contribution to the Second International Conference on Permafrost, Yakutsk, Siberia: Washington, D.C., National Academy of Sciences, p. 27–34.

Budd, W. F., and Smith, I. N., 1981, The growth and retreat of ice sheets in response to orbital radiation changes, *in* Allison, Ian, ed., Sea level, ice, and climatic change: International Associations of Hydrological Sciences Publication 131, p. 369–409.

—— , 1987, Conditions for growth and retreat of the Laurentide Ice Sheet: Géographie Physique et Quaternaire, v. XLI, no. 2, p. 279–290.

Craig, B. G., 1969, Late-glacial and postglacial history of the Hudson Bay region, *in* Hood, P. J., ed., Earth Science Symposium on Hudson Bay: Geological Survey of Canada Paper 68-53, p. 63–77.

Denton, G. H., and Hughes, T. J., 1981, The Arctic ice sheet: An outrageous hypothesis, *in* Denton, G. H., and Hughes, T. J., eds., The last great ice sheets: New York, Wiley-Interscience, p. 437–467.

—— , 1983, The Milankovitch thoery of ice ages: Hypothesis of ice-sheet linkage between regional insolation and global climate: Quaternary Research, v. 20, p. 125–144.

Denton, G. H., Hughes, T. J., and Karlen, W., 1986, Global ice-sheet system interlocked by sea level: Quaternary Research, v. 26, p. 3–26.

Dredge, L. A., and Cowan, W. R., 1989, Quaternary geology of the southwestern Canadian shield, *in* Fulton, R. J., ed., Quaternary geology of Canada and Greenland: Geological Survey of Canada, Geology of Canada, no. 1, p. 214–249.

Dredge, L. A., and Nielsen, E., 1985, Glacial and interglacial deposits in the Hudson Bay Lowlands: A summary of sites in Manitoba, *in* Current research, Part A, Geological Survey of Canada Paper 85-1A, p. 247–257.

Dredge, L. A., and Thorleifson, L. H., 1987, The middle Wisconsinan history of the Laurentide ice sheet: Géographie physique et Quaternaire, v. XLI, no. 2, p. 215–235.

Dredge, L. A., Nixon, F. M., and Richardson, R. J., 1986, Quaternary geology and geomorphology of northwestern Manitoba: Geological Survey of Canada Memoir 418, 38 p.

Dredge, L. A., Morgan, A. V., and Nielsen, E., 1990, Sangamon and pre-Sangamon interglaciations in the Hudson Bay Lowlands of Manitoba: Géographie physique et Quaternaire, v. 44, p. 319–336.

Dyck, W., and Fyles, J. G., 1963, Geological Survey of Canada radiocarbon dates II: Radiocarbon, v. 5, p. 39.

Dyke, A. S., and Prest, V. K., 1987, Late Wisconsinan and Holocene history of the Laurentide Ice Sheet: Géographie physique et quaternaire, v. XLI, no. 2, p. 237–263.

Dyke, A. S., Dredge, L. A., and Vincent, J. S., 1982, Configuration and dynamics of the Laurentide Ice Sheet during the late Wisconsin maximum: Géographie physique et Quaternaire, v. XXXVI, p. 5–14.

Eyles, N., Sladen, J. A., and Gilroy, S., 1982, A depositional model for stratigraphic complexes and facies superimposition in lodgement tills: Boreas, v. 11, p. 317–333.

Eyles, N., Eyles, C. H., and Miall, A. D., 1983, Lithofacies types and vertical profile models; an alternative approach to the description and environmental interpretation of glacial diamict and diamictite sequences: Sedimentology, v. 30, p. 393–410.

Flint, R. F., 1943, Growth of the North American ice sheet during the Wisconsin age: Geological Society of America Bulletin, v. 54, p. 325–362.

—— , 1971, Glacial and Quaternary geology: New York, John Wiley and Sons, 892 p.

Forman, S. L., Wintle, A. G., Thorleifson, L. H., and Wyatt, P. H., 1987, Thermoluminescence properties and age estimates for Quaternary raised marine sediments, Hudson Bay Lowland, Canada: Canadian Journal of Earth Sciences, v. 24, p. 2405–2411.

Harris, S. A., 1981, Climatic relationships of permafrost zones in areas of low winter snow cover: Arctic, v. 34, p. 64–70.

Hughes, O. L., 1965, Surficial geology of part of the Cochrane District, Ontario, Canada, *in* Wright, H. E., and Frey, D. G., eds., International studies on the Quaternary: Geological Society of America Special Paper 84, p. 535–565.

Ives, J. D., 1957, Glaciation of the Torngat Mountains: Geographical Bulletin, no. 10, p. 67–87.

Klassen, R. W., 1986, Surficial geology of north-central Manitoba: Geological Survey of Canada Memoir 419, 57 p.

Lee, H. A., 1960, Late glacial and post-glacial Hudson Sea episode: Science, v. 131, p. 1609–1611.

Lowdon, J. A., Fyles, J. G., and Blake, W., Jr., 1967, Geological Survey of Canada radiocarbon dates VI: Radiocarbon, v. 9, p. 156.

Martinson, D. G., Pisias, N. G., Hays, J. D., Imbrie, J., Moore, T. C., Jr., and

Shackleton, N. J., 1987, Age dating and the orbital theory of the ice ages: Development of a high-resolution 0 to 300,000-year chronostratigraphy: Quaternary Research, v. 27, p. 1–29.

McDonald, B. G., 1969, Glacial and interglacial stratigraphy, Hudson Bay Lowlands: Geological Survey of Canada Paper 68-53, p. 78–99.

—— , 1971, Late Quaternary stratigraphy and deglaciation in eastern Canada, *in* Turekian, K. K., ed., The late Cenozoic glacial ages: New Haven, Connecticut, Yale University Press, p. 331–353.

Miller, G. H., 1985, Aminostratigraphy of Baffin Island shell-bearing deposits, *in* Andrews, J. T., ed., Quaternary environments, Eastern Canadian Arctic, Baffin Bay and western Greenland: London, George Allen and Unwin, p. 394B–427B.

Miller, G. H., and Mangerud, J., 1985, Aminostratigraphy of European marine interglacial deposits: Quaternary Science Reviews, v. 4, p. 215–278.

Miller, G. H., Hearty, P. J., and Stravers, J. A., 1988, Ice-sheet dynamics and glacial history of southeasternmost Baffin Island and outermost Hudson Strait: Quaternary Research, v. 30, p. 116–136.

Mott, R. J., and DiLabio, R.N.W., 1990, Paleoecology of organic deposits of probable last interglacial age in northern Ontario: Géographie physique et Quaternaire, v. 44, no. 3, p. 309–318.

Netterville, J. A., 1974, Quaternary stratigraphy of the lower Gods River region, Hudson Bay Lowlands, Manitoba [M.S. thesis]: Calgary, University of Calgary, 79 p.

Nielsen, E., Morgan, A. V., Morgan, A., Mott, R. J., Rutter, N. W., and Causse, C., 1986, Stratigraphy, paleoecology, and glacial history of the Gillam area, Manitoba: Canadian Journal of Earth Sciences, v. 23, p. 1641–1661.

Nielsen, E., Churcher, C. S., and Lammers, G. E., 1988, A woolly mammoth (Proboscidea, Mammuthus primigenius) molar from the Hudson Bay Lowland of Manitoba: Canadian Journal of Earth Sciences, v. 25, p. 933–938.

Paterson, W.S.B., 1972, Laurentide Ice Sheet: Estimated volumes during late Wisconsin: Reviews of Geophysics and Space Physics, v. 10, no. 4, p. 885–917.

Prest, V. K., 1963, Red Lake–Landsdowne House area, northwestern Ontario—Surficial geology (parts of 42, 43, 52, 53): Geological Survey of Canada Paper 63-6, p. 23.

—— , 1966, Glacial studies, northeastern Ontario and northwestern Quebec, *in* Report of Activities, May to October 1965: Geological Survey of Canada Paper 66-1, p. 202–203.

—— , 1969, Retreat of Wisconsin and recent ice in North America: Geological Survey of Canada Map 1257A, scale 1:5,000,000.

—— , 1970, Quaternary geology of Canada, *in* Douglas, R.J.W., ed., Geology and economic minerals of Canada (fifth edition): Ottawa, Geological Survey of Canada Economic Geology Series, No. 1, p. 676–758.

Prest, V. K., Grant, D. R., and Rampton, V. N., 1968, Glacial map of Canada: Geological Survey of Canada Map 1253A, scale 1:5,000,000.

Prinsenberg, S. J., 1986, Salinity and temperature distributions of Hudson Bay and James Bay, *in* Martini, I. P., ed., Canadian inland seas: New York, Elsevier, p. 163–186.

Shackleton, N. J., and Opdyke, N. D., 1973, Oxygen isotope and paleomagnetic stratigraphy of equatorial Pacific core V28-238: Oxygen isotope temperatures and ice volumes on a 10^5 and 10^6 year scale: Quaternary Research, v. 3, p. 39–55.

Shilts, W. W., 1980, Flow patterns in the central North American ice sheet: Nature, v. 286, p. 213–218.

—— , 1982, Quaternary evolution of the Hudson/James Bay region: Le Naturaliste Canadien, v. 109, p. 309–332.

Shilts, W. W., and Wyatt, P. H., 1988, Aminostratigraphy of marine and associated nonglacial beds of the Hudson Bay Lowland: 17th Annual Arctic Workshop: Boulder, University of Colorado, Institute of Arctic and Alpine Research, p. 49.

Shilts, W. W., Cunningham, C. M., and Kaszycki, C. A., 1979, Keewatin Ice Sheet—Reevaluation of the traditional concept of the Laurentide Ice Sheet: Geology, v. 7, p. 537–541.

Skinner, R. G., 1973, Quaternary stratigraphy of the Moose River basin, Ontario: Geological Survey of Canada Bulletin 225, 77 p.

Stuiver, M., Deevey, S., and Rouse, I., 1963, Yale natural radiocarbon measurements VIII: Radiocarbon, v. 5, p. 312.

Stuiver, M., Heusser, C. J., and Yang, I. C., 1978, North American glacial history extended to 75,000 years ago: Science, v. 200, p. 16–21.

Sugden, D. E., 1977, Reconstruction of the morphology, dynamics, and thermal characteristics of the Laurentide Ice Sheet at its maximum: Arctic and Alpine Research, v. 9, p. 21–47.

Terasmae, J., 1958, Contributions to Canadian palynology: Part III, Nonglacial deposits along Missinaibi River, Ontario: Geological Survey of Canada Bulletin 46, p. 29–34.

Terasmae, J., and Hughes, O. L., 1960, A palynological and geological study of Pleistocene deposits in the James Bay Lowlands, Ontario: Geological Survey of Canada Bulletin 62, 15 p.

Thomas, R. H., 1979, The dynamics of marine ice sheets: Journal of Glaciology, v. 24, p. 167–177.

Thorleifson, L. H., 1989, Quaternary stratigraphy of the central Hudson Bay Lowland, northern Ontario, Canada [Ph.D. thesis]: Boulder, University of Colorado, 363 p.

Thorleifson, L. H., Wyatt, P. H., and Warman, T. A., 1992, Quaternary stratigraphy of the Severn and Winisk drainage basins, northern Ontario, Canada: Geological Survey of Canada Bulletin (in press).

Tyrrell, J. B., 1988, The glaciation of north-central Canada: Journal of Geology, v. 8, p. 147–160.

—— , 1914, The Patrician Glacier south of Hudson Bay: International Geological Congress, XII, Ottawa, p. 523–534.

Veilette, J. J., 1986, Former southwesterly ice flows in Abitibi-Temiscamingue: Implications for the configuration of the late Wisconsinan ice sheet: Canadian Journal of Earth Sciences, v. 23, p. 1724–1741.

Veillette, J. J., and Pomares, J. S., 1991, Older ice flows in the Mattagami-Chapais area, Quebec: Geological Survey of Canada Paper 91-1C, p. 143–148.

Veillette, J. J., Averill, S. A., LaSalle, P., and Vincent, J. S., 1989, Quaternary geology of Abitibi-Temiscamingue and mineral exploration: Geological Association of Canada Field Trip Guidebook, 112 p.

Walcott, R. I., 1972, Quaternary vertical movements in eastern North America: Quantitative evidence of glacio-isostatic rebound: Reviews of Geophysics and Space Physics, v. 10, p. 849–884.

Warman, T. A., 1987, The Quaternary sedimentology of the Severn River area, Hudson Bay Lowlands [B.S. thesis]: Hamilton, Ontario, McMaster University, 123 p.

Williams, L. D., 1978, Ice sheet initiation and climatic influences of expanded snow cover in arctic Canada: Quaternary Research, v. 10, p. 141–149.

—— , 1979, An energy-balance model of potential glacierization of northern Canada: Arctic and Alpine Research, v. 11, p. 443–456.

Wyatt, P. H., 1989, The stratigraphy and amino acid chronology of Quaternary sediments in central Hudson Bay Lowland [M.S. thesis]: Boulder, University of Colorado, 119 p.

—— , 1990, Amino acid evidence indicating two or more ages of pre-Holocene nonglacial deposits in Hudson Bay Lowland, northern Ontario: Géographie Physique et Quaternaire, v. 44, p. 389–393.

Geological Society of Canada Contribution 17589 (Nielsen)
Manuscript Accepted by the Society September 6, 1991

Geological Society of America
Special Paper 270
1992

Timing and character of the last interglacial-glacial transition in the eastern Canadian Arctic and northwest Greenland

Gifford H. Miller
Center for Geochronological Research, Institute of Arctic and Alpine Research, and Department of Geological Sciences, University of Colorado, Boulder, Colorado 80309-0450
Svend Funder
Geological Museum, Ostervoldgade 5-7, DK-1350, Copenhagen K, Denmark
Anne de Vernal
GEOTOP, Université du Québec à Montréal, C.P. 8888, Succursale "A," Montreal, Quebec H3C 3P8, Canada
John T. Andrews
Institute of Arctic and Alpine Research, and Department of Geological Sciences, University of Colorado, Boulder, Colorado 80309-0450

ABSTRACT

The records of glaciation and climate change preserved in sediments on the Canadian and northwest Greenland margins of Baffin Bay pertaining to the last interglacial-glacial transition are remarkably similar. In both regions, warmer than present terrestrial and nearshore marine facies of the last interglacial sensu stricto (s.s.) are overlain by glacial sediments that represent the most extensive advance of continental ice during the last glaciation. Chronometric controls (^{14}C, thermoluminescence, amino acids) indicate an isotope stage 5 age for this advance. Evidence for extensive high-latitude glacial erosion during stage 5 is recorded by abundant pre-Quaternary palynomorphs in Baffin Bay sediment cores, in contrast to a much reduced flux during the remainder of the last glaciation. Warm nearshore marine conditions (seasonally ice free) also occurred near the end of stage 5 along both the eastern Baffin Island and northwest Greenland coasts after the maximum glacial advance; surface water in central Baffin Bay apparently was dominated by meltwater at this time. Subsequently (isotope stages 4, 3, and 2), terrestrial conditions were colder and drier, sea-surface temperatures were lower, and ice margins were retracted. Minimum summer insolation at high latitudes, coupled with mild winters and vigorous meridional oceanic (and presumably atmospheric) circulation characterized the inception phase of the last glaciation during isotope stage 5. In contrast, the 20 ka B.P. (isotope stage 2) "last glacial maximum" was characterized by a zonal circulation regime that resulted in cold and dry conditions over Baffin Bay; the margins of the northwest Greenland and northeast Laurentide ice sheets did not extend beyond the fiords at this time.

INTRODUCTION

The pattern of Laurentide Ice Sheet retreat at the end of the last glaciation (Bryson and others, 1969; Dyke and Prest, 1987) reflects an ice sheet pinned along its northeastern margin, with the last remnant ice masses centered over the plateau (not the mountains) of northern Quebec and central Baffin Island. Indeed, the Barnes Ice Cap, north-central Baffin Island, apparently contains residual ice from the Laurentide Ice Sheet (Hooke, 1982). We argue that the pattern of ice-sheet retreat provides a first-order

Miller, G. H., Funder, S., de Vernal, A., and Andrews, J. T., 1992, Timing and character of the last interglacial-glacial transition in the eastern Canadian Arctic and northwest Greenland, *in* Clark, P. U., and Lea, P. D., eds., The Last Interglacial-Glacial Transition in North America: Boulder, Colorado, Geological Society of America Special Paper 270.

model of the inception phase of a glacial cycle, and that initial ice-sheet growth probably occurred over the high plateau regions of Baffin Island and northern Quebec.

Recent reconstructions of the Late Wisconsinan Laurentide Ice Sheet (e.g., Dyke and Prest, 1987) emphasize that the ice sheet probably always consisted of multiple dispersal centers, each of which responded somewhat independently to local control on accumulation and ablation. The Labradorean sector was the largest dome; domes in the Keewatin and Foxe-Baffin sectors account for most of the remainder of the ice sheet. The present Greenland Ice Sheet consists of two major domes and a number of minor ancillary dispersal centers; presumably a similar situation prevailed during the last glaciation in Greenland.

Despite the magnitude of the Laurentide and Greenland ice sheets, significant stretches of land along the coasts of eastern Baffin Island and western Greenland remained ice free throughout much of the last glacial cycle, and offer the potential of preserving direct field evidence of events during the transition from the last interglacial into the subsequent glaciation. Recognition of deposits formed during the last interglacial or during the earliest phase of the subsequent glaciation is hampered by problems inherent in dating deposits within this time range, and the lack of floral and/or faunal assemblages diagnostic of the last (as opposed to any other) interglaciation in the Arctic. Within these constraints, considerable progress has been made in the past 15 years in understanding the sequence of events during the earliest phase of the last glaciation.

The purpose of this chapter is to (1) review the stratigraphic and chronologic evidence for ice-sheet growth early in the last glacial cycle as recorded in terrestrial exposures on the Canadian and Greenland margins of Baffin Bay, in deep-sea sediment cores from Baffin Bay and the Labrador Sea, and in ice cores from the Greenland Ice Sheet and from ice caps in the Canadian Arctic; and (2) to compare the implications of this dataset pertaining to initial ice-sheet growth and the climate at the last glacial maximum (ca. 20 to 15 ka B.P.).

TERMINOLOGY

Last interglaciation

We rely on the marine sediment-core records from the North Atlantic and Norwegian Sea for a conceptual definition of the last interglaciation sensu stricto (s.s.), and we restrict our definition of the last interglacial to the interval spanned by oxygen-isotope substage 5e (e.g., CLIMAP Project Members, 1984; Mangerud and others, 1979). Based on this conceptual framework, we expect that the last interglacial was represented by a brief episode of warmer than present terrestrial summer temperatures at a time of near-modern sea level, correlative with substage 5e. The working definition for deposits of the last interglacial on eastern Baffin Island has been the youngest (stratigraphically) deposit close to present sea level that bears firm evidence in the form of plant macrofossils or pollen for terrestrial summer temperatures substantially above those of the present, and at least as warm as the Holocene optimum. This working definition is not applicable to northwest Greenland, where NORDQUA 86 Participants (1989) reported plant remains indicative of warmer than present terrestrial summer temperatures in deposits with thermoluminescence (TL) dates younger than the last interglacial (s.s.). Estimates of ocean surface-water temperature cannot readily be used to define the last interglacial (s.s.), because both inshore and open-marine surface waters were about as warm as present late in isotope stage 5. We recognize the inherent limitations of this definition due to the incompleteness of the terrestrial record and the possibility that older interglacial deposits may be closest to the present land surface; however, the lack of reliable absolute chronological control, and the absence of floral and/or faunal assemblages unique to the last interglacial, preclude a more rigorous definition. Amino-acid ratios help to identify older interglacials that may be mistaken for the most recent one (e.g., Miller, 1985), and TL dates have proven useful in subdividing nearshore marine deposits of stage 5 (e.g., Forman and others, 1987; Kronborg and others, 1990).

Last glaciation

We consider the last glaciation to span the interval from the end of the last interglacial (s.s.) to the final deglaciation of Foxe basin, about 6500 yr ago (Andrews, 1982). The dynamics and timing of advances over eastern Baffin Island may have been decoupled from events along the southern Laurentide border; consequently, we call the last glaciation of Baffin Island the Foxe Glaciation, because it reflects activity within the Foxe-Baffin sector (Prest, 1984) of the ice sheet. The last glaciation has been subdivided into Early, Middle and Late stades (Andrews and Ives, 1978a; Mangerud and others, 1974; Miller, 1985), the Early stade spanning the time period from about 115 to 70 ka ago (absolute ages follow Martinson and others, 1987). This differs from the standard Canadian definition that classifies all of isotope stage 5 as interglacial and the Early Wisconsinan beginning at the stage 5/4 transition (e.g., Fulton, 1989). On Bylot Island, the earliest phase of the last glaciation is known as the Eclipse glaciation (Klassen, 1985); it can be confidently correlated with the Early Foxe Glaciation (Ayr Lake advance) on the Clyde Foreland (Miller and others, 1977; Andrews and Miller, 1984). For Greenland, the chronostratigraphic terminology follows that advocated by Mangerud and others (1974), the Weichselian Glaciation commencing at the end of isotope substage 5e, and being essentially equivalent to the Foxe Glaciation. Local stadial names in the Thule area are the Saunders Ø interstade (last interglacial s.s.), Narssarssuk stade (Early Weichselian, equivalent to the Ayr Lake advance), and the Qarmat interstade (late stage 5).

LAST INTERGLACIAL-GLACIAL TRANSITION

Field evidence for extensive glaciation around Baffin Bay early in the last glacial cycle is found along the Qivitu and Clyde forelands of eastern Baffin Island, on Bylot Island, and in the

Thule area of northwest Greenland (Fig. 1). Inferential evidence comes from marine cores recovered from Baffin Bay and the Labrador Sea, and from the record preserved in ice cores around the Baffin Bay region. The most instructive field evidence records the interplay between glacial and marine environments. Ideally, field relations would include a complete transgressive/regressive marine cycle and deposits of the intervening glaciation (cf. Mode and others, 1983). However, the transgressive phase is seldom preserved, and most exposures document the deglacial hemicycle only, consisting of a till over bedrock or glacially eroded sediment, overlain by ice-proximal to ice-distal sediments that grade upward into littoral facies, on which a soil developed after isostatic readjustments elevated the site above the sea. In rare instances, the transgressive leg is preserved, recording a deepening of the water depth during ice buildup and a coarsening-upward sedimentary sequence reflecting the encroaching ice margin.

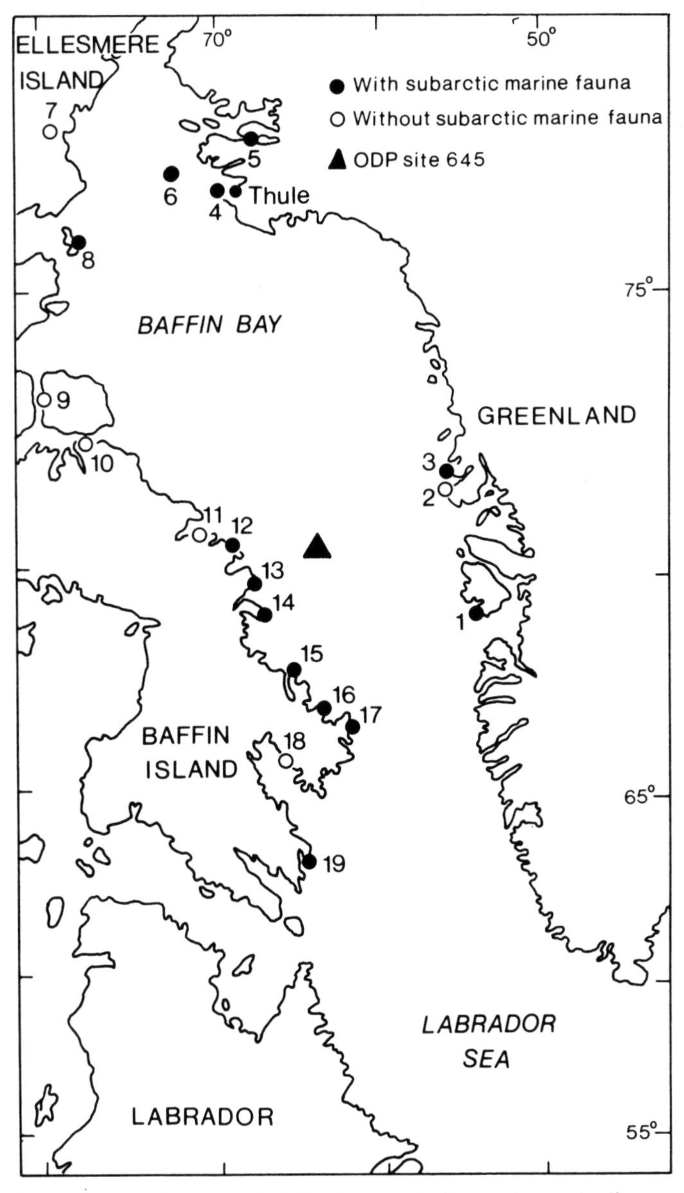

Figure 1. Location map of Baffin Bay and Labrador Sea and adjacent land masses with stage 5 (sensu lato) deposits (from Funder, 1990a). Sites discussed in the text include the Thule area (4), Coburg Island (8), Bylot Island (9), Pond Inlet (10), Clyde Foreland (12), Cape Henry Kater (13), Qivitu Peninsula (14), and ODP Baffin Bay core 645.

Baffin Island

Along northeastern Baffin Island, Miller and others (1977) and Miller (1985) recognized two distinct glacial episodes within the Foxe Glaciation: an extensive Early Foxe advance (Ayr Lake till of the Kogalu aminozone) and a more restricted glacial advance during the Late Foxe interval (Baffinland Drift; Andrews and Ives, 1978b). Isoleucine epimerization (D/L or alle/Ile) ratios are used to define and correlate the Early Foxe deposits. Well-preserved moraines of the Ayr Lake advance can be traced to ice-contact marine deltas from which in situ marine shells yield nonfinite radiocarbon ages (cf. Løken, 1966; Miller and others, 1977), and D/L ratios that fall within the Kogalu aminozone. Moraines correlative with the Ayr Lake advance have been mapped in the Qivitu region of northern Cumberland Peninsula (Nelson, 1982). Marine sediments associated with this advance contain molluscan fossils with Nelson's (1982) aminozone 2 D/L ratios; aminozone 2 can be correlated with the Kogalu aminozone at Clyde. On the basis of limiting radiocarbon, uranium series, and amino-acid dates, the Kogalu aminozone is ascribed a stage 5 age (Miller, 1985).

In a subsequent study of the glacial history of Bylot Island and surrounding regions, Klassen (1981, 1985) demonstrated that lateral moraines of the Eclipse glaciation were dominated by foreign erratic lithologies (dominantly carbonates, but other far-traveled lithologies are also present). Marine shell fragments within the moraines and associated drift have amino-acid D/L ratios similar to the Kogalu aminozone at Clyde. Geomorphically distinct raised marine deltas proximal to the terminus of this advance, and therefore postdating the advance, contain in situ molluscs with slightly lower D/L ratios than in Eclipse drift, but still within the range of the Kogalu aminozone at Clyde. Radiocarbon dates on the same collections are nonfinite. At the maximum of the Eclipse glaciation a major outlet glacier flowed out Lancaster Sound, along the east coast of Bylot Island, and calved into northern Baffin Bay. Klassen (1981, 1985) concluded that the Eclipse glaciation was of Early Wisconsin age, correlative to the Ayr Lake advance of the Clyde region, and that subsequent advances were restricted to channels around Bylot Island, and involved substantially smaller ice volumes.

A younger subdivision of the Kogalu aminozone is recognized by D/L ratios that are statistically lower than in other deposits of the Kogalu aminozone along the Clyde Foreland (Mode, 1985) and Cape Henry Kater Peninsula (Miller, 1985). Deposits containing the younger ratios are associated with deglaciation from an extensive ice advance and contain a marine molluscan fauna similar to that of the Holocene marine optimum.

The difference in D/L ratio is interpreted to indicate a late stage 5 age for the associated deposits, as opposed to an earlier stage 5 age for the bulk of the deposits of the Kogalu aminozone.

Collectively, these studies indicate an initial buildup of continental ice that advanced onto the continental shelf early during isotope stage 5; during deglaciation late in isotope stage 5, inshore marine conditions were similar to or warmer than the Holocene optimum. At both Clyde (Miller, 1976) and Bylot Island, it was demonstrated that many local glaciers have not been much larger than present since deglaciation from the Early Foxe glaciation, implying that precipitation was low throughout subsequent stages 4, 3, and 2. However, neither region provided evidence for conditions during the inception phase of the glacial advance.

Labrador/Ungava

Clear evidence of the earliest Wisconsin advance of the Labrador sector along the Labrador coast and around Ungava Bay has yet to be discovered, largely because subsequent advances obliterated most of the older deposits. Changes in rock-weathering characteristics mappable over much of northern Labrador (e.g., Clark, 1988, and references therein) are difficult to date, and no strong evidence has been presented to indicate a particular zone is correlative with the earliest Wisconsin advance.

Northwest Greenland

Quaternary deposits preserved in the Thule region of northwest Greenland provide the best record of the interglacial-glacial transition on the eastern side of Baffin Bay (NORDQUA 86 Participants, 1989; Funder, 1990a, 1990b). A transect of coastal exposures from near the edge of the Inland Ice to islands at the northern limit of Baffin Bay range from shallow-marine and beach facies to deep-water glacial-marine mud and diamicton. The sections include three superposed, generally coarsening-upward sequences, recording deglaciation followed by shallowing with regional isostatic readjustment. Thermoluminescence and amino-acid racemization analyses provide the chronologic framework; marine micro- and macro-faunas combined with palynology and macrofossils of terrestrial plants and insects provide paleoenvironmental information (Fig. 2).

The last interglacial (local name Saunders Ø interglacial) is represented by in situ and reworked marine sediment containing subarctic molluscan fossils, indicating oceanographic circulation in Baffin Bay similar to that of the Holocene. Marine deposits with nonfinite radiocarbon dates on Coburg Island (Fig. 1) that contain *Mytilus edulis* north of its maximum Holocene limit (Blake, 1973), are correlated with the Saunders Ø interglacial deposits at Thule (e.g., Funder, 1990b), and suggest a more vigorous meridional oceanographic circulation, with Atlantic water

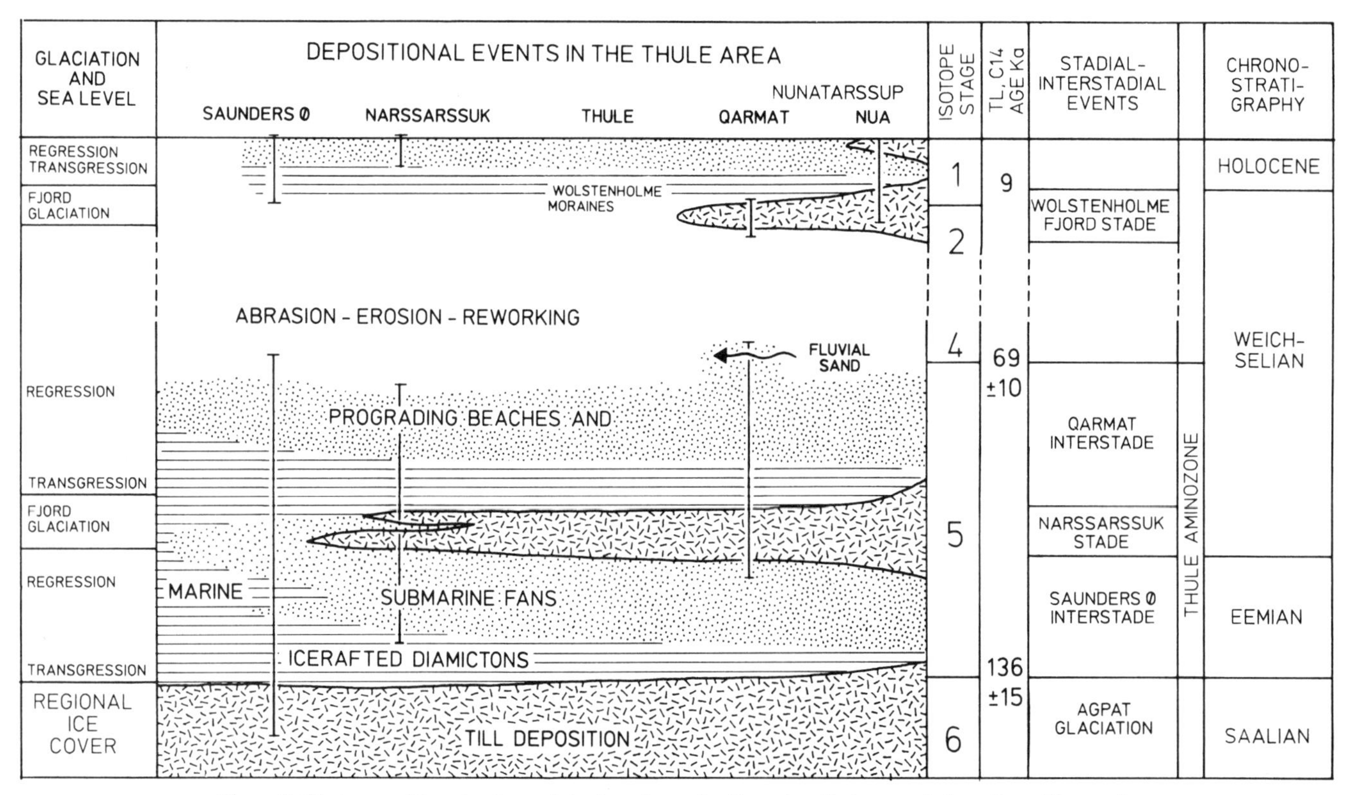

Figure 2. Thule area, Greenland, event stratigraphy and sedimentary facies associations (from Houmark-Nielsen, 1990). Vertical bars through the center portion of the diagram indicate the actual amount of sections exposed at each of the sites.

penetrating farther north during the Saunders Ø interglacial than in the present interglacial.

Last interglacial deposits near the inland ice in the Thule area were overrun by a subsequent advance of the Greenland Ice Sheet during the Narssarssuk stade. The ice margin did not reach Saunders Ø during this advance, the most extensive of the Weichselian. TL dates on sediment associated with this advance suggest a mid-stage 5 age. The Narssarssuk stade was terminated by the Qarmat interstade, when marine surface waters again warmed, and subarctic molluscs penetrated farther north than during the Holocene optimum. TL dates and amino-acid ratios indicate a late stage 5 age. Events in the Thule area can be correlated with those of eastern Baffin Island (Fig. 3).

Baffin Bay–Labrador Sea

Foraminifera, dinoflagellate cysts, and pollen extracted from marine sediments in cores (Fig. 4) recovered from Baffin Bay (Aksu and Piper, 1979; de Vernal and others, 1987; de Vernal and Mudie, 1989; Hillaire-Marcel and others, 1989, 1990) and the Labrador Sea (Fillon and Duplessy, 1980; de Vernal and Hillaire-Marcel, 1987) provide supplemental evidence for glaciation on the adjacent land masses during the earliest phases of the last glaciation. The primary age control for deep-sea cores spanning the last glacial cycle is by correlation of the oxygen-isotope variations to the global master isotope curve. However, the $\delta^{18}O$ signal in Baffin Bay and the northern Labrador Sea is complicated by the episodic addition of isotopically light meltwater from the adjacent Laurentide and Greenland ice sheets, which may obscure the global signal. This effect diminishes to the south. Although only a generalized isotopic framework has been established for Baffin Bay cores, a secure isotopic framework is available for most Labrador Sea cores.

The last interglacial is well represented only in cores from the Labrador Sea, where it is characterized by low $\delta^{18}O$ values and a diverse and productive dinoflagellate cyst assemblage indicative of temperate to subarctic surface waters. Pollen from cores recovered southwest of Greenland indicate substantial reduction of the Greenland Ice Sheet and development of a relatively rich terrestrial flora in southern Greenland during the last interglaciation. In Labrador Sea cores, the floral transition indicating a return to colder sea-surface temperatures at the end of the last interglacial is coincident with the isotope stage 5e/5d transition. Reworked pre-Quaternary palynomorphs are abundant throughout stage 5 sediment in Baffin Bay cores, supporting the interpretation that in the eastern Canadian Arctic and northwest Greenland, ice-sheet growth was initiated early in isotope stage 5, and active glacial erosion was maintained throughout stage 5, giving rise to the high flux of glacially eroded detritus. The dramatic decrease in pre-Quaternary palynomorphs in sediment of isotopic stages 4, 3, and 2 indicates a substantial decrease in high-latitude glacial activity relative to that of isotope stage 5, presumably because the locus of precipitation and ice-sheet activity shifted south. Rapid, high-amplitude $\delta^{18}O$ shifts in planktonic foraminifera from stage 5 sediment in Baffin Bay cores indicate discharge of large volumes of meltwater from adjacent ice sheets, suggesting substantial summer melt at this time. The absence of these fluctuations in the younger portions of the cores indicates less production of meltwater from ice sheets surrounding Baffin Bay. Baffin Bay cores do not record an influx of warm surface waters late in stage 5, as do the nearshore molluscan fauna of the late Kogalu aminozone on eastern Baffin Island and the Qarmat interstade in northwest Greenland. This discrepancy may reflect the influence of meltwater in determining surface-water characteristics of central Baffin Bay, as molluscan assemblages have proven themselves faithful watermass indicators by their present-day distribution and their changes within the Holocene (e.g., Andrews, 1972; Hjort and Funder, 1974). The decrease in pre-Quaternary palynomorphs and meltwater spikes during isotope stages 4, 3, and 2 suggests that the locus of maximum glacial activity shifted to more southerly latitudes at the end of stage 5.

Ice cores

Ice cores that span the last glacial cycle have been recovered from Greenland (Camp Century and Dye 3; Dansgaard and others, 1985; Reeh, 1989) and from the Devon and Agassiz ice caps (Koerner, 1989a) in the Canadian High Arctic. Although some uncertainty exists as to the precise chronologies of the ice cores near their base, the following general conclusions can be drawn. During the earliest phase of the last glaciation (probably throughout most of marine isotope stage 5) precipitation remained relatively high (i.e., similar to present; subsequently precipitation receipts were substantially reduced, and summer mean annual temperatures remained well below present until about the Pleistocene-Holocene transition (Dansgaard and others, 1982; Koerner, 1989a). This evidence implies that sufficient moisture was available to nourish high-latitude ice sheets throughout marine isotope stage 5, whereas maintenance of these ice sheets during the low-latitude glacial maximum may have been primarily by diminished summer temperatures during an interval of low precipitation. Koerner (1989b) recently argued that the ice-core records imply a substantially reduced Greenland Ice Sheet during the last interglacial.

REGIONAL SYNTHESIS

Broad similarities in reconstructed glacial histories from both sides of Baffin Bay and from deep-sea cores suggest regional climatic control on the inception phase of the last glaciation. Although deposits of the last interglacial are rare, deposits associated with the earliest advance of the last glaciation are well preserved around Baffin Bay. On both the Canadian and northwest Greenland margins, the Early Foxe/Weichselian advance is the most extensive advance of the last glaciation. Along Baffin Island, Early Foxe ice advanced onto the continental shelf, calving into Baffin Bay. To the north, outlet glaciers draining the north-central region of the ice sheet flowed out Lancaster Sound,

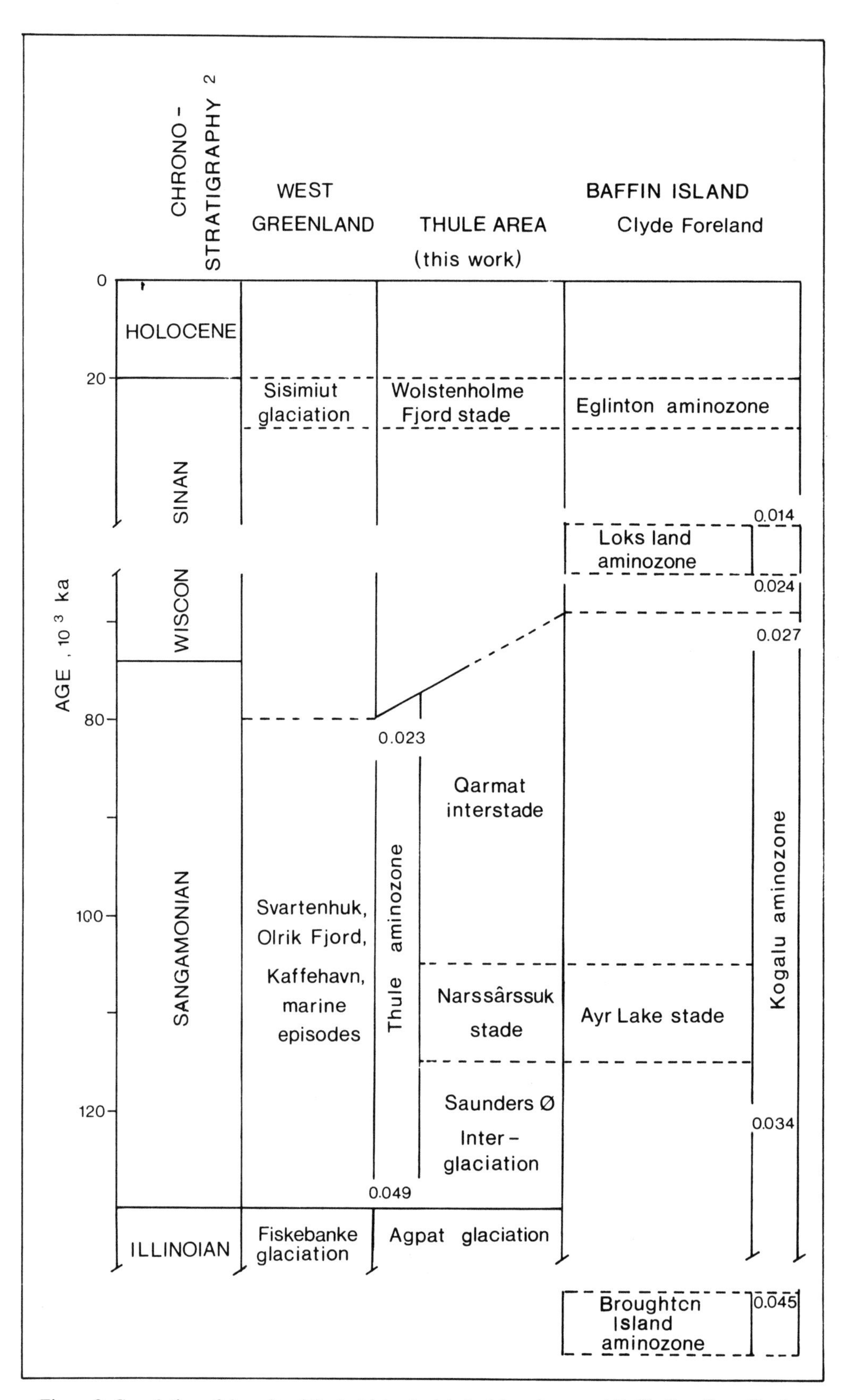

Figure 3. Correlation of deposits of the last interglacial-glacial cycle around Baffin Bay (from Funder, 1990a). West Greenland data from Kelly (1986); Baffin Island data from Andrews and others (1986).

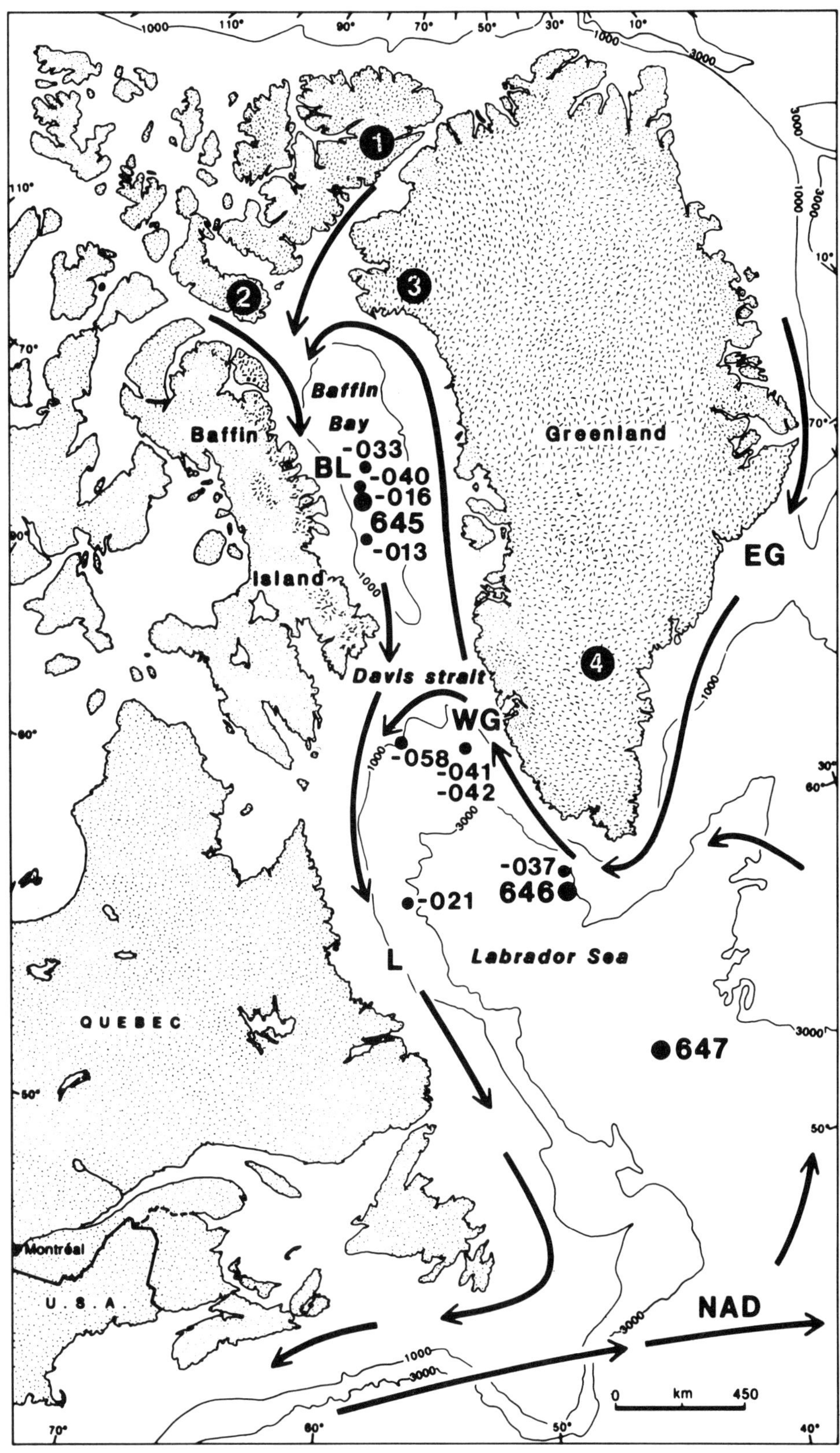

Figure 4. Location map for ODP coring sites 645, 646, and 647, a number of supplemental piston cores in Baffin Bay, and the location of ice cores from the Greenland Ice Sheet, and smaller ice caps in the Canadian Arctic. Modern ocean surface-water circulation is indicated by arrows; NAD=North Atlantic Drift; WG = West Greenland Current; EG = East Greenland Current; BL = Baffin Land Current; L = Labrador Current.

debouching directly into northern Baffin Bay (Klassen and Fisher, 1988). Both central and northern Baffin Island outlet glaciers delivered abundant carbonate erratics to Baffin Bay, and this lithology serves as a stratigraphic marker in Baffin Bay cores, linking the terrestrial and deep-sea records (e.g., Aksu and Piper, 1979). The extent of the advance on northwest Greenland was less dramatic, but in both Greenland and the Canadian Arctic, striated surfaces, reworking of older Quaternary sediment, and abundant meltwater deposits indicate that the ice was not frozen to its bed and that summer temperatures were high enough to produce substantial melt. Subarctic shallow-water marine mollusca and benthic foraminifera, and bones of large whales preserved in marine deposits that postdate the Early Foxe/Weichselian advance(s), indicate that Baffin Bay was at least seasonally ice free late in isotope stage 5. Although Baffin Bay sediment cores lack a clear signal for warm surface water at this time, the oxygen-isotope signal indicates abundant meltwater, and this may have precluded the establishment of a subarctic open-ocean planktonic foraminifera assemblage. That extensive high-latitude glaciation was restricted to the early phase of the last glaciation is supported by the abundance of pre-Quaternary palynomorphs in Baffin Bay stage 5 sediment, and their rarity in sediments from isotope stages 4, 3, and 2. The palynomorphs are derived from the Cretaceous-Tertiary sedimentary rocks that are common in the high Canadian Arctic and inter-island channels and on West Greenland, but are rare at lower latitudes over the Canadian Shield. Throughout isotope stage 5, sea-surface temperatures in the North Atlantic remained essentially "interglacial" (cf. CLIMAP Project Members, 1984), and precipitation over the Greenland (Dansgaard and others, 1982) and Canadian (Koerner, 1989b) ice caps was similar to present, in contrast to isotope stages 4, 3, and 2, when sea-surface temperatures in the North Atlantic were lower and precipitation in the Arctic was reduced.

On the basis of this regional synthesis, we reaffirm our earlier assertions (Miller and others, 1977; Andrews and others, 1986; Hillaire-Marcel and de Vernal, 1989; Funder, 1990a) that glaciation in the Canadian and Greenland Arctic adjacent to Baffin Bay is controlled primarily by oceanographic circulation (which controls winter accumulation) and summer insolation. Warm surface currents are required to advect sufficient moisture into the Arctic to grow continental ice on the Canadian side, and to cause the advance of the Greenland Ice Sheet beyond the inner fjord regions in northwest Greenland. Milankovitch minima in high-latitude summer insolation will be the intervals of minimum summer melting, although calving processes and advection are important variables as well. We postulate that the inception phase of the last glaciation occurred at the end of isotope substage 5e as summer insolation receipts at high latitudes diminished from the interglacial insolation maximum into the minimum centered on 115 ka B.P., the absolute minimum of the last glaciation (Berger, 1984), but while meridional oceanographic circulation was still strong. The warm winters and high precipitation produced by such a circulation, coupled with lowered summer temperatures, should have produced the most favorable conditions for initial ice-sheet growth over the plateau regions of Baffin Island and Labrador–New Quebec. Conditions in Arctic regions during the interval of maximum mid-latitude glaciation were, in contrast, very cold and much drier. Consequently, ice-sheet margins were reduced in the north and Baffin Bay presumably was covered by permanent sea ice throughout much of isotope stages 4, 3, and 2.

ACKNOWLEDGMENTS

Our understanding of conditions favorable for the growth of ice sheets at the end of the last interglacial have developed from a long period of field and laboratory studies of surficial geology exposed on land and sediment cores from adjacent deep-sea basins. We are indebted to numerous colleagues for stimulating discussions over the years and to our various funding agencies for financial support. We wish to acknowledge in particular the Surficial Processes program within the Division of Earth Science at the U.S. National Science Foundation for support of the Baffin Island program (Miller and Andrews), the Geological Surveys of Greenland and Denmark, and the Danish Research Council, as well as other agencies for support of the Thule program (Funder) and NSERC (Canada) and Fonds FCAR (Quebec) for support of the marine core analyses (de Vernal).

REFERENCES CITED

Aksu, A. E., and Piper, D.J.W., 1979, Baffin Bay in the past 100,000 years: Geology, v. 7, p. 245–248.

Andrews, J. T., 1972, Recent and fossil growth rates of marine bivalves, Canadian Arctic, and late Quaternary Arctic marine environments: Palaeogeography, Palaeoclimatology, Palaeoecology, v. 11, p. 157–176.

—— , 1982, Holocene glacier variations in the eastern Canadian Arctic: A review: Striae, v. 18, p. 9–14.

Andrews, J. T., and Ives, J. D., 1978a, "Cockburn" nomenclature and the late Quaternary history of the eastern Canadian Arctic: Arctic and Alpine Research, v. 10, p. 617–633.

—— , 1978b, Glacial inception and disintegration during the last glaciation: Annual Review of Earth and Planetary Sciences, v. 6, p. 205–228.

Andrews, J. T., and Miller, G. H., 1984, Quaternary glacial and nonglacial correlations for the Eastern Canadian Arctic, *in* Fulton, R. J., ed., Quaternary stratigraphy of Canada—A Canadian contribution to IGCP Project 24: Geological Survey of Canada Paper 84-10, p. 101–116.

Andrews, J. T., Aksu, A., Kelly, M., Klassen, R., Miller, G. H., Mode, W. N., and Mudie, P., 1986, Land/ocean correlations during the last interglacial/glacial transition, Baffin Island, northwestern North Atlantic: A review: Quaternary Science Reviews, v. 4, p. 333–355.

Berger, A., 1984, Accuracy and frequency stability of the earth's orbital elements during the Quaternary, *in* Berger, A. L., and others, eds., Milankovitch and climate: Boston, D. Reidel, p. 3–39.

Blake, W., Jr., 1973, Former occurrence of *Mytilus edulis* L. on Coburg Island, Arctic Archipelago: Naturaliste Canadien, v. 100, p. 51–58.

Bryson, R. A., Wendland, W. M., Ives, J. D., and Andrews, J. T., 1969, Radiocarbon isochrones on the disintegration of the Laurentide Ice Sheet: Arctic and Alpine Research, v. 1, p. 1–14.

Clark, P. U., 1988, Glacial geology of the Torngat Mountains, Labrador: Canadian Journal of Earth Sciences, v. 25, p. 1184–1198.

CLIMAP Project Members, 1984, The last interglacial ocean: Quaternary Research, v. 21, p. 123–224.

Dansgaard, W., Clausen, H. B., Gundestrup, N., Hammer, C. U., Johnsen, S. J., Kristinsdottir, P. M., and Reeh, N., 1982, A new Greenland deep ice core: Science, v. 218, p. 1273–1277.

Dansgaard, W., Clausen, H. B., Gundestrup, N., Johnsen, S. J., and Rygner, C., 1985, Dating and climatic interpretation of two deep Greenland ice cores: American Geophysical Union Geophysical Monograph 33, p. 71–76.

de Vernal, A., and Hillaire-Marcel, C., 1987, Paleoenvironments along the eastern Laurentide Ice Sheet margin and timing of the last ice maximum and retreat: Géographie Physique et Quaternaire, v. 41, p. 265–277.

de Vernal, A., and Mudie, P., 1989, Pliocene and Pleistocene palynostratigraphy at ODP sites 646 and 647, eastern and southern Labrador Sea, *in* Srivastava, S. P., Arthur, M. A., Clement, B., and others, Proceedings of the Ocean Drilling Program, Scientific Results, Volume 105: College Station, Texas A&M University, p. 401–422.

de Vernal, A., Hillaire-Marcel, C., Aksu, A. E., and Mudie, P. J., 1987, Palynostratigraphy and chronostratigraphy of Baffin Bay deep sea cores: Climatostratigraphic implications: Palaeogeography, Palaeoclimatology, Palaeoecology, v. 62, p. 97–105.

Dyke, A. S., and Prest, V. K., 1987, Late Wisconsinan and Holocene history of the Laurentide Ice Sheet: Géographie Physique et Quaternaire, v. 41, p. 237–263.

Fillon, R. H., and Duplessy, J. C., 1980, Labrador Sea bio-, tephro-, oxygen isotope stratigraphy and late Quaternary paleoceanographic trends: Canadian Journal of Earth Sciences, v. 17, p. 831–854.

Forman, S. L., Wintle, A. G., Thorleifson, H. L., and Wyatt, P. H., 1987, Thermoluminescence properties and age estimates of Quaternary marine sediments from the Hudson Bay Lowlands, Canada: Canadian Journal of Earth Sciences, v. 24, p. 2405–2411.

Fulton, R. J., 1989, Foreword to the Quaternary geology of Canada and Greenland, *in* Fulton, R. J., ed., Quaternary geology of Canada and Greenland: Geological Survey of Canada, Geology of Canada, no. 1, (also Geological Society of America, The Geology of North America, v. K-1), p. 1–11.

Funder, S., 1990a, The last interglacial/glacial cycle in Greenland: New perspectives: Striae.

——, 1990b, Late Quaternary stratigraphy and glaciology in the Thule area, northwest Greenland: Meddeleser om Grønland, Geoscience, no. 22, p. 1–63.

Hillaire-Marcel, C., and de Vernal, A., 1989, Isotopic and palynological records of the late Pleistocene in eastern Canada and adjacent ocean basins: Géographie Physique et Quaternaire, v. 43, p. 263–290.

Hillaire-Marcel, C., de Vernal, A., Aksu, A. E., and Macko, S., 1989, High-resolution isotopic and micropaleontological studies of upper Pleistocene sediments at ODP site 645, Baffin Bay, *in* Srivastava, S. P., Arthur, M. A., Clement, B., and others, Proceedings of the Ocean Drilling Program, Scientific Results, Volume 105: College Station, Texas A&M University, p. 599–616.

Hillaire-Marcel, C., de Vernal, A., and Aksu, A. E., 1990, High resolution palynostratigraphy and ^{18}O stratigraphy of upper Pleistocene sediments at ODP site 645, Baffin Bay, *in* Initial reports of the Ocean Drilling Project, Volume 105, part B: College Station, Texas A&M University.

Hjort, C., and Funder, S., 1974, The subfossil occurrence of *Mytilus edulis* L. in central East Greenland: Boreas, v. 3, p. 23–24.

Hooke, R.LeB., 1982, Wisconsin and Holocene delta ^{18}O variations, Barnes Ice Cap, Canada: Geological Society of America Bulletin, v. 93, p. 784–789.

Houmark-Nielsen, M., 1990, Local correlation and event stratigraphy, *in* Funder, S., ed., Late Quaternary stratigraphy in the Thule area, northwest Greenland: Meddeleser om Grønland, Geoscience.

Kelly, M., 1986, Quaternary, pre-Holocene, marine events of western Greenland: Grønlands Geologiske Undersøgelse, Report 13, 23 p.

Klassen, R. A., 1981, Aspects of the glacial history of Bylot Island, District of Franklin, *in* Current research, Part A: Geological Survey of Canada Paper 81-1A, p. 317–326.

——, 1985, An outline of glacial history of Bylot Island, District of Franklin, N.W.T., *in* Andrews, J. T., ed., Quaternary environments: Eastern Canadian Arctic, Baffin Bay and West Greenland: Boston, Allen and Unwin, p. 428–460.

Klassen, R. A., and Fisher, D. A., 1988, Basal flow condition at the northeastern margin of the Laurentide Ice Sheet, Lancaster Sound: Canadian Journal of Earth Sciences, v. 25, p. 1740–1750.

Koerner, F., 1989a, Queen Elizabeth Island glaciers, *in* Fulton, R. J., ed., Quaternary geology of Canada and Greenland: Geological Survey of Canada, Geology of Canada no. 1, p. 464–473.

——, 1989b, Ice core evidence for extensive melting of the Greenland Ice Sheet in the last interglacial: Science, v. 244, p. 964–968.

Kronborg, C., Mejdahl, V., and Sejrup, H. P., 1990, Thermoluminscence dating and amino acid analyses, *in* Funder, S., ed., Late Quaternary stratigraphy and glaciology in the Thule area, northwest Greenland: Meddeleser om Grønland, Geoscience.

Løken, O. H., 1966, Baffin Island refugia older than 54,000 years: Science, v. 153, p. 1378–3180.

Mangerud, J., Anderson, S. T., Berglund, B. E., and Donner, J. J., 1974, Quaternary stratigraphy of Norden, a proposal for terminology and classification: Boreas, v. 3, p. 109–128.

Mangerud, J., Sønstegaard, E., and Sejrup, H. P., 1979, Correlation of the Eemian (interglacial) Stage and the deep-sea oxygen-isotope stratigraphy: Nature, v. 277, p. 189–192.

Martinson, D. G., Pisias, N. G., Hays, J. D., Imbrie, J. I., Moore, T. C., and Shackleton, N. J., 1987, Age dating and the orbital theory of the ice ages: Development of a high-resolution 0 to 300,000 year chronostratigraphy: Quaternary Research, v. 27, p. 1–29.

Miller, G. H., 1976, Anomalous local glacier activity, Baffin Island, Canada: Paleoclimatic implications: Geology, v. 4, p. 503–504.

——, 1985, Aminostratigraphy of Baffin Island shell-bearing deposits, *in* Andrews, J. T., ed., Quaternary environments: Eastern Canadian Arctic, Baffin Bay and West Greenland: Boston, Allen and Unwin, p. 394–427.

Miller, G. H., Andrews, J. T., and Short, S. K., 1977, The last interglacial-glacial cycle, Clyde foreland, Baffin Island, N.W.T.: Stratigraphy, biostratigraphy and chronology: Canadian Journal of Earth Sciences, v. 14, p. 2824–2857.

Mode, W. N., 1985, Pre-Holocene pollen and molluscan records from eastern Baffin Island, *in* Andrews, J. T., ed., Quaternary environments: Eastern Canadian Arctic, Baffin Bay and West Greenland: Boston, Allen and Unwin, p. 502–519.

Mode, W. N., Nelson, A. R., and Brigham, J. K., 1983, A facies model of Quaternary glacial-marine cyclic sedimentation along eastern Baffin Island, N.W.T., Canada, *in* Molinia, B. F., ed., Glacial-marine sedimentation: New York, Plenum, p. 495–534.

Nelson, A. R., 1982, Aminostratigraphy of Quaternary marine and glaciomarine sediments, Qivitu Peninsula, Baffin Island: Canadian Journal of Earth Sciences, v. 19, p. 945–961.

NORDQUA 86 Participants, 1989, The Baffin Bay region during the last interglacial: Evidence from northwest Greenland: Géographie Physique et Quaternaire, v. 43, p. 255–262.

Prest, V. K., 1984, The late Wisconsinan glacier complex, *in* Fulton, R. J., ed., Quaternary stratigraphy of Canada—A Canadian contribution to IGCP Project 24: Geological Survey of Canada, Paper 84-10, p. 21–36, map 1584A, scale 1:7,500,000.

Reeh, N., 1989, Dynamic and climatic history of the Greenland Ice Sheet; *in* Fulton, R. J., ed., Quaternary geology of Canada and Greenland: Geological Survey of Canada, Geology of Canada, no. 1, p. 795–822.

MANUSCRIPT ACCEPTED BY THE SOCIETY SEPTEMBER 6, 1991

Printed in U.S.A.

Geological Society of America
Special Paper 270
1992

The Sangamonian and early Wisconsinan glacial record in the western Canadian Arctic

Jean-Serge Vincent
Geological Survey of Canada, 601 Booth Street, Ottawa, Ontario K1A 0E8, Canada

ABSTRACT

Widespread till sheets, glacial lake and glacial-marine sediments on Banks, Victoria, and Melville islands, and on the Beaufort Sea Coastal Plain of the Canadian mainland, may record a late Pleistocene glacial advance which extended to the area as early as the Sangamonian (broad sense) to early Wisconsinan. These sediments overlie beds of interglacial character and underlie in places nonglacial deposits, which have provided both nonfinite and finite ages, and glacial sediments of unquestionable late Wisconsinan age. In other places only a single till sheet is observed between the last interglacial and Holocene sediment suites. Although some workers have argued that the glacial units mentioned above are all late Wisconsinan, stratigraphic, paleoecologic, and chronologic data (^{14}C, Th/U, and amino acid analyses), from several localities, indicate that the glacial sediments are of likely Sangamonian (broad sense) to early Wisconsinan age and that the nonglacial beds underlying or overlying these date respectively from the Sangamonian and middle Wisconsinan. The dispersal centre during the ice advance was situated, as during other advances in northwestern Canada, west of Hudson Bay. The ice generally extended further during the Sangamonian (broad sense)/early Wisconsinan than the late Wisconsinan but not as far as it did during the early and middle Pleistocene. To help resolve apparent incongruities in interpretation of the late Pleistocene deposits and ice limits it is postulated that extensive Keewatin Sector Ice of the Laurentide Ice Sheet may have first advanced in northwestern Canada during the Sangamonian (broad sense)/early Wisconsinan and remained there until it finally disappeared in the late Wisconsinan.

INTRODUCTION

In this chapter I discuss late Pleistocene deposits thought to have been laid down as early as the Sangamonian–early Wisconsinan, on Banks, Melville, and Victoria islands in the southwestern Canadian Arctic Archipelago, and on the coastal plain of the District of Mackenzie and Yukon Territory. Previously published studies on this area, situated at the northwestern margin of the Laurentide Ice Sheet, are synthesized and assessed. Some new chronological data that support proposed age assignments are presented, and an attempt is made to reconcile apparently incongruous interpretations.

In the western Canadian Arctic (Fig. 1), as in all other areas of glaciated North America, it is difficult to identify unambiguously deposits of Sangamonian (132–80 ka) and early Wisconsinan (80–65 ka) age (Vincent and Prest, 1987). Although numerous exposures are available, and extremely well preserved and abundant fossils have been recovered in situ from units of varying ages, the absence of accurate numerical dating methods is a major stumbling block to establishing a well-delimited chronostratigraphy.

Most Quaternary scientists agree as to what are late Pleistocene (<132 ka) glacial deposits in the area, but disagree on their exact ages within that time interval. Some workers contend that the most extensive advance initially occurred during the Sangamonian (in the broad sense; 122–80 ka)–early Wisconsinan and that late Wisconsinan (23–10 ka) ice was less extensive (Rampton, 1988; Vincent, 1983, 1984, 1989). Others consider that the

Vincent, J.-S., 1992, The Sangamonian and early Wisconsinan glacial record in the western Canadian Arctic, *in* Clark, P. U., and Lea, P. D., eds., The Last Interglacial-Glacial Transition in North America: Boulder, Colorado, Geological Society of America Special Paper 270.

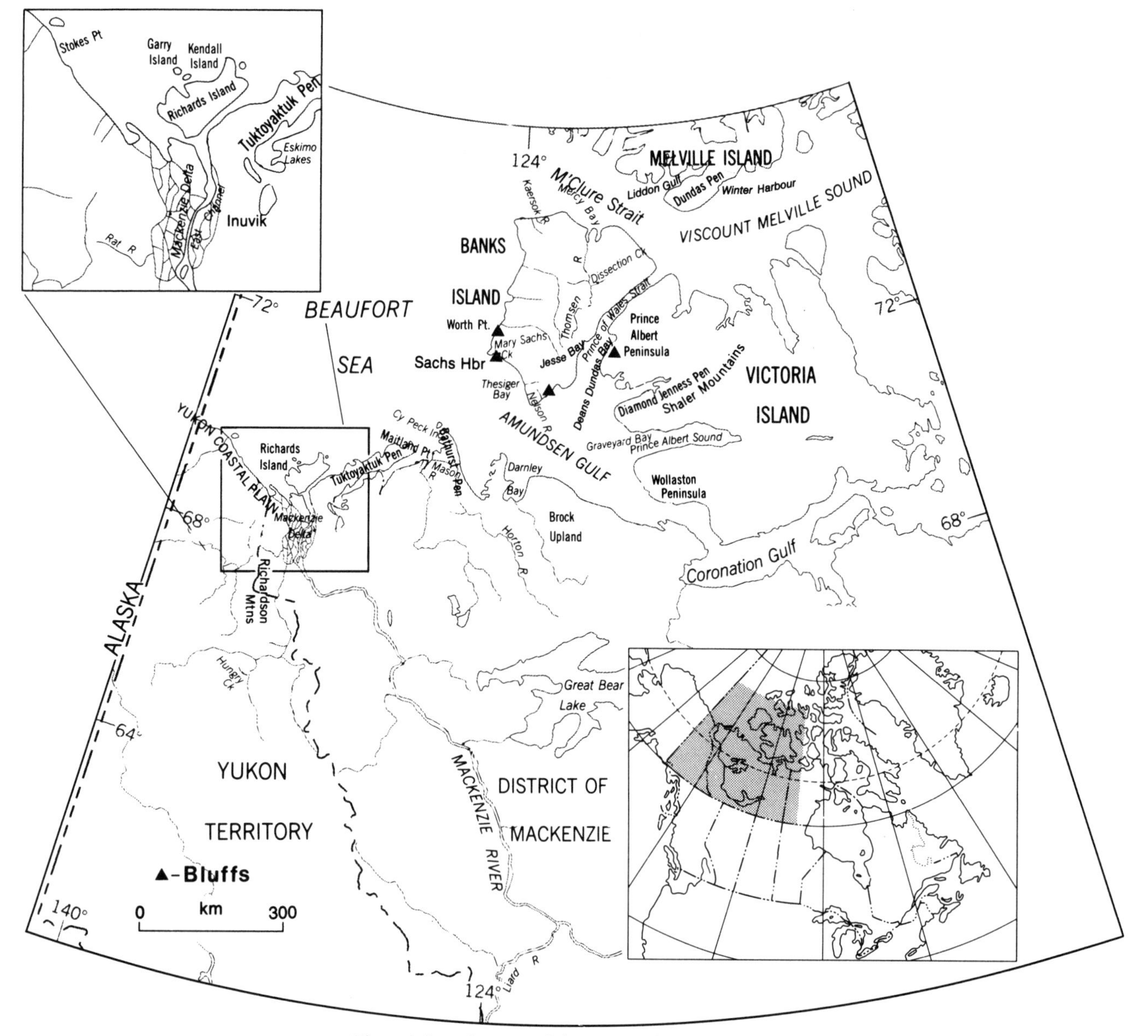

Figure 1. Location map of the western Canadian Arctic.

maximum Wisconsinan advance occurred during the late Wisconsinan (Mayewski and others, 1981; Denton and Hughes, 1983; Hughes, 1987; Dyke, 1987; Dyke and Prest, 1987a) and in some areas was followed by a local readvance (Hughes, 1987). Accordingly, these workers consider the Sangamonian–early Wisconsinan advance of the former researchers to be late Wisconsinan, and the younger advance, when recognized, to be a limited readvance.

LIMITS OF QUATERNARY GLACIERS

The extent of continental ice advances in the western Canadian Arctic have been mapped on Banks Island by Vincent (1980a, 1980b, 1982, 1983, 1984); on Melville and Victoria islands by Hodgson and others (1984), and Vincent (1982, 1983, 1984); on the coastal plain in the northwestern part of the District of Mackenzie by Fyles and others (1972), Hughes (1972, 1987), Klassen (1971), and Rampton (1980, 1981a, 1981b, 1981c, 1988); and on the Yukon Coastal Plain by Rampton (1982). Vincent (1984, 1989) attempted to correlate the various ice margins and to assign them ages.

Surficial geologic mapping and stratigraphic investigations in the area demonstrate that ice, originating west and northwest of Hudson Bay, advanced at least four times to the western Canadian Arctic (Fig. 2). Observed clasts and measured flow indica-

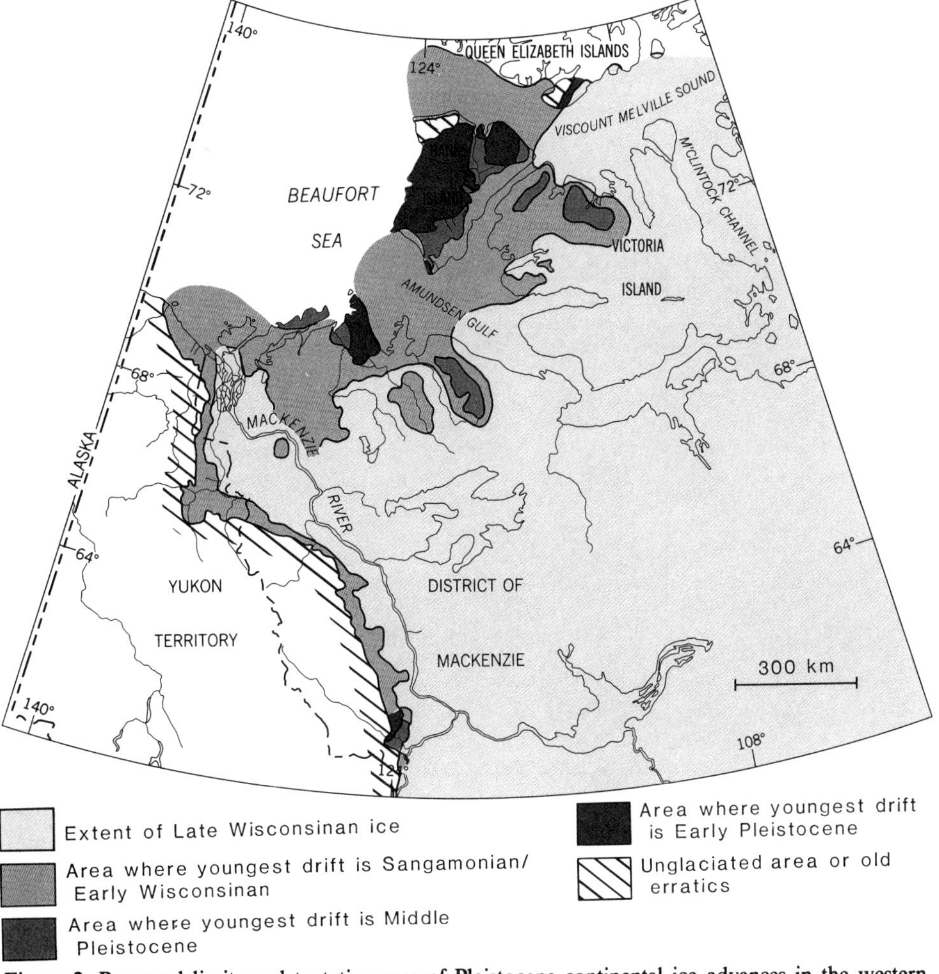

Figure 2. Proposed limits and tentative ages of Pleistocene continental ice advances in the western Canadian Arctic (modified from Vincent, 1989).

tors in tills of all ages support this assertion. To facilitate discussions, specific names were assigned to the various glaciations and stadial advances (Vincent, 1983). The oldest recognized advance (Banks glaciation; dark gray pattern in Fig. 2) is undoubtedly early Pleistocene in age (1.64–0.79 Ma); on southern Banks Island and on Bathurst Peninsula, paleomagnetic investigations (Vincent and others, 1984; Vincent, 1989; and Barendregt and Vincent, 1990) have shown that the tills and associated marine sediments deposited during this advance are magnetically reversed. The youngest advance (Russell stade of Amundsen glaciation; light gray pattern) is of late Wisconsinan age, based on a ^{14}C chronology.

The age assignment for the two intermediate ice advances (dark and light medium gray patterns) is more difficult to establish. Could one of these have occurred during the Sangamonian (broad sense) and/or the early Wisconsinan? The ages of these two intermediate glaciations, the Thomsen glaciation and the M'Clure stade of Amundsen glaciation, largely depends on the age assigned to the intervening nonglacial beds of the Cape Collinson Formation (Fig. 3). If these beds are Sangamonian, then the Thomsen glaciation occurred sometime during the middle Pleistocene (790–132 ka), and the youngest, the M'Clure stade of Amundsen glaciation, could have occurred in the Sangamonian (broad sense), the early Wisconsinan, the middle Wisconsinan (65–23 ka), or even the late Wisconsinan. However, if the nonglacial beds are older than the Sangamonian, or younger, the possibilities for alternate age assignments are numerous.

To resolve the chronological problem, attempts have been made to date wood and plant remains from interstratified nonglacial beds and shells from marine sediments associated with the ice advances. Radiocarbon ages have been obtained from various laboratories (Table 1). Although these laboratories are capable of providing ages older than 35 ka on wood and 20 ka on shells, the potential for contamination is high. I believe that reported ages in excess of these values should be looked at very critically. Thorium/uranium age estimates have also been obtained from

Géotop at the Université du Québec à Montréal (Table 1). The method used and the reliability of the age estimates was discussed by Causse and Vincent (1989). Generally, it seems that most of the uranium uptake occurs during the early diagenetic phase; i.e., shortly after the deposition of the sediments or burial of organics. However, secondary uptake of uranium may occur during subsequent episodes of water circulation. The calculated Th/U age estimates are therefore strictly related to the age of uranium uptake and must be treated as minimums. In addition, a few hundred amino-acid analyses have been performed on wood and

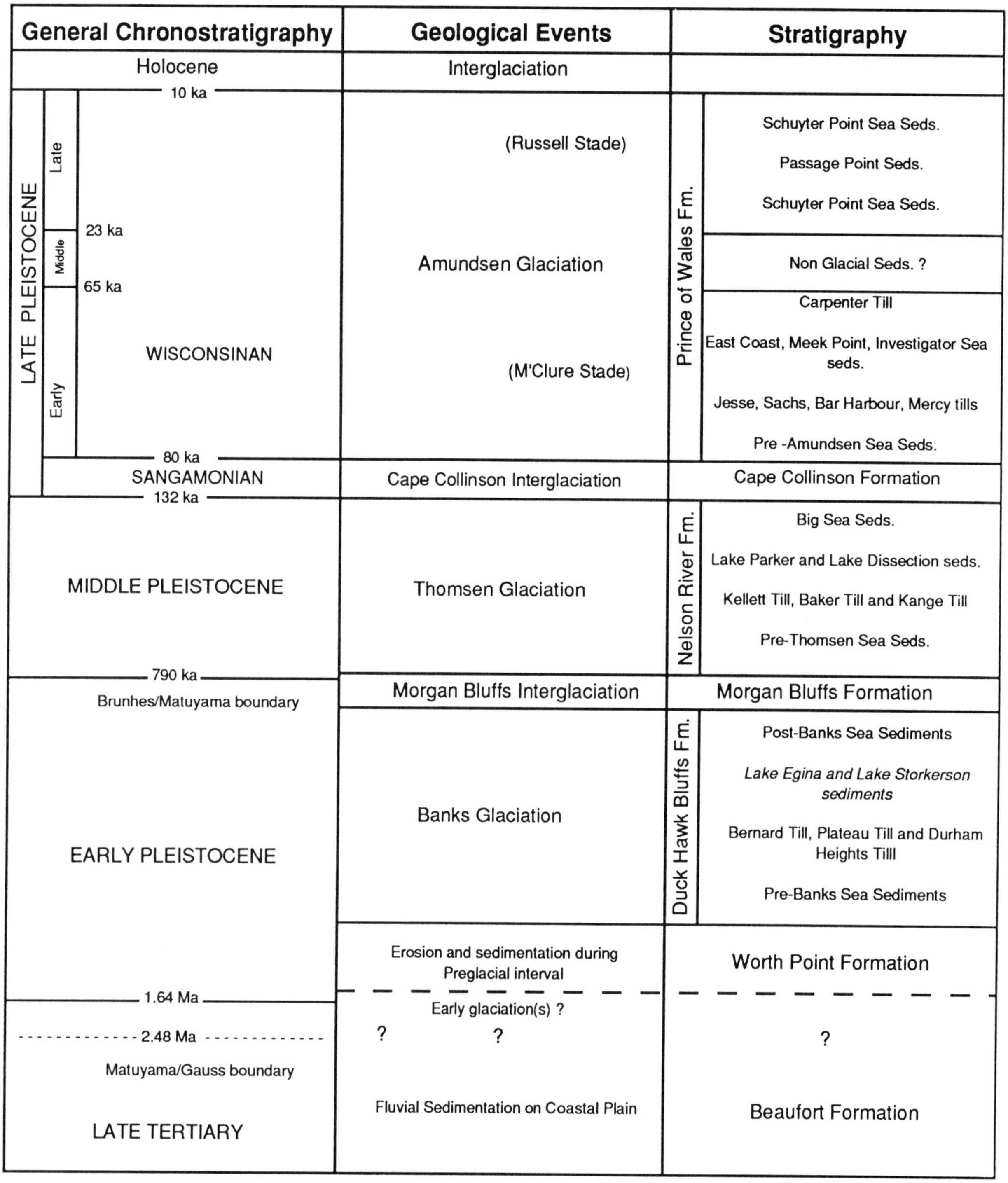

General Chronostratigraphy	Geological Events	Stratigraphy
Holocene	Interglaciation	
10 ka LATE PLEISTOCENE – WISCONSINAN Late 23 ka Middle 65 ka Early	(Russell Stade) Amundsen Glaciation (M'Clure Stade)	Prince of Wales Fm. Schuyter Point Sea Seds. Passage Point Seds. Schuyter Point Sea Seds. Non Glacial Seds. ? Carpenter Till East Coast, Meek Point, Investigator Sea seds. Jesse, Sachs, Bar Harbour, Mercy tills Pre -Amundsen Sea Seds.
80 ka SANGAMONIAN	Cape Collinson Interglaciation	Cape Collinson Formation
132 ka MIDDLE PLEISTOCENE	Thomsen Glaciation	Nelson River Fm. Big Sea Seds. Lake Parker and Lake Dissection seds. Kellett Till, Baker Till and Kange Till Pre-Thomsen Sea Seds.
790 ka Brunhes/Matuyama boundary EARLY PLEISTOCENE	Morgan Bluffs Interglaciation	Morgan Bluffs Formation
	Banks Glaciation	Duck Hawk Bluffs Fm. Post-Banks Sea Sediments *Lake Egina and Lake Storkerson sediments* Bernard Till, Plateau Till and Durham Heights Tilll Pre-Banks Sea Sediments
	Erosion and sedimentation during Preglacial interval	Worth Point Formation
1.64 Ma 2.48 Ma Matuyama/Gauss boundary LATE TERTIARY	Early glaciation(s) ? ? ? Fluvial Sedimentation on Coastal Plain	? Beaufort Formation

Figure 3. Correlation of Pleistocene lithostratigraphic units and geological events recognized on Banks Island (modified from Vincent, 1989).

shells by the facilities at the Institute of Arctic and Alpine Research of the University of Colorado and at the Department of Geology of the University of Alberta. Although these results help separate sediments into broad relative age groups, they do little to improve the numerical chronology.

RECORD ON BANKS ISLAND

On Banks Island, deposits assigned to the Cape Collinson Formation (Vincent, 1983; Vincent and others, 1983) are the youngest nonglacial sediments underlying till. At the type section (Figs. 4 and 5) along coastal bluffs east of the mouth of Nelson River (Fig. 1) and at another coastal section (Fig. 6) east of Mary Sachs Creek, near Sachs Harbour, the nonglacial deposits unconformably underlie marine sediments and till laid down during the M'Clure stade of the Amundsen glaciation. At Nelson River, an in situ tundra surface rests on marine sand and gravel and is overlain by M'Clure stade deposits. Wood, in growth position in the peat, has yielded a ^{14}C age of >61,000 yr B.P. (QL-1230) and a Th/U age estimate of 67,700 +11,700 / –9,900 yr B.P. (UQT-117). At Mary Sachs Creek, a thick in situ woody peat rests on marine sands and underlies glacial-marine sediments and till of the M'Clure stade. The peat is ^{14}C dated at >49,000 yr B.P. (GSC-3560-2). Abundant, paired, and extremely well preserved in situ *Portlandia arctica* shells from overlying glacial-marine sediments are dated at >37,000 yr B.P. (GSC-3698) by the ^{14}C method and at 85,900 +13,000 / –11,900 yr B.P. (UQT-143) by the Th/U method.

Faunal and floral remains of numerous insects and plants not living as far north as Banks Island today have been identified in the peats at the two sites, and indicate that the climate was distinctly warmer than that of the present interglaciation (Matthews and others, 1986). For example, dwarf birch (*Betula glandulosa*), which does not grow on the island today, has been found in growth position on the fossilized tundra surface at the site east of the mouth of Nelson River.

On the basis of the ^{14}C and Th/U age estimates, a middle Wisconsinan age for the Cape Collinson Formation can probably be ruled out. The Cape Collinson Formation could have been deposited at any time during the Sangamonian or could be older (middle Pleistocene). Because the beds represent the youngest known interval distinctly warmer than the present interglaciation, it is reasonable to assign the deposits to the Sangamonian (strict sense; 132–122 ka).

Assuming that the Cape Collinson beds are Sangamonian, deposits of the M'Clure stade of the Amundsen glaciation would necessarily be late Pleistocene, and an attempt can be made at assigning an age to them within that period. During this stade, lobes of continental ice from the Amundsen Gulf, Victoria Island, and Viscount Melville Sound advanced onto the coastal areas of Banks Island (Fig. 7) (Vincent, 1982, 1983, 1984, 1989). All lobes had low surface gradients and flow was controlled by the interisland channels.

The Prince of Wales lobe (Fig. 7) from the Amundsen Gulf and Victoria Island impinged on southern Banks Island and flowed northwestward in Prince of Wales Strait to lap onto the east coast of the island. The Jesse till, terminal moraines, and spectacular outwash plains that extend onto nonglaciated areas were deposited (Fig. 8). Glacial lakes were dammed and, following deglaciation, the eastern coastal areas below about 120 m were submerged by the East Coast Sea (Fig. 7). The marine limit cuts into the Jesse till and can be traced continuously for 350 km.

The Thesiger lobe (Fig. 7), flowed northwest from the Amundsen Gulf and impinged on the southwestern coast. The lobe left the Sachs till, formed moraines, and dammed glacial lakes in large valleys at its margin. During deglaciation, the area along Thesiger Bay and all the unglaciated western coast of Banks Island was submerged to about 20 m by the Meek Point Sea. The limit of this sea can be traced for 250 km. It clearly truncates the Sachs till surfaces in the south and the Bar Harbour till surfaces in the north. These relations indicate that the last recorded marine transgression on the west coast postdates the M'Clure stade. Southeast of Sachs Harbour, the Sand Hills moraine extends along the coast for 25 km and overlaps the Sachs till at its southeastern extremity. The moraine was probably constructed by a local ice readvance in Thesiger Bay after the initial retreat of the Thesiger lobe.

The Prince Alfred lobe (Fig. 7), probably an ice stream, impinged on the north coast of the island as it progressed westward into M'Clure Strait from Viscount Melville Sound. East of Mercy Bay, the lobe was confined to the marine channel by high cliffs, whereas to the west it overlapped the land, deposited the Mercy and Bar Harbour tills, and formed lateral moraines. Much of the north slope was also inundated by large glacial lakes. The Investigator Sea is thought to have flooded the north coast of Banks Island to about 30 m following ice retreat. The only shells discovered up to now on this coast, on the surface at 1.5 m above sea level, yielded a ^{14}C age of 11,300 ± 190 yr B.P. (GSC-5096).

The Prince of Wales and Thesiger lobes were contemporaneous; meltwaters of the first lobe built deltas into Glacial Lake Masik, which was dammed by the latter lobe. Similarly, the Prince of Wales and Prince Alfred lobes were contemporaneous, because the extent of Glacial Lake Ivitaruk in the Thomsen River basin was controlled by the simultaneous presence of the lobes in the M'Clure Strait and in the upper Thomsen River basin.

Age of the M'Clure stade

Mayewski and others (1981, p. 103–104), Denton and Hughes (1983, p. 134–135), Dyke (1987), and Dyke and Prest (1987a) have argued, in their reconstructions of events in northwestern Canada, that the M'Clure stade, as portrayed above, is late Wisconsinan. Vincent (1984, 1989), however, using radiometric dates on organics associated with or postdating the M'Clure stade, provided evidence that this ice advanced and in some cases clearly retreated before the late Wisconsinan.

In both stratigraphic sections (Figs. 5 and 6) discussed above, the Cape Collinson Formation underlies marine sediments and

TABLE 1. RADIOCARBON AND CORRECTED Th/U AGE ESTIMATES CITED IN THE TEXT

Laboratory Number*	Date (yr BP)	Locality	Reference or Collector	Material	Significance
		RADIOCARBON AGE ESTIMATES			
GSC-388	>32,400	Northwest of Graveyard Bay, Victoria Island	Blake, 1974; Vincent, 1984	Tundra plants	Interstadial deposits between early and late Wisconsinan tills?
GSC-481	17,860 ± 260	Ibyuk Pingo, Tuktoyaktuk District of Mackenzie	Lowdon and Blake, 1973; Rampton, 1988	Peat	Minimum age for Toker Point stade
GSC-562	>35,000	Garry Island, District of Mackenzie	Lowdon and others, 1971; Rampton, 1988	Marine shells	Minimum age for marine transgression after the Toker Point stade
GSC-690	>37,000	Richards Island, District of Mackenzie	Lowdon and Blake, 1968; Rampton 1988	Marine shells	Minimum age for marine transgression after the Toker Point stade
GSC-727	>33,000	West of Winter Harbour, Melville Island	Lowdon and Blake, 1968; Hodgson and others, 1984	Marine shells	Minimum age for marine transgression after deposition of Bolduc till
GSC-787	42,400 ± 1,900	West of Winter Harbour, Melville Island	Lowdon and Blake, 1968; Hodgson and others, 1984	Marine shells	Minimum age for marine transgression after deposition of Bolduc till
GSC-1088	>41,000	Kaersok River, Banks Island	Vincent, 1983, 1984	Moss	Minimum age for Prince Alfred lobe (M'Clure stade of Amundsen glaciation), Bar Harbour till, and lakes Ballast and Ivitaruk
GSC-1262	22,400 ± 240	Stokes Point, Yukon	Lowdon and Blake, 1976; Rampton, 1982	Peat	Minimum age for Buckland glaciation
GSC-1478	>19,000	South of Worth Point, Banks Island	Lowdon and Blake, 1973; Vincent, 1983, 1984	Marine shells	Minimum age for Meek Point Sea and M'Clure stade of Amundsen glaciation
GSC-1974	33,800 ± 880	Cy Peck Inlet, Bathurst Peninsula, District of Mackenzie	Lowdon and Blake, 1978; Rampton, 1988	Wood and bark	Minimum age for the Franklin Bay stade and the Toker Point stade
GSC-2422	36,900 ± 300	Hungry Creek, District of Mackenzie	Hughes and others, 1981; McNeely, 1989	Wood	Maximum age for the Hungry Creek glaciation
GSC-3371	21,300 ± 270	Rat River, District of Mackenzie	Catto, 1986; McNeely, 1989	Organic detritus	Minimum age for deglaciation of Upper Rat River
GSC-3560-2	>49,000	Coastal section east of Mary Sachs Creek, Banks Island	Vincent and others, 1983; Blake, 1987	Peat	Minimum age of Cape Collinson Formation deposits lying below M'Clure stade (Amundsen glaciation) deposits
GSC-3592	>38,000	Northwest of Graveyard Bay, Victoria Island	Vincent, 1984	Wood	Interstadial deposits between early and late Wisconsinan tills?
GSC-3613	>37,000	Northwest of Graveyard Bay, Victoria Island	Vincent, 1984	Willow leaves	Interstadial deposits between early and late Wisconsinan tills?
GSC-3698	>37,000	Coastal section east of Mary Sachs Creek, Banks Island	Vincent, 1984; Blake, 1987	Marine shells	Minimum age of Thesiger lobe (M'Clure stade of Amundsen glaciation)
GSC-3722	>39,000	Maitland Point, Bathurst Peninsula, District of Mackenzie	Vincent, 1989; Rampton, 1988	Wood	Driftwood postdating Mason River stade
GSC-3759	>38,000	South of Mason River mouth, District of Mackenzie	Vincent, 1989; Rampton, 1988	Wood	Driftwood postdating Mason River stade
GSC-3813	21,200 ± 240	Rat River, District of Mackenzie	Catto, 1986; McNeely, 1989	Organic detrius	Minimum age for deglaciation of Upper Rat River
GSC-4075	>36,000	South of Mason River mouth, District of Mackenzie	Vincent, 1989; Rampton, 1988	Wood	Driftwood postdating Mason River stade
GSC-4710	34,200 ± 400	Northeast of Worth Point, Banks Island	J.-S. Vincent	Marine Shells	Minimum age for Meek Point Sea and M'Clure stade of Amundsen glaciation

TABLE 1. RADIOCARBON AND CORRECTED Th/U AGE ESTIMATES CITED IN THE TEXT (continued)

Laboratory Number*	Date (yr BP)	Locality	Reference or Collector	Material	Significance
GSC-5095	>35,000	South of Worth Point, Banks Island	J.-S. Vincent and D. A. Hodgson	Marine shells	Minimum age for Meek Point Sea and M'Clure stade of Amundsen glaciation (same sample as TO-2206 and UQT-467-470)
QL-1230	>61,000	East of Nelson River mouth, Banks Island	Vincent, 1983, 1984	Wood	Minimum age of interglacial Cape Collinson Formation sediments
RIDDL-801	48,200 ± 1,100	South of Kendall Island, District of Mackenzie	Vincent, 1989	Marine shells	Minimum age for marine transgression after the Toker Point stade, should be considered nonfinite
TO-650	24,730 ± 260	South of Jesse Harbour, Banks Island	Vincent, 1989	Marine shells	Minimum age for East Coast Sea and M'Clure stade on eastern Banks Island
TO-796	43,550 ± 470	Garry Island, District of Mackenzie	Vincent, 1989	Marine shells	Minimum age for marine transgression after the Toker Point stade, should be considered nonfinite
TO-2206	41,090 ± 770	South of Worth Point, Banks Island	J.-S. Vincent and D. A. Hodgson	Marine shells	Minimum age for Meek Point Sea and M'Clure stade of Amundsen glaciation (same sample as GSC-5095 and UQT-467-470)
TO-2406	34,760 ± 330	West of Winter Harbour, Melville Island	D. A. Hodgson and J.-S. Vincent	Marine shells	Minimum age for marine transgression after deposition of Bolduc Till
		CORRECTED Th/U AGE ESTIMATES			
UQT-117†	$67,700^{+11,700}_{-9,900}$	East of Nelson River mouth, Banks Island	Causse and Vincent, 1989	Wood	Minimum age of interglacial Cape Collinson Formation sediments
UQT-142†	$31,800^{+3,800}_{-3,200}$	South of Jesse Harbour, Banks Island	Causse and Vincent, 1989	Marine shells	Minimum age for East Coast Sea and M'Clure stade on eastern Banks Island
UQT-143†	$85,900^{+13,000}_{-11,900}$	Coastal section east of Mary Sachs Creek, Banks Island	Causse and Vincent, 1989	Marine shells	Minimum age of Thesiger Lobe (M'Clure Stade of Amundsen glaciation)
UQT-207	$96,700^{+11,000}_{-10,000}$	East of Maitland Point, Bathurst Peninsula, District of Mackenzie	J.-S. Vincent	Wood	Driftwood postdating Mason River stade
UQT-230†	$61,000^{+8,800}_{-8,600}$	Northwest of Graveyard Bay, Victoria Island	Causse and Vincent, 1989	Wood	Interstadial deposits between early and late Wisconsinan tills?
UQT-233	23,700 ± 1,000	South of Mason River mouth, District of Mackenzie	J.-S. Vincent	Wood	Driftwood postdating Mason River stade
UQT-235	65,600 ± 3,600	South of Mason River mouth, District of Mackenzie	J.-S. Vincent	Wood	Driftwood postdating Mason River stade
UQT-467-470	$40,900^{+300}_{-500}$	South of Worth Point, Banks Island	J.-S. Vincent and D. A. Hodgson	Marine shells	Minimum age for Meek Point Sea and M'Clure stade of Amundsen glaciation (same sample as GSC-5095 and TO-2206)

*GSC = Radiocarbon Dating Laboratory, Geological Survey of Canada; QL = Quaternary Isotope Laboratory, University of Washington; RIDDL = Simon Fraser University RIDDL Group; TO = IsoTrace Laboratory, University of Toronto; UQT = Geotop Laboratory, Université du Québec à Montréal.
†Corrected Th/U age estimates differ from the uncorrected age estimates cited in Vincent (1989).

Figure 4. Section east of the mouth of Nelson River on Banks Island showing the Cape Collinson Formation deposits, presumably of Sangamonian age, underlain by middle Pleistocene Big Sea sediments (Nelson River Formation), and overlain by pre-Amundsen Sea sediments and Jesse till of the M'Clure stade of Amundsen glaciation (Prince of Wales Formation). Geological Survey of Canada photo 176161.

till of the M'Clure stade. The marine deposits were probably laid down in a glacioisostatic sea that submerged the land as it was progressively being depressed as the ice was advancing. At the site just east of the mouth of Mary Sachs Creek, on southwestern Banks Island, glacial-marine sediments overlie the interglacial peat and underlie the Sachs till laid down by the Thesiger lobe. This lobe reached its farthest extent just 1 km north of the bluffs shown in Figure 9. As mentioned previously, in situ shells from the glacial-marine sediments yielded a ^{14}C age of >37,000 yr B.P. (GSC-3698) and a Th/U age estimate of 85,900 +13,000/−11900 yr B.P. (UQT-143). Because the glacial-marine sediments grade into and are interstratified with the overlying Sachs till, ice must have reached its farthest extent well before the late Wisconsinan, perhaps during the Sangamonian (broad sense) if the Th/U age is correct. Dyke (1987, p. 599) did not dispute the relevance of this site in dating the age of the M'Clure stade advance, and correctly pointed out the following:

> That date clearly relates to a marine event prior to or during the last ice advance and does not contradict a Late Wisconsinan age for ice retreat.

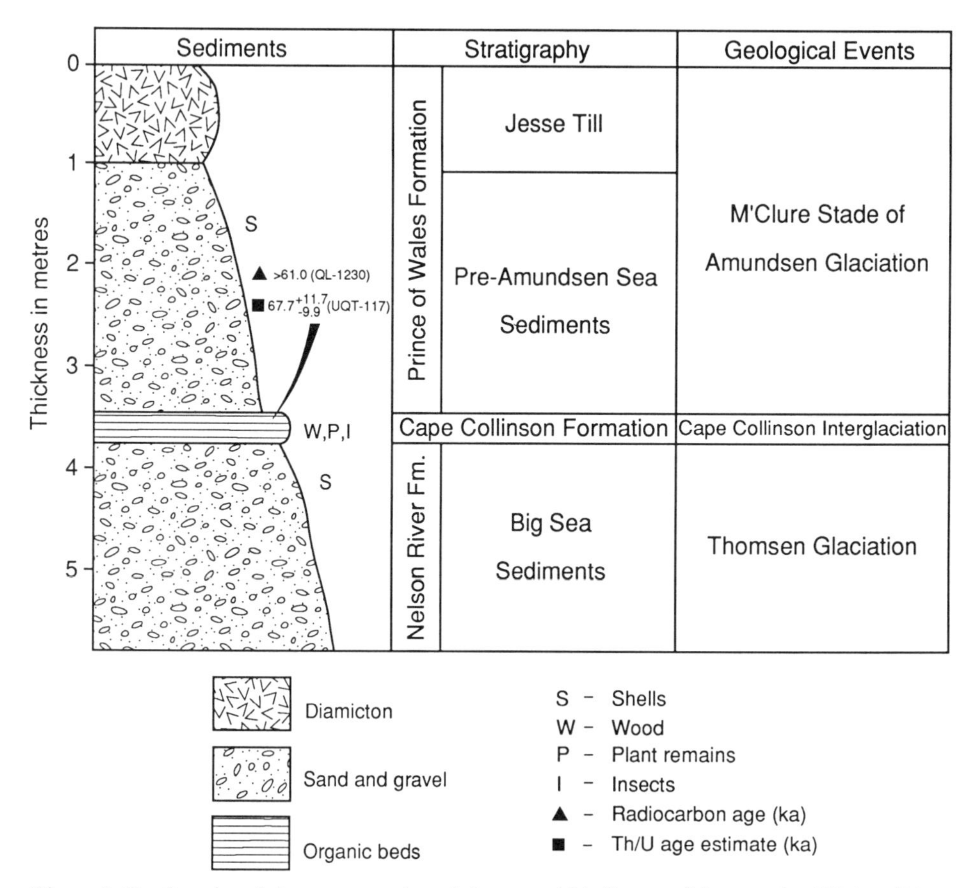

Figure 5. Stratigraphy of the upper portion of the coastal bluffs east of the mouth of Nelson River, southern Banks Island.

The sediment from which the shells came apparently does record ice-proximal sedimentation during the last major ice advance, which suggests that the Jesse Till and correlatives, including the Sachs Till, record a stable or quasi-stable ice cover throughout most of Late Wisconsinan and at least part of Middle Wisconsinan time.

On the basis of the now available Th/U age estimate, the period of stability of ice cover, proposed by Dyke, would have to be extended to the Sangamonian (broad sense) if actual retreat of the Thesiger lobe of the M'Clure stade only occurred in the late Wisconsinan.

South of Worth Point, on the west coast of Banks Island, shell fragments from raised nearshore Meek Point Sea deposits were initially ^{14}C at >19,000 yr B.P. (GSC-1478). Abundant in situ *Astarte borealis* were collected from the same deposit in 1990 and yielded ^{14}C ages of >35,000 yr B.P. (GSC-5095) and 41,090 ± 770 yr B.P. (TO-2206) and a Th/U age estimate of 40,900 +300/−500 yr B.P. (UQT-467-470). Fragile marine shells from other raised nearshore sediments of this sea, northeast of Worth Point, yielded a conventional ^{14}C age of 34,200 ± 1,400 yr B.P. (GSC-4710), which should be considered a minimum age. Obviously, the Meek Point Sea, which postdates or at least was coextensive with the Thesiger and Prince Alfred lobes of the M'Clure stade, predates the late Wisconsinan. It is unlikely that it submerged western Banks Island as late as 18–13 ka, as Dyke and Prest (1987b) have portrayed on their paleogeographic maps. The radiometric evidence points towards an older late Pleistocene age.

Vincent (1984, 1989) also argued that radiometric and other data, from the three sites discussed next, contribute to assigning an age to the M'Clure stade of the Amundsen glaciation. These data are admittedly more open to criticism than the previous data, but nevertheless may help in resolving the chronological problems.

In the Kaersok River Valley of northwestern Banks Island, a moss bed, ^{14}C dated at >41,000 yr B.P. (GSC-1088), occurs in the upper part of valley-fill deposits which were believed by Vincent (1984, 1989) to postdate the retreat of the Prince Alfred lobe in M'Clure Strait, because they are inset in deposits of that lobe. The site has not been studied in detail since it was discovered and sampled by J. G. Fyles in 1968. In particular, the genesis

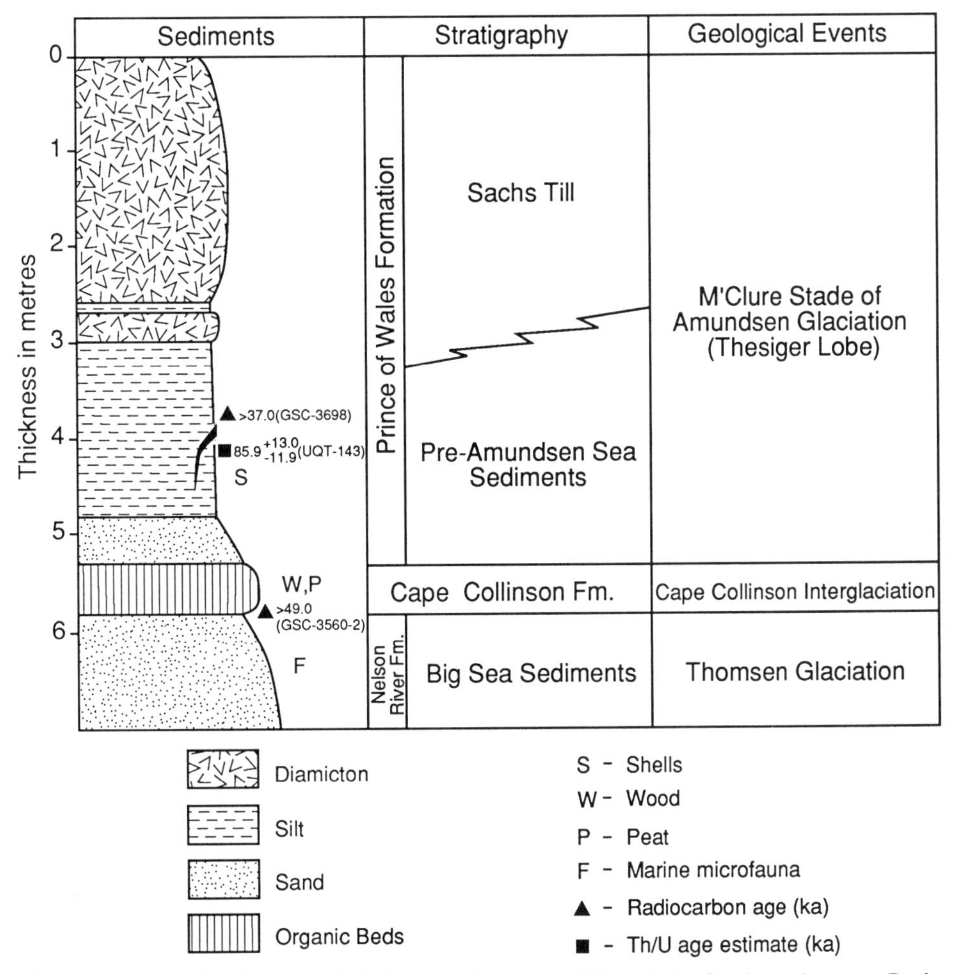

Figure 6. Stratigraphy of the coastal bluffs east of the mouth of Mary Sachs Creek, southwestern Banks Island.

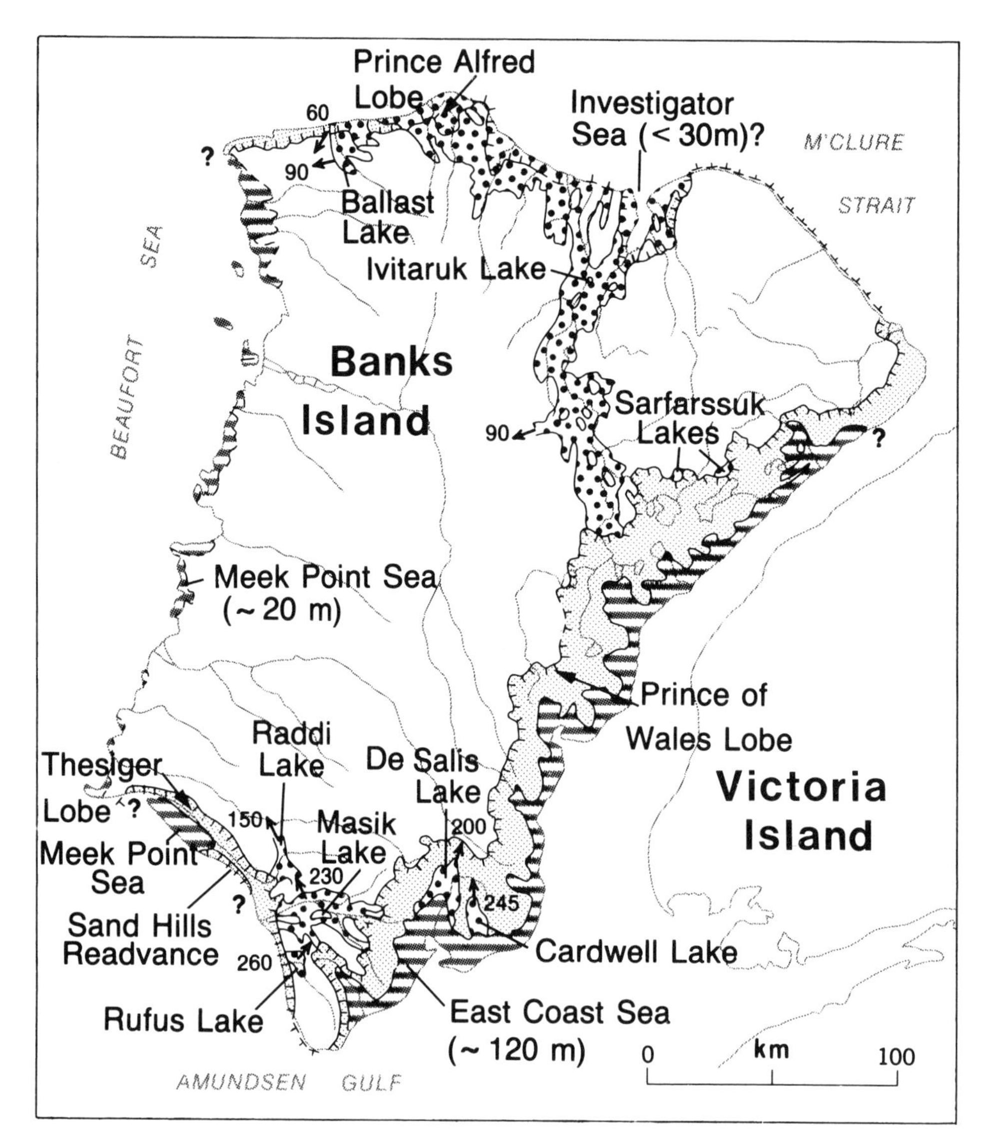

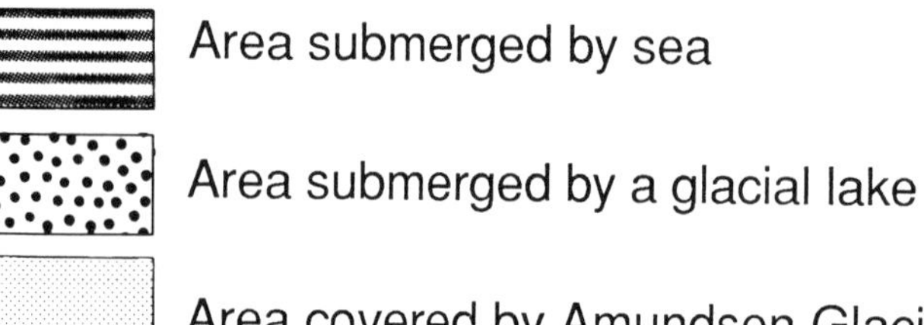

60 Location of lake outlet with altitude in metres, arrow indicates flow direction

Limit of ice advance (defined, assumed)

Figure 7. Paleogeographic map of Banks Island during the M'Clure stade of Amundsen glaciation (modified from Vincent, 1983).

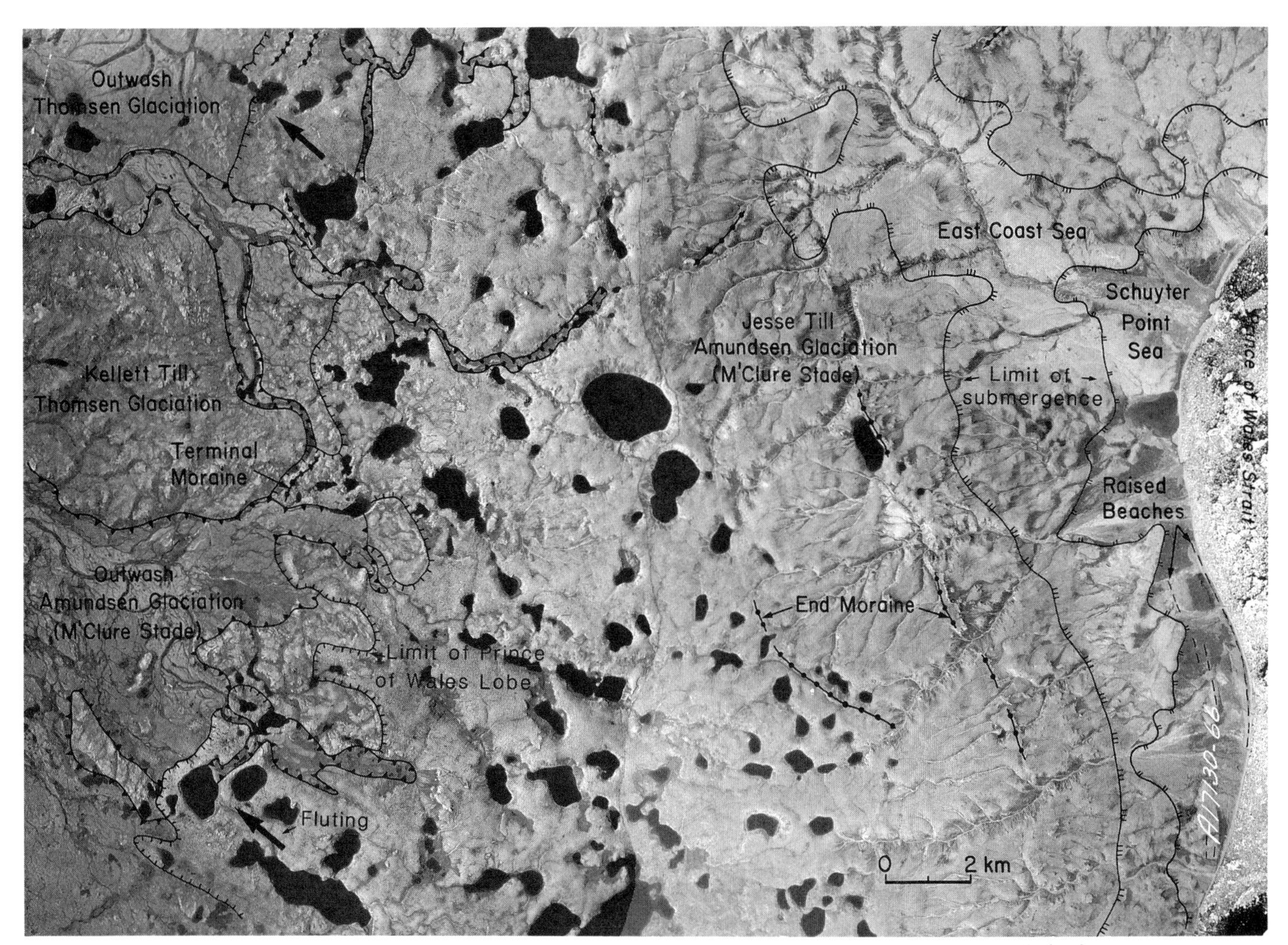

Figure 8. Areal view of the eastern slope of Banks Island, just south of Jesse Bay, showing the limit of extent of the Prince of Wales lobe during the M'Clure stade of Amundsen glaciation. Parts of air photographs A17057-93 and A17130-66, National Air Photo Library, Department of Energy, Mines and Resources, Canada (modified from Vincent, 1982).

of the deposits underlying and overlying the dated moss bed, and the stratigraphic and geomorphic setting of the site in relation to the Prince Alfred lobe, are still unclear. This is exemplified by comments in Dyke (1987) in which the sampled site is said to lie beyond the limit of the Prince Alfred lobe. Vincent's (1980a) surficial geology map shows that it lies ~6 km within it. Dyke (1987, p. 599–600) also argued, using a quote from Fyles (1969, p. 195), that the moss bed was glacially deformed and that "The date, therefore, more likely relates to an ice advance than to ice retreat." The comment of Fyles, as the study of his 1968 field notes indicates, was general in nature and did not relate to the dated moss site, but rather to another locality to the north. Further study is required before the site can be convincingly used, with the information provided by the Meek Point Sea dated sites, to assign a pre–late Wisconsinan age to the Prince Alfred lobe.

Radiocarbon-dated peats in fluvial deposits along Dissection Creek were also used by Vincent (1984, 1989) to argue for a pre–late Wisconsinan age of the Prince Alfred lobe. The stratigraphic and geomorphic setting of the site, in relation to the Prince Alfred lobe and the marine inundation that followed its retreat, cannot be unambiguously demonstrated, as argued by Dyke (1987). The uniqueness of the site and its possible bearing in helping to resolve the chronology warrant further investigation.

Amino-acid ratios of alloisoleucine / L-isoleucine (total) for six fragments of *Hiatella arctica* shells from an East Coast Sea delta overlying the Jesse till range between 0.04 (AAL-895A) and 0.09 (AAL-895C). These ratios differ markedly from those obtained for in situ shells of the likely middle Pleistocene Big Sea (0.16, 0.19, 0.21; AAL-527A, B, C) and those obtained for shells of the late Wisconsinan–Holocene Schuyter Point Sea (0.02, 0.03; AAL-533A, B, C). The ratios seems to indicate that the East Coast Sea, and the advance of the Prince of Wales lobe which it immediately postdates, are distinctly older than late Wisconsinan. An accelerator ^{14}C age on the East Coast Sea shell fragments

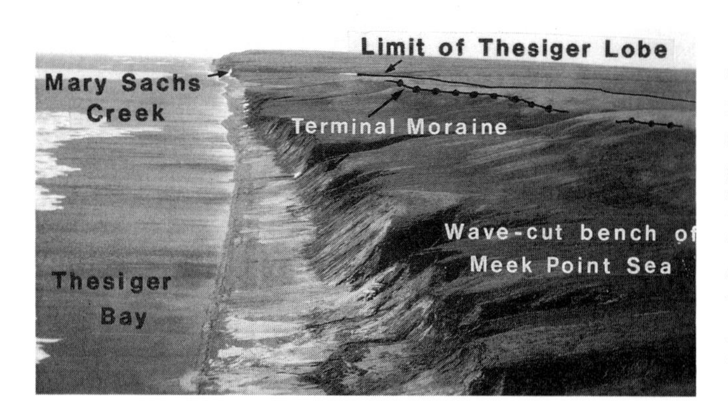

Figure 9. Coastal Bluffs east of the mouth of Mary Sachs Creek, southwestern Banks Island. Shown are the limit of the Thesiger lobe during the M'Clure stade of Amundsen glaciation and a wave-cut bench and scarp marking the limit of Meek Point Sea. The bluffs depicted in Figure 5 are in the foreground. Geological Survey of Canada photo 202775-T.

gave an age of 24,730 ± 260 yr B.P. (TO-650), whereas a Th/U determination yielded an age estimate of 31,800 +3,800/–3200 yr B.P. (UQT-142). Both these determinations should probably be considered minimum ages. Dyke (1987, p. 599) rightly pointed out that "The sample of shells utilized consisted only of fragments from the delta surface and could represent redeposited, ice-transported shells or shells eroded from older marine deposits, which are widespread on Banks Island." It should nevertheless be emphasized that the amino-acid ratios obtained are distinct from late Wisconsinan ones, and, more important, from all those obtained from Big Sea sediments, thick and widespread sequences of which immediately underlie M'Clure stade deposits on eastern Banks Island. The Big Sea sediments should have been the logical source of reworked shells in the East Coast Sea delta, but the ratios obtained contradict this.

Notwithstanding problems related to the last mentioned sites, the above data can be used to argue that M'Clure stade ice advanced into northwestern Canada well before the late Wisconsinan and sometime after a period distinctly warmer than the present, perhaps the Sangamonian (strict sense). An early Wisconsinan age for the ice advance is possible, but some of the Th/U age estimates point to a Sangamonian (broad sense) age. The age of ice retreat is still uncertain. On the north coast it could have been as early as >41,000 yr B.P. (GSC-1088), if the determination of Kaersok River mosses is accepted, and was certainly before 11,300 ± 190 yr B.P. (GSC-5096). On the east coast it could have been as early as 24,730 ± 260 yr B.P. (TO-650), if the age of shells attributed to the East Coast Sea is valid, and was certainly before 11,200 ± 100 yr B.P. (GSC-2545). Whatever the case, it should not be forgotten that a glacial (stadial) limit of clearly late Wisconsinan age can be traced to the east of Banks Island, on Melville and Victoria islands, and to the south on the mainland (Vincent, 1984, 1989).

RECORD ON MELVILLE ISLAND

Beds of possible Sangamonian age have not been observed on Melville Island, but deposits laid down by a lobe of ice flowing west from Viscount Melville Sound are present along the southern slope of the island (Hodgson and others, 1984; Hodgson, 1991). This lobe is probably the same as the Prince Alfred lobe, which impinged on the north coast of Banks Island (Vincent, 1984, 1989). It moved onto southern Dundas Peninsula, depositing the Bolduc till; onto areas bordering the western portion of Liddon Gulf, depositing the Liddon till; and at least as far as the southwestern extremity of the island (Fig. 2). The till sheets left by this ice are distinct from the younger Winter Harbour till, present on southern Dundas Peninsula, which has been associated with a major surge of continental ice at about 10 ka in the area bordering Viscount Melville Sound (Hodgson and Vincent, 1984). On Melville Island, it has not yet been possible to date the age of the initial ice advance down M'Clure Strait, but it is likely that ice retreat occurred late in the late Wisconsinan. Several ^{14}C age determinations on shells, including one from marine sediments conformably overlying till in the southwest of the island, have provided ages no older than 11,700 ± 110 yr B.P. (GSC-4167) (Hodgson, 1991). Some retreat may have occurred earlier in certain areas. *Hiatella arctica* valves and fragments collected on the surface of the Bolduc till gave ^{14}C ages of >33,000 yr B.P. (GSC-727) and 42,400 ± 1900 yr B.P. (GSC-787), best regarded as a minimum age. A single valve of *Hiatella arctica* from a relatively abundant third collection of shells overlying the Bolduc till also yielded a ^{14}C age of 34,760 ± 330 yr B.P. (TO-2406). If all these shells were laid down in a glacioisostatic sea that postdated the ice advance, then at least this margin of the ice sheet advanced and retreated well before the late Wisconsinan on Melville Island. However, it is possible that all these shells are erratics. Dyke (1983) and Dyke and others (1992), for example, have reported the presence of abundant old shells on the surface of late Wisconsinan tills on Somerset and Prince of Wales islands.

RECORD ON VICTORIA ISLAND

On Victoria Island, alluvial beds of possible Sangamonian age underlie M'Clure stade till in two localities near Deans Dundas Bay on Prince Albert Peninsula. Plant and insect remains in the beds indicate slightly warmer conditions than today (Matthews and others, 1986).

During the M'Clure stade, generally northwest-flowing ice probably covered most of Victoria Island, except for the central portion of Prince Albert Peninsula and perhaps part of the Shaler Mountains (Vincent, 1984, 1989). This ice advance deposited an unnamed till sheet and associated sediments that clearly predate an ice advance (or readvance?), the limit of which can be tentatively traced on northwestern Victoria Island (Fig. 2).

The age of the two distinct advances can be debated. A critical section (Fig. 10) near Graveyard Bay north of Prince Albert Sound exposes plant-bearing terrestrial sediments that

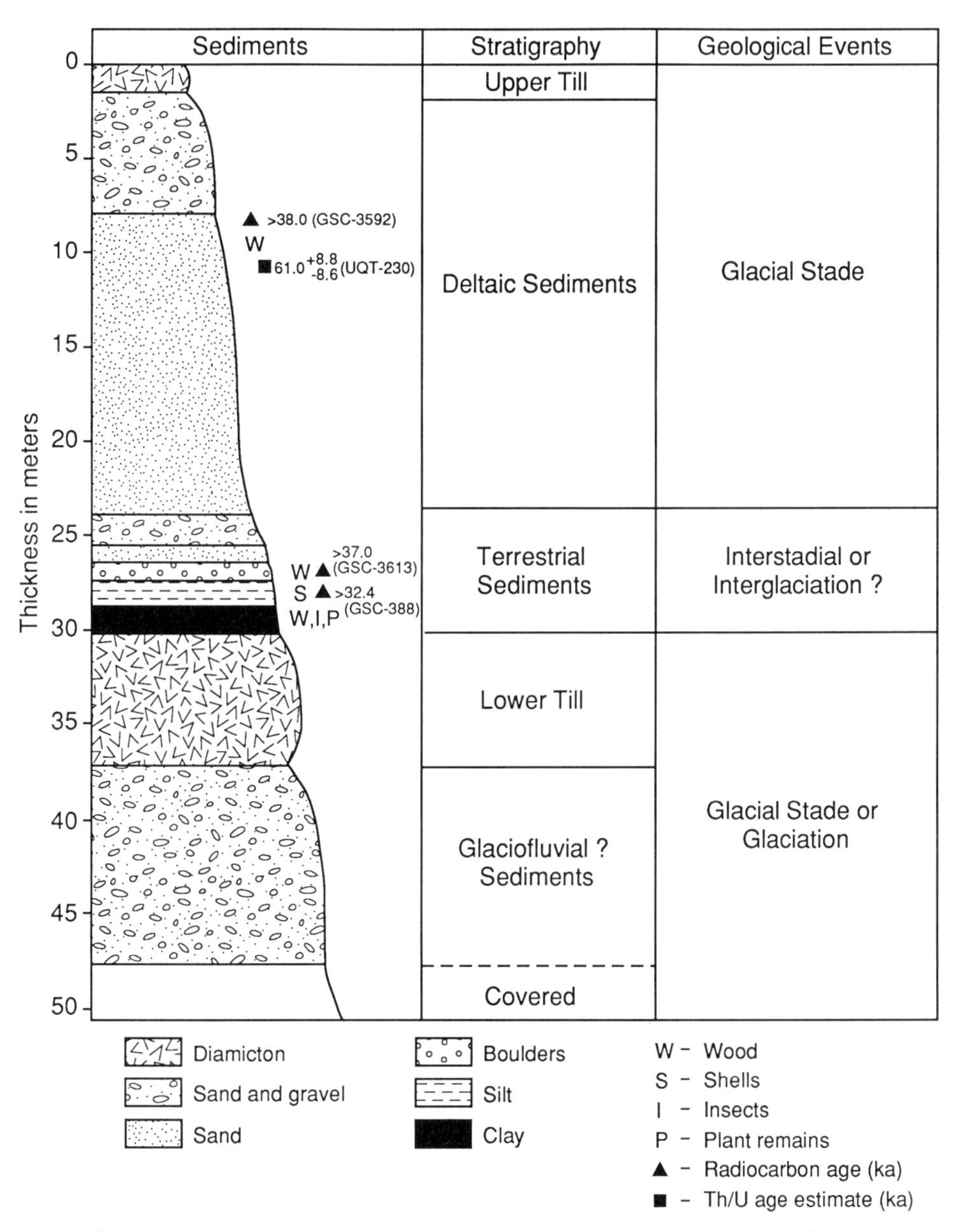

Figure 10. Stratigraphy of a river section near Graveyard Bay, on Diamond Jenness Peninsula, western Victoria Island (modified from Vincent, 1989).

separate the late Wisconsinan till from an older till. The plant remains yielded ^{14}C ages of >32,400 yr B.P. (GSC-388), >37,000 yr B.P. (GSC-3613), and >38,000 yr B.P. (GSC-3592), and a Th/U age estimate of 61,000 +8,800/–8,600 yr B.P. (UQT-230). The dated organic remains, including detrital wood and an homogeneous pocket of willow leaves, record environmental conditions similar to the present. The older till is thought to be the same till that elsewhere on northwestern Victoria Island lies on the surface beyond the late Wisconsinan glacial limit. If this correlation is correct, then the older till could well date from the Sangamonian (broad sense) or the early Wisconsinan, and a site where two late Pleistocene tills, separated by nonglacial fossil-bearing middle Wisconsinan deposits, would have been recognized for the first time in the southwestern Arctic archipelago. Alternatively, the organic-bearing beds could be older (Sangamonian?), and the site of little use in determining the chronology of late Pleistocene ice advances.

RECORD OF THE COASTAL PLAIN: NORTHWESTERN DISTRICT OF MACKENZIE AND YUKON TERRITORY

On the coastal plain of the mainland, from Bathurst Peninsula in the east to the Yukon-Alaska border in the west, organic-rich nonglacial deposits are widespread and clearly predate well-recognized late Pleistocene ice advances (Rampton, 1982, 1988; Vincent, 1989). In most instances it is still difficult to separate nonglacial Sangamonian deposits from middle Pleistocene and sometimes even middle Wisconsinan and Holocene ones.

The most likely Sangamonian deposits occur on low benches located on the west side of Bathurst Peninsula. The benches are cut into probable middle Pleistocene glacial deposits of the Mason River glaciation (Rampton, 1988). Rampton assigned sediments on the benches (Fig. 11) to the interglacial Ikpisugyuk Formation. They consist of sand and driftwood mats interpreted as being deposited in an intertidal beach. Nonfinite ^{14}C and Th/U age estimates were obtained on two *Picea* logs and one *Salix* log from different localities (>36,000 yr B.P.—GSC-4075 and 23,700 ± 1000 yr B.P.—UQT-233; >38,000 yr B.P.—GSC-3759 and 65,600 ± 3,600 yr B.P.—UQT-235; >39,000 yr B.P.—GSC-3722 and 96,700 +10,000/−11,000 yr B.P.—UQT-207; Table 1). The benches and the sediments which rest on them record a period when relative sea level was 5 to 10 m higher than present and when the Mackenzie River or Horton River basins provided an ice-free source of driftwood. The chronological and paleoecological data, together with the lack of evidence for emergence in the late Wisconsinan–Holocene (i.e., the sea has been progressively transgressing over the area since the late middle Wisconsinan—Hill and others, 1985), indicate that the deposits described above likely date from the Sangamonian (strict sense) a period during which relative sea level is known to have been higher than present. Numerous other deposits underlying till could also be Sangamonian. On the Yukon Coastal Plain, for example, fluvial sand and gravel with ice-wedge casts and paleosols and other sequences of nonglacial deposits are widespread (Rampton, 1982; Vincent, 1989). Like the Cape Collinson Formation on Banks Island, they represent the youngest interval distinctly warmer than the present interglaciation.

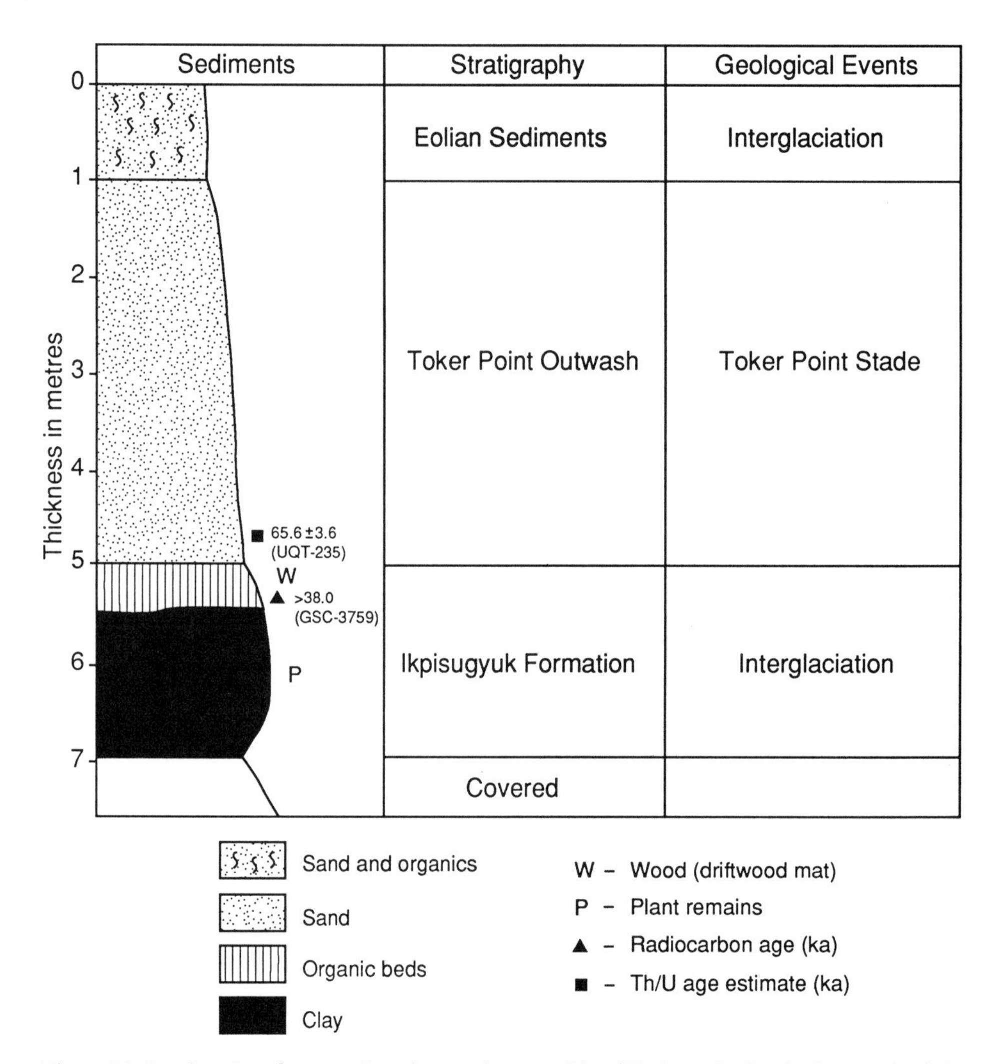

Figure 11. Stratigraphy of a coastal section on the west side of Bathurst Peninsula, just south of the Mason River mouth, District of Mackenzie.

Work by Fyles and others (1972), Hughes (1972, 1987), Klassen (1971), Rampton (1982, 1988), and Vincent (1989) on the coastal plain and adjacent areas has demonstrated that continental ice advanced at least twice in the area during the late Pleistocene following deposition of nonglacial deposits such as those described above. The older advance was correlated by Vincent (1989) to the Hungry Creek glaciation (Hughes and others, 1981), Buckland glaciation (Rampton, 1982), Toker Point stade, and Franklin Bay stade (Rampton, 1988). The younger advance is correlated to the Tutsieta Lake phase (Hughes, 1987), and Sitidgi stade (Rampton, 1988). During the older advance, Vincent (1989) proposed that the ice covered most of the coastal plain, except for higher areas of the Brock upland and parts of Bathurst and Tuktoyaktuk peninsulas, and that it extended well onto the Cordilleran mountain front, the Beaufort Sea Shelf, and the Yukon Coastal Plain (Fig. 2). During the younger stade, the ice is believed to have been much less extensive and to have not reached the Beaufort Sea except at the eastern end of the Amundsen Gulf and at the mouth of Mackenzie River.

As in the southwestern Arctic archipelago, the ages of the two distinct ice advances are the subject of much debate (see discussions in Vincent, 1989). Data from the area bordering the Beaufort Sea and from the Cordilleran mountain front are seemingly difficult to reconcile.

Along the Richardson Mountains, Hughes and others (1981) and Catto (1986) have provided evidence that the older and most extensive of the two late Pleistocene ice advances reached its maximum extent about 25,000 years ago. In the upper Rat River valley, organic material in deposits of a lake dammed by ice of this advance has provided ^{14}C ages of 21,300 ± 270 yr B.P. (GSC-3371) and 21,200 ± 240 yr B.P. (GSC-3813) (Table 1), thus indicating that Laurentide ice may have reached its maximum extent in the very early late Wisconsinan. On Hungry Creek (Fig. 12), allochthonous wood underlying till of this advance gave nonfinite ^{14}C ages and a finite age of 36,900 ± 300 yr B.P. (GSC-2422), providing evidence, if the single finite age is accepted, that the advance that deposited the till is necessarily late middle Wisconsinan or late Wisconsinan (Hughes and others,

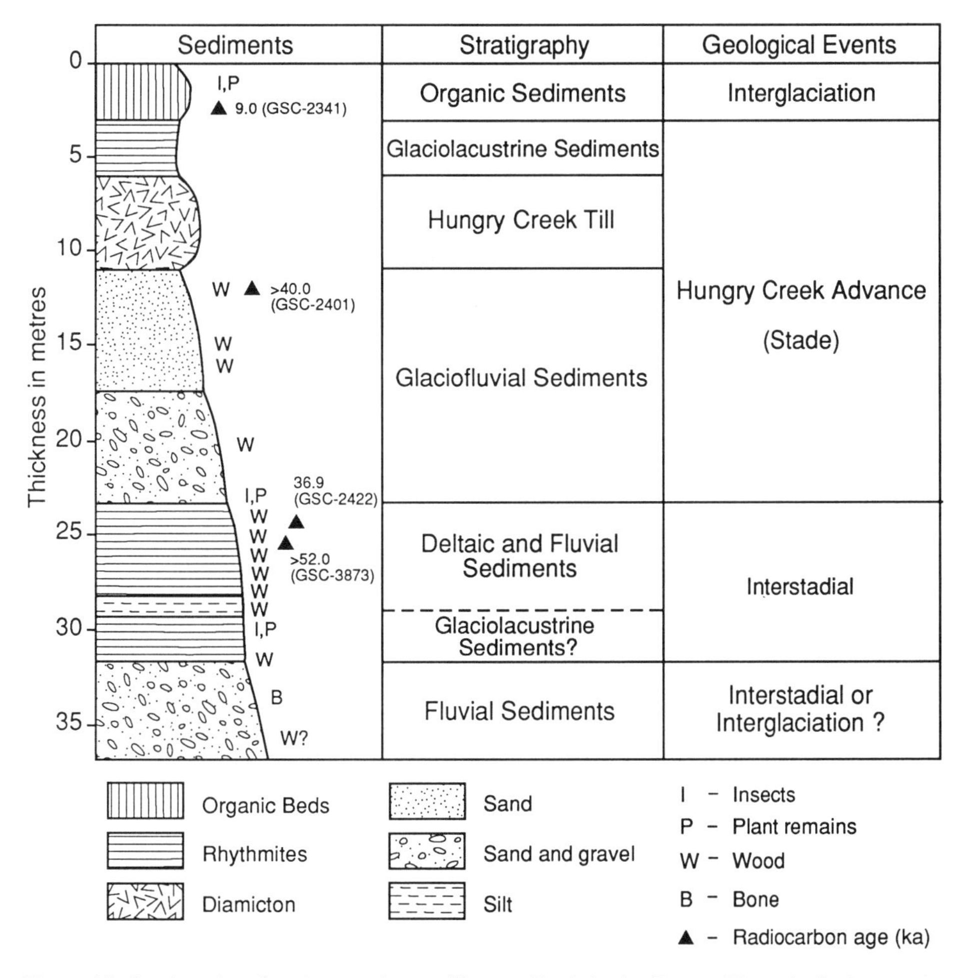

Figure 12. Stratigraphy of a river section on Hungry Creek in the Bonnet Plume basin (composite section drawn from information in Hughes and others, 1981).

1981). The 36.9 ka age is the only finite ^{14}C determination on material below till in the western Canadian Arctic. Everywhere else, ages from a similar stratigraphic situation have all been nonfinite.

However, other lines of evidence from the coastal plain indicate that the oldest of the two late Pleistocene ice advances predates the late Wisconsinan. Among these are ^{14}C age determinations on organic material that is assumed to postdate this glacial event. On Bathurst Peninsula, fragile but allochthonous willow twigs collected by Rampton (1988) in terrace sand postdating the Franklin Bay stade (= M'Clure/Toker Point stades) have been ^{14}C dated at 33,800 ± 880 yr B.P. (GSC-1974). On Tuktoyaktuk Peninsula, an autochthonous peat overlying till of the Toker Point stade that was tilted during growth of the Ibyuk pingo has provided a ^{14}C age of 17,860 ± 260 yr B.P. (GSC-481). On the Yukon Coastal Plain, allochthonous plant remains collected within pond deposits overlying till surfaces of the Buckland glaciation have provided ^{14}C ages as old as 22,400 ± 240 yr B.P. (GSC-1262) (Rampton, 1982).

Perhaps the best evidence for a pre–late Wisconsinan age assignment for the earlier ice advance on the coastal plain is provided by limiting ages on marine shells. On Garry Island, at the mouth of the Mackenzie River delta, two different collections of relatively abundant shells, gathered in nearshore marine deposits overlying till of the Toker Point stade (Figs. 13 and 14), were ^{14}C dated at >35,000 yr B.P. (GSC-562) and 43,550 ± 470 yr B.P. (TO-796), an age which should be considered nonfinite. Similarly, on Richards Island just south of Kendall Island, two different shell samples in similar deposits were ^{14}C dated at >37,000 yr B.P. (GSC-690) and 48,200 ± 1100 yr B.P.

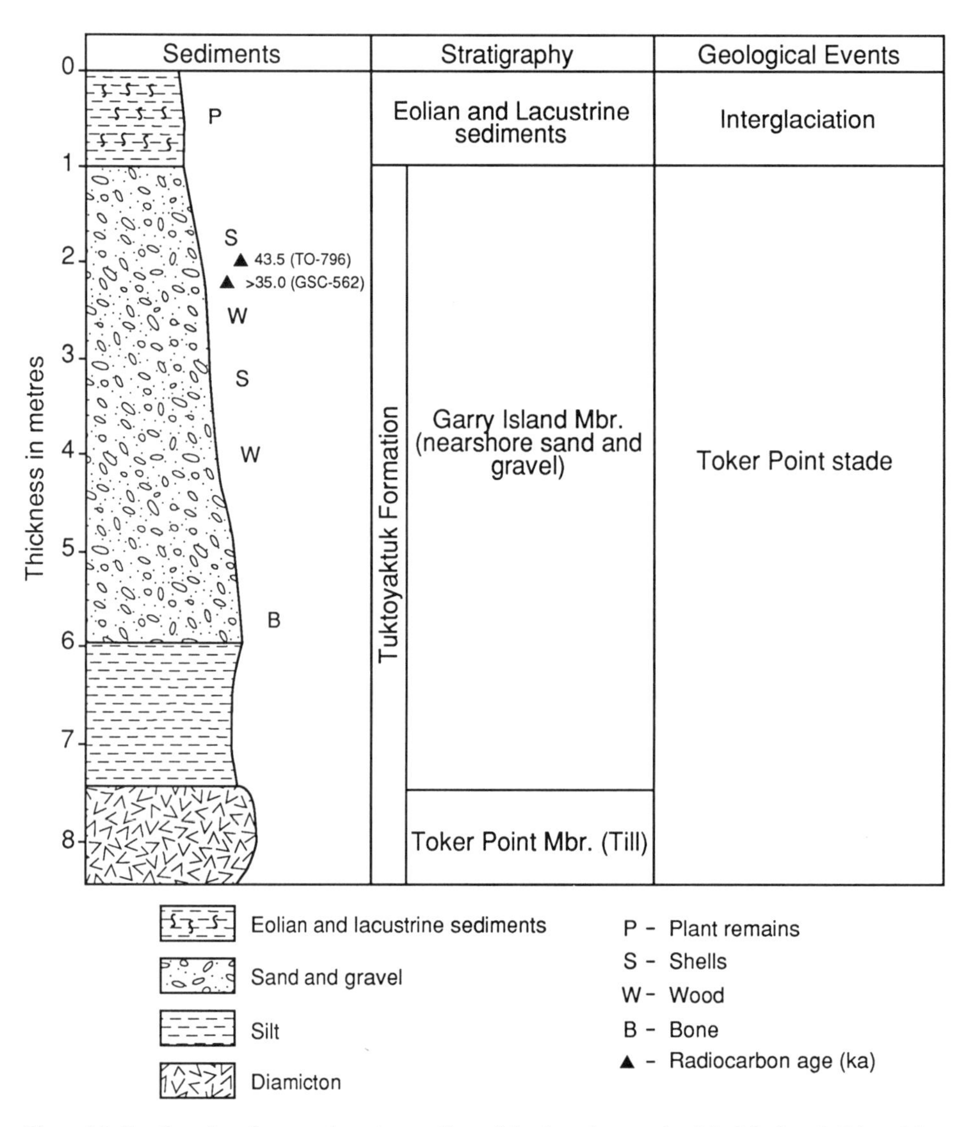

Figure 13. Stratigraphy of a coastal section on Garry Island, at the mouth of the Mackenzie River delta, District of Mackenzie.

Figure 14. Section on the north shore of Garry Island, at the mouth of the Mackenzie River delta, showing fossiliferous nearshore marine sand and gravel. The Toker Point till underlies the marine sediments in a section immediately to the northwest (Geological Survey of Canada photo 204939-A).

(RIDDL-801). The shells from these four collections were not observed in growth position in the nearshore sediments. Their abundance and state of preservation, compared to those found in older marine deposits on the coastal plain, leave little doubt that they are of the same age as the enclosing sediment. The dated marine deposits are therefore thought to be clearly associated with the presence of a sea that inundated the coastal areas during retreat of Toker Point stade ice. They provide strong evidence of an interval of local retreat that predates the late Wisconsinan.

The history of sea-level fluctuations on the coastal plain of the mainland must also be taken into account when assigning ages to the two glacial advances. During the older Toker Point stade/ Buckland glaciation advance, the ice extended well onto the Beaufort Sea Shelf (Fig. 2), and perhaps was sufficiently thick to depress part of the coastal plain below sea level of the time (Vincent, 1989). Following retreat of the ice, a shallow glacio-isostatic sea covered newly deglaciated areas, leaving behind marine deposits, such as those found on Garry Island and adjacent regions. The sea-level record is thus one of emergence. During the Sitidgi stade, the ice was much less extensive, forming one relatively thin lobe (Rampton, 1988) centered in the low-lying Mackenzie River Valley (Fig. 2). No record of emergence exists west of Darnley Bay in the southwestern Amundsen Gulf. Rather, the coastal plain has been submerging (Hill and others, 1985). On the basis of the sea-level record, one can therefore argue that the two late Pleistocene glacial advances are distinct stadial events well separated in time. During the Toker Point stade/Buckland glaciation, ice was very extensive, the coastal plain was isostatically depressed, and emergence occurred following deglaciation. During the younger Sitidgi stade, ice was much less extensive and the coastal plain was not depressed, but rather has been progressively submerging during the late Wisconsinan and Holocene.

It is tempting to associate the Toker Point stade/Buckland glaciation on the coastal plain and Beaufort Sea Shelf with the deposition of the Flaxman Member of the Gubik Formation on the coastal plain of northern Alaska. The glacial-marine deposits of the Flaxman Member (Fig. 15) can be traced from Cape Barrow at the western end of the Alaskan Coastal Plain to 20 km west of the Buckland glacial limit in the Yukon. They are thought to be 70–80 ka, on the basis of extensive thermoluminescence dating in Alaska (Carter and others, 1986). Clasts found in the glacial-marine clay are of dominantly Canadian origin and of similar provenance to clasts found in tills on the southwestern Arctic archipelago and the coastal plain of the mainland. The Flaxman Member logically represents a suite of distal glacial-marine sediments derived from ice located immediately to the east. On the basis of the age assignment in Alaska (Carter and others, 1986), a late Sangamonian (broad sense)–early Wisconsinan age for the ice advance would be plausible. Although the age assigned to the Flaxman Member supports an early Wisconsinan age for ice at the Buckland limit, the regionally extensive high sea level necessary to account for the Flaxman deposits makes it unnecessary to invoke a glacial-isostatic mechanism to

Figure 15. Coastal section on the Yukon Coastal Plain, just east of the Yukon-Alaska border, showing deposits of the Flaxman Member of the Gubik Formation. Note the clasts in the glacial-marine sediments (Geological Survey of Canada photo 191126).

explain the Garry Island and Richards Island raised marine deposits. The Flaxman deposits in Alaska extend well beyond the limit of any area that could have reasonably been subjected to crustal depression. The only obvious mechanism to explain the record would be a broad and more or less uniform regional uplift of the Canadian and Alaskan Beaufort Sea Coastal Plain.

CONCLUSION

In this chapter, chronological and other data from the coastal plain of the Canadian mainland and the southwestern Arctic archipelago were presented to support the postulate that ice originating west of Hudson Bay advanced toward northwestern Canada during the earlier part of the late Pleistocene. The assumed overall extent of this ice (Fig. 16) is substantially larger than the extent of the younger proposed late Wisconsinan ice (Fig. 2). The records on Banks, Melville, and Victoria islands, and on Bathurst Peninsula of the mainland, also indicate that early and middle Pleistocene glaciers were more extensive than late Pleistocene advances (Vincent, 1989).

Assigning a definite age to these pre–late Wisconsinan ice advances is difficult because of the lack of unquestionable geochronological data for deposits older than the limit of the ^{14}C method. Evidence was presented that seems to indicate that late Pleistocene Laurentide ice first advanced and covered extensive areas of the southwestern Arctic archipelago and the coastal plain of the mainland certainly before the late Wisconsinan, and possibly as early as the Sangamonian (broad sense). On southwestern Banks Island, a Th/U age estimate on shells indicates that M'Clure stade ice reached its fullest extent at a minimum of 86 ka. On the coastal plain of the mainland, Toker Point stade ice had already retreated from northern Richards Island >35 ka.

Whatever the precise age of the initial late Pleistocene ice advance in the area, authors, who have generally argued for a late Wisconsinan age for the ice advances discussed here, agree that ice was already present at its stadial limit in the very early late Wisconsinan, if not in the late middle Wisconsinan. For example, Dyke and Prest (1987b) showed ice of the M'Clure, Franklin Bay, and Toker Point stades, and Buckland glaciation, at their maximum position by at least 18 ka. The reconstruction of Hughes and others (1981) and of Catto (1986) imply that ice had reached its maximum position on the Cordilleran Mountain front

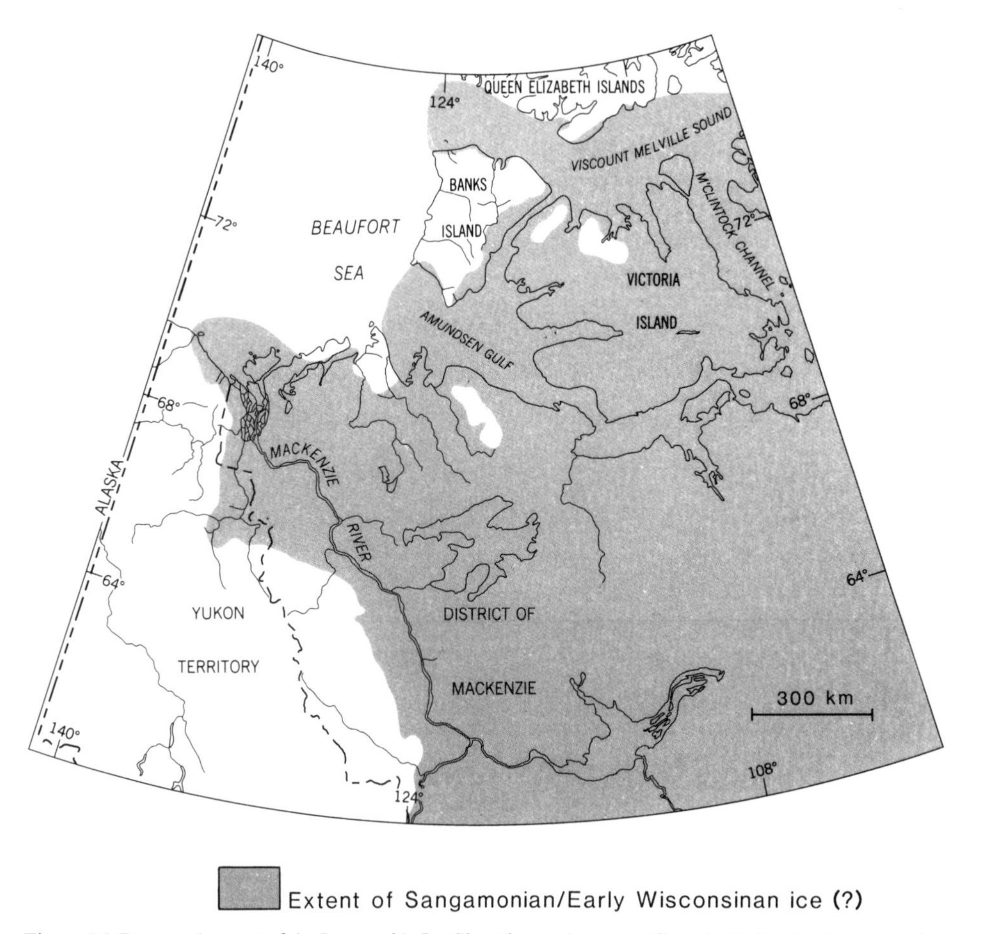

Figure 16. Proposed extent of the Laurentide Ice Sheet in northwestern Canada during the Sangamonian (broad sense)/early Wisconsinan (modified from Vincent, 1989).

by about 25 ka. It is important to realize that for ice to be present on Banks Island, at the edge of the Cordillera, and on the Yukon Coastal Plain at that time, it must have also covered much of northwestern Canada during the latter half of the middle Wisconsinan. Vincent and Prest (1987, Fig. 4) have estimated that >20 ka would be needed to generate enough Keewatin ice to extend to these distant margins of the Laurentide Ice Sheet.

For promoting discussion it can be argued that if ice covered large areas of northwestern Canada, ca. 40 to 45 ka, it could have done so much earlier in the late Pleistocene. Until now, as in many other areas at the periphery of the Laurentide Ice Sheet (Vincent and Prest, 1987), researchers have often considered that continental ice underwent two major Wisconsinan advances separated by a long nonglacial middle Wisconsinan interval. However, it is possible that in northwestern Canada, Keewatin sector ice of the Laurentide Ice Sheet persisted over much of the area, from perhaps as early as the Sangamonian (broad sense) through to the late Wisconsinan. Having stable or quasi-stable ice present in northwestern Canada during most of the late Pleistocene would perhpas help reconcile the diverging views on the age of the ice advances and the incongruities in the interpretation of the age of events; e.g., the overlap of eastern Banks Island by the East Coast Sea. In the southwestern Arctic archipelago, M'Clure stade ice may have advanced to its maximum extent during the Sangamonian (broad sense) or early Wisconsinan, as indicated by the dated sediments of the pre-Amundsen Sea/Meek Point Sea. This ice could have then substantially retreated in contact with the East Coast Sea before it readvanced during the latest late Wisconsinan (Russell stade). In other areas of northwestern Canada, such as near the Richardson Mountains, the ice of the Keewatin sector could have advanced to its farthest position during the late middle Wisconsinan, if Hughes and others (1981) interpretation of the Hungry Creek data is accepted, or much earlier if not.

Although many uncertainties remain, the record in northwestern Canada provides much information on the extent, dynamics, and timing of continental ("Laurentide") ice sheets. Information can also be acquired on ice-volume and sea-level relations and the response of the Earth's crust to loading. Interpretation of the paleoenvironments of Arctic areas also closely depend on our understanding of the timing of ice advances and the length of time ice sheets were present in the area.

ACKNOWLEDGMENTS

This is Contribution 25789 of the Geological Survey of Canada (GSC). Thanks are extended to Julie Brigham-Grette of the University of Massachusetts, Arthur S. Dyke, John G. Fyles, and Douglas A. Hodgson of the GSC, Darrell Kaufman of the University of Colorado, and Peter D. Lea, the coeditor of this volume, who offered valuable comments and suggestions that have improved this chapter. I am also indebted to Lucie Maurice and Tracy Barry, who prepared illustrations. Much of the field work in northwestern Canada was made possible thanks to the support, since 1974, of the Polar Continental Shelf Project.

REFERENCES CITED

Barendregt, R. W., and Vincent, J.-S., 1990, Late Cenozoic paleomagnetic record of Duck Hawk Bluffs, Banks Island, Canadian Arctic Archipelago: Canadian Journal of Earth Sciences, v. 27, p. 124–130.

Blake, W., Jr., 1974, Studies of glacial history in Arctic Canada. II. Interglacial peat deposits on Bathurst Island: Canadian Journal of Earth Sciences, v. 11, p. 1025–1042.

—— , 1987, Geological Survey of Canada radiocarbon dates XXVI: Geological Survey of Canada, Paper 86-7, 60 p.

Carter, L. D., Brigham-Grette, J., and Hopkins, D. M., 1986, Late Cenozoic marine transgressions of the Alaskan Arctic Coastal Plain, *in* Heginbottom, J. A., and Vincent, J.-S., eds., Correlations of Quaternary deposits and events around the margin of the Beaufort Sea: Contributions from a joint Canadian-American workshop, April 1984: Geological Survey of Canada Open-File 1237, p. 21–26.

Catto, N. R., 1986, Quaternary sedimentology and stratigraphy, Peel Plateau and Richardson Mountains, Yukon and N.W.T. [Ph.D. thesis]: Edmonton, University of Alberta, 728 p.

Causse, C., and Vincent, J.-S., 1989, Th/U disequilibrium dating of middle and late Pleistocene wood and shells from Banks and Victoria islands, Arctic Canada: Canadian Journal of Earth Sciences, v. 26, p. 2718–2723.

Denton, G. H., and Hughes, T. J., 1983, Milankovitch theory of ice ages: Hypothesis of ice-sheet linkage between regional insolation and global climate: Quaternary Research, v. 20, p. 125–144.

Dyke, A. S., 1983, Quaternary geology of Somerset Island: Geological Survey of Canada Memoir 404, 32 p.

—— , 1987, A reinterpretation of glacial and marine limits around the northwestern Laurentide Ice Sheet: Canadian Journal of Earth Sciences, v. 24, p. 591–601.

Dyke, A. S., and Prest, V. K., 1987a, The late Wisconsinan and Holocene history of the Laurentide Ice Sheet: Géographie physique et Quaternaire, v. 41, p. 237–263.

—— , 1987b, Paleogeography of northern North America, 18 000-5000 years ago: Geological Survey of Canada Map 1703A, scale 12,500,000.

Dyke, A. S., Morris, T. F., Green, D.E.C., and England, J., 1992, Quaternary geology of Prince of Wales Island: Geological Survey of Canada Memoir 433 (in press).

Fyles, J. G., 1969, Northwestern Banks Island, District of Franklin, *in* Report of activities, Part A: Geological Survey of Canada Paper 69-1A, p. 194–195.

Fyles, J. G., Heginbottom, J. A., and Rampton, V. N., 1972, Quaternary geology and geomorphology, Mackenzie Delta to Hudson Bay: International Geological Congress, 24th, Montréal, Canada, Guidebook A-30, 23 p.

Hill, P. R., Mudie, P. J., Moran, K., and Blasco, S. M., 1985, A sea-level curve for the Canadian Beaufort Shelf: Canadian Journal of Earth Sciences, v. 22, p. 1383–1393.

Hodgson, D. A., 1991, Quaternary geology of western Melville Island, Northwest Territories: Geological Survey of Canada Paper 89-21, 35 p.

Hodgson, D. A., and Vincent, J.-S., 1984, A 10 000 yr. B.P. extensive ice shelf over Viscount Melville Sound, Arctic Canada: Quaternary Research, v. 22, p. 18–30.

Hodgson, D. A., Vincent, J.-S., and Fyles, J. G., 1984, Quaternary geology of central Melville Island, Northwest Territories: Geological Survey of Canada Paper 83-16, 25 p.

Hughes, O. L., 1972, Surficial geology of northern Yukon Territory and northwestern District of Mackenzie, Northwest Territories: Geological Survey of Canada Paper 69-36, 11 p.

—— , 1987, The late Wisconsinan Laurentide glacial limits of northwestern Canada: The Tutsieta Lake and Kelly Lake phases: Geological Survey of Canada Paper 85-25, 19 p.

Hughes, O. L., and 6 others, 1981, Upper Pleistocene stratigraphy, paleoecology, and archaeology of the northern Yukon interior, Eastern Beringia 1. Bonnet Plume Basin: Arctic, v. 34, p. 329–365.

Klassen, R. W., 1971, Surficial geology, Franklin Bay (97C) and Brock River

(97D): Geological Survey of Canada Open-File 48, scale 1:250,000.
Lowdon, J. A., and Blake, W., Jr., 1968, Geological Survey of Canada radiocarbon dates VII: Radiocarbon, v. 10, p. 207–245.
——, 1973, Geological Survey of Canada radiocarbon dates XIII: Geological Survey of Canada Paper 73-7, 61 p.
——, 1976, Geological Survey of Canada radiocarbon dates XVI: Geological Survey of Canada Paper 76-7, 21 p.
——, 1978, Geological Survey of Canada radiocarbon dates XVIII: Geological Survey of Canada Paper 78-7, 20 p.
Lowdon, J. A., Robertson, I. M., and Blake, W., Jr., 1971, Geological Survey of Canada radiocarbon dates XI: Radiocarbon, v. 13, p. 255–324.
Matthews, J. V., Jr., Mott, R. J., and Vincent, J.-S., 1986, Preglacial and interglacial environments of Banks Island: Pollen and macrofossils from Duck Hawk Bluffs and related sites: Géographie physique et Quaternaire, v. 40, p. 279–298.
Mayewski, P. A., Denton, G. H., and Hughes, T. J., 1981, Late Wisconsin ice sheets of North America, *in* Denton, G. H., and Hughes, T. J., eds., The Last great ice sheets: New York, John Wiley and Sons, p. 67–178.
McNeely, R., 1989, Geological Survey of Canada radiocarbon dates XXVIII: Geological Survey of Canada Paper 88-7, 93 p.
Rampton, V. N., 1980, Surficial geology, Aklavik, District of Mackenzie: Geological Survey of Canada Map 31-1979, scale 1:250,000.
——, 1981a, Surficial geology, Malloch Hill, District of Mackenzie: Geological Survey of Canada Map 30-1979, scale 1:250,000.
——, 1981b, Surficial geology, Mackenzie Delta, District of Mackenzie: Geological Survey of Canada Map 32-1979, scale 1:250,000.
——, 1981c, Surficial geology, Stanton, District of Mackenzie: Geological Survey of Canada Map 33-1979, scale 1:250,000.
——, 1982, Quaternary geology of the Yukon Coastal Plain: Geological Survey of Canada Bulletin 317, 49 p.
——, 1988, Quaternary geology of the Tuktoyaktuk Coastlands, Northwest Territories: Geological Survey of Canada Memoir 423, 98 p.
Vincent, J.-S., 1980a, Surficial geology—Banks Island (north half), Northwest Territories: Geological Survey of Canada Map 16-1979, scale 1:250,000.
——, 1980b, Surficial geology—Banks Island (south half), Northwest Territories: Geological Survey of Canada Map 17-1979, scale 1:250,000.
——, 1982, The Quaternary history of Banks Island, N.W.T., Canada: Géographie physique et Quaternaire, v. 36, p. 209–232.
——, 1983, La géologie du Quaternaire et la géomorphologie de l'île Banks, Arctique canadien: Commission géologique du Canada, Mémoire 405, 118 p.
——, 1984, Quaternary stratigraphy of the western Canadian Arctic Archipelago, *in* Fulton, R. J., ed., Quaternary stratigraphy of Canada—A Canadian contribution to IGCP Project 24: Geological Survey of Canada Paper 84-10, p. 87–100.
——, 1989, Quaternary geology of the northern Canadian Interior Plains, *in* Fulton, R. J., ed., Quaternary geology of Canada and Greenland: Geological Survey of Canada, Geology of Canada, no. 1, p. 100–137.
Vincent, J.-S., and Prest, V. K., 1987, The early Wisconsinan history of the Laurentide Ice Sheet: Géographie physique et Quaternaire, v. 41, p. 199–213.
Vincent, J.-S., Occhietti, S., Rutter, N. W., Lortie, G., Guilbault, J. P., and Boutray, B.de, 1983, The late Tertiary–Quaternary stratigraphic record of the Duck Hawk Bluffs, Banks Island, Canadian Arctic Archipelago: Canadian Journal of Earth Sciences, v. 20, p. 1694–1712.
Vincent, J.-S., Morris, W. A., and Occhietti, S., 1984, Glacial and nonglacial sediments of Matuyama paleomagnetic age on Banks Island, Canadian Arctic Archipelago: Geology, v. 12, p. 139–142.

MANUSCRIPT ACCEPTED BY THE SOCIETY SEPTEMBER 6, 1991

Printed in U.S.A.

Geological Society of America
Special Paper 270
1992

The Sangamonian and early Wisconsinan stages in western Canada and northwestern United States

John J. Clague
Geological Survey of Canada, 100 West Pender Street, Vancouver, British Columbia, V6B 1R8, Canada
D. J. Easterbrook
Department of Geology, Western Washington University, Bellingham, Washington 98225
O. L. Hughes
Geological Survey of Canada, 3303-33rd Street N.W., Calgary, Alberta T2L 2A7, Canada
J. V. Matthews, Jr.
Geological Survey of Canada, 601 Booth Street, Ottawa, Ontario K1A 0E8, Canada

ABSTRACT

Lithostratigraphic and geochronologic data from Yukon Territory indicate relatively limited glaciation in the northern Canadian Cordillera during the early Wisconsinan. If the Cordilleran Ice Sheet existed in south and central Yukon during the early Wisconsinan, it was less extensive than during either the Illinoian or late Wisconsinan. In contrast, ice cover during the early Wisconsinan in British Columbia and northern Washington may have been comparable to that of the late Wisconsinan, as suggested by the widespread occurrence of glacial deposits between middle Wisconsinan and presumed Sangamonian nonglacial strata.

Sediments of probable Sangamonian age have been studied for pollen and plant and animal macrofossils. Climate during deposition of these sediments was warmer and drier than today. Plant communities probably had different distributions than at present, and permafrost may have been absent or more restricted over some areas in which it currently occurs. Little is known about the transition from the Sangamonian to the early Wisconsinan in western Canada and Washington, although limited data suggest that during the early and middle Wisconsinan there were perturbations in climate ranging from full glacial to temperate.

INTRODUCTION

The Quaternary Period is noteworthy for major climatic perturbations that caused continental ice sheets in middle latitudes of the Northern Hemisphere to wax and wane. The Cordilleran Ice Sheet and smaller independent satellite glaciers repeatedly enveloped much of the Canadian Cordillera and adjacent areas (Fig. 1; Flint, 1971; Hamilton and Thorson, 1983; Waitt and Thorson, 1983; Clague, 1989b). In this region, episodes of ice-sheet glaciation alternated with intervals during which glaciers were restricted to mountain systems, and lowlands and plateaus were ice free.

Much is known about the last glacial maximum (late Wisconsinan = marine oxygen-isotope stage 2) in northwestern North America (Fulton, 1971, 1984; Clague, 1981, 1989a; Hamilton and Thorson, 1983; Waitt and Thorson, 1983; Easterbrook, 1986; Blunt and others, 1987). Sediments and landforms record in detail late Wisconsinan glacier growth and decay, and abundant organic material allows precise dating and paleoecologic reconstruction. Similarly, our knowledge of Holocene and middle Wisconsinan environments is considerable, although most middle Wisconsinan deposits are covered by younger sediments or have been removed by late Wisconsinan ice.

In contrast, considerable uncertainty and confusion exist

Clague, J. J., Easterbrook, D. J., Hughes, O. L., and Matthews, J. V., Jr., 1992, The Sangamonian and early Wisconsinan stages in western Canada and northwestern United States, *in* Clark, P. U., and Lea, P. D., eds., The Last Interglacial-Glacial Transition in North America: Boulder, Colorado, Geological Society of America Special Paper 270.

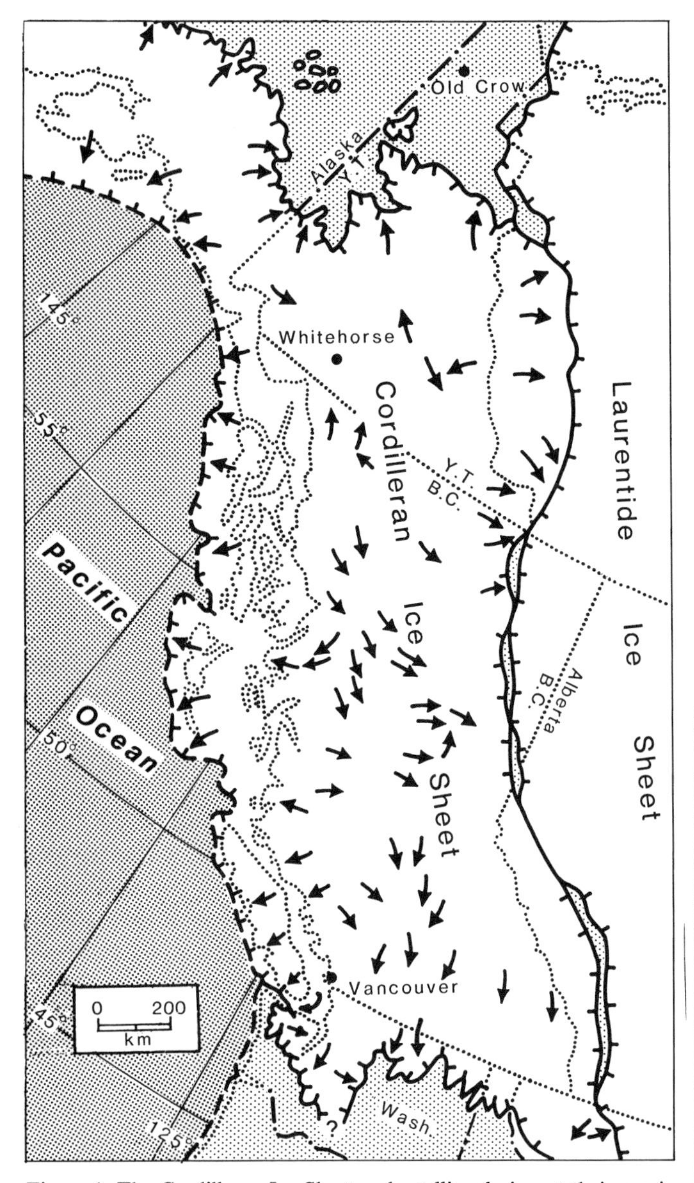

Figure 1. The Cordilleran Ice Sheet and satellite glaciers at their maximum extent (adapted from Flint, 1971, Fig. 18-1). Arrows indicate generalized directions of glacier flow. Nonglaciated coast is drawn at –100 m isobath.

about the character, chronology, and correlation of pre–middle Wisconsinan glacial and nonglacial events in northwestern North America, primarily because (1) sediments that are older than the middle Wisconsinan are sparse and poorly exposed in most areas, and (2) these sediments are beyond the range of radiocarbon dating, making reliable estimates of their age difficult to obtain. However, recent advances in amino-acid, fission-track, and thermoluminescence (TL) dating techniques and in the use of tephras as critical stratigraphic marker beds in some areas have yielded new insights on early Wisconsinan and older events.

This chapter summarizes geologic and paleoecologic information bearing on climatic conditions and ice cover in western Canada and northwestern Washington during the early Wisconsinan and Sangamonian (here correlated with marine oxygen-isotope stages 4 and 5, respectively). Pertinent stratigraphic, geomorphic, chronologic, and plant and animal fossil evidence are presented for Yukon Territory, British Columbia, and Washington (Fig. 2). No attempt is made to summarize similar information for the eastern sector of the Canadian Cordillera (western Alberta and western District of Mackenzie), because Pleistocene events in this region have not been dated and because correlations are tenuous at best.

A critical issue, which at present is not fully resolved, is the age of the penultimate phase of ice-sheet glaciation in western Canada. Past studies have shown that there was no ice sheet in the Canadian Cordillera during the middle Wisconsinan (Fulton, 1971, 1984; Clague, 1981). Drift that directly underlies middle Wisconsinan nonglacial sediments in British Columbia and drift located just beyond the late Wisconsinan glacial limit in Yukon Territory traditionally have been assigned to the early Wisconsinan (e.g., Fulton, 1971, 1984; Rampton, 1971; Clague, 1981;

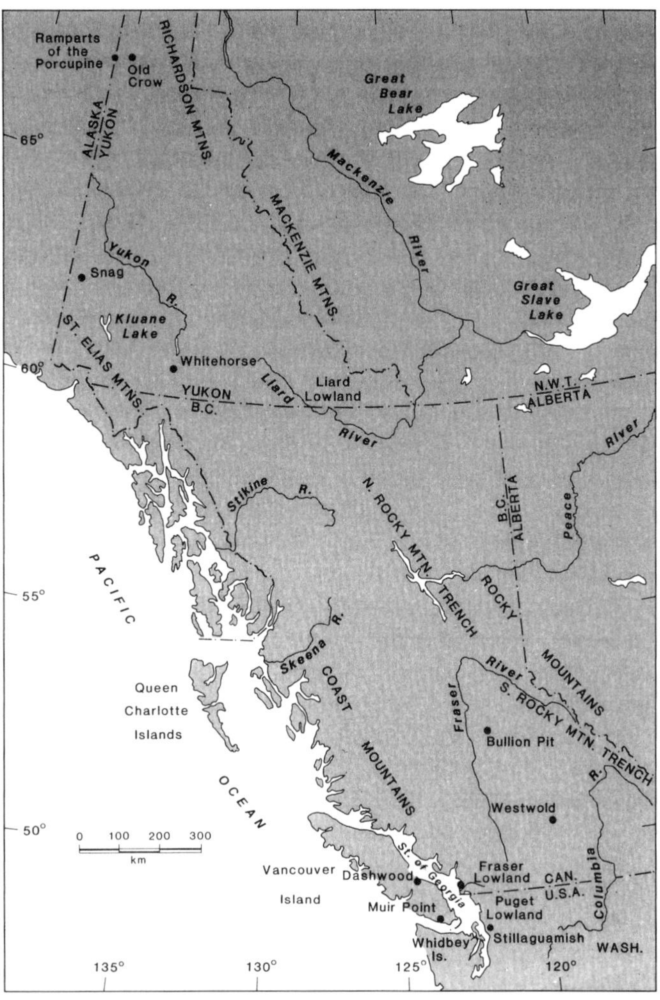

Figure 2. Locality index map.

Armstrong, 1981; Hicock and Armstrong, 1983). Chronologic control on these deposits is so poor, however, that a critical appraisal of their ages seems warranted. Furthermore, original age assignments were strongly influenced by a mid-continent chronostratigraphic framework that itself is being reappraised and extensively revised (Curry, 1989; Curry and Follmer, this volume).

YUKON TERRITORY

Stratigraphy

Deposits assigned to the Sangamonian and early Wisconsinan are known from glaciated and unglaciated areas of Yukon Territory. The best-documented examples are summarized in this section.

Sangamonian–early Wisconsinan sediments beyond glacial limits. Thick late Cenozoic sediments underlie the lowlands of Old Crow, Bluefish, and Bell basins far north and west of Pleistocene ice margins, near the present limit of boreal forest. These sediments were deposited mainly in fluvial, deltaic, and lacustrine environments. During the late Wisconsinan, the former eastward drainage in this region was blocked at the eastern front of the Richardson Mountains by the Laurentide Ice Sheet, and Old Crow, Bluefish, and Bell basins were inundated by a glacial lake that found an outlet to the west via the Ramparts of the Porcupine at the Alaska-Yukon border (Hughes, 1972; Thorson and Dixon, 1983). Subsequent downcutting of the Ramparts led to the incision of the late Cenozoic fill in these basins and the establishment of the present westerly drainage.

Organic-rich silt and sand of Sangamonian and Wisconsinan age are well exposed in steep bluffs along Porcupine, Bluefish, and Old Crow rivers. These sediments have been extensively studied, in part because bones thought to have been modified by humans were found on floodplains below some exposures (Morlan, 1980, 1986; Jopling and others, 1981).

A section at Ch'ijee's (Twelvemile) Bluff in Bluefish Basin (Fig. 3), 10 km southwest of the village of Old Crow, illustrates the complex history of sedimentation and erosion in this area during the Quaternary and provides the stratigraphic context for paleoecologic results presented later in this chapter (Hughes, 1972; Matthews and others, 1990b). Sediments low in the bluff (units 1 and 2) comprise oxidized and iron-cemented fluvial and deltaic silt, sand, and minor gravel of early Pleistocene or late Tertiary age. These are overlain by bedded, silty clay and clayey silt (unit 3), which probably accumulated in a large lake in Bluefish and Old Crow basins during early Pleistocene time (Hughes, 1972; Pearce and others, 1982). A major disconformity spanning most of the Pleistocene separates units 3 and 4. Unit 4, consisting of organic-rich fluvial and lacustrine(?) silt, sand, and gravel (Fig. 4), is of particular interest here because it probably encompasses the Sangamonian and much of the Wisconsinan. It can be divided into three subunits. Subunit 4a directly overlies unit 3 and consists of sand and gravel with lenses of organic detritus. It grades upward into interbedded fine sand, silt, and peat (subunit 4b). These sediments contain both a laterally continuous horizon of ice-wedge pseudomorphs and Old Crow tephra, which is widely distributed in northern Yukon Territory and adjacent Alaska (Westgate and others, 1985). The ice-wedge pseudomorphs in subunit 4b apparently developed after the tephra was deposited. Subunit 4c sharply and disconformably overlies subunit 4b and, like it, consists mainly of silt and sand with abundant woody and peaty zones. A conspicuous horizon of truncated ice-wedge pseudomorphs occurs within subunit 4c, above a zone of tightly folded cryoturbation structures. Unit 4 is sharply overlain by clay (unit 5) deposited in a glacial lake that covered this area during the late Wisconsinan. This clay, in turn, is overlain by peat and organic-rich fluvial silt of latest Pleistocene and Holocene age.

Sediments similar to those at Ch'ijee's Bluff are exposed along Old Crow River in Old Crow Basin (Fig. 5; Morlan, 1980, 1986; Jopling and others, 1981; Matthews and others, 1987). In situ and reworked lacustrine clay, correlative with unit 3 at Ch'ijee's Bluff, is overlain by thick fluvial silt, sand, and minor gravel containing large channel cut-and-fill structures, ice-wedge pseudomorphs, cryoturbation structures, paleosols, peat layers,

Figure 3. Ch'ijee's (Twelvemile) Bluff, northern Yukon Territory. Vertical bar indicates approximate location of section 3 in Figure 4. Bluff is ~115 m high. OC = Old Crow.

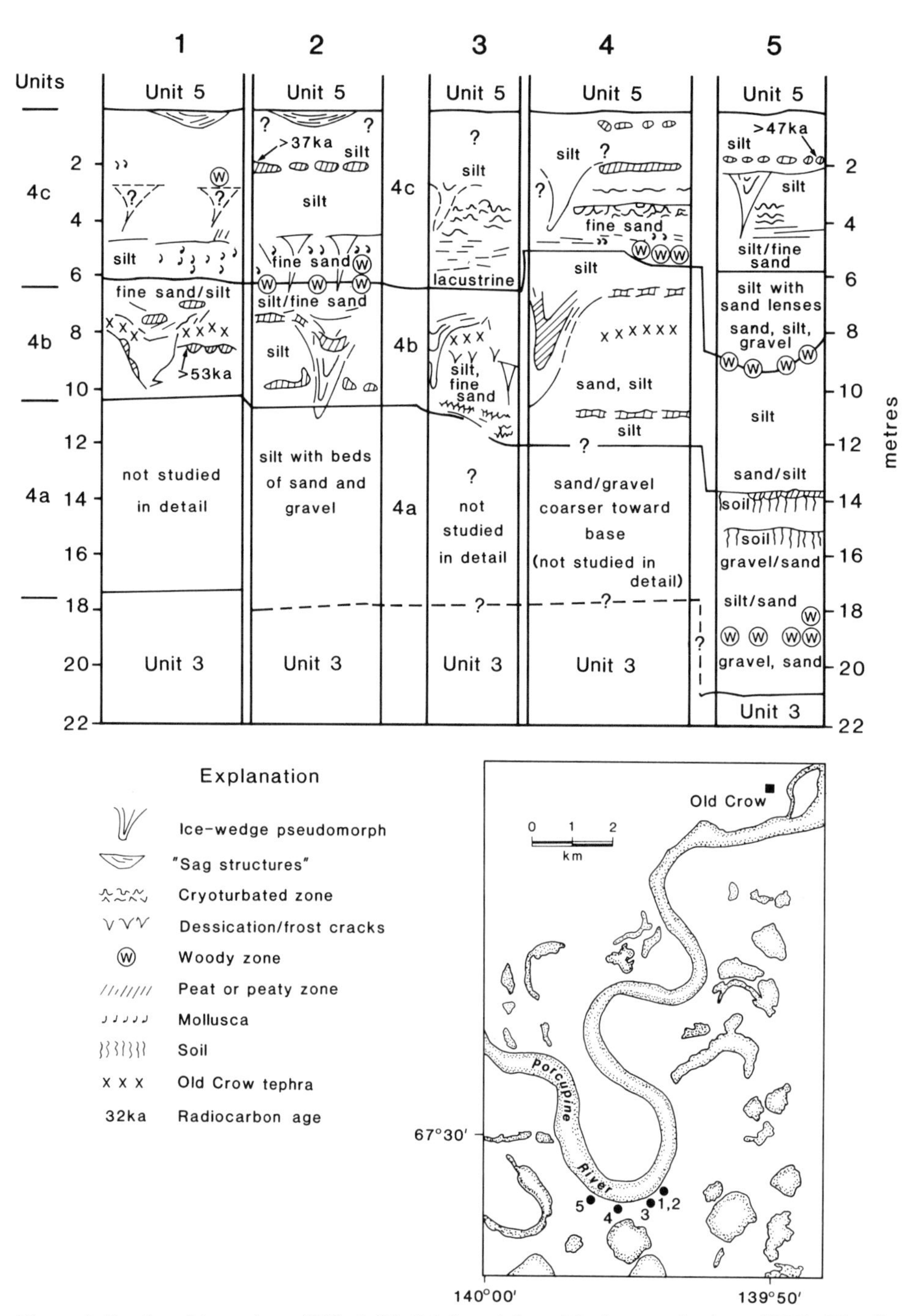

Figure 4. Stratigraphic sections, Ch'ijee's Bluff (adapted from Matthews and others, 1990b, Fig. 4). Datum is base of unit 5.

molluscs, and bone. These sediments are equivalent to unit 4 at Ch'ijee's Bluff. Unconformities, demarcated by concentrations of bone, truncated cryoturbation structures, and peat, are also present, especially in the upper part of the succession. The most conspicuous of these ("Disconformity A" of Morlan, 1980) merges with a paleosol and peat that formed on a coniferous forest floor. Old Crow tephra is present 1–2 m below Disconformity A. Glaciolacustrine clay, commonly having a capping layer of peat, sharply overlies this fluvial succession and forms the upper surface of much of Old Crow Basin. In places, however, the peat is underlain by fluvial silt and sand, which fill channels cut into the glaciolacustrine sediments.

Drift of the Cordilleran Ice Sheet. Bostock (1966) inferred four advances of the Cordilleran Ice Sheet in central

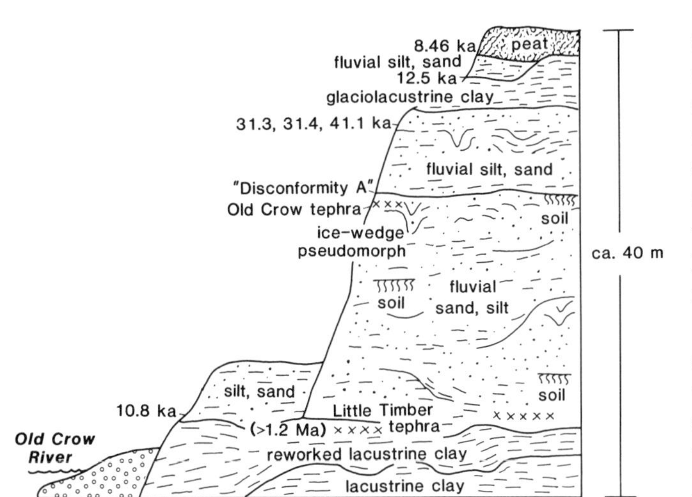

Figure 5. Composite stratigraphic section of Old Crow River bluffs (adapted from Morlan, 1980, Fig. 2.2).

Yukon Territory: Nansen (older), Klaza, Reid, and McConnell (youngest), each successive advance being less extensive than its predecessor (Fig. 6). Glacial features from the Nansen and Klaza advances are subdued, and the deposits of the respective glaciations have not been differentiated outside the area mapped by Bostock (1966). Ice-marginal features marking the limit of the Reid glaciation are moderately well preserved, and those of the late Wisconsinan McConnell glaciation are very well preserved, permitting airphoto interpretation of their limits across much of central Yukon Territory (Fig. 6; Hughes and others, 1969). Deposits of the pre-Reid, Reid, and McConnell glaciations are further distinguished by conspicuous differences in soil development (Foscolos and others, 1977; Rutter and others, 1978; Tarnocai and others, 1985; Smith and others, 1986). Soils on pre-Reid drift at well-drained sites are Luvisols with thick Bt horizons; soils on the Reid drift are moderately developed Luvisols, or Brunisols with thick Bm horizons; and soils on the McConnell drift are Brunsols with thin Bm horizons (see Canada Soil Survey Committee, 1978, for definition of soil terms).

In southwestern Yukon Territory, which was affected by successive advances of glaciers from the St. Elias Mountains, there are only two drift sheets at the surface. These have been assigned to the Mirror Creek and Macauley glaciations (Rampton, 1971), and are continuous with drift of the Reid and McConnell glaciations, respectively. Old Crow tephra overlies drift of the Mirror Creek glaciation beyond the late Wisconsinan (Macauley) limit.

Interdrift sediments. Sediments assigned to the nonglacial interval between the Reid (Mirror Creek) and McConnell (Macauley) glaciations have been identified at several sites in south and central Yukon Territory (Figs. 6, 7, 8). In sections along Silver Creek near the south end of Kluane Lake, the Icefield and Kluane drifts are separated by a weathering zone that formed during the Boutellier nonglacial interval (Fig. 7; Denton and Stuiver, 1967). The Kluane drift is equivalent to the Macauley drift (late Wisconsinan), and the Icefield drift probably correlates with the Mirror Creek drift (Denton and Stuiver, 1967; Rampton, 1971; Hughes and others, 1972). In the Snag-Klutlan area north of Kluane Lake, nonglacial beds beyond the range of radiocarbon dating are exposed beneath till of the Macauley glaciation

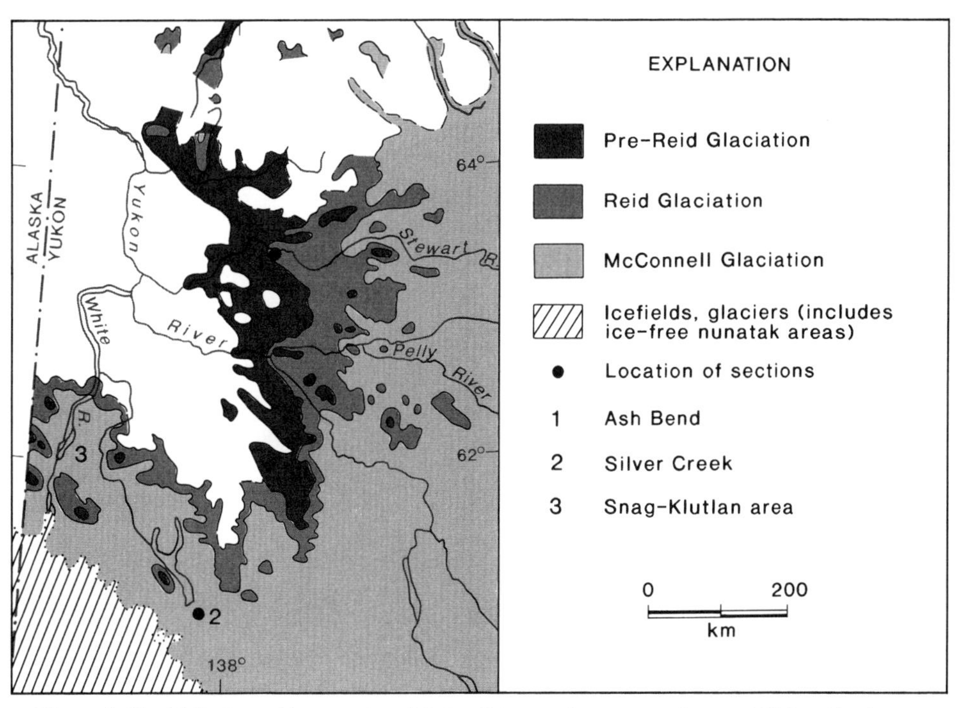

Figure 6. Glacial limits and key stratigraphic localities, southwestern and central Yukon Territory.

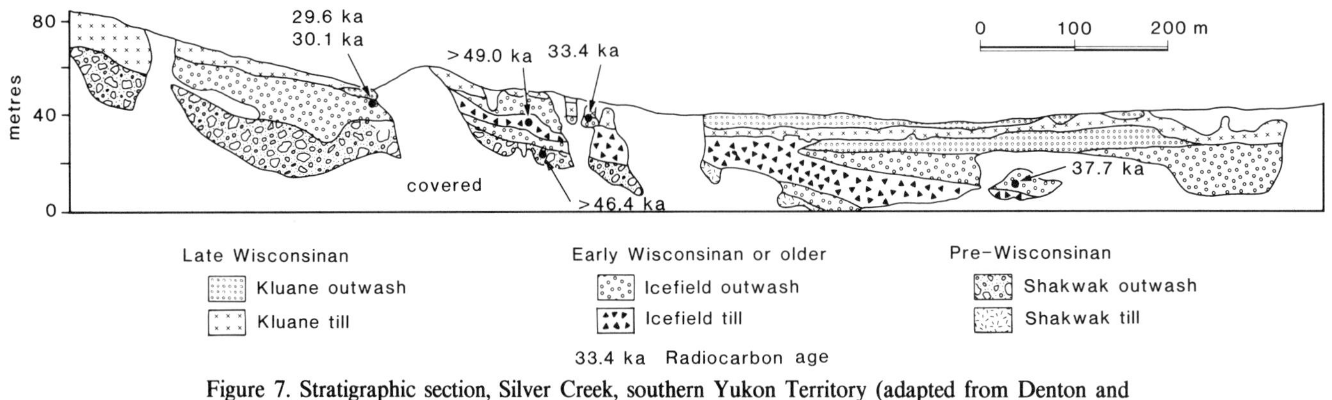

Figure 7. Stratigraphic section, Silver Creek, southern Yukon Territory (adapted from Denton and Stuiver, 1967, Plate 4A). The Boutellier nonglacial interval is recorded by thin sediments and by a paleosol developed on Icefield drift.

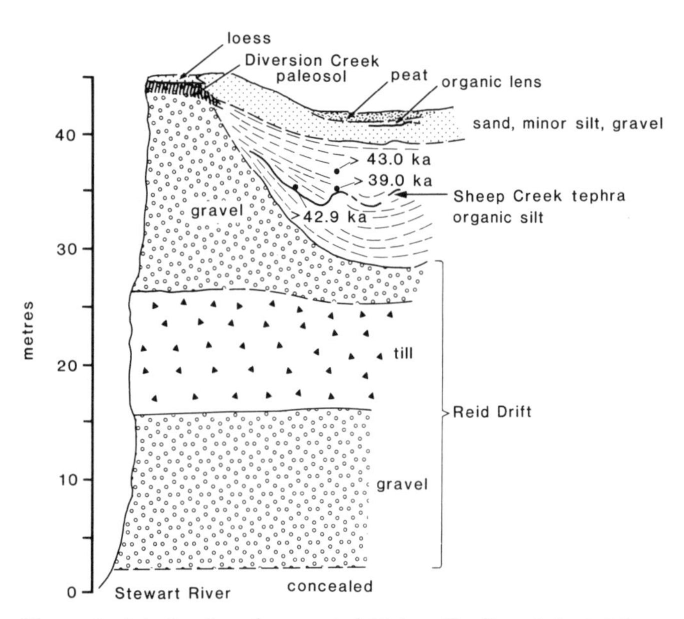

Figure 8. Ash Bend section, central Yukon Territory (adapted from Hughes and others, 1987, Fig. 28).

(Rampton, 1971; Hughes and others, 1972). In the Liard Lowland in southeastern Yukon Territory, organic-rich nonglacial beds occur between tills deposited during the last and penultimate advances of the Cordilleran Ice Sheet (Klassen, 1987). In central Yukon Territory, nonglacial beds underlie the McConnell till, and relatively old organic deposits overlie the Reid drift beyond the McConnell glacial limit (Hughes and others, 1989; Matthews and others, 1990a).

The Ash Bend section on Stewart River illustrates the stratigraphic context and character of pre-McConnell, post-Reid nonglacial deposits in central Yukon Territory (Fig. 8). At Ash Bend, peat and organic-rich silt containing wood, mammal bones, and a layer of volcanic ash (Sheep Creek tephra) fill a channel incised in the Reid drift (Hughes and others, 1987). A paleosol developed on Reid recessional outwash gravel, which is cut by this channel, is similar to the Diversion Creek paleosol, described from the type locality of the Reid glaciation (Smith and others, 1986). It shares affinities with other soils developed on the Reid drift and differs markedly from much more poorly developed soils that have formed on McConnell glacial deposits (Hughes and others, 1972; Foscolos and others, 1977; Rutter and others, 1978; Tarnocai and others, 1985; Smith and others, 1986). Blocks of oxidized sediment at the contact between the channel fill and the underlying outwash gravel suggest that the Diversion Creek paleosol may have begun to develop before the channel was eroded.

Chronology. Chronological control on late Pleistocene sediments in the Bluefish and Old Crow basins is provided by fission-track and TL ages on Old Crow tephra, radiocarbon ages on wood, peat, and bones, uranium-series ages on bone, paleomagnetic stratigraphy, and magnetic susceptibility measurements in loess.

Estimates of the age of Old Crow tephra range from about 80 to 200–250 ka. U-series ages on bones just above the tephra cluster around 60–80 ka, but do not provide tight age control because of reworking (Westgate and others, 1985). The tephra occurs in a core from Imuruk Lake in western Alaska, just above a paleomagnetic reversal thought to be the Blake event (100–125 ka; Smith and Foster, 1969; Shackleton, 1982; Verosub, 1982; Schweger and Matthews, 1985). The most recent TL age estimate for Old Crow tephra is 109 ± 14 ka (Berger, 1987), which is older than an earlier estimate of 86 ± 8 ka by Wintle and Westgate (1986). All of these determinations suggest an age younger than isotope substage 5e (ca. 125 ka; Martinson and others, 1987). However, Westgate (1988) reported an isothermal plateau fission-track age of 149 ± 13 ka on hydrated glass shards from Old Crow tephra, and Begét and Hawkins (1989) suggested, on the basis of magnetic-susceptibility variations in loess, that the tephra may be as old as 250 ka.

These results highlight the difficulty of accurately dating

Pleistocene events beyond the limit of the radiocarbon method. Nevertheless, they indicate that Old Crow tephra is older than early Wisconsinan (isotope stage 4) and probably older than Sangamonian (stage 5). This constrains the age of the upper part of the fluvial succession in the Bluefish and Old Crow basins (subunits 4b and 4c at Ch'ijee's Bluff and correlative sediments exposed in bluffs along Old Crow River). Subunit 4b, which contains Old Crow tephra, is probably pre-Sangamonian in age. Radiocarbon ages from subunit 4c at Ch'ijee's Bluff are all nonfinite, including one of >47,000 yr B.P. (GSC-3858) on wood fragments 1.5 m below unit 5. However, finite ages ranging from 31,300 ± 340 yr B.P. to 41,100 ± 1650 yr B.P. (GSC-1191 and GSC-2574, respectively) have been obtained on plant detritus, peat, and wood from the uppermost part of the fluvial succession in Old Crow Basin (Fig. 5). Evidence presented in the next section suggests that the warmest part of the Sangamonian (substage 5e) may be recorded by sediments at the base of subunit 4c.

A pre-Wisconsinan age for Old Crow tephra also has implications for the age of the Mirror Creek glaciation and its correlatives (Reid and Icefield glaciations). This glaciation must be older than early Wisconsinan because Old Crow tephra overlies the Mirror Creek drift. Although it could date to a cold substage of the Sangamonian (substage 5b or 5d), the Mirror Creek glaciation is more likely pre-Sangamonian in age (stage 6 or possibly even 8, if the magnetic-susceptibility record of Begét and Hawkins [1989] is correct).

Other data show that the nonglacial interval between the Reid glaciation and the late Wisconsinan McConnell glaciation was lengthy. Although no single section contains a complete record of this interval, the sections are thought to collectively encompass much of isotope stages 3, 4, and 5. The Boutellier nonglacial interval at Silver Creek in southwestern Yukon Territory began before 37,700 + 1700/-1300 yr B.P. (Y-1356; Denton and Stuiver, 1967). Twigs from nonglacial sediments lying between tills of the last and penultimate advances of the Cordilleran Ice Sheet have been radiocarbon dated at 23,900 ± 1140 yr B.P. and >30,000 yr B.P. (GSC-2811 and GSC-2949, respectively). Radiocarbon ages on organic material above the Mirror Creek drift and below the Mccauley drift in the Snag-Klutlan area are all nonfinite, except for one of 48,000 ± 1300 yr B.P. (GSC-732), for which contamination by modern rootlets is suspected (Rampton, 1971). Similarly, twigs and sedge fragments from channel-fill sediments overlying the Reid drift at the Ash Bend section have given infinite radiocarbon ages, ranging from >39,000 yr B.P. to >43,000 yr B.P. (GSC-2400 and GSC-2429, respectively; Fig. 8). Twigs within Sheep Creek tephra at Ash Bend have been dated at >42,900 yr B.P. (GSC-524). Vertebrate bones directly beneath the same tephra at Canyon Creek, Alaska, have given a U-series age of about 80 ka (Hamilton and Bischoff, 1984). At Canyon Creek, the tephra postdates outwash of the Delta glaciation, just as at Ash Bend it postdates Reid outwash. This supports the correlation of the Reid glaciation with the Mirror Creek and Delta glaciations, as suggested by Rampton (1971, p. 296), and shows that these events are pre-Wisconsinan in age.

Paleoecology

Detailed paleoecologic studies have been carried out on Quaternary sediments in the Old Crow and Bluefish basins (Lichti-Federovich, 1973, 1974; Morlan and Matthews, 1983; Schweger and Matthews, 1985; Matthews and others, 1990b). Of particular interest here are studies of pollen and plant and insect macrofossils in fluvial, pond, and organic sediments associated with and overlying Old Crow tephra (subunits 4b and 4c at Ch'ijee's Bluff, Fig. 4) (Matthews and others, 1990b). These studies show that the tephra and the sediments in subunit 4b were deposited in a shrub-tundra environment. Because the region is partly forested today, climate probably was colder than at present. Insect fossils just below the tephra include ants and such beetles such as *Helophorus oblongus* and *Agonum quinquepunctatum,* which do not range far beyond treeline today. Thus treeline probably was only slightly south of its present position just before tephra deposition.

Subsequently, climate ameliorated and permafrost in the Old Crow area became degraded. Sediments at the base of subunit 4c at Ch'ijee's Bluff contain abundant and diverse plant and insect macrofossils and pollen; some of the plants and insects presently have northern limits south of the Old Crow area (i.e., *Typha* (cattail) and the beetle families Elmidae and Heteroceridae) (Fig. 9). Pollen and macrofossils of *Picea* (spruce) are abundant, and other elements of a boreal forest association, including *Betula* (birch; both arboreal and shrub types) and *Alnus*

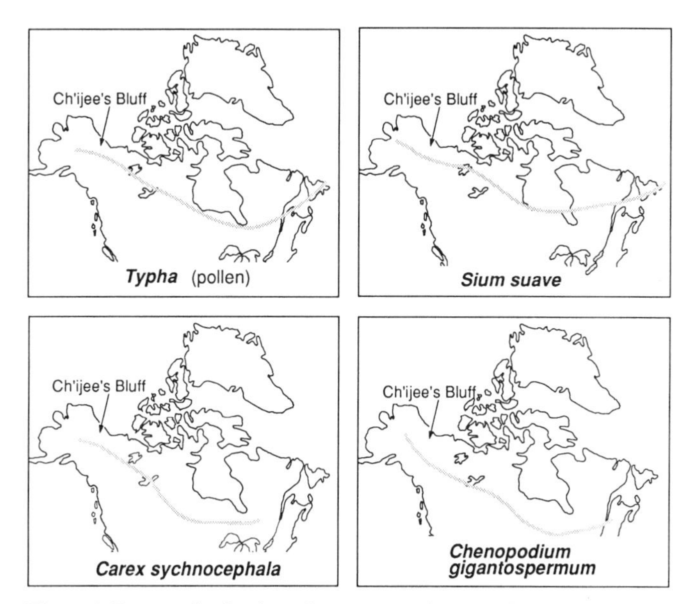

Figure 9. Present distribution of some plants found as fossils in the lower part of subunit 4c at Ch'ijee's Bluff near Old Crow (adapted from Matthews and others, 1990b, Fig. 6). Dotted lines indicate the northern limits of named taxa. Note that the present limits of these taxa are south of the site, suggesting that climate was warmer when this part of subunit 4c was accumulating. Distributional information modified from Porsild and Cody (1980).

(alder), are common. This evidence indicates that climate was warmer than at present during initial deposition of subunit 4c. This period, termed the Koy-Yukon thermal event or "interglaciation" (Schweger and Matthews, 1985; Matthews and others, 1990b), has been recognized at many sites throughout eastern Beringia and may correspond to the peak of the last interglaciation (isotope substage 5e). Alternative, but less likely, possibilities are that the Koy-Yukon thermal event represents a subsequent warm interval (substage 5a or 5c, or even stage 3) or an older interglaciation (stage 7).

The Koy-Yukon thermal event is recorded in Old Crow River exposures by "Disconformity A" (Morlan, 1980) and by sediments that immediately overlie it. This distinctive erosion surface passes laterally into a paleosol and peat that formed on a coniferous forest floor (Bombin, 1980; Morlan and Matthews, 1983). In places, it is associated with features that indicate regional thawing (Morlan, 1980; Matthews and others, 1987). Sediments directly above the unconformity contain abundant plant and insect fossils, some of which suggest a warmer summer climate than at present (Morlan and Matthews, 1983; Schweger and Matthews, 1985).

Subunit 4c contains a record of climatic oscillations postdating the Koy-Yukon thermal event. Pollen spectra from the lower part of the subunit just above the interglacial horizon provide evidence for alternating forest and tundra environments (C. E. Schweger, 1988, personal commun.). A peat layer located just above a zone of ice-wedged pseudomorphs in the middle of subunit 4c and dated >37,000 yr B.P. (GSC-2783) contains pollen and plant macrofossils indicative of forested conditions similar to the present. Sediments near the top of subunit 4c, which on the basis of radiocarbon ages are thought to date to the end of the middle Wisconsinan, show a trend toward cold-climate pollen assemblages. The vegetation in the Old Crow area during the early part of the late Wisconsinan was steppe tundra, with an abundance of grasses and herbs such as *Artemisia frigida* (McCourt, 1982).

It is difficult to correlate the climatic oscillations recorded in subunit 4c with those of the marine oxygen-isotope record. Some of the oscillations may correspond to cyclic cooling and warming linked to global ice-volume changes of isotope substages 5a–5d. Sediments of subunit 4c also record at least one interval (stage 4?) that was as cold as stage 2 (C. E. Schweger, 1988, personal commun.).

Paleoecologic studies have also been conducted on nonglacial sediments underlying late Wisconsinan drift and on similar sediments overlying drift of the Reid glaciation at a few sites in central and southern Yukon. These studies complement and supplement those from the Old Crow area. Pollen and bryophyte fossils from sediments underlying the Kluane drift at the Silver Creek section in southwestern Yukon indicate that this currently forested site was covered by tundra during the Boutellier nonglacial interval (Schweger and Janssens, 1980). In the Liard Lowland, an area currently covered by boreal forest, tundra conditions prevailed from before 30 ka until after 23.9 ka (Klassen, 1987). Preliminary pollen and macrofossil analysis of the channel-fill sediments at the Ash Bend section indicates that boreal forest existed in this part of central Yukon before deposition of Sheep Creek tephra, but that it gave way to tundra sometime thereafter. If this represents the transition from isotope stage 5 to 4, earlier soil formation and erosion of the channel may have taken place within stage 5.

BRITISH COLUMBIA

Stratigraphy

All surface sediments in British Columbia date to the late Wisconsinan and Holocene. Middle Wisconsinan and older sediments underlie late Wisconsinan drift in some valleys and coastal lowlands (Fulton, 1971; Clague, 1981) but, in general, are not well exposed. Detailed studies of middle Wisconsinan sediments have been conducted in a few areas and have provided considerable information on the lengthy "interstade" (Olympia nonglacial interval) that preceded late Wisconsinan glaciation. In contrast, little work has been done on older deposits, thus our knowledge of Sangamonian and early Wisconsinan events is poor.

Sangamonian (?) sediments. Possible Sangamonian deposits have been recognized at only a few localities in British Columbia. The Highbury sediments, identified at three sites in the Fraser Lowland in the southwest corner of the province, consist of fluvial, deltaic, and marine sediments (mainly silt and sand) and underlie pre–middle Wisconsinan drift and overlie older glacial sediments (Armstrong, 1975; Hicock and Armstrong, 1983). Highbury marine sediments grade upward into fluvial sediments, recording seaward progradation of deltas.

The Muir Point Formation on southern Vancouver Island is thought to correlate with the Highbury sediments on the basis of similar sediment types and stratigraphic position. It consists of silt, sand, gravel, and diamicton, deposited as alluvium and colluvium on a coastal floodplain (Hicock and Armstrong, 1983; Alley and Hicock, 1986; Hicock, 1990). At its type section, the Muir Point Formation is more than 30 m thick, is unconformably overlain by middle Wisconsinan nonglacial sediments and late Wisconsinan drift, and is underlain by till (Fig. 10).

Two well-defined stratigraphic units may record the transition from the Sangamonian to the Wisconsinan, although they could possibly be pre-Sangamonian in age (see below). The Mapleguard sediments, which underlie early Wisconsinan(?) till and glaciomarine sediments and are exposed at the base of some sea cliffs on eastern Vancouver Island (Fig. 11), consist of bedded silt, sand, and minor gravel of fluvial or perhaps deltaic or marine origin (Fyles, 1963; Hicock, 1980). Hicock and Armstrong (1983) proposed that the Mapleguard sediments are younger than, and stratigraphically separate from, the Muir Point Formation. The Mapleguard sediments may be strictly nonglacial in origin, although they are more likely outwash deposits laid down during an early Wisconsinan(?) advance (Hicock and Armstrong, 1983).

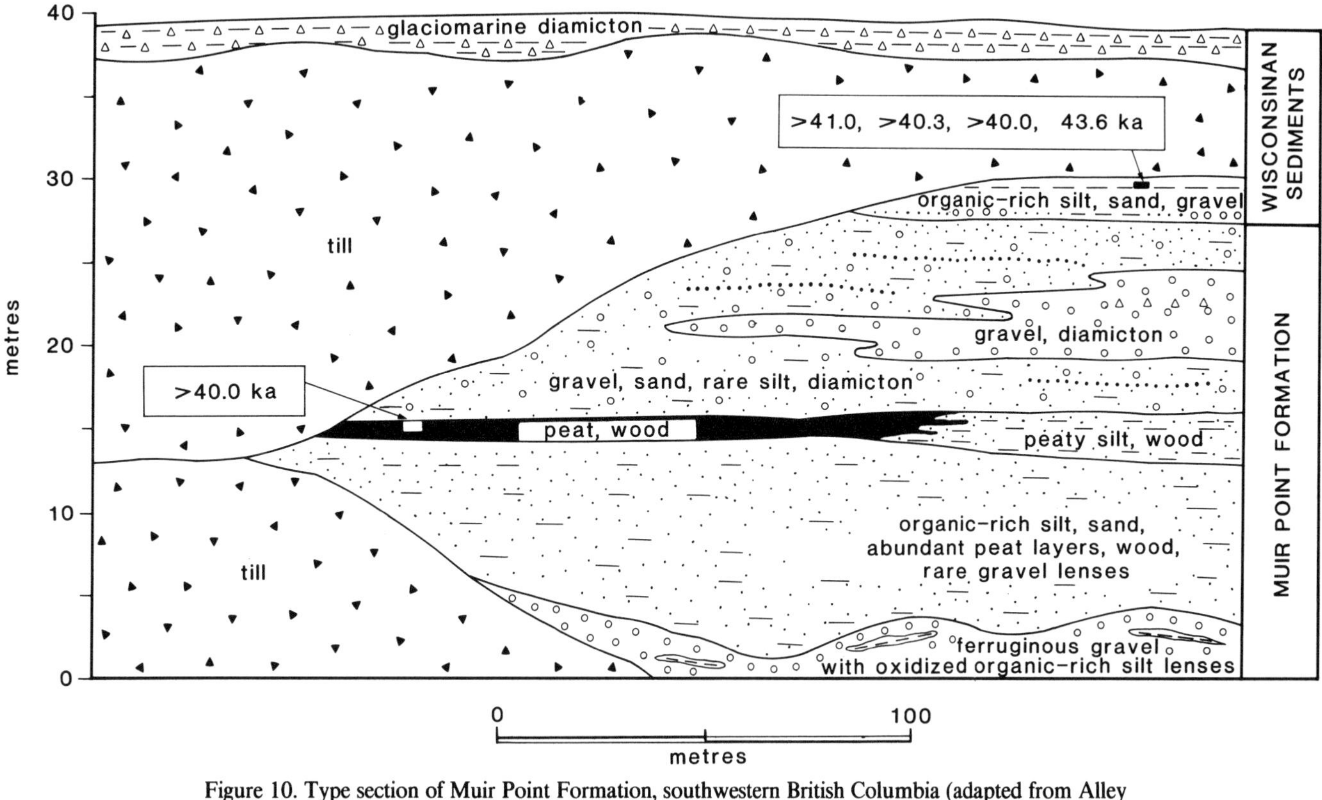

Figure 10. Type section of Muir Point Formation, southwestern British Columbia (adapted from Alley and Hicock, 1986, Fig. 2). Horizontal scale is approximate.

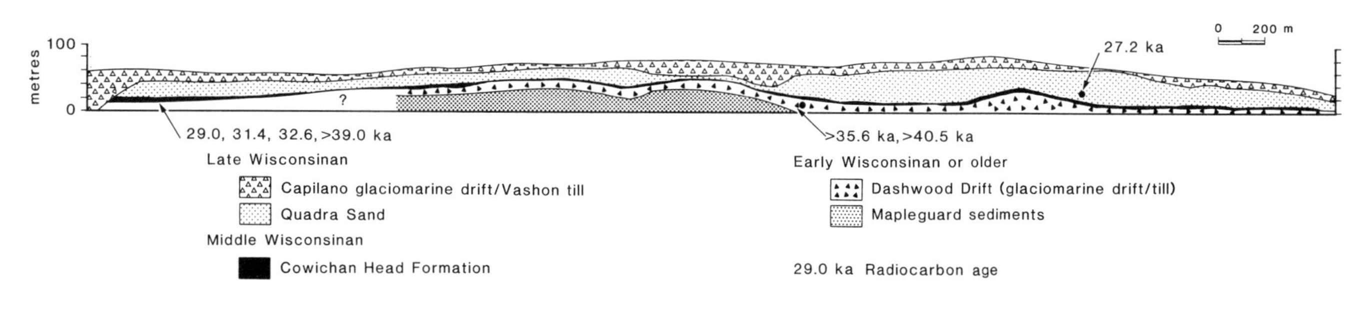

Figure 11. Stratigraphic section, Dashwood, southwestern British Columbia (adapted from Fyles, 1963, Map 1111A). The middle Wisconsinan Cowichan Head Formation appears to conformably overlie the Dashwood drift in this section.

Westwold Sediments (Fulton and Smith, 1978), identified at two sites in south-central British Columbia, may correlate with the Mapleguard sediments or part of the Muir Point Formation. At the type section, the Westwold sediments consist of 17 m of cross-bedded gravelly sand capped by about 2 m of clay, silt, and marl. Features resembling ice-wedge pseudomorphs are present near the top of the gravelly sand unit, indicating that permafrost may have been present. The thin, fine-grained upper unit, however, lacks such features and contains mollusc shells, plant impressions, and fragments of bison bones, fish, beetles, and rodents, suggesting subsequent climatic warming.

Early Wisconsinan(?) drift. Drift that has traditionally been assigned to the early Wisconsinan Substage has been identified in most parts of British Columbia where detailed stratigraphic work has been carried out, although it is largely restricted to major valleys and coastal lowlands.

In the interior of British Columbia and on Vancouver Island, this drift consists of a single till bounded by stratified sediments. A representative sequence from the Bullion placer gold pit in central British Columbia is shown in Figure 12. Early Wisconsinan(?) drift at this site consists of a till and overlying ice-contact gravel and sand. The drift is overlain by organic-rich, middle Wisconsinan nonglacial beds (Clague and others, 1990). The Okanagan Centre drift in south-central British Columbia comprises a till, an underlying unit of glaciolacustrine silt and glaciofluvial gravel, and an overlying unit of glaciolacustrine silt and minor glaciofluvial and beach gravel (Fulton and Smith, 1978). A similar succession has been found at other sites in the southern and central interior (Ryder, 1976, 1981) and in the Northern Rocky Mountain Trench (Rutter, 1976, 1977). On north-central Vancouver Island, the poorly exposed Muchalat River drift consists of till and overlying glaciolacustrine silt (Howes, 1981). Along the coast, the same till is overlain by fossiliferous glaciomarine mud (Howes, 1983). On eastern Vancouver Island south of Howes's study area, the Dashwood drift consists of till and an overlying unit of glaciomarine silt and silty sand (Fyles, 1963) (Fig. 11). Hicock and Armstrong (1983) included the previously mentioned Mapleguard sediments, which they consider to be glacial in origin, in the Dashwood drift.

The early Wisconsinan(?) Semiahmoo drift in the Fraser Lowland is more complex. It consists of two or more tills interlayered with glaciomarine, glaciofluvial, and possibly glaciolacustrine sediments (Armstrong, 1975; Hicock and Armstrong, 1983). Some of the "glaciofluvial" materials were deposited subaqueously as outwash fans and deltas. The complexity of this unit probably results from the fact that tidewater glaciers fluctuated in this area during the growth and decay phases of the Semiahmoo glaciation. The Semiahmoo and Dashwood drifts are similar in character and complexity to late Wisconsinan drift in the same area; thus, the pattern of glaciation in the Strait of Georgia region during the last two glacial periods probably was similar.

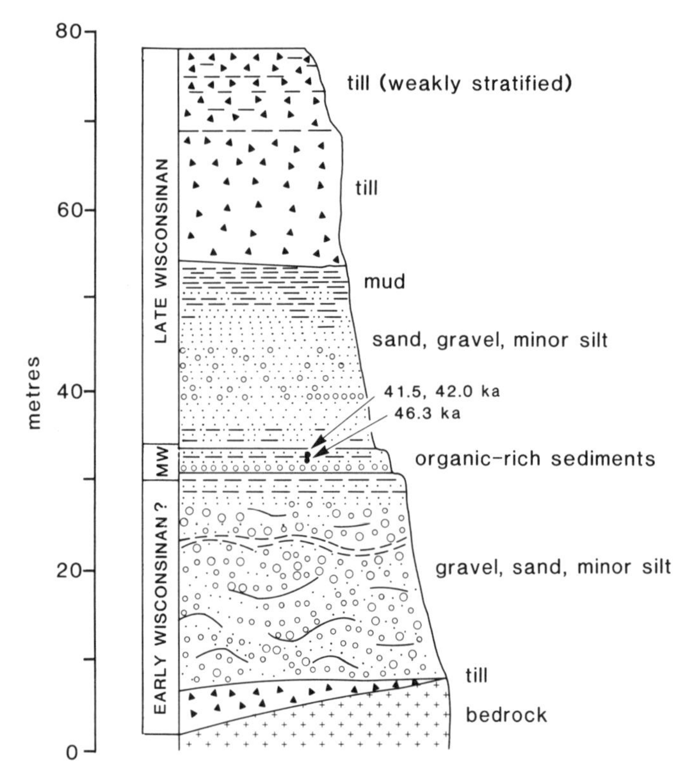

Figure 12. Stratigraphic section at the Bullion placer gold pit, central British Columbia (adapted from Clague and others, 1990, Fig. 6). The contact between middle Wisconsinan nonglacial sediments and underlying drift appears to be conformable.

Chronology

All radiocarbon ages from the Muir Point Formation and Dashwood and Semiahmoo drifts are infinite (for details on ages, see Clague, 1980; Hicock and Armstrong, 1983). It is clear from this and from stratigraphic considerations that the Dashwood and Semiahmoo drifts are no younger than early Wisconsinan, and that the Muir Point Formation and correlative units are no younger than Sangamonian. The radiocarbon ages themselves do not preclude the possibility that these deposits are pre-Sangamonian.

Amino-acid ratios have been determined for shells and wood from the Muir Point Formation and Dashwood and Semiahmoo drifts (Hicock and Rutter, 1986). Mean D/L ratios of aspartic acid in shells of *Clinocardium* (0.37), *Serripes* (0.35), *Nuculana* (0.44), *Mya* (0.44), *Macoma* (0.56), and *Hiatella* (0.44) from Dashwood and Semiahmoo glaciomarine sediments are greater than ratios for the same taxa in late Wisconsinan glaciomarine sediments (0.22, 0.23, 0.37, 0.28, 0.28, and 0.27, respectively), indicating that the two groups of deposits differ significantly in age. Numerical ages, however, have not been determined from these ratios because the temperature history of the fossils is poorly known. The interpretation of amino-acid ratios on fossil wood in this region is beset by the same problem (Hicock and Rutter, 1986).

The Muir Point Formation at its type locality is normally magnetized, indicating that it probably was deposited sometime during the Brunhes normal polarity chron (788 ka–present; Johnson, 1982). This is consistent with it being a correlative of the Whidbey Formation of northwestern Washington, which has been assigned to the Sangamonian on the basis of amino-acid age estimates, pollen, and stratigraphic position (Easterbrook, 1986; Blunt and others, 1987). If this correlation is correct, the Dashwood drift is most likely early Wisconsinan, and a correlative of the Possession Drift of northwestern Washington. The similarity in amino-acid ratios from the Dashwood and Possession drifts also implies that the two units correlate (Hicock and Rutter, 1986).

Correlation of penultimate glacial deposits in the British Columbia interior (e.g., Okanagan Centre drift) with those on the coast (Dashwood and Semiahmoo drifts) is based on similar stratigraphic position and the fact that all of these deposits are beyond the limit of radiocarbon dating. However, until the units are better dated, these correlations should be considered provisional.

Paleoecology

Fossil pollen assemblages from the Muir Point Formation provide a record of vegetation and climate during the last(?) interglaciation (Alley and Hicock, 1986; Hicock, 1990). *Pseudotsuga menziesii* (Douglas-fir) pollen is more abundant in these sediments than in the modern pollen rain in the same area. This, in conjunction with abundant *Alnus* (alder) and *Thuja plicata* (western red cedar), indicates that conditions were warmer and/or drier than at present on southern Vancouver Island. The uppermost organic-rich sediments of the Muir Point Formation record marked declines in *Pseudotsuga menziesii, Alnus,* and *Thuja plicata* pollen, and corresponding increases in *Tsuga heterophylla* (western hemlock) and *Picea* (spruce) pollen. This signifies a major change in forest structure, probably caused by cooler and/or moister climate. It is not known whether this climatic deterioration represents the onset of a glaciation or whether it was a cooler, moister phase during an interglaciation.

Pollen spectra of the Muir Point Formation differ markedly from those of middle Wisconsinan sediments in the same region. The latter, rich in *Picea, Tsuga heterophylla, Tsuga mertensiana* (mountain hemlock), and *Pinus* (pine) and with low amounts of *Pseudotsuga menziesii,* indicate a fluctuating, but generally cooler and perhaps moister climate than at present (Clague, 1978; Alley, 1979; Armstrong and others, 1985).

WASHINGTON

Stratigraphy

The Cordilleran Ice Sheet advanced into the northwestern United States on several occasions during the Pleistocene (Willis, 1898; Bretz, 1913; Crandell, 1965; Richmond and others, 1965; Waitt and Thorson, 1983; Easterbrook, 1986). Little is known of all but the latest of these advances in the eastern part of this region, but a detailed record of glacial and interglacial events and environments has been reconstructed for the Puget Lowland in northwestern Washington (Easterbrook, 1969, 1986; Hansen and Easterbrook, 1974; Blunt and others, 1987; Easterbrook and others, 1988, and references therein). This area was periodically covered by a large piedmont lobe of the Cordilleran Ice Sheet ("Puget lobe") that flowed south via the Strait of Georgia from source areas in the southern Coast Mountains. In this section, we briefly review the Sangamonian and early Wisconsinan part of this record.

Sangamonian(?) sediments. Interglacial sediments of the Whidbey Formation, which have been assigned a Sangamonian age on the basis of their stratigraphic position and pollen (Hansen and Mackin, 1949; Easterbrook and others, 1967; Heusser and Heusser, 1981; Easterbrook, 1986), are well exposed in sea cliffs in the central Puget Lowland (Fig. 13). These sediments underlie two drift sequences and consist of stratified clay, silt, sand, peat, and minor gravel deposited on broad floodplains and deltas (Hansen and Mackin, 1949; Easterbrook and others, 1967). At its type locality on Whidbey Island, the Whidbey Formation is 60 m thick, lies on the pre-Sangamonian Double Bluff drift, and is unconformably overlain by late Wisconsinan advance outwash (Esperance sand). At other sites, the Whidbey Formation is directly overlain by the Possession drift.

Early Wisconsinan(?) drift. The Possession drift, which is present in the northern Puget lowland, overlies the Whidbey Formation and underlies middle Wisconsinan nonglacial sediments and late Wisconsinan drift (Fig. 13; Easterbrook and others, 1967). At its type locality at Possession Point on Whidbey Island, the Possession drift consists of a single till, but elsewhere it includes glaciomarine sediments and outwash sand and gravel.

The southern extent of the Possession drift in the Puget Lowland is unknown, although the unit apparently does not extend beyond the late Wisconsinan glacial limit. Moderately weathered drift just south of the late Wisconsinan limit in the southernmost Puget Lowland is thought to be pre-Sangamonian in age and older than the Possession drift (see Summary and Discussion).

Chronology

All radiocarbon ages on wood and peat from the Whidbey Formation are nonfinite (Easterbrook, 1969, 1976; Blunt and others, 1987). Ages calculated from mean D/L ratios of leucine (0.35) and alloisoleucine/isoleucine (0.26) in *Saxidomus* shells from the Whidbey Formation at Admiralty Bay on Whidbey Island are on the order of 100 ± 40 ka (Blunt and others, 1987). Diagenetic temperatures used to calculate these ages were based on climatological data from a nearby site, adjusted for depth of burial and an average Pleistocene temperature reduction of 3.6 °C (the latter estimated from palynological records) (Kvenvolden and others, 1980). Racemization rate constants were also calibrated by determining D/L ratios of leucine and alloisoleucine/isoleucine in radiocarbon-dated, late Wisconsinan, *Saxidomus* shells (Blunt and others, 1987). Because amino-acid racemization is sensitive to temperature, significant errors in assigned diagenetic temperatures will result in large errors in age estimates. A gross age of 100 ka, however, is reasonable because paleoecologic evidence indicates that the Whidbey Formation was deposited at a time when climate was at least as warm as today (see below), most likely during the climatic optimum of the last interglaciation (isotope substage 5e). Unrealistically low diagenetic temperatures would be required to produce the measured amino-acid ratios if the Whidbey Formation dated to an earlier interglaciation (e.g., stage 7).

The Whidbey Formation is normally magnetized with declinations within 10° of north and inclinations of 46°–64° (Easterbrook and Othberg, 1976).

All radiocarbon ages in shells and wood from glaciomarine sediments of the Possession drift are nonfinite. Mean D/L ratios

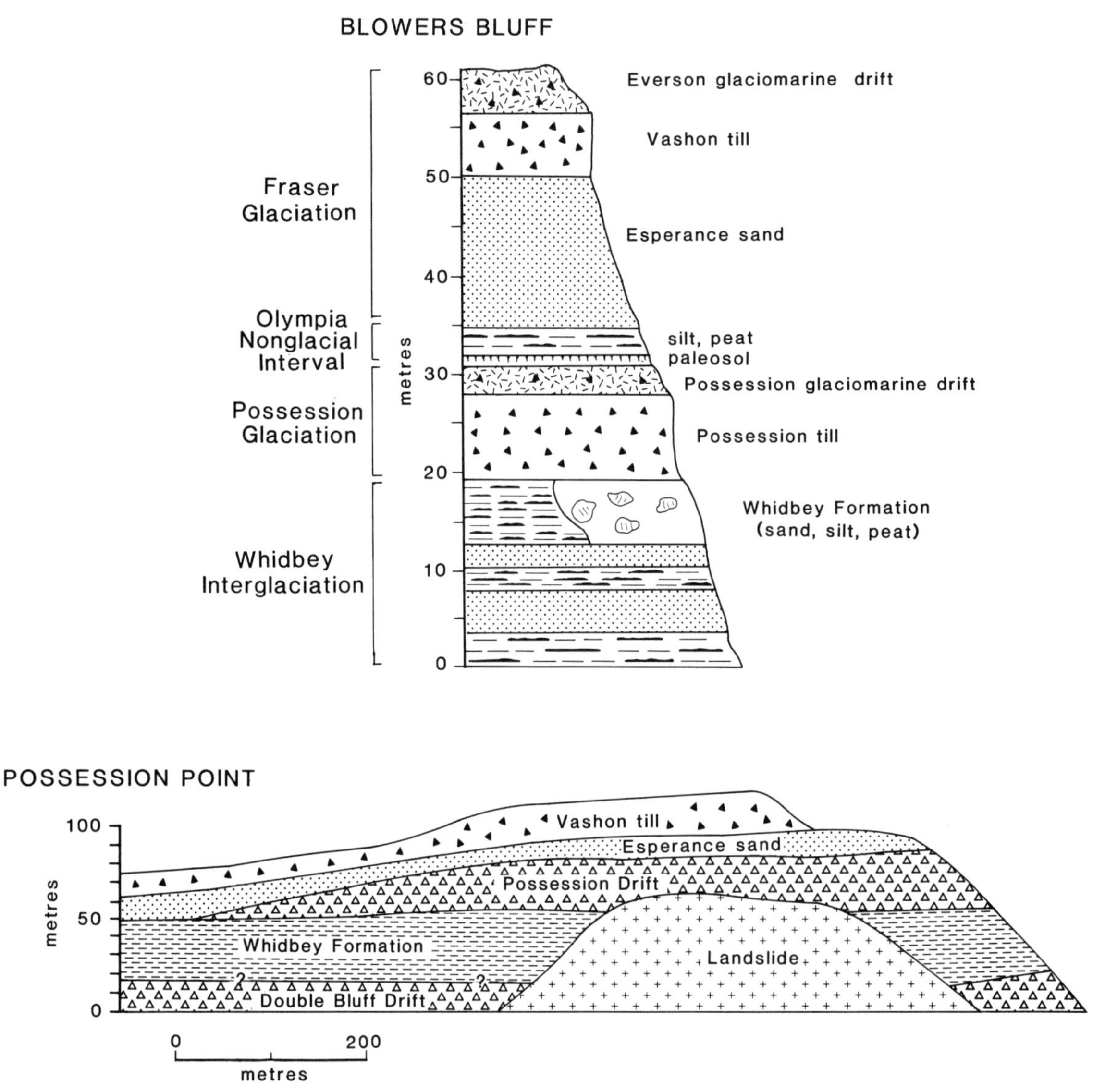

Figure 13. Stratigraphic sections at Blowers Bluff and Possession Point on Whidbey Island, Washington, showing the relation of the Whidbey Formation to the Possession drift. Horizontal scale is approximate.

of alloisoleucine/ isoleucine determined by Blunt and others (1987) on shells from the Possession drift at Stillaguamish, east of Whidbey Island, are 0.19 (*Saxidomus*), 0.20 (*Clinocardium*), 0.36 (*Nuculana*), and 0.16 (*Hiatella*). These authors calculated a mean age of 80 ± 22 ka from these ratios, using the same climatological data and assumptions about thermal history as above. Although it would be unwise to consider this an accurate age, given the possibility that the temperature assumptions of Blunt and others (1987) are in error, the data appear to preclude an Illinoian or older age for the Possession drift. The most likely possibility is that at least the early part of the Possession glaciation (Deception Pass stade of Easterbrook, 1976) is early Wisconsinan.

Preliminary TL data from the Whidbey Formation and Possession drift (G. W. Berger, 1989, personal commun.) are consistent with the amino-acid ages summarized above.

Paleoecology

Pollen analysis of the Whidbey Formation has provided information on the climate of northwestern Washington during the last interglaciation (Hansen and Mackin, 1949; Heusser and Heusser, 1981). Pollen spectra from this unit are rich in *Pseudotsuga menziesii* and *Alnus,* similar to the modern pollen rain in the Puget Lowland. This suggests that the Whidbey Formation was deposited under a climate similar to, or warmer than, the present. There is some pollen evidence, however, for cooler conditions at the beginning and end of Whidbey deposition.

Pollen spectra from the Whidbey Formation are similar to those from the Muir Point Formation in southwestern British Columbia (Alley and Hicock, 1986), lending support to the contention that the two units are equivalent (Hicock and Armstrong,

1983; Hicock, 1990). They differ, however, from middle Wisconsinan pollen spectra from both British Columbia and Washington. Middle Wisconsinan peat beds on Whidbey Island, for example, contain high percentages of *Pinus* and Gramineae (grass) pollen and moderate amounts of *Picea, Tsuga heterophylla,* and *Thuja plicata* pollen; *Pseudotsuga menziesii* is absent (Hansen and Easterbrook, 1974). This assemblage reflects an open landscape, dominated by grass and herbs, with scattered patches of pine and other trees, quite different from the mesic forests that covered this region during the warmest part of the Sangamonian. Climate was cooler and perhaps wetter than today.

SUMMARY AND DISCUSSION

Although progress is being made in deciphering Sangamonian and early Wisconsinan events and environments in western North America, efforts continue to be hampered by difficulties in satisfactorily dating critical deposits, leading to uncertainty in the correlation of sediments from one region to another. Because of these problems, many of the age assignments, correlations, and conclusions in this paper must be considered provisional; their acceptance (or rejection) requires a sounder chronologic framework than is currently available. Notwithstanding this, we can explore some of the implications of the stratigraphic, chronologic, and paleoecologic information summarized in preceding sections.

The presence of Old Crow tephra above the Mirror Creek drift in the Snag-Klutlan area of southwestern Yukon indicates that the Mirror Creek glaciation and its probable correlative, the Reid glaciation, are pre-Wisconsinan and probably Illinoian in age. Support for this conclusion is provided by the fact that the Diversion Creek paleosol (an interglacial soil) and Sheep Creek tephra postdate the Reid glaciation. Thus, at the northern margin of the Cordilleran Ice Sheet, no evidence exists for a glacial event between the Illinoian and late Wisconsinan. If an intervening advance occurred, it must have been less extensive than the late Wisconsinan. The Icefield drift in southwestern Yukon Territory (Fig. 7) may be a remnant of an intermediate age (i.e., early Wisconsinan) advance, although it could equally well be a product of the Reid glaciation.

In British Columbia, a major glaciation occurred just before a lengthy nonglacial period that spans at least the entire middle Wisconsinan. The extent of ice during this glaciation may have been similar to that of the late Wisconsinan. This glaciation is generally believed to be early Wisconsinan, but there are no numerical age determinations that conclusively demonstrate this.

At several sites in southern and central British Columbia, glacial deposits appear to conformably underlie nonglacial sediments that are, at least in part, middle Wisconsinan (Ryder and Clague, 1989) (Figs. 11, 12). This suggests that the glacial deposits may be early Wisconsinan. However, there could be a hiatus between the two sequences at these sites, or the nonglacial sequence could extend back into the early Wisconsinan or Sangamonian.

The strongest evidence for the existence of an ice sheet in the Cordillera during the early Wisconsinan comes from northwestern Washington where northern-provenance glacial deposits, which have yielded an amino-acid estimate of 80 ± 22 ka, underlie middle Wisconsinan sediments and overlie probable Sangamonian interglacial deposits.

The evidence for extensive early Wisconsinan glaciation in southern British Columbia and northwestern Washington seems, at first glance, difficult to reconcile with limited early Wisconsinan glacition in Yukon Territory. There are four possible explanations for this apparent enigma. (1) The Reid and Mirror Creek glaciations are, in fact, early Wisconsinan or late Sangamonian (stage 5b or 5d) in age, rather than Illinoian. (2) The Dashwood, Semiahmoo, and Possession glaciations are Illinoian. (3) The northern and southern sectors of the Cordilleran Ice Sheet responded differently during the early Wisconsinan than during the late Wisconsinan. (4) Differences in extent of glaciation in the north and the south are more apparent than real.

The first possibility seems very unlikely in view of the ages on Old Crow and Sheep Creek tephras and the relation of the Diversion Creek paleosol to the Reid drift. The second possibility also is unlikely because it requires that the amino-acid age estimates from the Possession drift and the Whidbey Formation are too young and that the nonglacial sequence between penultimate and last glacial deposits in British Columbia spans a longer period of time than is now thought to be the case.

Although neither of these possibilities can be ruled out with currently available data, the third and fourth explanations are more plausible. The Cordilleran Ice Sheet in Yukon Territory during the early Wisconsinan may have been only slightly less extensive than during the late Wisconsinan. This is compatible with evidence from the Puget Lowland indicating that the Puget lobe extended farther south during the late Wisconsinan than during the early Wisconsinan. A single drift with soils characterized by moderately well developed argillic B horizons with strong, blocky to prismatic structure lies 2–13 km beyond the late Wisconsinan glacial limit in southernmost Puget Lowland (Lea, 1983, 1984). Weathering rinds on basaltic stones in this drift average 1.01 ± 0.40 mm in thickness (Coleman and Pierce, 1981). In comparison, basaltic stones in Sangamonian marine terrace deposits on the Pacific coast of Washington are 0.79 ± 0.42 mm thick, on average. These observations suggest that this drift correlates with the Double Bluff drift in the northern Puget Lowland (Lea, 1983, 1984) and that there are no early Wisconsinan glacial deposits beyond the late Wisconsinan limit in this area.

Paleoecologic studies of sediments of the last interglaciation in Yukon Territory, British Columbia, and Washington provide evidence for a warmer and drier climate than today. In the Yukon, this period was marked by regional degradation of permafrost and a northward shift of treeline. The forests of south-coastal British Columbia and northwestern Washington contained more Douglas-fir and alder during the warmest part of the Sangamonian than today, and grasslands probably extended much farther north in interior British Columbia than at present.

Very little is known about the magnitude and character of climatic change during the transition from the Sangamonian to the early Wisconsinan. In the Old Crow area of Yukon Territory, the early Wisconsinan was a time of major climatic fluctuation characterized by alternation of tundra and forest environments. Detailed paleoecologic studies have not yet been conducted on presumed early Wisconsinan sediments in British Columbia and Washington, but middle Wisconsinan sediments in these areas record a much cooler and perhaps moister climate than that which prevailed during the Sangamonian.

REFERENCES CITED

Alley, N. F., 1979, Middle Wisconsin stratigraphy and climatic reconstruction, southern Vancouver Island, British Columbia: Quaternary Research, v. 11, p. 213–237.

Alley, N. F., and Hicock, S. R., 1986, The stratigraphy, palynology, and climatic significance of pre–middle Wisconsin Pleistocene sediments, southern Vancouver Island, British Columbia: Canadian Journal of Earth Sciences, v. 23, p. 369–382.

Armstrong, J. E., 1975, Quaternary geology, stratigraphic studies and revaluation of terrain inventory maps, Fraser Lowland, British Columbia (92G/1, 2, and parts of 92G/3, 6, 7, and H/4): Geological Survey of Canada Paper 75-1A, p. 377–380.

——, 1981, Post-Vashon Wisconsin glaciation, Fraser Lowland, British Columbia: Geological Survey of Canada Bulletin 322, 34 p.

Armstrong, J. E., Clague, J. J., and Hebda, R. J., 1985, Late Quaternary geology of the Fraser Lowland, southwestern British Columbia, *in* Tempelman-Kluit, D., ed., Field guides to geology and mineral deposits in the southern Cordillera (Geological Society of America, Cordilleran Section meeting guidebook): Vancouver, Geological Association of Canada, Cordilleran Section, p. 15-1–15-25.

Begét, J. E., and Hawkins, D. B., 1989, Influence of orbital parameters on Pleistocene loess deposition in central Alaska: Nature, v. 337, p. 151–153.

Berger, G. W., 1987, Thermoluminescence dating of the Pleistocene Old Crow tephra and adjacent loess, near Fairbanks, Alaska: Canadian Journal of Earth Sciences, v. 24, p. 1975–1984.

Blunt, D. J., Easterbrook, D. J., and Rutter, N. W., 1987, Chronology of Pleistocene sediments in the Puget Lowland, Washington: Washington Division of Geology and Earth Resources Bulletin 77, p. 321–353.

Bombin, M., 1980, Early and mid-Wisconsinan paleosols in the Old Crow Basin (Yukon Territory, Canada): American Quaternary Association, 6th Biennial Meeting, Abstracts and Program, p. 37–39.

Bostock, H. S., 1966, Notes on glaciation in central Yukon Territory: Geological Survey of Canada Paper 65-36, 18 p.

Bretz, J. H., 1913, Glaciation of the Puget Sound region: Washington Geological Survey Bulletin 8, 244 p.

Canada Soil Survey Committee, 1978, The Canadian system of soil classification: Canada Department of Agriculture Publication 1646, 164 p.

Clague, J. J., 1978, Mid-Wisconsinan climates of the Pacific Northwest: Geological Survey of Canada Paper 78-1B, p. 95–100.

——, 1980, Late Quaternary geology and geochronology of British Columbia, Part 1: Radiocarbon dates: Geological Survey of Canada Paper 80-13, 28 p.

——, 1981, Late Quaternary geology and geochronology of British Columbia, Part 2: Summary and discussion of radiocarbon-dated Quaternary history: Geological Survey of Canada Paper 80-35, 41 p.

——, compiler, 1989a, Quaternary geology of the Canadian Cordillera, *in* Fulton, R. J., ed., Quaternary geology of Canada and Greenland: Geological Survey of Canada, Geology of Canada, no. 1, p. 15–96.

——, 1989b, Cordilleran Ice Sheet, *in* Fulton, R. J., ed., Quaternary geology of Canada and Greenland: Geological Survey of Canada, Geology of Canada, no. 1, p. 40–42.

Clague, J. J., Hebda, R. J., and Mathewes, R. W., 1990, Stratigraphy and paleoecology of Pleistocene interstadial sediments, central British Columbia: Quaternary Research, v. 34, p. 208–226.

Coleman, S. M., and Pierce, K. L., 1981, Weathering rinds on andesitic and basaltic stones as a Quaternary age indicator, western United States: U.S. Geological Survey Professional Paper 1210, 56 p.

Crandell, D. R., 1965, The glacial history of western Washington and Oregon, *in* Wright, H. E., Jr., and Frey, D. G., eds., The Quaternary of the United States: Princeton, New Jersey, Princeton University Press, p. 341–353.

Curry, B. B., 1989, Absence of Altonian glaciation in Illinois: Quaternary Research, v. 31, p. 1–13.

Denton, G. H., and Stuiver, M., 1967, Late Pleistocene glacial stratigraphy and chronology, northeastern St. Elias Mountains, Yukon Territory, Canada: Geological Society of America Bulletin, v. 78, p. 485–510.

Easterbrook, D. J., 1969, Pleistocene chronology of the Puget Lowland and San Juan Islands, Washington: Geological Society of America Bulletin, v. 80, p. 2273–2286.

——, 1976, Quaternary geology of the Pacific Northwest, *in* Mahaney, W. C., ed., Quaternary stratigraphy of North America: Stroudsburg, Pennsylvania, Dowden, Hutchinson & Ross, p. 441–462.

——, 1986, Stratigraphy and chronology of Quaternary deposits of the Puget Lowland and Olympic Mountains of Washington and Oregon, *in* Šibrava, V., Bowen, D. Q., and Richmond, G. M., eds., Quaternary glaciations in the Northern Hemisphere: Quaternary Science Reviews, v. 5, p. 145–159.

Easterbrook, D. J., and Othberg, K., 1976, Paleomagnetism of Pleistocene sediments in the Puget Lowland, Washington, *in* Easterbrook, D. J., and Šibrava, V., eds., Quaternary glaciations in the Northern Hemisphere: IUGS-Unesco International Geological Correlation Program, Project 73-1-24, no. 3, p. 189–207.

Easterbrook, D. J., Crandell, D. R., and Leopold, E. B., 1967, Pre-Olympia Pleistocene stratigraphy and chronology in the central Puget Lowland, Washington: Geological Society of America Bulletin, v. 78, p. 13–20.

Easterbrook, D. J., Roland, J. L., Carson, R. J., and Naeser, N. D., 1988, Application of paleomagnetism, fission-track dating, and tephra correlation to lower Pleistocene sediments in the Puget Lowland, Washington, *in* Easterbrook, D. J., ed., Dating Quaternary sediments: Geological Society of America Special Paper 227, p. 139–165.

Flint, R. F., 1971, Glacial and Quaternary geology: New York, John Wiley & Sons, 892 p.

Foscolos, A. E., Rutter, N. W., and Hughes, O. L., 1977, The use of pedological studies in interpreting the Quaternary history of central Yukon Territory: Geological Survey of Canada Bulletin 271, 48 p.

Fulton, R. J., 1971, Radiocarbon geochronology of southern British Columbia: Geological Survey of Canada Paper 71-37, 28 p.

——, 1984, Quaternary glaciation, Canadian Cordillera, *in* Fulton, R. J., ed., Quaternary stratigraphy of Canada—A Canadian contribution to IGCP Project 24: Geological Survey of Canada Paper 84-10, p. 39–48.

Fulton, R. J., and Smith, G. W., 1978, Late Pleistocene stratigraphy of south-central British Columbia: Canadian Journal of Earth Sciences, v. 15, p. 971–980.

Fyles, J. G., 1963, Surficial geology of Horne Lake and Parksville map-areas, Vancouver Island, British Columbia: Geological Survey of Canada Memoir 318, 142 p.

Hamilton, T. D., and Bischoff, J. L., 1984, Uranium-series dating of fossil bones from the Canyon Creek vertebrate locality in central Alaska, *in* Reed, K. M., and Bartsch-Winkler, S., eds., The United States Geological Survey in Alaska: Accomplishments during 1982: U.S. Geological Survey Circular 939, p. 26–29.

Hamilton, T. D., and Thorson, R. M., 1983, The Cordilleran ice sheet in Alaska, *in* Porter, S. C., ed., Late Quaternary environments of the United States, Volume 1, the late Pleistocene: Minneapolis, University of Minnesota Press, p. 38–52.

Hansen, B. S., and Easterbrook, D. J., 1974, Stratigraphy and palynology of late Quaternary sediments in the Puget Lowland, Washington: Geological Society of America Bulletin, v. 85, p. 587–602.

Hansen, H. P., and Mackin, J. H., 1949, A pre-Wisconsin forest succession in the Puget Lowland, Washington: American Journal of Science, v. 247, p. 833–855.

Heusser, C. J., and Heusser, L. E., 1981, Palynology and paleotemperature analysis of the Whidbey Formation, Puget Lowland, Washington: Canadian Journal of Earth Sciences, v. 18, p. 136–149.

Hicock, S. R., 1980, Pre-Fraser Pleistocene stratigraphy, geochronology, and paleoecology of the Georgia depression, British Columbia [Ph.D. thesis]: London, University of Western Ontario, 230 p.

——, 1990, Last interglacial Muir Point Formation, Vancouver Island, British Columbia: Géographie physique et Quaternaire, v. 44, p. 337–340.

Hicock, S. R., and Armstrong, J. E., 1983, Four Pleistocene formations in southwest British Columbia: Their implications for patterns of sedimentation of possible Sangamonian to early Wisconsinan age: Canadian Journal of Earth Sciences, v. 20, p. 1232–1247.

Hicock, S. R., and Rutter, N. W., 1986, Pleistocene aminostratigraphy of the Georgia depression, southwest British Columbia: Canadian Journal of Earth Sciences, v. 23, p. 383–392.

Howes, D. E., 1981, Late Quaternary sediments and geomorphic history of north-central Vancouver Island: Canadian Journal of Earth Sciences, v. 18, p. 1–12.

——, 1983, Late Quaternary sediments and geomorphic history of northern Vancouver Island, British Columbia: Canadian Journal of Earth Sciences, v. 20, p. 57–65.

Hughes, O. L., 1972, Surficial geology of northern Yukon Territory and northwestern District of Mackenzie, Northwest Territories: Geological Survey of Canada Paper 69-36, 11 p.

Hughes, O. L., Campbell, R. B., Muller, J. E., and Wheeler, J. O., 1969, Glacial limits and flow patterns, Yukon Territory, south of 65 degrees north latitude: Geological Survey of Canada Paper 68-34, 9 p.

Hughes, O. L., Rampton, V. N., and Rutter, N. W., 1972, Quaternary geology and geomorphology, southern and central Yukon (northern Canada): International Geological Congress, 24th, Guidebook, Field Excursion A11, 59 p.

Hughes, O. L., Harington, C. R., Schweger, C., and Matthews, J. V., Jr., 1987, Stop 15: Ash Bend Section, *in* Morison, S. R., and Smith, C.A.S., eds., Guidebook to Quaternary research in Yukon: International Union for Quaternary Research Congress, 12th, Guidebook, Field Excursions A20a and A20b, p. 50–53.

Hughes, O. L., Rutter, N. W., and Clague, J. J., 1989, Yukon Territory (Quaternary stratigraphy and history, Cordilleran Ice Sheet), *in* Fulton, R. J., ed., Quaternary geology of Canada and Greenland: Geological Survey of Canada, Geology of Canada, no. 1, p. 58–62.

Johnson, R. G., 1982, Matuyama-Brunhes polarity reversal dated at 790 000 yr B.P. by marine-astronomical correlations: Quaternary Research, v. 17, p. 135–147.

Jopling, A. V., Irving, W. N., and Beebe, B. F., 1981, Stratigraphic, sedimentological and faunal evidence for the occurrence of pre-Sangamonian artefacts in northern Yukon: Arctic, v. 34, p. 3–33.

Klassen, R. W., 1987, The Tertiary-Pleistocene stratigraphy of the Liard Plain, southeastern Yukon Territory: Geological Survey of Canada Paper 86-17, 16 p.

Kvenvolden, K. A., Blunt, D. J., McMeanmin, M. A., and Straham, S. E., 1980, Geochemistry of amino acides in shells of the clam *Saxidomus, in* Douglas, A. G., and Maxwell, J. R., eds., Advances in organic geochemistry 1979: Physics and Chemistry of the Earth, v. 12, p. 321–332.

Lea, P. D., 1983, Glacial history of the southern margin of the Puget Lowland, Washington: Geological Society of America Abstracts with Programs, v. 15, p. 430.

——, 1984, Pleistocene glaciation of the southern margin of the Puget lobe, western Washington [M.S. thesis]: Seattle, University of Washington, 96 p.

Lichti-Federovich, S., 1973, Palynology of six sections of late Quaternary sediments from the Old Crow River, Yukon Territory: Canadian Journal of Botany, v. 51, p. 553–564.

——, 1974, Palynology of two sections of late Quaternary sediments from the Porcupine River, Yukon Territory: Geological Survey of Canada Paper 74-23, 6 p.

Martinson, D. G., Pisias, N. G., Hays, J. D., Imbrie, J., Moore, T. C., Jr., and Shackleton, N. J., 1987, Age dating and orbital theory of the ice ages: Development of a high-resolution 0 to 300,000-year chronostratigraphy: Quaternary Research, v. 27, p. 1–29.

Matthews, J. V., Jr., Harington, C. R., Hughes, O. L., Morlan, R. E., Rutter, N. W., Schweger, C. E., and Tarnocai, C., 1987, Stop 28: Schaeffer Mountain lookout and Old Crow Basin stratigraphy/paleontology, *in* Morison, S. R., and Smith, C.A.S., eds., Guidebook to Quaternary research in Yukon: International Union for Quaternary Research Congress, 12th, Guidebook, Field Excursions A20a and A20b, p. 75–83.

Matthews, J. V., Jr., Schweger, C. E., and Hughes, O. L., 1990a, Plant and insect fossils from the Mayo Indian Village Section (central Yukon): New data on middle Wisconsinan environments and glaciation: Géographie physique et Quaternaire, v. 44, p. 15–26.

Matthews, J. V., Jr., Schweger, C. E., and Janssens, J. A., 1990b, The last (Koy-Yukon) interglaciation in the northern Yukon: Evidence from Unit 4 at Ch'ijee's Bluff, Bluefish Basin: Géographie physique et Quaternaire, v. 44, p. 341–362.

McCourt, G. H., 1982, Quaternary palynology of the Bluefish Basin, northern Yukon Territory [M.S. thesis]: Edmonton, University of Alberta, 178 p.

Morlan, R. E., 1980, Taphonomy and archaeology in the upper Pleistocene of the northern Yukon Territory: A glimpse of the peopling of the New World: National Museums of Canada, National Museum of Man Mercury Series, Archaeological Survey of Canada Paper 94, 380 p.

——, 1986, Pleistocene archaeology in Old Crow Basin: A critical reappraisal, *in* Bryan, A. L., ed., New evidence for the Pleistocene peopling of the Americas: Orono, University of Maine, Center for the Study of Early Man, p. 27–48.

Morlan, R. E., and Matthews, J. V., Jr., 1983, Taphonomy and paleoecology of fossil insect assemblages from Old Crow River (CRH-15) northern Yukon Territory, Canada: Géographie physique et Quaternaire, v. 37, p. 147–157.

Pearce, G. W., Westgate, J. A., and Robertson, S., 1982, Magnetic reversal history of Pleistocene sediments at Old Crow, northwestern Yukon Territory: Canadian Journal of Earth Sciences, v. 19, p. 919–929.

Porsild, A. E., and Cody, W. J., 1980, Vascular plants of continental Northwest Territories, Canada: Ottawa, National Museums of Canada, 667 p.

Rampton, V. N., 1971, Late Pleistocene glaciations of the Snag-Klutlan area, Yukon Territory: Arctic, v. 24, p. 277–300.

Richmond, G. M., Fryxell, R., Neff, G. E., and Weis, P. L., 1965, The Cordilleran Ice Sheet of the northern Rocky Mountains, and related Quaternary history of the Columbia Plateau, *in* Wright, H. E., Jr., and Frey, D. G., eds., The Quaternary of the United States: Princeton, New Jersey, Princeton University Press, p. 231–242.

Rutter, N. W., 1976, Multiple glaciation in the Canadian Rocky Mountains with special emphasis on northeastern British Columbia, *in* Mahaney, W. C., ed., Quaternary stratigraphy of North America: Stroudsburg, Pennsylvania, Dowden, Hutchinson & Ross, p. 409–440.

——, 1977, Multiple glaciation in the area of Williston Lake, British Columbia: Geological Survey of Canada Bulletin 273, 31 p.

Rutter, N. W., Foscolos, A. E., and Hughes, O. L., 1978, Climatic trends during the Quaternary in central Yukon based upon pedological and geomorphological evidence, *in* Mahaney, W. C., ed., Quaternary soils: Norwich, Geo Abstracts, p. 309–359.

Ryder, J. M., 1976, Terrain inventory and Quaternary geology, Ashcroft, British Columbia: Geological Survey of Canada Paper 74-49, 17 p.

——, 1981, Terrain inventory and Quaternary geology, Lytton, British Columbia: Geological Survey of Canada Paper 79-25, 20 p.

Ryder, J. M., and Clague, J. J., 1989, British Columbia (Quaternary stratigraphy and history), *in* Fulton, R. J., ed., Quaternary geology of Canada and Greenland: Geological Survey of Canada, Geology of Canada, no. 1, p. 48–58.

Schweger, C. E., and Janssens, J.A.P., 1980, Paleoecology of the Boutellier nonglacial interval, St. Elias Mountains, Yukon Territory, Canada: Arctic and Alpine Research, v. 12, p. 309–317.

Schweger, C. E., and Matthews, J. V., Jr., 1985, Early and middle Wisconsinan environments of eastern Beringia: Stratigraphic and paleoecological implications of the Old Crow tephra: Géographie physique et Quaternaire, v. 39, p. 275–290.

Shackleton, J., 1982, Environmental histories from Whitefish and Imuruk lakes, Seward Peninsula, Alaska: Ohio State University, Institute of Polar Studies Report no. 76, 49 p.

Smith, C.A.S., Tarnocai, C., and Hughes, O. L., 1986, Pedological investigations of Pleistocene glacial drift surfaces in the central Yukon: Géographie physique et Quaternaire, v. 40, p. 29–37.

Smith, J. D., and Foster, J. H., 1969, Geomagnetic reversal in Brunhes Normal polarity epoch: Science, v. 163, p. 565–567.

Tarnocai, C., Smith, S., and Hughes, O. L., 1985, Soil development on Quaternary deposits of various ages in the central Yukon Territory: Geological Survey of Canada Paper 85-1A, p. 229–238.

Thorson, R. M., and Dixon, E. J., Jr., 1983, Alluvial history of the Porcupine River, Alaska: Role of glacial-lake overflow from northwest Canada: Geological Society of America Bulletin, v. 94, p. 576–589.

Verosub, K. L., 1982, Geomagnetic excursions: A critical assessment of the evidence as recorded in sediments of the Brunhes Epoch: Royal Society of London Philosophical Transactions, ser. A, v. 306, p. 161–168.

Waitt, R. B., Jr., and Thorson, R. M., 1983, The Cordilleran ice sheet in Washington, Idaho, and Montana, *in* Porter, S. C., ed., Late-Quaternary environments of the United States, Volume 1, the late Pleistocene: Minneapolis, University of Minnesota Press, p. 53–70.

Westgate, J. A., 1988, Isothermal plateau fission-track age of the late Pleistocene Old Crow tephra, Alaska: Geophysical Research Letters, v. 15, p. 376–379.

Westgate, J. A., Walter, R. C., Pearce, G. W., and Gorton, M. P., 1985, Distribution, stratigraphy, petrochemistry, and palaeomagnetism of late Pleistocene Old Crow tephra in Alaska and the Yukon: Canadian Journal of Earth Sciences, v. 22, p. 893–906.

Willis, B., 1898, Drift phenomena of Puget Sound: Geological Society of America Bulletin, v. 9, p. 111–162.

Wintle, A. G., and Westgate, J. A., 1986, Thermoluminescence age of Old Crow tephra in Alaska: Geology, v. 14, p. 594–597.

Manuscript Accepted by the Society September 6, 1991

Printed in U.S.A.

Geological Society of America
Special Paper 270
1992

Varied records of early Wisconsinan alpine glaciation in the western United States derived from weathering-rind thicknesses

Steven M. Colman
U.S. Geological Survey, Woods Hole, Massachusetts 02543
Kenneth L. Pierce
U.S. Geological Survey, Box 25046, MS 913, Denver Federal Center, Denver, Colorado 80225

ABSTRACT

Weathering-rind thicknesses were measured on volcanic clasts in sequences of glacial deposits in seven mountain ranges in the western United States and in the Puget lowland. Because the rate of rind development decreases with time, ratios of rind thicknesses provide limits on corresponding age ratios. In all areas studied, deposits of late Wisconsinan age are obvious; deposits of late Illinoian age (ca. 140 ka) also seem to be present in each area, although independent evidence for their numerical age is circumstantial. The weathering-rind data indicate that deposits that have intermediate ages between these two are common, and ratios of rind thicknesses suggest an early Wisconsinan age (about 60 to 70 ka) for some of the intermediate deposits. Three of the seven studied alpine areas (McCall, Idaho; Yakima Valley, Washington; and Lassen Peak, California) appear to have early Wisconsinan drift beyond the extent of late Wisconsinan ice. In addition, Mount Rainier and the Puget lowland, Washington, have outwash terraces but no moraines of early Wisconsinan age. The sequences near West Yellowstone, Montana; Truckee, California; and in the southern Olympic Mountains have no recognized moraines or outwash of this age. Many of the areas have deposits that may be of middle Wisconsinan age.

Differences in the relative extents of early Wisconsinan alpine glaciers are not expected from the marine oxygen-isotope record and are not explained by any simple trend in climatic variables or proximity to oceanic moisture sources. However, alpine glaciers could have responded more quickly and more variably than continental ice sheets to intense, short-lived climatic events, and they may have been influenced by local climatic or hypsometric effects. The relative sizes of early and late Wisconsinan alpine glaciers could also reflect differences between early and late Wisconsinan continental ice sheets and their regional climatic effects.

INTRODUCTION

Glacial deposits of early Wisconsinan age were once thought to be common beyond the limit of late Wisconsinan ice throughout North America (Flint, 1971, p. 520; Birkeland and others, 1971). As Quaternary dating methods have evolved over the past 20 years, many earlier correlations have been revised as age estimates based on relative dating methods have been replaced by those based on numerical or calibrated dating methods. In addition, the development of the marine oxygen-isotope record, a proxy measure of world-wide ice volume, has provided a firm framework for correlation of at least the major, first-order glaciations (Fullerton and Richmond, 1986). In general, the oxygen-isotope record suggests that glaciation during isotope stage 4 (early Wisconsinan) was somewhat less voluminous than that in either stage 2 (late Wisconsinan) or stage 6 (late Illinoian). As a result of these advances in dating and correlation, many deposits that were once considered early Wisconsinan have been

Colman, S. M., and Pierce, K. L., 1992, Varied records of early Wisconsinan alpine glaciation in the western United States derived from weathering-rind thicknesses, *in* Clark, P. U., and Lea, P. D., eds., The Last Interglacial-Glacial Transition in North America: Boulder, Colorado, Geological Society of America Special Paper 270.

reassigned to either younger (late Wisconsinan) or older (pre-Wisconsinan) glacial stages. Age estimates for glacial deposits in most areas of North America have followed this pattern, and many deposits that were once considered early Wisconsinan in age have been reassigned, mostly to pre-Wisconsinan ages (Table 1). Most of the deposits that remain assigned to the early Wisconsinan are less extensive than local late Wisconsinan deposits and are dated only by minimum-limiting radiocarbon ages (Richmond and Fullerton, 1986).

In the western United States, recent dating studies have tended to place deposits that were once considered early Wisconsinan into pre-Wisconsinan time (reviews in Richmond, 1986a, 1986b; Easterbrook, 1986; Fullerton, 1986). In general, we support this trend and have made several contributions to it (Table 2; Pierce and others, 1976; Pierce, 1979; Colman and Pierce, 1981, 1984). In contrast to recent trends, Richmond (1986a) suggested that, although some Bull Lake moraines in the Rocky Mountains have been dated to Illinoian time, others, especially those in the inner parts of Bull Lake moraine belts and their correlatives, may be early Wisconsinan in age.

In addition to these direct age estimates, other studies pertain to the relative extent of early Wisconsinan alpine glaciers. For example, Marchand and Allwardt (1981) inferred extensive early Wisconsinan glaciation in the Sierra Nevada on the basis of alluvial-fan stratigraphy in the northern San Joaquin Valley. In contrast, Atwater and others (1986) proposed relatively meager early Wisconsinan glaciation in that range, on the basis of indirect evidence for the volume of outwash deposited in the southern San Joaquin Valley.

Evidence for the relative extent of alpine glaciers is ordinarily preserved only in their terminal areas, and preservation of the evidence in turn depends on the sequence of glacial advances (Gibbons and others, 1984). An alpine glacier slightly more extensive than its predecessor commonly erodes or buries most or all of the evidence related to that predecessor. Thus, a late Wisconsinan valley glacier just slightly longer than its early Wisconsinan counterpart may override and bury all moraines of early Wisconsinan age. Alpine glaciers undoubtedly existed in many areas during early Wisconsinan time; features such as ice-contact lava flows in Yellowstone National Park (Richmond, 1986b) record evidence of those glaciers. However, except where deposits of those glaciers are preserved in the terminal areas, little or no direct record of their extent is normally preserved.

Throughout this chapter we refer to the marine oxygen-isotope record and its connection to continental and alpine glaciation (Mix, 1987). We use the composite isotope curves of Hays and others (1976) and Imbrie and others (1984) and the SPECMAP time scale (Imbrie and others, 1984; Martinson and others, 1987). We use the following generally accepted correlations between isotope stages and glaciations: 2—late Wisconsinan, 4—early Wisconsinan, 5 (or 5e)—Sangamonian (last interglaciation), and 6—late Illinoian.

The issue we examine here is the extent of early Wisconsinan glaciers relative to their late Wisconsinan and pre-

TABLE 1. SOME ALPINE GLACIAL DEPOSITS HISTORICALLY ASSIGNED AN EARLY WISCONSINAN AGE IN THE WESTERN UNITED STATES

Area	Deposit*	Present Assignment	Comments and References
W. Yellowstone	Bull Lake	Pre-Wisconsinan	Obsidian-hydration, ^{14}C, and K-Ar dating; Pierce and others (1976); Richmond (1986b)
McCall	Bull Lake	Pre-Wisconsinan	Multiple relative-age methods, correlation to W. Yellowstone Bull Lake; Colman and Pierce (1986); Richmond (1986a)
Yakima Valley	....	....	
Southern Olympic Mountains	Mobray	Pre-Wisconsinan	Weathering-rinds, stratigraphic relation to McCleary drift in Puget lowland; Colman and Pierce, unpublished data
Puget lowland†	Salmon Springs	Pre-Wisconsinan	Type section >1 m.y. by tephra correlation, surface drift (informally named McCleary) older than last interglaciation; Colman and Pierce (1984); Easterbrook (1986).
Lassen Peak	Tahoe	Early Wisconsinan	Multiple relative-age methods; Colman and Pierce (1981)
Truckee	Tahoe	Late Wisconsinan	Multiple relative-age methods, probably not typical of Tahoe in other parts of the Sierra Nevada; Colman and Pierce (1981); Fullerton (1986)

*See Colman and Pierce (1981) for references to historical age assignments.
†Deposits in the Puget lowland were deposited by the Puget Lobe of the Cordilleran ice sheet.

Wisconsinan counterparts in alpine areas of the western United States. Our identification of early Wisconsinan deposits is based primarily on studies of weathering rinds on volcanic clasts in these deposits and on related soil and stratigraphic studies. Many of these deposits are younger than deposits that were once considered early Wisconsinan in age, but are now thought to be pre-Wisconsinan. Although weathering rinds can be used to estimate numerical ages, such estimates involve several assumptions and uncertainties. By using weathering rinds simply as a high-resolution relative-age method, we will eliminate most of the assumptions and demonstrate that major age differences exist among Wisconsinan alpine glacial deposits, and that at least some of these deposits are probably early Wisconsinan in age. Major differences exist among the sequences of ages of Wisconsinan glacial deposits in alpine areas of the western United States, and we will speculate on the reasons for those differences.

METHODS AND AGE ESTIMATES

The thicknesses of weathering rinds on clasts in glacial deposits provides a simple, reproducible, widely applicable measure of weathering that can be used to estimate numerical ages (Colman and Pierce, 1981). This method has the advantage of being more widely applicable to alpine glacial deposits than traditional numerical-age methods and has quantitative and statistical advantages over most other relative-age methods.

The use of weathering rinds to estimate ages has been discussed in detail (Colman and Pierce, 1981, 1984, 1986), and only its salient aspects will be mentioned here. We measured the thickness of weathering rinds on basaltic and andesitic clasts in many glaciated areas of the western United States, concentrating on eight previously mapped areas. These areas and the deposits within them that were previously considered early Wisconsinan in age are listed in Table 1.

All rinds were measured in the field on clasts sampled from natural exposures, roadcuts, or soil pits. Clasts were taken from depths of 20 to 50 cm in the soil, from the upper part of the B horizon. The clasts were split and the rinds on 30 to 60 clasts of appropriate lithology were measured to the nearest 0.1 to 0.2 mm with a six-power comparator. Sampling sites were chosen on broad moraine crests or flat terrace surfaces.

A variety of factors, in addition to time, affect weathering-rind thickness; lithology and precipitation are particularly important (Colman and Pierce, 1981). However, we attempted to isolate time as a variable and to minimize variation in other factors. Within each area, rinds were measured only on specific types of basalt and andesite (Colman and Pierce, 1981) and sampling was confined to stable sites within restricted geographic areas in order to minimize local topographic and climatic effects. We used nested sampling designs with three or four levels (measurements, sites, landforms, and ages) to study sources of variation in each study area.

The original outer surfaces of the measured subsurface clasts appear to be preserved and we observed no evidence of the

TABLE 2. RIND THICKNESS (MEAN ± STANDARD DEVIATION) AND RATIOS FOR DEPOSITS OF WISCONSINAN AND LATE ILLINOIAN AGE*

Area (*Deposits*)	Late Wisconsinan Rind Thickness (mm)	Rind Ratio	Intermediate Rind Thickness (mm)	Rind Ratio	Pre-Wisconsinan Rind Thickness (mm)
West Yellowstone (WY) *(Deckard Flats, Pinedale, Bull Lake)*	0.10 ± 0.07	4.0	0.40 ± 0.22	0.51	0.78 ± 0.19
McCall, Idaho (MI) *(McCall, Williams Creek, Timber Ridge)*	0.35 ± 0.16	2.4	0.85 ± 0.24	0.53	1.61 ± 0.41
Yakima Valley (YV) *(Domerie, Ronald, Bullfrog, Indian John)*	0.25 ± 0.04 0.25 ± 0.04	2.1 2.8	0.52 ± 0.06 0.71 ± 0.12	0.50 0.68	1.05 ± 0.17 1.05 ± 0.17
Southern Olympic Mountains (OM) *(Inner Grisdale, outer Grisdale, Mobray)*	0.36 ± 0.24	2.6	0.94 ± 0.29	0.34	2.80 ± 0.66
Puget lowland (PL) *(Vashon, high Fraser Terrace, McCleary)*	0.22 ± 0.15	3.2	0.71 ± 0.30	0.70	1.01 ± 0.40
Mount Rainier (MR) *(Evans Creek, ..., Hayden Creek)*	0.19 ± 0.13	....	....	....	1.37 ± 0.43
Lassen Peak (LP) *(Tioga, early Tioga, Tahoe, pre-Tahoe)*	0.17 ± 0.09 0.17 ± 0.09	1.9 4.2	0.33 ± 0.15 0.72 ± 0.23	0.31 0.68	1.06 ± 0.31 1.06 ± 0.31
Truckee, California (TC) *(Tioga, ..., Donner Lake)*	0.16 ± 0.10	....	....	....	0.93 ± 0.24

*Data from Colman and Pierce (1981) and Colman and Pierce (unpublished). Map locations shown in Figure 5.

physical removal or destruction of the subsurface rinds. The rinds began to form shortly after deposition of the clasts, so the duration of weathering is nearly the same as the age of the deposit (Colman and Pierce, 1981). Clasts with asymmetric, inherited rinds are rare and few clasts in unweathered tills have rinds.

Weathering-rind thicknesses are effective relative-age indicators: they clearly differentiate deposits of different mapped ages (Fig. 1). Statistical analyses of the data show that deposit age is by far the most important variable controlling rind thickness within each sampling area (Colman and Pierce, 1981). For each area, mean rind thicknesses are significantly different among deposits of different mapped ages; in contrast, mean rind thicknesses on clasts from different landforms (moraines, terraces) within a given mapped age are not significantly different (Colman and Pierce, 1981).

All measurements for each mapped age of deposit within each area were combined to obtain mean rind thicknesses and standard deviations for each deposit. Data for selected deposits in each of the eight areas are given in Table 2. The complete data set of mean rind thicknesses formed the basis for our attempt to determine the relation between rind thickness and time. Unfortunately, because of differences in climate and lithology, such relations must be empirically determined for each area, and independent ages for calibrating these relations are rare. Only one of the areas studied, West Yellowstone, has sufficient independent age control to permit construction of a curve of rind thickness versus time. Near West Yellowstone, where the deposits have been dated by combined K-Ar, obsidian-hydration, and radiocarbon methods (Pierce and others, 1976; Pierce, 1979), we fit a logarithmic time function to the rind-thickness data (Fig. 2). Independently dated rind-thickness curves also have been derived for graywacke clasts in New Zealand (Chinn, 1981; Whitehouse and others, 1986; Knuepfer, 1988) and for basaltic clasts in Bohemia (Cernohouz and Solc, 1966). All of these studies clearly show that the rate of rind development decreases with time.

In each of the areas that lack independent numerical age controls, we correlated one deposit to the West Yellowstone Bull Lake on the basis of multiple independent relative-age data. The rind thickness for that deposit and the age of the West Yellowstone Bull Lake, about 140 to 150 ka (Pierce and others, 1976; Pierce, 1979; Richmond, 1986b), were used to calibrate the curve for that area (Fig. 3). The deposits used for calibration appear to be equivalent in age to late Illinoian drift in the midwestern United States and to deposits of marine oxygen-isotope stage 6 (Colman and Pierce, 1981). The deposits used for calibration all have soils, morphology, overlying loess and/or buried-soil stratigraphy, and other evidence that suggests that they are older than the last interglaciation, but younger than interglaciations that pre-

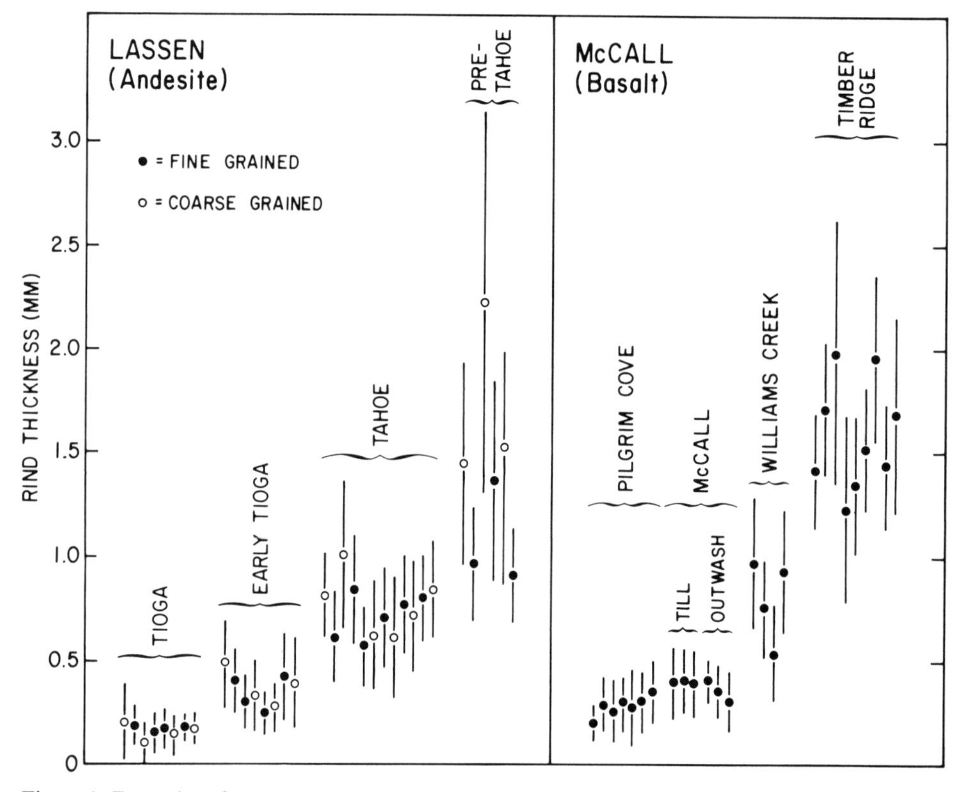

Figure 1. Examples of weathering-rind data for two of the study areas, Lassen Peak, California, and McCall, Idaho. Each point and bar represents the mean and standard deviation of all measurements at one sampling site. An early Wisconsinan age for Tahoe deposits at Lassen Peak and Williams Creek deposits at McCall is discussed in the text. Data from Colman and Pierce (1981).

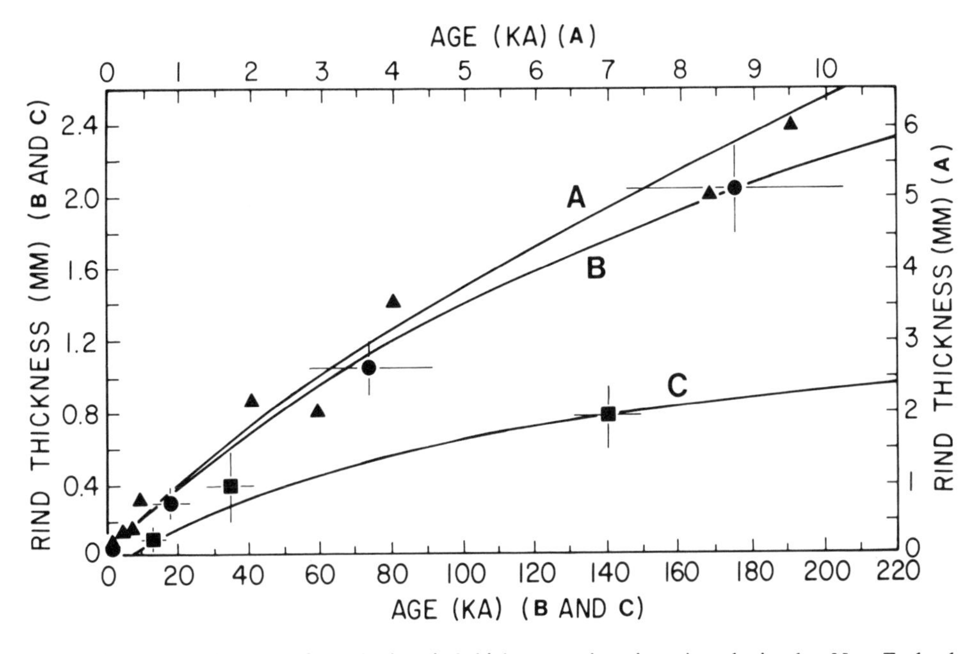

Figure 2. Calibrated curves of weathering-rind thickness against time. A and triangles, New Zealand (Chinn, 1981); B and circles, Bohemia (Cernohouz and Solc, 1966); C and squares, West Yellowstone, Montana (Colman and Pierce, 1981). From Colman (1981).

ceded the last (Colman and Pierce, 1981). Although the correlations are speculative, of the various possibilities, they seem to be the most consistent with the available data.

To construct rind-thickness vs. age curves for each of the study areas, we used the West Yellowstone curve and multiplied it by a constant for each area so that the curve passed through the 140 ka calibration point for that area (Fig. 3). The constant for each area is different because of rate differences resulting from variation in climate and lithology among the areas. The ages of other deposits in each area were determined by projecting the mean rind thicknesses for those deposits onto the appropriate curve. An age intercept of about 7 ka was used to account for the time necessary for weathering to reach the soil depths sampled (Figs. 2 and 3; Colman and Pierce, 1981).

The age estimates in Figure 3 involve a number of arguable assumptions (Colman and Pierce, 1981). Although they are useful as age hypotheses, they are *not* the basis for our arguments here. Nonetheless, these data suggest a complicated history of Wisconsinan glaciation in alpine areas in the western United States. Several age estimates for these glacial deposits fall in or close to the time of marine oxygen-isotope stage 4, whose boundaries are at about 59 and 71–74 ka (Imbrie and others, 1984; Martinson and others, 1987).

Several other aspects of the stratigraphies in our study areas are pertinent to the question of the extent of early Wisconsinan glaciation. One of the early Wisconsinan age estimates (Fig. 3) is represented by an outwash terrace in the Puget lowland that has no morainal counterpart, suggesting that early Wisconsinan ice, although clearly less extensive than late Wisconsinan ice, was at least well into the Puget lowland. In addition to the deposits represented in Figure 3, several outwash terraces occur along the Cowlitz River on the flank of Mount Rainier between the ones related to the pre-Wisconsinan Hayden Creek and the late Wisconsinan Evans Creek moraines. Thick loess on these terraces prevented the use of weathering-rind measurements to estimate their ages, but their number and height, and the degree of soil development on them (Dethier, 1988) suggest that one or more may be related to extensive early Wisconsinan ice. Near Truckee, California, sparse rind-thickness data for an unnamed glacial deposit suggest that it may be intermediate in age between late Wisconsinan and late Illinoian (Colman and Pierce, 1981; Fullerton, 1986). We concentrate here on deposits for which we have abundant weathering-rind data, but we note that additional evidence supports the existence of early Wisconsinan alpine glaciers comparable in extent to those of late Wisconsinan age.

RIND-THICKNESS RATIOS

Several assumptions and uncertainties encumber the numerical age estimates for glacial deposits in the western United States derived by the methods discussed above. However, rind-thickness data can be used in other ways that involve considerably fewer assumptions about actual calibrated rind-thickness curves, while yielding quantitative age information. The methods we use here employ the ratio of the rind thicknesses of two deposits compared with the ratio of their ages.

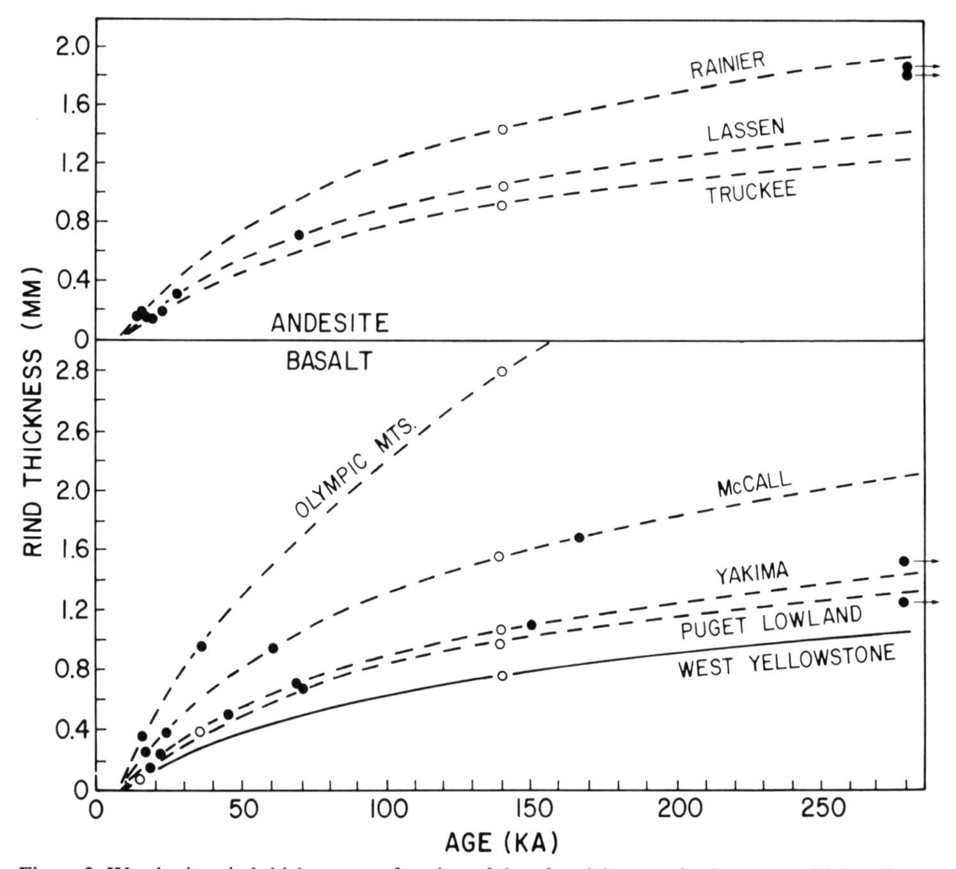

Figure 3. Weathering-rind thickness as a function of time for eight areas in the western United States. The curve for West Yellowstone appears as in Figure 2; the other curves are based on the form of the West Yellowstone curve calibrated by an inferred age (140–150 ka) of one deposit in each area (see text). Open circles plotted at 140 ka (except for West Yellowstone) are calibration points with inferred ages; other open circles are for dated deposits at West Yellowstone. Solid circles represent other glacial deposits and are plotted on the appropriate curve according to their mean rind thicknesses. Solid circles off the curves (with arrows, right) diagrammatically represent deposits that plot older than 300 ka on the closest curve. Modified from Colman and Pierce (1984).

Studies that have calibrated rates of rind development using independent numerical ages have employed different functions to model the rates, but all have shown that the rates decrease with time (Fig. 2; Colman, 1981; Colman and Dethier, 1986). Consequently, a linear relation between rind thickness and time is an absolute upper bound on the actual relation. Therefore, *the ratio of the ages of two deposits must be greater than the ratio of their rind thicknesses.* For example, if weathering rinds in one deposit are twice as thick as those in a second deposit, the first deposit *must* be more than twice as old as the second. We use this relation to place limits on possible early Wisconsinan deposits by comparing them to both older and younger deposits via rind ratios (Table 2). In the following discussion, we refer to deposits of possible middle or early Wisconsinan age as "intermediate." They are intermediate in age between the youngest comparable deposits in each area, which are readily identified as late Wisconsinan, and the next older comparable deposits, which we infer to be pre-Wisconsinan.

Several of the study areas contain two deposits of intermediate age (Table 2); of these, we consider the older of the two as a possible early Wisconsinan representative. The single intermediate deposits in two areas appear to be middle Wisconsinan in age: Pinedale deposits at West Yellowstone, which are dated at about 35 ka (Pierce and others, 1976), and outer Grisdale deposits in the southern Olympic Mountains, which have very low rind ratios (Table 2). The deposits that are candidates for an early Wisconsinan age are the Williams Creek at McCall, Idaho, the Bullfrog in the Yakima Valley, the high Fraser terrace in the Puget lowland, and the Tahoe at Lassen Peak.

Compared to late Wisconsinan deposits in each area, these four intermediate deposits have rind ratios of 2.4 to 4.2. Compared to the youngest pre-Wisconsinan deposits in each area, they have rind ratios of 0.53 to 0.70. Whatever the precise ages of the younger (late Wisconsinan) and older (pre-Wisconsinan) deposits, the intermediate deposits must be *at least* two to four times as old as the younger deposit, and *no more than* one-half to two-thirds as old as the next older deposit. These are strong constraints on the ages of the intermediate deposits and they require only two assumptions: (1) that variables other than time are relatively constant within each study area, and (2) that rind thickness increases systematically with time, at a rate that decreases with time. Our sampling procedures were designed to ensure that the first assumption is valid, and all available rate data suggest that the second assumption is true.

The youngest deposits of comparable extent to the intermediate deposits in each of the areas are clearly late Wisconsinan in age (Colman and Pierce, 1981, 1984). For any reasonable estimate of the age of the late Wisconsinan glacial maxima, a constraint of more than two to four times as old suggests that many of the intermediate deposits are early Wisconsinan in age. That weathering rinds within soil profiles begin to form a few thousand years after deposition (see the nonzero intercepts in Figs. 3 and 4; Colman and Pierce, 1981) argues for even greater age contrasts between the intermediate deposits and the late Wisconsinan deposits than the rind ratios imply.

The age of the next older glacial deposits in each area is more speculative, but we believe that they are late Illinoian in age, based on their correlation with the West Yellowstone Bull Lake drift, dated at about 140–150 ka (Pierce and others, 1976; Pierce, 1979), and on their soils, morphology, overlying loess and/or buried-soil stratigraphy, and other evidence that suggests that they predate the last interglaciation (Colman and Pierce, 1981). In addition, the oxygen-isotope record suggests that world-wide ice volumes during stage 6 were greater than those during either stage 2 or 4, so that representatives of stage 6 are likely to survive in moraine sequences. For any reasonable estimate of the age of the late Illinoian glacial maximum, a constraint of less than one-half to two-thirds as old (Table 2) suggests an early Wisconsinan age for the intermediate deposits. The rind-ratio constraints would allow an age within the time of oxygen-isotope stage 5, and the isotope record shows significant ice expansion during stages 5b and 5d. However, we consider an age equivalent to any part of stage 5 unlikely for deposits that indicate glaciation beyond the late Wisconsinan (stage 2) ice maximum.

These constraints are shown graphically in Figure 4, in which ages of 18 and 140 ka are assumed for late Wisconsinan and late Illinoian deposits, respectively. However, the conclusion that the intermediate deposits fit best in early Wisconsinan time is not sensitive to these exact ages. The same conclusion is reached for any reasonable age of deposits that correlate with oxygen-isotope stages 2 and 6. Only if the younger and older deposits are correlated with the wrong isotope stage entirely does this conclusion change.

Because late Wisconsinan deposits are readily identified and relatively well dated, the only realistic alternative to our interpretation of the early Wisconsinan ages of the intermediate deposits is that the older deposits are older than oxygen-isotope stage 6. This alternative has two objectionable consequences: (1) it places the older deposits, many of which were once considered early Wisconsinan (Table 1) and which are now thought to be late

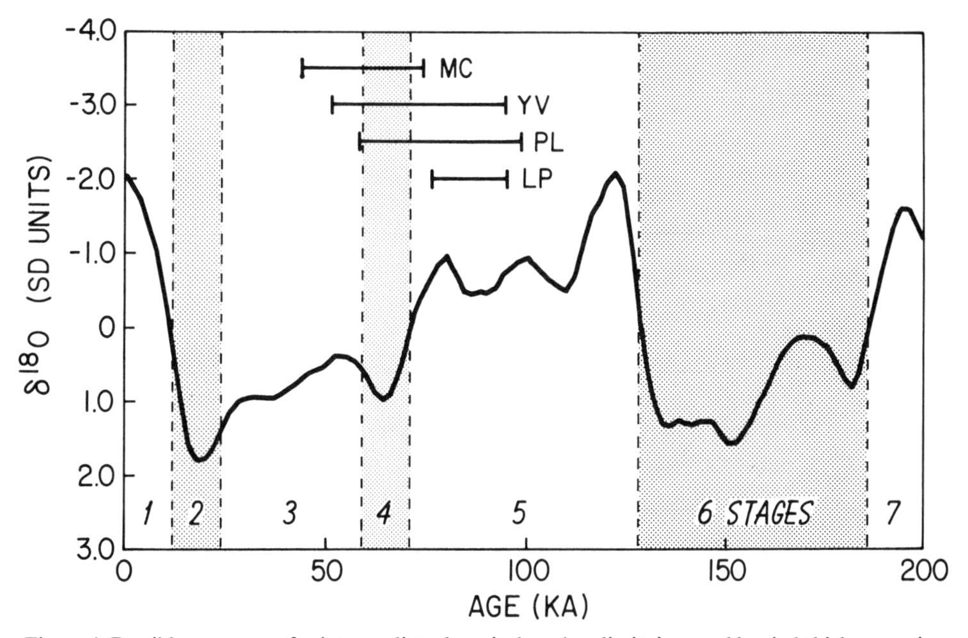

Figure 4. Possible age ranges for intermediate deposits based on limits imposed by rind-thickness ratios. Ages of 18 and 140 ka have been assumed for late Wisconsinan and late Illinoian deposits, respectively, in each study area. MC, Williams Creek deposits at McCall, Idaho; YV, Bullfrog deposits in Yakima Valley; PL, high Fraser Terrace in Puget lowland; LP, Tahoe at Lassen Peak. Oxygen-isotope data and stage boundaries from Imbrie and others (1984).

Illinoian (oxygen-isotope stage 6), into pre–stage 6 time; and (2) it makes the intermediate deposits the correlatives of oxygen-isotope stage 6, despite the fact that these deposits are found only in some areas and have relative-age data inconsistent with a pre-interglacial age.

In summary, we conclude that three of the seven alpine areas in the western United States that we examined contain early Wisconsinan till beyond the limit of late Wisconsinan ice. One additional alpine area (Mt. Rainier) and the Puget lowland contain outwash terraces, but no moraines of apparent early Wisconsinan age, suggesting that early Wisconsinan glaciers were comparable to, but somewhat less extensive than, their late Wisconsinan counterparts in these areas. In three alpine areas, no deposits of early Wisconsinan age have been clearly identified beyond the limit of late Wisconsinan ice. Even if we have miscorrelated deposits in one or two areas, our conclusion that early Wisconsinan glaciers were locally but variably extensive in alpine areas of the western United States remains unchanged.

SPECULATIONS ON THE REASONS FOR VARIATIONS IN THE EXTENT OF EARLY WISCONSINAN ALPINE GLACIERS

Weathering-rind data suggest that Wisconsinan glaciation in the western United States was complex. Early Wisconsinan glaciers in different areas varied in extent compared with their late Wisconsinan counterparts. The conclusion—that in about half of the areas studied, early Wisconsinan glaciers were more extensive than their late Wisconsinan counterparts—contrasts with both the marine oxygen-isotope record and with the stratigraphic record of continental ice sheets. The marine oxygen-isotope record suggests that world-wide ice volumes were less in stage 4 than in stage 2 (Imbrie and others, 1984; Martinson and others, 1987; Mix, 1990). The stratigraphic record of deposits beyond the limit of the late Wisconsinan Laurentide Ice Sheet shows that many of the deposits that were once assigned to the early Wisconsinan have been reassigned younger or older ages (Richmond and Fullerton, 1986); in many or most areas, the early Wisconsinan advance of continental ice appears to be less extensive than the late Wisconsinan advance (Richmond and Fullerton, 1986; Eyles and Westgate, 1987; Curry, 1989; Goldthwait, this volume; Oldale and Colman, this volume; Dreimanis, this volume; Eyles and Williams, this volume; Curry and Follmer, this volume).

Despite these indications that global ice volume was less during oxygen-isotope stage 4 than during stage 2, alpine glaciers could have responded more quickly than continental ice sheets to the climatic event associated with isotope stage 4. If glacial climatic conditions during stage 4 were as intense but of shorter duration relative to stage 2, or if the climatic conditions were more glacial in alpine areas than in the source areas of continental ice sheets, then early Wisconsinan alpine glaciers may have been more extensive, relative to their late Wisconsinan counterparts, than continental ice sheets. In addition to the smaller size and presumably faster response time of alpine glaciers, local threshold effects also may have dramatically affected the response of alpine glaciers. For example, some areas have hypsometric distributions such that a relatively small change in equilibrium-line altitude could place a large amount of terrain into the glacial source area, resulting in a disproportionate glacial response to a relatively small climatic change. Perhaps differences in size, response time, hypsometry, and local climatic conditions make correlation among deposits of alpine glaciers and between deposits of alpine glaciers and continental ice sheets more improbable than is commonly assumed.

The reasons for variations in extent among early Wisconsinan alpine glaciers in relation to their late Wisconsinan counterparts are not well understood. No simple trend in climatic variables or proximity to oceanic moisture sources is apparent in the geographic pattern of glacial extent, although the distribution of early Wisconsinan glaciers of similar relative extent could be related to distance from Pacific Ocean moisture and from the margins of the continental ice sheets (Fig. 5). Glaciation thresholds in the northwestern United States suggest that climatic gradients during the last glacial maxima were similar to those of today, and that the change in summer temperature was about 5.5 ± 1.5 °C (Porter, 1977; Porter and others, 1983), although Porter and others (1983) suggested a temperature depression of at least 10 °C for continental climates in the Rocky Mountains. Glaciation-threshold data from the Great Basin suggest a more complicated comparison between modern and Pleistocene climatic patterns (Zielinski and McCoy, 1987).

Recent computer simulations of climatic conditions at the height of the last glaciation suggest that the Laurentide Ice Sheet may have had a major effect on the atmospheric circulation patterns in the western United States (Kutzbach and Wright, 1985; Kutzbach and Guetter, 1986). Because the margins of the continental ice sheets in early Wisconsinan time were different than those in late Wisconsinan time, the effects on circulation patterns, especially storm tracks, may have been different. Specifically, the climatic modeling studies suggest that the jet stream split over the Laurentide Ice Sheet during the late Wisconsinan, producing an anticyclonic circulation cell over the northwestern United States, which resulted in easterly, rather than the present westerly, prevailing winds over parts of the western mountains. These patterns appear to be consistent with much of the vegetational data for the same period from the northwestern United States (Barnosky and others, 1987) and with geologic data from other areas (Kutzbach and Wright, 1985). The anticyclonic circulation cell over the northwestern United States is of regional scale and may have produced rather local effects. Differences in the position of the cell, or differences in its degree of development, between early Wisconsinan and late Wisconsinan time could have produced major differences in orographic effects or storm-track locations, thereby accounting for variability in the relative extent of Wisconsinan glaciers in the western United States. However, these climatic differences between early and late Wisconsinan time are speculative, because the configuration of early Wisconsinan Laurentide ice is not well known.

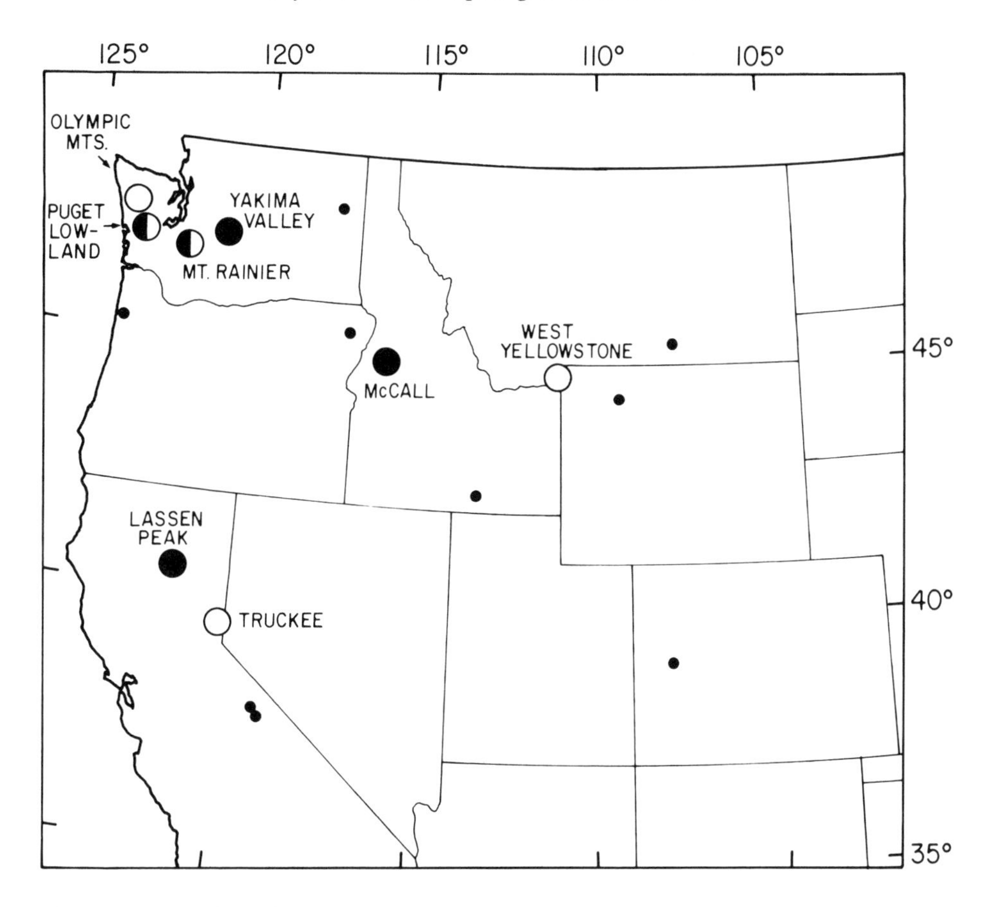

Figure 5. Geographic distribution of study areas with different sequences of Wisconsinan glacial deposits. Solid circles, early Wisconsinan moraines beyond the limit of late Wisconsinan ice; half-filled circles, early Wisconsinan outwash terrace but no moraine; open circle, no early Wisconsinan glacial deposits identified. Small circles represent other areas in which rinds were sampled by Colman and Pierce (1981).

If we are correct in our inference that the record of Wisconsinan alpine glaciation in the western United States is complex and varied, then the *lack* of correlation of preserved deposits among areas may be as important as correlation. It may be that the timings of major alpine glacial advances were correlative, but it appears that their relative extents, and thus the preservation of their age sequence of deposits, were different. General correlation trends, such as the depreciation of early Wisconsinan glaciation, are valid, but the more variable the response of glacial systems, the more likely it is that exceptions will exist. The geographic pattern of exceptions or variations, of which Figure 5 is an inadequate first attempt due to the small number of data points, may lead to important information about the dynamics of glacial climatic systems.

REFERENCES CITED

Atwater, B. F., Adam, D. P., Bradbury, J. P., Forester, R. M., Mark, R. K., and Lettis, W. R., 1986, A fan dam for Tulare Lake, California, and implications for the Wisconsin glacial history of the Sierra Nevada: Geological Society of America Bulletin, v. 97, p. 97–109.

Barnosky, C. W., Anderson, P. M., and Bartlein, P. J., 1987, The northwestern U.S. during deglaciation—Vegetation history and paleoclimatic implications, *in* Ruddiman, W. F., and Wright, H. E., Jr., eds., North America and adjacent oceans during the last deglaciation: Boulder, Colorado, Geological Society of America, Geology of North America, v. K-3, p. 289–321.

Birkeland, P. W., Richmond, G. M., and Crandell, D. R., 1971, Status of Quaternary stratigraphic units in the western conterminous United States: Quaternary Research, v. 1, p. 208–227.

Cernohouz, J., and Solc, I., 1966, Use of sandstone wanes and weathered basaltic crust in absolute chronology: Nature, v. 212, p. 806–807.

Chinn, T.J.H., 1981, Use of rock weathering rind thickness for Holocene absolute age-dating in New Zealand: Arctic and Alpine Research, v. 13, p. 33–45.

Colman, S. M., 1981, Rock-weathering rates as functions of time: Quaternary Research, v. 15, p. 250–264.

Colman, S. M., and Dethier, D. P., 1986, An overview of rates of chemical weathering, *in* Colman, S. M., and Dethier, D. P., eds., Rates of chemical weathering of rocks and minerals: New York, Academic Press, p. 1–20.

Colman, S. M., and Pierce, K. L., 1981, Weathering rinds on basaltic and andesitic stones as a Quaternary age indicator, western United States: U.S. Geological Survey Professional Paper 1210, 56 p.

——, 1984, Correlation of Quaternary glacial sequences in the western United States based on weathering rinds and related studies, *in* Mahaney, W. C., ed.,

Correlation of Quaternary sequences: Norwich, GeoBooks Ltd., p. 437–454.

——, 1986, The glacial sequence near McCall, Idaho: Weathering rinds, soil development, morphology, and other relative-age criteria: Quaternary Research, v. 25, p. 25–42.

Curry, B. B., 1989, Absence of Altonian glaciation in Illinois: Quaternary Research, v. 31, p. 1–13.

Dethier, D. P., 1988, The soil chronosequence along the Cowlitz River, Washington: U.S. Geological Survey Bulletin 1590-F, 47 p.

Easterbrook, D. J., 1986, Stratigraphy and chronology of Quaternary deposits in the Puget Lowland and Olympic Mountains of Washington and the Cascade Mountains of Washington and Oregon, *in* Sibrava, V., Bowen, D. Q., and Richmond, G. M., eds., Quaternary glaciations in the Northern Hemisphere: Quaternary Science Reviews, v. 5, p. 145–159.

Eyles, N. and Westgate, J. A., 1987, Restricted regional extent of the Laurentide Ice Sheet in the Great Lakes basins during early Wisconsin glaciation: Geology, v. 15, p. 537–540.

Flint, R. F., 1971, Glacial and Quaternary geology: New York, John Wiley and Sons, 892 p.

Fullerton, D. S., 1986, Chronology and correlation of glacial deposits in the Sierra Nevada, California, *in* Sibrava, V., Bowen, D. Q., and Richmond, G. M., eds., Quaternary glaciations in the Northern Hemisphere: Quaternary Science Reviews, v. 5, p. 161–169.

Fullerton, D. S., and Richmond, G. M., 1986, Comparison of the marine oxygen-isotope record, the eustatic sea-level record, and the chronology of glaciation in the United States of America, *in* Sibrava, V., Bowen, D. Q., and Richmond, G. M., eds., Quaternary glaciations in the Northern Hemisphere: Quaternary Science Reviews, v. 5, p. 197–200.

Gibbons, A. B., Megeath, J. D., and Pierce, K. L., 1984, Probability of moraine survival in a succession of glacial advances: Geology, v. 12, p. 327–330.

Hays, J. D., Imbrie, J., and Shackleton, N. J., 1976, Variations in the earth's orbit—Pacemaker of the ice ages: Science, v. 194, p. 1121–1132.

Imbrie, J., and eight others, 1984, The orbital theory of Pleistocene climate—Support from a revised chronology of the marine $\delta^{18}O$ record, *in* Berger, A., Imbrie, J., Hays, J., Kukla, G., and Saltzman, B., eds., Milankovitch and climate: Boston, D. Reidel Publishing Co., p. 269–305.

Knuepfer, P.L.K., 1988, Estimating ages of late Quaternary stream terraces from analysis of weathering rinds and soils: Geological Society of America Bulletin, v. 100, p. 1224–1236.

Kutzbach, J. E., and Guetter, P. J., 1986, The influence of changing orbital parameters and surface boundary conditions on climate simulations for the past 18,000 years: Journal of Atmospheric Sciences, v. 43, p. 1726–1759.

Kutzbach, J. E., and Wright, H. E., Jr., 1985, Simulation of the climate 18,000 years B.P.—Results for the North America/North Atlantic/European sector and comparison with the geologic record of North America: Quaternary Science Reviews, v. 4, p. 147–187.

Marchand, D. E., and Allwardt, A., 1981, Late Cenozoic stratigraphic units, northeastern San Joaquin Valley, California: U.S. Geological Survey Bulletin 1470, 70 p.

Martinson, D. G., Pisias, N. G., Hays, J. D., Imbrie, J., Moore, T. C., Jr., and Shackleton, N. J., 1987, Age dating and the orbital theory of the ice ages—Development of a high-resolution 0 to 300,000-year chronostratigraphy: Quaternary Research, v. 27, p. 1–29.

Mix, A. C., 1987, The oxygen isotope record of glaciation, *in* Ruddiman, W. F., and Wright, H. E., Jr., eds., North America and adjacent oceans during the last deglaciation: Boulder, Colorado, Geological Society of America, Geology of North America, v. K-3, p. 111–135.

Pierce, K. L., 1979, History and dynamics of glaciation in the northern Yellowstone Park area: U.S. Geological Survey Professional Paper 729-F, 90 p.

Pierce, K. L., Obradovich, J. D., and Friedman, I., 1976, Obsidian hydration dating and correlation of Bull Lake and Pinedale glaciations near West Yellowstone, Montana: Geological Society of America Bulletin, v. 87, p. 703–710.

Porter, S. C., 1977, Present and past glaciation threshold in the Cascade Range, Washington, U.S.A.—Topographic and climatic controls, and paleoclimatic implications: Journal of Glaciology, v. 18, p. 101–116.

Porter, S. C., Pierce, K. L., and Hamilton, T. D., 1983, Late Wisconsin mountain glaciation in the western United States, *in* Porter, S. C., ed., Late Quaternary environments of the United States, Volume 1, The late Pleistocene: Minneapolis, University of Minnesota Press, p. 71–114.

Richmond, G. M., 1986a, Stratigraphy and correlation of glacial deposits of the Rocky Mountains, the Colorado Plateau and the ranges of the Great Basin, *in* Sibrava, V., Bowen, D. Q., and Richmond, G. M., eds., Quaternary glaciations in the Northern Hemisphere: Quaternary Science Reviews, v. 5, p. 99–127.

——, 1986b, Stratigraphy and chronology of glaciations in Yellowstone National Park, *in* Sibrava, V., Bowen, D. Q., and Richmond, G. M., eds., Quaternary glaciations in the Northern Hemisphere: Quaternary Science Reviews, v. 5, p. 83–98.

Richmond, G. M., and Fullerton, D. S., 1986, Summation of Quaternary glaciations in the United States of America, *in* Sibrava, V., Bowen, D. Q., and Richmond, G. M., eds., Quaternary glaciations in the Northern Hemisphere: Quaternary Science Reviews, v. 5, p. 183–196.

Whitehouse, I. E., McSaveney, M. J., Knuepfer, P.K.L., and Chinn, T.J.H., 1986, Growth of weathering rinds on Torlesse Sandstone, Southern Alps, New Zealand, *in* Colman, S. M., and Dethier, D. P., eds., Rates of chemical weathering of rocks and minerals: New York, Academic Press, p. 419–435.

Zielinski, G. A., and McCoy, W. D., 1987, Paleoclimatic implications of the relationship between modern snowpack and late Pleistocene equilibrium-line altitudes in the mountains of the Great Basin, western U.S.A.: Arctic and Alpine Research, v. 19, p. 127–134.

Manuscript Accepted by the Society September 6, 1991

Printed in U.S.A.

Geological Society of America
Special Paper 270
1992

Early Wisconsin lakes and glaciers in the Great Basin, U.S.A

Charles G. Oviatt
Department of Geology, Kansas State University, Manhattan, Kansas 66506
William D. McCoy
Department of Geology and Geography, University of Massachusetts, Amherst, Massachusetts 01003

ABSTRACT

Little is known for certain about early Wisconsin (isotope stage 4) lakes and glaciers of the Great Basin. A moderate lake-level rise in the Bonneville basin is not well dated, but on the basis of amino-acid and radiocarbon ages, is thought to be early Wisconsin in age. A moderate rise of lakes in the basins of Lake Lahontan is dated as ca. 50 ka by U-series ages on tufa, but may have occurred earlier. In the southern Great Basin, Searles Lake fluctuated at levels below the threshold connecting it with Panamint Valley, and Panamint Valley apparently did not contain a large lake during the early Wisconsin. The glacial record is even less-well dated than the lacustrine record. The extent of glaciers in and around the Great Basin during the early Wisconsin is not known; ice extent was certainly greater than at present, but probably was less than the late Wisconsin maximum in most glaciated valleys. Further work is necessary to refine lacustrine and glacial chronologies, and to investigate the causes of lake vs. glacier expansion. Important clues to these questions will come from detailed studies of lacustrine and glacial sequences in different parts of the Great Basin.

INTRODUCTION

The Great Basin is a large region of internal drainage in western North America (Fig. 1). It consists of many individual closed basins separated by north-south–trending mountain ranges. High mountains on the western (Sierra Nevada) and eastern (Wasatch Range) margins of the Great Basin receive much more precipitation than the basin interior, and were extensively glaciated during the Pleistocene. Some of the high mountains within the Great Basin were also glaciated. Lakes formed in many of the closed basins in the Great Basin (Fig. 1) broadly coincident with glaciations, and some of these lakes overflowed to adjacent drainage basins. The late Quaternary stratigraphic records of a few individual lake basins and mountain ranges have been studied in detail, and the late Wisconsin history of the Great Basin is relatively well understood. However, the early Wisconsin history is not well known.

In this chapter, we review what is currently known about the sizes of lakes and glaciers in the Great Basin during the early Wisconsin. For the purposes of this chapter, we define the early Wisconsin in a broad sense as equivalent to marine oxygen-isotope substage 5d through stage 4, or ca. 115 to 59 ka (Martinson and others, 1987). However, because there are few reliable numerical ages for this time interval from the Great Basin, some amount of inference is necessarily involved in almost all correlations. Furthermore, the lack of age control precludes a discussion of the nature of the transition from the last major interglaciation (substage 5e) to the early Wisconsin, but certain characteristics of the culmination of the early Wisconsin glacial or lacustrine episodes can be summarized.

LACUSTRINE RECORDS

Bonneville basin

The Bonneville basin is the largest lake basin in the Great Basin, covering ~88,000 km^2 in the states of Utah, Wyoming, Idaho, and Nevada. Lakes in the Bonneville basin are fed primarily by large rivers that drain high mountains along its eastern margin.

Studies of the stratigraphic record of Lake Bonneville began with G. K. Gilbert (1890), who thought that there had been two

Oviatt, C. G., and McCoy, W. D., 1992, Early Wisconsin lakes and glaciers in the Great Basin, U.S.A., *in* Clark, P. U., and Lea, P. D., eds., The Last Interglacial-Glacial Transition in North America: Boulder, Colorado, Geological Society of America Special Paper 270.

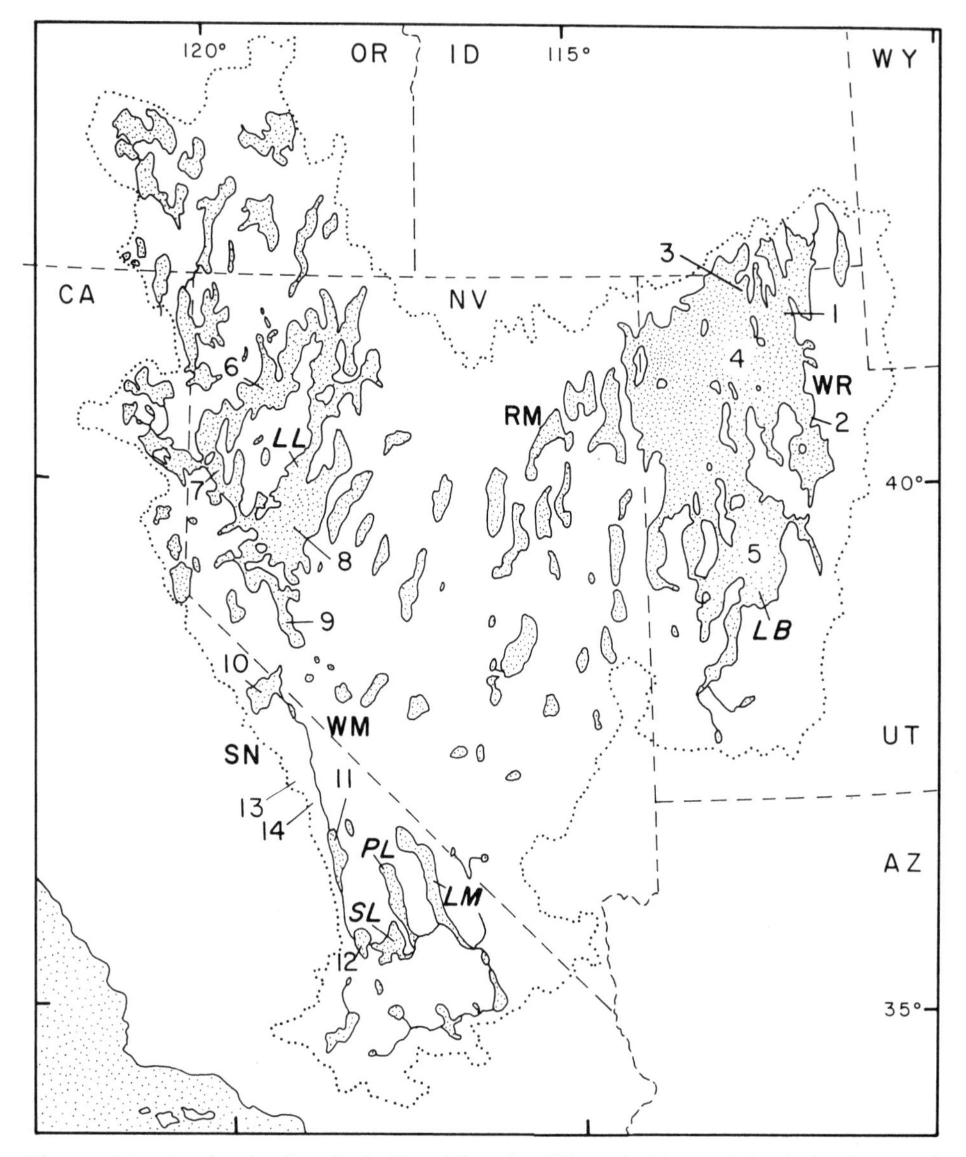

Figure 1. Map showing the Great Basin (dotted line), late Wisconsin lakes, and the glaciated mountain ranges discussed in the text. LB = Lake Bonneville; LL = Lake Lahontan; SL = Searles Lake; PL = Panamint Lake; LM = Lake Manly; WR = Wasatch Range; RM = Ruby–East Humboldt mountains; WM = White Mountains; SN = Sierra Nevada. Other localities mentioned in the text are numbered: 1 = Bear River; 2 = Little Cottonwood and Bells canyons; 3 = Hansel Valley; 4 = Great Salt Lake subbasin; 5 = Sevier Lake subbasin; 6 = Smoke Creek/Black Rock Desert subbasin; 7 = Pyramid Lake subbasin; 8 = Carson Desert subbasin; 9 = Walker Lake subbasin; 10 = Lake Russell (Mono Lake); 11 = Owens Lake; 12 = China Lake; 13 = Pine Creek; 14 = Sawmill Canyon. Modified from Morrison (1965a, Fig. 1) and Smith and Street-Perrott (1983, Fig. 10-1).

major rises in lake level separated by a long interval of low lake levels. Gilbert defined the yellow clay and the white marl as the stratigraphic units deposited in the two deep lakes. He interpreted an intervening unconformity and local gravel deposits as representing subaerial exposure during the period between the two major lake cycles. Subsequent workers (e.g., Antevs, 1948; Ives, 1951) largely accepted this interpretation. Hunt and others (1953) suggested new names for the stratigraphic units of Gilbert. The Alpine Formation was defined as the stratigraphic record of the earlier deep-lake cycle, and the Bonneville and Provo formations as the records of the later deep-lake cycle. All of these formations were considered Wisconsin in age. With some variations in interpretation, most workers during the 1950s, 1960s, and 1970s recognized at least one early Wisconsin deep-lake cycle in the Bonneville basin (e.g., Eardley and others, 1957; Bissell, 1963; Morrison, 1965b, 1965c, 1966; Morrison and Frye, 1965).

During the 1980s, the chronology of Lake Bonneville advanced by Morrison (1966), which had been widely accepted, was reinterpreted. On the basis of new radiocarbon ages, amino-acid-based correlations, and soil-stratigraphic studies, Scott and others (1983) suggested that the deposits in some areas that had been formerly mapped as the Alpine Formation were better interpreted as much older than their presumed early Wisconsin age, and in other areas were probably much younger. The Little Valley Alloformation and the Bonneville Alloformation were defined on the basis of their bounding unconformities in the attempt to avoid miscorrelations between units of similar lithology but vastly different ages (McCoy, 1987a; Oviatt, 1987). Based on ^{230}Th ages, and the amount and morphology of pedogenic carbonate that has accumulated in Little Valley deposits, the Little Valley Alloformation is probably Illinoian in age (oxygen-isotope stage 6; Scott and others, 1983). The Bonneville Alloformation is late Wisconsin in age (stage 2; Scott and others, 1983; Currey and Oviatt, 1985; McCoy, 1987a; Oviatt, 1987).

The stratigraphic studies of Scott and others (1983) indicated that early Wisconsin lakes in the Bonneville basin were confined below an altitude of 1380 m, or about 170 m below the altitude of the Bonneville shoreline, which marks the upper limit of late Wisconsin Lake Bonneville. Thus, by 1983 the hypothesis of a deep early Wisconsin lake in the Bonneville basin had been rejected by most, but not all workers (e.g., Varnes and Van Horn, 1984).

Deposits of a lake intermediate in age between late Wisconsin Lake Bonneville and the Little Valley lake cycle (about 140,000 yr B.P.; Scott and others, 1983) were discovered in northern Utah in 1983 and 1984 (Oviatt, 1986a, 1986b; Oviatt and others, 1987). At about the same time, McCalpin (1986; McCalpin and others, 1987; Robison and McCalpin, 1987) discovered lacustrine deposits in Hansel Valley, Utah, that had a similar intermediate stratigraphic position.

The intermediate aged lacustrine deposits in the Bonneville basin are called the Cutler Dam Alloformation (Oviatt and others, 1987). In exposures along the Bear River in northern Utah, the Cutler Dam Alloformation consists of both marginal and deeper-water lacustrine deposits, which contain ostracodes, plant macrofossils, mollusks, and vertebrates (Feduccia and Oviatt, 1986; Oviatt and others, 1987). In all exposures, the Cutler Dam Alloformation is found stratigraphically below the Bonneville Alloformation, and the two alloformations are separated by an unconformity and a buried soil named the Fielding Geosol (Oviatt and others, 1987). Along the Bear River, the upper limit of the Cutler Dam Alloformation (1340 m) is considerably lower than the upper limit (1610 m) of deposits formerly regarded as the early Wisconsin Alpine Formation (Morrison and Frye, 1965).

The age of the Cutler Dam Alloformation is limited by a radiocarbon age, by amino-acid analyses of mollusks, and, if correlation with the Hansel Valley deposits of McCalpin (1986) is correct, by thermoluminescence (TL) ages. The presence of the Fielding Geosol shows that the Cutler Dam Alloformation is unquestionably pre-Bonneville in age. A radiocarbon age of >36 ka (Beta-9845) on wood from the base of the Cutler Dam sediments places a limit on the age of the Cutler Dam lake—the lake was clearly pre-Bonneville in age, and at least 36,000 yr old (Oviatt and others, 1987).

Amino-acid analyses of mollusks place a poorly defined older limit on the age of the Cutler Dam deposits and yield only an estimate of its numerical age. The average alloisoleucine to isoleucine (aIle/Ile) ratio in the total hydrolysate of *Lymnaea* shells from Cutler Dam deposits (0.13 ± 0.01) is greater than that from Bonneville deposits in the same area (0.06), and significantly less than that from deposits of Little Valley age from around the basin (0.27 ± 0.03) (McCoy, 1987a; Oviatt and others, 1987). Therefore, the Cutler Dam shells and deposits are intermediate in age between the Bonneville and Little Valley alloformations.

Because of large uncertainties in postdepositional paleotemperatures in the northern Bonneville basin, the amino-acid data cannot be used to calculate precise numerical ages for the Cutler Dam Alloformation. If we assume that the effective temperature[1] for the Cutler Dam shells is the same as that for shells of the Little Valley Alloformation, and further assume that shells of the Little Valley Alloformation are about 140 to 150 ka, then an age of about 70–80 ka is obtained for the Cutler Dam Alloformation. If we assume an effective temperature equal to that for shells of the Bonneville Alloformation, then an age of 40–45 ka is obtained for the Cutler Dam Alloformation. Because the effective temperature for the Bonneville Alloformation shells is so dominated by the warm Holocene, we suggest that it provides an upper limit for possible effective temperatures for the Cutler Dam shells. The effective temperature calculated from Little Valley Alloformation shells is probably a better estimate of a long-term effective temperature for this region and is thus more applicable to the Cutler Dam shells. We therefore lean toward an older age, ca. 70 ka, for the Cutler Dam Alloformation.

The deposits in Hansel Valley described by McCalpin (1986; Robison and McCalpin, 1987) lie below an altitude of 1340 m, are stratigraphically between deposits believed to be correlative with the Little Valley and Bonneville alloformations, and have been dated at about 76 and 82 ka by TL (McCalpin, 1986). McCalpin (1986) also reported TL ages of about 138 ka on deposits of presumed Little Valley age, and about 13 ka on deposits of Bonneville age in Hansel Valley. These ages are con-

[1]Effective temperatures are calculated here by using independently known or assumed ages for shells of known aIle/Ile ratios in a kinetic equation derived for isoleucine epimerization in *Lymnaea* (McCoy, 1981) with ratios corrected for changes in preparation method. The absolute values of the effective temperatures discussed are not necessarily accurate and are not given here. It is their relative values that are important in these estimates, and which justify their use (McCoy, 1987b). The difference between the effective temperature for Little Valley Alloformation shells and that for Bonneville Alloformation shells is 4 K.

sistent with radiocarbon- and amino-acid-based estimates for the Little Valley and Bonneville alloformations, and suggest that the intermediate ages (76 and 82 ka) may also be reliable. Although allostratigraphic units cannot be correlated directly between the Cutler Dam area and Hansel Valley, the Cutler Dam Alloformation and the Hansel Valley deposits of McCalpin are likely to be equivalent, based on the similarity of the stratigraphic sequences and the altitudinal limits of the deposits. More work is necessary to test this correlation.

In summary, the Cutler Dam Alloformation is found at intermediate to low altitudes in the Bonneville basin. It is intermediate in age between the Little Valley Alloformation, which is probably correlative in time with marine oxygen-isotope stage 6, and the Bonneville Alloformation, which was deposited during isotope stage 2. Because the timing of the Bonneville lake cycle is well known to be broadly synchronous with isotope stage 2 (although lagging somewhat behind the glacier maximum), we infer that the Cutler Dam lake cycle was also broadly synchronous with an even-numbered isotope stage (most likely stage 4). Its precise timing, however, is not yet known. Although the Cutler Dam lake was large, over 60 m deep with a surface area of about 22,000 km^2, both the Little Valley lake and Lake Bonneville were much larger (44,000 and 51,000 km^2, respectively). Therefore, temperatures were probably not as cold and/or precipitation in the lake basin not as great, during Cutler Dam time (early Wisconsin?) as they were during the preceding or succeeding lacustrine episodes.

Lahontan basin

Based on detailed studies of Pleistocene deposits in the Lahontan basin, Russell (1885) concluded that there had been two major deep lakes that flooded the basin. Antevs (1945, 1948, 1952) also interpreted two major rises of lake level in the basin. On the basis of stratigraphic position and climatological arguments, he correlated the first rise with the Iowan and the second rise with the Mankato of the midwestern United States (Antevs, 1945). Morrison (1964) conducted much more detailed studies of the Carson Desert portion of the Lahontan basin and further refined the chronology of lake-level fluctuations. Morrison and Frye (1965) recognized two major lacustrine episodes within the late Pleistocene represented by deposits of the Eetza Formation (early Wisconsin) and the Sehoo Formation (late Wisconsin), separated by the Churchill soil and the subaerial deposits of the Wymeha Formation.

Several workers have contributed to the geochronology of lakes in the Lahontan basin over the past 30 years (Broecker and Orr, 1958; Broecker and Kaufman, 1965; Kaufman and Broecker, 1965, Benson, 1978; Thompson and others, 1986; Benson and Thompson, 1987a; Lao and Benson, 1988). Lao and Benson (1988) have synthesized the pre–late Wisconsin radiometric ages relating to lake levels. A U-series age on gastropod shells of about 120 ka (L-773L; Broecker and Kaufman, 1965; Kaufman and Broecker, 1965) collected from the upper part of the Eetza Formation suggests that the Eetza was deposited before the Wisconsin. Based on analogy with the last deep-lake cycle, we suggest that the uppermost Eetza lacustrine deposits are equivalent in age to isotope stage 6 rather than stage 5 (as would be indicated by a literal acceptance of the U-series age). The Churchill soil, which presumably formed (at least in part) during the last interglacial, is stratigraphically above the Eetza Formation.

Deposits of the Sehoo Formation appear to be mostly late Wisconsin in age (Kaufman and Broecker, 1965; Benson and Thompson, 1987a). However, Benson and Thompson (1987a) and Lao and Benson (1988) reported U-series ages of 49 ± 2 ka and 35 ± 10 ka for a tufa sample (PL-14) from the Pyramid Lake subbasin. Lao and Benson (1988) apparently preferred the former age, and indeed that sample has an acceptably high $^{230}Th/^{232}Th$ ratio (70 ± 14). We suggest that the age be considered a minimum for the sample, however, because the high level of uranium suggests the possibility of postdepositional uranium uptake. The altitude of the sample (1209 m) indicates that the western subbasins of the Lahontan basin were confluent, or at least that the Pyramid Lake subbasin spilled over Emerson Pass into the Smoke Creek–Black Rock Desert subbasin during the middle, or perhaps the early Wisconsin. The lake stood at a level approximately one-quarter of the distance between the historic high level of Pyramid Lake and the highest (late Wisconsin) shoreline of Lake Lahontan.

Lao and Benson (1988) reported a U-series age on a suite of tufa samples (WL84-7aI–WL84-7aIV and WL84-8a–WL84-8c) taken at altitudes of 1,315 and 1,317 m in the Walker Lake subbasin as 49 ± 5 ka. Because the $^{230}Th/^{232}Th$ ratios in those samples are low (3.6 to 5.8), the probability of contamination from detrital ^{230}Th is high. Therefore, Lao and Benson (1988) used the samples as a group to estimate the initial $^{230}Th/^{232}Th$ ratio and to calculate the age based on the isochron method. The extrapolated initial $^{230}Th/^{232}Th$ ratio is relatively high (2.64) compared to modern samples from the Walker Lake subbasin (~1.2 to 1.4). If the initial $^{230}Th/^{232}Th$ ratio for these samples was close to that of the modern samples, the U-series ages could be as much as 40% greater than 49 ka, or ca. 70 ka. (Lao and Benson [1988, Fig. 6] suggested a correlation of this moderate-sized lake with isotope stage 4.) The samples are from 7–9 m above the level of the sill at Adrian Pass, but may have been uplifted relative to the pass by faulting at the eastern base of the Wassuk Range. Therefore, the samples may have been deposited at a time when the relatively small, steep-sided Walker Lake subbasin overflowed into a much lower lake in the large, gently sloping Carson Desert subbasin. The samples may or may not be contemporaneous with sample PL-14 from the Pyramid Lake subbasin discussed above. It is possible that both samples were deposited in middle or early Wisconsin lakes. Lao and Benson (1988, Fig. 6) implied a correlation of these samples with isotope stage 4 of the deep-sea record, i.e., the early Wisconsin as defined here. Calculations by Benson and Thompson (1987b) and Benson and Paillet (1989) indicated that the coincidence of an overflowing lake at 1,308 m in the Walker Lake subbasin and a lake level of 1,209 m in the western subbasins is not climatically or hydrologically incongruous, but should be expected.

In summary, the evidence for moderate lake levels during the middle or early Wisconsin in the Lahontan basin is suggestive but not compelling. There remains no convincing stratigraphic evidence concerning early Wisconsin lake levels in the basin. The stratigraphic relations, chronology, and the nature of the transition from the last interglacial environments (Churchill soil) to the moderate early or middle Wisconsin lake levels remain unknown.

Searles Lake, Lake Panamint, and Lake Manly

Searles Lake, Lake Panamint, and Lake Manly are the last three lakes in a series of lakes connected by rivers during cooler and/or wetter periods of the Pleistocene (Fig. 1). The series of connected lakes, which also included Lake Russell, Owens Lake, and China Lake, was fed by runoff from the eastern slope of the Sierra Nevada (Smith, 1979).

Early studies by Gale (1914) established that overflow from China Lake to Searles Lake was probably coincident with glaciation in the Sierra Nevada. Detailed studies of sediment cores, and to a lesser extent surface outcrops, in the Searles Lake basin by Flint and Gale (1958), Stuiver (1964), Smith (1976, 1979, 1984), and Jannik and others (1991) over several decades of research have refined the chronology of lake fluctuations.

According to Smith (1976, 1979), Searles Lake filled and overflowed to Panamint Valley five times during the late Pleistocene. Many radiocarbon ages have been obtained for samples from the upper part of the sedimentary sequence at Searles Lake (Stuiver and Smith, 1979), and Smith (1979) used them to estimate sedimentation rates and to extrapolate ages for the lower parts of the sequence. Based on ages determined from extrapolated sedimentation rates and inferences about water depth from the geochemistry of the sediments, Smith (1979) suggested that Searles Lake may have overflowed during three long periods (3–11 ka) between about 90 and 55 ka. These three long periods of overflow were separated by two relatively short periods (2 ka) of intermediate lake levels.

G. I. Smith's (1979) chronology from Searles Lake is supported in part by the work of R. Smith (1975, 1978) in Panamint Valley. Based on extrapolated uplift rates of shorelines, R. Smith (1978) concluded that there were seven high lake stages in Panamint Valley between about 110 and 40 ka. Parts of the independently derived chronology for Panamint Lake are similar to the Searles Lake chronology, but other parts are in conflict. Because Panamint Valley is downstream from Searles Lake, and receives surface inflow from the Searles Lake basin, the two chronologies should be similar, or at least should show elevated lake levels in Panamint Valley when Searles Lake overflowed. Because it is unlikely that Panamint Valley could develop a deep lake without input from the Searles basin, however, conflicts arise where the Panamint Valley sequence indicates a high lake when Searles Lake did not overflow (R. Smith, 1978). Recently reported ^{36}Cl ages of salt layers, and an analysis of the salt budget in the waters of Searles and Panamint lakes (Jannik and others, 1991), indicate that Searles Lake overflowed into Panamint Valley about 150 ka and again between 24 and 10 ka. However, during the early Wisconsin interval (100 to 24 ka) Searles Lake fluctuated at levels below the threshold connecting the two lake basins.

Lake Manly was the Pleistocene lake in Death Valley, and received input from the overflowing Panamint Lake and from the Mojave and Amargosa rivers (Blackwelder, 1933, 1954; Hunt and Mabey, 1966; Hooke, 1972). Thus, the chronology of lake fluctuations in Death Valley should also be closely linked to those from Searles Lake and Panamint Lake. Blackwelder (1933) thought that Lake Manly was correlative with the Tahoe glaciation in the Sierra Nevada. Based on the weathering of pebbles in shoreline deposits of Lake Manly, Hunt and Mabey (1966) also suggested that Lake Manly was Tahoe in age (early Wisconsin in their view). Blackwelder (1954) recognized that Lake Manly rose twice in Wisconsin time, and that the early (Tahoe) lake was deeper than the late (Tioga) lake. Hooke (1972) named the higher stand the Blackwelder stand, and suggested that both lakes formed during the late Wisconsin. Dorn (1988) reported cation-ratio ages of rock varnish from shorelines of the Blackwelder stand that indicate that the shorelines were abandoned about 130 to 120 ka. Therefore, the Blackwelder stand shorelines are probably isotope stage 6 (Illinoian) in age, and at this point there is no convincing evidence for an early Wisconsin lake in Death Valley. Chlorine-36 studies in the Searles and Panamint lake basins indicate that Death Valley received inflow from these basins during stage 6, but not during any part of the Wisconsin (Jannik and others, 1991).

GLACIAL RECORDS

Very little is known about the early Wisconsin glacial record of the ranges of the Great Basin or of the major ranges in adjacent physiographic provinces that are hydrologically tributary to the Great Basin (Wasatch Mountains to the east and Sierra Nevada to the west). The most recent reviews of the Quaternary glaciations in the region were in Sibrava and others (1986). a brief review of the most relevant findings is given below.

Wasatch Range

Although some early workers assigned an early Wisconsin age to certain glacial deposits near the mouths of Little Cottonwood and Bells canyons, most workers today do not recognize deposits of that age in the area (Madsen and Currey, 1979; Richmond, 1986). A late Wisconsin age for the Bells Canyon till of Madsen and Currey (1979) is constrained by a radiocarbon age of 26,080 +1,200/−1,100 yr B.P. (GX-4737) on organic material from the B horizon of a paleosol developed in the underlying Dry Creek till, and by its relation to deposits of Lake Bonneville.

The age of the Dry Creek till is more equivocal, as it is not constrained by any radiometric ages other than the 26,000 yr B.P. age mentioned above. The relatively strong weathering characteristics (McCoy, 1977), fault offsets, and soil development (Scott, 1988) compared to those of the Bells Canyon till suggest to most recent workers that the Dry Creek till is pre-Wisconsin, and probably isotope stage 6 (Illinoian) in age (Madsen and Currey, 1979; Richmond, 1986; Scott, 1988). If that correlation is correct, there are no known exposed early Wisconsin glacial deposits in the Wasatch Range.

Sierra Nevada

Fullerton (1986) summarized the literature on the glacial chronology of the Sierra Nevada. Traditionally the Tahoe glaciation has been considered early Wisconsin and the Tioga late Wisconsin in age. Fullerton's (1986) discussion incorporated many of the recent data of Gillespie (1982, 1984), especially those concerning the ages of volcanic rocks interbedded with till in Sawmill Canyon of the southeastern Sierra Nevada. There, Gillespie has determined that two tills overlie a basalt that he dated using the $^{40}Ar/^{39}Ar$ method at 118 ± 7 and 119 ± 9 ka. These ages supported the inherently less-reliable K/Ar ages of Dalrymple (1964) and Dalrymple and others (1982), who suggested that the deposits overlying the basalt were probably younger than the last interglacial maximum of the marine record (isotope substage 5e). The younger of the two tills overlying the basalt is assigned a late Wisconsin age on the basis of its position and weathering characteristics. Fullerton (1986) assigned the older till an early Wisconsin age, although there remains considerable uncertainty as to its precise age.

Dorn and others (1987) have used cation-ratio dating of desert varnish formed on boulders along moraine crests to determine the age of deposits previously mapped as Tahoe moraines along Pine Creek (~75 km north of Sawmill Canyon in the southeastern Sierra Nevada) to be 143 to 156 ka. Therefore, they may be equivalent in age to isotope stage 6 (Illinoian) of the deep-sea record. It is important to reiterate the conclusions of Gillespie (1984) and Fullerton (1986) that not all moraines mapped as Tahoe are the same age—some are apparently early Wisconsin (in the broad sense), and some are decidedly pre-Wisconsin in age (see Colman and Pierce, 1981; Atwater and others, 1986; Phillips and others, 1990). Moraine exposure ages determined by the ^{36}Cl method indicate that at least some of the early Wisconsin valley glaciers in the Sierra Nevada were slightly more extensive than their late Wisconsin counterparts (Phillips and others, 1990).

Ruby and East Humboldt ranges

There are no numerical ages that delimit pre-late Wisconsin glacial deposits in the Ruby and East Humboldt ranges of north-central Nevada. Blackwelder (1931, 1934) and Sharp (1938) considered the extensive glacial deposits (Lamoille stage) at the mouths of canyons to be equivalent to the Tahoe moraines of the eastern Sierra Nevada and to the Iowan of the midwestern United States. Wayne (1984) has suggested that the Lamoille moraines are pre-Wisconsin, and probably Illinoian in age, based on relative weathering criteria and soil development. If Wayne is correct, there is no evidence concerning the relative extent of glaciers during the early or middle Wisconsin, except that they would appear to have been less extensive than the late Wisconsin Angel Lake glaciers.

White Mountains (California and Nevada)

Elliott-Fisk (1987) correlated moraines of the Perry Aiken glaciation in the White Mountains with the Tahoe moraines of the Pine Creek area, Sierra Nevada, which are thought to be stage 6 (Illinoian) in age by Dorn and others (1987; see above). Elliott-Fisk (1987) and Swanson and others (1989) argued that the uplift of the Sierra Nevada through Quaternary time has progressively enhanced its rain-shadow effect on the White Mountains to the east, thereby making successive glaciations less extensive, and resulting in a relatively complete record of glaciations in the White Mountains. If their interpretation is correct, there is no evidence of the extent of early Wisconsin glaciers.

DISCUSSION

A review of the evidence of early Wisconsin environments in the Great Basin reveals that very little can be said about that time period with any degree of certainty. Lakes may have enlarged moderately (though were still relatively low compared to their late Wisconsin maxima) in the Bonneville and Lahontan basins, but the dating of these events is very uncertain. Good age control in the Searles basin also indicates moderate lake rises in that basin.

Even less is known about the extent of glaciers in and around the Great Basin during the early Wisconsin. The existence of moraines intermediate in age between late Wisconsin and what may be Illinoian in certain ranges near the Great Basin such as the Sierra Nevada (Fullerton, 1986; Phillips and others, 1990) and the Salmon River Mountains of Idaho (Colman and Pierce, 1986) suggests that perhaps extensive glaciers did develop in the Great Basin ranges during the early Wisconsin, but that with some exceptions, their deposits were mostly obliterated by more extensive late Wisconsin glaciers.

A speculation, or working hypothesis, is that both lakes and glaciers were more extensive than at present during the early Wisconsin in the Great Basin. Neither were as extensive as their late Wisconsin successors, except that some glaciers of the Sierra Nevada may have approached, or locally exceeded, their late Wisconsin dimensions. This does not imply that lake and glacier expansion were in phase with each other, or that lake or glacier fluctuations in all parts of the Great Basin were in phase.

The more detailed record of lake and glacier fluctuations in the Great Basin during the late Wisconsin demonstrates a lack of synchroneity, especially in the Bonneville basin. Lake Bonneville attained its highest level about 15 ka, which was 5–7 ka later than the glaciers in the Wasatch Range had reached their maxima and had already retreated back upvalley (Scott and others, 1983). Lake Bonneville, however, was broadly contemporaneous with isotope stage 2, and it dropped very rapidly between about 13.5 and 12 ka, coincident with termination I of the deep-sea record. Therefore, it is likely that Lake Bonneville was responding in part to the same climatic factors that controlled glacier volume, but that other climatic factors were also influencing the size of the lake.

Lake level or lake surface area changes with changes in temperature and cloud cover, which control evaporation rate, and with changes in precipitation, and these variables are not always correlated with each other. For example, a shallow lake in the Sevier subbasin of the Bonneville basin overflowed continuously into the Great Salt Lake subbasin from about 12 to 10 ka, while the lake in the Great Salt Lake subbasin continued to decline and reached very low levels (Oviatt, 1988). The Sevier subbasin lake apparently was remaining full because of enhanced summer (monsoonal?) precipitation during that period, which did not reach the (northern) Great Salt Lake subbasin. Thus, at least in the Sevier subbasin, the lake was high at a time when summer temperatures were probably also high (COHMAP Members, 1988), and when mountain glaciers were probably retreating. This single example demonstrates that lakes and glaciers do not necessarily respond in the same way to changes in climate, and that a one-to-one correspondence between lake fluctuations and glacier fluctuations should not be expected. Spatial variations in lake-level maxima throughout the Great Basin on longer temporal scales are also possible, but are not yet documented.

In conclusion, lakes and glaciers may have generally expanded in the Great Basin during the early Wisconsin, but not all expansions were necessarily synchronous or of similar magnitude. This hypothesis could be tested by further work employing new methods for dating geomorphic surfaces (both glacial and lacustrine), mapping and stratigraphic studies in lake basins, and detailed analyses of cores of lake sediments from lakes throughout the Great Basin. A great deal remains to be learned about the late Pleistocene history of the Great Basin. Because the stratigraphic record of many lake basins is well preserved and inherently more complete than glacial records, studies of lake history will be important keys to understanding the paleoclimatic changes of this region. However, as we have pointed out, reliable reconstructions of glacier fluctuations (and other paleoenvironmental variables that we have not considered here) are equally important in painting a complete paleoclimatic picture.

ACKNOWLEDGMENTS

We are grateful to P. U. Clark, P. D. Lea, W. E. Scott, and R. S. Thompson for reviewing an earlier draft of this paper.

REFERENCES CITED

Antevs, E., 1945, Correlation of Wisconsin glacial maxima: American Journal of Science, v. 243-A, p. 1–39.

——, 1948, Climatic changes and pre-white man: University of Utah Bulletin, v. 38, p. 168–191.

——, 1952, Cenozoic climates of the Great Basin: Geologische Rundschau, v. 40, p. 94–108.

Atwater, B. F., and eight others, 1986, A fan dam for Tulare Lake, California, and implications for the Wisconsin glacial history of the Sierra Nevada: Geological Society of America Bulletin, v. 97, p. 97–109.

Benson, L. V., 1978, Fluctuation in the level of pluvial Lake Lahontan during the last 40,000 years: Quaternary Research, v. 9, p. 300–318.

Benson, L. V., and Paillet, F. L., 1989, The use of total lake-surface area as an indicator of climatic change: Examples from the Lahontan basin: Quaternary Research, v. 32, p. 262–275.

Benson, L. V., and Thompson, R. S., 1987a, Lake-level variation in the Lahontan basin for the past 50,000 years: Quaternary Research, v. 28, p. 69–85.

——, 1987b, The physical record of lakes in the Great basin, *in* Ruddiman, W. F., and Wright, H. E., Jr., eds., North America and adjacent oceans during the last deglaciation: Boulder, Colorado, Geological Society of America, Geology of North America, v. K-3, p. 241–260.

Bissell, H. J., 1963, Lake Bonneville—Geology of southern Utah Valley: U.S. Geological Survey Professional Paper 257-B, p. 101–130.

Blackwelder, E., 1931, Pleistocene glaciation in the Sierra Nevada and Basin Ranges: Geological Society of America Bulletin, v. 42, p. 865–922.

——, 1933, Lake Manly: An extinct lake in Death Valley: Geographical Review, v. 23, p. 464–471.

——, 1934, Supplementary notes on Pleistocene glaciation in the Great Basin: Washington Academy of Science Journal, v. 24, p. 217–222.

——, 1954, Pleistocene lakes and drainage in the Mojave region, southern California, Chapter 5, *in* Jahns, R. H., ed., Geology of southern California: California Division of Mines Bulletin 170, p. 35–40.

Broecker, W. S., and Kaufman, A., 1965, Radiocarbon chronology of Lake Lahontan and Lake Bonneville II, Great Basin: Geological Society of America Bulletin, v. 76, p. 537–566.

Broecker, W. S., and Orr, P. C., 1958, Radiocarbon chronology of Lake Lahontan and Lake Bonneville: Geological Society of America Bulletin, v. 69, p. 1009–1032.

COHMAP Members, 1988, Climatic changes of the last 18,000 years: Observations and model simulations: Science, v. 241, p. 1043–1052.

Colman, S. M., and Pierce, K. L., 1981, Weathering rinds on andesitic and basaltic stones as a Quaternary age indicator, western United States: U.S. Geological Survey Professional Paper 1210, 41 p.

——, 1986, Glacial sequence near McCall, Idaho: Weathering rinds, soil development, morphology and other relative age criteria: Quaternary Research, v. 25, p. 25–42.

Currey, D. R., and Oviatt, C. G., 1985, Durations, average rates, and probable causes of Lake Bonneville expansion, still-stands, and contractions during the last deep-lake cycle, 32,000 to 10,000 yrs ago, *in* Kay, P. A., and Diaz, H. F., EDS., Problems of and prospects for predicting Great Salt Lake levels (Proceedings, NOAA Conference, March 26–28, 1985): Salt Lake City, University of Utah, Center for Public Affairs and Administration, p. 9–24.

Dalrymple, G. B., 1964, Potassium-argon dates of three Pleistocene interglacial basalt flows from the Sierra Nevada, California: Geological Society of America Bulletin, v. 75, p. 753–757.

Dalrymple, G. B., Burke, R. M., and Birkeland, P. W., 1982, Concerning K-Ar dating of a basalt flow from the Tahoe-Tioga interglaciation, Sawmill Canyon, southeastern Sierra Nevada, California: Quaternary Research, v. 17, p. 120–122.

Dorn, R. I., 1988, A rock varnish interpretation of alluvial-fan development in Death Valley, California: National Geographic Research, v. 4, p. 56–73.

Dorn, R. I., Turrin, B. D., Jull, A.J.T., Linick, T. W., and Donahue, D. J., 1987, Radiocarbon and cation-ratio ages for rock varnish on Tioga and Tahoe morainal boulders of Pine Creek, eastern Sierra Nevada, California, and their paleoclimatic implications: Quaternary Research, v. 28, p. 38–49.

Eardley, A. J., Gvosdetsky, V., and Marsell, R. E., 1957, Hydrology of Lake Bonneville and sediments and soils of its basin: Geological Society of America Bulletin, v. 68, p. 1141–1201.

Elliott-Fisk, D. L., 1987, Glacial geomorphology of the White Mountains, California and Nevada: Establishment of a glacial chronology: Physical Geography, v. 8, p. 299–323.

Feduccia, A., and Oviatt, C. G., 1986, A trumpeter swan from the Pleistocene of Utah: Great Basin Naturalist, v. 46, p. 547–548.

Flint, R. F., and Gale, W. A., 1958, Stratigraphy and radiocarbon dates at Searles Lake, California: American Journal of Science, v. 256, p. 689–714.

Fullerton, D. S., 1986, Chronology and correlation of glacial deposits in the Sierra Nevada, California, *in* Sibrava, V., Bowen, D. Q., and Richmond, G. M., eds., Quaternary glaciations in the Northern Hemisphere: Quaternary Science Reviews, v. 5, p. 161–169.

Gale, H. S., 1914, Salines in the Owens, Searles, and Panamint basins, southeastern California: U.S. Geological Survey Bulletin 580-L, p. 251–323.

Gilbert, G. K., 1890, Lake Bonneville: U.S. Geological Survey Monograph 1, 438 p.

Gillespie, A. R., 1982, Quaternary glaciation and tectonism in the southeastern Sierra Nevada, Inyo County, California [Ph.D. thesis]: Pasadena, California Institute of Technology, 695 p.

—— , 1984, Evidence for both Wisconsin and Illinoian ages for the Tahoe glaciation, Sierra Nevada, California: Geological Society of America Abstracts with Programs, v. 16, p. 519.

Hooke, R. LeB., 1972, Geomorphic evidence for late-Wisconsin and Holocene tectonic deformation, Death Valley, California: Geological Society of America Bulletin, v. 83, p. 2073–2098.

Hunt, C. B., and Mabey, D. R., 1966, Stratigraphy and structure of Death Valley, California: U.S. Geological Survey Professional Paper 494-A, 162 p.

Hunt, C. B., Varnes, H. D., and Thomas, H. E., 1953, Lake Bonneville—Geology of northern Utah Valley, Utah: U.S. Geological Survey Professional Paper 257-A, 99 p.

Ives, R. L., 1951, Pleistocene valley sediments of the Dugway area, Utah: Geological Society of America Bulletin, v. 62, p. 781–797.

Jannik, N. O., Phillips, F. M., Smith, G. I., and Elmore, D., 1991, A ^{36}Cl chronology of lacustrine sedimentation in the Pleistocene Owens River system: Geological Society of America Bulletin, v. 103, p. 1146–1159.

Kaufman, A., and Broecker, W. S., 1965, Comparison of ^{230}Th and ^{14}C ages for carbonate materials from Lakes Lahontan and Bonneville: Journal of Geophysical Research, v. 70, p. 4039–4054.

Lao, Y., and Benson, L., 1988, Uranium-series age estimates and paleoclimatic significance of Pleistocene tufas from the Lahontan basin, California and Nevada: Quaternary Research, v. 30, p. 165–176.

Madsen, D. B., and Currey, D. R., 1979, Late Quaternary glacial and vegetational changes Little Cottonwood Canyon area, Wasatch Mountains, Utah: Quaternary Research, v. 12, p. 254–270.

Martinson, D. G., Pisias, N. G., Hays, J. D., Imbrie, J., Moore, T. C., Jr., and Shackleton, N. J.,1987, Age dating and the orbital theory of the ice ages: Development of a high resolution 0 to 300,000-year chronostratigraphy: Quaternary Research, v. 27, p. 1–29.

McCalpin, J., 1986, Thermoluminescence (TL) dating in seismic hazard evaluations: An example from the Bonneville basin, Utah, *in* Proceedings of the 22nd symposium on Engineering geology and soils engineering: Boise, Idaho, February 24-26, 1986: p. 156–176.

McCalpin, J., Robison, R. M., and Garr, J. D., 1987, Neotectonics of the Hansel Valley–Pocatello Valley corridor, northern Utah and southern Idaho, *in* Gori, P., and Hays, W. W., eds., Assessment of regional earthquake hazards and risk along the Wasatch Front, Utah: U.S. Geological Survey Open-File Report 87-585, v. 1, p. G1–G44.

McCoy, W. D., 1977, A reinterpretation of certain aspects of the late Quaternary glacial history of Little Cottonwood Canyon, Utah [M.A. thesis]: Salt Lake City, University of Utah, 84 p.

—— , 1981, Quaternary aminostratigraphy of the Bonneville and Lahontan basins, western U.S., with paleoclimatic implications [Ph.D. thesis]: Boulder, University of Colorado, 603 p.

—— , 1987a, Quaternary aminostratigraphy of the Bonneville basin, western United States: Geological Society of America Bulletin, v. 98, p. 99–112.

—— , 1987b, The precision of amino acid geochronology and paleothermometry: Quaternary Science Reviews, v. 6, p. 43–54.

Morrison, R. B., 1964, Lake Lahontan: Geology of southern Carson Desert, Nevada: U.S. Geological Survey Professional Paper 401, 156 p.

——, 1965a, Quaternary geology of the Great Basin, *in* Wright, H. E., Jr., and Frey, D. G., eds., The Quaternary of the United States: Princeton, New Jersey, Princeton University Press, p. 265–285.

—— , 1965b, Lake Bonneville: Quaternary stratigraphy of eastern Jordan Valley, Utah: U.S. Geological Survey Professional Paper 477, 80 p.

—— , 1965c, New evidence on Lake Bonneville stratigraphy and history from southern Promontory Point, Box Elder County, Utah: U.S. Geological Survey Professional Paper 525-C, p. 110–119.

—— , 1966, Predecessors of Great Salt Lake, *in* Stokes, W. L., ed., The Great Salt Lake: Utah Geological Society Guidebook to the Geology of Utah, no. 20, p. 77–104.

Morrison, R. B., and Frye, J. C., 1965, Correlation of the middle and late Quaternary successions of the Lake Lahontan, Lake Bonneville, Rocky Mountain (Wasatch Range), southern Great Plains, and eastern midwestern area: Nevada Bureau of Mines Report 9, 45 p.

Oviatt, C. G., 1986a, Geologic map of the Cutler Dam quadrangle, Box Elder and Cache counties, Utah: Utah Geological and Mineral Survey Map 91, scale 1:24,000.

—— , 1986b, Geologic map of the Honeyville quadrangle, Box Elder and Cache counties, Utah: Utah Geological and Mineral Survey Map 89, scale 1:24,000.

—— , 1987, Lake Bonneville stratigraphy at the Old River Bed, Utah: American Journal of Science, v. 287, p. 383–398.

—— , 1988, Late Pleistocene and Holocene lake fluctuations in the Sevier Lake basin, Utah, USA: Journal of Paleolimnology, v. 1, p. 9–21.

Oviatt, C. G., McCoy, W. D., and Reider, R. G., 1987, Evidence for a shallow early or middle Wisconsin-age lake in the Bonneville basin, Utah: Quaternary Research, v. 27, p. 248–262.

Phillips, F. M., Zreda, M. G., Smith, S. S., Elmore, D., Kubik, P. W., and Sharma, P., 1990, Cosmogenic chlorine-36 chronology for glacial deposits at Bloody Canyon, eastern Sierra Nevada: Science, v. 248, p. 1529–1532.

Richmond, G. M., 1986, Stratigraphy and correlation of glacial deposits of the Rocky Mountains, the Colorado Plateau and the ranges of the Great Basin, *in* Sibrava, V., Bowen, D. Q., and Richmond, G. M., eds., Quaternary glaciations in the Northern Hemisphere: Quaternary Science Reviews, v. 5, p. 99–127.

Robison, R. M., and McCalpin, J., 1987, Surficial geology of Hansel Valley, Box Elder County, Utah: Utah Geological Association Publication 16, p. 335–349.

Russell, I. C., 1885, Geologic history of Lake Lahontan: U.S. Geological Survey Monograph 11, 287 p.

Scott, W. E., 1988, Temporal relations of lacustrine and glacial events at Little Cottonwood and Bells canyons, Utah, *in* Machette, M. N., ed., In the footsteps of G. K. Gilbert—Lake Bonneville and neotectonics of the eastern Basin and Range province: Utah Geological and Mineral Survey Miscellaneous Publication 88-1, p. 78–81.

Scott, W. E., McCoy, W. D., Shroba, R. R., and Rubin, M., 1983, Reinterpretation of the exposed record of the last two lake cycles of Lake Bonneville, western United States: Quaternary Research, v. 20, p. 261–285.

Sharp, R. P., 1938, Pleistocene glaciation in the Ruby–East Humboldt Range, northeastern Nevada: Journal of Geomorphology, v. 1, p. 296–321.

Sibrava, V., Bowen, D. Q., and Richmond, G. M., eds., 1986, Quaternary glaciations in the Northern Hemisphere: Quaternary Science Reviews, v. 5, 514 p.

Smith, G. I., 1976, Paleoclimatic record in the upper Quaternary sediments of Searles Lake, California, U.S.A., *in* Horie, S., ed., Paleolimnology of Lake

Biwa and the Japanese Pleistocene: Kyoto, Japan, Kyoto University, v. 4, p. 577–604.

—— , 1979, Subsurface stratigraphy and geochemistry of late Quaternary evaporites, Searles Lake, California: U.S. Geological Survey Professional Paper 1043, 130 p.

—— , 1984, Paleohydrologic regimes in the southwestern Great Basin, 0–3.2 m.y. ago compared with other long records of "global" climate: Quaternary Research, v. 22, p. 1–17.

Smith, G. I., and Street-Perrott, F. A., 1983, Pluvial lakes of the western United States, *in* Wright, H. E., Jr., and Porter, S. C., eds., Late-Quaternary environments of the United States, Volume 1, The late Pleistocene: Minneapolis, University of Minnesota Press, p. 190–212.

Smith, R.S.U., 1975, Late-Quaternary pluvial and tectonic history of Panamint Valley, Inyo and San Bernardino counties, California [Ph.D. thesis]: Pasadena, California Institute of Technology, 295 p.

—— , 1978, Pluvial history of Panamint Valley, California: Unpublished guidebook for the Friends of the Pleistocene, Pacific Cell, 11 November, 1978, 36 p.

Stuiver, M., 1964, Carbon isotopic distribution and correlated chronology of Searles Lake sediments: American Journal of Science, v. 262, p. 377–392.

Stuiver, M., and Smith, G. I., 1979, Radiocarbon ages of stratigraphic units, *in* Smith, G. I., ed., Subsurface stratigraphy and geochemistry of late Quaternary evaporites, Searles Lake, California: U.S. Geological Survey Professional Paper 1043, p. 68–74.

Swanson, T. W., Elliot-Fisk, D. L., Dorn, R. I., and Phillips, F. M., 1989, Quaternary glaciation of the Chiatovich Creek basin, White Mountains, CA-NV: A multiple dating approach: Geological Society of America Abstracts with Programs, v. 20, no. 7, p. A209.

Thompson, R. S., Benson, L. V., and Hattori, E., 1986, A revised chronology for the last Pleistocene lake cycle in central Lake Lahontan: Quaternary Research, v. 25, p. 1–9.

Varnes, D. J., and Van Horn, R., 1984, Surficial geologic map of the Oak City area, Millard County, Utah: U.S. Geological Survey Open-File Report 84-115, 3 sheets.

Wayne, W. J., 1984, Glacial chronology of the Ruby Mountains–East Humboldt Range, Nevada: Quaternary Research, v. 21, p. 286–303.

Manuscript Accepted by the Society September 6, 1991

Printed in U.S.A.

Geological Society of America
Special Paper 270
1992

A pre–late Wisconsin paleolimnologic record from the Estancia Valley, central New Mexico

Frederick W. Bachhuber
Department of Geosciences, University of Nevada, Las Vegas, Nevada 89154-4010

ABSTRACT

Sedimentologic and paleontologic analyses of a 10.5 m subsurface section from the Estancia Valley provides a high-resolution paleolimnologic record of a pre–late Wisconsin section that is suspected to represent early through middle Wisconsin time. Relative abundances of ostracodes, foraminifers, *Ruppia,* and charophytes, along with sedimentologic aspects, depict a time interval characterized by alternating dry and wet playas and shallow saline lakes. The saline lakes exhibit marked evolution from brackish to saline to, presumably, hypersaline conditions. The overall paleoenvironmental conditions are in great contrast to those represented by the overlying full glacio-pluvial sediments of late Wisconsin age.

The pre–late Wisconsin section, as a whole, represents cold/dry paleoclimatic conditions. Minor climatic excursions from this norm are recognized, but the section is more notable for its quasistability. The lack of any fresh-water phase during this time implies that there was no regional climatic shift of the magnitude of the late Wisconsin. It is surmised here that if a cold/dry climate results in saline lakes, at the extreme, then full glacio-pluvial systems must require a significant increase in precipitation along with cold paleoclimatic conditions in order to maintain deep, fresh-water lakes in the currently arid southwestern United States.

INTRODUCTION

Pluvial environments throughout the world and especially in the southwestern United States provide a detailed record of climatic change. Although our knowledge of late Wisconsin pluvial systems is profuse, there are great uncertainties in regard to chronologies and paleoenvironmental conditions during the last interglacial/glacial transition. The Lahontan (Morrison and Davis, 1984) and Bonneville (Scott and others, 1983) basins, among others, apparently contain a deep-water pluvial record that predates the last interglacial, but the records are not continuous over the transition to early and middle Wisconsin time. Whereas the oceanic record provides a general framework of climate change during this time, the terrestrial record, for the most part, is scarce or poorly understood.

The Estancia Valley in central New Mexico contains a subsurface pre–late Wisconsin section, part of which, I believe, is of early and middle Wisconsin age. Although geochronologic uncertainties still exist, the sedimentologic and paleontologic data from the subsurface section imply limnologic and climatologic contrast to modern and late Wisconsin conditions. The paleoclimatologic inferences derived from the section may have regional significance.

Regional setting

The Estancia Valley watershed lies within a physiographically closed basin near the geographic center of the state of New Mexico (Fig. 1). The north-trending, elliptically shaped basin is ~80 km long and 40 km wide, with a total watershed area of about 5,200 km^2. The highest elevations within the watershed (up to 3,078 m) occur along the west-flanking Manzano and Sandia mountains. Basin closure to the east is provided by the Pedernal Hills (up to 2,312 m), and to the south by Juames Mesa (2,106 m). To the north, a broad saddle (~1,983 m) separates the Estancia Valley from the Galisteo watershed. The lowest elevations occur in the semiarid central portion of the watershed,

Bachhuber, F. W., 1992, A pre-late Wisconsin paleolimnologic record from the Estancia Valley, central New Mexico, *in* Clark, P. U., and Lea, P. D., eds., The Last Interglacial-Glacial Transition in North America: Boulder, Colorado, Geological Society of America Special Paper 270.

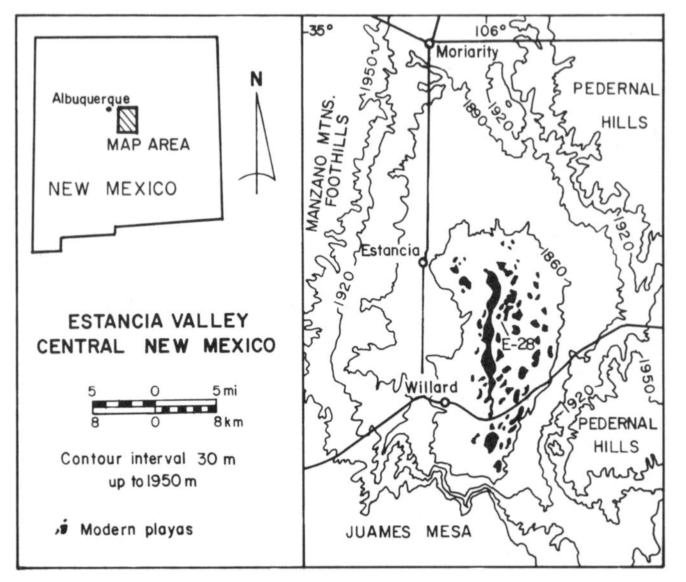

Figure 1. Index map of the Estancia Valley in central New Mexico indicating distribution of modern playas and location of playa E-28 (right center). Sandia Mountains 35 km northwest of map border. Original contour lines in feet, converted to meters.

where a complex of deflation basins are incised up to 10 m into the valley floor (~1,853 m). The deflation basins are floored by playas and ringed by stabilized parabolic dunes.

Present climate and hydrology

Today the mean annual temperature of the central part of the Estancia Valley is 10 °C, with a 43 year mean July temperature of 21 °C (warmest month) and a mean January temperature of 0 °C (coldest month) (Bachhuber, 1971). Annual precipitation within the central part of the valley is 12.3 cm (watershed average of 36.6 cm; Smith and Anderson, 1982) with 34% falling during July–August as high-intensity rainstorms. The modern temperature range and precipitation pattern translates to an annual water-budget deficit of 152 cm (Harbour, 1958). This deficit precludes the existence of perennial water bodies in the deflation basins of the valley, which remain as dry playas throughout the year. Each deflation basin, ringed by a parabolic dune, is a small drainage catchment isolated from high-elevational surface runoff, but not from regional ground-water recharge and flow from local perched water tables. The water table within the deflation basins fluctuates seasonally on the order of 0.5 m, the highest level occurring during the winter months. Playas contain ~1.5 m of playa sediment, but the basins today undergo neither net aggradation nor degradation.

Holocene

The Holocene history of the Estancia Valley is not well understood. However, there are at least two generations of dunes, the first deflation/dune-building event possibly having been terminated by a subpluvial episode (Bachhuber, 1982). The last deflation event, culminating in the modern deflation-basin/parabolic-dune complex, commenced after 4,660 ± 170 yr B.P. (GX-13321), an age derived from a charcoal horizon at the base of a parabolic dune. The modern deflation basin-dune complex is currently inactive, suggesting that, during dune formation, the Estancia Valley was more xeric (hotter and/or drier) than at present.

Late Wisconsin

The position and great number (more than 80) of deflation basins carved into the valley floor provide a unique opportunity to study a complex late Quaternary lithostratigraphic and biostratigraphic sequence. The flanks of the deflation basins expose an up to 10-m-thick, high-resolution, late Wisconsin paleolimnologic and paleoclimatic record, described by Bachhuber (1982, 1989). The record delineates two major pluvial episodes, late Lake Estancia and Lake Willard, separated by interpluvial sediment of the Estancia Playa complex (Fig. 2). Late Lake Estancia, the older and larger of the pluvial systems, culminated in a deep, oligotrophic, fresh-water phase, but water-level and salinity fluctuations are evident prior to and following this phase. The first of three fresh-water phases of late Lake Estancia is radiocarbon dated at 24,300 ± 560 yr B.P. (AA-1868) (all radiocarbon ages discussed in text are listed in Fig. 3). This age, obtained on salamander bones, marks the beginning of Wisconsin full glaciopluvial conditions in the Estancia Valley (Bachhuber, 1989).

Sediment of late Lake Estancia is overlain by the Estancia Playa complex, a unit characterized by subaerially deposited silt and gypsarenite (sand composed predominately of detrital gypsum). These interpluvial sediments represent deposition in a broad playa developed on the desiccated surface of late Lake Estancia. Following interpluvial conditions, the basin was again flooded, with subsequent deposition of Lake Willard sediment. Lake Willard did not attain the size or depth of late Lake Estancia, but its fresh-water nature is documented by Bachhuber (1989). The second of two high-water stages of the lake is radiocarbon dated at 12,460 ± 135 yr B.P. (Beta-25819, ETH-4155), a single accelerator date on ostracode valves. Lake Willard probably persisted into the early Holocene, at which time the lake desiccated. The desiccation event is marked by a 1-m-thick, indurated gypsarenite termed the Willard soil. Late Lake Estancia and Lake Willard contain abundant paleontologic evidence indicating that each had at least one fresh-water phase. These distinct fresh-water assemblages are unique for Wisconsin time.

Pre-late Wisconsin

In contrast to the highly detailed late Wisconsin record, the older subsurface record has remained poorly understood. Based on test-hole and outcrop data, Titus (1969) recognized (1) a lower alluvial unit that rests on bedrock, (2) a lower lacustrine clay, (3) the Medial sand of alluvial and eolian origin, and (4) an

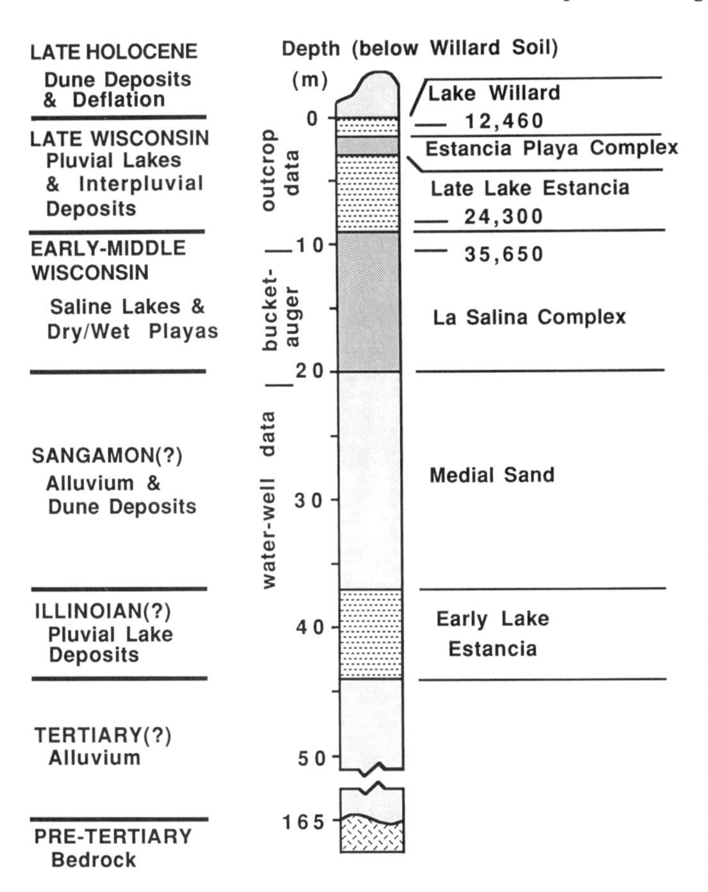

Figure 2. Generalized stratigraphic section of Quaternary and pre-Quaternary deposits of the Estancia Valley, derived from outcrop and subcrop data. Approximate stratigraphic location of radiocarbon ages on right (see Fig. 3 for radiocarbon age detail).

upper lacustrine clay. The upper lacustrine clay is correlative with the late Wisconsin pluvial stands established by Bachhuber (1971, 1982, 1989), but also includes a pre-late Wisconsin section, likely of early and middle Wisconsin age. The stratigraphic section between the Medial sand and the late Wisconsin sediments of late Lake Estancia is herein termed the La Salina complex (Fig. 2). In addition, Titus' Medial sand and the lower lacustrine clay (early Lake Estancia) are now thought to be no younger than Sangamon and no younger than Illinoian in age, respectively (Bachhuber, 1989) (Fig. 2).

Whereas the late Wisconsin section crops out along the flanks of most of the deflation basins, the uppermost portion of the La Salina complex is exposed only along the sides of the most deeply incised basins. These basins occur in the north-central portion of the deflation-basin complex, centered around playa E-28 (Fig. 1). Playa E-28 is one of the more deeply wind-excavated playas and thus contains one of the longest outcrop sections. Also, it is located in the center (deepest-water) portions of late Lake Estancia, Lake Willard, and, probably, the basins of pre-late Wisconsin time. The playas flanking E-28 step upward elevationally toward the basin margins, and therefore exhibit a shorter stratigraphic record. In addition, the sections exposed along the sides of these flanking playas have a more littoral aspect.

In earlier work (Bachhuber and McClellan, 1977; Bachhuber, 1982), the pre-late Wisconsin sediment exposed along the flanks of the centrally located playas was termed "pre-pluvial" and was believed to represent the early infilling stage of late Lake Estancia. Although this is partly the case, the lower portions of the exposures are distinctly pre-late Wisconsin in age with radiocarbon ages of 30,440 ± 520 yr B.P. (Beta-25542), >33,000 yr B.P. (A-1001), and 35,650 +3000/−2180 yr B.P. (A-4903). The last two dates could be considered as nonfinite minimum-limiting ages. The bottom of the exposed section as well as the upper part of the subsurface section are, therefore, at least of middle Wisconsin age. Age assignment of the lower part of the subsurface section, however, is speculative.

Sample collection. To better ascertain the nature of the La Salina complex, a trench was dug from the base of the E-28 outcrop section through the thin colluvium toward the edge of the playa. The base of the Willard soil was used as the zero datum. Approximately 10.5 m of sediment was exposed from the Willard soil to the bottom of the 1.5-m-deep trench. Below this point, bucket-auger samples were collected at 10 cm intervals, because the stiff clay and intercalated gypsarenite precluded the use of a Livingston piston corer. Although the sediment within the bucket auger was disturbed, stratigraphic integrity was preserved at least within 10 cm intervals.

During the course of three trips, a total of 12.03 m of subsurface section was collected from along the sides of trenches and with the bucket auger, to a depth of 21.03 m below the Willard soil. In March of 1986, the subsurface section was first sampled to a depth of 16.3 m below the Willard soil (9.0 m of exposure, 1.5 m trench, 5.8 m bucket auger) (Bachhuber, 1987). In March 1987, a new trench was dug adjacent to the 1986 trench. Duplicate bucket-auger samples were collected with stratigraphic integrity established by a distinctive marker lamina, and new samples were obtained to a depth of 19.8 m below the Willard soil (Bachhuber, 1988). Uppermost samples were collected with a 4½" (11.4 cm) diameter bucket auger, but progressively smaller-diameter augers were necessary below, ending with a 2" (5 cm) diameter bucket. The upper 5 m of the March 1987 hole was cased with 4" (10 cm) diameter PVC drainpipe, and reentered in June 1987, with no noticeable wall collapse at the greater uncased depths. At this time, an additional 1.23 m of section was obtained. Sample collecting ended at 21.03 m below the Willard soil when the hole began to drift in compacted gypsum sand, believed to be the top of the Medial sand. The March 1987 hole encountered flowing artesian water at a depth of 7.5 m below the playa surface (18 m below the Willard soil). Flow into the playa was still occurring in June 1987. Upon completion of sample collection, the hole was plugged.

Sample treatment. Trench and bucket-auger samples from the section and samples from the lower portion of the outcrop

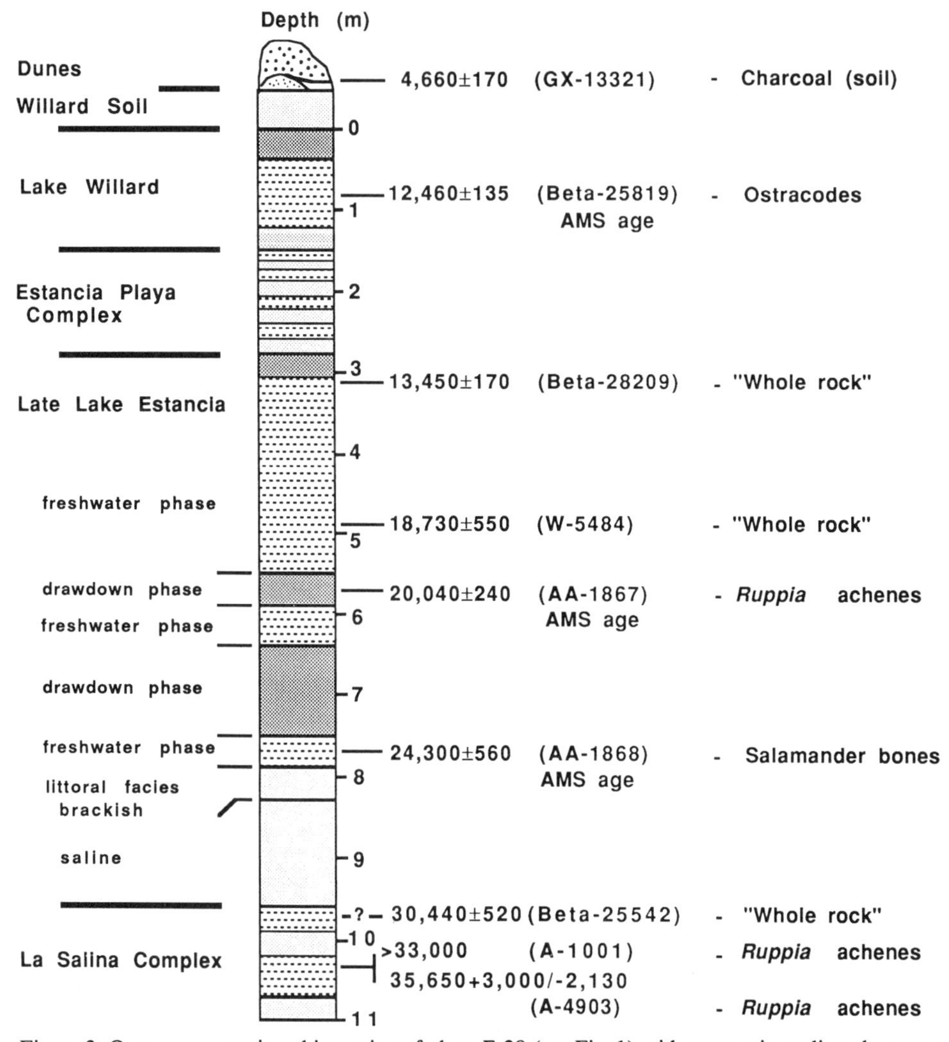

Figure 3. Quaternary stratigraphic section of playa E-28 (see Fig. 1) with composite radiocarbon ages and type of material dated. Datum for this and all subsequent figures is the base of the Willard soil. Not all radiocarbon ages are referenced in text (see Bachhuber, 1989, for Late Lake Estancia detail).

were air dried, and then placed in a beaker with water. The ~250 gm of sediment generally disaggregated rapidly. Floating material, consisting mainly of microfossils and minor organic debris, was decanted off onto a filter, and the remaining sediment was wet sieved through a series of three sieves. Because of the large number of samples processed, various-sized sieve stacks were used. The coarsest sieves were 2.0, 1.4, and 0.833 mm; medium sieves were 0.600, 0.417, and 0.300 mm; and the finest sieves 0.180, 0.150, and 0.124 mm. Sediment and fossils passing through the finest sieves were not saved. Typically, this finest fraction represented 50% to 60% of the original dry weight. The use of different-sized sieve stacks has created problems in paleontologic and sedimentologic analyses, and in retrospect, a more standardized procedure should have been used. The data, however, are relatively uniform and consistent with the overall lithostratigraphy and paleoclimatic history of the region.

SEDIMENTOLOGY AND COMPOSITION

The La Salina complex, as a whole, consists of clay, silty and sandy clay, and gypsarenite. All sediment has a high clay or silt component, varying in color from tan to flint gray. Even though sediment disaggregation occurs rapidly in water, a significant portion of the original clay-silt matrix remains after wet sieving as sand-sized pellets. In point counts made on the finest-sieve fraction, pellets of original-matrix material were counted as such, but the counts were excluded from the other grain-component percentages.

Percentages or relative abundances of seven grain components are plotted in Figure 4: gypsum crystals subdivided into (1) laths; and (2) selenite; (3) bassanite; (4) valves and tests of ostracodes and foraminifers; (5) charophyte casts; (6) silicate detritus; and (7) wood. Lath-shaped gypsum, valves and/or tests, charophyte casts, and silicate detritus are plotted as percentages of grains counted on the fine-sieve fraction exclusive of original-matrix pellets; gypsum as selenite, bassanite, and wood are plotted as relative abundances.

Gypsum lenses. Gypsum occurs virtually throughout the section in one of three dominant modes: discoid or lens-shaped crystals, elongated blades or lath-shaped crystals, and selenite crystals or fragments. Discoid or lens-shaped crystals, constituted by (111) prisms, are found in almost all samples of the La Salina complex and late Wisconsin sediment. Because of their ubiquitous distribution throughout the section, their occurrence is not plotted in Figure 4. The crystals occur individually and only rarely as rosettes of intergrown crystals. Individual crystals vary in diameter from less than 0.05 mm to more than 4 mm. The lenses occur as the dominant component of thin, intercalated gypsarenites or "embedded" in a clay or silty-clay matrix. In gypsarenites, especially, the lenses commonly show the effects of transport and, at the extreme, may be abraded to thin disk-shaped grains. Exsolution effects on lenses are common.

The gypsum lenses are interpreted as the product of secondary growth in unconsolidated sediment (cf. Carozzi, 1960). They precipitate initially in a clay or silty-clay matrix under sulfate-saturated phreatic and vadose conditions. They form primarily along the margins of wet playas and the interior of dry playas. Once formed, the lenses either remain in the sediment matrix, are reworked by flash floods, or are deflated from the playa surface and incorporated into surrounding dunes, a process that is currently occurring. Gypsum lenses are then eroded from the dunes and transported back to the playa surface, where they are reincorporated into the sediment. Their occurrence in the late Wisconsin pluvial sediment as well as interpluvial and subpluvial sediment of the La Salina complex attests to the efficient recycling of lenses under highly varied climates, even full-pluvial conditions. However, lenses are particularly numerous in the subsurface section of the La Salina complex, implying a limnologic boundary condition manifested by alternating dry and wet playas and shallow, saline-lake environments. These depositional environments are defined more fully in the section on paleolimnologic interpretation.

Elongated blades or lath-shaped gypsum crystals. Lath-shaped gypsum crystals, elongated parallel to [101] and (111), occur within a restricted stratigraphic interval in the subsurface section from 13 to 17 m. These crystals, where unbroken, are up to 5 mm long, with a length-to-width ratio of about 10:1. Elongation of the crystals is characteristic of precipitation from a

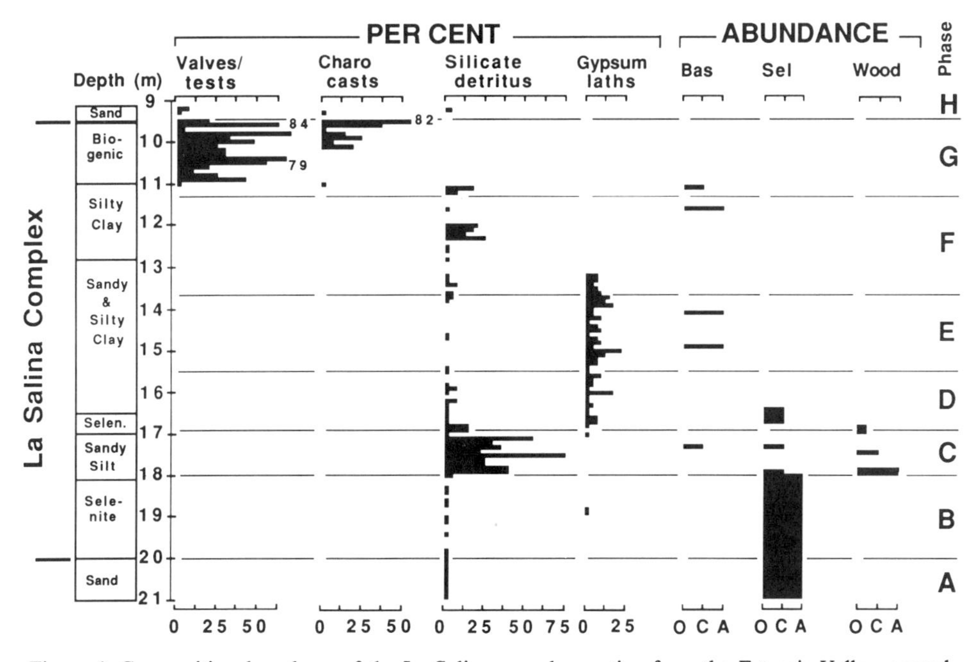

Figure 4. Compositional analyses of the La Salina complex section from the Estancia Valley, central New Mexico. Valves/tests = ostracode valves + foraminiferal tests; Charo casts = $CaCO_3$-encrusted, charophyte-stem casts; Silicate detritus = chert + lithic fragments + quartz + mica; Gypsum laths = lath-shaped gypsum crystals; Bas = bassanite; Sel = selenite; Wood = wood fragments. The first four categories are expressed as a percent of grains counted, exclusive of original-matrix pellets; the last three categories are expressed as a relative abundance: O = occasional, C = common, A = abundant. Phase is discussed in text.

sulfate-saturated, surface-water body (Ogniben, 1955), and, as such, the crystals are recognized as the only primary evaporite in either the La Salina complex or late Wisconsin sections of the Estancia Valley. The lath-shaped forms probably crystallized in shallow, isolated pools of saline water at the margins of wet playas or during the desiccation of dry playas, becoming incorporated in a clay or sand matrix during the next wet phase. The occurrence of lath-shaped gypsum between 13 and 17 m is interpreted to reflect alternating wet and dry playas, in contrast to more persistent, shallow, saline-lacustrine environments identified in other portions of the subsurface section.

Selenite. Selenite, coarsely crystalline gypsum, occurs as large crystals locally exceeding 1 cm in length or as fractured fragments. It occurs in abundance only in the lowermost portion of the subsurface section between 18 and 21 m, and in two discrete intervals between approximately the 16.5 to 17.5 m level. Samples to a depth of 20 m typically have 40%–50% of retained-sieve weight on the coarse sieve and a loss of 40%–50% of total sample weight through the fine sieve. The medium- and fine-sieve fractions also contain high percentages of fractured selenite. The flowing artesian water mentioned above originates predominantly from these selenite-bearing sediments. The selenite most likely represents secondary growth where sulfate-rich ground water (9360 mg/l) percolates through coarse- and fine-grained sediment under artesian conditions. The pervasive precipitation of selenite crystals has almost completely altered the original texture and composition of the host sediment. For this reason, selenite, a compositional term, is used in the stratigraphic columns of Figure 4 and other figures.

Below a depth of 20 m, the sediment is still dominated by selenite, but it becomes highly fractured and compacted. Average grain size also decreases significantly. The bulk of the material (up to 60% of total-sieve weight) is retained on the fine sieve, and less than 10% of total sample weight is lost through the fine sieve. I interpret the lowermost 1 m of section as detrital gypsum of unreworked gypsum-dune material, predating the La Salina complex.

Bassanite. Bassanite, a hemihydrate of gypsum ($2CaSO_4 \cdot H_2O$) (Sonnenfeld, 1984), is recorded at five stratigraphic levels (Fig. 4). In these horizons, bassanite, as determined by X-ray analysis, clearly replaces gypsum lenses and lath-shaped crystals. In some samples, macroscopic replacement of gypsum by milky bassanite varies from 0% to 100%.

Degens (1965) indicated that bassanite replaces gypsum only above 100 °C. This suggests that replacement of gypsum by bassanite normally occurs only at considerable burial depths. In the Estancia subsurface section, however, burial depth was never greater than 20 to 40 m, a depth where temperature was invariably less than 100 °C. In addition, bassanite occurs in the Estancia sediments in discrete horizons, separated by intervals where replacement is not in evidence. Therefore, it appears that gypsum can dehydrate at temperatures below 100 °C in surface environments.

Moiola and Glover (1965) documented gypsum replacement by bassanite on Clayton Playa, west-central Nevada. There, gypsum crystals were observed to grow in a silty-clay matrix following exposure of the sediment and subsequent drying. Many of the crystals were coated with a white "adherent material" that consisted of bassanite and anhydrite. The process of gypsum-crystal formation, followed by dehydration and the formation of a thin bassanite rind, occurred over 11 months of subaerial exposure. In general, I believe this is the process responsible for bassanite formation in the Estancia sediments. Here, a white coating on gypsum crystals does exist, but true pseudomorphs of bassanite after gypsum also occur. These relations seem to indicate periods of subaerial exposure substantially longer than a year. The occurrence of bassanite in the La Salina complex section is therefore interpreted as representing short-term climatic shifts to more xeric and hotter conditions.

Valves/tests. In the upper portion of the subsurface section and the lower portion of the outcrop section, ~9.5 to 11 m, ostracode valves and foraminferal tests compose a significant portion of the sediment retained on the fine sieve. The extremely high valve/test percentages of some samples classifies them as either ostracodal or foraminiferal sand. Although treated as sedimentologic entities in Figure 4, the valve/test category attains greater significance in terms of the paleoecology of the specific organisms, as discussed in a following section. In general terms, the large number of valves/tests is a reflection of the biologic productivity and quasistability of shallow, brackish to saline, lacustrine environments.

Charophyte casts. Charophytes are submerged thallophytic plants that commonly have $CaCO_3$ encrustations around thallus sections and the female fruiting structures (oogonia, or when calcified, gyrogonites). Remnants of the plant (oogonia, calcified and uncalcified mats of thalli, and impressions on clay) occur throughout the section, but only at the higher stratigraphic levels is it a significant sedimentologic component (Fig. 4). At these levels, 9.5 to 10.2 m, calcified thallus casts occur in abundance and, similar to valves/tests, indicate lacustrine stability and high biologic productivity relative to lower parts of the section.

Silicate detritus. Detrital silicate grains consisting of chert, lithic fragments, quartz, and mica occur sporadically throughout the section, forming a significant component of the sediment between 17 and 18 m (Fig. 4). Chert is typically black and commonly contains fusilinids, bryozoans, and other fossils. The most likely sources of the chert are the Paleozoic outcrops in the west-flanking Manzano Mountains and the south-flanking Juames Mesa. Lithic fragments, derived from various metamorphic and granitic rocks, as well as quartz and mica, were likely derived in part from the Pedernal Hills, a Precambrian metamorphic complex forming the eastern closure of the basin. The Precambrian core of the Manzano Mountains to the west is also a likely source for this material. The abundance of silicate detritus between 17 to 18 m may record increased alluviation due to flash floods or small, perennial streams draining from the surrounding highlands. Sedimentation occurred on dry playa surfaces or in shallow, wet playas.

Wood. Fragments of unidentified wood are common in the sediment containing the highest percentages of detrital silicate material (Fig. 4). The wood was probably derived from coniferous trees growing at high elevations along the margins of the basin and, together with the silicate detritus, washed into the basin by flash floods or small, perennial streams.

PALEONTOLOGY

Sieved and dried samples from the La Salina complex were placed in individual 100 × 15 mm plastic petri dishes, which proved to be expedient for efficiency of sample storage and reexamination, and for their weak electrostatic properties. Sediment samples with low fossil densities could be gently shaken in the dishes, where electrostatic forces held the fossils to the dish surface when the sediment was poured off. This procedure facilitated the recording of very low densities. All three sieve sizes and the floating fraction for each sample were microscopically examined for their fossil content.

Ostracodes. The relative abundance of ostracode valves was determined by counting all juvenile and adult valves and valve fragments of 0.1 gm (samples with high density) to 1 gm of sediment retained on the fine sieve. Adult ostracodes found on the coarse sieves were also noted and factored in the fine-sieve counts. Due to the significant number of fragments present in many of the samples, counting of broken valves was deemed necessary. In many cases, fragments could be identified to the generic level. In turn, based on the specific identification of complete valves within a particular sample and recognizing that ostracode diversity is low, fragments were assigned to a specific level. On the basis of crushing experiments of complete valves into different-sized fractions, and noting the number of the fragments of various size from complete crushed valves, the valve-fragment tabulation was extrapolated to complete-valve equivalents. These data, along with complete-valve counts, are recorded as number of valves per gram on Figure 5. Because relative abundance of ostracodes varies from one valve per gram (smallest unit plotted is <5/gm) to more than 13,000 valves per gram, the figure contains a scale change. The smallest subdivision plotted in Figure 5 includes samples that had no valves encountered in the one gram sample, but did have observable valves in the sample as a whole.

Foraminifers. The relative abundance of foraminifers was determined by the same procedure used for ostracodes. All juvenile and adult tests within 0.1 or 1 gm samples of the finest-sieve fraction were tabulated and plotted as number of tests per gram (Fig. 5). Foraminifers varied from 1 test per gram to more than 8,000 per gram, again necessitating a change of scale in the figure.

Ruppia. *Ruppia,* a rooted, submerged, aquatic plant, is rep-

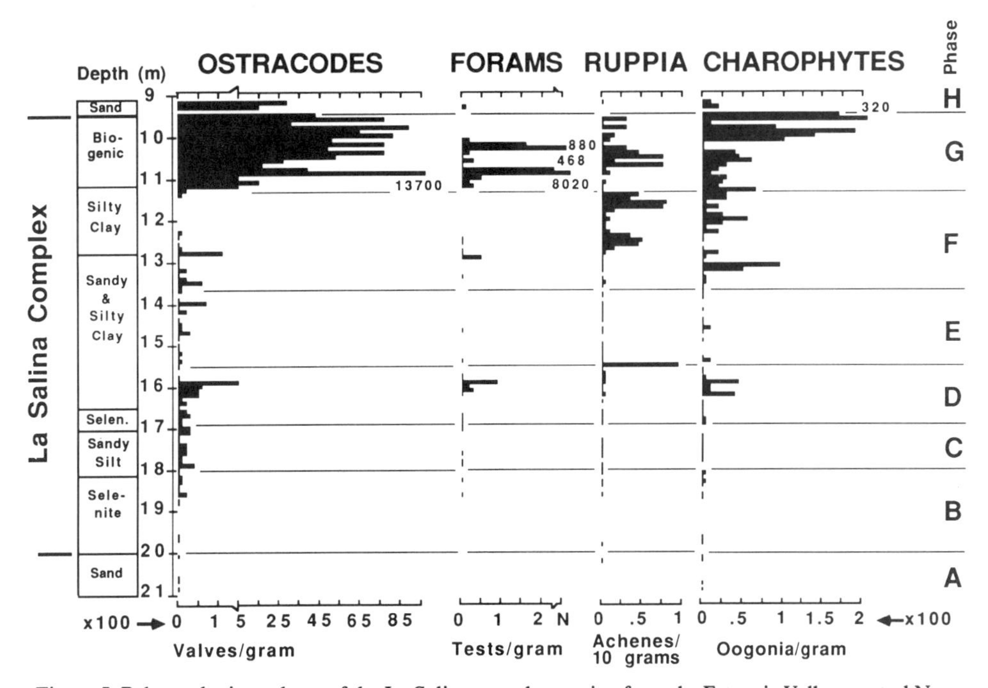

Figure 5. Paleontologic analyses of the La Salina complex section from the Estancia Valley, central New Mexico. Data are expressed as actual counts/gram or counts/10 gram. Ostracodes = total of all valves and full-valve equivalents; Forams = total of all foraminiferal tests; Ruppia = total of all *Ruppia* achenes; Charophytes = total of all oogonia. Note scale change of Valves/gram and Tests/gram. All scaled values ×100. Phase is discussed in text.

resented in the Estancia section by achenes (seeds) and fruiting-stem attachments (podogyns and peduncles). Two fossil/sedimentologic associations are recognized. Ovoid achenes, with an average length of 2.5 mm, are found in gypsarenite and silt, whereas achenes with attached fruiting-stems are found in clay and silty clay. In the latter association, the achenes and filamentous fruiting stems form relatively continuous mats up to 2 cm in thickness. The persistent attachment of the achenes and fruiting stems suggests little or no transport and, hence, deposition of the material essentially at the growing site. Current knowledge of the plant's ecology (Verhoeven, 1979) suggests a water depth of no greater than 2 m during formation of this biocoenesis.

The achenes and stem attachments are retained on the medium and coarse sieves or occur within the float. Even though many samples contain an abundance of stem attachments, only the number of achenes was tabulated. Data in Figure 5 are expressed as number of achenes per 10 gm of total sediment weight retained on all three sieves. *Ruppia* achenes vary in abundance from 0.5/10 gram to over 75/10 gram. The higher abundances are invariably associated with large numbers of peduncles and podogyns, many of which remain attached to the achenes. The smallest subdivision plotted (<2/10 gram) represents sparse occurrences based on achene and/or stem-attachment fragments.

Charophytes. Stem casts of charophytes are an important sedimentologic component in the upper stratigraphic portions of the La Salina complex sediment (9.5–10.2 m), but other remains of the algae are found throughout the section (see Fig. 4). Thick mats of uncalcified or partially calcified thallus segments are associated with the high-abundance stem-cast horizons. In addition, oogonia occur throughout the section as significant components of gypsarenite or scattered within clay and silty clay. These cylindrically whorled fruiting structures, both calcified (gyrogonites) and uncalcified, are retained on either the fine or medium sieve. The tabulation of oogonia is based on counting 1 gm samples of the sieve fraction containing the fruiting structures. Data are expressed as number of oogonia per gram (Fig. 5). The majority of oogonia do not have a carbonate encrustation, likely owing to the greater solubility of the high Mg calcite typical of the structures. No stratigraphic trends were observed in terms of uncalcified vs. calcified oogonia.

PALEOECOLOGY

The high abundances of various fossils in the subsurface section greatly influence sediment structure and texture. Sedimentologic interpretations, therefore, must take into account the relation between the sediment and the organisms living on the substrate or in the overlying water column. The main value of the organisms, however, is the understanding of their paleoecologic requirements. Because all organisms identified in the La Salina complex are extant, paleoecologic inferences can be drawn based on their modern distributions and ecology.

Ostracoda

Six species of ostracodes have been identified in the La Salina complex of the Estancia Valley, and an additional species coincides with the transition to the first, late Wisconsin, pluvial episode (Fig. 6). All of the species are also found in the late

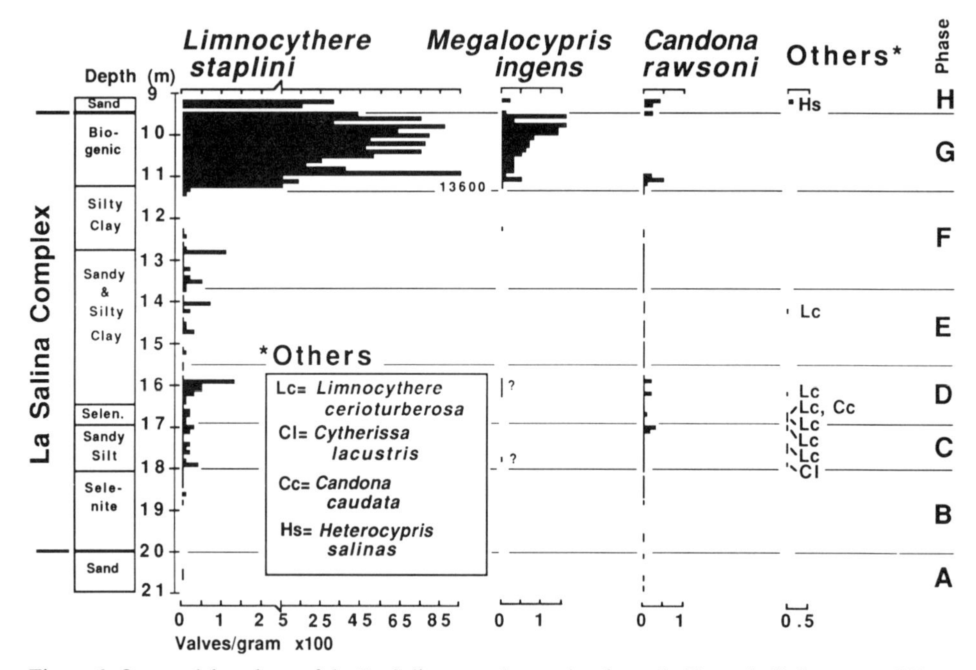

Figure 6. Ostracodal analyses of the La Salina complex section from the Estancia Valley, central New Mexico. Data are expressed as actual valve counts/gram of all identified species. Tabulations include full-valve equivalents. Note scale change of *Limnocythere staplini* graph. Phase is discussed in text.

Wisconsin pluvial section, which spans the time from the early infilling stage of late Lake Estancia through Lake Willard. The late Wisconsin ostracodes, originally described by Bachhuber (1971), have been reassessed based on examination of the material by R. M. Forester (1988, personal commun.).

***Limnocythere staplini* Gutentag and Benson, 1962.** *Limnocythere staplini* is the most-common ostracode in the Estancia Valley stratigraphic record. The ostracode is sporadically and, at times, abundantly found in the La Salina complex (Fig. 6), as well as having an almost universal distribution throughout the late Wisconsin section (Bachhuber, 1971). Although initially considered to be specifically a saline-water ostracode, *L. staplini* has now been identified in deposits that represent infrahaline conditions of deep, oligotrophic late Lake Estancia, as well as brackish and saline phases. Delorme (1969a) verified the apparent euryhaline nature of *L. staplini,* based on collections of the ostracode from modern Canadian lakes varying in salinity from 149 to 199,000 ppm total dissolved salts (TDS).

The euryhaline nature of *L. staplini* does not permit an evaluation of specific salinity conditions per se, although low relative abundances may be characteristic of fresh water. Whereas total salinity may not strongly influence the distribution of *L. staplini,* the ostracode may be sensitive to specific-anion composition (Forester, 1983). The modern occurrences of *L. staplini* plotted on an anion trilinear diagram (Forester, 1986) indicate that the species is restricted to alkaline-depleted, Cl- or SO_4-enriched water in saline environments. This interpretation is advanced for the occurrences of *L. staplini* within the La Salina complex.

***Candona rawsoni* Tressler, 1957.** *Candona rawsoni* occurs throughout the La Salina complex, but relative abundances are typically low (Fig. 6). Many occurrences, especially in the lower portions of the section, are based on valve fragments only. In contrast, high abundances of *C. rawsoni* occur in many stratigraphic horizons of late Lake Estancia and Lake Willard (Bachhuber, 1971), where infrahaline phases have been documented (Bachhuber, 1989). In the earlier work, the species was reported as *C. nyensis,* a junior synonym of *C. rawsoni,* and *C.* aff. *C. wanlessi,* now recognized as a juvenile form of *C. rawsoni.* The variations in relative abundances (i.e., productivity) may be a function of salinity, but, similar to *L. staplini*, *C. rawsoni* exhibits a high degree of salinity tolerance. Delorme (1969a) reported its occurrence in modern lakes varying from 141 to 42,770 ppm TDS, i.e., infrahaline to saline. However, even though the occurrence of the ostracode may not be specifically controlled by TDS, its occurrence is a reflection of anion composition. Plotting the modern occurrences of *C. rawsoni* on an anion trilinear diagram, Forester (1986) noted that the ostracode distribution falls into two main anion-compositional areas: one of alkaline-rich water and the other of alkaline-depleted water. The area of alkaline-depleted water partially coincides with that of *L. staplini.* It is believed here that the cooccurrence of the two ostracodes in the La Salina complex section indicates water of relatively high salinity that is enriched in the sulfate anion. This interpretation does not apply to the high abundances of *C. rawsoni* in late Lake Estancia and Lake Willard sediment. In these paleolakes, the high abundance of *C. rawsoni* relative to *L. staplini* indicates lower salinity and alkaline-enriched water.

***Megalocypris ingens* Delorme, 1967.** *Megalocypris ingens* is a common ostracode in the upper portion of the La Salina complex (9.5–11 m) (Fig. 6). Delorme (1967) described the species as *Cypriconcha ingens,* but later determined that the genus *Cypriconcha* was congeneric with *Megalocypris,* the senior synonym (Delorme, 1969b). The species was originally reported in the late Wisconsin section in the Estancia Valley as *Eucypris* sp. in what was believed to represent shallow, brackish to saline conditions (Bachhuber, 1971). Although there are few published ecologic data on *M. ingens,* Delorme (1967) reported the species from two lakes in south-central Saskatchewan, including Waldsea Lake, described by Last and Schweyen (1985) as a saline, meromictic lake with mixolimnion of 29,000 ppm TDS and monimolimnion of 53,000 ppm TDS. In addition, R. M. Forester (1989, personal commun.) observed that the species typically lives in highly ephemeral environments, including the fluctuating littoral zone of lakes. These observations support the original paleoecologic interpretation of Bachhuber (1971).

In the late Wisconsin section within the Estancia Valley, high abundances of *M. ingens* are found in gypsarenite. In these occurrences, the valves are usually convex upward with a strong perferred long-axis orientation, indicating reworking of littoral sand by gentle currents. In the La Salina complex, however, complete valves are rarely encountered. *M. ingens* has a large (up to 3 mm in length), fragile valve that is extremely susceptible to breakage. Its occurrence as disseminated fragments in the La Salina complex, therefore, indicates moderately strong wave action and redeposition in the littoral zone. Data plotted in Figure 6 represent whole-valve equivalents based on the counting of fragments, many of which are extremely small. As a result, it is suspected that *M. ingens* is moderately overrepresented relative to the other species. Nonetheless, the occurrence of *M. ingens* in the upper portion of the La Salina complex reinforces the interpretation of dynamic saline environments.

Other ostracodes. Three additional species of ostracodes are identified in the La Salina complex (Fig. 6). All three exhibit extremely low relative abundances and restricted stratigraphic distribution. Ecologic data suggest that the three species, *Limnocythere ceriotuberosa* Delorme, 1967, *Cytherissa lacustris* Sars, 1863, and *Candona caudata* Kaufmann, 1900, have a strong preference or exclusivity for fresh-water conditions. *C. lacustris,* especially, is found living in deep, cold-water, limnetic environments (Delorme, 1969a). All three species are reported from fresh-water phases of late Lake Estancia and Lake Willard, although *L. ceriotuberosa* was identified as *L.* aff. *L. verrucosa* (Bachhuber, 1971). The occurrence of these species in the La Salina complex, therefore, might indicate a fresh-water lake stand. Other evidence, however, indicates that this was not the situation. The extremely low relative abundances (one valve of *C. caudata,* one valve fragment of *C. lacustris*), cooccurrence with foraminifers, and a high percentage of detrital silicate material (see Fig. 4) all point to a redeposited assemblage. I believe that these fresh-water forms have been reworked from elevationally high, basin-margin sediment of early Lake Estancia, interpreted as Illinoian(?) in age (Fig. 2) (Bachhuber, 1989). Early Lake Estancia has been documented only in subcrop (Titus, 1969), but it is conceivable that basin-margin exposures once ringed much of the valley. The redeposition of these ostracodes may be the first paleolimnologic evidence of the fresh-water nature of early Lake Estancia, as inferred by Bachhuber (1989).

One additional ostracode, *Heterocypris salinas* Brady, 1969 (*Cyprinotus salinas* in Bachhuber, 1971), occurs at a depth of about 9.2 m, within the lower portion of a 1.5-m-thick, poorly bedded, poorly sorted, gravel-bearing, sandy clay and clayey sand (only the base of which is depicted at the top of the section in Figs. 4–7). Within the unit, thin gypsarenite laminae are common, and charophytes, foraminifers, and other ostracodes are abundant. R. M. Forester (1989, personal commun.) reports that *H. salinas* inhabits basins fed by ground-water discharge and can tolerate rapid fluctuations in environmental parameters such as TDS and pH. The ostracode is most common in Ca-Mg-HCO_3-enriched fresh water to Ca-enriched, saline-water systems.

In the Estancia Valley, *H. salinas*-bearing sediment is overlain by the first flint-gray clay of late Lake Estancia (24,300 yr B.P.). The *H. salinas* zone is interpreted as representing a saline, littoral facies of embryonic late Lake Estancia of late Wisconsin age. This interpretation is supported by the cooccurrence with other saline-indicator organisms. In places, however, the *H. salinas* zone is characterized by a red-stained, blocky clay reminiscent of a geosol. The true character of the blocky clay cannot be determined because of poor exposures; it is found only in the few deepest deflation basins in a zone of relatively high ground-water seepage. If it is truly a geosol, the unit is superimposed on uppermost pre–late Wisconsin sediment and would represent an hiatus prior to establishment of late Lake Estancia ca. 24,300 yr B.P. At this time, however, such an interpretation is speculative.

Foraminifera

Marine foraminifers were originally reported from Quaternary sediment of the Estancia Valley by Bachhuber and McClellan (1977). Two species, *Protelphidium orbiculare* (Brady) and *Cribroelphidium selseyense* (Heron-Allen and Earland), are found in abundance in the sediment representing the early embryonic phases of late Lake Estancia and Lake Willard. These occurrences are used to delineate salinity conditions (25,000–35,000 ppm TDS) for the early phases of pluvial-lake growth and significantly depressed summer temperature by about 10 °C from that of the present (Bachhuber and McClellan, 1977). The occurrence of these same foraminifers in the La Salina complex suggests similar paleoenvironmental conditions.

Abundance peaks of foraminifers occur in the upper part of the section (10–11 m) and at about 13 m and 16 m (Fig. 5). These horizons are interpreted to represent relatively permanent, saline lakes of moderate depth. In contrast, low abundances of foraminifers in other portions of the section may represent shorter-term, wet-playa conditions. The occurrence of foraminifers in the early saline phase of both late Lake Estancia and Lake Willard, separated by a full-pluvial (i.e., fresh-water) climate, implies avian reintroduction of the organisms into the valley (Bachhuber and McClellan, 1977). The sporadic occurrence of foraminifers within the La Salina complex, however, may also be a function of organism survival in an encysted state during short-term evaporative episodes, such as occur in ephemeral saline lakes of Australia (Cann and De Deckker, 1981).

Ruppia

Ruppia achenes (Fig. 5) and fruiting stalks occur in great abundance in many stratigraphic horizons of the La Salina complex of the Estancia Valley. The plant remains are identified as *Ruppia maritima* (L.), but the taxonomy of North American *Ruppia* is not well known (Verhoeven, 1979), and a revision may be necessary. Regardless, as a genus, *Ruppia* can be a significant paleoecologic indicator because all modern representatives have an association with saline waters. *Ruppia* not only tolerates saline conditions but requires them for maximum growth and reproduction (Husband and Hickman, 1985). Although the plant does occur in low-salinity systems, it does not flower and fruit under this condition. The large numbers of achenes in the Estancia sediment, therefore, argue against a low-salinity depositional environment. At the high end of the salinity range, annuals, such as *R. maritima,* occur in waters of up to 217,000 ppm TDS (Brock, 1979). Perennials are somewhat more restricted, with the reported upper limit at 46,000 ppm TDS (Brock, 1982). Because the abundance peaks of *Ruppia* coincide with low abundances of other organisms in the Estancia section, the plant is generally believed to represent the more-extreme salinity conditions.

Charophytes

Charophytes, in terms of paleoecologic value, are one of the more perplexing fossil occurrences in the Estancia Valley. The remains of charophyte vegetation exhibit a stratigraphic distribution that implies paleoenvironmental control. Unfortunately, the environmental tolerances and the taxonomy of the charophytes found in Estancia sediments are not well constrained, which is a problem for the plant group as a whole (see Proctor, 1980). The charophytes, in general, are algae thought to thrive in fresh-water ponds and lakes, but some brackish-water and saline forms are recognized.

The Estancia charophytes were originally identified as *Chara canescens* (Bachhuber, 1971), a brackish-water form, but this identification is now in question. R. M. Forester (1988, personal commun.) has identified both *C. hornemanni* and *Lamprothamnium* sp. in the uppermost La Salina complex sediment, with the more-abundant *Lamprothamnium* being equivalent to *C. canescens* of Bachhuber (1971). Of the two forms, *Lamprothamnium*, especially, can be found in saline ponds with up to 75,000 ppm TDS (Bisson and Kirst, 1980). Bisson and Kirst noted that *Lamprothamnium* is not only capable of thriving in a wide range of salinities, but it is also adaptable to wide salinity fluctuations. Even though the identification of the Estancia charophytes is still questionable, their association with other organisms is consistent with brackish to saline environments during the deposition of the La Salina complex. Similar charophytes occur in modern ephemeral saline lakes of Australia (Burne and others, 1980).

PALEOLIMNOLOGIC INTERPRETATION

The La Salina complex contains a unique stratigraphic record within the Estancia Valley. The ~12 m of sediment, exposed along the flanks of deflation basins or tested with a bucket auger, reflect paleolimnologic trends that are in great contrast to those of the radiocarbon-dated late Wisconsin section, as well as those of the present dry-playa depositional environment. Lake Lake Estancia and Lake Willard, of late Wisconsin age, are typified by a fresh-water flora and fauna (Bachhuber, 1989), whereas the La Salina complex is defined and limited by its abundance and diversity of saline-indicator organisms, none of which are found in the valley today.

On the basis of sedimentologic and paleontologic trends, the sampled subsurface section and basal portion of the exposed section of the Estancia Valley is subdivided into eight paleolimnologic phases (Fig. 7), six of which compose the La Salina complex. The six phases of the La Salina Complex share certain similarities, and the paleolimnologic interpretations advanced herein for the various phases are not markedly different from each other. A case could be made that observed trends are more a function of local environmental facies than of basin-wide paleolimnologic conditions. This is especially true in the context of attempting to make basin-wide reconstructions on a single section. I believe, however, that the sampled section of the La Salina complex is representative of deposition within the most-interior basin of the Estancia Valley. This contention is drawn by analogy with sediments of late Lake Estancia and Lake Willard at playa E-28, which are the thickest and have the most-profundal aspect of all sections within the valley. Therefore, the paleolimnologic trends, depicted in the La Salina complex, are interpreted to represent mainly climatic change through time. In the following sections, no attempt is made to quantify the degree of climatic change. Paleoclimatologic inferences are relative, compared to the modern climate defined as warm/dry.

The paleolimnologic interpretations that follow are based on the recognition of three main paleoenvironmental conditions: dry playa, wet playa, and saline lake. The dry playa is a system in which ground-water support is lacking, and the valley floor is flooded only during and immediately after a large precipitation event. With termination of surface runoff, evaporation reduces the water level, and within days or weeks the playa surface is subaerially exposed. Although the temporary water column undergoes salinity change during drawdown, water quality may be relatively fresh to only moderately brackish throughout much of its existence. Water quality may permit certain organisms to

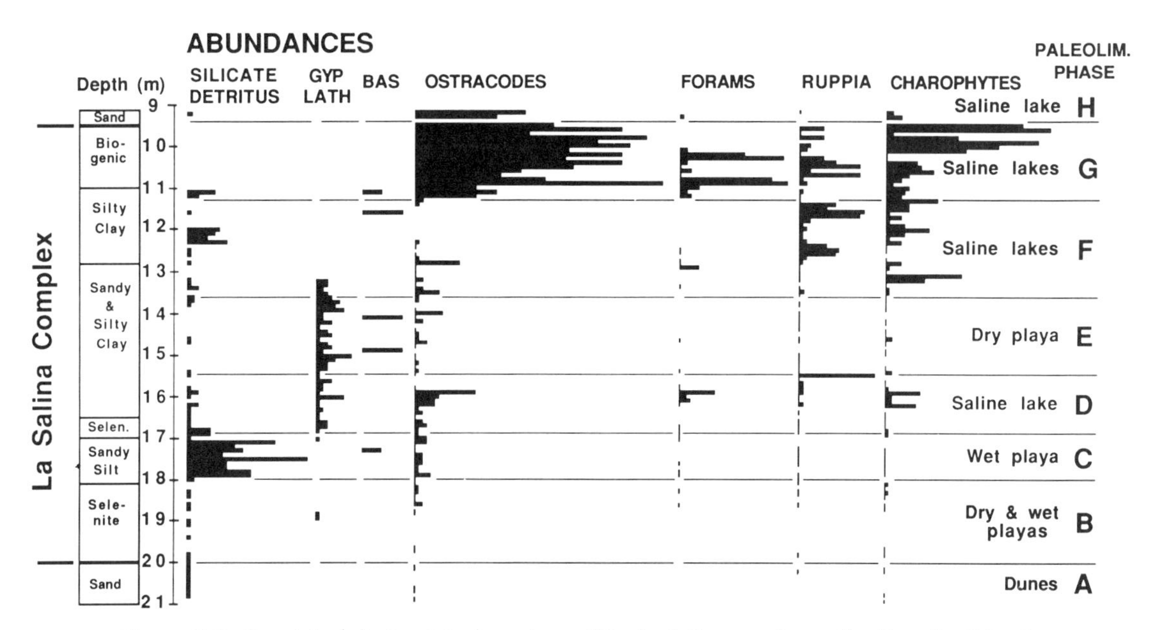

Figure 7. Sedimentologic/paleontologic analyses of the La Salina complex section from the Estancia Valley, central New Mexico. Data are expressed as relative abundances within individual categories (refer to Figures 4, 5, and 6 for percent, count, and abundance tabulations). GYP LATH = lath-shaped gypsum crystals; BAS = bassanite; FORAMS = foraminiferal tests.

"bloom," but the short residence time usually does not allow the growth and reproduction of complex organisms such as ostracodes. Dry-playa sediment is, therefore, biologically impoverished. In addition, much of the sediment (or precipitated minerals) is removed by deflation when the playa surface is subaerially exposed. As a result, thin sections of dry-playa sediment may represent long periods of time.

In contrast, the wet playa contains water more often than not. The wet playa, like the dry playa, is controlled by precipitation events and evaporation, but it has a more permanent groundwater support system. Therefore, the interior of the basin is subaerially exposed only infrequently. However, dramatic seasonal water-level fluctuations produce a broad ephemeral mudflat in which precipitation of minerals (such as gypsum lenses in the unconsolidated sediment or soluble salts on the surface) can occur. Salinity varies as water level and volume fluctuates seasonally, but the ratio of specific anions may remain essentially the same. Because anion ratios could have a greater influence on biologic occurrence and productivity than salinity per se (Forester, 1983, 1986), the wet playa could support organisms during seasonal changes in TDS. Within its longer time frame, a wet playa can be a biologically productive system that contains viable populations of complex organisms. But this system periodically "crashes" with resultant elimination of all organisms. The sedimentologic/paleontologic record of a wet playa should then be marked by rapid changes in abundance and occurrence of the various taxa.

The saline lake represents a permanent body of water of variable depth in which water chemistry evolves slowly. Although seasonal increases and decreases in salinity occur, specific-anion composition remains relatively unchanged (Forester, 1987). Water-chemistry stability permits high biologic productivity and a long-term reproductive potential that favors complex plant and animal groups. Within the context of stability, nonetheless, water chemistry does evolve over a long period of time. This permits the sequential establishment of populations that have significantly different ecologic requirements. In a saline lake, the paleontologic record should record these broader sequential changes and population replacements.

An additional concept introduced here is that of a saline-lake sequence. A saline-lake sequence records the evolution of a wet playa, typified by widely fluctuating water level and salinity, into a persistent shallow, saline lake. The saline lake also exhibits water-level and salinity fluctuations, but these occur over a longer time frame and with a gradual progression toward increased salinity and termination in a hypersaline condition. The concept being advanced is that some paleoclimatic threshold is passed which allows the wet playa to expand into a saline lake. Upon reaching this condition, continued paleolimnologic evolution occurs even within a fixed set of paleoclimatic parameters.

The hydrology of the three depositional systems discussed above varies only slightly, but I believe there are important limnologic thresholds between dry playa–wet playa, wet playa–saline lake, and saline lake–fresh-water lake. Movement across these thresholds, most often, is climate-forced.

Phase A. Sediments deposited during phase A (the bottom of the section, from 20 to 21 m) are distinguished by ~1 m of compacted gypsum sand, comprising well-sorted, highly fractured selenite. Whereas gypsarenites are common in both the pre–late Wisconsin and late Wisconsin sections, the gypsum sand of phase A is unique. In other intervals, the gypsarenites may be well sorted, but they invariably consist of reworked gypsum lenses, instead of fractured selenite (an exception being phase B sediment, discussed below). The sediment of phase A is remarkable for its thickness, uniform consistency, low loss of material through the finest sieve, and for its extremely low abundance of fossils.

Phase A sediment is interpreted as primary gypsum-dune material that accumulated on a broad, low-relief alluvial surface, with isolated playas and interdune areas serving as the source of fractured selenite. The selenite precipitated in these low-relief environments where ground water intersected the surface. Subsequently, the selenite, along with the few organisms occupying the intermittently flooded playas, was deflated into gypsum dunes. The depositional environment outlined here is analogous to the modern playa/dune complex of White Sands National Monument, about 240 km south of the Estancia Valley. Climatic conditions at White Sands are warm/dry, the interpretation advanced for phase A.

Phase A sediment occurs at the approximate subsurface level of the top of the Medial sand (Fig. 2), a 17-m-thick alluvial and eolian unit described by Titus (1969). Until a deeper core is obtained, correlation of phase A sediment with the top of the Medial sand will remain tentative. Phase A sediment, however, is distinct from that of the La Salina complex, and is considered an older unit.

Phase B. The sediments of phase B (18–20 m) mark the beginning of the La Salina complex. They are characterized by the occurrence of water-saturated, poorly sorted sandy clay and gypsarenites. The gypsarenites consist predominantly of selenite fragments of the fine and medium sieve range (0.124–0.6 mm). All sediment has an abundance of large selenite crystals and crystal fragments (>0.833 mm). The large selenite crystals, which have extensively altered the original sedimentary texture, are likely of secondary origin, related to the artesian flow of high-sulfate water through the sediment. Owing to the pervasive growth of selenite, the phase is defined more on postdepositional hydrologic properties than on paleolimnologic characteristics. Note, however, that the sand fraction of the sediment is similar to that of phase A, and as such may represent reworked dune material. Foraminifers first appear in the sediment of phase B, but the biostratigraphy is also defined by its impoverished flora and fauna. These aspects suggest either a dry playa or short-term wet playas, perhaps in association with eroding basin-margin gypsum dunes. The occurrence of playa sediment and foraminifers in phase B is interpreted as representing a regional rise in the ground-water table from that of phase A, and mean summer temperature of about 10 °C, or a depression of 10 °C from that of the present (Bachhuber and McClellan, 1977). The phase A/phase B transition is envisioned to be from a landscape dominated by gypsum dunes (phase A) to one dominated by playas (phase B).

Phase C. Paleolimnologic phase C (~17–18 m) is delineated on the basis of abundant detrital silicate material within a silty and sandy clay. The abundance of silicate detritus and the only occurrence of wood (see Fig. 4) suggests sediment input by flash floods or small perennial streams draining the surrounding highlands. Deposition probably occurred on a broad, subaerially exposed playa surface and/or the interior of shallow, wet playas. Minor climatic instability is reflected in the first occurrence of bassanite, which is believed to represent a subaerially exposed playa surface under hot and dry conditions, and minor ostracode peaks, mainly of *Limnocythere staplini* (Fig. 6), indicative of wet playas. Although phase C records the low-abundance occurrence of a number of fresh-water ostracodes (Fig. 6), these taxa are believed to be redeposited from basin-margin exposures of early Lake Estancia sediment of Illinoian(?) age (Bachhuber, 1989). The sedimentologic and paleontologic evidence indicates that phase C reflects more mesic conditions than phases A and B, attributed to an increase in watershed precipitation. If there was a significant increase in precipitation during phase C and summer temperature was as low as that inferred for phase B, however, the wet playas should have expanded into persistent saline lakes. The lack of a saline lake record in phase C, therefore, suggests that mean summer temperature and evaporation were higher than during phase B.

Phase D. Phase D (15.5–17 m) is the first saline-lake sequence recognized in the La Salina complex. The silty clay of phase D records a gradual increase in ostracodal abundance, terminating in a major abundance peak. The dominant ostracode is *L. staplini,* but *Candona rawsoni* is also present. The occurrence of either of these two ostracodes could reflect salinity conditions from infrahaline (if the waters are enriched with respect to alkalinity) to a maximum of 200,000 ppm TDS for *L. staplini* and 45,000 ppm TDS for *C. rawsoni* (if the waters are enriched with respect to sulfate). The cooccurrence of the two ostracodes, however, suggests a sulfate-dominated, Ca-enriched saline system (Forester, 1986). Although the absolute salinity is not known, it must have been moderately high, thus precluding the appearance of taxa with a lower salinity tolerance.

Coinciding with the ostracodal-abundance peak is the first major occurrence of foraminifers. The foraminifers may be indicative of dramatic temperature lowering, but they also constrain salinity to a range of 25,000–35,000 ppm TDS (Bachhuber and McClellan, 1977). The initial occurrence of the ostracodes in the absence of foraminifers indicates that the early developmental phase of the saline-lake sequence was brackish (upper limit of 25,000 ppm TDS) and enriched in sulfate. The stratigraphically higher cooccurrence with foraminifers constrains salinity within the 25,000–35,000 ppm TDS range, probably still enriched in sulfate.

As ostracodes and foraminifers flourished during phase D (an indication of productivity related to the stability of the saline system), charophytes reached their first abundance peak. If the charophytes are correctly identified as *Lamprothamnium,* their occurrence is consistent within the salinity range depicted by the other organisms. Their paleoecologic significance, however, is not so much as a salinity indicator, but rather as a possible competitor for limited habitation space. Charophyte abundance peaks, as well as those of other organisms, precede those of *Ruppia* in phase D and stratigraphically higher portions of the La Salina complex section. *Ruppia* is recognized as a saline indicator, but it could coexist with the other organisms in the lower salinity range, as well as at much higher salinities (up to about 200,000 ppm TDS), well beyond the tolerance of most ostracodes, foraminifers, and charophytes. The relative lag of *Ruppia* abundance peaks is probably a function of competition from charophytes, which are more successful at colonizing the same substrate and water depths (about 2 m) required by *Ruppia.* The inability of *Ruppia* to compete with other macrophytes in lower-salinity environments is well documented (Verhoeven, 1975, 1980). In the saline-lake sequence of phase D, it appears that *Ruppia* did not thrive until the upper tolerance limit for charophytes was reached. This limit is placed at about 75,000 ppm TDS, whereupon charophytes are eliminated from the lake with subsequent rapid colonization of the substrate by *Ruppia.*

Continued evolution of the saline-lake sequence to a hypersaline stage eventually resulted in the elimination of all organisms from the system. There is no mineralogic record of this final evolutionary stage in the form of evaporitic sequences, possibly due to deflation or by dissolution during a subsequent wet phase.

The paleontologic evidence of phase D exemplifies the concept of the saline-lake sequence. Changing abundances of the various fossil groups trace lake evolution from wet playas to, presumably, hypersaline conditions. Intermediate stages with salinity <25,000 ppm TDS, 25,000–35,000 ppm TDS, 35,000–200,000 ppm TDS, and >200,000 ppm TDS are recognized. Phase D paleolimnologic conditions may have been initiated and maintained by dramatically lowered summer temperatures (about 10 °C below present, as evidenced by foraminifers). In addition, I believe the annual precipitation regimen was also low, because evolution from a saline lake to a fresh-water system did not occur in phase D.

Phase E. Phase E (13.5–15.5 m) is defined by the relatively high abundance of lath-shaped gypsum crystals and the low abundances of fossils. In addition, bassanite, as pseudomorphs of gypsum lenses and lath-shaped crystals, occurs in two discrete horizons. The abundance of lath-shaped gypsum is suggestive of primary gypsum precipitation from shallow pools of saturated water. Although this could occur along the margins of saline lakes, this interpretation is not supported by the general absence of foraminifers, charophytes, and *Ruppia.* Minor ostracodal-abundance peaks of *L. staplini* (Fig. 6) may reflect short-term wet playas, but the paleolimnologic phase, as a whole, appears to represent a dry-playa depositional environment. The gypsum laths, bassanite, and impoverished biota suggest a lowered regional water table from that of phase D, with subaerial exposure of the basin interior. Furthermore, the occurrence of bassanite implies higher temperatures and reduced effective moisture.

Phase F. The paleontologic trends of a saline-lake sequence noted in phase D are again in evidence in phase F (11.2–13.5 m). Abundance peaks of ostracodes, foraminifers, and charophytes are followed by a *Ruppia* peak. The lower portion of phase F is indicative of a persistent, saline lake that evolved from initial brackish condition to higher salinities. The saline lake of phase F does not appear to terminate in a hypersaline condition. Instead, as *Ruppia* decreases, charophyte abundance again increases. The increase in charophytes, along with a pronounced increase in detrital silicate deposition, may reflect a time of higher precipitation and consequent decrease in salinity of the lake. Because foraminifers and/or ostracodes do not show a corresponding increase in abundance, I speculate that, as salinity decreased, charophytes reoccupied the saline lake, but high specific-anion composition (probably sulfate) prevented the return of other organisms. The inferred lower-salinity excursion is followed by a return to high *Ruppia* abundance, lower charophyte abundance, and termination of detrital silicate input. All reflect a general decrease in precipitation and a corresponding increase in salinity level of the lake system.

The sedimentologic and paleontologic associations of phase F indicate a cold/dry climate with one pronounced higher-precipitation excursion. Paleoclimatic conditions were similar to those of phase D.

Phase G. Paleolimnologic phase G (9.2–11.2 m) marks the development of the most persistent and biologically productive saline-lake sequence identified in the Estancia Valley section. The interval is characterized by high abundances of all taxa, as reflected in the dominance of biogenic sediment, consisting of charophyte-stem casts and extremely high abundances of ostracodal valves and foraminiferal tests. In addition, the ostracode assemblage contains *Megalocypris ingens* as an important constituent (Fig. 6). The consistently high ostracode abundances suggest that during phase G, the Estancia Valley contained a large saline lake. However, variations in foraminiferal and *Ruppia* abundances attest to fluctuations of salinity and water depth. Such fluctuations, in turn, indicate a certain degree of climatic instability, but not to the point where the lake evolved into a fresh-water body or, at the other extreme, desiccated.

I infer that paleoclimatic conditions during phase G were cold and dry, essentially a continuation of the pattern initiated during phase F. Phase G sediment represents the upper part and termination of La Salina complex sedimentation. On the basis of radiocarbon ages (>33,000 yr B.P.; 35,650 +3000/–2130 yr B.P.) on *Ruppia* achenes and a "whole-rock" sample (30,440 ± 520 yr B.P.) from phase G, the upper part of the La Salina complex is clearly pre–late Wisconsin in age, and probably middle Wisconsin age.

Phase H. Paleolimnologic phase H is recognized as the earliest developmental stage of late Lake Estancia, a late Wisconsin pluvial system that culminated in a deep, oligotrophic, fresh-water body. Phase H contains many of the paleontologic elements found throughout the underlying parts of the section, but also records the first appearance of the ostracode *Heterocypris salinas*. This species may reflect the only significant increase in regional ground-water discharge into the Estancia Valley since the beginning of La Salina complex deposition. The phase H water body eventually grew under the influence of full glacio-pluvial climatic conditions that are inferred to have been cold/wetter, with resultant replacement of saline-indicator organisms by the fresh-water biota of late Lake Estancia. The late Lake Estancia section is characterized by fresh-water gastropods, pelecypods, ostracodes, algae, and cutthroat trout (Bachhuber, 1989), an assemblage vastly different in character from those in underlying sediments. Although late Lake Estancia culminated in a deep, oligotrophic body, lake-level and salinity fluctuations are evident. During times of increased aridity or temperature, the paleontologic elements typical of the La Salina complex return to the stratigraphic record.

PALEOCLIMATIC SYNTHESIS

It is evident that paleolimnologic conditions during deposition of the La Salina complex were significantly different from those of much of the late Wisconsin, and also different from those of the present. Paleontologic and sedimentologic data strongly support the contention that the La Salina complex records generally more xeric conditions than fresh-water late Lake Estancia, and was characterized by a system of alternating dry and wet playas and shallow saline lakes. The paleoclimatic inferences drawn from the proxy record can be compared with the modern climate and hydrology of the Estancia Valley, as well as hydrologic models previously purposed for late Lake Estancia (Leopold, 1951; Antevs, 1954; Harbour, 1958; Lyons, 1969; Galloway, 1970; Brakenridge, 1978; Smith and Anderson, 1982).

Modern dry playas. Present climatic conditions result in a dry playa system throughout the year. The dry playa/wet playa threshold is much closer to being crossed, however, during the six winter months (October through March, 1.7 °C mean temperature and 4.1 cm mean precipitation) than during the six summer months (April through September, 15.5 °C mean temperature and 8.3 cm mean precipitation). Theoretically, a transition into a wet playa during the winter period would require an increase in precipitation and/or a temperature reduction. A reduction in temperature below the winter mean of 1.7 °C, however, would reduce surface runoff, thus producing no immediate change in the hydrology of the system. Conversely, an increase in winter precipitation of 4 to 5 cm would probably move the dry playa into a wet playa.

In contrast, it is much more difficult to produce a wet playa during the summer period. Under the high evaporation of June through August (20 °C mean temperature, which is mainly responsible for the 152 cm annual water-budget deficit), a five-fold increase in precipitation would not result in a crossing of the dry playa/wet playa boundary. This transition would require an even more unrealistic increase in rainfall. If the summer period versus the winter period is viewed as a basis for predicting long-term climatic change, the crossing of a limnologic/hydrologic threshold will likely occur under conditions of reduced temperature (but not below a mean of 1 °C), even with annual precipitation as low as that of the present. Hydrologic models of the Estancia Valley support this contention.

Hydrologic modeling. In many respects, the Estancia Valley is the birthplace of hydrologic modeling. Leopold (1951) and Antevs (1954) used the valley and the strandlines of late Lake Estancia as a model for their pioneering work on glacio-pluvial climates. This pioneering work did not have the benefit of the paleolimnologic record, radiocarbon ages, and age relation of the strandline sequences. Although detailed work in the valley has increased our knowledge of the pluvial system, even today the strandline-age and basin-overflow relations are still questionable. This has lead hydrologic modelers to contrasting interpretations.

Hydrologic models for the Estancia basin, in general, fall into three groups: those that invoke (1) drastically depressed mean annual temperatures with no change in precipitation (Galloway, 1970; Brakenridge, 1978), (2) a dramatic increase in mean annual precipitation with no temperature change (Harbour, 1958), or (3) a combination of mean annual temperature reduction and an increase in precipitation (Leopold, 1951; Antevs, 1954; Lyons, 1969; Smith and Anderson, 1982). Smith and Anderson (1982) summarized the various models, and calculated the hydrologic balance in the closed-lake system required to stabilize the lake level at various strandline elevations. Their summary is presented in Figure 8, with modifications discussed below.

In the calculation of equilibrium volumes, Smith and Anderson (1982) relied on work by Bachhuber (1971, 1977, 1982) that interpreted basin overflow occurring during late Lake Estancia at about 12,000 yr B.P. The highest interior strandline was interpreted as the high level of Lake Willard, previously believed to be of Holocene age. Although these age relations and the timing of lake overflow have been reinterpreted (Bachhuber, 1989), the equilibrium volumes calculated by Smith and Anderson (1982) are still of paleoclimatic significance. They concluded that maintenance of an overflow lake stage, now believed to be early Lake Estancia of Illinoian(?) age (Bachhuber, 1989), likely required a mean annual temperature reduction of 5.5–7.5 °C and about a 40%–75% increase in mean annual precipitation. Maintenance of a lake level at the highest interior strandline, now believed to be of late Lake Estancia age and dated as late Wisconsin, likely required a mean annual temperature reduction of 4–5 °C with an increase in precipitation of about 40% (Fig. 8). These paleoclimatic conditions approximate those proposed by Leopold (1951) and Antevs (1954), but temperature depression is not as extreme as proposed by Galloway (1970) and Braken-

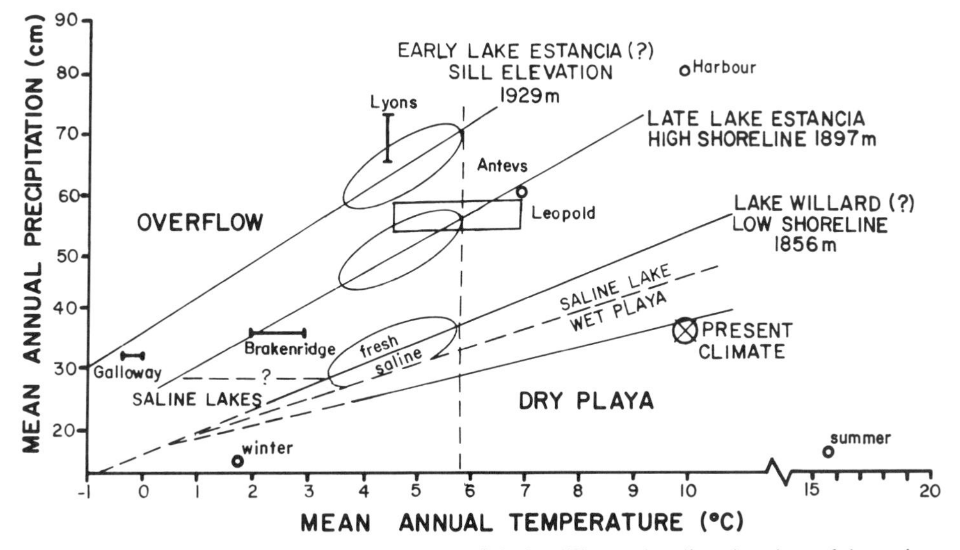

Figure 8. Graph showing climatic parameters possible for different shoreline elevations of the various pluvial stands, and saline lakes, wet playas, and dry playas. Ellipses represent probable climate during early Lake Estancia overflow, highest late Lake Estancia, and Lake Willard. Note the range of previous climatic estimates of pluvial conditions and the climate of modern dry playas during the winter (October–March) and summer (April–September) months. Vertical dashed line at 5.75 °C is inferred maximum mean annual temperature derived from foraminifer data. Redrawn and modified from Smith and Anderson (1982).

ridge (1978), whose models did not invoke a corresponding increase in precipitation.

On the basis of the occurrence of foraminifer *Protelphidium orbiculare*, which is believed to have an upper thermal-tolerance limit of 10 °C, Bachhuber and McClellan (1977) inferred a lowering of mean August air temperature of 9.7 °C from that of the present. They concluded that a summertime temperature depression of this magnitude would have assured the growth and development of late Lake Estancia and Lake Willard into fresh-water bodies, even without an increase in precipitation. This conclusion appears to support the paleoclimatic models of Galloway (1970) and Brakenridge (1978), but the 9.7 °C temperature depression is inferred only for the warmest summer month, and does not necessarily represent a mean annual temperature reduction of this magnitude. I now believe that summertime temperatures are more critical for the maintenance of a lake system in the Estancia Valley than winter or mean annual temperatures. If all months that currently have a mean temperature above 10 °C (May through August) underwent a temperature depression to 10 °C (the assumed lethal limit of the foraminifer) and the winter month averages remained the same, mean annual temperature for the valley would be 5.75 °C, a temperature higher than that proposed by Smith and Anderson (1982). This value, shown as a vertical dashed line in Figure 8, is considered to be a maximum mean annual temperature, because it is unlikely that significant temperature depression would occur only during the warmest months. Slight to moderate reduction in winter temperatures would be expected, thus reducing mean annual temperature below 5.75 °C, but probably not to the extreme proposed by Galloway (1970) and Brakenridge (1978).

Figure 8 depicts the hydrologic equilibrium line at various strandline elevations, as calculated by Smith and Anderson (1982). I have moved their ellipse of "probable climate during overflow" up the equilibrium line to coincide with a maximum mean annual temperature of 5.75 °C, derived from the foraminiferal data. For comparative purposes, similar ellipses are constructed for the highest interior strandline (late Lake Estancia) and the lowest strandline (probably a Lake Willard recessional stand). The position and size of the ellipses are conjecture, but should represent a reasonable range of paleoclimatic conditions during each lake phase.

The hydrologic model of Smith and Anderson (1982) was designed to provide insight into paleoclimatic conditions during high lake stands, but I believe it also represents the lower threshold conditions of the La Salina complex. The lowest equilibrium line of Smith and Anderson (1982), representing their "empty lake basin," is depicted in Figure 8 as the threshold between dry playa/wet playa. If the interpretation of the foraminifer data is correct, i.e., mean annual temperature maximum of 5.75 °C, the saline lake/fresh-water lake threshold can be partially established. The elevation of the lowest strandline approximately coincides with a foraminifer-bearing littoral facies of embryonic Lake Willard. The strandline and littoral facies probably are not time synchronous, but the relation suggests that a lake standing at 1856 m was probably saline. The equilibrium line at 1856 m (Fig. 8) then approximates the saline lake/fresh-water lake threshold. Consequently, the wet playa/saline lake threshold must fall between the 1856 m equilibrium line and the dry playa/wet playa threshold (Fig. 8).

A simple projection of the equilibrium lines/thresholds into the lower temperature range is probably an unrealistic representation of the three depositional environments. Instead, in the region of lower temperatures, I have extended the saline lake/fresh-water threshold line horizontally. The delimited region of a saline lake is for illustrative purposes only, but could represent a reasonable range of climatic conditions. It is clear that in the region of reduced mean annual temperature, where boundaries are closely spaced, a relatively minor change in precipitation will force a crossing of a threshold. Therefore, under cold/dry conditions, climatic quasistability can result in alternating dry playas, wet playas, and saline lakes, similar to the paleolimnologic sequence recorded in the La Salina complex. A crossing of the saline lake/fresh-water lake threshold, however, would require a greater degree of climatic change, i.e., precipitation comparable to that suggested by Smith and Anderson (1982).

La Salina complex paleoclimate and hydrologic conditions. The hydrologic model along with the comparison of the modern dry playa environment supports the climatic inferences derived from the La Salina complex paleolimnologic record. Data derived from the hydrologic model and the paleoclimatic inferences are summarized in Figure 9.

The most important climatic implication is that much of the La Salina complex was deposited during a time of quasistability, generally characterized by cold/dry conditions. A cold/dry climate was initially believed to be the impetus for the development and maintenance of late Lake Estancia under the influence of full glacio-pluvial conditions (Bachhuber, 1982), a hypothesis supported by other researchers (Brakenridge, 1978; Galloway, 1983). The quasiuniformity of the Estancia Valley La Salina complex record, however, now leads me to believe that cold/dry climates in the southwest are manifested by a system of dry and wet playas and saline lakes. Furthermore, it is now believed that growth and maintenance of large pluvial lakes, such as late Lake Estancia, requires a significant increase (40%–70%) in watershed precipitation. Therefore, full glacio-pluvial paleolimnologic systems in the southwest require cold/wet climatic conditions.

GEOCHRONOLOGY

The radiocarbon ages of phase G (Fig. 3) demonstrate the pre–late Wisconsin age of the La Salina complex. Finite ages of 30,440 ± 520 (Beta-25542) and 35,650 ± 3000/–2130 (A-4903) indicate a middle Wisconsin age for the upper part of the section, provided that such ages do not reflect contamination by younger carbon. Although chronologic uncertainties exist below the dated horizons, I believe that the La Salina complex is a relatively continuous, early through middle Wisconsin section. Major unconformities were not observed, petrocalcic horizons were not

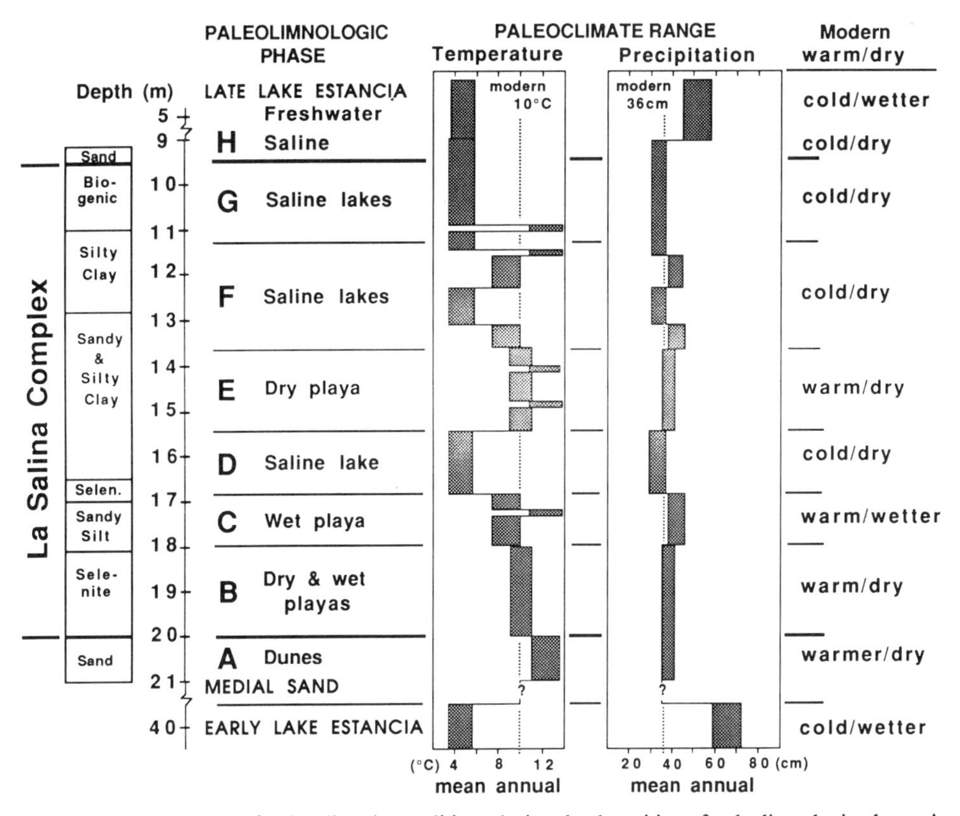

Figure 9. Inferred range of paleoclimatic conditions during the deposition of paleolimnologic phases A through H, the highest stand of late Lake Estancia, and early Lake Estancia.

encountered, and no major breaks occur in the paleontologic record. The section appears to be one of relatively continuous aggradation, but perhaps with some section thinning by deflation during dry-playa episodes.

If the complete subsurface section is one of stratigraphic continuity, phase A sediment, at the base of the La Salina complex, may be correlative with the uppermost part of the Medial sand (Titus, 1969), a 16-m-thick, well-sorted quartzose sand and gravel unit. Titus (1969) believed that the Medial sand was deposited by alluvial and eolian activity following the desiccation of early Lake Estancia, also known only in subcrop. Bachhuber (1989) suggested that the Medial sand is of Sangamon(?) age and the underlying clay of early Lake Estancia is of Illinoian(?) age. Although these age designations are speculative and should be considered as minimal ages, they appear to be reasonable in view of present knowledge of the subsurface section. An amino-acid racezimation study, being conducted on *Ruppia* achenes from the La Salina complex and late Wisconsin section, may help to resolve some of the geochronologic problems.

If the Estancia Valley was paleoclimatically in phase with other paleolake basins in the western United States, the following regional correlations are apparent. Sediments of early Lake Estancia may be equivalent in age to marine oxygen-isotope stage 6 (Shackleton and Opdyke, 1973). As such, they are probably correlative to the Little Valley alloformation of the Bonneville basin (Scott and others, 1983) and the Etza alloformation of the Lahonton basin (Morrison and Davis, 1984). The La Salina complex may correlate with oxygen-isotope stages 4 and 3, and may compare, in part, with the Cutler Dam alloformation of the Bonneville basin (Oviatt and others, 1987) and the Wyemaha alloformation (Morrison and Davis, 1984) of the Lahontan basin. Late Lake Estancia is clearly correlative with oxygen-isotope stage 2 and the Bonneville alloformation (Currey and others, 1984) of Lake Bonneville and the Sehoo alloformation (Morrison and Davis, 1984) of Lake Lahontan.

CONCLUSIONS

It is recognized (Bachhuber, 1989) that the Estancia Valley contains a high-resolution, late Wisconsin paleolimnologic and paleoclimatic record. In this chapter I demonstrate that a high-resolution pre–late Wisconsin record also exists. Sedimentologic and paleontologic analyses of a 10.5 m subsurface section leads to the following conclusions.

1. The La Salina complex of the Estancia Valley is bracketed by overlying late Wisconsin sediment, predominantly of pluvial-lakes origin, and is underlain by gypsum-dune material believed to be of Sangamon(?) age. Therefore, the minimum age

of the La Salina complex is middle Wisconsin, and the complex may record more-or-less continuous deposition throughout the early and middle Wisconsin.

Since completion of this chapter, a radiocarbon age has been obtained from the *Ruppia* abundance peak of phase D (15.5–17 m). *Ruppia* achenes at the top of the interval (Fig. 7) are dated at >48,800 yr B.P. (AA-6330). The nonfinite age further constrains the age of phase D sediment as no younger than lower-middle Wisconsin. This supports my contention that the La Salina complex spans early through middle Wisconsin time.

2. The La Salina complex exhibits sedimentologic and paleontologic trends that are believed to represent paleolimnologic conditions that vary from dry playa to wet playa to shallow, saline lakes. These trends permit a subdivision of the section into relatively distinct paleolimnologic phases, including saline-lake sequences in phases D, F, and G.

3. I believe that saline-lake sequences were initiated and maintained by cold/dry paleoclimatic conditions. Summer temperatures up to 10 °C lower than present and mean annual temperature at least 4.25 °C lower than present are inferred. Watershed mean annual precipitation approximated or was lower than the present (36.6 cm). Cold/dry paleoclimatic conditions were the norm during deposition of the La Salina complex.

4. Paleolimnologic trends suggest that paleoclimatic excursions from the norm did occur. Development of dry playas may reflect an increase in aridity or, more likely, an increase in summer temperatures, whereas, wet playas containing high silicate detritus may reflect a moderate increase in annual precipitation or runoff due to occasional large storms. Although minor temperature and/or precipitation excursions from the norm are noted, the La Salina complex paleoclimatic record is more remarkable for its quasistability.

5. Assuming that the La Salina complex represents a relatively continuous early through middle Wisconsin section, there are no paleolimnologic trends that permit a subdivision of the sediment into a distinct early Wisconsin and a distinct middle Wisconsin section. This suggests that there is no paleoclimatic difference between the two. Conversely, either the early or the middle Wisconsin section may be largely missing.

6. There is nothing even remotely suggestive of a fresh-water system in the La Salina complex sediment, despite inferred cold/dry climates. This is in great contrast to the overlying late Wisconsin section (late Lake Estancia and Lake Willard) and, presumably, early Lake Estancia of Illinoian(?) age. The inferred cold/dry conditions of full glacio-pluvial times, as proposed by other researchers (Galloway, 1983; Brakenridge, 1978), is then brought into question. Such climates may produce, at the extreme, saline lakes. Transition into pluvial conditions apparently requires the colder temperatures, at least during the summer months, but, in addition, there must be an increase in precipitation. Full glacio-pluvial climatic condition in the southwestern United States, therefore, are believed to have been cold/wet.

7. If large pluvial systems are time synchronous with major glacial advances and if pluvial systems are maintained by the same degree of climatic change responsible for glacial advances, the lack of a fresh-water lake in the La Salina complex suggests that there was no major glacial event during the time of deposition of the section.

ACKNOWLEDGMENTS

I thank Dann Halverson, Kurt Goebel, Allan Scott, Ed Thomas, and Todd Oppenborn for their assistance in the field. Their physical bearing provided the means for extracting the bucket-auger section. I thank R. M. Forester for examinaing the Estancia charophytes and ostracodes. His influence is felt well beyond the identification of a few organisms, for his meticulous research provides the basis of much paleolimnologic interpretation. Special thanks are due my daughter, Rachel Bachhuber, who spent untold hours picking "critters." I welcome her to the often-mundane world of geologic science. Funding for three radiocarbon ages was provided by the University of Nevada, Las Vegas Research Council. Radiocarbon ages Beta-25542 and Beta-28209 are used with permission of Roger Y. Anderson. Vera Markgraf provided radiocarbon age W-5484. Water analysis of the artesian system was provided by John Hess, Desert Research Institute. Much of the research done for this paper was completed at the University of Alberta while on sabbatical leave from the University of Nevada, Las Vegas. I gratefully acknowledge the financial support, use of facilities, and encouragement offered by both universities. The quality of this paper was greatly enhanced by reviews from Richard Forester, an anonymous reviewer, and the volume editors, Peter Lea and Peter Clark.

REFERENCES CITED

Antevs, E., 1954, Climate of New Mexico during the last glacial-pluvial: Journal of Geology, v. 62, p. 182–191.

Bachhuber, F. W., 1971, Paleolimnology of Lake Estancia and the Quaternary history of the Estancia Valley, central New Mexico [Ph.D. thesis]: Albuquerque, New Mexico, University of New Mexico, 238 p.

—— , 1982, Quaternary history of the Estancia Valley, central New Mexico, *in* Grambling, J. A., and Wells, S. G., eds., Albuquerque county II: New Mexico Geological Society, 33rd Field Conference, Guidebook, p. 343–346.

—— , 1987, An early to mid-Wisconsin climatic record from the Estancia Valley, central New Mexico: Geological Society of America Abstracts with Programs, v. 19, no. 7, p. 577.

—— , 1988, Paleontologic analysis of an early through middle Wisconsin section from the Estancia Valley, central New Mexico: Geological Society of America Abstracts with Programs, v. 20, no. 7, p. A206.

—— , 1989, The occurrence and paleolimnologic significance of cutthroat trout (*Oncorhynchus clarki*) in pluvial lakes of the Estancia Valley, central New Mexico: Geological Society of America Bulletin, v. 101, p. 1543–1551.

Bachhuber, F. W., and McClellan, W. A., 1977, Paleoecology of marine Foraminifera in the pluvial Estancia Valley, central New Mexico: Quaternary Research, v. 7, p. 254–267.

Bisson, M. A., and Kirst, G. O., 1980, *Lamprothamnium*, a euryhaline charophyte; I. Osmotic relations and membrane potential at steady state: Journal of Experimental Botany, v. 31, p. 1223–1235.

Brakenridge, G. R., 1978, Evidence for a cold dry full-glacial climate in the American Southwest: Quaternary Research, v. 9, p. 22–40.

Brock, M. A., 1979, Accumulation of proline in a submerged aquatic halophyte,

Ruppia L.: Oecologia, v. 51, p. 217–219.

——, 1982, Biology of the salinity tolerant genus *Ruppia* L. in saline lakes in S. Australia. I. Morphological variation within and between species and ecophysiology: Aquatic Botany, v. 13, p. 219–248.

Burne, R. V., Bauld, J., and De Deckker, P., 1980, Saline lake charophytes and their geological significance: Journal of Sedimentary Petrology, v. 50, p. 281–293.

Cann, J. H., and De Deckker, P., 1981, Fossil Quaternary and living Foraminifera from athalassic (non-marine) saline lakes, southern Australia: Journal of Paleontology, v. 55, p. 660–670.

Carozzi, A. V., 1960, Microscopic sedimentary petrography: New York, John Wiley and Sons, 485 p.

Currey, D. R., Oviatt, C. G., and Czaromski, J. E., 1984, Late Quaternary geology of Lake Bonneville and Lake Waring: Utah Geological Association Publication 13, p. 227–237.

Degens, E. T., 1965, Geochemistry of sediments: Englewood Cliffs, New Jersey, Prentice-Hall, 342 p.

Delorme, L. D., 1967, New freshwater Ostracoda from Saskatchewan, Canada: Canadian Journal of Zoology, v. 41, p. 357–363.

——, 1969a, Ostracodes as Quaternary paleoecological indicators: Canadian Journal of Earth Sciences, v. 6, p. 1471–1476.

——, 1969b, On the identity of the ostracode genera *Cypriconcha* and *Megalocypris*: Canadian Journal of Zoology, v. 47, p. 271–281.

Forester, R. M., 1983, Relationship of two lacustrine ostracode species to solute composition and salinity: Implications for paleohydrochemistry: Geology, v. 11, p. 435–438.

——, 1986, Determination of the dissolved anion composition of ancient lakes from fossil ostracodes: Geology, v. 14, p. 796–798.

——, 1987, Late Quaternary paleoclimate records from lacustrine ostracodes, *in* Ruddiman, W. F., and Wright, H. E., Jr., eds., North America and adjacent oceans during the last deglaciation: Boulder, Colorado, Geological Society of America, The Geology of North America, v. K-3, p. 261–276.

Galloway, R. W., 1970, The full-glacial climate in the southwestern United States: Annals of the Association of American Geographers, v. 60, p. 245–256.

——, 1983, Full-glacial southwestern United States: Mild and wet or cold and dry: Quaternary Research, v. 19, p. 236–248.

Harbour, J., 1958, Microstratigraphy and sedimentational studies of an early man site near Lucy, New Mexico [M.S. thesis]: Albuquerque, New Mexico, University of New Mexico, 111 p.

Husband, B. C., and Hickman, M., 1985, Growth and biomass allocation of *Ruppia occidentalis* in three lakes, differing in salinity: Canadian Journal of Botany, v. 63, p. 2004–2014.

Last, W. M., and Schweyen, T. H., 1985, Late Holocene history of Waldsea Lake, Saskatchewan, Canada: Quaternary Research, v. 24, p. 219–234.

Leopold, L. B., 1951, The Pleistocene climate in New Mexico: American Journal of Science, v. 249, p. 152–168.

Lyons, T. R., 1969, A study of the Paleo-Indian and desert culture complexes of the Estancia valley area, New Mexico [Ph.D. thesis]: Albuquerque, New Mexico, University of New Mexico, 335 p.

Moiola, R. J., and Glover, E. D., 1965, Recent anhydrite from Clayton Playa, Nevada: American Mineralogist, v. 50, p. 2063–2069.

Morrison, R. B., and Davis, J. O., 1984, Quaternary stratigraphy and archeology of the Lake Lahontan area: A re-assessment (field trip 13), *in* Lintz, J., Jr., ed., Western geological excursions (Geological Society of America, annual meeting guidebook: Reno, Nevada, Mackay School of Mines, v. 1, p. 252–281.

Ogniben, L., 1955, Inverse graded bedding in primary gypsum of chemical deposition: Journal of Sedimentary Petrology, v. 25, p. 273–281.

Oviatt, C. G., McCoy, W. D., and Reider, R. G., 1987, Evidence for a shallow early or middle Wisconsin-age lake in the Bonneville basin, Utah: Quaternary Research, v. 27, p. 248–262.

Proctor, V. W., 1980, Historical biogeography of *Chara* (Charophyta); an appraisal of the Braun-Wood Classification plus a falsifiable alternative for future consideration: Journal of Phycology, v. 16, p. 218–233.

Scott, W. E., McCoy, W. D., Shroba, R. R., and Rubin, M., 1983, Reinterpretation of the exposed record of the last two cycles of Lake Bonneville, western United States: Quaternary Research, v. 20, p. 261–285.

Shackleton, N. J., and Opdyke, N. D., 1973, Oxygen-isotope and paleomagnetic stratigraphy of equatorial Pacific core V28-238: Oxygen-isotope temperatures and ice volumes on a 10^5 year and 10^6 year scale: Quaternary Research, v. 3, p. 39–55.

Smith, L. N., and Anderson, R. Y., 1982, Pleistocene-Holocene climate of the Estancia basin, central New Mexico, *in* Grambling, J. A., and Wells, S. G., eds., Albuquerque country II: New Mexico Geological Society, 33rd Field Conference, Guidebook, p. 347–350.

Sonnenfeld, P., 1984, Brines and evaporites: Orlando, Florida, Academic Press, 613 p.

Titus, F. B., 1969, Late Tertiary and Quaternary hydrogeology of the Estancia basin, central New Mexico [Ph.D. thesis]: Albuquerque, New Mexico, University of New Mexico, 179 p.

Verhoeven, J.T.A., 1975, *Ruppia*-communities in the Camargue, France. Distribution and structure in relation to salinity and salinity fluctuations: Aquatic Botany, v. 1, p. 217–241.

——, 1979, The ecology of *Ruppia*-dominated communities in western Europe. I. Distribution of *Ruppia* representatives in relation to their autecology: Aquatic Botany, v. 6, p. 197–268.

——, 1980, The autecology of *Ruppia*-dominated communities in western Europe. III. Aspects of production, consumption and decomposition: Aquatic Botany, v. 8, p. 209–253.

MANUSCRIPT ACCEPTED BY THE SOCIETY SEPTEMBER 6, 1991

Index

[Italic page numbers indicate major references]

Typeset by WESType Publishing Services, Inc., Boulder, Colorado
Printed in U.S.A. by Malloy Lithographing, Inc., Ann Arbor, Michigan